Anthropoid Origins

ADVANCES IN PRIMATOLOGY

Current Volumes in the Series:

ANTHROPOID ORIGINS
Edited by John G. Fleagle and Richard F. Kay

COMPARATIVE BIOLOGY AND EVOLUTIONARY RELATIONSHIPS OF TREE SHREWS
Edited by W. Patrick Luckett

EVOLUTIONARY BIOLOGY OF THE NEW WORLD MONKEYS AND CONTINENTAL DRIFT
Edited by Russell L. Ciochon and A. Brunetto Chiarelli

NEW INTERPRETATIONS OF APE AND HUMAN ANCESTRY
Edited by Russell L. Ciochon and Robert S. Corruccini

NURSERY CARE OF NONHUMAN PRIMATES
Edited by Gerald C. Ruppenthal

PRIMATES AND THEIR RELATIVES IN PHYLOGENETIC PERSPECTIVE
Edited by Ross D. E. MacPhee

SENSORY SYSTEMS OF PRIMATES
Edited by Charles R. Noback

SIZE AND SCALING IN PRIMATE BIOLOGY
Edited by William L. Jungers

SPECIES, SPECIES CONCEPTS, AND PRIMATE EVOLUTION
Edited by William H. Kimbel and Lawrence B. Martin

A Continuation Order Plan is available for this series. A continuation order will bring delivery of each new volume immediately upon publication. Volumes are billed only upon actual shipment. For further information please contact the publisher.

Anthropoid Origins

Edited by

JOHN G. FLEAGLE
State University of New York at Stony Brook
Stony Brook, New York

and

RICHARD F. KAY
Duke University Medical Center
Durham, North Carolina

Springer Science+Business Media, LLC

Library of Congress Cataloging-in-Publication Data

Anthropoid origins / edited by John G. Fleagle and Richard F. Kay.
p. cm. -- (Advances in primatology)
Revised papers from a conference held at Duke University, May 1992.
Includes bibliographical references and index.

1. Primates--Evolution--Congresses. 2. Primates, Fossil--Congresses. I. Fleagle, John G. II. Kay, Richard F. III. Series.
QL737.P9A64 1994
599.80438--dc20 94-43006
CIP

DOI 10.1007/978-1-4757-9197-6

Originally published by Plenum Press, New York in 1994.
MyCopy version of the original edition 1994

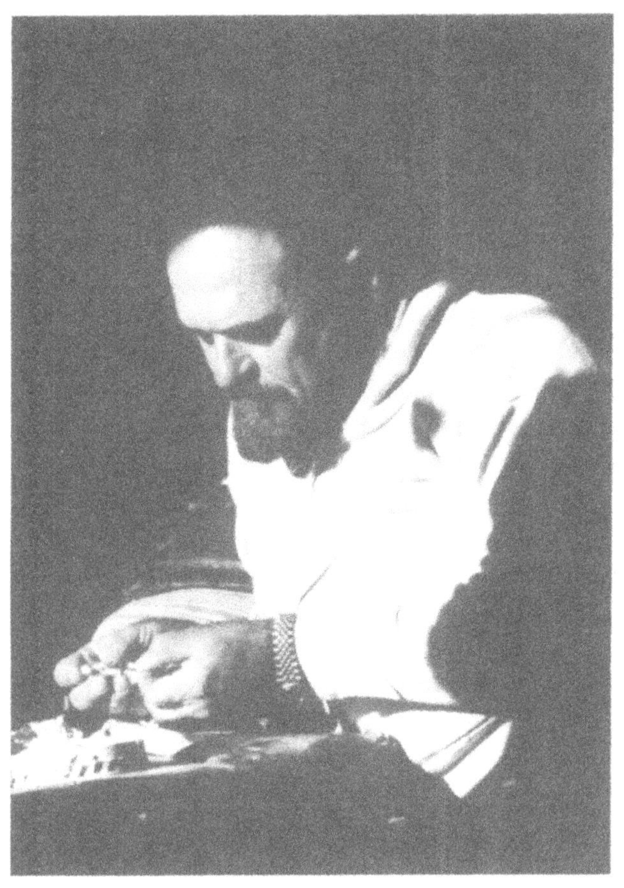

Dedicated to Elwyn L. Simons
Who over the past thirty-five years
invented paleoprimatology as a
discipline, and whose ongoing
discoveries in Egypt continue
to set the agenda for studies of
anthropoid origins

Contributors

K. Christopher Beard
Section of Vertebrate Paleontology
Carnegie Museum of Natural History
Pittsburgh, Pennsylvania 15213

Thomas M. Bown
Paleontology and Stratigraphy Branch
U.S. Geological Survey
Denver, Colorado 80225

Matt Cartmill
Department of Biological Anthropology and Anatomy
Duke University
Durham, North Carolina 27710

Prithijit S. Chatrath
Duke University Primate Center
Durham, North Carolina 27706

Russell L. Ciochon
Departments of Anthropology and Pediatric Dentistry
University of Iowa
Iowa City, Iowa 52242

Herbert H. Covert
Department of Anthropology
University of Colorado
Boulder, Colorado 80309-0233

Marian Dagosto
Department of Cell and Molecular Biology
Northwestern University Medical School
Chicago, Illinois 60611, and
Department of Mammalogy
American Museum of Natural History
New York, New York 10024

John G. Fleagle
Department of Anatomical Sciences
School of Medicine
Health Sciences Center
State University of New York
Stony Brook, New York 11794

Susan M. Ford
Department of Anthropology
Southern Illinois University
Carbondale, Illinois 62901

Jens Lorenz Franzen
Forschungsinstitut Senckenberg
D-60325 Frankfurt am Main, Germany

Daniel L. Gebo
Department of Anthropology
Northern Illinois University
DeKalb, Illinois 60115

Philip D. Gingerich
Museum of Paleontology
University of Michigan
Ann Arbor, Michigan 48109

Marc Godinot
Laboratoire de Paléontologie
Institut des Sciences de l'Evolution
Université Montpellier II
F-34095 Montpellier Cédex 5, France

Patricia A. Holroyd
Department of Biological Anthropology and Anatomy
Duke University
Durham, North Carolina 27710, and
U.S. Geological Survey
Denver, Colorado 80225

William L. Hylander
Department of Biological Anthropology and Anatomy
Duke University Medical Center
Durham, North Carolina 27710

Richard F. Kay
Department of Biological Anthropology and Anatomy
Duke University
Durham, North Carolina 27710

Mary C. Maas
Department of Biological Anthropology and Anatomy
Duke University
Durham, North Carolina 27710

R. D. E. MacPhee
Department of Mammalogy
American Museum of Natural History
New York, New York 10024

D. Tab Rasmussen
Department of Anthropology
Washington University
St. Louis, Missouri 63130

Matthew J. Ravosa
Department of Biological Anthropology and Anatomy
Duke University Medical Center
Durham, North Carolina 27710
Present address:
Department of Cell and Molecular Biology
Northwestern University Medical School
Chicago, Illinois 60611-3008, and
Department of Zoology
Division of Mammals
Field Museum of Natural History
Chicago, Illinois 60605-2496

Kenneth D. Rose
Department of Cell Biology and Anatomy
Johns Hopkins University School of Medicine
Baltimore, Maryland 21205

Callum Ross
Department of Biological Anthropology and Anatomy
Duke University
Durham, North Carolina 27705
Present address:
Department of Anatomical Sciences
School of Medicine
Health Sciences Center
State University of New York
Stony Brook, New York 11794

Elwyn L. Simons
Duke Primate Center
Duke University
Durham, North Carolina 27705

Blythe A. Williams
Department of Anthropology
University of Colorado
Boulder, Colorado 80309
Present address:
Department of Biological Anthropology and Anatomy
Duke University
Durham, North Carolina 27710

Preface

This volume brings together information about recent discoveries and current theories concerning the origin and early evolution of anthropoid primates—monkeys, apes, and humans. Although Anthropoidea is one of the most distinctive groups of living primates, and the origin of the group is a frequent topic of discussion in the anthropological and paleontological literature, the topic of anthropoid origins has rarely been the focus of direct discussion in primate evolution. Rather, discussion of anthropoid origins appears as a major side issue in volumes dealing with the origin of platyrrhines (Ciochon and Chiarelli, 1980), in discussions about the phylogenetic position of *Tarsius,* in descriptions of early anthropoid fossils, and in descriptions and revisions of various fossil prosimians. As a result, the literature on anthropoid origins has a long history of argument by advocacy, in which scholars with different views have expounded individual theories based on a small bit of evidence at hand, often with little consideration of alternative views and other types of evidence that have been used in their support. This type of scholarship struck us as a relatively unproductive approach to a critical issue in primate evolution. We hoped that by bringing together scholars with expertise in different areas of primate evolution and often divergent views of this issue, we could generate a common focus and broader awareness, in order to sort out areas of general agreement, areas of great disagreement, and areas of acknowledged ignorance. In doing so we would be in a better position to either resolve the question of anthropoid origins or at least set the stage for more fruitful discussion of the topic in future years.

To this end, in May 1992, we joined with Professor Elwyn Simons, Scientific Director of the Duke University Primate Center, to organize a conference and workshop on Anthropoid Origins at Duke University. Through four days of presenting papers, discussing anthropoid origins, examining new fossils, and watching lemurs, all of the participants gained a better knowledge of both the available fossil material, much of it unpublished at the time, and the views of the other participants. Following the conference, many of the participants wrote up and submitted manuscripts on various aspects of anthropoid origins; these were sent to other participating scientists as well as "outside" scholars for peer review. The manuscripts were revised, often drastically, before they were accepted for publication in the volume. As a result the papers contained here reflect the combined input of many researchers and demonstrate real progress in addressing the question of anthropoid origins.

The problem of anthropoid origins is not a single issue, but a wide range of issues. These obviously include the phylogenetic origin of anthropoids (What is the sister taxon of anthropoids?), the geographical origin of the group (Where did anthropoids first evolve?), and the less frequently addressed problem of the adaptations of the earliest anthropoids (Did the origin of anthropoids involve a major adaptive shift in primate evolution, or some "keystone adaptation"?). But, in addition, the question includes sorting out the phylogenetic relationships (and adaptations) of an increasing diversity of extinct species that appear to be near the border between prosimians and anthropoids. While the origin of anthropoids may appear to be a single point in a cladogram of living primates, the stem between the earliest anthropoid and the subsequent divergence of modern anthropoid groups could, and almost certainly did, contain dozens of extinct taxa possessing a variety of combinations of "anthropoid" features. The anthropoid tree is much bushier at the base than the diversity of modern taxa would indicate.

This volume addresses these many aspects of anthropoid origins from a variety of perspectives. Some chapters consider anthropoid origins from the perspective of the prosimians that preceded and perhaps gave rise to higher primates. Some take an anthropoid perspective to describe the early anthropoid radiation and "look back" to see where the group may have come from. They examine both prosimians and anthropoids using cranial, dental, or postcranial features. The authors have been encouraged to compare and contrast their results with those of others in the volume so that both individually and collectively they strive for some degree of synthesis.

This volume has benefited from the contributions of many persons and institutions. Funding for the conference was provided in part by the LSB Leakey Foundation, The Wenner-Gren Foundation, The National Science Foundation, Duke University, and the State University of New York. The conference and the volume would not have been possible without the many contributions of Dr. Elwyn Simons, who as scientist, host, and mentor to many of the participants laid the base for most of the research included in the volume. Ken Glander, Prithijit Chatrath, Friderun Ankel-Simons, Todd Rae, Mario Gagnon, Irene Lofstrum, Gay Gaster, and Rachael Hougam were instrumental in making the conference a success. Todd Rae, Walter Hartwig, Joan Kelly, and Kaye Reed made herculean efforts in putting together the chapters that make up the volume. Many of our colleagues aided in reviewing the manuscripts. Most of all, however, this volume reflects the willingness of our colleagues to share their time, their thoughts, their diverse skills, and their sense of humor, in a common effort to make some progress in understanding this critical aspect of primate evolution.

John G. Fleagle
Richard F. Kay

Stony Brook, New York
Durham, North Carolina

Contents

The Early Radiation of Euprimates and the Initial Diversification of Omomyidae

1

KENNETH D. ROSE, MARC GODINOT, and THOMAS M. BOWN

Introduction

The source of euprimates (Hoffstetter, 1977) or "true primates"—the taxonomic assemblage including extant lemurs, lorises, tarsiers, and anthropoids and their unquestioned extinct relatives—remains one of the paramount unresolved questions in primate evolution. Any discussion of the origin and initial radiation of euprimates in general, and Omomyidae in particular, must begin with a broader consideration of the origin, relationships, and composition of the order Primates.

KENNETH D. ROSE • Department of Cell Biology and Anatomy, Johns Hopkins University School of Medicine, Baltimore, Maryland 21205. MARC GODINOT • Laboratoire de Paléontologie, Institut des Sciences de l'Evolution, Université Montpellier II, F-34095 Montpellier Cédex 5, France. THOMAS M. BOWN • Paleontology and Stratigraphy Branch, U.S. Geological Survey, Denver, Colorado 80225.

Anthropoid Origins, edited by John G. Fleagle and Richard F. Kay. Plenum Press, New York, 1994.

Primate Origins and the Status of Plesiadapiformes

The origin of primates has commonly been considered to lie among the Insectivora (s.l.). In recent years Erinaceomorpha have been viewed as the most probable source, partly because of dental resemblance between certain early erinaceomorphs and early Paleocene *Purgatorius* (Figs. 1A and 2A; Clemens, 1974; Schwartz *et al.*, 1978; Szalay and Delson, 1979), a putative plesiadapiform that has been widely held to be the oldest and most primitive primate (e.g., Van Valen and Sloan, 1965; Simons, 1972; Szalay and Delson, 1979; Fleagle, 1988; Hoffstetter, 1988). Dental similarities often cited (and usually considered to be derived traits) include relatively broad, low-crowned molars with broad, basined talonids, reduced paraconids, and less acute cusps than in other contemporary "insectivores." But these characters are also particularly plesiadapiform- or primate-like in European Eocene Amphilemuridae. Until fairly recently this led to the frequent misallocation of amphilemurids, now recognized as erinaceomorphs (Szalay, 1971; Koenigswald and Storch, 1983; Novacek *et al.*, 1985), to the primates—as evidenced by the names *Amphilemur, Alsaticopithecus,* and *Gesneropithex* (see, e.g., Heller, 1935; Hürzeler, 1946, 1947; Simons, 1962; Russell *et al.*, 1967, 1975). Erinaceomorphs are also said to share with primates specific basicranial traits, including a significant petrosal contribution to the bulla and a well-developed promontory branch of the internal carotid artery (ICA) (MacPhee and Cartmill, 1986). An erinaceomorph–primate alliance has never been demonstrably very close, however, and new basicranial evidence from primitive Eocene erinaceomorphs contradicts the latter two features, suggesting that putative synapomorphies were independently acquired (i.e., convergent) and that erinaceomorphs in fact have no special alliance with primates (Novacek *et al.*, 1983; MacPhee *et al.*, 1988).

If primates are not closely related to erinaceomorphs, to what other high-level taxa might they be allied? Other extant groups that have been considered to be particularly close to primates, either individually or collectively, include tree shrews (Scandentia), bats (particularly Megachiroptera), and flying lemurs (Dermoptera)—that is, essentially, Gregory's (1910) Archonta. With the possible exception of Dermoptera (discussed further below), no fossil evidence has yet been found that would substantiate any of these hypothetical sister-group relationships.

Earlier in this century, tree shrews were regarded as primates (e.g., Simpson, 1945), largely as a result of detailed anatomic studies by Carlsson (1922) and Le Gros Clark (e.g., 1924a,b, 1925, 1933). More recently, the consensus of diverse research on tree shrews is that they constitute a primitive eutherian clade with no special relationship to primates (e.g., Martin, 1968, 1986; Luckett, 1980); i.e., similarities to primates are symplesiomorphous or convergent features. However, reevaluation of cranial traits over the last few years has once again raised the possibility that tree shrews and euprimates form a clade (Wible and Covert, 1987; Novacek, 1989; Kay *et al.*, 1992).

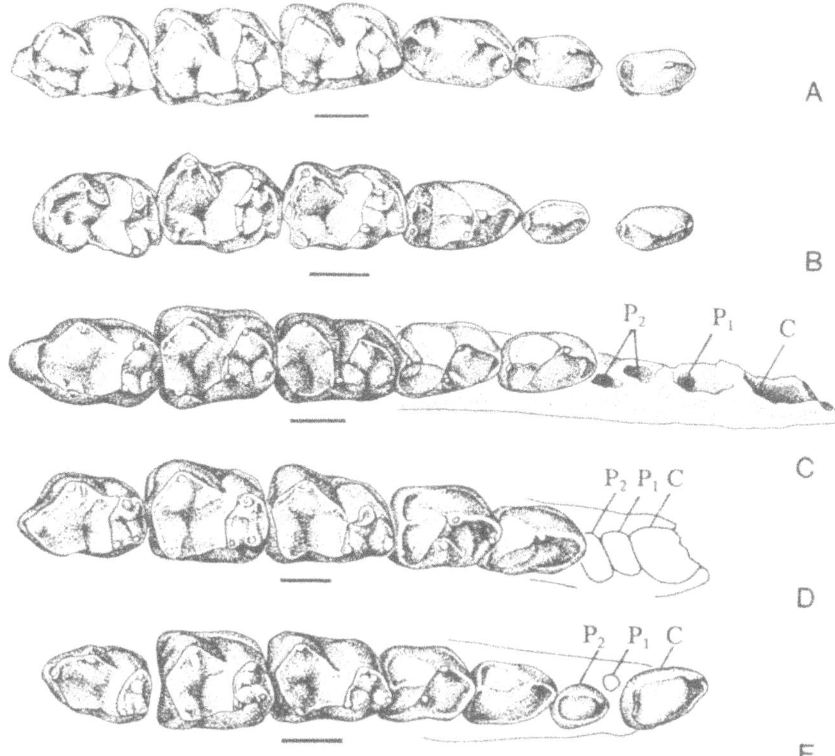

Fig. 1. Occlusal views of left lower dentitions of primitive plesiadapiforms and euprimates, drawn to approximately unit length of M_1. A, *Purgatorius unio*, P_2–M_3, UCMP 107406 (reversed). B, *Palaechthon alticuspis*, P_2–M_3, based on AMNH 35488 and USNM 9532 (reversed). C, *Donrussellia provincialis*, P_3–M_3 and alveoli of canine and P_{1-2}, based on MNHN numbers RI-170 and RI-430. D, *Steinius vespertinus*, P_2–M_3 and alveoli of canine and P_{1-2}, based on AMNH 16835, USGS 28325, and USGS 28326. E, *Teilhardina americana*, canine, P_2–M_3, and alveolus of P_1, UW 6896. Scales are 1 mm. Acronyms for institutions housing specimens illustrated in this and subsequent figures are as follows: AMNH, American Museum of Natural History, New York; CL, Wouters collection, in Royal Natural History Museum, Brussels; MNHN, Muséum National d'Histoire Naturelle, Paris; PSS, Geological Institute Paleontology and Stratigraphy Section, Mongolian Academy of Sciences, Ulan Bator; UCMP, Museum of Paleontology, University of California at Berkeley; UM, University of Michigan Museum of Paleontology, Ann Arbor; USGS, U.S. Geological Survey, Denver; USNM, National Museum of Natural History, Smithsonian Institution, Washington, D.C.; UW, University of Wyoming Geological Museum, Laramie; WL, Wouters collection, in Royal Natural History Museum, Brussels; YPMPU, Princeton University collection, at Yale Peabody Museum, New Haven, Connecticut.

Molecular studies generally have not supported either a sister-group relationship between tree shrews and primates or a monophyletic Archonta (e.g., Goodman *et al.*, 1985; Miyamoto and Goodman, 1986; Wyss *et al.*, 1988; Adkins and Honeycutt, 1991; Bailey *et al.*, 1992; Ammerman and Hillis, 1992). An exception is Shoshani's (1986) immunodiffusion analysis, which upheld a weak link between Primates and Scandentia as well as a larger clade

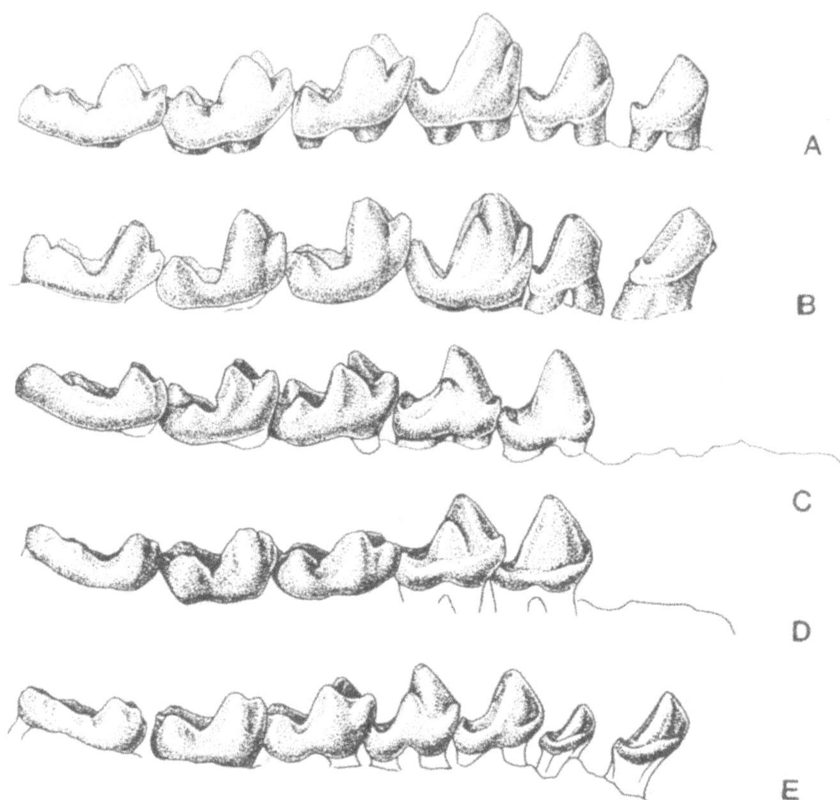

Fig. 2. Medial views of left lower dentitions of the same plesiadapiforms and euprimates as in Fig. 1 (same scales as in Fig. 1). A, *Purgatorius unio*. B, *Palaechthon alticuspis*. C, *Donrussellia provincialis*. D, *Steinius vespertinus*. E, *Teilhardina americana*.

composed of Primates, Scandentia, and Chiroptera (but not Dermoptera). More recent molecular studies have found evidence for a clade composed of Primates, Scandentia, and Dermoptera, but not Chiroptera (Adkins & Honeycutt, 1991; Ammerman and Hillis, 1992). Thus, the question remains unresolved, but it seems fair to say that there is evidence—both molecular and morphological—that primates belong to a monophyletic group comprising at least some of the higher taxa traditionally included in the Archonta, but whose precise composition and interrelationships are still being debated (e.g., McKenna, 1975; Szalay, 1977; Novacek, 1992; Beard, 1993). This "modified Archonta," which may also include Plesiadapiformes and some other extinct groups, is unified by particular adaptations for arboreal life.

Recent studies of retinotectal pathways (optic tracts) in Chiroptera indicate that megabats share an otherwise unique crossover pattern with Primates and Dermoptera, leading to provocative speculation that Primates, Dermoptera, and Megachiroptera constitute a clade to the exclusion of Microchiroptera and Scandentia (Pettigrew, 1986, 1991; Pettigrew *et al.*, 1989). The preponderance of other evidence (both morphological and molecular), however, seems overwhelmingly contrary to such an alliance and still favors bat monophyly (e.g., Novacek and Wyss, 1986; Miyamoto and Goodman, 1986; Wible and Novacek, 1988; Adkins and Honeycutt, 1991; Thewissen and Babcock, 1991; Greenwald, 1991; Baker *et al.*, 1991; Bailey *et al.*, 1992; Ammerman and Hillis, 1992).

Considerations of primate origins have conventionally been based on comparisons between nonprimate groups and Paleocene members of Plesiadapiformes rather than later-appearing Euprimates. This comparison has been predicated on the belief that plesiadapiforms are unquestionably primates (e.g., Szalay *et al.*, 1987), a presumption that has been increasingly disputed in recent years (e.g., Martin, 1968, 1986, 1990; Cartmill, 1974; MacPhee *et al.*, 1983; Wible and Covert, 1987). Plesiadapiforms include several families of so-called "archaic primates"—Plesiadapidae, Carpolestidae, Saxonellidae, Paromomyidae, Micromomyidae, Palaechthonidae, and probably Picrodontidae and Microsyopidae. They were common and diverse in the Paleocene of North America and (less so) Europe, accounting for upwards of 20% of the mammalian assemblage at most Paleocene sites in western North America (Rose, 1981a,b). Postcranial evidence (where known) indicates that plesiadapiforms were arboreal (Szalay *et al.*, 1975; Beard, 1989, 1991).

Most of the 30 or so genera of plesiadapiforms are known only from dentitions, which are characterized by enlarged central incisors, specialization and/or loss of premolars and lower canines, and conservative low-crowned molars. Many of these modifications appear to have occurred in parallel in different lineages. The most obvious specialization—the enlarged, procumbent incisors—although commonly considered homologous and therefore a synapomorphy of plesiadapiforms, has never been so demonstrated. Actually, plesiadapiform incisors vary widely in form (compare the broad, lanceolate I_1 of *Microsyops* with the long and narrow I_1 of *Phenacolemur* and *Carpolestes* or the stout chisel-like I_1 of *Plesiadapis* and *Chiromyoides*). Picrodontids and microsyopids are, in fact, fundamentally different in some respects and may not be closely related to other plesiadapiforms. Moreover, Beard (1993) limited Plesiadapiformes to the Plesiadapidae, Saxonellidae, Carpolestidae, Paromomyidae, and Galeopithecidae (see further discussion below). Thus, the composition and the monophyly of Plesiadapiformes are still controversial issues.

Allocation of plesiadapiforms to Primates has been based largely on strikingly close dental (mostly molar) resemblance to euprimates (particularly early Eocene representatives), but it has long been known that the most euprimate-like teeth are found in younger, more derived plesiadapiforms that cannot be directly related to later primates (for instance, compare late Tiffanian *Plesiadapis* molars to those of Wasatchian *Cantius*); more primitive plesiadapiforms are less like euprimates. In fact, the only plesiadapiform that is not obviously too derived dentally to be ancestral to euprimates is early

Paleocene *Purgatorius!* Despite its very primitive overall dental structure, it should not be assumed that all characters of *Purgatorius* (Figs. 1A and 2A) are necessarily primitive for primates. For example, traits such as the prominent paraconid on P_4 and the presence of both a paracone and a metacone on P^4 (a feature more reminiscent of palaechthonids than euprimates) could be apomorphic relative to the euprimate morphotype. Moreover, *Purgatorius* differs from all euprimates and plesiadapiforms (except *Berruvius lasseroni*) in having crestiform molar paraconids. Still, its close correspondence in other details of dental anatomy indicates that *Purgatorius* is a primitive plesiadapiform, probably most closely related to palaechthonids (Figs. 1B and 2B; Clemens, 1974; Gunnell, 1989).

Close correspondence in auditory anatomy—notoriously controversial because of difficulty in interpretation—has also been claimed: for instance, a petrosal bulla, inflated bullar floor, degree of promontory exposure within the middle ear, and "shield" over fenestra cochleae, are inferred characters of *Plesiadapis* allegedly shared with euprimates (Gingerich, 1976; MacPhee, 1981; Szalay *et al.*, 1987); but bullar composition cannot be unequivocally determined, and the other characters occur in Miocene erinaceids. Moreover, *Plesiadapis* evidently lacked a significant internal carotid system (MacPhee and Cartmill, 1986; but see Szalay *et al.*, 1987, for an opposing view), in contrast to euprimates but in common with extant Dermoptera (Hunt and Korth, 1980). Indeed, diagnostic basicranial traits of Primates are elusive: most characteristic traits are either primitive or known to occur in many other orders (MacPhee and Cartmill, 1986). And plesiadapiforms uniformly lacked a postorbital bar, opposable hallux, and nails on most digits—traits considered diagnostic of Primates, or at least Euprimates, by most authors. It seems increasingly probable that the close resemblances between plesiadapiforms and euprimates could have resulted from convergence.

The most stimulating new evidence on the relationships of plesiadapiforms pertains to early Eocene paromomyids. Recently discovered postcranial skeletal elements of *Ignacius* and *Phenacolemur* have derived traits—particularly phalangeal structure and proportions and details of the carpus and tarsus—suggesting that paromomyids are functionally similar and phylogenetically related to extant *Cynocephalus* (Dermoptera) [e.g., longer middle than proximal phalanges, inferentially associated with interdigital webbing of a patagium; lunate distal to scaphoid; aspects of the sustentaculum tali and calcaneocuboid joints (Beard, 1989, 1990, 1993)]. *Plesiadapis* shares several of these traits but lacks the specialized phalangeal proportions (Beard, 1990, 1993). Largely on the basis of these new postcranial fossils, Beard (1993) considered Plesiadapoidea (consisting of Plesiadapidae, Carpolestidae, and Saxonellidae) to be the sister group of Eudermoptera (composed of Paromomyidae and Galeopithecidae) and grouped both of these in the order Dermoptera. He further concluded that Dermoptera and Primates are sister groups, for which he proposed the new mirorder Primatomorpha (Beard, 1991, 1993).

Krause (1991) challenged the existence of convincing evidence for longer intermediate phalanges in paromomyids, and thus for a patagium and gliding capability, citing ambiguity in the identification and association of these isolated elements. Whether or not Krause's criticism is justified, the postcranial traits cited by Beard, though suggestive, are not yet demonstrative of special paromomyid–dermopteran relationship. The possibility remains that these characters (and the inferred gliding capability) in the two taxa evolved convergently. Further evidence is needed to make a compelling case for plesiadapiform–galeopithecid ties.

Basicranial evidence from a new skull of *Ignacius* also provides weak support for a plesiadapiform–dermopteran clade: the auditory bulla was apparently composed of the entotympanic (rather than the petrosal as in all euprimates), which contacts the basioccipital as in the extant dermopteran *Cynocephalus* (Hunt and Korth, 1980; Kay *et al.*, 1990). Furthermore, the principal endocranial blood supply must have come through the vertebral arteries, not via the internal carotid artery or the ascending pharyngeal branch of the external carotid as in euprimates (Kay *et al.*, 1990, 1992). The absence of both the stapedial and the promontory arteries is based on absence of grooves or tubes and on the caliber of the carotid foramen, which is too small to have transmitted a significant internal carotid artery. But other craniodental specializations of *Ignacius* and *Phenacolemur* exclude them from direct ancestry of *Cynocephalus*, and cranial and postcranial details of the most primitive paromomyids (needed to corroborate cited traits as synapomorphic with those of *Cynocephalus*) are unknown. Moreover, what Kay *et al.* (1992) defended as an entotympanic/petrosal suture in the skull of *Ignacius* could be merely a fracture line. Thus, despite these intriguing new fossils, it is quite possible that resemblances to *Cynocephalus* are convergent or less exact than has been proposed. In any case, a latest common ancestor of paromomyids or plesiadapiforms and *Cynocephalus* could not have been younger than early Paleocene.

Still, no other early Tertiary mammals have so closely approached euprimates in certain diagnostic skeletal traits as plesiadapiforms. The postcranial skeletons of plesiadapiforms are highly arboreally adapted (Szalay *et al.*, 1975; Beard, 1989, 1991) and in some cases seem to foreshadow euprimate adaptations (e.g., *Plesiadapis* apparently had a mobile, but not habitually abducted, hallux capable of grasping: Szalay *et al.*, 1987; Szalay and Dagosto, 1988). For these reasons, plesiadapiforms will continue to be crucial to the question of primate origins, either as an archaic sister-group (alone or with *Cynocephalus*) of euprimates still lacking many of the key primate traits and with many of their own specializations, or as a primate-like adaptive group occupying a similar ecological niche but more primitive in some cranial and skeletal features than living or fossil euprimates.

Both the composition and the phylogenetic position of microsyopoids are controversial. Some authors retain all or some of them in Plesiadapiformes, but others exclude them. They are noteworthy here because, according to MacPhee and Cartmill (1986), their basicranial anatomy except for the auditory bulla (not preserved, but some variant of ecto- or entotympanic rather than petrosal is considered most likely) is consistent with the predicted basal primate condition. However, Szalay *et al.* (1987) have contended that the basicranium of microsyopids shares unique features with extant Dermoptera. Other claims of ties with euprimates or Scandentia have not been substantiated in print.

The Earliest Supposed Euprimates

Euprimates are united by several derived traits—including a postorbital bar, opposable hallux (and tarsal modifications), and nails—not present in any known plesiadapiform (see Szalay et al., 1987, and Dagosto, 1988, for recent lists of euprimate synapomorphies). It should be noted that the ungual phalanges of the most primitive euprimates, though nail bearing, were more claw-like than those of extant primates (Godinot, 1992b). The source of euprimates is as obscure as the origin of plesiadapiforms, and, so far, the fossil record has provided few significant clues. It would help, of course, to have a more precise understanding of the euprimate morphotype, and considerable new fossil evidence has promoted progress toward that end.

We shall first consider several poorly known Paleocene–early Eocene taxa that have been advanced as the "oldest" or "most primitive" known euprimates. None of them preserves any of the (nondental) diagnostic euprimate characters, so assessments must be based entirely on dental anatomy (which some authorities have already dismissed as a result of homoplasy for the much better known Plesiadapiformes).

Szalay and Li (1986) proposed *Decoredon anhuiensis* (Xu, 1976) from the middle Paleocene of Anhui Province, south China, as the oldest euprimate, possibly an omomyid or representative of a more primitive basal stock. As used by Szalay and Li, the species included not only the holotype maxilla (originally described by Xu, 1976, as an anagalid in the genus *Diacronus*) but also dentary fragments that constituted the holotype of *Decoredon elongetus* (Xu, 1977) (originally allocated to Hyopsodontidae). P_4 and M_3 do bear some resemblance to those of Eocene euprimates, but it is difficult to say how precisely because the specimens are so poorly preserved. The lower cheek teeth, however, are much more elongate, with narrower talonids and more buccally displaced paraconids, than in early Eocene euprimates; and the talonid basin is open lingually through a deep lingual notch, especially evident on M_3, unlike primitive euprimates (although *Donrussellia gallica* approaches this condition). The assigned upper dentition is badly preserved and occludes poorly with the lower teeth, rendering conspecific assignment suspect and making details almost impossible to decipher. Thus, convincing evidence that *Decoredon* represents a euprimate is currently lacking.

In a better state of preservation but also very incomplete is a maxilla purportedly from the late Paleocene of Guangdong Province, China, designated as the holotype of *Petrolemur brevirostre*. Its describer (Tong, 1979) referred it to Adapidae, which it resembles somewhat in molar morphology, but its molariform P^4 and elongate P^3 are unlike those of early euprimates. Subsequent authors believed it to be nonprimate (Russell and Gingerich, 1980; Szalay, 1982), and, if the age assessment is accurate, it is unlikely that it belongs to either the adapiform or omomyid radiations. Szalay and Li (1986) hinted that new material may reaffirm its primate status, but documentation has yet to be published. This intriguing species merits further study to reassess its possible relationship to primates.

Dentally more euprimate-like is *Altanius orlovi* from the early Eocene of Mongolia (Fig. 3; Dashzeveg and McKenna, 1977), which has been widely accepted as an omomyid. Initially based on a dentary with P_4–M_3, its dentition has recently become much more completely known (Gingerich *et al.*,

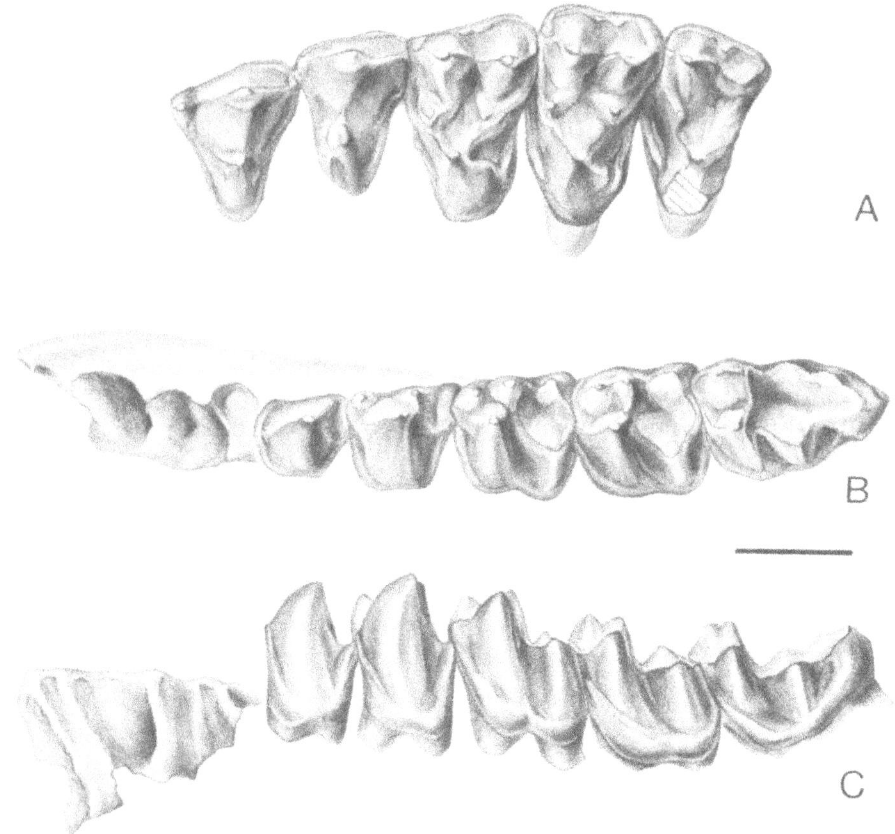

Fig. 3. *Altanius orlovi.* A, left P^3–M^3 in occlusal view, PSS 20-61. B, left P_3–M_3, and alveoli of I_{1-2}, C, and P_{1-2}, in occlusal view, PSS 20-58. C, lateral view of PSS 20-58. Scale is 1 mm.

1991), and it displays a curious combination of probable specializations [very high lingual cusps, short talonids with very steep entocristids (especially on M_1), pronounced exodaenodonty, and broad anteroposteriorly compressed P_{3-4}] coupled with extremely primitive features (lower dental formula of 2-1-4-3; small, nearly vertical incisors; large canine; P_{2-4} two-rooted; unreduced third molars; rudimentary hypocones). Other characters, such as high paraconids, well-developed paraconid and metaconid on P_4, buccolingually compressed trigonid on M_1, and strong conules, are more difficult to polarize, but their distribution among early primates (*sensu lato*) suggests that all of them could be derived. Still, it should be recognized that the polarity of many of these characters is uncertain. Although championed as a primitive omomyid by Dashzeveg and McKenna (1977) and Gingerich *et al.* (1991) (because of its very small size and overall resemblance to certain early Eocene omomyids), it is at once more primitive than any known omomyid in having a two-rooted P_2 and unreduced P_1, but more derived than primitive omomyids in its high lingual cusps, short talonids, and degree of exodaenodonty. Gingerich *et al.* (1991) acknowledged the very primitive state of *Altanius,* and their own cladistic analysis in fact placed it as an unambiguous sister-group of other euprimates, not as a cladistic omomyid.

The specialized traits of *Altanius* recall those of the primitive carpolestid plesiadapiform *Elphidotarsius* (Fig. 4) and also the plesiadapid *Pronothodectes* and were the basis for Rose and Krause's (1984) alternative hypothesis that *Altanius* might be a plesiadapiform rather than a euprimate. However, several previously unknown features revealed in the recently described dentitions may seem at first glance more characteristic of primitive euprimates: the retention of four premolars (compacted posteriorly but otherwise unmodified), a large canine, and small incisors. All known plesiadapiforms differ in having an enlarged central incisor (it is even somewhat enlarged in *Purgatorius;* Kielan-Jaworowska *et al.,* 1979), and all except *Purgatorius* have reduced or lost the canine and have lost P_1. But these are unquestionably apomorphies (i.e., all plesiadapiforms except *Purgatorius* are derived in these traits relative to *Altanius*). At the same time, these resemblances of *Altanius* to euprimates are almost certainly primitive and cannot be used as evidence of special relationship to euprimates. The upper molars of *Altanius* have prominent, projecting conules with conule cristae, and a strong *Nannopithex* fold running from the protocone to the postcingulum (Fig. 5). These derived(?) traits are particularly characteristic of carpolestid and plesiadapid upper molars, whereas the conules are small and low and the *Nannopithex* fold is weakly developed or absent in the primitive euprimates *Donrussellia* and *Teilhardina.*

Thus, we believe that the relationships of *Altanius* remain open to debate and that possible plesiadapiform affinity should not be discounted. In fact, there seem to be more potentially synapomorphic traits supporting plesiadapiform affinity than euprimate affinity (see also Godinot, Chapter 10, this volume). What is undeniable and most intriguing is that *Altanius* bears

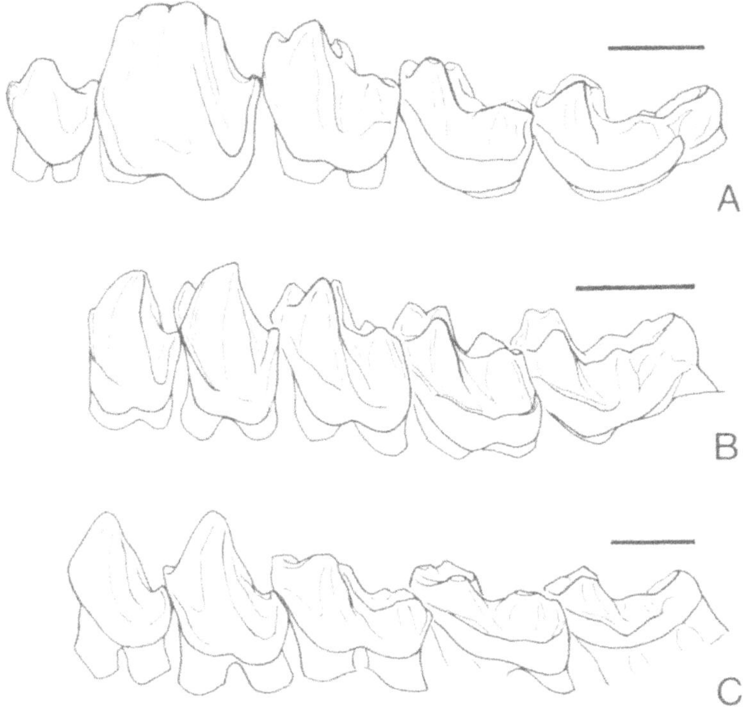

Fig. 4. Left P_3–M_3 in lateral view of (A) *Elphidotarsius*, cf. *E. florencae*, YPMPU 14791 (reversed); (B) *Altanius orlovi*, PSS 20-58; and (C) *Teilhardina belgica*, CL 192. Note high-crowned molars with short talonids in *Elphidotarsius* and *Altanius*. Scales are 1 mm.

close resemblance to certain primitive representatives of both euprimates and plesiadapiforms. Based on its dental anatomy, it is reasonable to conclude that whether euprimate or plesiadapiform, it represents a very early (primitive) offshoot from the basal stock, collateral to other known lines; it is unlikely to belong to any recognized family. It is even conceivable that *Altanius*, with its close resemblances to both groups, is the first fossil evidence of a common origin for plesiadapiforms and euprimates (see also Gingerich *et al.*, 1991).

Arguably the most tantalizing recent discovery bearing on the origin of euprimates is a small number of isolated teeth from the late Paleocene (Thanetian) of Morocco, described as *Altiatlasius koulchii* and allocated to Omomyidae by Sigé *et al.* (1990). *Altiatlasius* is the oldest primate from Africa—indeed, from anywhere—if its late Paleocene age (based on selachians; Capetta *et al.*, 1987) can be corroborated. Sigé *et al.* compared it most closely with omomyids, particularly *Omomys*, citing resemblances such as its diminutive size, bunodont cusps, large protocone, poorly developed conules and hypocone, absent *Nannopithex* fold, lingual cingulum on M^2, and lingual

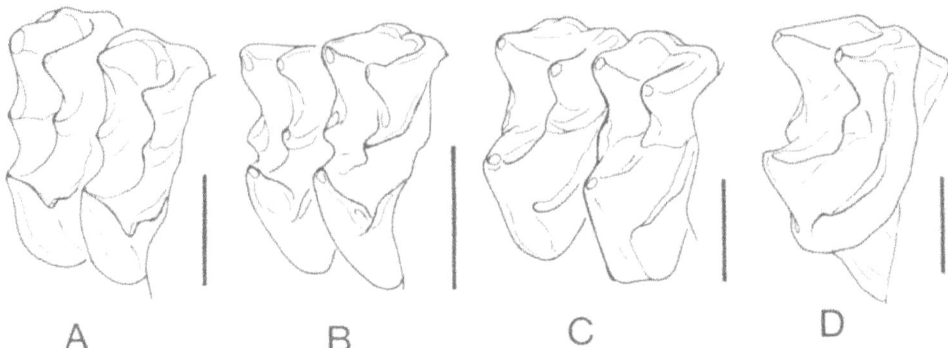

Fig. 5. Left upper molars (M^1 or M^{1-2}) in posterior oblique view. A, *Elphidotarsius*, cf. *E. florencae*, YPMPU 17439 (reversed); B, *Altanius orlovi*, PSS 20-61; C, *Teilhardina belgica*, WL 1398 (reversed); D, *Donrussellia gallica*, MNHN Av 4830. Scales are 1 mm.

paraconid, among others. They acknowledged that many of these traits are, or could be, primitive and listed them also as close resemblances to certain plesiadapiforms—microsyopids (*Berruvius*), micromomyids—and various other primates. In fact, few if any of these characters are demonstrably synapomorphic with omomyids, and character polarities are often unclear. *Altiatlasius* has a higher paraconid and smaller talonids than in the most primitive omomyids and adapiforms (primitive traits?) as well as a small pericone, bunodont cusps, and lack of a postmetaconule crista (more characteristic of adapiforms than of omomyids, and presumably derived relative to the primitive euprimate condition). Its P_3, if properly allocated, is lower-crowned and broader than would be expected in a generalized adapiform or omomyid.

These teeth appear to represent a very small and primitive euprimate (perhaps even a "protosimiiform," *sensu* Hoffstetter, 1982), but it is difficult to be more specific, for diagnostic characters of omomyids versus adapiforms are very difficult to enumerate at this early stage, especially from isolated teeth. Moreover, the phylogenetic position of *Altiatlasius* is necessarily vague because we lack knowledge of its dental formula and the relative size, morphology, and root configuration of the antemolar teeth (none of which are known from serially associated dentitions). Study of *Altiatlasius* (by M.G.) in the light of the new north African anthropoids suggests that it might be more closely related to them than to omomyids plus adapiforms. If this proves true, *Altiatlasius* would require the addition of new, more primitive characters in the euprimate morphotype and would suggest an earlier split between protosimiiforms and adapiforms plus omomyids (see Godinot, Chapter 10, this volume). However, because the material is still very fragmentary, and because some of its characters are unexpected for a euprimate, it would be premature to introduce it into our reconstruction of the euprimate morphotype.

Early Euprimate Radiations

The oldest undisputed euprimates come from the beginning of the Eocene (early Wasatchian and early Sparnacian) of North America and Europe. They belong to the taxa Omomyidae (often included in Tarsiiformes, sometimes in Omomyiformes; some authors separate Microchoeridae at the family level) and Adapiformes (often called simply Adapidae, but better divided into three families, Notharctidae, Adapidae *sensu stricto,* and the later-appearing Sivaladapidae). Adapiforms are strepsirhine and lemur-like but resemble lemuriforms largely in primitive characters. Omomyids and adapiforms are demonstrably euprimates because, when adequately preserved, they possess a series of derived anatomic features of the skull and skeleton otherwise found together only in extant primates (e.g., postorbital bar, petrosal bulla, opposable hallux, nails). Many species, however, are known only from dentitions.

Adapiforms are generally considered to be more primitive dentally (at least in the antemolar dentition); the more plesiomorphic ones are characterized by a dental formula of 2-1-4-3 above and below, larger canines, less mesiodistally compacted premolars, and two-rooted P_2. However, most of the derived antemolar traits that have conventionally differentiated omomyids from adapiforms (e.g., loss of premolars, reduction of canines, enlargement of central incisors) characterize the most primitive omomyids to a much lesser degree. Moreover, primitive adapiform molars lack a postmetaconule crista and show subtle traits in the lower molars (a tendency for decreasing depth of the protocristid notch and more widely separated protoconid–metaconid posteriorly, a labially shifted hypoconulid, a relatively straight postcristid, and an anteriorly shifted entoconid, variably set off by a notch) that may be derived relative to omomyid molars (Godinot, 1992a). Although these may seem to be relatively minute details, they represent a morphological shift toward typical adapiform dental specializations. Nonetheless, it is important to note that as the fossil record has improved, the most primitive known adapiforms and omomyids have more closely converged toward the same dental morphology, making it increasingly difficult to distinguish the two groups using dental criteria. The principal distinctions at the most archaic stage now known pertain to relative compaction of the antemolar series: omomyids differ from adapiforms in showing slight size reduction of the canine and P_1, loss of one root of P_2, and anteroposterior compression of P_{3-4} (resulting in a crown base that is more oblique in lateral perspective and projects mesiad over the anterior root; Figs. 1 and 2).

Euprimate postcranial synapomorphies found in omomyids and adapiforms, besides an opposable hallux and nails, include modified distal femoral and overall tarsal anatomy and joint specialization for highly developed forearm supination. Recent studies of isolated postcranial remains of early euprimates indicate that adapiforms resemble extant strepsirhines (lemurs and lorises) in some apparently derived features of the manus [divergent pollex, ulnar-deviated hand posture (e.g., Godinot and Beard, 1991)] and tarsus [fibular facet on astragalus sloping gently laterad, not more vertical as in omomyids, *Tarsius*, and anthropoids; flexor hallucis longus groove on astragalus laterally displaced relative to tibial articulation; broader cubonavicular contact (Gebo, 1986, 1988; Dagosto, 1988; Covert, 1988)]. This evidence not only supports a natural strepsirhine clade but also indicates that Adapiformes as well as Omomyidae are characterized by derived traits. Other anatomic features (distal elongation of calcaneus and navicular, modified navicular/entocuneiform joint), however, appear to be more derived in omomyids (Szalay and Dagosto; 1988; Dagosto, 1988).

Primitive Adapiforms

Most students regard *Donrussellia* (Figs. 1C and 2C), from the early Sparnacian of western Europe, as the most primitive adapiform. The genus includes the oldest euprimates reported from Europe (Godinot, 1978, 1981; Godinot *et al.*, 1987; see also Antunes and Russell, 1981; Estravis and Russell, 1989, on the age of beds producing *Donrussellia*). It is very small (M_1 length 2.2–3.4 mm) and has sharper (less bunodont) and more peripheral cusps and a less reduced paraconid than most other adapiforms. Its dentary also may be more primitive than that of any other known euprimate in having a small coronoid process and a shallow anterior horizontal ramus (Godinot *et al.*, 1987). Despite its overall primitive structure, *Donrussellia* shares certain probably derived dental traits with undoubted adapiforms [e.g., loss of the postmetaconule crista, a more or less continuous crest from the postprotocrista to the metacone (see Szalay, 1976), and a labial hypoconulid; see Godinot, 1992a]. Thus, what is known of the jaw and dentition of *Donrussellia* indicates that it is the most primitive known adapiform and that it lies in or near the ancestry of all other adapiforms.

Notharctidae as viewed here includes two subfamilies, Notharctinae and Cercamoniinae (see also Franzen, 1987, for a similar arrangement), whose initial differentiation may be discernible within the paraphyletic genus *Donrussellia*. *D. magna* shares probably derived features of M_3 with the basal notharctine *Cantius* (Godinot, 1992a), whereas another unidentified species of *Donrussellia* probably gave rise to *Protoadapis*, a primitive cercamoniine. The precise origin of the dentally more specialized Adapidae is much less clear. Where preserved, alveoli indicate that primitive adapiforms had small, essentially vertical incisors (the oldest genus in which this is known is *Cantius*).

Cantius (which includes most of the species formerly placed in *Pelycodus;* Fig. 6) first occurs in both Europe and North America at the beginning of the Eocene and is a common constituent of early Eocene faunas. The oldest European records are from the early Ypresian of Rians and Meudon (Godinot, 1981; Russell *et al.*, 1988). The oldest North American species is *C. torresi*

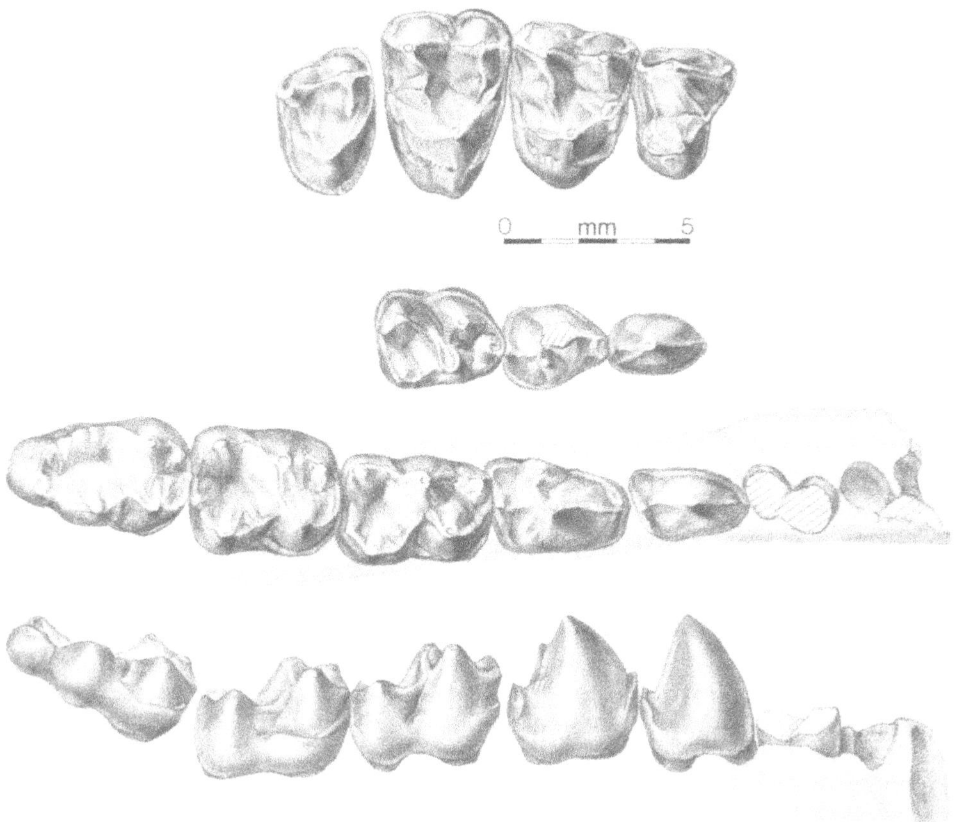

Fig. 6. Cheek teeth of *Cantius*. A, *Cantius ralstoni*, holotype, AMNH 16089, right P^4–M^3. B, *Cantius torresi*, holotype, UM 83470, left dentary with P_3–M_1, crown view. C, D, *Cantius ralstoni*, YPM 23317, right P_3–M_3 and alveoli of P_{1-2} in crown and lateral views (P_3 partly restored from AMNH 16093).

from the basal Wasatchian of Wyoming, which Gingerich (1986) suggested might be ancestral to all other adapiforms including *Donrussellia*. Such a relationship, however, seems contradicted by several probably derived features: larger size, more bunodont and inflated cusps, more mesiodistally compacted lower premolars, more molarized P_4, and more quadrate upper molars than in *Donrussellia* (Godinot *et al.*, 1987; Godinot, 1988a, 1992a; Rose and Bown, 1991). Furthermore, *C. torresi* may even be derived relative to the slightly younger *C. ralstoni* in having shorter and broader P_{3-4} and more squared upper molars (although its weaker *Nannopithex* fold is probably primitive relative to *C. ralstoni*). This suggests the possibility that there were at least two immigrations of the genus into North America (Godinot, 1992a). All North American adapiforms except late Eocene *Mahgarita* (a cercamoniine) are notharctines derivable ultimately from species of *Cantius*.

The European adapiform radiation was more diverse and of greater duration. It comprised two distinct episodes: an initial radiation of primitive forms, succeeded by a second group of apparently foreign origin. The first group includes *Donrussellia, Protoadapis, Europolemur,* and three slightly later genera of smaller body size: *Anchomomys, Periconodon,* and *Buxella* (these three genera constitute the Anchomomyini of Szalay and Delson, 1979; see also Godinot, 1988b). *Protoadapis* and *Europolemur* (included in Protoadapini by Szalay and Delson, 1979) were rooted in *Cantius* by Franzen (1987), but an origin in some species of *Donrussellia* appears more probable.

This initial radiation of primitive adapiforms was informally called "protoadapines" by Gregory (1920), but a formal taxon was evidently not proposed until more than a half-century later. Szalay and Delson (1979) proposed the tribe Protoadapini for *Protoadapis, Europolemur,* and three other genera, and Franzen (1987) elevated it to a subfamily Protoadapinae. However, because these family-level taxa include *Protoadapis (Cercamonius) brachyrhynchus,* for which Gingerich (1975) earlier coined the subfamily Cercamoniinae, the latter term has priority for any subfamily or tribe that includes this species (unless an earlier formal definition of Protoadapinae exists). Hence, we use the subfamily Cercamoniinae for these primitive European adapiforms. Being descended from a species of *Donrussellia,* Cercamoniinae are the sister group of Notharctinae, and the two subfamilies appear to be close in morphology and adaptation (Franzen, 1987). For these reasons they are best united together in a family Notharctidae.

The second radiation of European adapiforms includes *Adapis, Leptadapis, Cryptadapis,* the more primitive *Microadapis,* and probably also *Caenopithecus.* Gingerich (1977a: fig. 8) showed the latter genus as a descendant of *Protoadapis,* but this is difficult to confirm because *Caenopithecus* is very autapomorphic and not easily linked with other taxa. *Adapis* and *Leptadapis* display very peculiar dental and postcranial adaptations (Stehlin, 1912; Dagosto, 1983). Hence, this group of related genera is best distinguished at the family level as Adapidae, *sensu stricto.* These "true" Adapidae do not appear until the middle Eocene (MP 13–14 levels of the European mammalian scale; see Schmidt-Kittler, 1987). *Anchomomys* has been suggested as a possible source of this group (e.g., Franzen, 1987); however, although there are some similarities in lower molar morphology between *Anchomomys* and Adapidae *sensu stricto* (e.g., the paracristid), specialized characters of *Anchomomys* indicate that they must be convergences. This leaves the Adapidae without a clear origin in the known European lower Eocene record. Since they appear as an already diversified group, an allochthonous origin is probable (Franzen, 1987). Franzen suggested an African source, but no undoubted adapiform is yet known from Africa. *Donrussellia* is primitive enough to provide a remote ancestral morphology on the European continent; hence, the differentiation of Adapidae in an unsampled area neighboring northwestern Europe (e.g., to the south or east) seems more likely.

Primitive Omomyids

The oldest known omomyid is *Teilhardina* from the basal Sparnacian of Europe and early Wasatchian of North America, nearly contemporaneous with the oldest adapids. Recently collected specimens indicate that the oldest and most primitive representatives on both continents—European *T. belgica* and the slightly more derived *T. americana* (Figs. 1E and 2E) from Wyoming—typically retained a tiny P_1, often situated lateral to the long axis of the tooth row. They also had small, relatively vertical incisors (based on alveoli known only in *T. belgica*) and a large canine (primitive euprimate traits), coupled with somewhat broader and more mesiodistally compressed lower cheek teeth, reduced third molars, and more elevated molar protocones (with longer lingual slope) compared with *Donrussellia* (Gingerich 1977b; Bown and Rose, 1987). The hypocone is small or absent, and the *Nannopithex* fold tends to be weakly developed or absent. The overall dental anatomy and stratigraphic position of *Teilhardina* place it very near the base of the omomyid radiation (at least for Anaptomorphinae; Bown and Rose, 1987), but it already appears to be more advanced in degree of basal crown inflation and third molar reduction than, for example, *Omomys*.

Fossils of the rare Wasatchian omomyid *Steinius vespertinus* (Figs. 1D and 2D) found in Wyoming over the last 2 years show that in most features this species is dentally as primitive as *Teilhardina* or more so (Rose and Bown, 1991): its canine alveolus indicates a tooth relatively larger than in other omomyids except for *Teilhardina,* and it is the only other omomyid known to retain four lower premolars. In addition, its P_{3-4} are tall and relatively elongate compared to those of other omomyids, P_3 and the third molars are unreduced, and the cheek teeth show minimal basal inflation (the molars have relatively peripheral cusps), all presumably primitive traits relative to *Teilhardina.* Judging from alveoli, however, *Steinius vespertinus* apparently had a moderately enlarged central incisor, probably a derived condition relative to the most primitive *Teilhardina.* But except for its enlarged I_1, *Steinius* possesses a combination of extremely primitive dental features. The fact that *S. vespertinus* comes from about 400 m stratigraphically higher than *T. americana,* and is therefore about 2 MA younger, implies that an older substantial segment of the *Steinius* lineage is unknown.

Steinius and *Teilhardina* thus are close relatives near the base of Omomyidae. The existence of two such primitive omomyids in the early Wasatchian of North America, neither clearly derived from the other, raises the possibility that North American omomyids, like notharctids, may have evolved from more than one omomyid immigrant. It is likely that their common ancestor would be recognized as an omomyid rather than as an adapiform or some more generalized primitive euprimate, but these primitive omomyids cannot be far—at least in dental structure—from the euprimate stem.

Recent studies of omomyids suggest that the early diversification of the family was complex. Several—probably at least six or seven—separate lineages or clades originated from species that would probably be recognized as belonging in the genera *Teilhardina* or *Steinius*. Many of these are best represented in Wasatchian strata of the Bighorn Basin, Wyoming. It is now clear that antemolar dentitions usually provide more detailed phylogenetic information than molars in this family. Incisor size and orientation have been given considerable weight in assessing relationships. Existing evidence favors relatively small (i.e., I_1 and I_2 approximately equal in size, I_1 not enlarged), more or less vertical incisors as the primitive state for euprimates and suggests that hypertrophy of I_1 occurred repeatedly in different lineages of omomyids. Reversion of a procumbent, hypertrophied I_1 to the primitive condition seems much less probable (although there is evidence that this may have occurred in at least one lineage; see below). If these conclusions are accepted, several lineages characterized by small incisors (tribe Washakiini, *Absarokius*, *Anaptomorphus*) presumably emerged from the omomyid stem very early. Lower premolar size and structure are also significant data for deciphering omomyid phylogeny, whereas molars are often less useful because they have been conservative in many early Eocene omomyid lineages, usually differing only in minute details (Bown and Rose, 1987).

As a working hypothesis, lineages characterized by relatively longer and narrower P_{3-4} (less modified from the primitive euprimate state) may be traced to a species of *Steinius*, whereas those with more mesiodistally compressed or more molarized P_{3-4} probably originated from a species of *Teilhardina*. These criteria and molar similarity suggest that Bridgerian *Omomys* and its close relatives evolved from a species like *S. vespertinus* (Matthew, 1915; Bown and Rose, 1984; Honey, 1990; Rose and Bown, 1991). Similarly, molar details of the primitive uintaniin *Jemezius* suggest that the tribe Uintaniini may be a separate clade derived from a species of *Steinius* (Beard, 1987; Honey, 1990). In both cases, however, much of the close resemblance is in presumed primitive characters.

Other early lineages are probably descendants of *Teilhardina americana* or a close relative. Like *T. americana*, the anaptomorphine genera *Tetonius* and *Absarokius* exhibit mesiodistal compression of P_{3-4}, reduced third molars, and basal inflation of lower molar crowns. Both are derived relative to *T. americana* in having lost P_1 and reduced or lost P_2, and *Tetonius* is further derived in having an enlarged I_1 and more reduced canine. (Additional apomorphies characterize *Pseudotetonius ambiguus* and *Tatmanius szalayi*, more advanced members of the *Tetonius* clade.) Although geologically younger, later Wasatchian *Absarokius* is more primitive than *Tetonius* in having small lower incisors, a larger P_2, and a larger lower canine (Bown and Rose, 1987). Therefore, *Absarokius* cannot be derived from any known species of *Tetonius* (as was long believed), and its ultimate source must have been a very primitive species of *Teilhardina*. The synapomorphies of *Absarokius* and *Tetonius*, however, suggest that they share a more recent common ancestor, but they must have diverged very early in the Wasatchian.

Wasatchian *Chlororhysis* and Bridgerian *Anaptomorphus,* though poorly known, seem to belong to two additional independent lines emerging from a primitive species of *Teilhardina. Chlororhysis* is very similar to *T. americana* in premolar structure and canine size but is larger, lacks P_1, and has a weaker metaconid on P_4. It probably had a relatively small I_1. *Anaptomorphus* is more derived than *T. americana* in inflation of P_4 and the molars, in loss of P_{1-2}, and in larger size (about the size of *Tetonius* and *Absarokius*). It contrasts with *Tetonius* and resembles *Absarokius,* however, in having small incisors, which suggests an origin from a species of *Teilhardina* at least as primitive as *T. americana* and perhaps more so.

Late Wasatchian *Loveina* is the probable stem taxon of washakiin omomyids (Simpson, 1940; Szalay, 1976, Honey, 1990). Its mesiodistally short, stout P_3 and P_4, the latter with a well-developed trigonid, and molar characters suggest that it, too, evolved from a species of *Teilhardina* (Bown and Rose, 1984; Honey, 1990). I_1 was not significantly enlarged in *Loveina* (Simpson, 1940; Szalay, 1976) and was slightly smaller than I_2 in the Bridgerian washakiin *Washakius* (Covert and Williams, 1991) and in the Lostcabinian washakiin *Shoshonius* (K. C. Beard, *personal communication*). These facts indicate that the last common ancestor of Washakiini probably had relatively small incisors.

Relationships among the smallest Wasatchian omomyids are particularly difficult to resolve. Small size, an enlarged I_1, and molarization of P_{3-4} suggest that *Anemorhysis* (here considered to include *Tetonoides*) evolved from a species of *Teilhardina* more derived than *T. americana* (Bown and Rose, 1987). Recently described *Arapahovius advena* is closely similar to *Teilhardina crassidens, T. tenuiculus,* and *Anemorhysis pearcei;* hence, the source of *Arapahovius* seems to be from a species in the *Teilhardina–Anemorhysis* complex (Savage and Waters, 1978; Bown and Rose, 1991; Rose, in press). A possible obstacle to these proposed relationships, however, is the relatively large third molars of *Anemorhysis* and *Arapahovius.* If these genera descended from *Teilhardina,* their third molars must be secondarily enlarged compared to the somewhat reduced third molars of *Teilhardina.* Alternatively, the large third molars could be plesiomorphic, which would preclude origin from *Teilhardina,* implying descent from an unknown more primitive omomyid (not *Steinius,* which differs in dental structure) and substantial parallelism to derived species of *Teilhardina.* Such a hypothesis is implicit in the recent proposal by Beard *et al.* (1992), based on molar morphology, to expand the tribe Trogolemurini to include *Anemorhysis, Tetonoides,* and *Arapahovius* in addition to *Trogolemur.* At present it is uncertain which, if any, of these scenarios is correct, and the relationships of these small omomyids remain moot. Nonetheless, it is encouraging that considerable new fossil evidence of small omomyids has been discovered very recently, so clarification of their relationships may not be far off.

From the previous discussion the following conclusions emerge. The initial diversification of Omomyidae was more complex than is commonly believed, and the widely held division of nonmicrochoerine omomyids into two subfamilies, Anaptomorphinae and Omomyinae, is probably overly simplistic. Instead, this basal radiation is best represented at present as two unresolved polytomies (the *Steinius* and *Teilhardina* clades) that do not correspond to the traditional subfamilies as currently construed (e.g., Szalay, 1976; Fleagle, 1988). If Washakiini evolved from a species of *Teilhardina,* as is now believed, then Omomyinae in the current concept is at least diphyletic, since other lineages of Omomyinae more likely descended from species of *Steinius* (Bown and Rose, 1991; Rose and Bown, 1991).

Within Omomyidae, character reversal and widespread parallelism confound attempts to decipher relationships. Particular derived characters that must have evolved independently in two or more lineages include enlargement of I_1 (usually associated with crowding of anterior teeth), loss of one or more lower premolars, hypertrophy of P_4, molarization of P_4 (involving more distinct metaconid and paraconid or development of a talonid basin), reduction or enlargement of third molars, crenulation of enamel, and presence of a mesostyle. It is also possible that in one or more instances small size of I_1 was a derived state (either from a somewhat enlarged primitive state, as in *Purgatorius,* or as a reversal from a derived hypertrophied state).

The preceding remarks have been limited to early euprimates of Europe and North America. Primitive Omomyidae and Adapidae have also been reported from the early or middle Eocene of Pakistan (Russell and Gingerich, 1980, 1987), but the remains are so fragmentary that they provide little more than the oldest uncontested Asian records of euprimates. Other questionable records of primitive euprimates from Asia and Africa were alluded to earlier.

As we have mentioned, euprimates appear in the holarctic record abruptly at the beginning of the Eocene, their phylogenetic and geographic source largely speculative at present. Krause and Maas (1990) recently proposed that primates and some other "modern" orders may have originated in isolation in Africa or, more likely, on the Indian subcontinent following dispersal of the primitive eutherian stock from Africa to India before India rifted from Madagascar in the Late Cretaceous. According to this scenario, primates and other mammals of modern aspect would have spread into Holarctica after the Indian plate collided with Asia near the beginning of the Eocene. Appealing as this hypothesis may be, the discovery of *Altiatlasius* in the late Paleocene—if its age and identity are confirmed—seems to refute it and to support Africa as a more probable source, at least for euprimates.

The Origin of Anthropoidea

Recently described skulls of the washakiin omomyid *Shoshonius* may provide new evidence of a close relationship between omomyids and Tarsiiformes, which would have important implications for the early diversification of euprimates (Beard *et al.*, 1991; Beard and MacPhee, Chapter 3, this volume). According to Beard *et al.*, the skulls show that *Shoshonius*, like *Tarsius*, has very large orbits (the largest relative to skull size known among omomyids, but still smaller than in *Tarsius*), a very short snout, and several unique basicranial traits (ventrolateral location of posterior carotid foramen, basioccipital flange overlapping bulla, presence of suprameatal foramen). Based on these synapomorphies, Beard *et al.* (1991) concluded that *Shoshonius* is more closely allied with *Tarsius* than is any other known omomyid or anthropoid. (Skulls of *Loveina*, *Teilhardina*, and *Steinius* are unknown.) Assuming a monophyletic Haplorhini (= omomyids, tarsiids, and anthropoids), a close relationship between the omomyid *Shoshonius* and Tarsiidae implies that anthropoids must already have diverged from the tarsiiform clade by late Wasatchian time (Beard *et al.*, 1991).

Until very recently, however, the oldest known fossil anthropoids were from late Eocene deposits of North Africa and Burma (e.g., Simons, 1989), some 15 MA younger than *Shoshonius*. Sigé *et al.* (1990) suggested that late Paleocene *Altiatlasius* was the sister group of Anthropoidea, which implies that anthropoids were in existence by the late Paleocene; but the phylogenetic position of *Altiatlasius* is ambiguous. The recent discovery of diminutive bunodont euprimate teeth from the early or middle Eocene of Algeria, representing a primitive new anthropoid named *Algeripithecus minutus* (Godinot and Mahboubi, 1992), is the first demonstrable evidence that early Anthropoidea were penecontemporaneous with omomyids such as *Shoshonius*.

Not all workers accept the monophyly of Haplorhini; some consider adapiforms to be the sister group of Anthropoidea (e.g., Gingerich, 1981; Simons and Rasmussen, 1989; Rasmussen, 1990). But the search for anthropoidean ancestry should not necessarily be restricted to adapiforms or omomyids. The discovery of *Algeripithecus* in North Africa—strengthened by recent reports of a diversity of late Eocene anthropoids from North Africa (de Bonis *et al.* 1988; Simons, 1989, 1992)—is strongly suggestive that during the earliest episode of euprimate evolution, at least three collateral radiations were under way: Tarsiiformes, Adapiformes, and Anthropoidea. Should *Altanius* prove to be a euprimate, it could represent a fourth line. The common ancestor of all of these clades may have been a primitive euprimate that is not unequivocally referable to any of the named higher taxa.

Conclusions

Early Paleocene *Purgatorius* is a primitive plesidapiform and is the oldest and dentally most primitive mammal that has been regarded as a primate. The phylogenetic position of Plesiadapiformes is uncertain. Although they were long considered primates, evidence has increased that the closest dental resemblances between plesiadapiforms and the earliest euprimates must have evolved independently. Whether plesiadapiforms eventually prove to be the sister group of primates or more distantly related, they represent the most primate-like mammals known from the Paleocene.

Early Eocene *Altanius orlovi* from Mongolia is extremely primitive dentally and cannot with certainly be assigned to either Euprimates or Plesiadapiformes, although it possesses apparent synapomorphies with plesiomorphic members of both groups. Whatever its true affinities, *Altanius* must represent a primitive line that diverged very early from other members of its clade.

Altiatlasius koulchii is based on several isolated teeth from the late Paleocene of Morocco. *Altiatlasius* may well be the oldest known euprimate, but its precise affinities will remain uncertain until more complete specimens become available.

The oldest or most primitive undoubted euprimates are the adapiforms *Donrussellia* and *Cantius* and the omomyids *Teilhardina* and *Steinius,* all from the early Eocene. The early diversification of Omomyidae involved as-yet unresolved branching of five or more omomyid lineages or clades from species that would probably be included in *Teilhardina* and at least two from species that would be recognized as *Steinius*. Although known Anaptomorphinae appear to be monophyletic in origin (from a primitive species of *Teilhardina*), Omomyinae as currently conceived are at least diphyletic; monophyly in origin might be maintained for both subfamilies by transferring Washakiini from the Omomyinae to the Anaptomorphinae. Recent discovery of anthropoidean teeth (*Algeripithecus minutus*) in late early or middle Eocene sediments of North Africa suggests that Anthropoidea may represent a clade separate from both Omomyidae and Adapiformes. *Altiatlasius* is possibly related to this clade.

An assessment of the dental characters found in or shared by the most primitive adapiforms and omomyids leads to the following features likely to characterize the euprimate dental morphotype: small, more or less vertical incisors associated with a shallow anterior dentary; unreduced canine; four premolars, P_{2-4} two-rooted; P_{3-4} transversely narrow and moderately elongate, P_3 simple, P_4 with small, low paraconid (or no paraconid) and metaconid; P_3 and third molars not appreciably reduced; uninflated cheek teeth with peripheral cusps and smooth enamel; molar cusps moderately acute, paraconids lower than metaconids; *Nannopithex* fold and hypocone weak or absent; conules with pre- and postconule cristae; and upper molars with pre- and postcingula, probably incomplete lingually. This hypothesis will be tested by future discoveries. When primitive anthropoids are better known, modification of this hypothetical basal euprimate condition may become necessary.

Acknowledgments

We thank the organizers, John Fleagle, Richard Kay, and Elwyn Simons, for inviting us to participate in the symposium on which this volume is based.

We are grateful to Philip Gingerich for providing casts of the newly described specimens of *Altanius orlovi* and *Cantius torresi*. Specimens and casts of *Teilhardina belgica* were kindly made available by P. Gigase, G. Wouters, and the Royal Natural History Museum (Brussels); many other casts were generously supplied by D. E. Savage. The figures were expertly prepared by Elaine Kasmer. Commentary on an earlier draft by Chris Beard, David Krause, and two anonymous reviewers led to substantial improvements in the manuscript. This research has been supported by NSF grant DEB-8918755.

References

Adkins, R. M., and Honeycutt, R. L. 1991. Molecular phylogeny of the superorder Archonta. *Proc. Natl. Acad. Sci. USA* **88:**10317–10321.

Ammerman, L. K., an D. M. Hillis. 1992. A molecular test of bat relationships: Monophyly or diphyly? *Syst. Biol.* **41:**222–232.

Antunes, M. T., and Russell, D. E. 1981. Le gisement de Silveirinha (Bas Mondego, Portugal): La plus ancienne faune de vertébrés Éocènes connue en Europe. *C. R. Acad. Sci. Paris* **293:**1099–1102.

Bailey, W. J., Slightom, J. L., and Goodman, M. 1992. Rejection of the "flying primate" hypothesis by phylogenetic evidence from the ϵ-globin gene. *Science* **256:**86–89.

Baker, R. J., Novacek, M. J., and Simmons, N. B. 1991. On the monophyly of bats. *Syst. Zool.* **40:**216–231.

Beard, K. C. 1987. *Jemezius*, a new omomyid primate from the early Eocene of northwestern New Mexico. *J. Hum. Evol.* **16:**457–468.

Beard, K. C. 1989. Postcranial anatomy, locomotor adaptations, and paleoecology of early Cenozoic Plesiadapidae, Paromomyidae, and Micromomyidae (Eutheria, Dermoptera). Ph.D. dissertation, Johns Hopkins University, Baltimore.

Beard, K. C. 1990. Gliding behaviour and palaeoecology of the alleged primate family Paromomyidae (Mammalia, Dermoptera). *Nature* **345:**340–341.

Beard, K. C. 1991. Vertical postures and climbing in the morphotype of Primatomorpha: Implications for locomotor evolution in primate history. In: *Origine(s) de la Bipédie chez lez Hominidés*, pp. 79–87. Cahiers de Paléoanthropologie, CNRS, Paris.

Beard, K. C. 1993. Phylogenetic systematics of the Primatomorpha, with special reference to Dermoptera. In: F. S. Szalay, M. J. Novacek, and M. C. McKenna (eds.), *Mammal Phylogeny, Vol. 2, Placentals*, pp. 129–150. Springer-Verlag, New York.

Beard, K. C., Krishtalka, L., and Stucky, R. K. 1991. First skulls of the early Eocene primate *Shoshonius cooperi* and the anthropoid–tarsier dichotomy. *Nature* **349:**64–67.

Beard, K. C., Krishtalka, L., and Stucky, R. K. 1992. Revision of the Wind River faunas, early Eocene of central Wyoming. Part 12. New species of omomyid primates (Mammalia: Primates: Omomyidae) and omomyid taxonomic composition across the early–middle Eocene boundary. *Ann. Carnegie Mus.* **61:**39–62.

Bonis, L. de, Jaeger, J.-J., Coiffait, B., and Coiffait, P.-E. 1988. Découverte du plus ancien primate Catarrhinien connu dans l'Éocène supérieur d'Afrique du Nord. *C. R. Acad. Sci. Paris* **306**(ser. II):929–934.

Bown, T. M., and Rose, K. D. 1984. Reassessment of some early Eocene Omomyidae, with description of a new genus and three new species. *Fol. Primatol.* **43:**97–112.

Bown, T. M., and Rose, K. D. 1987. Patterns of dental evolution in Early Eocene anaptomorphine primates (Omomyidae) from the Bighorn Basin, Wyoming. *Paleont. Soc. Mem.* **23:**1–162.

Bown, T. M., and Rose, K. D. 1991. Evolutionary relationships of a new genus and three new species of omomyid primates (Willwood Formation, Lower Eocene, Bighorn Basin, Wyoming). *J. Hum. Evol.* **20:**465–480.

Capetta, H., Jaeger, J.-J., Sigé, B., Sudre, J., and Vianey-Liaud, M. 1987. Compléments et précisions biostratigraphiques sur la faune Paléocène à mammifères et sélaciens du Bassin d'Ouarzazate (Maroc). *Tertiary Res.* **8:**147–157.

Carlsson, A. 1922. Über die Tupaiidae und ihre Beziehungen zu den Insectivore und den Prosimiae. *Acta Zool.* **3:**227–270.

Cartmill, M. 1974. Rethinking primate origins. *Science* **184:**436–443.

Clemens, W. A. 1974. *Purgatorius,* an early paromomyid primate (Mammalia). *Science* **184:**903–905.

Covert, H. H. 1988. Ankle and foot morphology of *Cantius mckennai:* Adaptations and phylogenetic implications. *J. Hum. Evol.* **17:**57–70.

Covert, H. H., and Williams, B. A. 1991. The anterior lower dentition of *Washakius insignis* and adapid–anthropoidean affinities. *J. Hum. Evol.* **21:**463–467.

Dagosto, M. 1983. Postcranium of *Adapis parisiensis* and *Leptadapis magnus* (Adapiformes, Primates). *Fol. Primatol.* **41:**49–101.

Dagosto, M. 1988. Implications of postcranial evidence for the origin of euprimates. *J. Hum. Evol.* **17:**35–56.

Dashzeveg, D., and McKenna, M. C. 1977. Tarsioid primate from the early Tertiary of the Mongolian People's Republic. *Acta Palaeont. Pol.* **22:**119–137.

Estravis, C., and Russell, D. E. 1989. Découverte d'un nouveau *Diacodexis* (Artiodactyla, Mammalia) dans l'Éocène inférieur de Silveirinha, Portugal. *Palaeovertebrata* **19:**29–44.

Fleagle, J. G. 1988. *Primate Adaptation and Evolution.* Academic Press, New York.

Franzen, J. L. 1987. Ein neuer Primate aus dem Mitteleozän der Grube Messel (Deutschland, S-Hessen). *Cour. Forsch. Inst. Senckenberg* **91:**151–187.

Gebo, D. L. 1986. Anthropoid origins—the foot evidence. *J. Hum. Evol.* **15:**421–430.

Gebo, D. L. 1988. Foot morphology and locomotor adaptation in Eocene primates. *Fol. primatol.* **50:**3–41.

Gingerich, P. D. 1975. A new genus of Adapidae (Mammalia, Primates) from the late Eocene of southern France, and its significance for the origin of higher primates. *Contrib. Mus. Paleont. Univ. Michigan* **24:**163–170.

Gingerich, P. D. 1976. Cranial anatomy and evolution of early Tertiary Plesiadapidae (Mammalia, Primates). *Univ. Michigan Pap. Paleont.* **15:**1–140.

Gingerich, P. D. 1977a. New species of Eocene primates and the phylogeny of European Adapidae. *Fol. Primatol.* **28:**60–80.

Gingerich, P. D. 1977b. Dental variation in early Eocene *Teilhardina belgica,* with notes on the anterior dentition of some early tarsiiformes. *Fol. Primatol.* **28:**144–153.

Gingerich, P. D. 1981. Early Cenozoic Omomyidae and the evolutionary history of tarsiiform primates. *J. Hum. Evol.* **10:**345–374.

Gingerich, P. D. 1986. Early Eocene *Cantius torresi*—oldest primate of modern aspect from North America. *Nature* **320:**319–321.

Gingerich, P. D., Dashzeveg, D., and Russell, D. E. 1991. Dentition and systematic relationships of *Altanius orlovi* (Mammalia, Primates) from the early Eocene of Mongolia. *Geobios* **24:**637–646.

Godinot, M. 1978. Un nouvel adapidé (primate) de l'Éocène inférieur de Provence. *C. R. Acad. Sci. Paris* **286:**1869–1872.

Godinot, M. 1981. Les mammifères de Rians (Éocène inférieur, Provence). *Palaeovertebrata* **10:**43–126.

Godinot, M. 1988a. Dental morphology of *Donrussellia,* the most primitive known adapiform (Primates). *J. Vert. Paleont.* **8**(3, suppl.):15A–16A.

Godinot, M. 1988b. Les primates adapidés de Bouxwiller (Éocène Moyen, Alsace) et leur apport à la compréhension de la faune de Messel et à l'évolution des Anchomomyini. *Cour. Forsch. Inst. Senckenberg* **107:**383–407.

Godinot, M. 1992a. Apport à la systématique de quatre genres d'Adapiformes (Primates, Éocène). *C. R. Acad. Sci. Paris* **314**(ser. II):237–242.

Godinot, M. 1992b. Early euprimate hands in evolutionary perspective. *J. Hum. Evol.* **22:**267–283.

Godinot, M., and Beard, K. C. 1991. Fossil primate hands: A review and an evolutionary inquiry emphasizing early forms. *Hum. Evol.* **6**:307–354.

Godinot, M., and Mahboubi, M. 1992. Earliest known simian primate found in Algeria. *Nature* **357**:324–326.

Godinot, M., Crochet, J.-Y., Hartenberger, J.-L., Lange-Badré, B., Russell, D. E., and Sigé, B. 1987. Nouvelles données sur les mammifères de Palette (Éocène inférieur, Provence). *Munch. Geowiss. Abh.* **10**:273–288.

Goodman, M., Czelusniak, J., and Beeber, J. E. 1985. Phylogeny of Primates and other eutherian orders: a cladistic analysis using amino acid and nucleotide sequence data. *Cladistics* **1**:171–185.

Greenwald, N. S. 1991. Primate–bat relationships and cladistic analysis of morphological data. *Am. J. Phys. Anthrop.* [*Suppl.*] **12**:82.

Gregory, W. K. 1910. The orders of mammals. *Bull. Am. Mus. Nat. Hist.* **27**:1–524.

Gregory, W. K. 1920. On the structure and relations of *Notharctus*, an American Eocene primate. *Mem. Am. Mus. Nat. Hist.* [*new ser.*] **3**:51–243.

Gunnell, G. F. 1989. Evolutionary history of Microsyopoidea (Mammalia, ?Primates) and the relationship between Plesiadapiformes and Primates. *Univ. Michigan Pap. Paleont.* **27**:1–157.

Heller, F. 1935. *Amphilemur eocaenicus* n. g. et n. sp., ein primitiver Primate aus dem Mitteleozän des Geiseltales bei Halle a. S. *Nova Acta Leopoldina N.F.* **2**:293–300.

Hoffstetter, R. 1977. Phylogénie des primates. Confrontation des résultats obtenus par les diverses voies d'approche du problème. *Bull. Mem. Soc. Anthrop. Paris* **4**:327–346.

Hoffstetter, R. 1982. Les primates simiiformes (= Anthropoidea) (compréhension, phylogénie, histoire biogéographique). *Ann. Paleont.* **68**:241–290.

Hoffstetter, R. 1988. Origine et évolution des primates non humains du nouveau monde. In: *L'Évolution dans sa Réalité et ses Diverses Modalités. Colloque international, Paris, 1985, Fondation Singer-Polignac*, pp. 133–170. Masson, Paris.

Honey, J. G. 1990. New washakiin primates (Omomyidae) from the Eocene of Wyoming and Colorado, and comments on the evolution of the Washakiini. *J. Vert. Paleont.* **10**:206–221.

Hunt, R. M., Jr., and Korth, W. W. 1980. The auditory region of Dermoptera: Morphology and function relative to other living mammals. *J. Morphol.* **164**:167–211.

Hürzeler, J. 1946. *Gesneropithex peyeri* nov. gen. nov. spec., ein neuer Primate aus dem Ludien von Gosden (Solothurn). *Ecologae Geol. Helv.* **39**:354–361.

Hürzeler, J. 1947. *Alsaticopithecus leemanni* nov. gen. nov. spec., ein neuer Primate aus dem unteren Lutétien von Buchsweiler im Unterelsass. *Eclogae Geol. Helv.* **40**:343–356.

Kay, R. F., Thorington, R. W., Jr., and Houde, P. 1990. Eocene plesiadapiform shows affinities with flying lemurs not primates. *Nature* **345**:342–344.

Kay, R. F., Thewissen, J. G. M., and Yoder, A. D. 1992. Cranial anatomy of *Ignacius graybullianus* and the affinities of the Plesiadapiformes. *Am. J. Phys. Anthrop.* **89**:477–498.

Kielan-Jaworowska, Z., Bown, T. M., and Lillegraven, J. 1979. Eutheria. In: J. A. Lillegraven, Z. Kielan-Jaworowska, and W. A. Clemens (eds.), *Mesozoic Mammals*, pp. 221–258. University of California Press, Berkeley.

Koenigswald, W. V., and Storch, G. 1983. *Pholidocercus hassiacus*, ein Amphilemuride aus dem Eozän des "Grube Messel" bei Darmstadt (Mammalia, Lipotyphla). *Senckenbergiana Lethaea* **64**:447–495.

Krause, D. W. 1991. Were paromomyids gliders? Maybe, maybe not. *J. Hum. Evol.* **21**:177–188.

Krause, D. W., and Maas, M. C. 1990. The biogeographic origins of late Paleocene–early Eocene mammalian immigrants to the Western Interior of North America. In: T. M. Bown and K. D. Rose (eds.), *Dawn of the Age of Mammals in the northern part of the Rocky Mountain Interior, North America*, pp. 71–105. *Geol. Soc. Am. Spec. Pap.* 243.

Le Gros Clark, W. E. 1924a. The myology of the tree-shrew (*Tupaia minor*). *Proc. Zool. Soc. Lond.* **1924**:461–497.

Le Gros Clark, W. E. 1924b. On the brain of the tree-shrew (*Tupaia minor*). *Proc. Zool. Soc. Lond.* **1924**:1053–1074.

Le Gros Clark, W. E. 1925. On the skull of *Tupaia*. *Proc. Zool. Soc. Lond.* **1925**:559–567.

Le Gros Clark, W. E. 1933. The brain of the Insectivora. *Proc. Zool. Soc. Lond.* **1933**:975–1013.

Luckett, W. P. (ed.). 1980. *Comparative Biology and Evolutionary Relationships of Tree Shrews*. Plenum Press, New York.

MacPhee, R. D. E. 1981. Auditory regions of primates and eutherian insectivores. *Contrib. Primatol.* **18:**1–282.

MacPhee, R. D. E., and Cartmill, M. 1986. Basicranial structures and primate systematics. In: D. Swindler and J. Erwin (eds.), *Comparative Primate Biology, Vol. 1, Systematics, Evolution, and Anatomy,* pp. 219–275. Alan R. Liss, New York.

MacPhee, R. D. E., Cartmill, M., and Gingerich, P. D. 1983. New Palaeogene primate basicrania and the definition of the order Primates. *Nature* **301:**509–511.

MacPhee, R. D. E., Novacek, M. J., and Storch, G. 1988. Basicranial morphology of early Tertiary erinaceomorphs and the origin of primates. *Am. Mus. Novit.* **2921:**1–42.

Martin, R. D. 1968. Towards a new definition of Primates. *Man* **3:**377–401.

Martin, R. D. 1986. Primates: A definition. In: B. Wood, L. Martin, and P. Andrews (eds.), *Major Topics in Primate and Human Evolution,* pp. 1–31. Cambridge University Press, Cambridge.

Martin, R. D. 1990. *Primate Origins and Evolution.* Princeton University Press, Princeton.

Matthew, W. D. 1915. A revision of the Lower Eocene Wasatch and Wind River faunas. Part IV. Entelonychia, Primates, Insectivora (part). *Bull. Am. Mus. Nat. Hist.* **34:**429–483.

McKenna, M. C. 1975. Toward a phylogenetic classification of the Mammalia. In: W. P. Luckett and F. S. Szalay (eds.), *Phylogeny of the Primates,* pp. 21–46. Plenum Press, New York.

Miyamoto, M. M., and Goodman, M. 1986. Biomolecular systematics of eutherian mammals: Phylogenetic patterns and classification. *Syst. Zool.* **35:**230–240.

Novacek, M. J. 1989. Higher mammal phylogeny: The morphological–molecular synthesis. In: B. Fernholm, K. Bremer, and H. Jörnvall (eds.), *The Hierarchy of Life,* pp. 421–435. Elsevier, Amsterdam.

Novacek, M. J. 1992. Mammalian phylogeny: Shaking the tree. *Nature* **356:**121–125.

Novacek, M. J., and Wyss, A. R. 1986. Higher-level relationships of the recent eutherian orders: Morphological evidence. *Cladistics* **2:**257–287.

Novacek, M. J., McKenna, M. C., Neff, N. A., and Cifelli, R. L. 1983. Evidence from earliest known erinaceomorph basicranium that insectivorans and primates are not closely related. *Nature* **306:**683–684.

Novacek, M. J., Bown, T. M., and Schankler, D. 1985. On the classification of the early Tertiary Erinaceomorpha (Insectivora, Mammalia). *Am. Mus. Novit.* **2813:**1–22.

Pettigrew, J. D. 1986. Flying primates? Megabats have the advanced pathway from eye to midbrain. *Science* **231:**1304–1306.

Pettigrew, J. D. 1991. Wings or brain? Convergent evolution in the origin of bats. *Syst. Zool.* **40:**199–216.

Pettigrew, J. D., Jamieson, B. G. M., Robson, S. K., Hall, L. S., McNally, K. I., and Cooper, H. M. 1989. Phylogenetic relations between microbats, megabats and primates (Mammalia: Chiroptera and Primates). *Phil. Trans. R. Soc. Lond. [Biol.]* **325:**489–559.

Rasmussen, D. T. 1990. The phylogenetic position of *Mahgarita stevensi:* Protoanthropoid or lemuroid? *Int. J. Primatol.* **11:**439–469.

Rose, K. D. 1981a. The Clarkforkian Land-Mammal Age and mammalian faunal composition across the Paleocene–Eocene boundary. *Univ. Michigan Pap. Paleont.* **26:**1–197.

Rose, K. D. 1981b. Composition and species diversity in Paleocene and Eocene mammal assemblages: An empirical study. *J. Vert. Paleont.* **1:**367–388.

Rose, K. D. 1994. Anterior dentition and relationships of the early Eocene omomyids *Arapahovius advena* and *Teilhardina demissa,* sp. nov. *J. Human Evol.,* in press.

Rose, K. D., and Bown, T. M. 1991. Additional fossil evidence on the differentiation of the earliest euprimates. *Proc. Natl. Acad. Sci. USA* **88:**98–101.

Rose, K. D., and Krause, D. W. 1984. Affinities of the primate *Altanius* from the early Tertiary of Mongolia. *J. Mammal.* **65:**721–726.

Russell, D. E., and Gingerich, P. D. 1980. Un nouveau primate omomyide dans l'Éocene du Pakistan. *C. R. Acad. Sci. Paris* **291:**621–624.

Russell, D. E., and Gingerich, P. D. 1987. Nouveaux primates de l'Éocène du Pakistan. *C. R. Acad. Sci. Paris* **304**(II):209–214.

Russell, D. E., Louis, P., and Savage, D. E. 1967. Primates of the French early Eocene. *Univ. Calif. Publ. Geol. Sci.* **73:**1–46.

Russell, D. E., Louis, P., and Savage, D. E. 1975. Les Adapisoricidae de l'Éocene Inférieur de France. Réévaluation des formes considérées affines. *Bull. Mus. Nat. Hist. Nat. ser. 3* (327), *Sci. Terre* **45:**129–193.

Russell, D. E., Galoyer, A., Louis, P., and Gingerich, P. D. 1988. Nouveaux vertébrés sparnaciens du Conglomérat de Meudon à Meudon, France. *C. R. Acad. Sci. Paris* **307:**429–433.

Savage, D. E., and Waters, B. T. 1978. A new omomyid primate from the Wasatch Formation of southern Wyoming. *Fol. Primatol.* **30:**1–29.

Schmidt-Kittler, N. (ed.). 1987. International symposium on mammalian biostratigraphy and paleoecology of the European Paleogene—Mainz, February 18th–21st 1987. Münchner *Geowiss. Abh. A* vol. 10, 311 pp.

Schwartz, J. H., Tattersall, I., and Eldredge, N. 1978. Phylogeny and classification of the primates revisited. *Yearb. Phys. Anthrop.* **21:**95–133.

Shoshani, J. 1986. Mammalian phylogeny: Comparison of morphological and molecular results. *Mol. Biol. Evol.* **3:**222–242.

Sigé, B., Jaeger, J.-J., Sudre, J., and Vianey-Liaud, M. 1990. *Altiatlasius koulchii* n. gen. et sp., primate omomyidé du Paléocène supérieur du Maroc, et les origines des Euprimates. *Palaeontographica* [*A*] **214:**31–56.

Simons, E. L. 1962. A new Eocene primate genus, *Cantius,* and a revision of some allied European lemuroids. *Bull. Br. Mus. Nat. Hist.* **7:**1–36.

Simons, E. L. 1972. *Primate Evolution.* Macmillan, New York.

Simons, E. L. 1989. Description of two genera and species of late Eocene Anthropoidea from Egypt. *Proc. Natl. Acad. Sci. USA* **86:**9956–9960.

Simons, E. L. 1992. Diversity in the early Tertiary anthropoidean radiation in Africa. *Proc. Natl. Acad. Sci. USA* **89:**10743–10747.

Simons, E. L., and Rasmussen, D. T. 1989. Cranial morphology of *Aegyptopithecus* and *Tarsius* and the question of the tarsier–anthropoidean clade. *Am. J. Phys. Anthrop.* **79:**1–23.

Simpson, G. G. 1940. Studies on the earliest primates. *Bull. Am. Mus. Nat. Hist.* **77:**185–212.

Simpson, G. G. 1945. The principles of classification and a classification of the mammals. *Bull. Am. Mus. Nat. Hist.* **85:**1–350.

Stehlin, H. G. 1912. Die Säugetiere des schweizerischen Eocaens. Critischer Catalog der Materialen. *Abh. Schweiz. Palaont. Ges.* **38:**1165–1298.

Szalay, F. S. 1971. Relationships of the alleged primate *Gesneropithex peyeri* Hürzeler, 1946. *J. Mammal.* **52:**824–826.

Szalay, F. S. 1976. Systematics of the Omomyidae (Tarsiiformes, Primates). Taxonomy, phylogeny, and adaptations. *Bull. Am. Mus. Nat. Hist.* **156:**157–450.

Szalay, F. S. 1977. Phylogenetic relationships and a classification of the eutherian Mammalia. In: M. K. Hecht, P. C. Goody, and B. M. Hecht (eds.), *Major Patterns in Vertebrate Evolution,* pp. 315–374. Plenum Press, New York.

Szalay, F. S. 1982. A critique of some recently proposed Paleogene primate taxa and suggested relationships. *Fol. Primatol.* **37:**153–162.

Szalay, F. S., and Dagosto, M. 1988. Evolution of hallucial grasping in the primates. *J. Hum. Evol.* **17:**1–33.

Szalay, F. S., and Delson, E. 1979. *Evolutionary History of the Primates.* Academic Press, New York.

Szalay, F. S., and Li, C.-K. 1986. Middle Paleocene euprimate from southern China and the distribution of primates in the Paleogene. *J. Hum. Evol.* **15:**387–397.

Szalay, F. S., Tattersall, I., and Decker, R. L. 1975. Phylogenetic relationships of *Plesiadapis*—postcranial evidence. *Contrib. Primatol.* **5:**136–166.

Szalay, F. S., Rosenberger, A. L., and Dagosto, M. 1987. Diagnosis and differentiation of the order Primates. *Yearb. Phys. Anthrop.* **30:**75–105.

Thewissen, J. G. M., and Babcock, S. K. 1991. Distinctive cranial and cervical innervation of wing muscles: New evidence for bat monophyly. *Science* **251:**934–936.

Tong, Y.-S. 1979. A late Paleocene primate from S. China. *Vert. Palas.* **17:**65–70.

Van Valen, L., and Sloan, R. E. 1965. The earliest primates. *Science* **150:**743–745.

Wible, J. R., and Covert, H. H. 1987. Primates: Cladistic diagnosis and relationships. *J. Hum. Evol.* **16:**1–22.

Wible, J. R., and Novacek, M. J. 1988. Cranial evidence for the monophyletic origin of bats. *Amer. Mus. Novit.* **2911:**1–19.

Wyss, A. R., Novacek, M. J., and McKenna, M. C. 1988. Amino acid sequence versus morphological data and the interordinal relationships of mammals. *Mol. Biol. Evol.* **4:**99–116.

Xu, Q. 1976. New materials of Anagalidae from the Paleocene of Anhui (B). *Vert. Palas.* **14:**242–251.

Xu, Q. 1977. Two new genera of old Ungulata from the Paleocene of Qianshan Basin, Anhui. *Vert. Palas.* **15:**119–125.

Recently Recovered Specimens of North American Eocene Omomyids and Adapids and Their Bearing on Debates about Anthropoid Origins

2

HERBERT H. COVERT
and BLYTHE A. WILLIAMS

Introduction

Most recent reviews of primate evolution recognize that modern primates can be divided into two monophyletic suborders, the Strepsirhini (lemuriforms and lorisiforms) and the Haplorhini (tarsiers, monkeys, apes, and humans). It is also accepted by most researchers that the radiation of modern primate taxa

HERBERT H. COVERT • Department of Anthropology, University of Colorado, Boulder, Colorado 80309-0233. BLYTHE A. WILLIAMS • Department of Anthropology, University of Colorado, Boulder, Colorado 80309. *Present address:* Department of Biological Anthropology and Anatomy, Duke University, Durham, North Carolina 27710.
Anthropoid Origins, edited by John G. Fleagle and Richard F. Kay. Plenum Press, New York, 1994.

dates to at least the beginning of the Eocene. The earliest occurring animals that appear to be seeded within this radiation belong to the families Adapidae and Omomyidae, the Eocene euprimates (Hoffstetter, 1974; Szalay and Delson, 1979; Covert, 1986; MacPhee and Cartmill, 1986; Martin, 1986; Wible and Covert, 1987; and Szalay *et al.*, 1987).

It has been noted for decades that adapids resemble extant lemurids and indriids (and have been termed lemur-like), and omomyids resemble tarsiers (and have been termed tarsier-like) (e.g., Gregory, 1915, 1920; Simons, 1961). It has also been argued that these resemblances indicate broad phylogenetic relationships. However, the phylogenetic significance of these resemblances is frequently debated. For example, Cartmill and Kay (1978) offered tentative support for an omomyid–tarsiid link based primarily on basicranial data. They also argued that extant tarsiids and anthropoids are probably more closely related to one another than either is to any known omomyid. They found no evidence linking adapids with extant strepsirhines. In contrast, Gingerich and Schoeninger (1977) suggested that, on the basis of basicranial and dental features, adapids are ancestral to both anthropoids and strepsirhines. They accepted the omomyid–tarsiid relationship. More recently, Rasmussen (1986) has argued that omomyids are in the ancestry of tarsiids, that adapids are in the ancestry of anthropoids, and that there are no known derived characteristics linking adapids with extant strepsirhines to the exclusion of other taxa. However, several other authors (e.g., Gebo, 1987, 1988; Beard *et al.,* 1988 Covert, 1988; Dagosto, 1988; Dagosto and Gebo, Chapter 17, this volume) have noted that there are indeed postcranial synapomorphies that unite adapids and extant strepsirhines. Beard *et al.* (1991; Beard and MacPhee, Chapter 3, this volume) have argued that the early Eocene North American omomyid *Shoshonius* shares a number of basicranial traits uniquely with tarsiers and that these traits indicate that these animals are more closely related to one another than either is to anthropoids.

In this chapter we address the pertinent dental and postcranial characters that have been used to (1) link adapids with anthropoids, (2) link adapids with extant strepsirhines, and (3) link tarsiids with anthropoids. We also discuss the implications of newly recovered material that may shed light on the question of anthropoid origins.

Utility of the Terms Lemur-like and Tarsier-like

As noted above, adapids have often been referred to as the "lemur-like" Eocene primates, and omomyids have often been referred to as the "tarsier-like" Eocene primates. Research on both extant and extinct species during the past 20 years or so has rendered these terms essentially meaningless and even misleading. As reviewed by Fleagle (1988), extant lemuroids show a great deal of adaptive diversity. These animals range in size from about 600 g (*Lepilemur*) to about 10,000 g (*Indri*). They include nocturnal, diurnal, and cathemeral forms. Some are frugivores, others folivores, and one is a specialized insectivore (*Daubentonia*). The vast majority of these creatures are arboreal, with *Lemur catta* being the only species that spends much time on the ground. Locomotor adaptations include the hindlimb-dominated acrobatic leaping of the indriids and *Lepilemur,* the quadrupedal branch-running and leaping of the members of *Lemur,* and the deliberate arboreal quadrupedalism of *Daubentonia.* A consideration of the adaptive diversity shown by these creatures makes the term "lemur-like" such a broad category that it provides no specific information about the adaptive profile of any single organism. To call something "tarsier-like" creates a different sort of problem. As reviewed by Fleagle (1988), modern tarsiers are highly specialized creatures. They are small-bodied (80 to 150 g), nocturnal, faunivorous, and vertical-clinging and leaping creatures. Thus, labeling an extinct organism tarsier-like implies that it has a similar highly specialized adaptive profile.

Recent research on the omomyids and adapids has yielded a great deal of information about their adaptations. For example, *Smilodectes gracilis* was a moderate-sized (approximately 2000 g), diurnal, folivorous, hindlimb-dominated acrobatic leaping middle Eocene North American adapid (Covert, 1986). Although each of these adaptive attributes fits comfortably within the lemur-like category, simply describing *S. gracilis* as lemur-like does not accurately reflect what we know about its anatomy and inferred adaptations. Dagosto (1983) demonstrated that *Adapis parisiensis,* a late Eocene European adapid, had locomotor specializations most similar to extant lorisines. Thus, even with the realization that the term lemur-like includes a wide range of locomotor adaptations, some adapids fall beyond these limits. In addition, Gebo *et al.* (1991) have provided a list of features of the North American adapid tarsals that distinguish them from all lemurs, indicating that these animals did not grasp with their feet in the same fashion as do the lemurs and that these Eocene creatures are not particularly closely related to the extant lemurs. At present, adaptive profiles are known for only a handful of the omomyids, and none are like that of *Tarsius.* The late Eocene European *Necrolemur* comes closest to being tarsier-like in being nocturnal and sharing a number of leaping specializations in the hindlimb and foot with *Tarsius.* It was, however, probably frugivorous instead of faunivorous (Covert, 1986). *Absarokius abbotti,* an early Eocene North American omomyid, was a small arboreal animal (probably around 125 g) that was adept at climbing, running, and leaping. It was most likely frugivorous, and its overall adaptive profile was probably much more similar to that of a mouse lemur than to that of a tarsier (Covert and Hamrick, 1994).

In sum, as our knowledge about the adaptive diversity of extant and extinct primates has grown, it has become clear that the terms "lemur-like" and "tarsier-like" are inappropriate to apply to adapids and omomyids. The lemuroids are an adaptively diverse group, and thus "lemur-like" is a vague term to apply to an extinct organism. Although some adapids do fall within this broad adaptive category, at least one, *Adapis parisiensis,* does not. *Tarsius* is a highly specialized creature, and thus "tarsier-like" is a very specific term and one that does not aptly describe any known omomyid. Probably the most pertinent aspect of this information for questions of anthropoid origins is the realization that omomyids are not characterized by the hindlimb or anterior dentition specializations shown by *Tarsius.* Omomyids cannot be argued to be too specialized to be within the ancestry of anthropoid primates based on features of these systems.

New Early Eocene Primate Material

Since 1987 we have been collecting early Eocene vertebrate remains in the Washakie Basin of south central Wyoming for the University of Colorado Museum. These deposits preserve fossils dating to the early, early Eocene (Graybullian subage of the Wasatchian), to the middle early Eocene (late Graybullian or early Lysitean subage), and to the late early Eocene (Lostcabinian subage). Omomyid and adapid dental and postcranial remains have been recovered from each of these stratigraphic levels (Table I). This postcranial material is of special interest here because the skeletons of most early Eocene primates are extremely poorly known and because it appears that omomyids and adapids differ significantly in a number of skeletal features. Additionally, these features appear to be of phylogenetic importance in debates about primate sub- and infraordinal relationships.

None of our newly recovered primate postcranial material was found in direct association with dental remains. We have followed traditional methods of allocating this material to a given taxon. At each of our localities primate dental remains are more common than skeletal remains. We have thus assumed that the skeletal remains can be attributed to one of the taxa represented by dental remains at a given locality. At the locality level this decision is based on size factors. For example, at Lostcabinian localities we have recovered dentitions of at least four omomyid taxa (*Absarokius abbotti, Trogolemur* cf. *myodes, Loveina minuta,* and *Chlorohysis* n. sp.), with *Absarokius abbotti* being significantly larger than the other three species. Omomyid skeletal elements in these deposits come in two sizes. Thus, the larger material is attributed to *Absarokius,* and we are unable to attribute the smaller specimens with any degree of certainty to any one of the three smaller taxa. Unless dental and postcranial material are collected in association with one another, we cannot be absolutely sure of the taxonomic attribution at the species or even the genus level of specific skeletal elements. But in this analysis this is a moot point as long as we have correctly distinguished adapids and omomyids. Because the members of these families differ significantly in size from one another in the Wasatch of Wyoming, there can be no doubt about our attribu-

Table I. Primate Postcranial Specimens from the Washakie Basin

Lower Graybull specimens
Cantius mckennai
1 talus UCM 60940
2 distal phalanges UCM 56856, 56857
Tetonoides n.sp.
3 tali UCM 56853, 56854, 56855
3 calcanei UCM 56852
Late Graybull—Early Lysite specimens
Copelemur praetutus and/or *Cantius trigonodus*
3 tali UCM 62605, 67910, 67911
2 calcanei UCM 60944, 67912
1 fifth metatarsal UCM 60945
Tetonoides pearcei
2 tali UCM 56892, 60901
1 navicular UCM 56893
Lysite
Copelemur australotutus
2 tali UCM 65075, 65394
1 calcaneus USM 60937
Adapid
1 calcaneus UCM 60941
1 hallucial metatarsal UCM 60941
1 distal radius UCM 60941
Arapahovius gazini
1 proximal femur UCM 60946
1 distal femur uncatalogued
1 calcaneus UCM 67850
Lostcabin
Cantius venticolus
3 tali UCM 60918, 60920, 62671
1 capitate UCM 59783
1 cuboid UCM 60919
1 entocuneiform UCM 59781
1 distal humerus UCM 65214
Small adapid
1 calcaneus UCM 62662
2 entocuneiforms UCM 59782, 62659
1 capitate UCM 62661
1 distal phalanx UCM 60948
Absarokius abbotti
1 talus UCM 62681
2 calcanei UCM 58954, 67907
2 naviculars UCM 60934, 67908
1 tibia UCM 62673
1 distal humerus UCM 62672
Small omomyid
1 talus uncataloged
3 calcanei UCM 58658, 62698
1 distal humerus uncatalogued

tions at the family level. It should also be noted here that few, if any, Wasatchian primates have been argued to be in the direct ancestry of anthropoids or strepsirhines in the recent literature, and we certainly do not argue otherwise. This new material is important, however, because it provides us with a much better understanding of the distribution of a number of morphological traits that have been argued to be of taxonomic importance among the adapids and omomyids.

The Adapidae–Omomyidae Paradox

The following phylogenies concerning anthropoid origins have been suggested in the literature: an omomyid origin for anthropoids and tarsiers (Fig. 1A), an adapid origin for anthropoids and an omomyid origin for tarsiers (Fig. 1B), and the origin of anthropoids from an unsampled primate group (Fig. 1C).

Gingerich (1981a) has noted that combined employment of the various lines of evidence (ontogenetic, biomolecular, comparative anatomy, and paleontology) to understand anthropoid origins results in a paradox. This paradox has been outlined by Rasmussen (1986) as follows. (1) Comparative anatomy and biomolecular studies of modern primates provide strong evidence for linking tarsiids with anthropoids in a monophyletic group versus the "toothcombed prosimians" (referred to in this paper as the extant strepsirhines). (2) Paleontological evidence suggests that omomyid primates dating from the Eocene are probably ancestral to the living tarsiids. (3) Paleontological evidence suggests that adapid primates are probably ancestral to living strepsirhines. (4) Paleontological evidence suggests that adapids are probably ancestral to living anthropoids. As Gingerich (1981a) and Rasmussen (1986) noted, some of these ideas are contradictory, and if each is accepted, it is not possible to develop an internally consistent evolutionary scenario. Specifically, if tarsiids and anthropoids form a monophyletic group, and omomyids are in the ancestry of tarsiids, adapids cannot be ancestral to both anthropoids and living strepsirhines.

Rasmussen (1986) accepted the evidence outlined by neontologists that tarsiids and anthropoids are a monophyletic group. He argued (citing Cartmill and Kay, 1978) that the evidence for a close relationship between adapids and extant strepsirhines is lacking. He further argued (citing Gingerich and Schoeninger, 1977) that there is good evidence for a close relationship between adapids and extant anthropoids.

Among the traits cited as evidence of an exclusive Adapidae–Anthropoidea relationship are seven dental and postcranial characteristics (Table II) (Gingerich and Schoeninger, 1977; Gingerich, 1980, 1981a; Rasmussen, 1986). These traits have been argued to be absent among the Omomyidae.

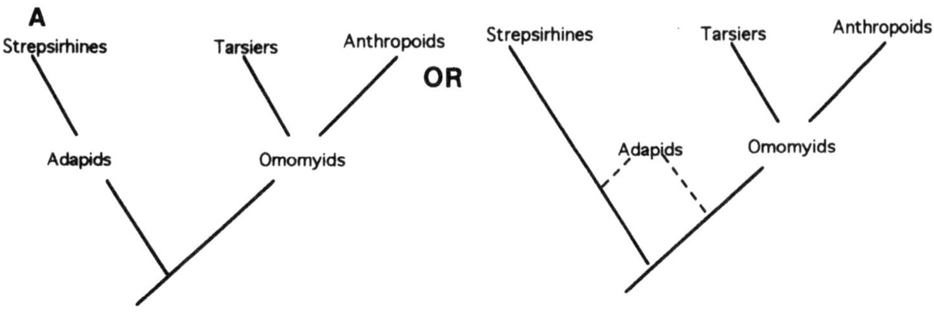

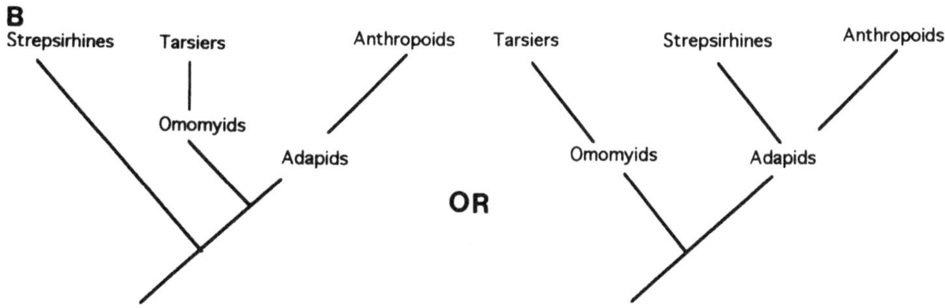

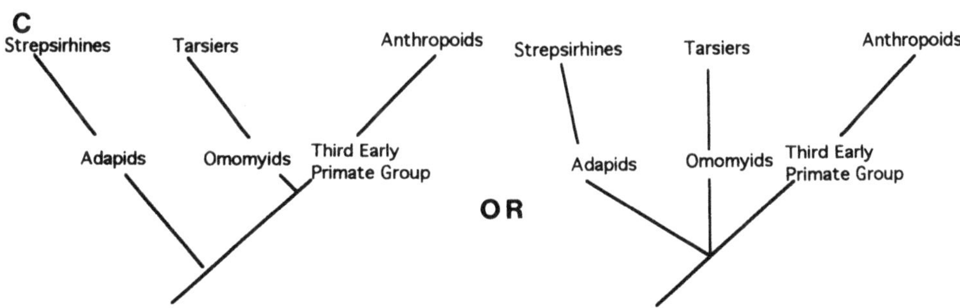

Fig. 1. Alternative views about primate relationships. A: These phylogenies view omomyids to be stem haplorhine primates and represent the position held by many researchers including Szalay and Delson (1978) and Martin (1990). B: These phylogenies view adapids to be the ancestral group to anthropoids. The phylogeny on the left is that of Rasmussen (1986), and that on the right is that of Gingerich and Schoeninger (1977). C: These phylogenies argue that anthropoids are descended from a third group of early primates. The phylogeny on the left divides primates into strepsirhines and haplorhines, whereas the one on the right recognizes an ancient trichotomy within the order primates. Note that in each case adapids and omomyids are paraphyletic taxa.

Table II. Traits That Have Been Used to Link Adapids with Anthropoids

1. Small, vertically implanted, and spatulate lower incisors
2. Lateral lower incisor larger than central lower incisor
3. Upper canine with a honing wear facet against an enlarged anterior-most lower premolar
4. Fusion of the mandibular symphysis
5. Sexual dimorphism in body and canine size
6. Calcaneus and navicular not elongated
7. Unfused tibia and fibula

Evaluation of Proposed Adapid–Anthropoid Similarities

We evaluate these proposed similarities in two ways. First, we consider their distribution among primates to see if they are actually limited to adapids and anthropoids. If the trait is more widespread within the order, then it is not considered to be of significant importance in phylogeny reconstruction. Second, if they are limited to these groups among primates, we consider the polarity of the trait. Here we follow a simple outgroup comparison with other mammals such as members of Plesiadapiformes and Scandentia. If the adapid–anthropoid condition appears to be primitive, we do not accept it as evidence of a close phylogenetic relationship between these groups. If the adapid–anthropoid condition appears to be derived, we accept it as potential evidence of a close phylogenetic relationship between these two groups.

1. Small, Vertically Implanted, Spatulate Lower Incisors

2. I_2 Larger Than I_1

Many omomyids have large, procumbent, and lanceolate central lower incisors (particularly the derived anaptomorphines) and greatly reduced lateral incisors. These characteristics are often cited by researchers as among the most diagnostic for distinguishing between adapids and omomyids (e.g., Gingerich, 1977; Simons, 1989; Rose and Bown, 1991).

The lower incisors of the adapids *Notharctus, Smilodectes,* and *Cantius* are all small and spatulate. They are not, however, vertically implanted. Rather, as has been noted by Rosenberger *et al.* (1985), they are slightly procumbent (Fig. 2). The incisors of *Adapis parisiensis* also are slightly procumbent; however, they differ from the above-mentioned adapids in being relatively wider teeth.

Cartmill and Kay (1978) suggested that certain omomyids also had small,

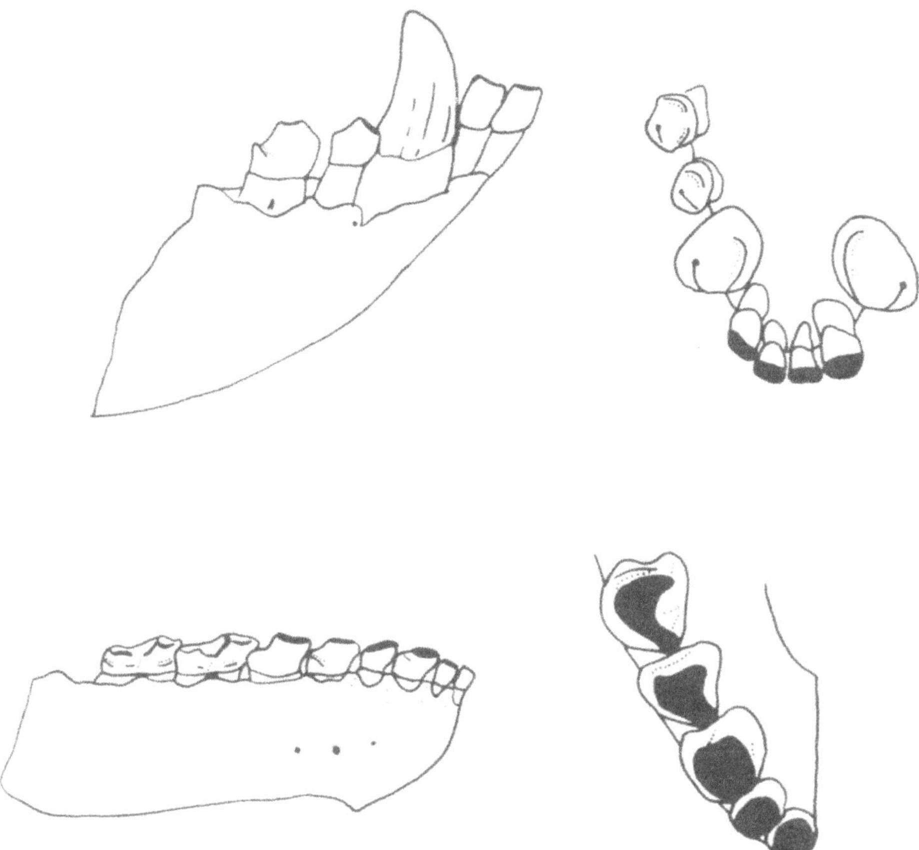

Fig. 2. Top left: *Notharctus tenebrosus* (YPM 12956) lateral view of mandible with I_1–P_2. Top right: Occlusal view of same. Bottom left: *Washakius insignis* (UW 12580) lateral view of mandible with I_1–M_2. Bottom right: Occlusal view of same. Note that incisors are only slightly procumbent and are of similar proportions in both specimens.

nearly vertically implanted incisors on the basis of alveolar size and orientation. Newly described material of the omomyid *Washakius insignis* with incisors and canine preserved intact clearly demonstrates this condition (Covert and Williams, 1991). In *Washakius*, the lateral incisor is slightly larger than the central incisor (Fig. 2). Heavy wear on the incisors precludes determination of whether or not they are spatulate. Gingerich (1977) noted that one specimen preserving the gingival portion of this tooth has a faint ridge of enamel running along its medial surface. Because the lower incisor of *Tarsius* has a similar enamel ridge and is pointed, Gingerich suggested that the central incisor of *Washakius* was also pointed. We prefer to be slightly more conservative and state that the occlusal crown morphology of *Washakius* is presently unknown. It is worth noting, however, that the incisors of a number of adapids are clearly spatulate and more closely resemble those of anthropoids in this feature than do known unworn incisors of any omomyid. Gingerich *et al.* (1991) described the anterior alveoli of the Asian omomyid *Altanius orlovi* and noted that the lower incisors of *Altanius* appear to have been small, vertically implanted, and of equal size. Gingerich *et al.* concluded that *Altanius* is one of the most primitive primates known to date. They concluded that small, vertically implanted, and nearly equal-sized incisors are primitive for primates.

Differences in incisor morphology and orientation are clearly more complex than has been previously argued. Adapids are not identical to anthropoids in incisor orientation, and certain omomyids resemble adapids and anthropoids in the relative size of their incisors. We argue that if either outgroup comparison or commonality is used to analyze these features, it appears that small, nearly vertically implanted incisors of approximately equal size were primitive for primates. Although incisor crowns are known for several adapid and early anthropoid taxa, the incisor crown morphology of omomyids is poorly known, and is unknown for any of the primitive taxa (e.g. *Altanius* and early *Teilhardina*). Thus, features relating to crown morphology cannot, at present, be used for deciphering anthropoid origins.

In sum, given the present evidence, incisor morphology does not provide compelling support for models evolving anthropoids from adapids. Further, when one considers the incisor specializations seen among some derived omomyid groups, it is clear that if anthropoids are descended from this group the ancestor would have to be one of the more primitive omomyids such as the washakiins, or *Teilhardina*.

3. Upper Canine with Honing Wear Facet against Enlarged Anterior Lower Premolar

A number of adapids (*Cantius, Notharctus, Smilodectes, Adapis, Leptadapis,* and *Caenopithecus*) do exhibit a lower premolar hone for the upper canine. However, unlike anthropoids, the premolar serving this function is small (Fig. 3). In addition, the upper canine of members of *Lemur* often hones across the anteriormost lower premolar (Cartmill and Kay, 1978). Thus, this is a feature that some adapids share not only with anthropoids but with some strepsirhines as well.

This character is also fairly complex. The lower premolar that the upper canine hones across varies among primates. For most adapids it is the first premolar, but in the adapine *Caenopithecus* it is the second premolar. For platyrrhines it is the second premolar, and for catarrhines it is the third premolar. As noted by Kay (1980), an anterior lower premolar with a canine hone appears to be a homoplastic rather than a shared-derived similarity and, thus, uninformative for determining anthropoid origins.

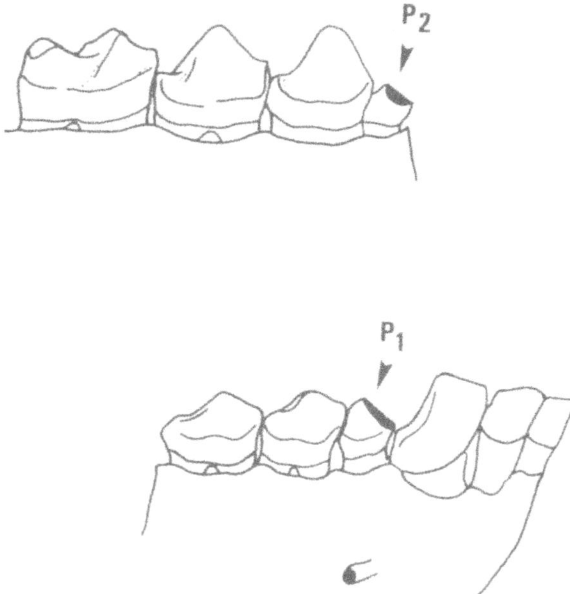

Fig. 3. Anterior lower premolar serving as a canine hone: top, *Caenopithecus lemuroides;* bottom, *Adapis parisiensis.* Note that the premolar serving as a honing surface is small.

4. *Fusion of the Mandibular Symphysis*

This trait evolved at least three times within the Adapidae: once for notharctines (*Notharctus*) and twice for adapines (in *Mahgarita* and for the common ancestor of *Adapis, Leptadapis,* and *Caenopithecus*). This trait has also apparently evolved at least three times among more recent strepsirhines; *Archaeolemur, Paleopropithecus,* and *Megaladapis* each have developed a fused mandibular symphysis independently (Ravosa and Hylander, Chapter 14, this volume).

The distribution of this trait among strepsirhines demonstrates that it is not unique to adapids and anthropoids among primates. This is a feature that has evolved time and again among mammals. It has evolved at least five times among Carnivora, probably in response to increasing body size (M. J. Ravosa, *personal communication*). Cartmill and Kay (1978) noted that fusion of the mandibular symphysis has evolved independently in many herbivorous groups of mammals.

Further, according to Rosenberger *et al.* (1985), the mandibular symphyseal morphology of adapids that show fusion is fundamentally different from that of anthropoids. The fused mandibular symphysis of adapids, when viewed in cross section, appears to be elongate and obliquely oriented. The long axis of the fused bone nearly parallels the occlusal surface of the postcanine dentition. In contrast, the fused symphysis of anthropoids, when viewed in cross section, is much more vertical in orientation and is nearly perpendicular to the occlusal plane of the postcanine teeth. Although it would be possible to evolve the anthropoid condition from that of adapids, we agree with the assessment by Rosenberger *et al.* (1985) that the mandibular morphology of adapids and anthropoids are not strikingly similar to one another.

Finally, not all anthropoid primates have a fused mandibular symphysis. *Catopithecus browni* and some specimens of *Parapithecus grangeri,* primitive anthropoids from the Fayum region of Egypt, apparently lacked symphyseal fusion (Kay and Simons, 1983; Simons, 1990). A fused mandibular symphysis is clearly not a unique resemblance between adapids and anthropoids.

5. Sexual Dimorphism in Body and Canine Size

Many of the earliest known anthropoids from the Fayum of Egypt appear to have exhibited sexual dimorphism (Fleagle *et al.*, 1980), as do many modern anthropoids. Some adapids also appear to have been sexually dimorphic in canine size. Krishtalka *et al.* (1990) provide striking evidence for this within *Notharctus venticolus,* and a number of the other North American adapids probably also had dimorphic canines. In the past it appeared that the European genera *Adapis* and *Leptadapis* also were sexually dimorphic (Gingerich, 1981b), but the recent analysis by Laneque (1992) raises doubt about this. Rather, Laneque concludes that what had been argued to be male and female members of a single species of *Adapis* are better considered members of different species. Extant strepsirhines lack significant sexual dimorphism in body size, apparently as did omomyids.

Within modern anthropoid primates, body size dimorphism ranges from none (e.g., callitrichines, gibbons) to great (baboons, gorillas). As Rosenberger *et al.* (1985, p. 29) have noted, "dimorphic features vary greatly at the species level and are difficult to establish as homologous states across larger taxonomic boundaries." The distribution among extant anthropoids and the possible parallel evolution among European and North American adapids indicate that sexual dimorphism in body size and canine size does not provide compelling evidence for an adapid origin for Anthropoidea. If, on the other hand, the European and North American adapids had sexually dimorphic canines retained from a last common ancestor (presumably a primitive *Cantius* or *Donrussellia*), and if the earliest anthropoids are documented to have sexually dimorphic canines, then one could suggest that this trait links adapids to anthropoids.

6. Calcaneus and Navicular Not Elongated

Moderate elongation of the tarsus is a shared-derived character linking all euprimates (Dagosto, 1988). Adapids and anthropoids are similar to one another in lacking the dramatic anterior calcaneal and navicular elongation seen in at least one omomyid, *Necrolemur.* However, the vast majority of omomyids for which the calcaneus and/or navicular are known show only moderate elongation of these elements (Szalay, 1976; Gebo, 1988; Dagosto, 1988; Covert and Hamrick, 1994). Among extant strepsirhines there is a great deal of variation in calcaneal elongation: the large-bodied Malagasy primates and lorisids have a slightly elongated calcaneus and navicular like those of adapids and anthropoids; cheirogaleines have moderate elongation of these elements; and galagids have extreme elongation approaching the state shown by *Tarsius* (Gebo, 1988) (Table III). Our recent field work in the Washakie Basin of Wyoming has yielded a number of new adapid and omomyid calcanei, and these specimens add significant information to our knowledge

Table III. Calcaneal Measurements and Ratios following Gebo (1988)[a]

Taxon	Specimen number	CL	ACR	MCR	PCR
Absarokius	UCM 58904	9.98	0.54	0.20	0.26
Absarokius	UCM 67907	10.75	0.56	0.20	0.24
Omomyidae	UCM 58658	8.20	0.54	0.24	0.22
Arapahovius	UCM 67850	8.60	0.53	0.26	0.21
Omomys	UCM 67878	14.85	0.53	0.18	0.29
Cantius	UCM 63110	17.70	0.43	0.32	0.25
Cantius or *Copelemur*	UCM 67912	20.09	0.40	0.31	0.29
Cantius	UCM 62662	21.10	0.41	0.28	0.31
Notharctus	UCM 65984	24.90	0.36	0.32	0.32
Lemuridae			0.43 0.39–0.47	0.30 0.23–0.36	0.27 0.21–0.32
Cheirogalidae			0.54 0.49–0.61	0.24 0.19–0.29	0.22 0.19–0.26
Galaginae			0.68 0.64–0.71	0.14 0.10–0.18	0.18 0.13–0.20
Tarsiidae			0.76 0.76–0.77	0.10 0.08–0.12	0.13 0.12–0.15
Omomyidae			0.52 0.49–0.54	0.25 0.22–0.31	0.23 0.16–0.28
Adapidae			0.37 0.29–0.43	0.33 0.30–0.36	0.30 0.25–0.36

[a]All University of Colorado (UCM) early Eocene specimens are from the Washakie Basin with the exception of the *Omomys* and *Notharctus* (middle Eocene specimens from the Bridger Basin). Comparative data from Gebo (1988) with mean (above) and range (below) representing seven lemurid species, four cheirogaleid species, six galagine species, three tarsiid species, five omomyid species, and ten adapid species. CL, calcaneal length; ACR, anterior calcaneal ratio; MCR, midcalcaneal ratio; PCR, posterior calcaneal ratio.

about the variation in the elongation of this element in early Eocene primates. For example, new specimens of the adapid genera *Cantius* and *Copelemur* show elongation similar to previously described members of this family (Fig. 4). New specimens of the omomyid genera *Tetonoides, Arapahovius,* and *Absarokius* show moderate calcaneal elongation, as do previously described North American omomyids, a degree of elongation that is most similar to that in cheirogalines (Table III, Fig. 5). Some debate exists about the primitive calcaneal morphology (in terms of anterior elongation) for euprimates. Gebo (1988) suggested that the condition shown by the majority of omomyids (i.e., moderate elongation) is primitive, whereas we believe that it is just as likely that the slight elongation shown by adapids is the primitive condition. In either case, the lack of extreme elongation is not useful for determining phylogenetic relationships between Eocene taxa and anthropoids.

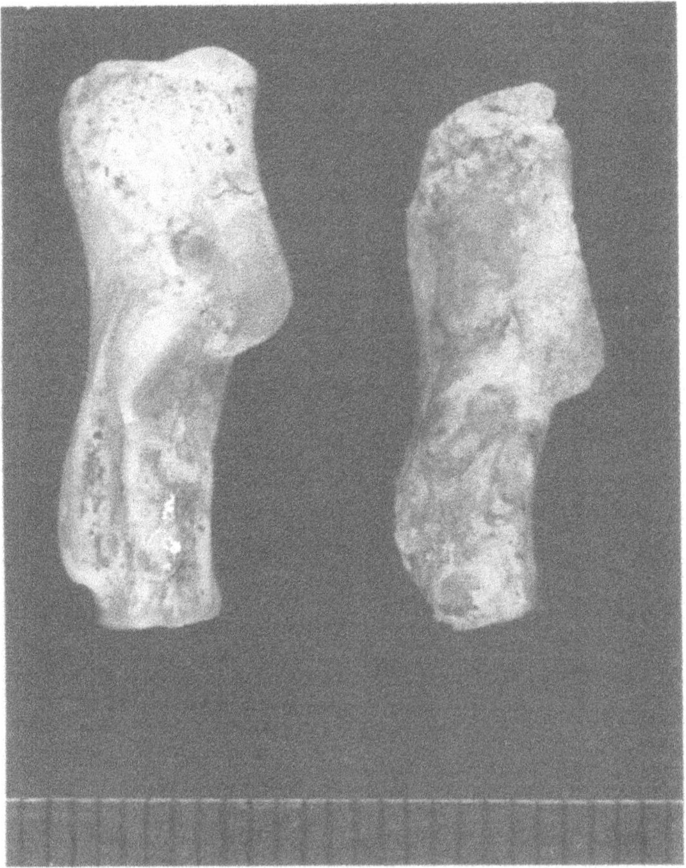

Fig. 4. Medial views of calcaneus of *Lemur* (left) and *Cantius,* UCM 62662 (right). Note similar overall shape and proportion.

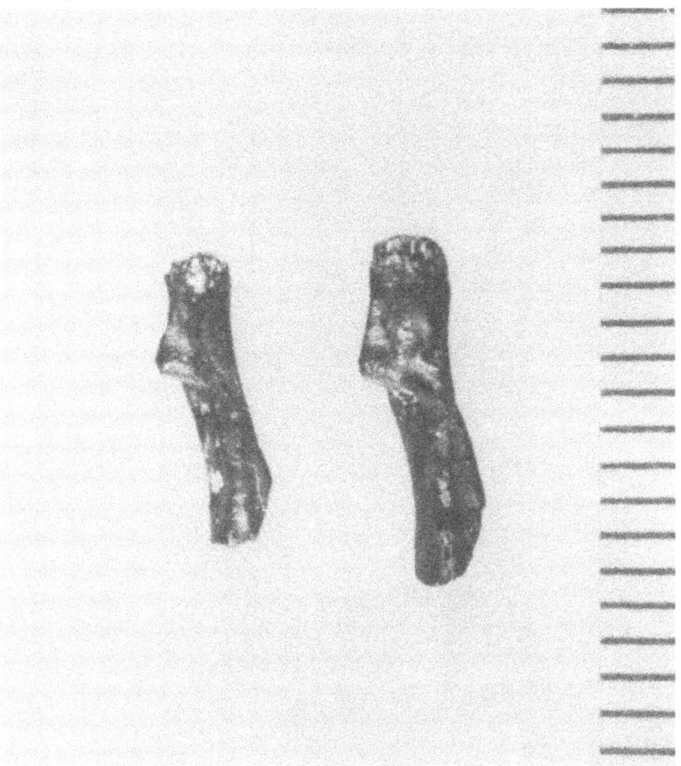

Fig. 5. Medial view of two late Wasatchian omomyid calcanei: genus indeterminate UCM 58658 (left) and *Absarokius* UCM 58904 (right). Note similar overall shape and proportion. The anterior elongation of these specimens is characteristic of most omomyids and resembles that of cheirogalines, not that of *Tarsius*. UCM 58658 calcaneal length, 8.20 mm; UCM 58904 calcaneal length, 9.98 mm.

7. *Unfused Tibia and Fibula*

Modern primates exhibit a great range of variation in distal tibiofibular articulation. Whereas most show a small and somewhat mobile articulation, a few primates (e.g., *Microcebus, Saimiri*) show a more extensive opposition of these elements, and only *Tarsius* exhibits tibiofibular fusion (Fleagle and Simons, 1983). Dagosto (1985) has found that only one very late-occurring omomyid, *Necrolemur,* exhibited tibiofibular fusion. Omomyids are actually more similar to anthropoids in some morphological aspects of the distal tibia (such as the orientation of the medial malleolus) than are adapids (Dagosto, 1985). Recently recovered material of the omomyid *Absarokius abbotti* (Fig. 6) demonstrates an extensive distal articulation between the tibia and fibula, yet they remained separate (Covert and Hamrick, 1994). This morphology is

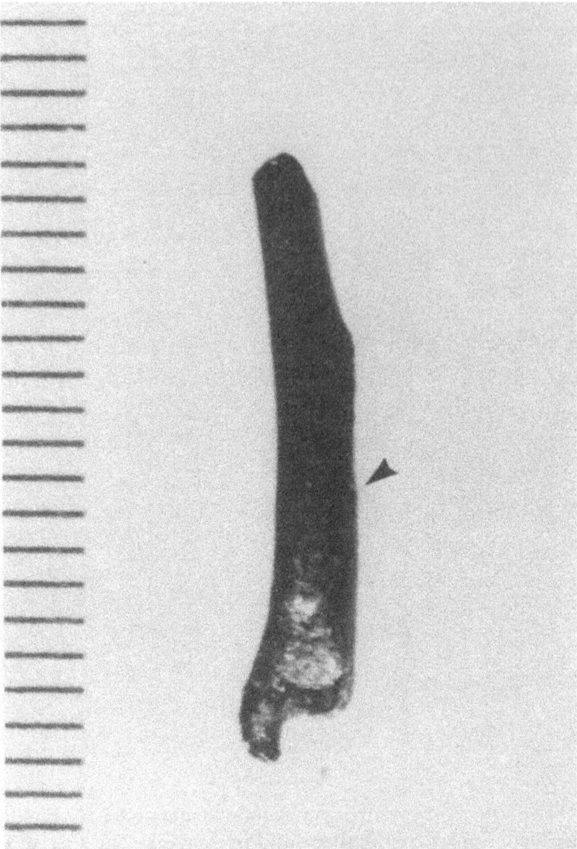

Fig. 6. Tibia of *Absarokius* (UCM 62673). Note the extensive facet for articulation with the fibula (arrow marks proximal extent of facet).

most similar to that seen in the Fayum anthropoid *Apidium phiomense* and is somewhat intermediate between the morphology of *Saimiri* and *Tarsius*.

An unfused tibia and fibula is the primitive state for primates, with at least one omomyid exhibiting the derived condition of tibiofibular fusion. Therefore, an unfused distal tibiofibular articulation is of no use for demonstrating a close relationship between adapids and anthropoids, as each group simply displays the primitive condition. Further, some omomyids are not so specialized in the anatomy of this region that they could not have been within the ancestry of anthropoids.

Summary of Characteristics

The newly recovered material of Eocene primates and recently published detailed morphological analyses of extant and fossil taxa have provided crucial information that has allowed us to examine critically many of the postcranial and dental characteristics used to unite anthropoids with adapids. We have found that these traits provide little, if any, support for a close relationship between these primates. Some are clearly shared primitive features that are characteristic of primates in general (characters 1, 2, 6, and 7). Others are widespread throughout the order (characters 3, 4, and 5) and so do not uniquely link adapids with anthropoids.

Basicranial traits have also been argued to link adapids and anthropoids (e.g., Gingerich, 1981a; Rasmussen, 1986; Simons and Rasmussen, 1989). Many of these traits have been critically evaluated (e.g., MacPhee and Cartmill, 1986). Further discussion of the primate basicrania and its phylogenetic importance is provided by Ross (Chapter 15, this volume) and Beard and MacPhee, (Chapter 3, this volume).

Similarities among Adapids and Living Strepsirhines

There are several postcranial shared-derived traits that might link adapids uniquely with extant strepsirhines (Table IV). Each trait is discussed below to evaluate its usefulness for supporting an exclusive phylogenetic relationship among these primates. Whereas previous reports about these features are intriguing, the primary problem with interpretation of their importance is that most of the omomyid and adapid postcranial elements are known only from extremely small samples. Four of the traits linking adapids with extant strepsirhines occur in the tarsus, and one on the tibia. The following discussion includes information on new specimens that significantly increase the number of taxa known by these elements. Thus, we are able to improve our understanding of the distribution of the traits among early primates.

Table IV. Presumed Shared-Derived Traits Uniting Adapids with Extant Strepsirhines

1. Well-developed talar posterior trochlear shelf.
2. Anterolateral orientation of the tibial medial malleolus.
3. Talofibular facet projects laterally.
4. Groove for flexor hallucis longus positioned lateral to the posterior trochlear facet.
5. Navicular cuboid articulation contiguous with that of the mesocuneiform and the ectocuneiform.
6. Deeply excavated facet for ulnar styloid process in carpus (adapines, not notharctines).

1. Well-Developed Talar Posterior Trochlear Shelf

At present, the talus is known for only a handful of adapid and omomyid taxa, and the majority of these taxa are either middle or late Eocene in age. Early Eocene taxa for which this element is known include the adapid *Cantius* and the omomyids *Teilhardina belgica*, cf. *Tetonius*, and *Arapahovius* (e.g., Covert, 1985, 1988; Gebo, 1988; Szalay and Delson, 1979). Anthropoids, most omomyids, and plesiadapiforms lack or have a weakly developed posterior trochlear shelf. Adapids, in contrast, usually have a well-developed posterior trochlear shelf. This distribution suggests that the adapid–extant-strepsirhine condition is derived (Beard *et al.*, 1988; Covert, 1988; Dagosto, 1988). Our recent field work in the Washakie Basin, Wyoming, has yielded approximately 16 new early Eocene primate tali. Each of the adapid tali (*Cantius* and *Copelemur*) has a well-developed posterior trochlear shelf (Fig. 7). Each of the omomyid tali (*Tetonoides, Anemorhysis, Absarokius,* and genus indeterminate) has a poorly developed posterior trochlear shelf (Fig. 8). Thus, these specimens strengthen previous suggestions about the distribution of talar posterior trochlear shelf variation and demonstrate that purported adapid and

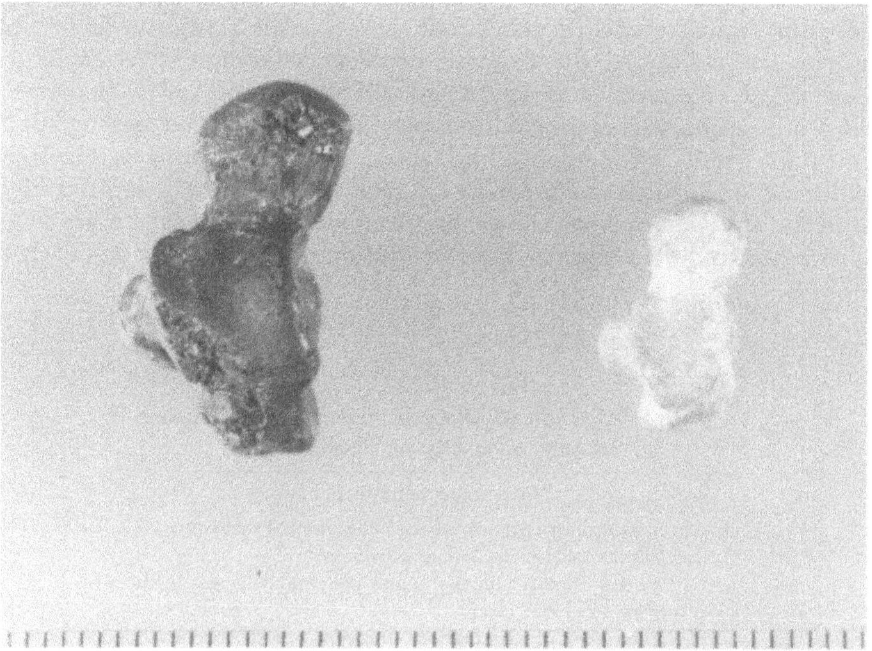

Fig. 7. Relative posterior trochlear shelf size. Dorsal view of two notharctine tali (UCM 62671 on left, UCM 62605 on right). Although these specimens differ significantly in size (UCM 6271 talar length, 21.2 mm; UCM 62605 talar length, 13.5 mm), both have well-developed posterior trochlear shelves, and the talofibular facet projects laterally from the talar body.

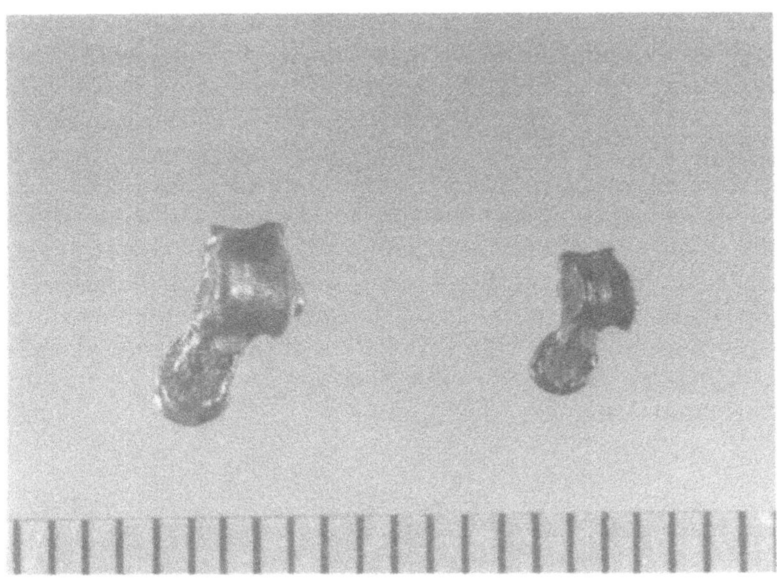

Fig. 8. Omomyid talus morphology. Dorsal view of *Absarokius* talus, UCM 62681 (left), and *Tetonoides* talus, UCM 60901 (right). Note weak posterior trochlear shelf development and lack of talofibular facet projection. Although these specimens differ significantly in size (*Tetonoides* talar length, 4.9 mm; *Absarokius* talar length, 6.15 mm), both are quite similar in morphology.

omomyid conditions appear to be widespread and nearly uniform within the respective families. We state nearly uniform here because there are counterexamples within both the adapids and omomyids. Dagosto (1988) noted that the talus of *Adapis parisiensis* has only a weakly developed posterior trochlear shelf, and Godinot and Dagosto (1983) noted that the talus of *Necrolemur antiquus* has a fairly well-developed posterior trochlear shelf. We believe that the condition shown by *Adapis parisiensis* is secondarily derived because the talus of *Leptadapis magnus* and those of the Messel adapids have a well-developed posterior trochlear shelf that closely resembles that of notharctines. We further assume that the talus of *Necrolemur* has converged toward the adapid condition from the common omomyid condition.

2. *Anterolateral Orientation of the Tibial Medial Malleolus*

The tibia was known previously for only two or three Eocene omomyids—taxa that were of middle and late Eocene age. Dagosto (1985) noted that primates differ from primitive eutherians in having a tibial medial malleolus that is rotated in such a way that its articular surface is oriented somewhat anterolaterally. Dagosto also reported that extant primates can be subdivided on the basis of this trait: strepsirhines usually show more rotation than do haplorhines. Adapids resemble extant strepsirhines in having a fairly strong rotation of the medial malleolus, whereas omomyids resemble extant haplorhines in showing only moderate rotation of the medial malleolus. The variation seen among primitive mammals, haplorhines, and strepsirhines may be interpreted as a morphocline in which strepsirhines, including adapids, show the most derived condition (Covert, 1988). Our recent field work has yielded new distal tibial fragments of early Eocene primates. The morphology of the distal tibia of our new adapid specimens (*Cantius*) shows fairly strongly rotated medial malleoli, whereas our new omomyid specimen (*Absarokius*) shows only moderate rotation of the medial malleolus. These new specimens strengthen an inferred morphocline for the orientation of the tibial medial malleolus in demonstrating that at least one early Eocene omomyid shows the purported haplorhine condition.

3. Talofibular Facet Morphology

Gebo (1988) has noted that the adapid talus is like that of extant strepsirhines and differs from that of all other primates in the morphology of the talofibular facet. In the adapid–extant-strepsirhine condition, the talofibular facet slopes obliquely outward rather than being vertical as in omomyids, anthropoids, and most other eutherians. Because the morphology of haplorhines more closely resembles that of primitive mammals, and the adapid–extant-strepsirhine state is unique among primates and all other eutherians, Gebo (1988) and Beard *et al.* (1988) have argued that the adapid–strepsirhine condition is derived. Our recently collected early Eocene primate tali (see above) expand our sample of taxa known by the talus and reinforce the suggestions of Gebo about the phylogenetic importance of the talofibular facet morphology (Figs. 7 and 8).

4. Position of the Flexor Hallucis Longus Groove

A second adapid–strepsirhine shared-derived trait identified by Gebo (1988) is the position of the flexor hallucis longus groove. Among anthropoids and omomyids, the groove is in the midline of the posterior trochlear facet of the talus, whereas in adapids and living strepsirhines it is positioned lateral to the posterior trochlear facet. Again, the morphology of haplorhines appears to resemble that of other eutherians, and, as has been argued by Gebo (1988) and by Beard *et al.* (1988), the adapid–extant-strepsirhine condition is apparently derived. Our recently collected early Eocene primate tali both increase the sample of taxa known by this element and reinforce the suggestions of Gebo about the phylogenetic importance of this trait.

5. Morphology of the Naviculocuboid Articulation

Dagosto (1988) and Dagosto and Gebo (Chapter 17, this volume) have noted that primates vary in the morphology of the naviculocuboid articulation. Omomyids, *Tarsius,* and anthropoids resemble most mammals in having a cuboid facet on the navicular that is contiguous solely with that of the ectocuneiform. In adapids and living strepsirhines, the cuboid facet is contiguous with that of both the mesocuneiform and ectocuneiform. The condition in adapids and strepsirhines is apparently the derived one. As is the case with the other tarsals, the navicular is extremely poorly known. We have collected four additional omomyid naviculars: one is attributable to *Tetonoides,* two to *Absarokius,* and the fourth's generic placement is indeterminate. Each shows a naviculocuboid articulation like that described by Dagosto for other omomyids.

6. Articulation between the Ulna and Carpus

Beard *et al.* (1988) noted that adapines (but not notharctines) have a strepsirhine-like articulation between the ulna and carpus. Further, the morphology shown by the notharctines resembles that of extant haplorhines and other eutherians. Thus, it appears that the adapine–strepsirhine morphology is derived.

Summary of Adapid–Extant-Strepsirhine Similarities

These traits, taken together, appear to provide stronger evidence for an adapid–extant-strepsirhine relationship than do the previously discussed traits proposed to link adapids and anthropoids (also see Dagosto and Gebo, Chapter 17, this volume). Further, recently recovered early Eocene primate postcranial remains from the Washakie Basin have added substantially to our understanding of the general nature of the distribution of a number of these traits for both adapids and omomyids, thus strengthening this argument. For example, we have documented that both large and small adapids show the presumed derived traits that link them with strepsirhines. Likewise, both large and small omomyids are homogeneous in morphology. The homogeneous nature of both the adapid and omomyid traits is also interesting in that the earliest omomyid and adapid tarsals do not appear to be noticeably more similar to one another than those from the middle Eocene. This contrasts with the dental situation, where it has been noted by numerous researchers that the earliest-occurring omomyids and adapids are very similar to one another in dental morphology, so much so that there has been much debate about the omomyid or adapid status of these creatures (e.g., Gingerich *et al.*, 1991; Rose *et al.*, Chapter 1, this volume). Finally, one could question whether a number of the ankle features used to link adapids to extant strepsirhines are functionally linked to one another. This is a possibility, but we believe it is a weak one. The distinctive rotation of the adapid tibial medial malleolus does not articulate with the talofibular facet and has only a slight contact with the well-developed posterior trochlear shelf.

Potential Shared-Derived Traits Uniting Tarsiids with Anthropoids

Numerous authors have noted interesting and presumably derived similarities linking tarsiers with anthropoids (Martin, 1990). For example, Pocock (1918) noted that tarsiers grouped with anthropoids in lacking a rhinarium. Luckett (1975) detailed a number of traits of the placenta shared by these taxa to the exclusion of other extant primates. Goodman (1992) argued that the most recent biomolecular information available links tarsiers to anthropoids. Basicranial synapomorphies have been identified by MacPhee and Cartmill (1986) and Ross (Chapter 15, this volume), and shared derived dental features by Kay and Williams (Chapter 13, this volume). We are unaware of any postcranial tarsiid–anthropoid synapomorphies.

We agree with Martin's (1990) view that the various lines of evidence linking tarsiers with anthropoids provides substantial support for dividing extant primates into two groups: strepsirhines and haplorhines.

Potential Shared-Derived Traits Uniting Omomyids and Tarsiids and/or Anthropoids

According to a number of recent reviews, many of the similarities shown by omomyids, tarsiers, and anthropoids are shared primitive traits. Beard *et al.* (1988) and Covert (1988) noted that the similar morphological traits shared by omomyids and extant haplorhines in the ankle and tarsus more closely resemble those of other eutherians than those of adapids and extant strepsirhines and are thus probably primitive. Rosenberger *et al.* (1985) failed to identify any omomyid–extant-haplorhine shared-derived traits in mandibular or anterior dental anatomy. Likewise, MacPhee and Cartmill (1986) were unable to identify compelling basicranial evidence linking omomyids to extant haplorhines. In fact, MacPhee and Cartmill (1986) suggested that Omomyidae may not be a monophyletic group but rather a grouping of early euprimates that lack any diagnostic adapid features.

Beard *et al.* (1991) recently argued that the early Eocene North American omomyid *Shoshonius* shares a number of derived traits with *Tarsius* to the exclusion of anthropoids and other omomyids known from cranial material. Unfortunately, only three other omomyids (*Necrolemur, Rooneyia,* and *Tetonius*) are known from sufficiently complete cranial material to be included in their analysis, so it cannot be determined if this evidence suggests a specific relationship between *Shoshonius* and tarsiers or a more general one between a more substantial group of omomyids with tarsiers.

Thus, evidence grouping omomyids with tarsiids and/or anthropoids does not appear to be as strong as that grouping adapids with strepsirhines or tarsiids with anthropoids.

Conclusions

Much debate remains about the relationships among the adapids, omomyids, and more recently occurring strepsirhines, tarsiids, and anthropoids. We believe that recent discoveries and analyses of North American adapids and omomyids have strengthened the hypothesis that extant strepsirhines are descended from adapids. The tarsal differences between omomyids and adapids are particularly interesting in that they appear to delineate these groups more clearly than does dental evidence in the earliest Eocene. Additionally, many of the traits that have been argued to support an adapid–anthropoid relationship are not supported by new evidence. The above information coupled with substantial evidence linking extant anthropoids with tarsiers and some, albeit poor, evidence linking tarsiers with omomyids lead us to consider it most plausible that the anthropoid ancestor may be either an unknown (or poorly known) omomyid or a member of a presently poorly sampled third group of basal primates. Given this situation, new specimens described by Godinot and Mahboubi (1992; Godinot, Chapter 10, this volume) and Beard *et al.* (1994) are of critical importance for understanding anthropoid origins. We believe that as this new group becomes better known, it is likely to be seen as more closely related to omomyids than to adapids.

The Eocene of North America and Europe saw extensive adaptive radiations for both Omomyidae and Adapidae. These animals document an interesting chapter of primate evolutionary history and one that is much richer than indicated by terms such as "lemur-like" and "tarsier-like." As reviewed in many of the chapters in this volume, it is difficult to identify plausible anthropoid ancestors within these radiations. Rather, what we appear to be sampling are successful early radiations of primates that were cousins to the anthropoid ancestors. As noted by Godinot (Chapter 10, this volume) and Beard *et al.*, (1994), the search for anthropoid ancestors is now shifting to early Cenozoic deposits of Africa and Asia.

ACKNOWLEDGMENTS

We thank P. Robinson, J. Wible, T. Bown, D. Gebo, M. Dagosto, C. Beard, R. Anemone, C. Harrisville-Wolff, D. Ayers-Darling, M. Hamrick, M. Anthony, P. Holroyd, and E. Strasser for informative discussions and comments on the morphology of early primates. H.H.C. thanks R. F. Kay, E. L. Simons, M. Cartmill, and R. D. E. MacPhee for numerous insightful discussions about primate relationships. Special thanks go to Dr. J. A. Lillegraven of the University of Wyoming Geological Museum for permission to study and describe the *Washakius* material. Finally, we thank the following persons and institutions for access to skeletal material of extant and extinct primates in their collections: the American Museum of Natural History, Drs. G. Musser and R. H. Tedford; National Museum of Natural History (Smithsonian Institution), Drs. R. Thorington and R. Emry; Yale Peabody Museum, Dr. J. H. Ostrom; Museum of Comparative Zoology at Harvard University, Drs. F. A. Jenkins, Jr., and M. Rutzmoser; U.S. Geological Survey (Denver), Dr. T. M. Bown; the University of Wyoming Geological Museum, Dr. J. A. Lillegraven; and the Duke University Primate Center, Dr. E. L. Simons.

We especially thank Drs. J. G. Fleagle, R. F. Kay, and E. L. Simons for inviting us to participate in the Anthropoid Origin Conference on which this volume is based.

References

Beard, K. C., Dagosto, M., Gebo, D. L. and Godinot, M. 1988. Interrelationships among primate higher taxa. *Nature* **331**:712–714.

Beard, K. C., Krishtalka, L. and Stucky, R. K. 1991. First skulls of the early Eocene primate *Shoshonius cooperi* and the anthropoid-tarsier dichotomy. *Nature* **349**:64–67.

Beard, K. C., Qi, T., Dawson, M. R., and Wang, B. 1994. A diverse new primate fauna from middle Eocene fissure-fillings in southeastern China. *Nature* **368**:604–609.

Cartmill, M., and Kay, R. F. 1978. Craniodental morphology, *Tarsius* affinities, and primate suborders. In: D. J. Chivers and K. A. Joysey (eds.), *Recent Advances in Primatology, Vol. 3, Evolution,* pp. 205–214. Academic Press, London.

Covert, H. H. 1985. Adaptations and evolutionary relationships of the Eocene primate family Notharctidae. Ph.D. dissertation, Duke University.

Covert, H. H. 1986. Biology of early Cenozoic primates. In: D. R. Swindler and J. Erwin (eds.), *Comparative Primate Biology, Vol. 1: Systematics, Evolution, and Anatomy,* pp. 335–359. Alan R. Liss, New York.

Covert, H. H. 1988. Ankle and foot morphology of *Cantius mckennai:* Adaptations and phylogenetic implications. *J. Hum. Evol.* **17**:57–70.

Covert, H. H., and Hamrick, M. W. 1993. Description of new skeletal remains of the early Eocene anaptomorphine primate *Absarokius* (Omomyidae) and a discussion about its adaptive profile. *J. Hum. Evol.* **25**:351–362.

Covert, H. H., and Williams, B. A. 1991. The anterior lower dentition of *Washakius insignis* and adapid–anthropoidean affinities. *J. Hum. Evol.* **21**:463–467.

Dagosto, M. 1985. The distal tibia of primates with special reference to the Omomyidae. *Int. J. Primatol.* **6**:45–75.

Dagosto, M. 1988. Implications of postcranial evidence for the origin of euprimates. *J. Hum. Evol.* **17**:35–56.

Fleagle, J. G. 1988. *Primate Adaptation & Evolution.* Academic Press, San Diego.

Fleagle, J. G., and Simons, E. L. 1983. The tibio-fibular articulation in *Apidium phiomense,* an Oligocene anthropoid. *Nature* **301**:238–239.

Fleagle, J. G., Kay, R. F., and Simons, E. L. 1980. Sexual dimorphism in early anthropoids. *Nature* **287**:328–330.

Gebo, D. L. 1987. Anthropoid origins—the foot evidence. *J. Hum. Evol.* **15**:421–430.

Gebo, D. L. 1988. Foot morphology and locomotor adaptation in Eocene primates. *Fol. Primatol.* **50**:3–41.

Gebo, D. L., Dagosto, M., and Rose, K. D. 1991. Foot morphology and evolution in early Eocene *Cantius. Am. J. Phys. Anthropol.* **86**:51–73.

Gingerich, P. D. 1977. Dental variation in early Eocene *Teilhardina belgica* with notes on the anterior dentition of some early Tarsiiformes. *Fol. Primatol.* **28**:144–153.

Gingerich, P. D. 1980. Eocene Adapidae, paleobiogeography, and the origin of South American Platyrrhini. In: R. L. Ciochon and A. B. Chiarelli (eds.), *Evolutionary Biology of New World Monkeys and Continental Drift,* pp. 123–138. Plenum Press, New York.

Gingerich, P. D. 1981a. Early Cenozoic Omomyidae and the evolutionary history of the Tarsiiform primates. *J. Hum. Evol.* **10**:345–374.

Gingerich, P. D. 1981b. Cranial morphology and adaptations in Eocene Adapidae. I. Sexual dimorphism in *Adapis magnus* and *Adapis parisiensis. Am. J. Phys. Anthropol.* **56**:217–234.

Gingerich, P. D., and Schoeninger, M. 1977. The fossil record and primate phylogeny. *J. Hum. Evol.* **6**:483–505.

Gingerich, P. D., Dashzeveg, D., and Russell, D. E. 1991. Dentition and systematic relationships of *Altanius orlovi* (Mammalia, Primates) from the early Eocene of Mongolia. *Geobios* **24**:637–646.

Godinot, M., and Dagosto, M. 1983. The astragalus of *Necrolemur* (Primates, Microchoerinae). *J. Paleontol.* **57**:1321–1324.

Godinot, M., and Mahboubi, M. 1992. Earliest known simian primate from Algeria. *Nature* **357**:324–326.

Goodman, M. 1992. Molecular systematics of the Primates. *Am. J. Phys. Anthropol.* [*Suppl.*] **14**:82–83.

Gregory, W. K. 1915. I. On the relationships of the Eocene lemur *Notharctus* to the Adapidae and to other primates. *Bull. Geol. Soc. Am.* **26**:419–446.

Gregory, W. K. 1920. On the structure and relations of *Notharctus,* an American Eocene primate. *Mem. Am. Mus. Nat. Hist. N.S.* **3**(2):49–243.

Hoffstetter, R. 1974. Phylogeny and geographical deployment of the primates. *J. Hum. Evol.* **3**:327–350.

Kay, R. F. 1980. Platyrrhine origins: A reappraisal of the dental evidence. In: R. L. Ciochon and A. B. Chiarelli (eds.), *Evolutionary Biology of New World Monkeys and Continental Drift,* pp. 159–188. Plenum Press, New York.

Kay, R. F., and Simons, E. L. 1983. Dental formulae and dental eruption patterns in Parapithecidae (Primates, Anthropoidea). *Am. J. Phys. Anthropol.* **62**:363–375.

Krishtalka, L., Stucky, R. K., and Beard, K. C. 1990. The earliest fossil evidence for sexual dimorphism in primates. *Proc. Natl. Acad. Sci. USA* **87**:5223–5226.

Laneque, L. 1992. Analyse de matrice de distance euclidienne de la region du museau chez *Adapis* (Adapiformes, Eocene). *C.R. Acad. Sci. Paris* [*Ser. II*] **314**:1387–1393.

Luckett, W. P. 1975. Ontogeny of the fetal membranes and placenta. In: W. P. Luckett and F. S. Szalay (eds.), *Phylogeny of the Primates,* pp. 157–182. Plenum Press, New York.

MacPhee, R. D. E., and Cartmill, M. 1986. Basicranial structures and primate systematics. In: D. R. Swindler and J. Erwin (eds.), *Comparative Primate Biology, Vol. 1: Systematics, Evolution, and Anatomy,* pp. 219–276. Alan R. Liss, New York.

Martin, R. D. 1986. Primates: A definition. In: B. A. Wood, L. B. Martin, and P. Andrews (eds.), *Major Topics in Primate and Human Evolution,* pp. 1–31. Cambridge University Press, Cambridge.

Martin, R. D. 1990. *Primate Origins and Evolution: A Phylogenetic Reconstruction.* Princeton University Press, Princeton.

Pocock, R. I. 1918. On the external characters of the lemurs and *Tarsius. Proc. Zool. Soc. Lond.* **1918:**19–53.

Rasmussen, D. T. 1986. Anthropoid origins: A possible solution to the Adapidae–Omomyidae paradox. *J. Hum. Evol.* **15:**1–12.

Rose, K. D., and Bown, T. M. 1991. Additional fossil evidence on the differentiation of the earliest euprimates. *Proc. Natl. Acad. Sci. USA* **88:**98–101.

Rosenberger, A. L., Strasser, E., and Delson, E. 1985. Anterior dentition of *Notharctus* and the adapid–anthropoid hypothesis. *Fol. Primatol.* **44:**15–39.

Simons, E. L. 1961. Notes on Eocene tarsioids and a revision of some Necrolemurinae. *Bull. Br. Mus. Nat. Hist. Geol.* **5**(3):43–69.

Simons, E. L. 1990. Discovery of the oldest known anthropoidean skull from the Paleogene of Egypt. *Science* **247:**1567–1569.

Simons, E. L., and Rasmussen, D. T. 1989. Cranial morphology of *Aegyptopithecus* and *Tarsius* and the question of the tarsier–anthropoidean clade. *Am. J. Phys. Anthropol.* **79:**1–25.

Szalay, F. S. 1976. Systematics of the Omomyidae (Tarsiiformes, Primates): Taxonomy, phylogeny, and adaptations. *Bull. Am. Mus. Nat. Hist.* **156**(3):157–450.

Szalay, F. S., and Delson, E. 1979. *Evolutionary History of the Primates.* Academic Press, New York.

Szalay, F. S., Rosenberger, A. L., and Dagosto, M. 1987. Diagnosis and differentiation of the Order Primates. *Yearb. Phys. Anthropol.* **30:**75–105.

Wible, J. R., and Covert, H. H. 1987. Primates: Cladistic diagnosis and relationships. *J. Hum. Evol.* **16:**1–22.

Cranial Anatomy of *Shoshonius* and the Antiquity of Anthropoidea

3

K. CHRISTOPHER BEARD
and R. D. E. MACPHEE

Introduction

Although most, if not all, modern systematists agree that the living anthropoid or simiiform primates (New and Old World monkeys, apes, and humans) constitute a natural, monophyletic assemblage, there is widespread disagreement over whether or not certain fossil forms should be included in this taxon. Unfortunately, not all of this disagreement reflects the morphological ambiguity of such poorly known fossil taxa as *Hoanghonius stehlini, Amphipithecus mogaungensis, Pondaungia cotteri,* and *Afrotarsius chatrathi.* Rather, it is at least partly a result of different notions regarding how the taxon Anthropoidea should be defined. Obviously, a major prerequisite for understanding the origin and early evolutionary history of any group of organisms is a theoretically sound definition of what actually constitutes the taxon under investigation. Without this type of formal definition, a great deal of time and

K. CHRISTOPHER BEARD • Section of Vertebrate Paleontology, Carnegie Museum of Natural History, Pittsburgh, Pennsylvania 15213. R. D. E. MACPHEE • Department of Mammalogy, American Museum of Natural History, New York, New York 10024.

Anthropoid Origins, edited by John G. Fleagle and Richard F. Kay. Plenum Press, New York, 1994.

effort can be wasted by different investigators talking past one another about semantic issues that are largely irrelevant to the actual topic of debate.

Because we are inherently interested in the phylogenetic origin of the anthropoid clade, we define the primate suborder Anthropoidea vertically. That is, we will call any primate an anthropoid if it can be shown to be either (1) a nested member of the clade that includes all living anthropoids (the "crown group" of Ax, 1985) or (2) an animal that lies outside this clade but that shares more recent common ancestry with it than with any other living primates (the "stem lineage" of Ax, 1985). To our knowledge, the only viable alternative to this vertical definition of Anthropoidea is a horizontal one based on the presence or absence of some key character, such as the postorbital septum. We strongly prefer the vertical solution because (1) we have little knowledge of the sequence whereby key anthropoid characters evolved; (2) there is no consensus in any case concerning which character should serve as the basis for defining Anthropoidea horizontally; and (3) any horizontal definition would, by design, exclude the most significant taxa for understanding the origin and early history of the anthropoid clade (i.e., basal taxa).

The focus of this study, *Shoshonius cooperi,* is not an anthropoid by either of the definitions just presented. However, *Shoshonius* does appear to be an early member of the clade that many workers consider to be the sister group of Anthropoidea, the Tarsiiformes. As such, the anatomy and age of *Shoshonius* place constraints on hypotheses of anthropoid origins, topics that we will discuss at some length later in this chapter. Since the publication of an earlier, preliminary study of the cranial anatomy of *Shoshonius* (Beard *et al.,* 1991), field parties from the Carnegie Museum of Natural History have been successful in recovering additional skulls pertaining to this species (Table I). In some cases these new specimens clarify details of anatomy that were not observable on the original specimens. Our goal here is to describe key aspects

Table I. Cranial Specimens of *Shoshonius cooperi* from the Lost Cabin Member, Wind River Formation, Wyoming

Specimen No.	Locality	Description
CM 31366	K-6, Q-11	Skull, neurocranium dorsally crushed but basicranium almost intact
CM 31367	K-6, Q-11	Skull, laterally crushed; right side moderately distorted, left side badly distorted and incomplete
CM 60492	K-6, Q-6	Partial skull, very crushed; endocranial surface of left auditory region detached and separately prepared
CM 60493	K-6, Q-6	Skull, slightly dorsoventrally crushed; palate, orbits well preserved; both auditory regions missing
CM 60494	K-6, Q-6	Skull, slightly laterally crushed; auditory regions complete but distorted
CM 60495	K-6, Q-6	Skull, very crushed dorsoventrally, but with little distortion; bullae broken, but auditory regions complete; parts of posterior skull roof separately prepared

Table II. Abbreviations Used in Text and Figures

ASF	alisphenoid flange
B	bulla
BOF	basioccipital flange
ChS	character state
CS	central stem, character state
EAM	external acoustic meatus
E	ectotympanic
EPL	external pterygoid lamina
FM	foramen magnum
FO	foramen ovale
H	hyoid (part)
ICC	internal carotid canal
IPL	internal pterygoid lamina
IPS	inferior petrosal sinus
JF	jugular foramen
MR	mastoid region
OC	occipital condyle
OTU	operational taxonomic unit (nominal taxon)
PAR	parietal
PC	promontorial canal
PCF	posterior carotid foramen
PF	parotic fissure
PGF	postglenoid foramen
PGP	postglenoid process
POP	paroccipital process
PTP	posttympanic process
SC	stapedial canal
SF	suprameatal foramen
SMF	stylomastoid foramen
SQ	squamosal
SQE	squamosal entoglenoid
AMNH	American Museum of Natural History
CM	Carnegie Museum of Natural History
MCZ	Museum of Comparative Zoology, Harvard University
TMM	Texas Memorial Museum, University of Texas

of the cranial anatomy of *Shoshonius* in the broader framework of a phylogenetic analysis of cranial traits among primate higher-level taxa and certain critical fossils. The abbreviations we use are defined in Table II.

Materials and Methods

Materials

We have examined almost all of the critical fossil cranial specimens discussed in this paper, and in preparing this paper we have had the benefit of access to the extensive neontological collections of the AMNH. Our morphological evaluations are brief if existing literature on the structures concerned is adequate. Disputed or questionable interpretations are referenced and discussed when required. As a matter of practice, we have noted significant polymorphisms wherever they occur.

Much of our character list and discussion concern features of the posterior basicranium, a traditional source of cranial characters in systematic primatology. The anatomy of the ear region is not necessarily harder to interpret than other parts of the skull, but interpretation of fossils is often complicated by the fact that structures are usually small (and therefore easily damaged during fossilization or preparation) and packed within narrowly confined spaces (and therefore difficult to expose). In introducing their discussion of the morphology of basicranial structures in extinct primates, MacPhee and Cartmill (1986, p. 249) observed that the sample available to them for study was on the whole "a poor one despite the prominence given to basicranial traits in primate systematics." Ironically, poor material can become a fertile field for overinterpretation if researchers attempt to squeeze out more information than the material is capable of offering up (e.g., misinterpretation of the basicranial arterial network in *Ignacius* by MacPhee *et al.*, 1983; see Kay *et al.*, 1992). The problem is most severe when the basicranial sample size is $N = 1$, because there is no way of controlling for undetected distortion or polymorphism (see Table III). This is not to say that we should wait until large samples of a particular fossil taxon accrue before embarking on an effort to try to interpret what the evidence means. However, we emphasize that fossils, such as *Shoshonius,* that are represented by several individuals are probably more reliable as morphological documents than those known only from single specimens, for the simple reason that any interpretation has to be consistent with all of the evidence. Consistency need not always imply accu-

Table III. Examples of Difficulties in Morphological Information Retrieval for Specimens of *Shoshonius cooperi*[a]

	CM 60493	CM 60494	CM 60495	CM 31366	CM 31367
Posterior carotid foramen	ND	D	P	D	P
Stylomastoid foramen	ND	D	ND	P	P
Anular bridge/crista tympanica	ND	P	P	ND	ND
Stapedial canal	ND	ND	P	?P	D
Promontorial canal	ND	D	D	D	D
"Suprameatal foramen"	P	D	P	ND	P

[a]On side showing the best-preserved example of the feature, the feature is: P, present and undistorted (feature is preserved and surrounding context is not at all or only slightly disrupted); D, present but distorted (feature is preserved, but context is so damaged that original morphological relations are unclear); ND, not describable (feature not preserved, not exposed, or uninterpretable).

racy, but it is surely the best tool that we have to guard against fundamental errors of anatomic interpretation.

In the character analysis, we separately list and score taxa conventionally placed in Omomyidae that are sufficiently well represented by cranial material to make the exercise worthwhile. The taxa in question include *Shoshonius, Necrolemur, Rooneyia,* and *Tetonius.* The first two of these taxa are represented by multiple specimens, and their cranial anatomy can now be considered to be comparatively well known. *Rooneyia* finds a place on this list because the single skull attributable to this genus is unusually well preserved. Unfortunately, the same cannot be said of the only known skull of *Tetonius,* and it is certainly debatable whether its inclusion adds anything to the analysis. As it is now known, *Tetonius* does not differ significantly from *Shoshonius* with respect to the few characters for which it can be scored.

In contrast to our separate treatment of the preceding omomyid taxa, we considered adapiforms as a single OTU because we found that the four taxa complete enough for scoring (*Notharctus, Smilodectes, Adapis,* and *Leptadapis*) display the same states for the characters under consideration. *Mahgarita stevensi* has been left out of consideration because much controversy now attends its morphological interpretation, and we have not yet had the opportunity to examine all of the material assigned to this taxon. We do, however, comment on some of Rasmussen's (1990) evaluations of character states of *Mahgarita.* Except for *Pronycticebus,* also not separately considered here, no other adapiform is represented by well-preserved (or well-prepared) cranial remains.

Certain higher-level primate taxa have implied different taxonomic assemblages to different authors over the past several years. Our usage of these taxa is as follows: Strepsirhini = Adapiformes + Lemuriformes; Lemuriformes = Lemuroidea + Lorisoidea; Tarsiiformes = Tarsiidae + Omomyidae.

Methods

Here it is only necessary to point out that we utilized the most recent version of PAUP available to us (version 3.0s; Swofford, 1990) in attempting to discover the most parsimonious replicate of phylogeny given our character/taxon matrix. We have taken special pains to provide adequate information on the morphology and distributions of certain characters; our readers may not always agree with our polarity decisions, but we hope that the level of detail will make it clear why we reached the conclusions we did.

Perusal of the literature on the primate basicranium generated over the last 25 years amply demonstrates that investigations have tended to follow the same ruts along well-traveled avenues. Thus, in all such studies it has become mandatory to discuss whether or not tubes for stapedial arteries are larger than ones for promontorial arteries, whether ectotympanics are covered by a lip of the bulla or perched on its margin, whether the bullae are separated by a suture from the cochlear housing or continuous with it, and so on. Some of this may be justified, on the argument that well-researched and well-performing characters are preferable to ones of less certain status. However, the value of new material (such as that of *Shoshonius*) is that it permits one not only to hunt for potentially valuable additional characters but also to "test" prior assumptions about the taxonomic distribution of characters that were already known.

Characters and Transformations

The characters (transformation series) and polarity inferences presented in this section form the basic data used for phylogeny reconstruction in the next section (see Table IV). Further discussion of the anatomy of most of these characters can be found in papers by Cartmill (1980), Cartmill and MacPhee (1980), Rosenberger (1985), MacPhee and Cartmill (1986), and references cited by these authors. Where pertinent, conditions in *Shoshonius cooperi* are reported in detail.

1. Bulla, Structure of: 0, Suture Separating Bulla from Petrosal Bone, or Developmental Evidence for Presence of More Than One Element in Non-ectotympanic Portion of Bulla; 1, Suture Absent, No Developmental Evidence for Elements Additional to Petrosal

The petrosal bulla is the only currently recognized cranial synapomorphy of extant primates that is absent (or indeterminable) in all their probable sister groups (MacPhee and Cartmill, 1986). The observational evidence indicating uniform bullar structure in primates is of two types. First, no extant primate possesses a suture in the bulla (excluding, of course, sutures related to the ectotympanic in species that have tympanopetrosal bullae). Second, in young specimens there is never any developmental independence of the bullar process, which always arises in continuity with the underside of the petrosal bone. Alleged cases of bullar polymorphism within living primate taxa (e.g., alleged entotympanic involvement in the bullae of marmosets and tarsiers; see Starck, 1975; Schwartz, 1978) have not held up under more detailed scrutiny (MacPhee and Cartmill, 1986).

Likewise, unequivocal sutures in the nonectotympanic portion of the bulla have not been demonstrated in any fossil taxon currently accepted as (eu)primate. This may be regarded as a necessary, but not a sufficient, condition for the assertion that the bullae of fossil primates are compositionally

Table IV. Taxon–Character Matrix

Plesiadapidae	1 0	1	0 0 0 0 2 ? 0 0 0 0 0 0 0	0	0 0
Paromomyidae	0 0	1	0 0 0 0 2 0 0 0 1 0 0 0 0	0	0 0
Galeopithecidae	0 0	1	2 0 1 0 2 1 0 0 0 0 0 0 0	0	0 0
Lorisoidea	1 0	0	2 0 1 0 1 1 0 0 0 0 0 0 0	0	1 1
Cheirogaleidae	1 1	0	1 0 0 0 1 1 0 0 0 0 0 0 0	0	1 1
Lemuroidea	1 1	0	1 1 0 0 0 1 0 0 0 0 0 0 0	0	1 1
Adapiformes	1 1	0	1 1 0 0 0 1 0 0 0 0 0 0 0	0	0 1
Shoshonius	1 0	0	0 1 0 0 0 0 1 0 1 1 0 1	1	0 1
Tetonius	1 ?	?	? 1 ? ? 0 ? ? ? ? 1 0 1	1	0 1
Necrolemur	1 0	1	0 1 1 1 0 0 1 0 0 1 0 1	0	0 1
Rooneyia	1 0	1	0 1 1 ? 0 ? 0 0 0 0 0 0 0	0	0 1
Tarsius	1 0	1	2 1 2 1 0 0 1 1 1 1 0 1	1	0 1
Callitrichidae	1 0	0	2 1 1 0 0 1 0 1 0 0 1 0	1	0 1
Cebidae	1 0	0	2 1 1 0 0 1 0 1 0 0 1 0	0,1	0 1
Catarrhini	1 0	0,1	2 1 1 0 0 1 0 1 0 0 1 0	0,1	0 1

equivalent to those of extant primates. (Sufficiency would require ontogenetic confirmation, unlikely to be forthcoming for extinct taxa.) The paromomyid dermopteran *Ignacius* (Kay *et al.*, 1990, 1992) and perhaps its close relative *Phenacolemur* apparently fail to meet this standard, but the plesiadapid *Plesiadapis* had a bulla that was continuous with the promontory in the adult—a fact previously regarded by many observers as *prima facie* evidence for its petrosal origin (e.g., MacPhee, 1979). By contrast, all extant nonprimate archontans have at least one entotympanic in their bullae, usually in addition to other structures (Cartmill and MacPhee, 1980; Wible and Martin, 1993). We regard possession of at least one entotympanic as primitive for archontans; it is possibly primitive for Eutheria (MacPhee and Novacek, 1993) or even Theria (Norris, 1993).

Because the issue of bullar composition in fossil primates probably cannot be resolved, appropriate criteria for differentiating character states have to be found if one still chooses to use bullar structure in transformation series. Presence or absence of an edge-to-edge suture in the part of the bulla that is not ectotympanic in derivation is the criterion we have used here. Scoring requires no additional comment, except to repeat the point that, because there is no suture between the bulla and the cochlear housing in adult *Plesiadapis,* it must be scored in the same way as true primates.

The bean-shaped auditory bulla of *Shoshonius,* seen in its least deformed condition in CM 31366 (Fig. 1), bears a slight depression on its external surface that corresponds to the limits of the ectotympanic (as judged by the position of the anular bridge in other specimens). This depression probably, but not certainly, marks the obliterated tympanopetrosal suture. There is no indication of a true suture on the medial side of the bulla (see Character 10), although it must be noted that in this and other specimens the bulla is crushed along its junction with the central stem.

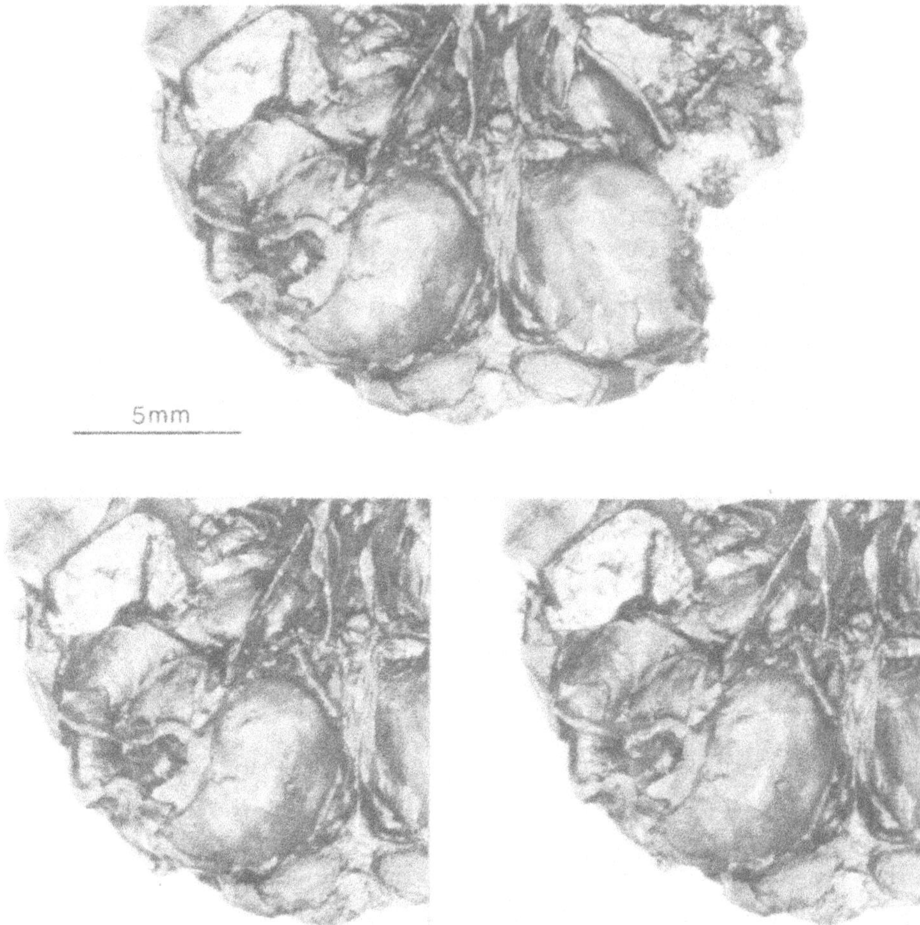

Fig. 1. Complete basicranium (top) and closeup of right auditory region (bottom, stereo pair) of *Shoshonius cooperi* CM 31366, ventral aspect. Key to structures is provided on facing page. Comparison with other specimens indicates that the posterior carotid foramen would have been situated high up on the bulla, in the damaged area (asterisk) adjacent to the posterior lip of the external acoustic meatus (EAM). A small portion of the internal carotid canal (ICC) is visible through EAM.

2. *Ectotympanic, Form of: 0, Crura and Body Expanded Relative to Ontogenetically Early Condition; 1, Crura and Body Not Expanded Relative to Ontogenetically Early Condition*

A narrow, ring-like ectotympanic—i.e., one that is little more than a sill embracing the tympanic membrane—is widely believed to be primitive for

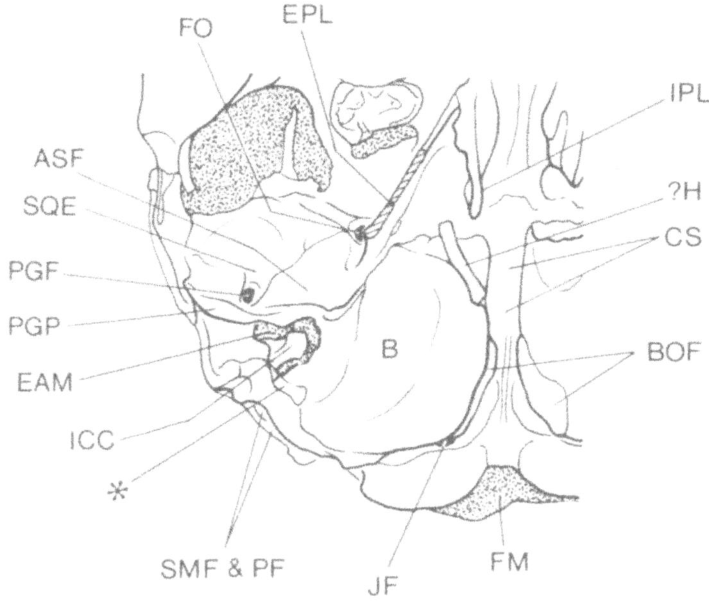

mammals because this character state is encountered in some extant eutherians (e.g., soricomorph lipotyphlans, *Orycteropus*), some metatherians, monotremes, and a disparate selection of early Tertiary mammals (Novacek, 1977). As usually analyzed, departures from an anular shape from expansion of the ectotympanic's medial or lateral borders are regarded as derived. However, MacPhee (1981, 1987) and MacPhee and Cartmill (1986) have argued that most "ring-like" ectotympanics are in fact slightly expanded and that the truly annular ectotympanic is a specialized retention of a fetal character limited to adapiforms and extant Malagasy lemurs (including Cheirogaleidae) among primates and rarely encountered elsewhere among mammals. Conditions in large-bodied Malagasy primates are assumed to be autapomorphous within Lemuroidea (MacPhee, 1987); the data matrix therefore does not reflect polymorphism for this character within this taxon.

Rasmussen (1990) questioned whether there was sufficient evidence to establish the nature of the ectotympanic in the adapiform *Mahgarita* but nevertheless went on to propose a solution (i.e., ectotympanic probably "extrabullar") grossly at variance with previous interpretations of ectotympanic anatomy in Adapiformes. We are unconvinced by Rasmussen's discussion on this point and refer the reader to Ross's evaluation of the relevant material (Chap-

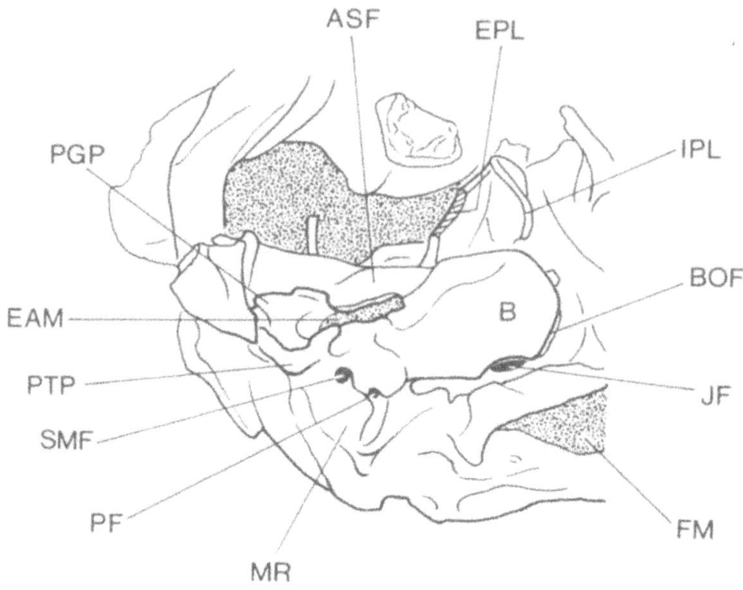

ter 15, this volume). It may be briefly noted that the transversely fractured basicranium (TMM 41715-1) referred to *M. stevensi* may be the key to the problem, if it is as well-preserved as Rasmussen's (1990) extensive discussion of its anatomy would imply. In any mammal, transverse sections of the cranium that pass through the central portion of the tympanic cavity must *by necessity* include segments of the ectotympanic, because of the absolutely small size and proximity of the structures involved. Accordingly, we predict that *in situ* remnants of the ectotympanic are probably represented in this specimen of *Mahgarita* and suggest that someone take up the task of properly preparing TMM 41715-1 so that this can be demonstrated.

3. External Acoustic Meatus, Shape of: 0, Not Extended as a Tube; 1, Extended as a Tube

Although the tubular EAM is traditionally considered to be a derived character (e.g., Kay *et al.*, 1992), with our outgroups it is actually parsimonious to conclude that it is primitive. However, successive plausible additions to the outgroup pool (e.g., megachiropterans, microsyopids) tend to reverse this polarity, and perhaps this is the solution most workers would prefer. Our strong suspicion is that tubular EAMs have been invented independently on

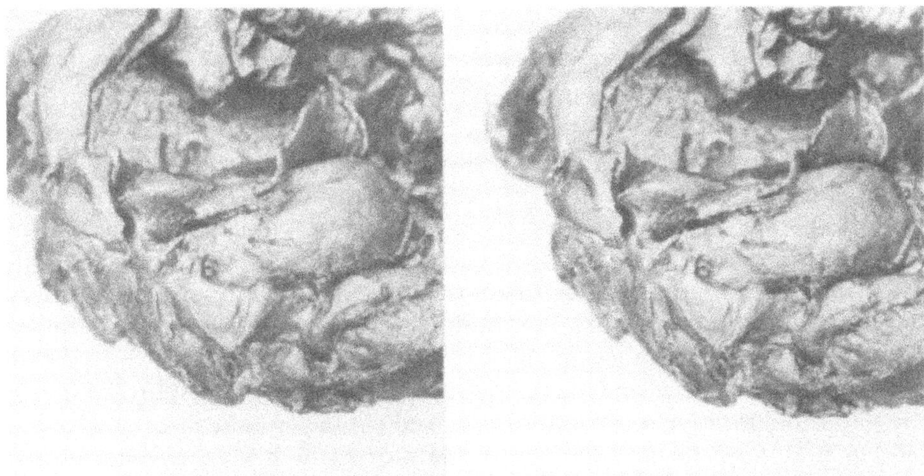

Fig. 2. Right auditory region of *Shoshonius cooperi* CM 31366, oblique posterior view with basicranium facing toward top of page (stereo pair). Relationship of posttympanic process (PTP), stylomastoid foramen (SMF), and parotic fissure (PF) is clear in this aspect (see key to structures, facing page).

several occasions during primate evolution, making this a character of low phylogenetic valence.

It is rare in primates for the meatus to terminate flush with its porus, and there is usually at least a slight degree of eversion of the ventral meatal lip (as in most extant lemurs, for example). Although contrasts in relative EAM development are not stark when primates are considered as a group, certain length differences are patent enough to permit useful character state discrimination.

Lorisoids could be scored as either polymorphic, to account for differing conditions in lorises and galagos, or as primitive, on the argument that the quasitubular meatus of *Loris* is not exclusively formed by the ectotympanic and is therefore autapomorphous. We prefer the latter resolution and consider the tubeless state to be primitive for lorisoids as a group. Other scorings follow usual practice. The meatus was nontubular in early anthropoids (Cartmill *et al.*, 1981; Simons, 1990), and it is still so in extant New World monkeys. We score catarrhines as polymorphic for this character in order to encompass the different character states expressed by fossil (e.g., *Aegyptopithecus*) and extant taxa.

It was noted in the original report (Beard *et al.*, 1991) that *Shoshonius* lacks a tubular EAM. We confirm this on the basis of a new specimen (Figs. 1 and 2). In its general conformation, the EAM of *Shoshonius* resembles that of small *Galago*, in which the ventral lip of the meatus is very slightly everted.

4. Annular Bridge, Completeness of: 0, Bridge Present, Complete (Recessus Dehiscence Absent); 1, Bridge Present, Incomplete (Recessus Dehiscence Present); 2, Bridge Absent

Among primate taxa in which the ectotympanic bone is intrabullar or aphaneric, the annular bridge has come to mean any sheet of bone that bridges the gap between the ectotympanic's crista tympani and the internal aspect of the bullar wall (cf. MacPhee and Cartmill, 1986). Obviously, if the ectotympanic is extrabullar or phaneric, there is no annular bridge. In adapiforms and extant lemurs, the annular bridge is present and definitely petrosal in origin, because in all known taxa there is a gap, the recessus dehiscence, between it and the ectotympanic ring. In some subfossil lemurs the gap is obliterated, a secondarily derived condition (MacPhee, 1987). *Necrolemur* and *Rooneyia* are similar to adapiforms in having a wide annular bridge but crucially differ in lacking the recessus dehiscence. This difference is conventionally interpreted as indicating that in omomyids the ectotympanic forms all of the annular bridge. This is not known for certain, however, because no young stages of omomyids are available to document ectotympanic ontogeny in this group (Conroy, 1980; MacPhee and Cartmill, 1986). Moreover, in *Megaladapis* the annular bridge is demonstrably formed by both the ectotympanic and the petrosal (MacPhee, 1987).

The polarity of this character is problematic. All living primates except lemurs display ChS 4.2. In outgroups other than Galeopithecidae and Paromomyidae, there is always at least a small annular bridge but never a recessus dehiscence. This suggests that the condition found in lemurs and adapiforms (ChS 4.1), in which the recessus dehiscence is present, is derived, but which of the two "absence" states is primitive is difficult to decide. *Plesiadapis* has a wide anular bridge without a recessus dehiscence, but Kay *et al.* (1992, p. 484) state that in *Ignacius* "the tympanic ring is fused with the bulla, and no subtympanic recess is apparent." In this respect *Ignacius* resembles modern *Cynocephalus,* suggesting that ChS 4.2 may represent a synapomorphy for Eudermoptera (*sensu* Beard, 1993). We tentatively consider ChS 4.0 as primitive, based on its occurrence in tupaiids, plesiadapids, and omomyids.

Because the bridge is on the internal face of the bulla, it is only exposed in those specimens in which the bulla is severely damaged. Small fragments of the bridge exposed in CM 60494 and 60495 reveal that it is quite narrow (<1 mm in width) and supported by a few tiny struts. Beard *et al.* (1991) stated that the bridge in *Shoshonius* is "similar" to that of *Necrolemur,* but new specimens confirm that it is actually much narrower, and the septae that support it are minute. Hence, the annular bridge in *Shoshonius* is morphologically intermediate between the conditions found in *Necrolemur* and *Rooneyia* on the one hand and *Tarsius* on the other. Conservatively, we score *Shoshonius* as primitive for this character, thus assuming no morphocline with *Tarsius.*

5. *Arterial Canals, Proximal Divisions of Internal Carotid Enclosed in: 0, Arterial Canals Absent; 1, Arterial Canals Present*

Production of intratympanic bony tubes around divisions of the internal carotid is rare in mammals other than primates and some of their relatives (MacPhee, 1981). On this basis we consider possession of tubes to be derived relative to the eutherian morphotype. Some primatomorphs lack tubes, but for different reasons. Absence of canals in cheirogaleids and lorisoids is almost certainly secondarily derived, consequent upon the loss of the internal carotid and its functional replacement by the ascending pharyngeal (see Character 8). Absence of canals in plesiadapids, paromomyids, and galeopithecids is possibly secondary as well, but in these cases absence cannot be linked to the formation of a pharyngeocarotid anastomosis (Kay *et al.,* 1992).

Promontorial and stapedial canals are not separately distinguished in the definition of this character, and it is therefore important to point out that the stapedial artery is absent in the adult stage of all anthropoids (a derived trait). There are doubts concerning the presence and size of stapedial tubes in some fossil primates (see Ross, Chapter 15, this volume). In the discussion below we attempt to clarify the situation in *Tetonius* and *Shoshonius.*

In AMNH 4194, the only known cranial specimen of *Tetonius,* the right auditory capsule is exposed, but poor preparation has destroyed many surfaces. There is a cast of the lumen of a large canal on the ventral aspect of the promontory that, by its position and relations, must be the promontorial canal. Unfortunately, the tube was broken distal to the origin of the stapedial canal, no part of which remains on the lateral aspect of the promontory. However, on the adjacent part of the roof of the tympanic cavity there is a small cylinder—in the correct position to be a stapedial canal—that bends sharply to the anterolateral. The facial canal never displays such an angulation, and we conclude that this apparent canal is for the stapedial.

The arterial tubes are badly damaged in all specimens of *Shoshonius* (Table III). Conditions in two of the newly collected crania, CM 31366 and 31367, indicate that in *Shoshonius* the internal carotid canal divides into two daughter canals, one for the stapedial artery, the other for the promontorial artery (Fig. 3). This point was not possible to ascertain in the original description (Beard *et al.,* 1991). In the new specimens the promontorial canal's distal section is intact where it leaves the promontory for the roof of the tympanic cavity, but the point at which it enters the endocranium is not exposed. In none of the specimens was it possible to recognize the track of the endocranial part of the stapedial artery (as ramus superior or middle meningeal).

In primate systematics much continues to be made of the relative sizes of the promontorial and stapedial tubes (e.g., Rasmussen, 1990) despite the fact that adapiforms are certainly, and omomyids are probably, internally variable

Table V. Position of Posterior Carotid Foramen Relative to Other Structures in *Shoshonius* and Some Other Primates[a]

Taxon[b]	PCF-SMF	PCF-PTP	PCF-EAM	M2W
Shoshonius cooperi	2.0 (0.63)[c]	1.0 (0.31)[c,d,e]	1.0 (0.31)[e]	3.2
Eulemur fulvus	2.6 (0.44)	3.2 (0.54)	3.8 (0.64)	5.9
Microcebus murinus	2.0 (0.90)	2.0 (0.90)	2.2 (1.0)	2.2
Avahi laniger	4.0 (1.08)	4.5 (1.22)	3.5 (0.95)	3.7
Galago senegalensis	4.7 (1.34)	4.8 (1.37)	4.8 (1.37)	3.5
Alouatta caraya	8.3 (1.12)	11.4 (1.51)	6.7 (0.91)	7.4
Tarsius syrichta	5.9 (1.51)	6.0 (1.54)	5.4 (1.38)	3.9

[a]PCF, posterior carotid foramen; SMF, stylomastoid foramen; PTP, posttympanic process; EAM, lip of external acoustic meatus; M2W, buccolingual width of M^2. In each column, numbers in parentheses represent average value of variate divided by M2W. All measurements are in mm; see Fig. 4 for explanation.
[b]Sample size for each extant species, $N = 3$.
[c]CM 31367.
[d]CM 60495.
[e]CM 60494.

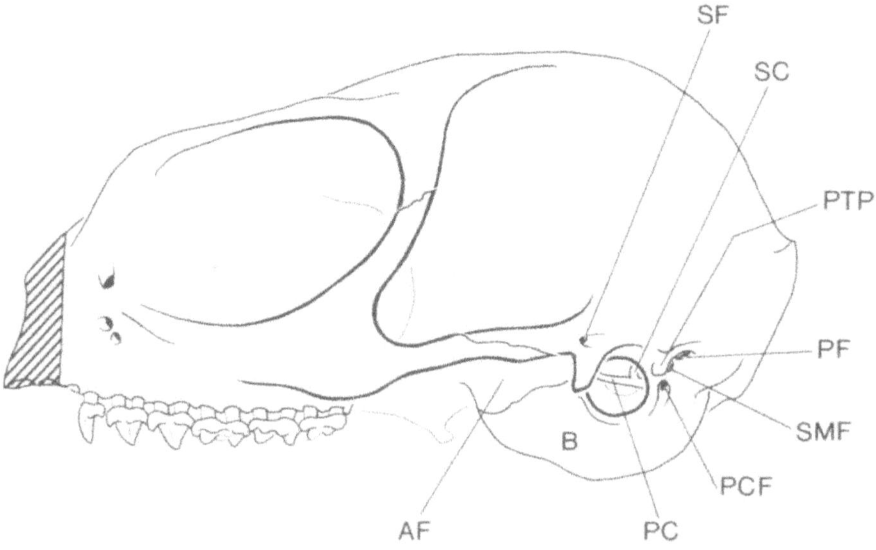

Fig. 3. Reconstruction of the skull of *Shoshonius*. Shape of the dorsal part of the neurocranium is conjectural, but the size of the orbits and the placement of the suprameatal foramen (SF), parotic fissure (PF), posterior carotid foramen (PCF), and alisphenoid flange (AF) closely correspond to dispositions in CM 60494, 31367, and 31366, after correction for distortion. Premaxilla (shaded area) is not known.

for arterial caliber (MacPhee and Cartmill, 1986). In view of the uncertainties expressed above, we will not hazard a guess as to the comparative size of arterial tubes in *Shoshonius*. (Size relationships as shown in Fig. 3 are for illustrative purposes only.)

Beard *et al.* (1991, Fig. 2) identified an apparent bony tube lying on the ventrolateral surface of the cochlear promontory as the "facial canal." The canal is depicted in a location in which the facial nerve is never found in mammals, and it must therefore represent some other structure. Reexamination of *Shoshonius* fossils indicates that the structure is not hollow and that it is separated from the promontory proper by a thin seam of matrix. Since the arterial tubes and facial canals of primate middle ears are derived from the petrosal bone during development (MacPhee, 1981), the "facial canal" must be part of some other bone. The likeliest candidate is the malleus (hammer), one of the three auditory ossicles. This ossicle typically possesses an elongated lever arm, the manubrium, which is embedded in the tympanic membrane. Parts of the malleus can be detected in several skulls, and it is clear that *Shoshonius* possessed an elongated malleolar manubrium. CM 31367, one of the two newly collected skulls, also possesses a "canal" plastered onto one side of the promontory, but in this case the structure emerges from the epitympanic recess. Because the "canal" is therefore inconstant in orientation as well, there is no reason to believe it represents anything other than a preservational artifact. *Shoshonius* had only one tube crossing its promontory.

6. *Posterior Carotid Foramen, Position of: 0, Posterolateral; 1, Posteromedial; 2, Anterolateral*

The PCF position has been thought to be comparatively constant within major primate groups, and for decades it has figured prominently in systematic investigations of primates (e.g., Gregory, 1920; Le Gros Clark, 1934; Simons and Russell, 1960; Szalay, 1975; Cartmill *et al.*, 1981). We mention this because *Shoshonius* does not conform to expectations as defined by this literature: its PCF location is radically different from the conditions found in other omomyids and *Tarsius* but is rather similar to that of adapiforms and lemurs. Details concerning the distribution of these character states among primatomorphs can be found elsewhere (MacPhee and Cartmill, 1986).

Beard *et al.* (1991, p. 65) originally described and illustrated the PCF of *Shoshonius* as being positioned "ventrolaterally, just posterior to the external acoustic meatus. This location, derived among primates, most closely approximates the PCF position in *Tarsius*, which is also ventrolateral but just anterior to the external acoustic meatus." On the basis of additional but still imperfect evidence, it now appears that the true position of this foramen is not only posterior to the external acoustic meatus but high up on the bullar surface, close to the position of the posttympanic process and the stylomastoid foramen. (The SMF is also out of position in Fig. 2 of Beard *et al.*, 1991.) As we stress below, however, *no single skull preserves both the bulla and the PCF in an undistorted state,* and thus a thorough morphological analysis is required in order to remove ambiguities.

The right auditory region of CM 31366 (Fig. 2) is the best preserved in the available sample. The middle and anterior parts of the bulla in this specimen are smoothly contoured and free of large cracks, in contrast to CM 60494, the bullae of which are extensively damaged. The two also differ in that the apparent center of the bulla of CM 60494 bears a large foramen (PCF) at the lateral end of a deep trough. Neither the trough nor the foramen is found in an equivalent position in CM 31366. Indeed, in this specimen no PCF is visible at all, because it was located in the posterior, less well-preserved part of the bulla. This inference is based on conditions in CM 60495 and 31367, in which the foramen is preserved even though the bulla is mostly destroyed.

In CM 60495, the PCF is located at the end of a trough incised into the bullar wall, as in CM 60494. The two specimens differ in that in CM 60495 the tube conducting the internal carotid to the posterior pole of the promontory is still in continuity with the PCF, thus demonstrating that the original relations of the portion of the bulla bearing this foramen have not been greatly distorted. In CM 60495 the PCF is located approximately 1 mm ventral to the lip of the posttympanic process, which is also intact. However, the SMF is not in its original location, and the margin of the EAM is broken, so the position of the PCF with respect to these structures cannot be reliably measured on this specimen. Conditions in CM 31367 confirm that the PCF is located high up on the posterior bullar wall, just beneath the posttympanic process. The SMF, the original location of which is preserved in this specimen, is situated about 2 mm dorsal and posterior to the PCF. Even though the bulla as a whole has been plastically deformed anteriorly in CM 60494, the closest approach of the margin of the EAM to the PCF (shortest gap, ~1 mm) can be reliably measured because there are no distorting breaks between these features. Triangulating with these figures, we find that the undistorted position of the PCF of *Shoshonius* lay on average within 1 mm of the posttympanic process and the posterior margin of the external acoustic meatus and within 2 mm of the stylomastoid foramen. This conclusion is depicted in our reconstruction of the skull of *Shoshonius* (Fig. 3).

It now remains to discuss how the PCF position should be scored in *Shoshonius* in relation to character states in other primates. Locative terms such as "posterolateral," "posterior," "ventral," and so forth have occasioned misunderstanding and debate because different authors have used these terms in different ways and because a rather large amount of interpretive baggage has been attached to PCF position in the past (see Ross, Chapter 15, this volume). The use of measurements may avoid some difficulties. Table V compares linear distances between the PCF and three other structures in *Shoshonius* and representative extant primates (see also Fig. 4). Each entry in each column lists the absolute value and, in parentheses, a corrected value that

Table VI. Relative Size of Central Stem in *Shoshonius* and Some Other Primates[a]

Taxon[b]	CSW	M2W	RW
Shoshonius cooperi CM 60493	1.1	3.3	0.33
Avahi laniger	2.8	3.5	0.80
Microcebus murinus	1.4	2.2	0.64
Tarsius syrichta	0.9	3.9	0.23
Alouatta caraya	8.5	7.4	1.15
Galago senegalensis	2.8	3.5	0.80

[a]CSW, width of central stem at closest approach of bullae; M2W, buccolingual width of M^2; RW, relative width of central stem (CSW/M2W). All measurements are in mm.
[b]Sample size for each extant species, $N = 3$.

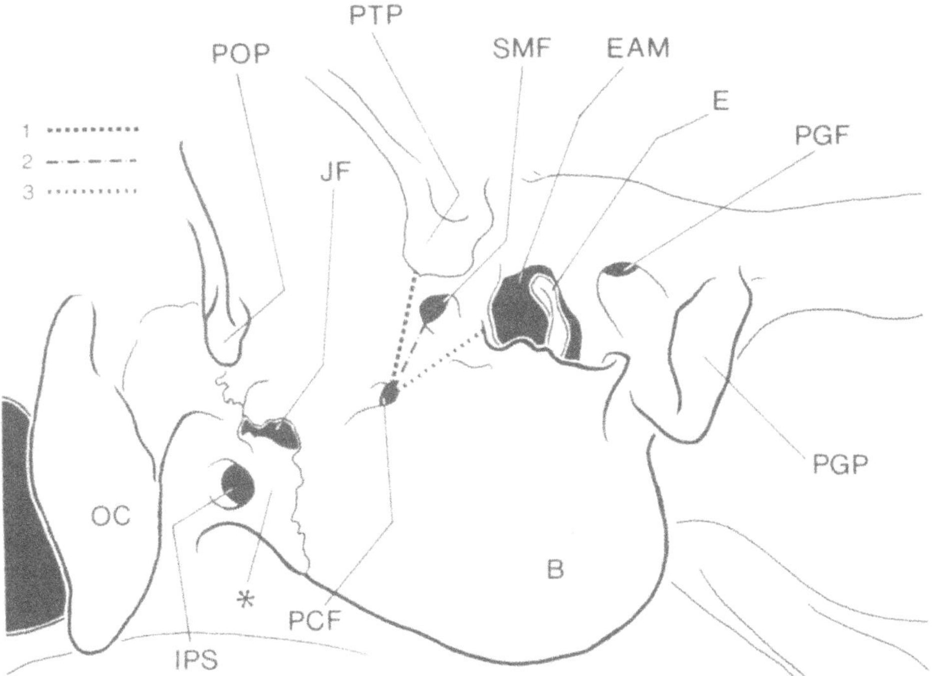

Fig. 4. Measurements used to define the position of the posterior carotid foramen on the bullar surface, in this case in *Eulemur fulvus*. Key: 1, PCF-PTP; 2, PCF-SMF; 3, PCF-EAM (see Table V). Asterisk identifies small basioccipital excrescence on bulla.

attempts to control for differences in body size (variable/M2W). The lower the corrected value, the relatively closer the PCF is to the compared structure. Note that, within this sample, *Shoshonius* exhibits the shortest relative distance between the PCF and either the posttympanic process or the posterior margin of the external acoustic meatus and the second shortest distance between the PCF and the SMF. *Galago, Tarsius,* and *Alouatta* have higher values for nearly all measurements, but lemurs (especially *Eulemur fulvus*) are quite close. Indeed, according to these figures, not only is PCF position in *Shoshonius* unlike that of *Tarsius,* either descriptively or metrically, it is possibly more exaggeratedly posterolateral than in most extant lemurs. We do not have comparable measurements for *Necrolemur, Microchoerus,* or *Rooneyia,* but in each of these taxa PCF position is descriptively rather like that of *Galago* and is therefore different from both *Shoshonius* and *Tarsius.* Conditions in *Tetonius* are unknown, because (*contra* Wortman, 1903, p. 407) nothing regarding PCF position can be legitimately interpreted from the type skull.

For scoring purposes, we relied primarily on the relative (corrected) distance between the PCF and SMF. The values in the table permit a reasonable, if arbitrary, discrimination among posterolateral (0.44–1.10), posteromedial (1.11–1.40), and anterolateral (>1.41) conditions, although more taxa and specimens ought to be measured before these character states are related to specific ranges.

The relatively posterolateral position of the PCF, combined with a posterior septum and bony tube for the internal carotid, are standard features of the adapiform ear region as defined by MacPhee and Cartmill (1986). These features are now seen to co-occur in *Shoshonius,* which tends to suggest that this complex should be regarded as primitive for primates (*contra* Archibald, 1977; MacPhee, 1987). The nature of the arterial network of plesiadapids and paromomyids continues to be controversial. Kay *et al.* (1992) have identified an aperture on the posterolateral side of the bulla in *Ignacius* that may have given passage to a tiny internal carotid artery (or perhaps only to the internal carotid nerve). A similar arrangement has been inferred for *Plesiadapis* (MacPhee and Cartmill, 1986). In *Cynocephalus* the internal carotid nerve enters the middle ear posteromedially and runs in the transpromontorial position across the cochlear housing (Wible, 1993). Although the proximal part of the internal carotid is completely involuted, it would appear on this evidence that *Cynocephalus* is derived from an ancestor in which carotid routing was much as it is in a loris. Nonprimate primatomorphs thus provide no clear signal of the primitive position of the PCF. We resolve the issue by regarding the posteromedial position of galeopithecids as derived, since plesiadapids and paromomyids apparently show the same, posterolateral, state. Another approach leading to the same conclusion would be to regard conditions in the outgroups as void for uncertainty and assume that the inferred primitive primate condition is primitive at a higher level. With this in mind, *Shoshonius* can be regarded as simply plesiomorphous for a trait for which *Tarsius* is autapomorphously derived. If Cheirogaleidae and lorisoids are more closely related to each other than the former is to (other) lemurs, a possibility most recently reviewed by Yoder (1992), then lorises and galagos are also autapomorphous for PCF position.

7. *Suprameatal Foramen, Presence of: 0, Absent; 1, Present*

There is no definition of the suprameatal foramen that is widely agreed on by comparative morphologists. Worse, judging from the nonconstancy of foraminal position in therians in which it is said to occur, the term "suprameatal foramen" has been applied to foramina that are probably nonhomologous. Although a complete history of the use of this term will not be offered here, it is critical to point out that some of the current confusion is a result of identifications made in two papers that were published in 1986. In his influential paper on the skull morphology of *Leptictis,* Novacek (1986) used the term "suprameatal foramen" to refer to an aperture situated immediately posterior to the morphological root of the zygomatic process of the squamosal, just above the position of the EAM. Previously, this foramen had been identified as the "subsquamosal foramen" by Cope (1880), Gregory (1910), Butler (1956), and other authors. No justification for name transference was provided by Novacek (1986), although he may have based the change on an unsupported inference that the foramen was pierced by the "suprameatal vein." In the same year, MacPhee and Cartmill (1986) drew attention to a peculiarity of cranial circulation in *Tarsius:* using serially sectioned material, they demonstrated that the intracranial stapedial ("middle meningeal") circulation is connected to the extracranial posterior auricular artery via an anastomosis that travels through a very large (and also partly venous) aperture in the cranial sidewall—their suprameatal foramen. In hindsight, the choice of this anatomic nomen was probably not propitious, although the positive outcome was an intelligible account of vascular organization on the sidewall of the tarsier skull.

Here, we are expressly interested in suprameatal foramina of the sort encountered in *Tarsius.* This foramen, which is constantly present in species of *Tarsius* (including *T. pumilus*), is a comparatively enormous aperture situated above the EAM in the interval bounded by the postglenoid and posttympanic processes (Fig. 5). As far as is now known, the relative size and vascular contents of the tarsier's suprameatal foramen are unique features of this genus among living mammals. By contrast, the subsquamosal foramen of Cope (1880), first identified in certain marsupials, has always been assumed to be an exclusively venous port linked to the transverse sinus and its dependencies. We acknowledge that assumptions do not constitute effective evidence and urge that a proper investigation of vasculature within marsupial subsquamosal foramina be undertaken with appropriate material. However, we repeat that *Tarsius* is the only mammal in which an arterial suprameatal foramen has been identified so far.

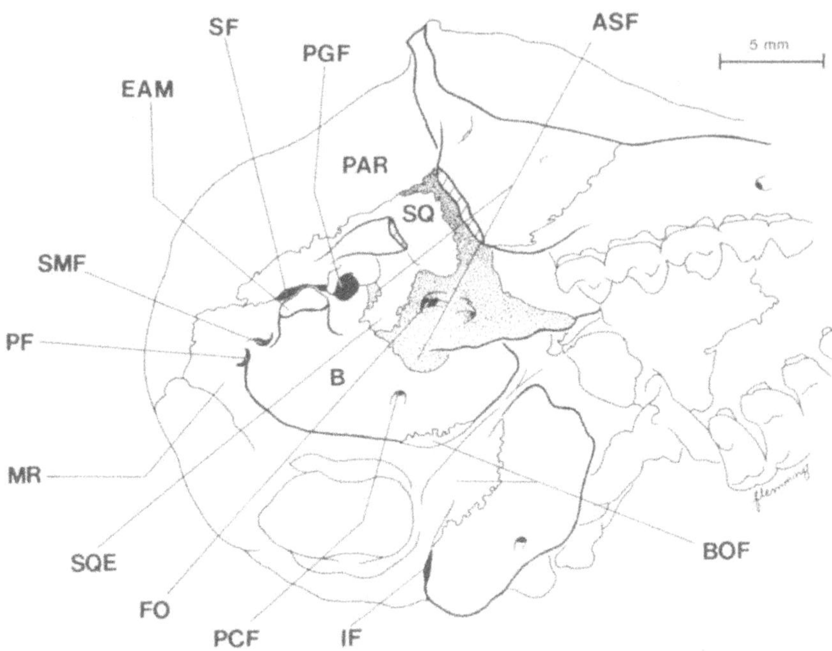

Fig. 5. Right auditory region of a juvenile of *Tarsius spectrum* AMNH 196478, oblique ventral view, showing suprameatal foramen (SF) and basioccipital and alisphenoid flanges (BOF, ASF). Most cranial sutures are open in this specimen, and the full extent of the right alisphenoid bone (in stipple) can be demonstrated.

Notwithstanding any morphological significance these facts may have, from a systematic standpoint they only have meaning if the tarsier condition can be placed on a morphocline that includes at least one other character state demonstrable in a real taxon. MacPhee and Cartmill (1986) noted that, of the taxa known to them, only *Necrolemur* possessed an aperture of the right size and in the right place to be homologized with the suprameatal foramen of *Tarsius*. We stand by this assessment on criteria of position and size (see also Ross, Chapter 15, this volume); a more definitive argument for homology would require knowledge of the vascular contents of the foramen in *Necrolemur,* which we do not have.

It remains to consider two other cases in which suprameatal foramina have been identified, both of which fail to convince using available criteria. *Shoshonius* exhibits a small foramen (double in some specimens) in the interval between the postglenoid and posttympanic processes, and Beard *et al.* (1991) regarded the similarity to the suprameatal foramen of *Tarsius* as sufficient to warrant a supposition of homology. Against this *prima facie* case at least two objections can be raised. First, the foramen in *Shoshonius* is positioned somewhat further anteriorly than that in *Tarsius,* on the morphological root of the zygomatic process of the squamosal (Fig. 3). Because of the configuration of the sidewall of the skull in *Shoshonius* (best seen in CM 60493), it is probable that this foramen led into diploe within the squamosal root of the zygoma and not into the endocranium. Hence, there is little likelihood that it carried a vessel connected with the stapedial circulation. Second, and again unlike *Tarsius,* the foramen in *Shoshonius* is not in broad communication with the postglenoid foramen behind an intervening projection of the tubular EAM. The end effect is therefore not tarsier-like in detail, which is all that is necessary to decide here. It may eventually prove useful to reclassify the aperture in *Shoshonius* as a subsquamosal foramen, although other kinds of comparisons and a meaningful definition of the latter aperture would be required to take this step.

In our view, the same conclusion—nonhomology with the condition expressed by *Tarsius*—must apply to the identification of a suprameatal foramen in *Ignacius* by Kay *et al.* (1992). As described and illustrated by these workers, the suprameatal foramen of *Ignacius* is positioned far out on the root of the zygomatic arch and thus even further away from the auditory region (and the postglenoid foramen) than the similar foramen of *Shoshonius.* Whatever this foramen is in a homological sense, it is not in the position or size range of the suprameatal foramen of *Tarsius,* which is all that is of interest here.

It remains to make some useful sense of these observations. The subsquamosal foramen as usually understood is considered primitive, possibly for therians generally (Gregory, 1910; Novacek, 1986). The amount of variation within placentals (cf. Gregory's illustration of this foramen in *Solenodon,* his Fig. 18) for this feature and its uncertain relationship to the postsquamosal foramen strongly suggest to us that emissary venous foramina on or near the root of the zygomatic arch have been developed on several occasions. Whatever one's preference for the homology of the aperture above the EAM in *Tarsius,* this foramen's enormous size is unquestionably derived on any comparative criterion, even if the aperture is itself primitive. The sidewalls of the skulls of *Rooneyia* and *Tetonius* are not preserved well enough to decide what they carried, and they are scored accordingly in the data matrix. *Plesiadapis* possesses an unusual foramen (*trou post-glenoïdien secondaire* of Saban, 1963), which we cannot homologize with anything in primates. Accordingly, character scoring in the data matrix shows derived presence of the suprameatal foramen *sensu* MacPhee and Cartmill (1986) in *Tarsius* and *Necrolemur* only.

8. *Internal Carotid Artery, Size of: 0, Unreduced; 1, Reduced (Function Assumed by Ascending Pharyngeal); 2, Reduced (Function Assumed by Vertebrobasilar System or Vessels Other Than the Ascending Pharyngeal)*

There is widespread agreement that the primitive pattern of blood supply to the brain involves the internal carotid artery, assisted by the vertebrobasilar system. We score as "unreduced" any group in which the stem of the internal carotid is large as judged by dissected specimens or arterial canal caliber. We recognize two derived patterns, the difference between them being which other major cranial vessel(s) assumes the function of supplying the brain. Plesiadapids, paromomyids, and galeopithecids all lack evidence of functional internal carotids and ascending pharyngeals in the adult stage (MacPhee and Cartmill, 1986; Kay *et al.*, 1990, 1992; Wible, 1993). In the case of *Cynocephalus,* it is now known that the circulus arteriosus is supplied by the orbital rete via an arteria anastomotica (Wible, 1993), but in fossil dermopterans the method of supply will probably never be satisfactorily resolved. Morphological conditions in cheirogaleids and lorisoids are undisputed, although the question of homology versus analogy for their ascending pharyngeals remains vexing (Yoder, 1992).

9. Parotic Fissure, Presence of: 0, Present (Exposed on Sidewall of Skull); 1, Absent (Incorporated into Middle Ear Cavity by Growth of Posterior Part of Bulla)

Anatomic correlates of expressions of this character were discussed by MacPhee (1981) in connection with exposure of the fossa for the origin of the stapedius muscle. In mammalian embryos, the stapedius fossa is located adjacent to the crista parotica, a low ridge on the otic capsule ventral to the eminence of the lateral semicircular canal. This is a highly conservative relationship that is rarely modified. Ontogenetic studies reveal that in extant strepsirhines (and evidently in anthropoids also), the stapedius fossa is progressively sealed within the middle ear cavity as a result of the pattern of growth of the caudal part of the bulla, a section of which originates from the crista parotica. By contrast, in lipotyphlans, elephant shrews, tree shrews, and probably most other mammals, the stapedius fossa is never completely incorporated into the middle ear and remains at least partly exposed on the sidewall of the skull, normally in a furrow or trench (hereafter, parotic fissure).

Another aspect of this conservative morphological pattern is that the parotic fissure is situated in close proximity to the distalmost part of the intratympanic course of the facial nerve, the course of which is essentially invariant in eutherians. When there is no complete stylomastoid foramen or facial canal, a distinction between the parotic fissure/stapedius fossa and the furrow for the facial nerve is often hard to make. Authors not making the kinds of distinctions reviewed here often refer to this whole complex as the "primitive stylomastoid foramen."

Tarsius is the only extant primate investigated to date that expresses the ancient pattern of stapedius fossa exposure (MacPhee and Cartmill, 1986). As far as we can tell, the parotic fissure is never exposed in adapiforms, but conditions are less clear in omomyids. Saban's (1963) depiction of the auditory region of *Necrolemur* strongly implies that this omomyid possessed an unenclosed parotic fissure. The fissure also appears to be present in *Shoshonius,* judging from the condition in CM 31366 (Fig. 2), the only specimen in which the area is intact enough for interpretation. In this specimen, the aperture of the putative stylomastoid foramen is continued posteriorly as a narrow furrow that arcs beneath a slight bulge on the sidewall of the skull (probably the eminence formed by the lateral semicircular canal). The fissure is not as wide as that in *Tarsius,* but its shape and position are similar. We cannot determine character states for *Rooneyia* and *Tetonius* and therefore score this character as unknown for these taxa. Conditions in Plesiadapidae and Paromomyidae are also uncertain, although it seems more likely than not that the stapedius fossa is outside the tympanic cavity in *Phenacolemur* (MacPhee, 1981). In *Cynocephalus,* however, the stapedius fossa is evidently enclosed by the bulla.

It is interesting that possibly as many as three tarsiiform taxa possess the parotic fissure, while no other primates do.

10. Basioccipital Flange, Presence of: 0, Absent; 1, Present

In all extant species of *Tarsius,* the entire lateral aspect of the basioccipital and exoccipital, from the sphenooccipital synchondrosis to the jugular foramen, gives origin to a thin shell of bone that broadly overlaps the medial aspect of the bulla (Fig. 5). Its free margin is digitated and thus seems to mimic a suture, but in a young disarticulated skull (AMNH 196623) it is clear that the flange is actually a squamous overgrowth that does not directly face onto the tympanic cavity. Young stages of *Tarsius* affirm that the bulla originates from the petrosal and that there is no *intrabullar* suture on the medial side of the bulla at any stage of development (MacPhee and Cartmill, 1986).

Among known omomyids, the basioccipital flange is also well developed in *Shoshonius* and *Necrolemur.* It is absent or insignificant in *Rooneyia.* We distinguish as nonhomolgous the condition seen in occasional specimens of lemurids and adapiforms in which there is a low, thickened rise on the basioccipital (?crest for prevertebral musculature) that overlaps the bulla to a small extent (Fig. 4).

11. Apical Aditus (of Anterior Accessory Cavity), Presence of: 0, Absent; 1 Present

In extant tarsiers and anthropoids, there is a narrow conduit between the tympanic cavity *per se* and the expansion in front of it (cellule-filled in anthropoids), identified by MacPhee and Cartmill (1986) as the "anterior accessory cavity." The existence of this conduit—the apical aditus (of the anterior accessory cavity)—has provoked less controversy than the cavity into which it leads.

MacPhee and Cartmill (1986) demonstrated that the apical aditus is not a primary feature of the middle ear cavity but develops as a result of intense pneumatization within the anterior margin of the petrosal plate.

According to the evidence reviewed by MacPhee and Cartmill (1986), the apical aditus does not occur in primates other than tarsiers and anthropoids. The only detailed objection that has been raised to this generalization is Rasmussen's (1990), whose views are based on his understanding of the morphology of *Mahgarita*. In *Mahgarita* the cochlear housing and promontorial canal are situated approximately in the middle of the uncomplicated middle ear cavity. Rasmussen (1990) regards the position of these structures as marking a boundary between the "tympanic cavity/[auditory] tube" and his "anteromedial cavity." The latter is described as "a large inflated intrabullar chamber" (p. 448). The promontorial canal bears a fragment of transversely directed septum, of unknown size, that circumscribed a "perforating foramen" that permitted communication between the tympanic cavity proper and the anteromedial cavity. The conclusion, developed in detail elsewhere in his paper, is that the septum, anteromedial cavity, and perforating foramen correspond closely to conditions seen in *Aegyptopithecus* and that for this and other reasons anthropoids evolved from adapiform ancestors not unlike *Mahgarita*.

In our view, Rasmussen's terminology clarifies nothing and instead adds an unnecessary layer of obscurity. Calling the medial and lateral portions of the middle ear cavity the "tympanic cavity/tube" and "anteromedial cavity" is no worse—but certainly no better—than calling the same volumes the "tympanic cavity" and "hypotympanic sinus," as did Gregory (1920), or the "cavité tympanique" and "diverticulum D2," which were Saban's (1963) names for the same volumes. MacPhee and Cartmill (1986) avoid these names and instead concentrate on a processual argument for distinguishing certain kinds of pneumatization that seem to have systematic significance. They utilize the expression "intrabullar cavity" to mean one that actually develops within the petrosal plate, and they extensively document two examples in which such development occurs (lorises and tarsiers). By contrast, Rasmussen (1990) treats any pocketing on the anteromedial side of the cochlear housing as an anteromedial cavity, no matter how it is formed, thereby removing any basis for making a distinction. This is warranted if the point to be made is that pneumatic spaces are useless as characters, but this would appear to go against his intention. Since the ontogeny of the "anteromedial cavity" cannot be assessed from fossils, nothing is added by additionally defining this space as "intrabullar" (unless all that is meant by intrabullar is that the cavity is inside the bulla, in which case the adjective is otiose).

Rasmussen calls the communication between the tympanic cavity proper and the anteromedial cavity a "foramen." It is true that, in a topographic sense, the projecting cochlear housing and transverse septum pinch the volume of the middle ear cavity in such a way that an internal constriction is created. The "foramen" thus produced is said to be the same as the apical foramen defined by MacPhee and Cartmill (1986). We think Rasmussen's terminology and the assessment on which it is based are odd. Why not, for example, call the internal constriction created by the posterior border of the middle cranial fossa a "foramen" that permits communication between the anterior and posterior cranial fossae? The real distinction here is between apertures produced by point bursts of bone remodeling (pneumatic foramina, of which the apical aditus and the mastoid aditus are examples) and ones produced by the mere conjunction of bone territories or bony prominences.

There is no recognizable equivalent of the apical aditus in nonprimate primatomorphs.

Shoshonius possessed an anterior septum (Beard *et al.*, 1991), as did, in fact, all Eocene primates for which we have a relevant record. The septum in these primates is actually best described as the medial wall of a canal or semicanal for the auditory tube, and its developmental origin is to be explained as a result of the inflating bulla leaving a passageway for the only channel that connects the middle ear to the nasopharynx. Here it may simply be said that the short, recumbent, unadorned septum of *Shoshonius* is very like that of omomyids, adapiforms, and Recent lemurs, all of which inflate the front portion of their middle ears in an uncomplicated fashion. Very likely, then, the anterior septum as seen in *Shoshonius* is merely primitive.

12. Central Stem, Size of: 0, Broad; 1, Narrow

"Central stem" refers to the midline keel of the posterior basicranium, normally composed of the basisphenoid and basioccipital bones. In the primitive condition the central stem is comparatively broad, but in some taxa the keel has become so narrow that the bullae are almost in midline contact. Relative narrowness is expressed in Table VI as the ratio of the width of the central stem to the buccolingual width of M^2. *Shoshonius* and *Tarsius* are at a derived extreme within primates (Figs. 1 and 5). The paromomyid *Ignacius* evidently possessed a very narrow central stem (Kay *et al.*, 1992), but this feature was clearly broad in Plesiadapidae.

13. Alisphenoid Flange, Bullar Onlap of: 0, Nonexistent or Trivial; 1, Extensive

Although many mammals, including some scandentians and primates, display a narrow connection between the external pterygoid lamina and the rostral surface of the bulla, it is extremely rare for the latter to overlap the bulla extensively. The connection in *Tupaia* (MacPhee, 1981, p. 159) and *Galago*, for example, consists of a hair-fine bridge, often broken in prepared skulls, which presumably represents a structurally antecedent condition. In *Tarsius,* the overlap is so extensive that a prong of the alisphenoid reaches almost as far as the anterior margin of the postglenoid foramen laterally, thereby nearly excluding the entoglenoid portion of the squamosal from contacting the bulla. Medially, it covers much of the part of the bulla housing the anteromedial cavity. It seems to us useful to refer to this posterior extension of the external pterygoid process by a separate name, alisphenoid flange (a deliberate parallel to the basioccipital flange, which also overlaps the bulla extensively without participating in the bullar wall *per se;* see Character 10).

AMNH 196478 is a young adult specimen of *Tarsius spectrum* in which most cranial sutures are still open (Fig. 5). In this specimen it is possible to determine the full extent of the alisphenoid flange because on the right side most of the squamosal is broken off and on the left side the suture between the squamosal entoglenoid and the alisphenoid flange is patent. The relationship of the alisphenoid and the squamosal is complicated. In essence, the alisphenoid actually forms most of the posterior wall of the infratemporal fossa, but much of it is occluded by squamosal overgrowth. The tiny entoglenoid portion of the squamosal sends down a narrow tongue that overlaps part of the alisphenoid flange, which in turn overlaps the bulla.

As noted by Rosenberger (1985), *Necrolemur* and *Microchoerus* resemble *Tarsius* in the degree of alisphenoid overlap. Presence of a tarsier-like condition for this character in *Necrolemur* is confirmed by C. Ross (*personal communication*) on the basis of his study of MCZ 8879. Beard *et al.* (1991) reported a similar condition in *Shoshonius cooperi,* an assessment that is fully supported by additional material. The least deformed example of the alisphenoid flange in this species is seen in CM 31366 (Fig. 2), in which the lamina (on the right side) can be seen to extend almost as far as the EAM. We interpret a line passing through the area in question as a suture separating the entoglenoid region of the squamosal and the flange *per se.* Conditions in *Tetonius* are less certain because the basicranium is severely damaged. However, we noted that on the left side of the unique cranial specimen, there is a large process in the correct position to be the alisphenoid flange. Its identity as such is corroborated by the fact that its inner aspect appears to have been an articular surface. (Because of its position, this surface cannot be a contact facet for the ectotympanic, a small part of which is preserved inside the middle ear as Szalay, 1976, noted.) There is no indication of an alisphenosquamosal suture, and we assume that it is obliterated in the available specimen. By contrast, bullar overlap is nonexistent or negligible in adapiforms (Rosenberger, 1985), with the possible exception of *Mahgarita.*

14. Postorbital Septum, Presence of: 0, Absent; 1, Present

There is no dispute concerning the polarity of this character, but there is disagreement regarding the homology of the character state expressed by

Tarsius. Based on morphological differences between the complete postorbital septum of anthropoids and the periorbital flanges of *Tarsius* (cf. Simons and Rasmussen, 1989), we doubt the homology of these structures in these taxa (*contra* Cartmill, 1980; Ross, Chapter 15, this volume). On the other hand, *Tarsius* obviously differs from every other taxon in our comparative set. By scoring it as "primitive", we signify only that it does not bear the anthropoid character state. The alternative is to treat the tarsier condition as autapomorphous, which we do not follow here. *Shoshonius* resembles living and fossil strepsirhines and other omomyids in retaining the primitive version of this character (Beard *et al.*, 1991).

15. Choanae, Shape of: 0, Broad; 1, Very Narrow and "Peaked"

Rosenberger (1985) distinguished the narrow and dorsally peaked choanae of *Tarsius, Necrolemur,* and *Tetonius* as a derived condition relative to the much broader and more gently arching choanal roof found in other living and fossil primatomorphs. *Rooneyia* is once again distinct (and primitive) within known omomyids. We agree with Rosenberger's assessment of the polarity and distribution of these character states and add that the derived condition also occurs in *Shoshonius*. Ross (Chapter 15, this volume) dismisses choanal shape as a useful character for phylogenetic analysis on the grounds that it is functionally correlated with orbital hypertrophy. To this objection we offer two rejoinders: (1) we have not incorporated orbital hypertrophy as a character in this analysis, so its possible functional correlation with choanal shape bears no role in assessing whether or not the latter character should be included here; (2) since *Necrolemur* has relatively much smaller orbits than either *Shoshonius* or *Tarsius* (Beard *et al.*, 1991; Ross, Chapter 15, this volume), yet all three taxa show the same choanal morphology, we question the intimate functional relationship between these characters alleged by Ross (Chapter 15, this volume).

16. Snout, Reduction of: 0, Unreduced; 1, Reduced

Living and fossil dermopterans and strepsirhines retain a pronounced snout relative to the condition in *Tarsius*. Polarization of character states seems uncontroversial on the basis of distributional criteria and the severe reduction of the nasal fossa and olfactory bulbs in *Tarsius* (cf. Cartmill, 1980, and references cited therein). We scored *Necrolemur* and *Rooneyia* as primitive for this character, because both taxa exhibit relatively unreduced, galago-like snouts (Stehlin, 1916; Wilson, 1966). Although none of the *Shoshonius* skulls currently available preserves the facial region completely intact, in the best-preserved examples (e.g., CM 60493) all that is missing is the premaxilla and the floor of the nasal aperture. Alveoli for anterior teeth indicate that the tooth row was sharply U- or V-shaped at the front of the palate. As there is no possibility that the snout extended much beyond the alveolar border of the anteriormost teeth, we feel justified in scoring *Shoshonius* as derived. A similar logic can be applied to the interpretation of the condition in *Tetonius*. Most living anthropoids display snout reduction, although there are obvious exceptions. The morphotypic state for anthropoids is difficult to glean from available fossils (e.g., Fayum anthropoids). Accordingly, we score cebids and catarrhines as polymorphic. Callitrichids invariably show the derived state for this character.

17. Toothcomb, Presence of: 0, Absent; 1, Present

The toothcomb of lemuriform primates remains one of the most diagnostic synapomorphies of that clade (e.g., Szalay and Seligsohn, 1977; Rosenberger and Strasser, 1985). Equivalents to the lemuriform toothcomb are not found elsewhere among primatomorphs, including galeopithecids (the "comb" of which is obviously nonhomologous with that of lemuriforms). Although only the alveoli of the lower incisors and canines are known in *Shoshonius*, these demonstrate, unsurprisingly, that a toothcomb was absent. The lower anterior dentition of *Shoshonius* probably resembled that of its close relative *Washakius insignis*, recently described by Covert and Williams (1991).

18. Postorbital Bar, Presence of: 0, Absent; 1, Present

The polarity and distribution of this character among primatomorphs is undisputed and requires no further discussion. Although it is not directly relevant to this analysis, we do not accept the claimed homology between the postorbital bar of primates and that of tupaiids (e.g., Kay *et al.*, 1992) for reasons elaborated elsewhere (Beard, 1993).

Parsimony Analyses

Most Parsimonious Trees

The results of our character analysis were coded for parsimony assessment using PAUP 3.0s (Table IV). The branch-and-bound algorithm, which guarantees discovery of all most-parsimonious trees (MPTs), yielded 492 MPTs for our data matrix, each having a tree length (TL) of 36 and a consistency index (CI) of 0.583. Obviously, given this number of MPTs, our data set provides relatively little resolution of the phylogenetic relationships among primatomorphs. Indeed, a strict consensus of these 492 MPTs (Fig. 6)

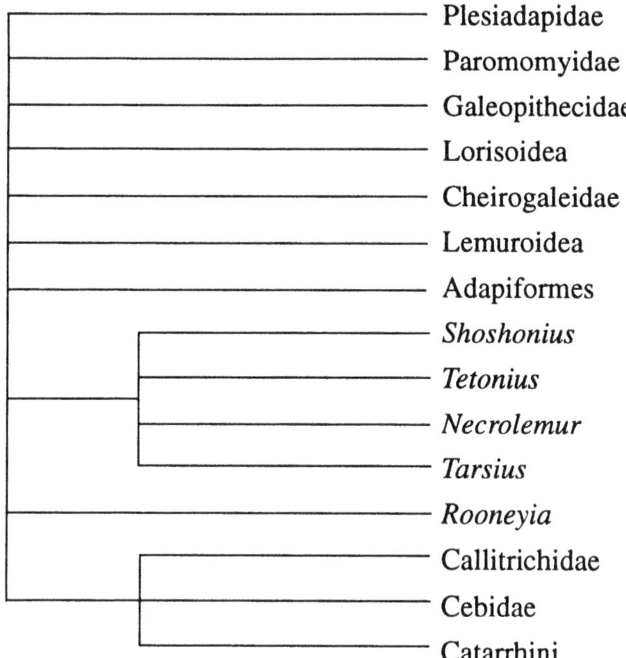

Fig. 6. Strict consensus tree for 59 MPTs, with no constraints and no weighting scheme imposed.

supports the monophyly of only two clades of primates. The first of these approximates the classical notion of Tarsiiformes by including *Shoshonius, Tetonius, Necrolemur,* and *Tarsius* (but not *Rooneyia*). The second group includes the three anthropoid taxa. Relationships within these two clades, as well as all other relationships among Primatomorpha, are left unresolved by this analysis.

Two general observations regarding these results merit discussion here. First, we found it surprising that this data set yielded such a low level of resolution of phylogenetic relationships. For example, primate monophyly is not supported by our strict consensus tree despite the fact that such diagnostic primate characters as the postorbital bar (Character 18) were included. Closer examination of the data revealed that, in cases such as this, further phylogenetic resolution was impeded by the presence of "crossing synapomorphies" (MacPhee, 1993)—characters that show taxonomic distributions that conflict with those of characters such as the postorbital bar, which would otherwise provide clear-cut support for the monophyly of primates. This phenomenon led us to resort to the *a posteriori* reweighting option available in PAUP, whereby our characters were weighted according to the maximum value of their rescaled consistency indices (see below). By this means, characters that show taxonomic distributions consistent with the strict consensus tree (such as Character 18) are heavily weighted, whereas characters that show a high level of homoplasy on the strict consensus tree (e.g., Character 3) are weighted less.

On the other hand, the low level of phylogenetic resolution yielded by our data set may simply reflect the paucity of clear-cut, shared-derived cranial traits for certain higher-level primate taxa. For example, aside from the potential cranial synapomorphies that have been cited in support of a strict *Tarsius* + Anthropoidea clade (e.g., Cartmill, 1980; Cartmill *et al.*, 1981; Ross, Chapter 15, this volume), we are unaware of well-documented derived cranial characters that would link anthropoids with *any* other known clade of primates. As Fig. 6 illustrates, conflicting evidence provided by different cranial characters of *Shoshonius* and other omomyids effectively overwhelms these putative tarsier–anthropoid cranial synapomorphies, which (together with other considerations) leads us to question their validity. If our skepticism on this point is legitimate, the cranial anatomy of anthropoids serves more to distinguish them from all other primates than to aid in the identification of their nearest relatives.

Interestingly, character support for the *Shoshonius* + *Tetonius* + *Necrolemur* + *Tarsius* clade is stronger than that for the clade comprising the three anthropoid taxa. The four tarsiiforms are united by their common possession of basioccipital and alisphenoid flanges that overlap the bulla and choanal shape (ChS 10.1, 13.1, and 15.1), which are each unique and unreversed characters. In addition, under the DELTRAN optimization regime of PAUP 3.0s, the monophyly of these taxa is supported by their possession of a parotic fissure (ChS 9.0), which is interpreted as a secondary reversal on the most parsimonious character walk. In contrast, the monophyly of Anthropoidea is unequivocally supported only by the presence of a postorbital septum and apical aditus (ChS 11.1 and 14.1), of which only the postorbital septum is unique and unreversed (the apical aditus evolved convergently in *Tarsius*). Under the ACCTRAN optimization regime of PAUP 3.0s, snout reduction (ChS 16.1) provides further character support for anthropoid monophyly.

A Posteriori Reweighting

In an effort to overcome the problem of crossing synapomorphies noted above, we undertook a second analysis employing an *a posteriori* weighting scheme in which characters were weighted according to the maximum value of their rescaled consistency indices based on an initial parsimony run (Swofford, 1990). The individual weights applied to each character under this procedure are given in Table VII. A branch-and-bound search of this reweighted data matrix yielded five MPTs having TL = 22,212 and CI = 0.693. Unsurprisingly, a strict consensus of these five MPTs shows a high level of resolution of the phylogenetic relationships among primatomorphs (Fig. 7). Of greater interest is the fact that the more highly resolved topology in Fig. 7

Table VII. Character Weights Based on Maximum Rescaled Consistency Indices[a]

1(1000); 2(1000); 3(400); 4(556); 5(375); 6(300); 7(1000); 8(1000); 9(1000); 10(1000); 11(333); 12(250); 13(1000); 14(1000); 15(1000); 16(333); 17(1000); 18(1000)

[a]Character number given outside parentheses; character weight given inside parentheses.

conforms with a number of widely accepted higher-level clades of primatomorphs, including Eudermoptera, Primates, Strepsirhini, Lemuriformes, and Cheirogaleidae + Lorisoidea. Curiously, however, *Rooneyia* comprises the sister group of all other primates, a finding that reinforces its distinctiveness with respect to other putative tarsiiforms.

Despite the fact that the topology of the consensus tree illustrated in Fig. 7 corresponds to several of our preconceived notions regarding primate phylogeny, we place relatively little significance on this result. As Novacek (1993) has pointed out, there is an element of circularity involved in using the consistency indices of individual characters from a first parsimony run as the means for weighting the same set of characters in successive runs. However, we argue that the weights accorded to individual characters in Table VII provide insight into their utility for reconstructing phylogenetic relationships among primatomorphs. As is the case with any limited data set, inferences

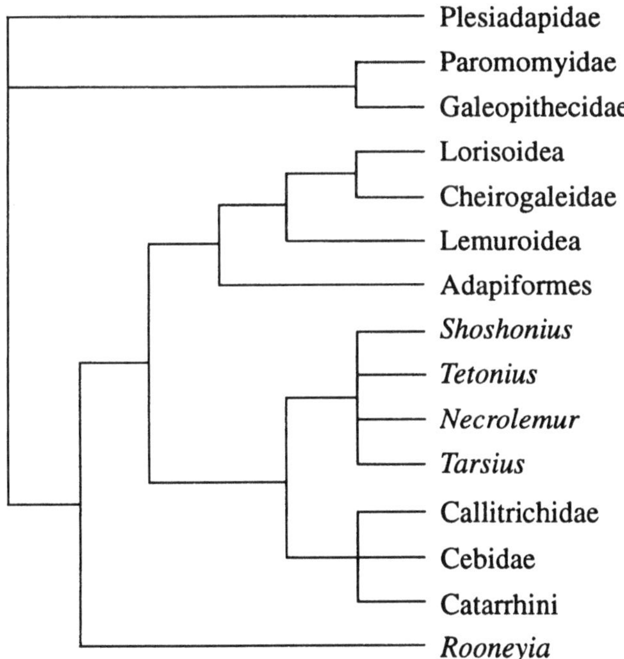

Fig. 7. Strict consensus tree for five MPTs, with *a posteriori* weighting imposed.

concerning the reliability of phylogenetic results must be based on comparisons with similar results derived from other, preferably unrelated, data sets (see below).

Discussion of Phylogenetic Results

Monophyly and Relationships of Tarsiiformes

Among the most persistent groupings we encountered in our parsimony analyses was one in which *Tarsius* was united with one or more of the extinct taxa that are traditionally classified as omomyids. This strong association between *Tarsius* and omomyids reflects their common possession (in whole or in part) of a number of derived cranial characters, including presence of basioccipital and alisphenoid flanges that overlap the bulla posteromedially and anterolaterally, respectively; narrow, peaked choanae and narrow central stem of basicranium; and presence of a reduced snout, parotic fissure, and suprameatal foramen. We interpret our findings as additional support for the monophyly of Tarsiiformes (Tarsiidae + Omomyidae), a view that has been endorsed, at least implicitly, by most (e.g., Wortman, 1903; Gregory, 1915; Simons, 1961; Szalay, 1976; Hoffstetter, 1977; Gingerich, 1981; Rosenberger, 1985; Simons and Rasmussen, 1989; Beard *et al.,* 1991; Beard and Wang, 1991), but not all (e.g., Simpson, 1940, 1955; Hürzeler, 1948; Cartmill and Kay, 1978; Cartmill, 1980; Cartmill *et al.,* 1981; Schmid, 1981, 1982; MacPhee and Cartmill, 1986) students of the primate fossil record.

Within Tarsiiformes, there is substantial evidence that individual omomyid taxa share more recent common ancestry with *Tarsius* than they do with each other. Hence, the traditional concept of Omomyidae (e.g., Szalay, 1976; Gingerich, 1981) probably consists of a paraphyletic assemblage of primitive tarsiiform taxa united solely on the basis of symplesiomorphy (Rosenberger, 1985; Beard *et al.,* 1991). Perhaps the most problematic of these putative fossil tarsiiform taxa is *Rooneyia viejaensis* (Wilson, 1966), which consistently fell outside of a *Tarsius* + *Shoshonius* + *Tetonius* + *Necrolemur* clade in our parsimony analyses. We do not attempt to resolve the phylogenetic position of *Rooneyia* here. However, because virtually all aspects of its cranial anatomy differ so strongly from the conditions found in *Shoshonius,* we find it extremely unlikely that this enigmatic animal is derived from a washakiin ancestry, as was tentatively suggested by Wilson (1966) and Szalay (1976, Fig. 138). Hence, we agree with Rosenberger (1985) that *Rooneyia* must represent an extremely basal tarsiiform clade, if that is what it is. In our opinion, alternative possibilities concerning the phylogenetic position of *Rooneyia* must also be considered, including the chance that *Rooneyia* bears some special relationship with anthropoids, as originally suggested by Wilson (1966) and Simons (1968,

p. 11). Obviously, further material of this interesting taxon could go a long way toward resolving these issues.

On strictly philosophical grounds, we would not hesitate to break with tradition and divide the longstanding concept of Omomyidae into assemblages of taxa that might more closely approximate actual clades. At present, the primary obstacle to this solution is our ignorance of the cranial anatomy of *Omomys* and its close allies in the tribe Omomyini. Because *Omomys* is the type genus of the family Omomyidae Trouessart, 1879, we have no way of knowing whether or not this family should be considered a junior synonym of Tarsiidae Gray, 1825 in a phylogenetic classification. The phylogenetic relationships among *Tarsius, Shoshonius,* and other omomyines (i.e., Omomyini and Uintaniini) are not clearly resolved, so we refrain from formally referring *Shoshonius* and its close relatives in the tribe Washakiini to the Tarsiidae here. Nevertheless, of the extinct tarsiiforms that are currently represented by relatively complete cranial remains, *Shoshonius* appears to share most recent common ancestry with *Tarsius.*

Congruence with Other Data Sets

If our phylogenetic reconstruction based on cranial characters is accurate, it follows that extant *Tarsius* diverged from the clades that ultimately gave rise to all other living primates by at least 50.5 Ma, which is the age of *Shoshonius* (Beard *et al.,* 1991). If the strict monophyly of Tarsiiformes is accepted, *Tarsius* must have diverged from all other extant primate clades by at least the basal Wasatchian/Sparnacian (about 57 Ma), which marks the first appearance of omomyids in the fossil record (e.g., Krause and Maas, 1990). Aside from the large body of neontological evidence supporting the monophyly of Haplorhini, how does this phylogenetic conclusion conform with those derived from other data sets?

Dental and postcranial evidence pertaining to omomyids provides very little support for the monophyly of Tarsiiformes at present, but to our knowledge these data sets cannot be interpreted as refuting tarsiiform monophyly either. The dentition of extant *Tarsius,* although derived in numerous details, is generally characterized by very simple, tribosphenic molar morphology. Among extinct primate taxa, only certain omomyids and *Afrotarsius chatrathi* from the Fayum of Egypt approach the molar morphology of *Tarsius* (e.g., Gingerich, 1981; Simons and Bown, 1985), but none of these taxa exhibits clear dental synapomorphies with *Tarsius* (e.g., Ginsburg and Mein, 1987). Modern tarsiers also display numerous postcranial autapomorphies that are not found in omomyids, which retain many postcranial characters that are plausibly interpreted as primitive for primates (e.g., Simpson, 1940; Dagosto, 1985; Gebo, 1987, 1988). Most of the derived features of the omomyid postcranium appear to be synapomorphies for Haplorhini rather than Tarsiiformes (e.g., Szalay and Dagosto, 1980, 1988; Dagosto, 1985), although we stress that the taxonomic representation of postcranial characters across Omomyidae remains poor at best.

A large and diverse assemblage of neontological data sets bears on the phylogenetic position of *Tarsius* with respect to other extant primates. The vast majority of these data support the monophyly of Haplorhini (e.g., Hoffstetter, 1988; Martin, 1990), as we mentioned previously. Although it is impossible to test the monophyly of Tarsiiformes directly on the basis of neontological data sets (since all tarsiiform taxa aside from *Tarsius* itself are extinct), several frequent results of these studies are consistent with our conclusion that *Tarsius* shares a geologically remote (*ca.* 57 Ma) common ancestry with other extant primates. For example, even though the nucleotide sequence data of Koop *et al.* (1989a,b) corroborates the monophyly of Haplorhini, the level of support for the *Tarsius* + Anthropoidea grouping in their MPTs is weak compared with that for certain other primate clades. Specifically, their data require 230 additional steps (mutations) before the monophyly of Anthropoidea is broken and 91 before the monophyly of Strepsirhini dissolves, whereas only 28 additional steps are required to break the association between *Tarsius* and anthropoids in favor of a monophyletic Prosimii. The phylogenetic position of *Tarsius* based on data from the COII gene is equally problematic. Adkins and Honeycutt (1993) found in their analysis of transversion data that *Tarsius* would not group with primates unless constrained to do so. Adding primate monophyly as a constraint produced two MPTs in which *Tarsius* formed the sister group either of all other primates or of strepsirhines. Constraining haplorhine monophyly required five steps additional to that for the shortest tree. These different, but generally low, levels of support for linking *Tarsius* with other major primate clades may reflect a relatively short interval of common ancestry between *Tarsius* and other extant primate clades. This possibility was suggested earlier by Dutrillaux and Rumpler (1988, p. 132) on the basis of their study of the karyotype of *Tarsius syrichta:* "The *Tarsius* lineage was isolated very early during primate evolution and either did not share any common trunk with strepsirhines or simians or shared only a very short trunk."

Alternative Phylogenetic Reconstructions

The hypothesis that *Tarsius* represents the sole living survivor of a great radiation that flourished during the Eocene has been widely favored by paleontologists since Cope's time. However, there is another tradition, of almost equal age, that primarily focuses on the evaluation of neontological evidence for ascertaining the phylogenetic position of *Tarsius.* In recent decades workers have attempted to retrieve phylogenetically useful information from both of these data sets but have not necessarily agreed on the interpretation of their results. For example, most paleoprimatologists continue to see merit in arrangements that group Tarsiiformes (*Tarsius* + Omomyidae) and Anthropoidea within a larger assemblage, Haplorhini. However, there have been some recent challenges to this view. For example, Schwartz and Tattersall (1987) concluded that the linkage between tarsiiforms and anthropoids was dubious because *Tarsius* and extant lorisoids (including cheirogaleids) share a variety of allegedly homologous derived states not seen in other primates. In agreement with other commentators (Dutrillaux and Rumpler, 1988; Hoffstetter, 1988; Koop *et al.*, 1989b), we find the probability that the Schwartz–Tattersall hypothesis is correct (or even parsimonious, except in the most contrived circumstances) to be vanishingly small, and we will not consider it further here.

A possibly more plausible argument, largely developed by Cartmill and his colleagues (Cartmill and Kay, 1978; Cartmill, 1980; Cartmill *et al.*, 1981; MacPhee and Cartmill, 1986; see also Schmid, 1981, 1982; Ross, Chapter 15, this volume), takes the position that *Tarsius* and anthropoids uniquely share certain character states (thereby supporting Haplorhini) that are not known to occur in any omomyid investigated to date (contrary to the conventional concept of Tarsiiformes). This argument has been challenged by several authors, most forcefully by Rasmussen (Simons and Rasmussen, 1989; Rasmussen, 1990). Many of Rasmussen's criticisms are addressed by Ross (Chapter 15, this volume), who cautiously adopts the conclusion that *Tarsius* is indeed most closely related to Anthropoidea. However, his character support for this conclusion is very meager and, like its *locus classicus,* significantly depends on the interpretation of certain basicranial characters. For example, Ross (Chapter 15, this volume) notes that the anterior accessory cavity and transverse septum *sensu* MacPhee and Cartmill (1986) are limited to tarsiers and anthropoids among vertebrates and are "therefore probably unique to them in the history of life." Without endorsing this particular statement, we generally agree that narrowly distributed apomorphies are highly desirable for defining natural groups. Difficulties arise, however, when (1) the apomorphous character state of interest is limited to the target taxon and its imputed sister; (2) no developmental or acceptable morphoclinal evidence links this state with any other; and (3) the morphology of the character state is not exactly the same in the two taxa alleged to be sister groups. This last point is crucial, since perceived anatomic differences in the tarsier and anthropoid versions of these structures have engendered most of the controversy concerning their interpretation (cf. Simons and Rasmussen, 1989).

One may nevertheless agree with Ross (Chapter 15, this volume) that the differences are in fact minor, that the anterior accessory cavity and associated structures develop in a similar manner in tarsiers and anthropoids, and that they therefore pass some sort of pretest of homology. However, just as in all other phylogenetic analyses, a pretest is not definitive. The possibility cannot be discounted that the tarsier and anthropoid versions of the anterior accessory cavity and transverse septum, despite their "uniqueness," are still the product of convergence. It is sometimes argued that this age-old problem can be sidestepped by taking as the arbiter of homology the most parsimonious arrangement of character states on the most parsimonious tree(s). Thus, despite the fact that we operationally homologized the tarsier and anthropoid conditions of the apical aditus (ChS 11.1), the most parsimonious distribution of character states on our MPTs (Figs. 6 and 7) required that the apical aditus be gained twice during primate evolution—once on the terminal stem leading to *Tarsius* and again at the base of the anthropoid radiation. Yet Ross's (Chapter 15, this volume) data set permits a different conclusion, because his similar character state is unambiguously placed on the stem that supports a monophyletic grouping limited to *Tarsius* + Anthropoidea, in the context, of course, of a rather different and more elaborate character set than the one we used.

Which of these solutions is "better"? In answering this question, we think that it is important to note that very little new information supportive of a tarsier–anthropoid dyad has come to light during the past 15 years. Ross's (Chapter 15, this volume) study, for example, ably reassesses data in the published literature but adds little that is absolutely new. On the other hand, the new chromosomal and molecular evidence cited in the preceding section clearly seems to weaken the case for a close evolutionary relationship between tarsiers and anthropoids. Similarly, we note that paleontological evidence accrued over the past decade—particularly the new skulls of *Shoshonius*—has failed to corroborate Cartmill's (1980, p. 261) prediction that "in some respects the last common ancestor of *Tarsius* and anthropoids was more like a small monkey than a modern tarsier." One may quibble whether schemes of relationships that fail to garner additional support when placed under scrutiny are actually "falsified," but the practical result is the same. We suggest that not only has the "tarsiphile" hypothesis of anthropoid origins failed to withstand further scrutiny, but also that its only other competitors, the "lemurphile" and "omomyophile" hypotheses, have fared no better. Perhaps we have missed something fundamental in our efforts to decipher the earliest phases of anthropoid evolution, as the next section posits.

Higher-Level Primate Phylogeny and Anthropoid Origins

Over the past several decades, one of the most enduring concepts among students of primate evolution has been the idea that anthropoids evolved relatively recently from a prosimian ancestry that has been the subject of dispute (e.g., Gingerich, 1980; Rosenberger and Szalay, 1980; Cartmill, 1980). Two factors seem to have contributed greatly to this near consensus model of anthropoid origins. First, for many years the stratigraphic distribution of anthropoids in the fossil record was restricted to rocks no older than the earliest Oligocene and/or latest Eocene Jebel Qatrani Formation in the Fayum Province of Egypt (e.g., Simons, 1989, 1990, 1992). Second, this stratigraphic distribution conformed with influential *scala naturae* notions that posited that, because anthropoids are "more advanced" than prosimians, they must have taken longer to evolve. As a result of these considerations, it was assumed that a "prosimian–anthropoid transition" occurred near the Eocene–Oligocene boundary, and the only point of contention that remained was to settle the ancestral prosimian source for anthropoids. This notion of a very late origin for anthropoids has been endorsed most recently and explicitly by Rasmussen and Simons (1992, p. 503), who predicted "that as more is learned about Old World primate faunas *during the latest Eocene*, more 'protoanthropoid' prosimians and prosimian-like anthropoids will be discovered, which will not be easily and unambiguously classified into either suborder" (emphasis added).

Perhaps the most problematic corollary of these traditional hypotheses concerning anthropoid origins is that they are forced to deny the strict monophyly of whatever prosimian stock is favored as the ancestral source for higher primates. For example, Rasmussen (1986) rejected the monophyly of Strepsirhini (adapiforms + lemuriforms) while advocating an adapiform ancestry for anthropoids. Similarly, workers who favor an omomyid origin for anthropoids (e.g., Rosenberger and Szalay, 1980) as well as those who prefer a strict *Tarsius* + Anthropoidea clade (e.g., Cartmill and Kay, 1978; Cartmill, 1980; Cartmill *et al.*, 1981; MacPhee and Cartmill, 1986) are forced to reject the monophyly of Tarsiiformes. Unsurprisingly, advocates of an adapiform ancestry for anthropoids have been among the most vocal supporters of tarsiiform monophyly in recent years (Gingerich, 1981; Rasmussen, 1986; Simons and Rasmussen, 1989). In the same vein, workers who have favored an omomyid ancestry for anthropoids have been among the most strident supporters of strepsirhine monophyly (e.g., Rosenberger *et al.*, 1985).

Two sets of data have been accumulating over the past several years that effectively refute the traditional hypothesis that anthropoids evolved from a prosimian stock near the Eocene–Oligocene boundary. The first of these data sets consists of recent research that corroborates the strict monophyly of both Strepsirhini and Tarsiiformes, thereby stripping away from Anthropoidea any viable ancestral source among the well-known Eocene adapiforms and omomyids of North America and Europe. Clear synapomorphies, especially in the ankle region, linking adapiforms with extant lemuriforms have been identified only recently (Gebo, 1986, 1988; Beard *et al.*, 1988; Dagosto, 1988; Covert, 1988). Additional derived features of dental and carpal anatomy link certain adapiforms (especially *Adapis* and its close allies) with lemuriforms, thus reinforcing the view that adapiforms are the paraphyletic stem lineage for lemuriforms (Gingerich, 1975; Beard *et al.*, 1988; Beard and Godinot, 1988). Likewise, Rosenberger's (1985) reassessment of the cranial anatomy of *Necrolemur* and other omomyids and the new evidence regarding the cranial anatomy of *Shoshonius cooperi* described here and elsewhere (Beard *et al.*, 1991) strongly support the monophyly of Tarsiiformes. Hence, these new phylogenetic studies of strepsirhines and tarsiiforms suggest that both of these primate clades were established as monophyletic categories by their first, synchronous appearance in the fossil record about 57 Ma. The logical corollary is that Anthropoidea must be a clade of similar antiquity, since the sister taxon of Anthropoidea must be Tarsiiformes, Strepsirhini, or a Tarsiiformes + Strepsirhini clade.

More direct evidence supporting the antiquity of Anthropoidea is derived from recent paleontological discoveries. Among the most significant of these new findings is the recent description of *Algeripithecus minutus*, a diminutive anthropoid represented only by isolated teeth from middle Eocene strata at the locality of Glib Zegdou in western Algeria (Godinot and Mahboubi, 1992). As originally noted by Godinot and Mahboubi (1992), we can deduce that Anthropoidea as a clade is considerably older than *Algeripithecus* because of the nested position of this taxon well within the anthropoid radiation. Of perhaps even greater antiquity are anthropoid specimens from Chambi in central Tunisia, a locality that is thought to be of early Eocene age (Hartenberger *et al.*, 1985). Isolated upper molars from Chambi that are referable to Anthropoidea have been described by Hartenberger and Marandat (1992), who assigned them to the new genus and species *Djebelemur martinezi*. The holotype of this species is a left mandible preserving P_3–M_3 that we believe represents a species distinct from the referred upper molars. More problematic than these Algerian and Tunisian fossils is the possibility of anthropoid affinities for *Altiatlasius koulchii*, a much more primitive animal that is currently represented by isolated teeth from the late Paleocene of Morocco (Sigé *et al.*, 1990). Although we do not formally endorse the anthropoid affinities of *Altiatlasius* here, we agree with Sigé *et al.* (1990) and Godinot and Mahboubi (1992) that certain of its dental traits can be interpreted as evidence for such a relationship. More nearly complete fossils of this important taxon are required to reach more definitive conclusions. Nevertheless, all of these recent paleontological discoveries from the early Cenozoic of North Africa suggest a much greater age for the anthropoid clade than many workers had previously suspected. This conclusion is corroborated by the high diversity of anthropoids in younger Paleogene localities in North Africa and Oman (Thomas *et al.*, 1988; de Bonis *et al.*, 1988; Simons, 1989, 1990, 1992).

The recent discovery of a new radiation of basal anthropoids, the Eosimiidae, in middle Eocene fissure-fillings in southeastern China, emphatically underscores the wide geographical range and high taxonomic diversity of early Cenozoic anthropoids (Beard *et al.*, 1994).

Given the preceding lines of evidence that the anthropoid clade is roughly as ancient as the two major "prosimian" clades that have long been considered its potential ancestors, what conclusions can be drawn about the phylogenetic position of Anthropoidea with respect to other primates? We suggest that a definitive answer to this question is not yet within our grasp, although the possibilities are limited, and certain phylogenetic hypotheses have more support than others. In essence, we recognize three major clades of primates, Strepsirhini, Tarsiiformes, and Anthropoidea, each of which appears to date to the basal Eocene if not earlier. For these three taxa, there are only three cladograms that depict the range of possibilities for interrelationships among them. In decreasing order of their current level of support, these three cladograms are characterized by the monophyly of Haplorhini (Tarsiiformes + Anthropoidea), Prosimii (Tarsiiformes + Strepsirhini), and Simiolemuriformes (Strepsirhini + Anthropoidea), respectively. Hence, the new evidence regarding the great antiquity of Anthropoidea is still consistent with the monophyly of Haplorhini, which we tentatively support. At the same time, however, the likelihood that Anthropoidea was actually the first major clade to diverge from the primitive primate stock has been dramatically increased by provocative new fossil discoveries. Ironically, if this latter phylogeny ultimately proves to be true, it would vindicate a prosimian/anthropoid dichotomy within Primates on cladistic criteria, although this notion was originally proposed for phenetic reasons that have long been derided by cladists.

ACKNOWLEDGMENTS

We are grateful to the organizers of the "Anthropoid Origins" symposium on which this book is based, Drs. J. G. Fleagle, R. F. Kay, and E. L. Simons, for inviting us to participate. The cranial specimens of *Shoshonius* forming the basis of this study were collected over the duration of a long-standing field program in the Wind River Formation that was initiated by our colleagues, Drs. L. Krishtalka and R. K. Stucky, to whom we express our deep appreciation. Thanks also go to A. R. Tabrum (CM) for his skilled preparation of the *Shoshonius* skulls and for help during all phases of field work in the Wind River Formation. Some material of *Shoshonius*, including the excellent new skull CM 31366, was prepared by Clare Flemming (AMNH), who also drew most of the illustrations. The outstanding stereopairs in Figs. 1 and 2 were produced by Lorraine Meeker (AMNH). We also thank Rich Kay, Veronica M. MacPhee, Tab Rasmussen, and Callum Ross for reading and improving the manuscript version of this chapter. Financial support for this study has been provided by NSF BSR 8709242, NSF BSR 9020276, and the M. Graham Netting Research Fund, Carnegie Museum of Natural History.

References

Adkins, R. M., and Honeycutt, R. L. 1993. A molecular examination of archontan and chiropteran monophyly. In: R. D. E. MacPhee (ed.), *Primates and Their Relatives in Phylogenetic Perspective*, pp. 227–249. Plenum Press, New York.

Archibald, J. D. 1977. Ectotympanic bone and internal carotid circulation of eutherians in reference to anthropoid origins. *J. Hum. Evol.* **6**:609–622.

Ax, P. 1985. Stem species and the stem lineage concept. *Cladistics* **1**:279–287.

Beard, K. C. 1993. Phylogenetic systematics of the Primatomorpha, with special reference to Dermoptera. In: F. S. Szalay, M. J. Novacek, and M. C. McKenna (eds.), *Mammal Phylogeny: Placentals*, pp. 129–150. Springer-Verlag, New York.

Beard, K. C., and Godinot, M. 1988. Carpal anatomy of *Smilodectes gracilis* (Adapiformes, Notharctinae) and its significance for lemuriform phylogeny. *J. Hum. Evol.* **17**:71–92.

Beard, K. C., and Wang, B. 1991. Phylogenetic and biogeographic significance of the tarsiiform primate *Asiomomys changbaicus* from the Eocene of Jilin Province, People's Republic of China. *Am. J. Phys. Anthropol.* **85**:159–166.

Beard, K. C., Dagosto, M., Gebo, D. L., and Godinot, M. 1988. Interrelationships among primate higher taxa. *Nature* **331**:712–714.

Beard, K. C., Krishtalka, L., and Stucky, R. K. 1991. First skulls of the early Eocene primate *Shoshonius cooperi* and the anthropoid–tarsier dichotomy. *Nature* **349**:64–67.

Beard, K. C., Qi, T., Dawson, M. R., Wang, B., and Li, C. 1994. A diverse new primate fauna from middle Eocene fissure-fillings in southeastern China. *Nature* **368**:604–609.

Butler, P. M. 1956. The skull of *Ictops* and the classification of the Insectivora. *Proc. Zool. Soc. Lond.* **126**:453–481.

Cartmill, M. 1980. Morphology, function, and evolution of the anthropoid postorbital septum. In: R. L. Ciochon and A. B. Chiarelli (eds.), *Evolutionary Biology of the New World Monkeys and Continental Drift,* pp. 243–274. Plenum Press, New York.

Cartmill, M., and Kay, R. F. 1978. Cranio-dental morphology, tarsier affinities, and primate suborders. In: D. J. Chivers and K. A. Joysey (eds.), *Recent Advances in Primatology, Vol. 3. Evolution,* pp. 205–214. Academic Press, London.

Cartmill, M., and MacPhee, R. D. E. 1980. Tupaiid affinities: the evidence of the carotid arteries and cranial skeleton, pp. 95–132. *In* W. P. Luckett (ed.), *Comparative Biology and Evolutionary Relationships of Tree Shrews.* Plenum Press, New York.

Cartmill, M., MacPhee, R. D. E., and Simons, E. L. 1981. Anatomy of the temporal bone in early anthropoids, with remarks on the problem of anthropoid origins. *Am. J. Phys. Anthropol.* **56**:3–21.

Conroy, G. C. 1980. Ontogeny, auditory structures, and primate evolution. *Am. J. Phys. Anthropol.* **52**:443–451.

Cope, E. D. 1880. On the foramina perforating the posterior part of the squamosal bone of the Mammalia. *Proc. Am. Phil. Soc.* **18**:452–461.

Covert, H. H. 1988. Ankle and foot morphology of *Cantius mckennai:* Adaptations and phylogenetic implications. *J. Hum. Evol.* **17**:57–70.

Covert, H. H., and Williams, B. A. 1991. The anterior lower dentition of *Washakius insignis* and adapid–anthropoidean affinities. *J. Hum. Evol.* **21**:463–467.

Dagosto, M. 1985. The distal tibia of Primates with special reference to the Omomyidae. *Int. J. Primatol.* **6**:45–75.

Dagosto, M. 1988. Implications of postcranial evidence for the origin of euprimates. *J. Hum. Evol.* **17**:35–56.

de Bonis, L., Jaeger, J.-J., Coiffat, B., and Coiffat, P.-E. 1988. Découverte du plus ancien primate catarrhinien connu dans l'Éocène supérieur d'Afrique du Nord. *C. R. Acad. Sci., Paris* [*Ser. II*] **306**:929–934.

Dutrillaux, B., and Rumpler, Y. 1988. Absence of chromosomal similarities between tarsiers (*Tarsius syrichta*) and other primates. *Fol. Primatol.* **50**:130–133.

Gebo, D. L. 1986. Anthropoid origins—the foot evidence. *J. Hum. Evol.* **15**:421–430.

Gebo, D. L. 1987. Functional anatomy of the tarsier foot. *Am. J. Phys. Anthropol.* **73**:9–31.

Gebo, D. L. 1988. Foot morphology and locomotor adaptation in Eocene primates. *Fol. Primatol.* **50**:3–41.

Gingerich, P. D. 1975. Dentition of *Adapis parisiensis* and the evolution of lemuriform primates. In: I. Tattersall and R. W. Sussman (eds.), *Lemur Biology,* pp. 65–80. Plenum Press, New York.

Gingerich, P. D. 1980. Eocene Adapidae, paleobiogeography, and the origin of South American Platyrrhini. In: R. L. Ciochon and A. B. Chiarelli (eds.), *Evolutionary Biology of the New World Monkeys and Continental Drift,* pp. 123–138. Plenum Press, New York.

Gingerich, P. D. 1981. Early Cenozoic Omomyidae and the evolutionary history of tarsiiform primates. *J. Hum. Evol.* **10**:345–374.

Ginsburg, L., and Mein, P. 1987. *Tarsius thailandica* nov. sp., premier Tarsiidae (Primates, Mammalia) fossile d'Asie. *C. R. Acad. Sci. Paris* [*Ser. II*] **304:**1213–1215.

Godinot, M., and Mahboubi, M. 1992. Earliest known simian primate found in Algeria. *Nature* **357:**324–326.

Gregory, W. K. 1910. The orders of mammals. *Bull. Am. Mus. Natl. Hist.* **27:**1–524.

Gregory, W. K. 1915. I. On the relationship of the Eocene lemur *Notharctus* to the Adapidae and to other Primates. II. On the classification and phylogeny of the Lemuroidea. *Bull. Geol. Soc. Am.* **26:**419–446.

Gregory, W. K. 1920. On the structure and relations of *Notharctus*, an American Eocene primate. *Mem. Am. Mus. Nat. Hist.* **3:**49–243.

Hartenberger, J.-L., and Marandat, B. 1992. A new genus and species of an early Eocene primate from North Africa. *Hum. Evol.* **7:**9–16.

Hartenberger, J.-L., Martinez, C., and Ben Said, A. 1985. Découverte de mammifères d'âge Éocène inférieur en Tunisie centrale. *C. R. Acad. Sci. Paris* [*Ser. II*] **301:**649–652.

Hoffstetter, R. 1977. Phylogénie des Primates: confrontation des résultats obtenus par les diverses voies d'approche du problème. *Bull. Mem. Soc. Anthropol. Paris* [*Ser. XIII*] **4:**327–346.

Hoffstetter, R. 1988. Relations phylogéniques et position systématique de *Tarsius:* Nouvelles controverses. *C. R. Acad. Sci. Paris* [*Ser. II*] **307:**1837–1840.

Hürzeler, J. 1948. Zur Stammesgeschichte der Necrolemuriden. *Schweiz. Palaeontol. Abh.* **66:**1–46.

Kay, R. F., Thorington, R. W., Jr., and Houde, P. 1990. Eocene plesiadapiform shows affinities with flying lemurs not primates. *Nature* **345:**342–344.

Kay, R. F., Thewissen, J. G. M., and Yoder, A. D. 1992. Cranial anatomy of *Ignacius graybullianus* and the affinities of the Plesiadapiformes. *Am. J. Phys. Anthropol.* **89:**477–498.

Koop, B. F., Siemienick, D., Slightom, J. L., Goodman, M., Dunbar, J., Wright, P. C., and Simons, E. L. 1989a. *Tarsius* δ- and β-globin genes: Conversions, evolution, and systematic implications. *J. Biol. Chem.* **264:**68–79.

Koop, B. F., Tagle, D. A., Goodman, M., and Slightom, J. L. 1989b. A molecular view of primate phylogeny and important systematic and evolutionary questions. *Mol. Biol. Evol.* **6:**580–612.

Krause, D. W., and Maas, M. C. 1990. The biogeographic origins of late Paleocene–early Eocene mammalian immigrants to the western interior of North America. In: T. M. Bown and K. D. Rose (eds.), *Dawn of the Age of Mammals in the Northern Part of the Rocky Mountain Interior, North America,* pp. 71–105. Geological Society of America Special Paper 243, Boulder.

Le Gros Clark, W. E. 1934. On the skull structure of *Pronycticebus gaudryi. Proc. Zool. Soc. Lond.* **1934:**19–27.

MacPhee, R. D. E. 1979. Entotympanics, ontogeny and primates. *Fol. Primatol.* **31:**23–47.

MacPhee, R. D. E. 1981. Auditory regions of primates and eutherian insectivores: Morphology, ontogeny, and character analysis. *Contrib. Primatol.* **18:**1–282.

MacPhee, R. D. E. 1987. Basicranial morphology and ontogeny of the extinct giant lemur *Megaladapis. Am. J. Phys. Anthropol.* **74:**333–355.

MacPhee, R. D. E. 1993. Summary. In: R. D. E. MacPhee (ed.), *Primates and Their Relatives in Phylogenetic Perspective,* pp. 363–373. Plenum Press, New York.

MacPhee, R. D. E., and Cartmill, M. 1986. Basicranial structures and primate systematics. In: D. R. Swindler and J. Erwin (eds.), *Comparative Primate Biology, Vol. 1: Systematics, Evolution, and Anatomy,* pp. 219–275. Alan R. Liss, New York.

MacPhee, R. D. E., and Novacek, M. J. 1993. Definition and relationships of Lipotyphla. In: F. S. Szalay, M. J. Novacek, and M. C. McKenna (eds.), *Mammal Phylogeny: Placentals,* pp. 13–31. Springer-Verlag, New York.

MacPhee, R. D. E., Cartmill, M., and Gingerich, P. D. 1983. New Palaeogene primate basicrania and the definition of the order Primates. *Nature* **301:**509–511.

Martin, R. D. 1990. *Primate Origins and Evolution: A Phylogenetic Reconstruction.* Princeton University Press, Princeton.

Norris, C. A. 1993. Changes in the composition of the auditory bulla in southern Solomon Islands populations of the grey cuscus, *Phalanger orientalis breviceps* (Marsupialia, Phalangeridae). *Zool. J. Linn. Soc.* **107:**93–106.

Novacek, M. J. 1977. Aspects of the problem of variation, origin and evolution of the eutherian auditory bulla. *Mammal Rev.* **7:**131–149.

Novacek, M. J. 1986. The skull of lepticid insectivorans and the higher-level classification of eutherian mammals. *Bull. Am. Mus. Nat. Hist.* **183:**1–112.

Novacek, M. J. 1993. Reflections on higher mammalian phylogenetics. *J. Mamm. Evol.* **1:**3–30.

Rasmussen, D. T. 1986. Anthropoid origins: a possible solution of the Adapidae–Omomyidae paradox. *J. Hum. Evol.* **15:**1–12.

Rasmussen, D. T. 1990. The phylogenetic position of *Mahgarita stevensi:* Protoanthropoid or lemuroid? *Int. J. Primatol.* **11:**439–469.

Rasmussen, D. T., and Simons, E. L. 1992. Paleobiology of the oligopithecines, the earliest known anthropoid primates. *Int. J. Primatol.* **13:**477–508.

Rosenberger, A. L. 1985. In favor of the necrolemur–tarsier hypothesis. *Fol. Primatol.* **45:**179–194.

Rosenberger, A. L., and Strasser, E. 1985. Toothcomb origins: Support for the grooming hypothesis. *Primates* **26:**73–84.

Rosenberger, A. L., and Szalay, F. S. 1980. On the tarsiiform origins of Anthropoidea. In: R. L. Ciochon and A. B. Chiarelli (eds.), *Evolutionary Biology of the New World Monkeys and Continental Drift,* pp. 139–157. Plenum Press, New York.

Rosenberger, A. L., Strasser, E., and Delson, E. 1985. Anterior dentition of *Notharctus* and the adapid–anthropoid hypothesis. *Fol. Primatol.* **44:**15–39.

Saban, R. 1963. Contribution à l'étude de l'os temporal des primates. Description chez l'homme et les prosimiens. Anatomie comparée et phylogénie. *Mem. Mus. Natl. Hist. Nat.* [*Ser. A*] **29:**1–378.

Schmid, P. 1981. Comparison of Eocene nonadapids and *Tarsius.* In: A. B. Chiarelli and R. S. Corruccini (eds.), *Primate Evolutionary Biology,* pp. 6–13. Springer-Verlag, Berlin.

Schmid, P. 1982. *Die systematische Revision der europäischen Microchoeridae Lydekker, 1887 (Omomyiformes, Primates).* Juris Druck + Verlag, Zurich.

Schwartz, J. H. 1978. If *Tarsius* is not a prosimian, is it a haplorhine? In: D. J. Chivers and K. A. Joysey (eds.), *Recent Advances in Primatology, Vol. 3: Evolution,* pp. 195–202. Academic Press, London.

Schwartz, J. H., and Tattersall, I. 1987. Tarsiers, adapids and the integrity of Strepsirhini. *J. Hum. Evol.* **16:**23–40.

Sigé, B., Jaeger, J.-J., Sudre, J., and Vianey-Liaud, M. 1990. *Altiatlasius koulchii* n. gen. et sp., primate omomyidé du Paléocène supérieur du Maroc, et les origines des euprimates. *Paleontographica* [*A*] **214:**31–56.

Simons, E. L., 1961. Notes on Eocene tarsioids and a revision of some Necrolemurinae. *Bull. Br. Mus. Nat. Hist. Geol.* **5:**43–69.

Simons, E. L. 1968. Early Cenozoic mammalian faunas, Fayum Province, Egypt. Part 1. African Oligocene mammals: introduction, history of study, and faunal succession. *Bull. Peabody Mus. Nat. Hist. Yale Univ.* **28:**1–21.

Simons, E. L. 1989. Description of two genera and species of late Eocene Anthropoidea from Egypt. *Proc. Natl. Acad. Sci. USA* **86:**9956–9960.

Simons, E. L. 1990. Discovery of the oldest known anthropoidean skull from the Paleogene of Egypt. *Science* **247:**1567–1569.

Simons, E. L. 1992. Diversity in the early Tertiary anthropoidean radiation in Africa. *Proc. Natl. Acad. Sci. USA* **89:**10743–10747.

Simons, E. L., and Bown, T. M. 1985. *Afrotarsius chatrathi,* first tarsiiform primate (?Tarsiidae) from Africa. *Nature* **313:**475–477.

Simons, E. L., and Rasmussen, D. T. 1989. Cranial anatomy of *Aegyptopithecus* and *Tarsius* and the question of the tarsier–anthropoidean clade. *Am. J. Phys. Anthropol.* **79:**1–23.

Simons, E. L., and Russell, D. E. 1960. Notes on the cranial anatomy of *Necrolemur. Breviora* **127:**1–14.

Simpson, G. G. 1940. Studies on the earliest primates. *Bull. Am. Mus. Nat. Hist.* **77:**185–212.

Simpson, G. G. 1955. The Phenacolemuridae, new family of early primates. *Bull. Am. Mus. Nat. Hist.* **105:**411–442.

Starck, D. 1975. The development of the chondrocranium in Primates. In: W. P. Luckett and F. S. Szalay (eds.), *Phylogeny of the Primates: A Multidisciplinary Approach,* pp. 127–155. Plenum Press, New York.

Stehlin, H. G. 1916. Die Säugetiere des schweizerischen Eocaens. Critischer Catalog der Materialen. VII(2): *Caenopithecus, Necrolemur–Microchoerus, Nannopithex, Anchomomys, Periconodon, Amphichiromys, Heterochiromys,* Nachträge zu *Adapis,* Schlussbetrachtungen zu den Primaten. *Abh. Schweiz. Palaontol. Ges.* **41:**1299–1552.

Swofford, D. L. 1990. PAUP: Phylogenetic Analysis Using Parsimony, Version 3.0. Computer program distributed by the Illinois Natural History Survey, Champaign.

Szalay, F. S. 1975. Phylogeny of primate higher taxa: the basicranial evidence. In: W. P. Luckett and F. S. Szalay (eds.), *Phylogeny of the Primates: A Multidisciplinary Approach,* pp. 91–125. Plenum Press, New York.

Szalay, F. S. 1976. Systematics of the Omomyidae (Tarsiiformes, Primates): Taxonomy, phylogeny, and adaptations. *Bull. Am. Mus. Nat. Hist.* **156:**157–450.

Szalay, F. S., and Dagosto, M. 1980. Locomoter adaptations as reflected on the humerus of Paleogene primates. *Fol. Primatol.* **34:**1–45.

Szalay, F. S., and Dagosto, M. 1988. Evolution of hallucial grasping in the Primates. *J. Hum. Evol.* **17:**1–33.

Szalay, F. S., and Seligsohn, D. 1977. Why did the strepsirhine tooth comb evolve? *Fol. Primatol.* **27:**75–82.

Thomas, H., Roger, J., Sen, S., and Al-Sulaimani, Z. 1988. Découverte des plus anciens "anthropoides" du continent arabo-africain et d'un primate tarsiiforme dans l'Oligocène du Sultanat d'Oman. *C. R. Acad. Sci., Paris [Ser. II]* **306:**823–829.

Wible, J. R. 1993. Cranial circulation and relationships of the colugo *Cynocephalus* (Dermoptera, Mammalia). *Am. Mus. Novit.* **3072:**1–27.

Wible, J. R., and Martin, J. R. 1993. Ontogeny of the tympanic floor and roof in archontans. In: R. D. E. MacPhee (ed.), *Primates and Their Relatives in Phylogenetic Perspective,* pp. 111–148. Plenum Press, New York.

Wilson, J. A. 1966. A new primate from the earliest Oligocene, west Texas: Preliminary report. *Fol. Primatol.* **4:**227–248.

Wortman, J. L. 1903. Studies of Eocene Mammalia in the Marsh Collection, Peabody Museum. Part II. Primates [Classification of the Primates]. *Am. J. Sci. [Ser. 4]* **15:**399–414.

Yoder, A. D. 1992. The applications and limitations of ontogenetic comparisons for phylogeny reconstruction: The case of the strepsirhine internal carotid artery. *J. Hum. Evol.* **23:**183–195.

The Messel Primates and Anthropoid Origins

4

JENS LORENZ FRANZEN

Introduction

The fossil site of Grube Messel is situated about 25 miles southeast of Frankfurt am Main (Germany). Dating from the early middle Eocene, Messel is famous for its preservation of generally complete and articulated skeletons of mammals and other vertebrates, sometimes even including soft body contours and gut contents (Schaal and Ziegler, 1992). During Eocene times carcasses of the animals would float into a freshwater lake and be deposited within clayish sediments on its bottom. The cadavers did not ascend to the surface again because decompositional inflation was prevented by hydrostatic pressures exceeding 10 psi (Elder, 1985). This corresponds to a lake depth over 10 m. Eocene Lake Messel was meromictic (Goth, 1990). Thus, still-water conditions, lack of oxygen and benthonic scavengers, and the metabolism of decomposing anaerobic bacteria near the bottom resulted in an extraordinary quality of preservation (Franzen, 1985, 1990).

Since it was surrounded in Eocene times by a dense rain forest of tropical to subtropical character, it is surprising that Messel has up to now yielded only a few primate skeletons, and these are largely fragmented. This could be because the primates, living high in the trees, were safe from any floods that would normally cause the death of animals and their immediate transport into the lake (Franzen, 1985). So, in the case of primates, the combination of

JENS LORENZ FRANZEN • Forschungsinstitut Senckenberg, D-60325 Frankfurt am Main, Germany.
Anthropoid Origins, edited by John G. Fleagle and Richard F. Kay. Plenum Press, New York, 1994.

death, transport, and fragmentation resulted only from rare accidents with crocodiles, which lived in large numbers in the lake and its surroundings (Franzen, 1987, 1988; J. L. Franzen and E. Frey, 1993).

In any case, Messel has produced the only articulated primate material from the whole Paleogene of Eurasia except for that from the Geiseltal locality. Therefore, the investigation of the Messel primates is crucial for understanding the early evolution of that order including the problem of anthropoid origins.

The Messel Primates

To date six primate specimens have been recovered from Grube Messel. They represent at least three different taxa, all of which belong to the Adapiformes.

The first discovery was made by the Hessisches Landesmuseum Darmstadt in 1975. The specimen, HLD-Me 7430, was described by Koenigswald in 1979 as an adapid indeterminable on the genus and species level. It consists of the rear part of a skeleton comprising most of the pelvis, both hindlimbs, and a large baculum (Fig. 1). Remarkable are the opposable hallux, the scutiform end phalanges, and particularly a toilet claw on the second toe.

The second discovery was made by a team from the Forschungsinstitut und Naturmuseum Senckenberg in 1982 (Franzen, 1987). The specimen, SMF-ME 1228, consists of the anterior half of a skeleton extending distally just to the os sacrum (Fig. 2). It is evidently a juvenile and includes an almost complete skull with most of the dentition (Fig. 3). The completely reduced metaconule of the upper molars characterizes the genus *Europolemur,* and a hypocone in *statu nascendi* indicated a new species, *Europolemur koenigswaldi* (Franzen, 1987).

The third primate specimen from Messel was another, albeit disarticulated, rear part of a large adapiform (Fig. 4). Discovered by a team from the Staatliches Museum für Naturkunde Karlsruhe in 1984, LNK Me 684 corresponds almost exactly with the Darmstadt specimen except for being less complete and about 10% smaller. It was described by Koenigswald in 1985.

The fourth primate was recovered from the Messel oilshale by a team from the Forschungsinstitut und Naturmuseum Senckenberg, again in 1982. This specimen was wrapped with the label "limb of an unknown mammal" and was recognized as a primate only after being prepared. SMF-ME 1683 is a primary fragment of an anterior right limb consisting of radius, ulna, and the manus lacking only a few distal phalanges (Fig. 5). It was described by Franzen in 1988 and 1993. In the absence of any pertinent dentition it is not classifiable for the moment, although it corresponds with the Darmstadt and the Karlsruhe specimens in being about 50% larger than *Europolemur koenigswaldi* (Franzen, 1988). The quality of its preservation is really remarkable. The

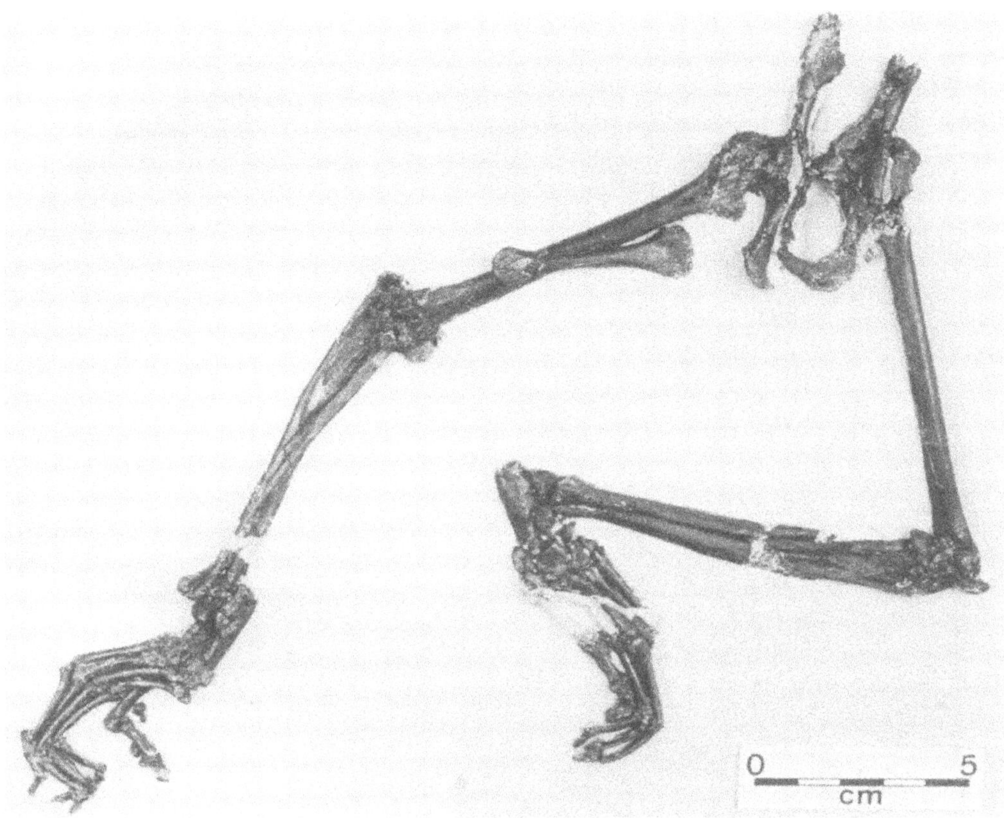

Fig. 1. First discovery of a Messel primate in 1975. Rear part of a skeleton of an as yet unidentified large adapiform (Hessisches Landesmuseum Darmstadt, HLD-Me 7430). Photo: Hessisches Landesmuseum Darmstadt (M. Schwab).

thumb is clearly opposable and bears a scutiform terminal phalanx. Its articulation with the trapezium is more or less hidden. Obviously it is rather flat, either a ball (trapezium) and socket (mtI) joint (Franzen, 1988, p. 281) or slightly saddle-shaped (Franzen, 1992). In contrast to all lemuriforms, but corresponding with the Anthropoidea, there is no contact of the centrale with the uncinatum, but there is with the trapezium on the radial side of the carpus. The specimen was also recently discussed by Godinot and Beard (1991, 1993).

The fifth Messel specimen was discovered in 1990 by a team from the Staatliches Museum für Naturkunde Karlsruhe. It was presented by Franzen and Frey at the International Messel Conference in November 1991 at Darmstadt (Franzen and Frey, 1993). The specimen, LNK Me 1125, displays the rear part of a skeleton comprising a fragmentary pelvis, both hindlimbs, and a long tail of 29 vertebrae (Fig. 6). Size and proportions are indicative of an

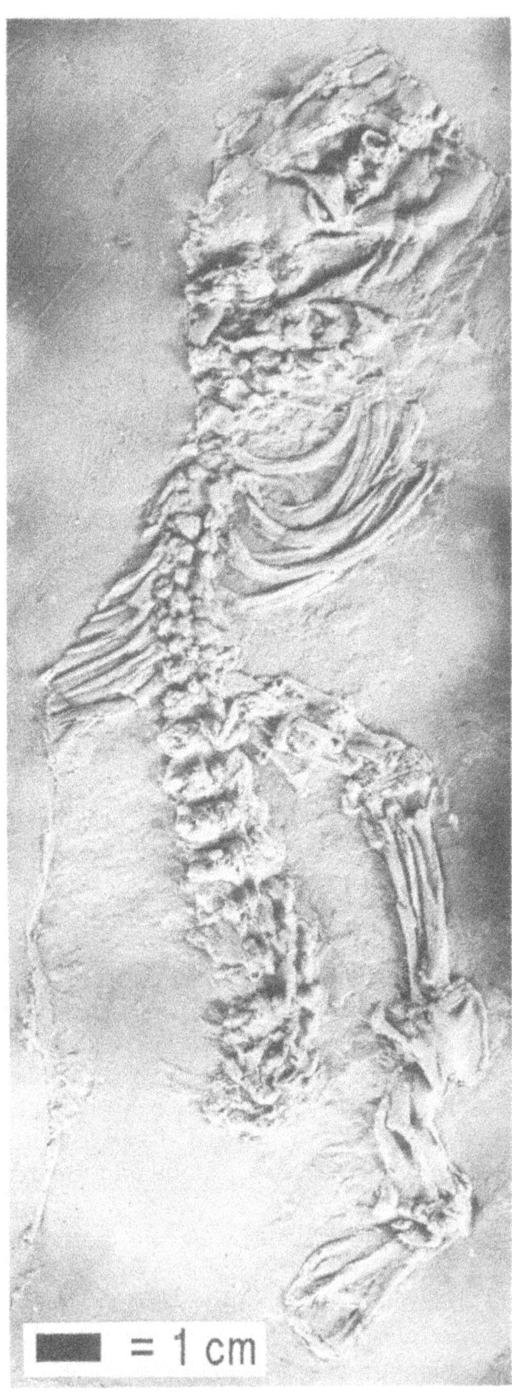

Fig. 2. Second Messel primate. Anterior part of a skeleton of *Europolemur koenigswaldi* (Franzen, 1987), discovered by the Forschungsinstitut und Naturmuseum Senckenberg in 1982 (SMF-ME 1228 B), coated with ammonium chloride. Photo: Forschungsinstitut Senckenberg Frankfurt am Main (E. Pantak-Wein).

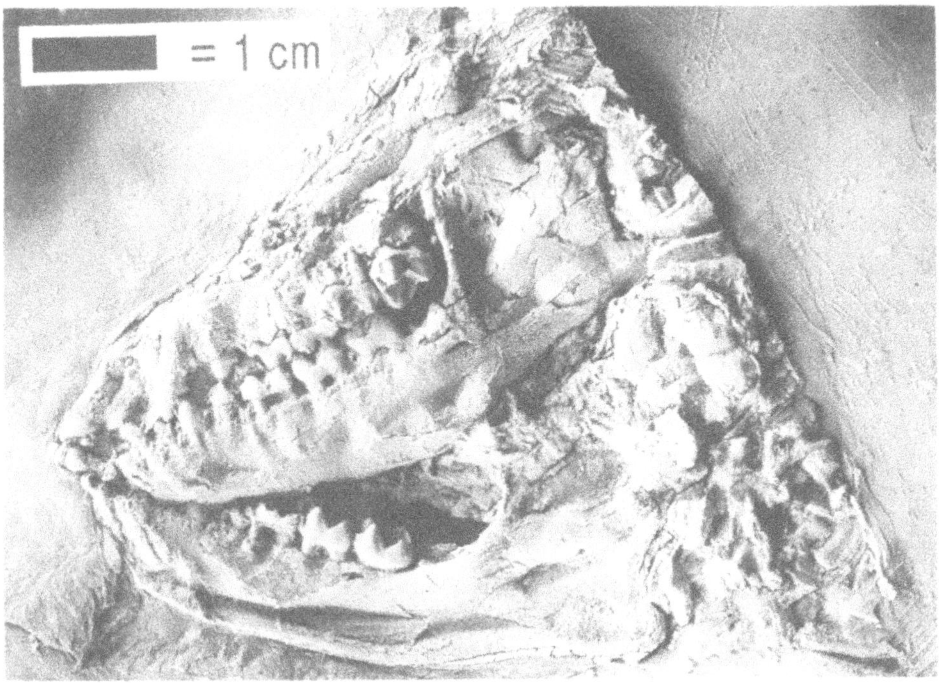

Fig. 3. Second Messel primate. Type specimen of *Europolemur koenigswaldi* (Franzen, 1987), left side of the skull (SMF-ME 1228 A), coated with ammonium chloride. Photo: Forschungsinstitut Senckenberg Frankfurt am Main (E. Pantak-Wein).

affiliation with *Europolemur koenigswaldi* (Table I), which thus becomes the first primate species from Messel known by its entire osteology (J. L. Franzen and E. Frey, 1993). At a first glance one has the impression that the tail could have been prehensile. But looking at the weak processes of the proximal caudal vertebrae and comparing the length proportions of all the vertebrae of the tail, Franzen and Frey arrived at the conclusion that the tail of this specimen does not correspond with that of *Ateles* but more closely resembles that of *Callithrix*. The same holds true for its other proportions and for the body size as a whole. This comparison suggests that *Europolemur koenigswaldi* had a body weight of 167–360 g (Napier and Napier, 1967, p. 79), whereas Gingerich's body mass software (Gingerich, 1990) gives a figure of about 400 (217–1416) g. In any case, *Europolemur koenigswaldi* falls below "Kay's threshold" (Kay, 1975; Gingerich, 1981, p. 356), indicating an insectivorous diet.

As in *Notharctus* and in the first Messel primate the distal part of the calcaneus is longer relative to the whole length of this bone than in true adapids, whereas the trochlea of the talus is wedge-shaped and does not display parallel ridges as in the case of *Adapis*. The neck of the talus is curved medially. Taken together with short metatarsals and long phalanges, this

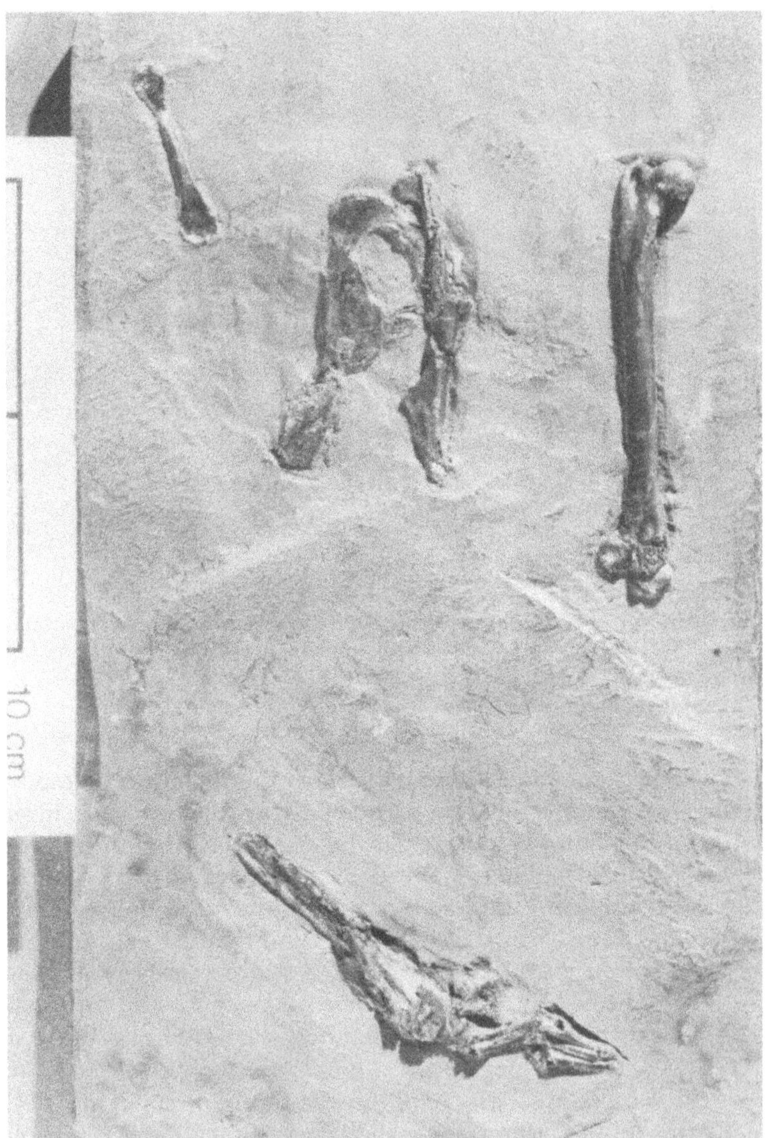

Fig. 4. Third Messel primate. Rear part of the skeleton of a large adapiform discovered by the Staatliches Museum für Naturkunde Karlsruhe in 1984 (LNK Me 684). Photo: Forschungsinstitut Senckenberg Frankfurt am Main (E. Haupt).

Fig. 5. Fourth Messel primate. Right zeugopodium (radius and ulna) and manus of a large adapiform discovered by the Forschungsinstitut und Naturmuseum Senckenberg in 1982 (SMF-ME 1683). Photo: Forschungsinstitut Senckenberg Frankfurt am Main (E. Pantak-Wein).

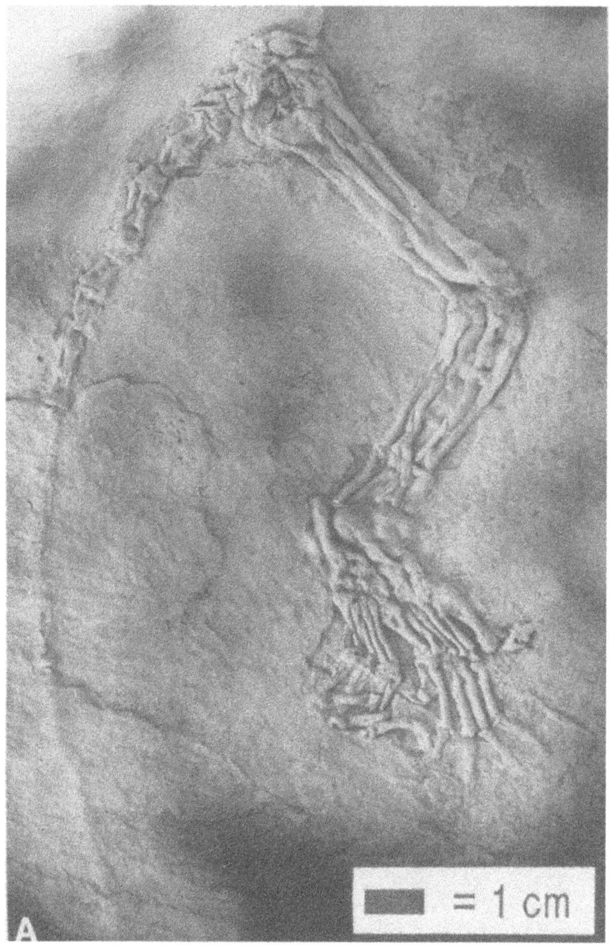

points to quadrupedal climbing closely adherent to broad branches (Fig. 7). Such locomotion would be consistent with narrow and pointed terminal phalanges, which, except for the hallux, appear rather claw- than nail-bearing.

The sixth primate specimen from Messel fell into our hands only recently. It was found by an unknown person, who sold it by means of an agent to a private collector in Switzerland, who in turn made it available to Frey and Franzen for investigation.

The composition of this specimen is as obscure as its origin. There can be no doubt that it is a primate and that it comes from Messel. But there is also no doubt that it is a falsification combining original parts with a cast of another mammal (mostly of the tail), and the distal parts of the limbs and some vertebrae are carved out of polyester (Fig. 8).

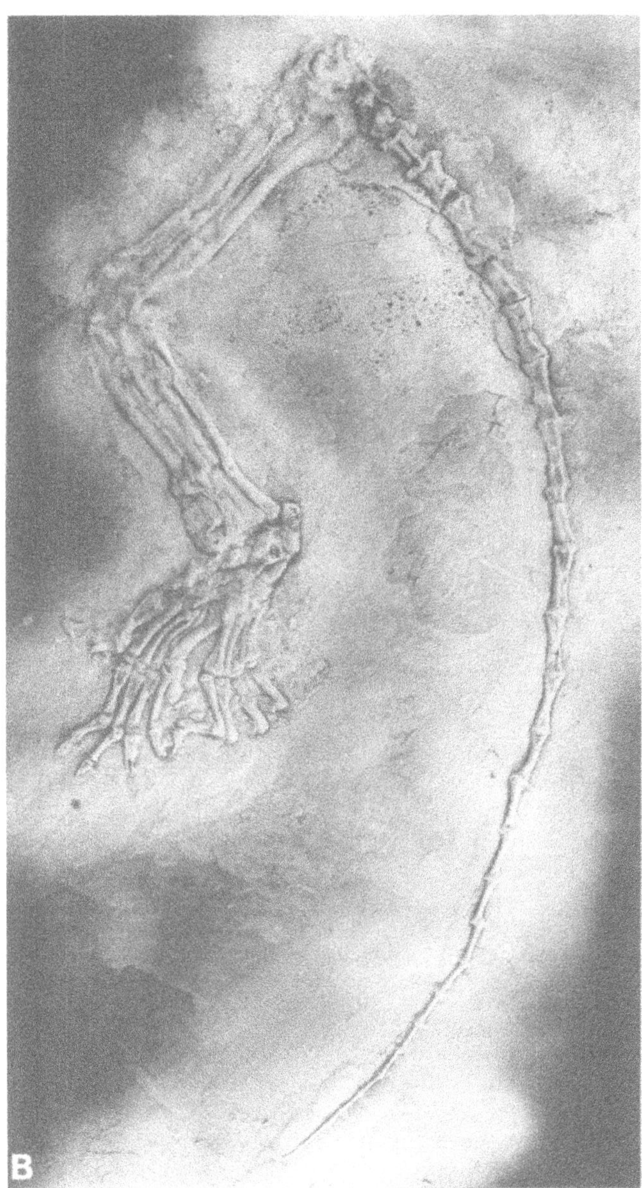

Fig. 6. Fifth Messel primate. Rear part of the skeleton of *Europolemur koenigswaldi* discovered by the Staatliches Museum für Naturkunde Karlsruhe in 1990. (A) Left part of the specimen (LNK Me 1125a). (B) Right part of the specimen (LNK Me 1125b), coated with ammonium chloride. Photos: Forschungsinstitut Senckenberg Frankfurt am Main (E. Pantak-Wein).

Table I. Measured and Calculated Limb Bone Lengths of *Europolemur koenigswaldi* Specimens Based on *Notharctus osborni* and *Callithrix jacchus* Specimens[a]

	Humerus	Radius	Femur	Tibia
	Lengths based on *Notharctus osborni*			
Notharctus osborni	71.0	67.0	122.0	101.0
Europ. koenigswaldi Holotype Humerus	45	36	**77**[b]	**64**
Europ. koenigswaldi Holotype Radius	45	36	**66**	**54**
Europ. koenigswaldi New specimen Femur	**35**	**33**	60.5	51.5
Europ. koenigswaldi New specimen Tibia	**36**	**34**	60.5	51.5
	Lengths based on *Callithrix jacchus*			
Callithrix jacchus	34.4	29.8	42.2	45.0
Europ. koenigswaldi Holotype Humerus	45	36	**55**	**59**
Europ. koenigswaldi Holotype Radius	45	36	**51**	**54**
Europ. koenigswaldi New specimen Femur	**50**	**43**	60.5	51.5
Europ. koenigswaldi New specimen Tibia	**39**	**34**	60.5	51.5

[a]Lengths of humerus, radius, femur, and tibia of the Frankfurt (SMF-ME 1228) and Karlsruhe (LNK-Me 1125) specimens of *Europolemur koenigswaldi* compared with the lengths of these bones calculated on the basis of comparison with *Notharctus osborni* (AMNH 11474) and *Callithrix jacchus* specimens (Franzen, 1988). The measured and calculated lengths correspond so well that they confirm the association of both specimens in a single species.
[b]Lengths in boldface type are calculated.

The original parts, however, are interesting enough. First, there is the anterior half of a skeleton comprising skull, cervicals, the proximal part of the thorax, and the forelimbs down to the middle and distal end of radius and ulna, respectively. Also, there is the rear part of a skeleton, presumably of the same individual, comprising three lumbar vertebra, the os sacrum, part of the pelvis, caudal vertebrae 1–6, and both hindlimbs down to the middle of tibia and fibula. All original parts evidently derive from a juvenile as indicated by the dentition as well as unfused epiphyses of the vertebra and the limb bones.

Fig. 7. Reconstruction of *Europolemur koenigswaldi* sitting on a horizontal branch of a tree, about half natural size. Drawing by Dr. Eberhard ("Dino") Frey, Staatliches Museum für Naturkunde Karlsruhe.

Of particular interest is the skull. It is almost complete as seen from its left side (Fig. 9). Its facial part, which is rather short and steep, displays the deciduous canine and three milk molars of the upper and the lower jaw as well as the lower first molar of the permanent dentition. The latter is damaged at its labiodistal corner. The formula of the deciduous dentition was ?-1-3/?-1-3. The most anteriorly situated milk molars, the $dP^2{}_2$, are very small and only one-rooted. Also, the next deciduous molars, the $dP^3{}_3$, are rather small. They are unicuspid but two-rooted, whereas the $dP^4{}_4$ are more or less molarized.

The only primate taxa from the Paleogene of Europe that correspond to such dentition are "*Pronycticebus*" *neglectus,* described by Thalmann, Haubold,

Fig. 8. Sixth Messel primate. Two original parts of a skeleton of *Caenopithecus neglectus* (Thalmann *et al.*, 1989) discovered by an unknown private collector. The original parts are embedded in polyester together with artifacial distal parts of the limbs and some vertebrae as well as a cast of the tail vertebrae of another mammal in order to fake a complete skeleton. Boundaries between original and faked parts are marked by short lines. The polyester is solidified by fiber material that obstructs the analysis of the radiograph. Photo: Forschungsinstitut Senckenberg Frankfurt am Main (E. Haupt).

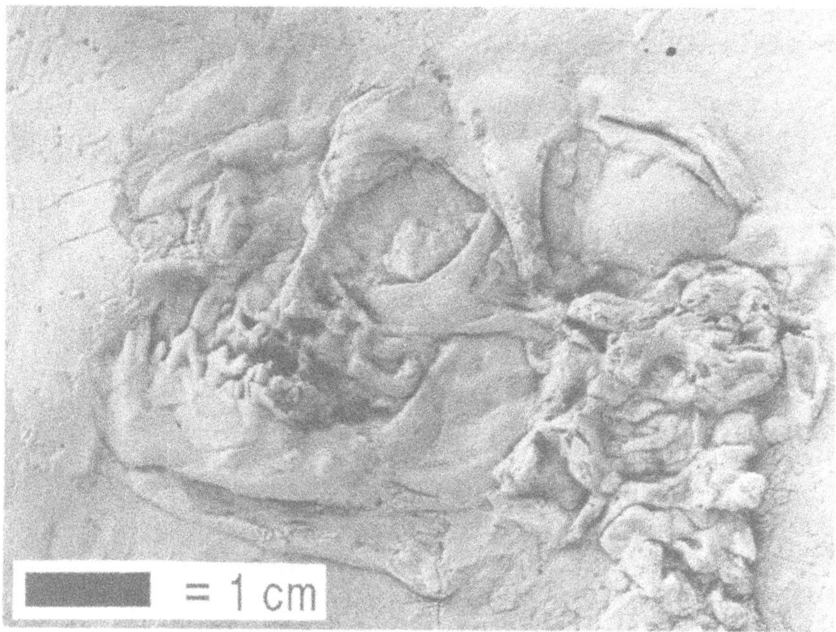

Fig. 9. Skull of *Caenopithecus neglectus* (Pohl Collection) seen from its left side, coated with ammonium chloride. Photo: Forschungsinstitut Senckenberg Frankfurt am Main (E. Pantak-Wein).

and Martin (1989) from the Geiseltal locality, and *Caenopithecus lemuroides,* described by Rütimeyer (1862) and Stehlin (1916) from the Egerkingen locality in Switzerland. *Cercamonius brachyrhynchus* (Stehlin, 1912) has also P_1 completely reduced and a P_2 that is only one-rooted, but P_3 is larger than P_4, and the metaconid of M_1 is more distally situated, whereas the difference in the ventral contour line of the mandible may be a reflection of different stages of ontogenetic development. *Mahgarita* from the late middle Eocene (Uintan) from Texas also displays such dentition (Szalay and Delson, 1979, pp. 125–126, Fig. 58), but its M_1 does not have such a sharp crest between proto- and metaconid, and it is also paleogeographically too far away to be considered in this context. *Mahgarita* may be the result of parallel development.

In my opinion the species from the Geiseltal is much better placed in the genus *Caenopithecus* (Rütimeyer, 1862) than in *Pronycticebus* (Grandidier, 1904). The specimen from the Geiseltal differs from *Pronycticebus* not only in its dental formula, comprising only three instead of four premolars, but also in its highly reduced P^2_2 and in the absence of a metaconule in the upper molars. All these characters are typical for *Caenopithecus* but are absent in *Pronycticebus.* The characters, however, that led Thalmann *et al.* (1989, p. 117) to regard the specimen from the Geiseltal as "most similar to *Pronycticebus gaudryi,*" namely "the outline of the molars" and an M^3 that "is distolingually slightly expanded and has a small swelling in the pericone region," appear comparatively weak. *Caenopithecus neglectus* from the Geiseltal differs from the only other species of that genus, *Caenopithecus lemuroides* from the upper middle Eocene of Egerkingen gamma, Bouxwiller, and Egerkingen Huppersand by its smaller size and in lacking a buccal cingulum in its cheek teeth (Thalmann *et al.,* 1989, p. 117).

Concerning the determination of the Messel specimen the hypodigm of *Caenopithecus* does not include any deciduous teeth. The configuration of the milk dentition of the Messel specimen, however, corresponds exactly with that of the permanent dentition of *Caenopithecus* in the lack of any $dP^1{}_1$ and in $dP^2{}_2$ that are very tiny, unicuspid, and uniradical. Thus, the degree of reduction of the deciduous dentition of the Messel specimen reflects that of the premolars of *Caenopithecus neglectus.* Also, the first lower molar, which is the only tooth directly comparable, corresponds in size and morphology exactly with that of *Caenopithecus neglectus.* A buccal cingulid is present in the trigonid. Therefore, I refer the Messel specimen to this species, although the type material of *Caenopithecus neglectus* comes from the "untere Mittelkohle" of the Geiseltal, which is stratigraphically just a little bit younger than the Messel formation (MP 12 instead of MP 11). Hence, the temporal range of the genus *Caenopithecus* is extended from the end to the beginning of the Geiseltalian, the middle Eocene of the terrestric stratigraphy of Europe (Franzen and Haubold, 1986).

Although the orbital region of the Messel specimen is somewhat deformed, its diameter can be estimated at about 14 mm (anteroposteriorly). This also corresponds very well with *Caenopithecus neglectus* from the Geiseltal, where "the diameter of the orbit has been calculated . . . to be about 14–15 mm" (Thalmann *et al.,* 1989, p. 118). Brought into relation with its basal cranial length of about 47.5 mm (Fig. 10) the Messel specimen falls near *Microcebus coquereli* and *Phaner furcifer,* both of which are known to be "fully nocturnal" (Gingerich and Martin, 1981, p. 243). In this respect *Caenopithecus neglectus* resembles *Pronycticebus gaudryi,* which is considered by Simons (1972, p. 129) also to be nocturnal.

The Systematic Position of the Messel Primates

All the Messel specimens differ considerably morphologically from the Adapidae *sensu strictu,* namely, *Adapis* (Cuvier, 1821) and *Leptadapis* (Gervais, 1876). Differences exist (1) in the shape of the skull, which neither in *Europolemur koenigswaldi* (specimen 2) nor in *Caenopithecus neglectus* (specimen 6) displays such a prominent sagittal crest and such a massive zygomatic arch as in the case of "true" adapids; (2) in the dentition (Franzen 1987, p. 174), which in *Europolemur koenigswaldi* is characterized by nonmolarized P4/4 and a sharp crest running from the proto- to the metacone ("crista obliqua"); and (3)

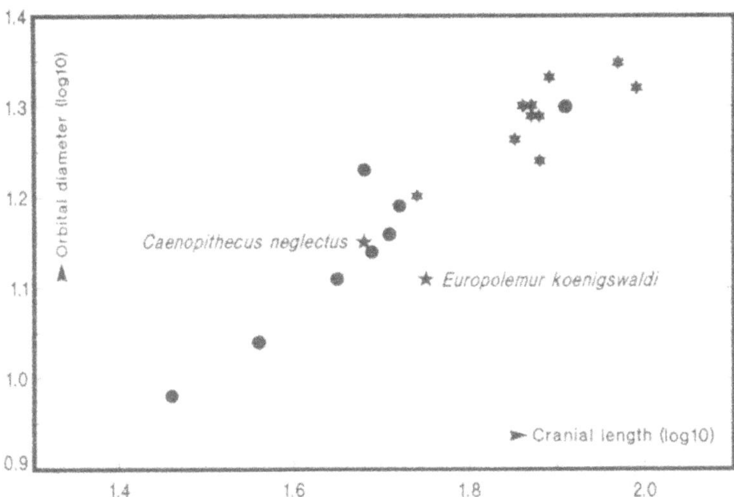

Fig. 10. Relationship between orbital diameter and cranial length of living lemurs (black dots, nocturnal; asterisks, diurnal), and of *Europolemur koenigswaldi* and *Caenopithecus neglectus* from Messel (data on living lemurs from Gingerich and Martin, 1981, Table 4).

in limb proportions and morphology, particularly an anteriorly elongated calcaneus and astragalar neck, a trochlea that is not so long and more wedge-shaped in dorsal view, a posterior astragalar shelf that is larger than that in *Adapis,* and in the distal extent of the intratrochanteric fossa of the femur, which is long in the Messel adapiforms and short in *Adapis* (Koenigswald, 1979; Dagosto, 1983). The cuboid is longer than wide, and the basal phalanges of the manus are considerably longer than the metacarpals.

Dentally the Messel primates belong to a group that, except for *Europolemur* and *Caenopithecus,* comprises genera such as *Agerinia* (Crusafont-Pairó and Golpe-Posse, 1973), *Anchomomys* (Stehlin, 1916), *Cantius* (Simons, 1962), *Cercamonius* (Gingerich, 1975), *Donrussellia* (Szalay, 1976), *Periconodon* (Stehlin, 1916), *Pronycticebus* (Grandidier, 1904), *Protoadapis* (Lemoine, 1878), and eventually *Mahgarita* (Wilson and Szalay, 1976). They are quite near to the North American notharctines, whereas the "true" adapids, *Adapis* and *Leptadapis,* are clearly different. Therefore, in 1987 I suggested distinguishing this group from the "true" adapids not only on a subfamily level (Franzen, 1987), but shifting it from the Adapidae (Trouessart, 1879) to the Notharctidae (Trouessart, 1879).

For priority reasons, however, I now prefer to name this group the Cercamoniinae (Gingerich, 1975) instead of the Protoadapinae (Szalay and Delson, 1979). When Gingerich (1975) defined this new subfamily, he based it on the genus *Cercamonius* alone. Although he regarded the Cercamoniinae and the Notharctinae as being "more closely related than either is to the Adapinae," he retained the new subfamily within the Adapidae (Gingerich, 1975, p. 169). Szalay and Delson (1979, p. 107), however, regarded *Cercamonius* as

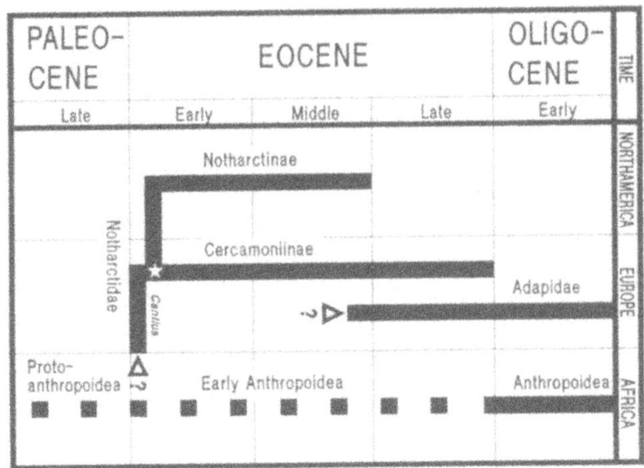

Fig. 11. Phylogeny, paleobiogeography, and a new classification of the Adapiformes.

"morphologically indistinguishable from *Protoadapis*," thereby creating the Protoadapini as a new tribe of the Adapidae, subfamily Adapinae, comprising the genera *Protoadapis, Europolemur, Mahgarita, Pronycticebus,* and *Agerinia.* But, as Gingerich pointed out, *Cercamonius* is clearly different from *Protoadapis,* particularly in lacking P_1 and "having P_2 reduced to a small single-rooted tooth." Although "*Cercamonius* resembles most closely *Protoadapis* and *Notharctus* in molar structure" (Gingerich, 1975, p. 165), it is principally different from "true" adapids by its nonmolarized P_4 as well as its relatively low-crowned and less crested molars. These characters are also typical for the other genera referred to here as being in the subfamily Cercamoniinae, whereas the reduced anterior premolars occur in *Caenopithecus* and *Mahgarita* only.

The Cercamoniinae obviously derive from *Donrussellia* by way of *Cantius,* and immigrants of the latter group gave rise to the North American Notharctinae (Godinot, 1988). Hence, these two subfamilies have a more recent ancestor in common than either of them with the Adapidae (Fig. 11). No antecedents are known of the Adapidae from sites earlier than the late middle Eocene of Europe [upper Geiseltalian *sensu* Franzen and Haubold (1986) = MP 13 of Schmidt-Kittler (1987)]. Therefore, the Adapidae are regarded as immigrants suddenly arriving together with other mammals (e.g., the palaeotheres and many artiodactyls) in central Europe from elsewhere at the end of the middle Eocene.

Classification of the Adapiformes

The Cercamoniinae are the European sister group of the North American Notharctinae. Therefore, both are united in the same family, the Notharctidae, and the Adapidae consist of the Adapinae only. The North American Notharctinae diverge phylogenetically from the European Cercamoniinae mainly by the emergence of a hypocone from the postprotocone crista and not from the cingulum as in the case of their European relatives, and by the development of strong mesostyles in the upper molars. *Magharita,* which developed a true hypocone, is either the result of a parallel development with the European Cercamoniinae or a member of that subfamily arriving late in North America. *Caenopithecus,* on the other hand, develops molar mesostyles within the Cercamoniinae.

In order to reflect the morphological, stratigraphic, and paleogeographic relationships as closely as possible, the following classification of the Adapiformes is suggested (Fig. 11).

Notharctidae (Trouessart, 1879)

Included subfamilies: Notharctinae (Trouessart, 1879), Cercamoniinae (Gingerich, 1975).

Diagnosis: Adapiformes with sagittal crest of the skull low or lacking, gracile zygomatic arch, and fused mandibular symphysis. In upper molars crest between proto- and metacone ("crista obliqua") sharp and continuous. P^4_4 not molarized. Hindlimbs longer than forelimbs. Calcaneus elongated anteriorly. Astragalar neck elongated. Trochlea of astragalus in dorsal view not so long and more wedge-shaped. Posterior astragalar shelf large. Intratrochanteric fossa of the femur long. Cuboid in dorsal view longer than wide. Basal phalanges considerably longer than metacarpals.

Distribution: Early to late Eocene of North America and Europe.

Notharctinae (Trouessart, 1879)

Type: Notharctus (Leidy, 1870).

Included genera: Pelycodus (Cope, 1875), *Smilodectes* (Wortman, 1903), *Copelemur* (Gingerich and Simons, 1977).

Diagnosis: Notharctidae with small orbita. Upper molars developing a mesostyle; hypocone arising from postprotocone crista ("*Nannopithex* fold"). P^1_1 present. Posterior astragalar shelf larger than in the Cercamoniinae.

Distribution: Early to middle Eocene of North America.

Cercamoniinae (Gingerich, 1975)

Type: Cercamonius (Gingerich, 1975).

Included genera: Agerinia (Crusafont-Pairó and Golpe-Posse, 1973), *Anchomomys* (Stehlin, 1916), *Caenopithecus* (Stehlin, 1916), *Cantius* (Simons, 1962),

Donrussellia (Szalay, 1976), *Huerzeleris* (Szalay, 1974), *Periconodon* (Stehlin, 1916), *Pronycticebus* (Grandidier, 1904), *Protoadapis* (Lemoine, 1878), and eventually *Mahgarita* (Wilson and Szalay, 1904).

Diagnosis: Notharctidae with small to large orbita. Upper molars usually without mesostyle (exception: *Caenopithecus*); hypocone developing from cingulum. $P^1{}_1$ present or lacking. $P^2{}_2$ in some cases very tiny (*Caenopithecus, Cercamonius, Mahgarita*). Posterior astragalar shelf smaller than in Notharctinae.

Distribution: Early to late Eocene of Europe (with the exception of *Cantius* and *Mahgarita,* which respectively occur also and only in North America).

Adapidae (Trouessart, 1879)

Included subfamily: Adapinae only.

Diagnosis: Adapiformes with skull displaying a high sagittal crest and very massive zygomatic arch. Orbita small. Upper molars always lacking mesostyle. Crest between proto- and metacone ("crista obliqua") in upper molars weak or lacking; hypocone developing from cingulum. $P^1{}_1$ present, $P^2{}_2$ well developed, and $P^4{}_4$ molarized. Fore- and hindlimbs about equally long. Calcaneus anteriorly short. Astragalar neck short. Trochlea of astragalus proximodistally long; ridges more parallel. Posterior astragalar shelf small. Intratrochanteric fossa of the femur short. Cuboid in dorsal view more square-shaped. Metacarpals about as long as basal phalanges.

Relationships and Anthropoid Origins

In my opinion it is only the Notharctidae (Cercamoniinae plus Notharctinae) that can be considered a sister group of possibly African early anthropoids. The Adapidae are something else. The problem is, all important character complexes of early Notharctidae, such as skull, dentition, limb bones, manus, and pes, appear still plesiomorphic enough to permit the derivation of the Anthropoidea (Franzen, 1987, pp. 175–176), but there is almost no structure that is complex and apomorphic enough to be regarded as a well-founded synapomorphy of Notharctidae and Anthropoidea together. This is not astonishing considering the large time interval of about 15 MA that exists between the cercamoniins from Messel and the earliest anthropoids from the Fayum.

A possible synapomorphy of Anthropoidea and Notharctidae could exist in the centrale being separated from the uncinatum by the capitate on one side while articulating with the trapezium on the other (Franzen, 1988). This feature is common to Notharctidae and Anthropoidea to the exclusion of Lemuriformes. But the question is whether it is plesiomorphic or apomorphic for euprimates (Beard *et al.,* 1988; Beard and Godinot, 1988; Godinot and Beard, 1993; Franzen, 1993).

Another synapomorphy could exist in the incisors of the Notharctidae. Their morphology closely approaches the condition seen in the Anthropoidea. This refers not only to the spatulate morphology, the shovel shape of the lower central incisors, and the size relations of I_1 and I_2 ($I_1 < I_2$) but also to the ovoid outline of their lateral upper and lower incisors (Fig. 13). On the other hand, all the incisors of notharctids are more brachyodont and labiolingually relatively narrower than those of the anthropoids, but this can easily be understood as a plesiomorphy. Even Rosenberger *et al.* (1985, p. 25), who denied any relationship between the incisor complexes of notharctids and anthropoids, could "essentially agree that the lower incisors of notharctines are anthropoid-like in appearance." It was mainly in view of "the more complex shapes of *Adapis–Leptadapis* as morphologically highly distinct" that they refused any sister-group relationship with the anthropoids. But this argument at best pertains to the "true" adapids alone. It does not refer to the Notharctidae. They also considered the morphologies of the upper incisors of anthropoids and what they called adapids as "distinctly different and probably nonhomologous," and pointed in this context particularly to a strong mesial prong of I^1 in all notharctines and adapines. In their opinion this feature is lacking in anthropoids, but this is not the case. For example, *Aegyptopithecus* among the earliest anthropoids as well as *Saimiri, Macaca,* and *Cercopithecus* among the recent ones still display this trait (compare Simons, 1967, p. 323, Fig. 5A; see Fig. 12). And although the lower lateral incisors of the Notharctidae are larger than the central ones, this difference of size is by far smaller than that in recent lemurs, and it exists also in platyrrhines, e.g., *Saimiri* (Fig. 12). So the incisor complex of the Notharctidae can be regarded as potentially ancestral for the Anthropoidea, but that of the lemuriforms is extremely different. This refers to the toothcomb, to the size reduction of the upper incisors, and to the very large interincisal diastema (Fig. 12). It cannot be excluded that the incisor morphotype of the Notharctidae was ancestral for recent lemurs as well. But their transformational distance from the ancestral morphotype is much larger than that of the Anthropoidea.

In any case, two possible synapomorphies are not enough to solve the problem of the phylogenetic position of the Messel primates with respect to anthropoid origins. But the Messel primates as well as the Notharctidae as a whole appear still plesiomorphic enough to permit the derivation of anthropoids. They seem to be much closer to a common ancestor with anthropoids than the Omomyidae, which are already too specialized in a tarsiiform direction, considering the disproportional enlargement of their central incisors, the beginning elongation of calcaneus and navicular, and the distal fusion of tibia and fibula (Gingerich, 1981).

Therefore, we should not look any further for direct ancestor–descendant relationships between Omomyidae and Anthropoidea or Notharctidae and Anthropoidea, but rather for the derivation of the Anthropoidea and the

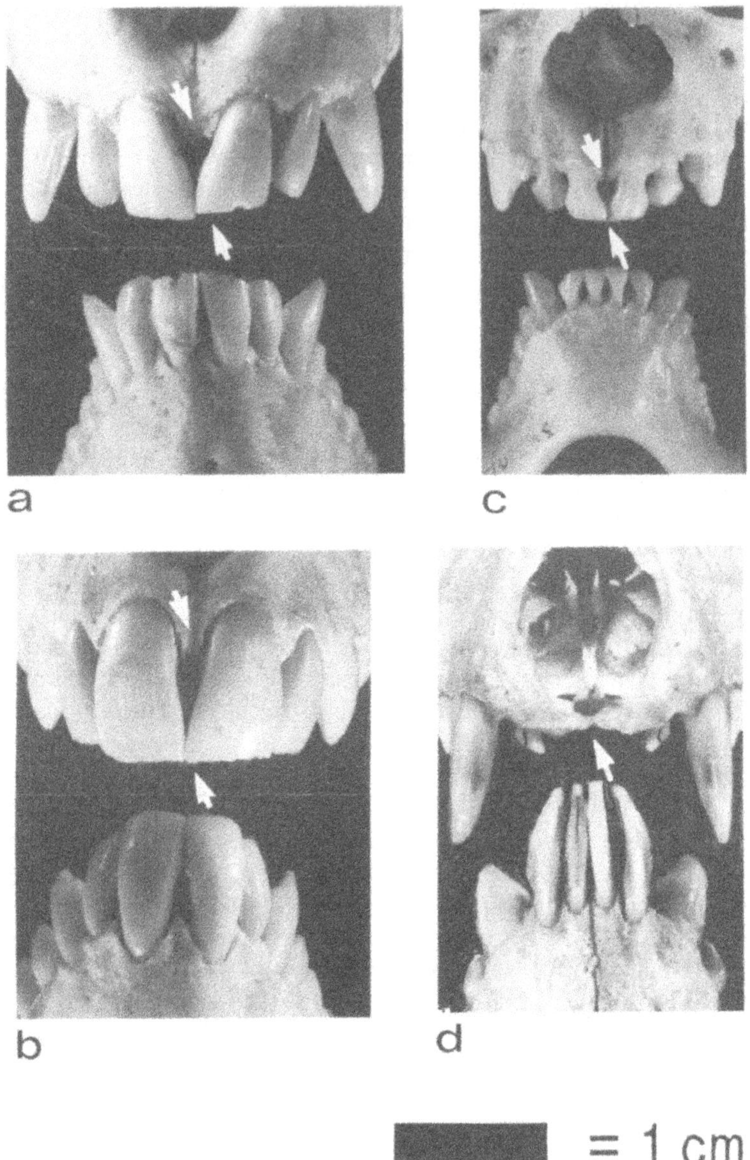

Fig. 12. Labial view of upper and lower incisors of (a) *Cercopithecus aethiops*, (b) *Macaca fascicularis*, (c) *Saimiri sciureus*, and (d) *Lemur mongoz*, scale indicated. Arrows are pointing to mesial prong of I^1 and to interincisal diastema. Photos: Forschungsinstitut Senckenberg Frankfurt am Main (E. Pantak-Wein).

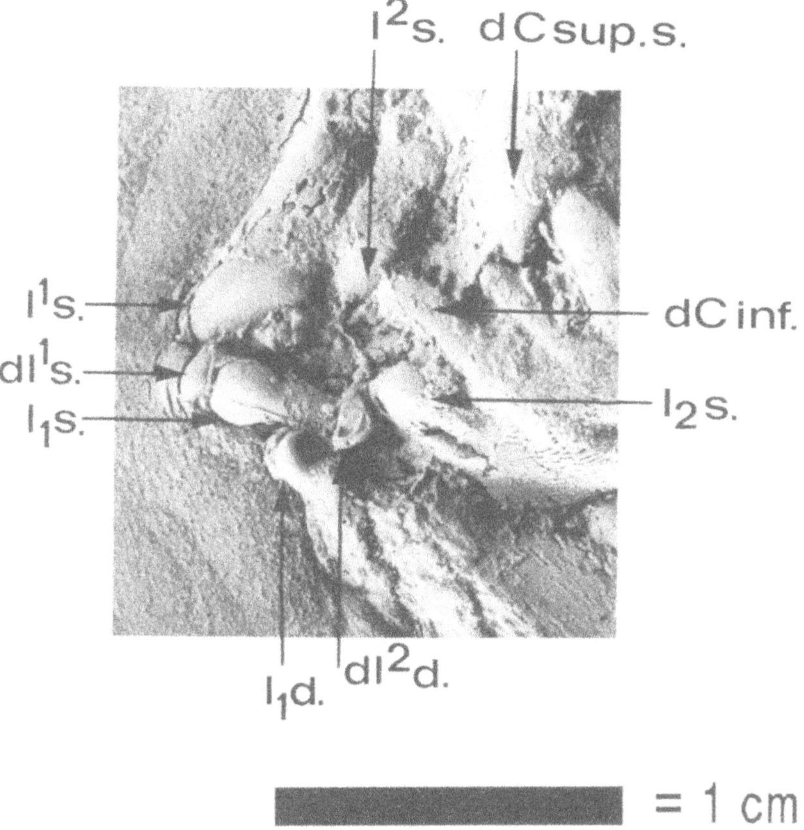

Fig. 13. Frontal dentition of *Europolemur koenigswaldi* (type specimen, SMF-ME 1228A), labial view, coated with ammonium chloride, individual teeth and scale indicated. Photo: Forschungsinstitut Senckenberg Frankfurt am Main (E. Pantak-Wein).

Notharctidae from an as yet unknown group of Protoanthropoidea that may have existed in Africa during the late Paleocene (Franzen, 1987, p. 175) and that was morphologically close to the Notharctidae, particularly to *Donrussellia*. What we still need to solve the problem of anthropoid origins, however, are more and better-preserved specimens of Messel quality as well as primate discoveries from the Eocene and Paleocene of subsaharan Africa.

Conclusions

1. The infraorder Adapiformes (Szalay and Delson, 1979) has to be structured into the families of the Notharctidae (Trouessart, 1879) and the Adapidae (Trouessart, 1879) in order to reflect their morpho-

logical relationships. The Notharctidae comprise the subfamilies of the Notharctinae (Trouessart, 1879) and the Cercamoniinae (Gingerich, 1975), while the Adapidae are restricted to the Adapinae (Trouessart, 1879) only. The classification of the Asian representatives of the Adapiformes is not relevant to this context. Their phylogenetic position and classification should be discussed separately.

2. All discussions of an "adapid" origin of anthropoids fail to meet the relevant arguments as long as they do not differentiate between Notharctidae (including Cercamoniinae) and "true" Adapidae.
3. The Notharctidae are not regarded as direct ancestors but as a possible sister group of early Anthropoidea that may have existed in (subsaharan?) Africa during the Eocene (Franzen, 1987, p. 175).

Acknowledgments

I thank the hosts of the conference on "Anthropoid Origins," Richard Kay (Durham) and John Fleagle (Stony Brook), and all those who helped them for the excellent organization of this extraordinary opportunity to meet most the relevant fossils and their investigators at the same time. My sincere thanks go to Elizabeth Culotta and Hans Thewissen for their generous and most amicable hospitality while housing me during that meeting. I am also very grateful to the LSB Leakey Foundation and the Wenner-Gren Foundation for Anthropological Research, who supported my travel to Durham.

The photographs in this chapter were prepared by Erwin Haupt and Elke Pantak-Wein (Forschungsinstitut Senckenberg) as well as Marisa Schwab (Hessisches Landesmuseum Darmstadt). The drawing of a *Europolemur* sitting in a tree (Fig. 7) is a product of the skills of Dr. Eberhard ("Dino") Frey (Staatliches Museum für Naturkunde Karlsruhe).

References

Beard, K. C., and Godinot, M. 1988. Carpal anatomy of *Smilodectes gracilis* (Adapiformes, Notharctinae) and its significance for lemuriform phylogeny. *J. Hum. Evol.* **17:**71–92.

Beard, K. C., Dagosto, M., Gebo, D. L., and Godinot, M. 1988. Interrelationships among primate higher taxa. *Nature* **331:**712–714.

Dagosto, M. 1983. Postcranium of *Adapis parisiensis* and *Leptadapis magnus* (Adapiformes, Primates). *Fol. Primatol.* **41:**49–101.

Elder, R. 1985. Principles of Aquatic Taphonomy with Examples from the Fossil Record. UMI Dissertation Information Service, Ann Arbor, Michigan.

Franzen, J. L. 1985. Exceptional preservation of Eocene vertebrates in the lake deposit of Grube Messel (West Germany). *Phil. Trans. R. Soc. Lond.* [*B*] **311:**181–186.

Franzen, J. L. 1987. Ein neuer Primate aus dem Mitteleozän der Grube Messel (Deutschland, S-Hessen). *Cour. Forsch. Inst. Senckenberg* **91:**151–187.

Franzen, J. L. 1988. Ein weiterer Primatenfund aus der Grube Messel bei Darmstadt. *Cour. Forsch. Inst. Senckenberg* **107:**275–289.

Franzen, J. L. 1990. Grube Messel, In: D. E. G. Briggs and P. R. Crowther (eds.), *Paleobiology. A Synthesis,* pp. 289–294. Blackwell Science Publishers, Oxford.

Franzen, J. L. 1993. The oldest primate hands: Additional remarks and observations. In: H. Preuschoft and D. J. Chivers (eds.), *Hands of Primates.* Springer-Verlag, Vienna.

Franzen, J. L. and Frey, E. 1993. *Europolemur* completed. *Kaupia* **3:**113–130.

Franzen, J. L. and Haubold, H. 1986. The European middle Eocene of mammalian stratigraphy. Definition of the Geiseltalian. *Mod. Geol.* **10:**159–170.

Gingerich, P. D. 1975. A new genus of Adapidae (Mammalia, Primates) from the late Eocene of southern France, and its significance for the origin of higher primates. *Contrib. Mus. Paleontol. Univ. Michigan* **24**(15)**:**163–170.

Gingerich, P. D. 1981. Early Cenozoic Omomyidae and the evolutionary history of tarsiiform primates. *J. Hum. Evol.* **10:**345–374.

Gingerich, P. D. 1990. Prediction of body mass in mammalian species from long bone lengths and diameters. *Contrib. Mus. Paleontol. Univ. Michigan* **28**(4)**:**79–92.

Gingerich, P. D., and Martin, R. D. 1981. Cranial morphology and adaptations in Eocene Adapidae. II. The Cambridge skull of *Adapis parisiensis. Am. J. Phys. Anthropol.* **56:**235–257.

Godinot, M. 1988. Dental morphology of *Donrussellia,* the most primitive known Adapiform (Primates). *J. Vert. Paleontol.* **8**(3)**:**15A.

Godinot, M., and Beard, K. C. 1991. A survey of fossil primate hands: a review and an evolutionary inquiry emphasizing early forms. *Human Evolution* **6**(4)**:**307–354.

Godinot, M., and Beard, K. C. 1993. A survey of fossil primate hands. In: H. Preuschoft and D. J. Chivers (eds.), *Hands of Primates.* pp. 335–378. Springer-Verlag, Vienna.

Goth, K. 1990. Der Messeler Ölschiefer—ein Algenlaminit. *Cour. Forsch. Inst. Senckenberg* **131:**1–143.

Haubold, H. 1987. Geiseltalium: ein neues laudsäugetier-zeitalter im paläogen. *Hall. Jb. Geowiss.* **12:**120–121.

Kay, R. F. 1975. The functional adaptations of primate molar teeth. *Am. J. Phys. Anthropol.* **43:**195–216.

Koenigswald, W. von 1979. Ein Lemurenrest aus dem eozänen Ölschiefer der Grube Messel bei Darmstadt. *Palaontol. Z.* **53**(1/2)**:**63–76.

Koenigswald, W. von 1985. Der dritte Lemurenrest aus dem mitteleozänen Ölschiefer der Grube Messel bei Darmstadt. *Carolinea* **42:**145–148.

Napier, J. R., and Napier, P. H. 1967. *A Handbook of Living Primates. Morphology, Ecology and Behaviour of Nonhuman Primates.* Academic Press, London.

Rosenberger, A. L., Strasser, E., and Delson, E. 1985. Anterior dentition of *Notharctus* and the Adapid–Anthropoid hypothesis. *Fol. Primatol.* **44:**15–39.

Schaal, S., and Ziegler, W. (eds.) 1992. *Messel. An Insight into the History of Life and of Earth.* Clarendon Press, Oxford.

Schmidt-Kittler, N. (ed.) 1987. International Symposium on Mammalian Biostratigraphy and Paleoecology of the European Paleogene, Mainz, February 18th–21st 1987. *Munch. Geowiss. Abh.* **A(10):**1–312.

Simons, E. L. 1967. The significance of primate paleontology for anthropological studies. *Am. J. Phys. Anthropol.* **27**(3)**:**307–332.

Simons, E. L. 1972. *Primate Evolution. An Introduction to Man's Place in Nature.* Macmillan, New York; Collier-Macmillan, London.

Stehlin, H. G. 1912. Die Säugetiere des schweizerischen Eocaens. Critischer Catalog der Materialien, T. 7(1): *Adapis. Abh. Schweiz. Palaontol. Ges.* **37:**1163–1298.

Stehlin, H. G. 1916. Die Säugetiere des schweizerischen Eocaens. Critischer Catalog der Materialien, T. 7(2): *Caenopithecus—Necrolemur—Microchoerus—Nannopithex—Anchomomys—*

Periconodon—Amphichiromys—Heterochiromys—Nachträge zu *Adapis*—Schlussbetrachtungen zu den Primaten. *Abh. Schweiz. Palaontol. Ges.* **41:**1297–1552.

Szalay, F. S., and Delson, E. 1979. *Evolutionary History of the Primates.* Academic Press, New York.

Thalmann, U., Haubold, H., and Martin, R. D. 1989. *Pronycticebus neglectus*—an almost complete Adapid primate specimen from the Geiseltal (GDR). *Palaeovertebrata* **19**(3)**:**115–130.

Trouessart, E.-L. 1879. *Catalogue Systématique, Synonymique et Géographique des Mammifères Vivants et Fossiles. I: Primates. (Simiae, Prosimiae, Chiroptera); II: Insectivora.* E. Deyrolles, Paris.

Relative Ages of Eocene Primate-Bearing Deposits of Asia

5

PATRICIA A. HOLROYD
and RUSSELL L. CIOCHON

Introduction

Paleontologists have often looked to Asia as a center of origin for anthropoid primates (e.g., Pilgrim, 1927; Colbert, 1937, 1938; Ba Maw *et al.*, 1979; Gingerich, 1980; Ciochon and Chiarelli, 1980; Ciochon *et al.*, 1985; Ciochon and Etler, 1994). In part, this notion has been founded in the accepted earlier occurrence of putative anthropoids or "protoanthropoids" in the Asian faunal record. In particular, the occurrence of characters similar to those found in African anthropoids in the Pondaung primates *Amphipithecus* and *Pondaungia* from presumed upper Eocene deposits in the Pondaung Hills of Burma (now Myanmar) has led to conjecture that this area represented the center of origin for higher primates. However, the criterion of earlier occurrence has been eliminated by the redating of the earliest Fayum anthropoid fauna to the late Eocene (Kappelman, 1992; Rasmussen *et al.*, 1992) such that the Fayum anthropoids would be penecontemporaneous with the Asian primates *Am-*

PATRICIA A. HOLROYD • Department of Biological Anthropology and Anatomy, Duke University, Durham, North Carolina 27710, and U.S. Geological Survey, Denver, Colorado 80225. RUSSELL L. CIOCHON • Departments of Anthropology and Pediatric Dentistry, University of Iowa, Iowa City, Iowa 52242.

Anthropoid Origins, edited by John G. Fleagle and Richard F. Kay. Plenum Press, New York, 1994.

phipithecus, Pondaungia, Hoanghonius, and *Rencunius* (see Chapter 7, this volume) based on current age assessments.

Amphipithecus mogaungensis and *Pondaungia cotteri* from the Pondaung fauna of Burma and *Hoanghonius stehlini* and *Rencunius zhoui* from the Heti fauna of Shaanxi and Shanxi Provinces, China, have generally been considered late Eocene in age, and these faunas are considered part of the Sharamurunian Land Mammal Age (LMA) (see, e.g., Delson, 1977; Russell and Zhai, 1987). Recently, a primate molar from Thailand has been classified as ?*Hoanghonius* and is also believed to be late Eocene in age (Ducrocq *et al.,* 1992). Other Asian primates—*Asiomomys changbaicus* from the Huadian fauna, Jilin Province, China; *Lushius qinlinensis* from the Lushi fauna, Henan Province, China; and the undescribed primates from Jiangsu Province, China—have been regarded as middle Eocene in age (Russell and Zhai, 1987; Beard and Wang, 1991; Qi *et al.,* 1991).

The three primate taxa from the Kuldana Formation of Pakistan—*Panobius afridi, Kohatius coppensi,* and cf. *Agerina* sp.—have been regarded as earliest middle Eocene or latest early Eocene in age (Russell and Gingerich, 1980, 1987; Gingerich and Russell, 1990). The only Asian primate of certain early Eocene age is *Altanius orlovi* from the Tsagan Khushu beds, Naran Bulak Formation, Mongolia (Dashzeveg and McKenna, 1977; Gingerich *et al.,* 1991). These localities and others discussed in the text are shown in Fig. 1.

The age of some of these primates, especially those typically considered to pertain to late Eocene faunas, is now questioned as a result of changes in the worldwide correlation of Eocene epoch and subepoch boundaries. Because these primates and the relative ages have been used in hypotheses of an Asian origin of anthropoids, it is important that they be placed in a current temporal framework. Here we summarize recent changes in Eocene biochronology, reexamine the evidence for the age of Asian primates and their associated faunas, and present a revised conception of Asian Eocene biochronology that is consistent with current evidence.

Eocene Biostratigraphy and Correlation

The dating of Asian continental Eocene deposits has traditionally been accomplished by faunal correlations. Currently, no radiometric dates or paleomagnetic reversal data are available for Asian mammal-bearing deposits of Eocene age. The only radiometric dates presently available for the *entire* Paleogene of Asia are dates of 31.3 and 32.0 Ma for a basalt in the Hsanda Gol Formation in Mongolia (Evernden *et al.,* 1964), indicating an early Oligocene age for this deposit. Hence, faunal correlations remain the only method by which to judge the age of Asian Eocene mammals. In China and Mongolia age assessments have been based principally on a limited number of mammalian correlations with the North American fossil record. In Southeast Asia, the

Fig. 1. Map showing locations of primate-bearing deposits and other mammal faunas of Asia discussed in this chapter.

correlation of the terrestrial and marine sequences in the Chindwin–Irrawaddy Basin of central Burma to standard stages in India and Europe has been used for age determination in that region of Asia.

These earlier faunal correlations are still valid. However, changes in the correlation of standard reference ages in Europe and North American LMAs relative to subepoch boundaries have not yet been integrated with the Asian record. These shifts, beginning with the placement of the Bartonian Stage in the middle Eocene by Berggren *et al.* (1978) and most recently with the redating of the entire Chadronian LMA to the late Eocene (Swisher and Prothero, 1990; Prothero and Swisher, 1992), suggest that the conventional age assessments for middle and upper Eocene primate-bearing deposits in Asia should be revised. These shifts are reviewed, and their effects on the dating of the Asian primate record are examined below. The assignments of an early Eocene age to *Altanius orlovi* and of a late early Eocene to early middle Eocene age for the Kuldana Formation primates are not affected by these changes, but all late Eocene age assignments for Asian primates are.

The first change in the placement of subepoch boundaries that has an impact on the interpretation of Asian primates was a realignment of the Bartonian Standard Stage in Europe. In the late 1970s debate erupted over the age of the Bartonian. Although the Bartonian had long been considered contemporaneous with the late Eocene Ludian/Priabonian, Cavalier (1979) and Cavelier and Pomerol (1976) demonstrated that it was older than the Priabonian. Berggren *et al.* (1978) summarized this debate and suggested an expanded concept of the middle Eocene that included not only the Lutetian Stage but also the Bartonian. This change in the antiquity of the Bartonian was adopted by a majority of workers, but it was never applied explicitly to the deposits of the Chindwin–Irrawaddy Basin that had been correlated with the Bartonian.

A second change that affects age interpretations of Asian primates is the realignment of North American LMAs relative to subepoch boundaries based on a revised magnetostratigraphy and radiometric dating of key stratigraphic sequences encompassing the early Oligocene and terminal Eocene (Swisher and Prothero, 1990; Berggren and Prothero, 1992; Prothero and Swisher, 1992). Prothero and Swisher (1992) demonstrate that the Uintan LMA, previously considered to be late middle through early late Eocene in age, is now placed entirely within the middle Eocene subepoch and that the Chadronian LMA occurs entirely within the late Eocene, with its upper limit being approximately equivalent to the Eocene–Oligocene boundary. They also retained the entire Duchesnean LMA within the late Eocene subepoch. However, Holroyd and Maas (Chapter 11, this volume) explain that at least the early Duchesnean lies within the middle Eocene subepoch based on magnetostratigraphic correlation with the European Bartonian, with the middle–late Eocene boundary probably occurring within the youngest half of the Duchesnean LMA in North America. This revised placement of subepoch boundaries, contrasted with previous correlations, is shown in Fig. 2.

These changes in North American biostratigraphy are so recent that their impact on the correlation of Asian sediments has not yet been examined in detail. Berggren and Prothero (1992) present a brief case based on faunal turnover and the radiometric dates of Evernden *et al.* (1964) for a realignment of the Asian "middle Oligocene" Hsanda Gol fauna to the early Oligocene and the "early Oligocene" Ergilin fauna to the late Eocene subepoch. Ducrocq (1993) has also suggested, based on analysis of cenograms of the "middle" Oligocene Hsanda Gol and Ergilin faunas, that these sites are older than previously thought. Dashzeveg (1992) has independently suggested a similar recorrelation, in part subdividing the Ergilin fauna and suggesting that the older part of the fauna is late Eocene. These realignments appear to push the Sharamurunian "late Eocene" faunas (those including *Hoanghonius*) back to the middle Eocene. This suggestion and the faunal evidence to support it are examined below.

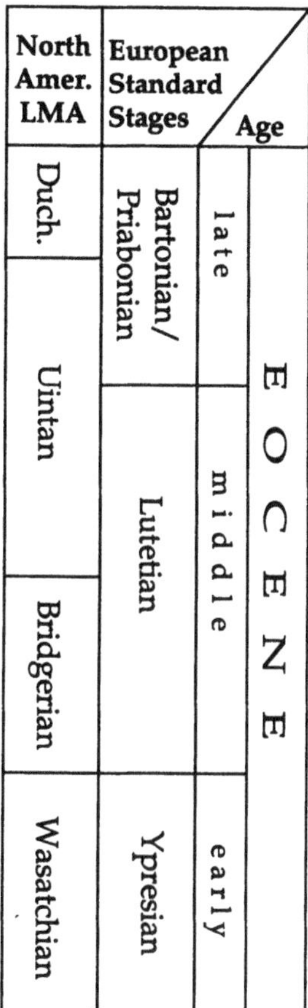

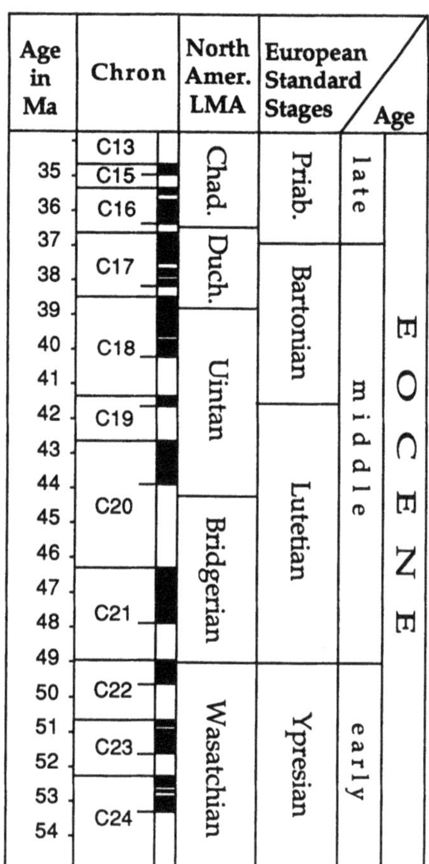

Fig. 2. Correlation chart showing differences in age concepts for the European standard stages and North American LMAs before (on the left) and after (on the right) the realignments of subepoch boundaries by Berggren *et al.* (1978), Prothero and Swisher (1992) and Holroyd and Maas (Chapter 11, this volume). Radiometric ages (in Ma) from Cande and Kent (1992).

Correlation of the Chinese Sites

The most recent review for the age of the primate-bearing sites of China was presented by Russell and Zhai (1987). The localities of Lushi and Heti were assigned to subepochs based on their correlation to the Inner Mongolian Irdin Manha and Shara Murun faunas. The combined faunas of Irdin Manha and Lushi were used by them to exemplify the middle Eocene Irdinmanhan LMA and those of Shara Murun and Heti to typify a late Eocene Sharamurunian LMA. The faunas were correlated to the worldwide time scale by virtue of their faunal similarities to North American land mammal assemblages.

The Asian LMAs discussed here and shown in Fig. 3 are those most

Asian (Chinese) LMA	Included local faunas	Age	
Sharamurunian	Shara Murun Heti (includes *Hoanghonius & Rencunius)*	late	EOCENE
Irdinmanhan	Irdin Manha Lushi (includes *Lushius*) Jiangsu (includes undescribed adapines, tarsiiformes, and putative anthropoids) Huadian (includes *Asiomomys)* Arshanto	middle	
Bumbanian	Tsagan Khushu (includes *Altanius)*	early	

Fig. 3. Asian Land Mammal Ages, their traditional correlations to standard stages and/or subepochs, and selected included local faunas.

commonly used for subdividing this continent's faunas. Recently, efforts have been directed to subdivide these LMAs further (e.g., Dashzeveg, 1992; Wang, 1992; Qi, 1987) or to propose additional LMAs (Tong, 1989). Both Dashzeveg (1992) and Wang (1992) dealt with restricted geographic areas and time periods in which primates are unknown, thus their work has little bearing on the discussion here. We concur with Qi's (1987) recognition of an older Arshantan element of the Irdinmanhan LMA, and we recognize this distinction within the more classically formulated Irdinmanhan LMA. However, we do not (at this time) recognize an additional, distinct Naduan LMA (Tong, 1989), intermediate in age between the Sharamurunian and Ergilin LMA and based on the Naduo fauna from Guangxi Province. Although the Naduo fauna contains no primates, it is relevant to a discussion of primate-bearing sites because this fauna has been correlated with the Pondaung fauna (Ding *et al.*, 1977; Qiu, 1977; Tang, 1978). However, as discussed by Russell and Zhai (1987), there are serious problems in typifying the Naduo fauna, primarily based on unclear stratigraphic provenance of finds as well as taxonomic identifications. The Naduo fauna was correlated with that of the Pondaung based on identification of Pondaung anthracothere genera at the Guangxi site. However, the Guangxi anthracotheres represent new genera (Holroyd and Ciochon, 1991), and work on other taxa from the area is still in progress. Thus, until further study of this fauna is completed, we do not feel the evidence is sufficiently strong to recognize a distinct Naduan LMA or to place it with accuracy within the regional biostratigraphic framework.

Below, intraregional correlations between the Chinese faunas are examined, as are their intercontinental correlations.

Intraregional Correlations

Lushi and Heti

The primate-bearing sites of Lushi and Heti have been, respectively, correlated with the Irdin Manha and Shara Murun faunas based on the number of shared species (Russell and Zhai, 1987). Examination of faunal resemblances between the Chinese faunas does support, albeit weakly, a correlation of the Heti and Shara Murun faunas and that of Lushi with the Irdin Manha fauna. Tables I and II present data on the number of shared taxa and generic and species faunal resemblance indices (FRI) for the two Chinese and two Inner Mongolian faunas. The FRIs are quite low for any pairwise comparison. The species FRI between the Irdin Manha and Lushi fauna of 17.9% is lower than species FRIs between temporally grouped composite faunas within the North American early Eocene reported by Flynn (1986), and the generic FRIs are lower than those between Europe and North America in the early Eocene (see Flynn, 1986; Krause and Maas, 1990). However, the fact that some species are shared between each pair of sites (with the exception of Irdin Manha and Heti) suggest that they do not differ substantially in age.

These comparisons fail to support the notion of marked faunal turnover between the Irdinmanhan and Sharamurunian LMAs as suggested by Russell and Zhai (1987); several species, notably *Honanodon hebetis* and *Propterodon morrisi,* occur in both LMAs. None of these individual faunas shows marked similarity to any other at either the specific or generic level, so it is difficult at this time to typify coherent land mammal ages or turnover between them.

Low overall resemblances between faunas may be a combined function of poor sampling and/or latitudinal, temporal, and environmental differences (Flynn, 1986). Certainly, the faunas considered here are still poorly sampled. However, because the number of shared species and genera is the only evidence available for the interpretation of the age of these faunas, the fact that

Table I. Number of Taxa Shared between North Asian Faunas Discussed Here

	Species/genera[a]	Lushi	Irdin Manha	Heti	Shara Murun
Lushi	*30*/28	—	8	4	5
Irdin Manha	*28*/25	*5*	—	0	5
Heti	*29*/25	*1*	*0*	—	6
Shara Murun	*31*/29	2	*1*	4	—

[a]Number of genera in each fauna are shown in roman type; numbers of species in each fauna appear in *italics*.

Table II. Faunal Resemblance Indices[a] for Species and Genera Shared between North Asian Mammalian Faunas

	Lushi	Irdin Manha	Heti	Shara Murun
Lushi	—	32.0	16.0	17.9
Irdin Manha	*17.9*	—	0.0	20.0
Heti	*3.4*	*0.0*	—	24.0
Shara Murun	*6.7*	*3.6*	*13.8*	—

[a]The FRI is the total number of taxa shared divided by the total number of taxa in the smaller of the two faunas; this ratio is multiplied by 100 and expressed as a percentage. Species FRIs are shown in *italic* below the diagonal, generic FRIs above. Counts are compiled from faunal lists in Russell and Zhai (1987).

several genera and a limited number of species are shared between any pair of these faunas (with the exception of Irdin Manha and Heti) suggests that none are separated from one another by a significant temporal gap. Thus, if the Irdin Manha fauna (and the Lushi fauna by its similarity) is early Uintan and/or late Bridgerian in age, based on correlations to North America (Qi, 1987; see below), the Shara Murun and Heti faunas are probably only slightly younger in age and should also be placed in the middle Eocene subepoch, probably equivalent to the late Uintan or early Duchesnean. Otherwise we would have to accept the notion that some species and genera persisted for more than 5 million years with little morphological change.

Member III of the Huadian Formation, Jilin Province, China

Asiomomys changbaicus derives from Member III of the Huadian Formation, Jilin Province, northeastern China. The fauna, described by Wang and Li (1990), is largely endemic and difficult to correlate to other Chinese sites. On the basis of general evolutionary stages of the mammals with their closest relatives in Asia and North America, the age of the fauna has been estimated as equivalent to the late Uintan (Wang and Li, 1990; Beard and Wang, 1991). Under current concepts of North American subepoch boundaries, this would place *Asiomomys* in the middle Eocene.

Jiangsu

A diverse assemblage of undescribed new primate genera (including adapids, tarsiers, and putative anthropoids) has been reported from fissure fillings in Liyang county, Jiangsu Province (Qi *et al.*, 1991; Rae, 1993). The two mammalian species currently described from this locality, *Lushilagus lohoensis* and *Miacis lushiensis,* are conspecific with species from Lushi. Thus, the Jiangsu primates are considered to be Irdinmanhan in age and approximately equivalent in age to *Lushius.*

Correlation to North America

Faunal evidence regarding the contemporaneity of Asian and North American faunas has primarily been based on evolution in helaletid and amynodontid perissodactyls. Russell and Zhai (1987) reviewed this evidence and concluded that the Irdin Manha perissodactyls showed greatest similarities with late Bridgerian/early Uintan faunas, noting in particular the cooccurrence of *Forstercooperia grandis* at both Irdin Manha and earliest Uintan localities in the Western Interior of North America. They therefore placed the Irdin Manha fauna in the middle Eocene, rejecting a previously proposed early late Eocene age. The Shara Murun perissodactyls apparently arose from endemic Irdin Manha forms and show no close similarity to those of North America.

Qi (1987) provided evidence for a slightly finer correlation of the Asian and North American faunas by suggesting that the Irdin Manha faunas are close in age to the underlying Arshanto fauna because the two faunas share three species and nine genera in common. Qi (1987) correlated the Arshanto with the late Bridgerian, early middle Eocene time. This correlation supports Russell and Zhai's (1987) conclusions and suggests that the Irdin Manha fauna has closer equivalence to the earliest Uintan in age.

Russell and Zhai (1987) placed the boundary between the middle and late Eocene subepochs in northern Asia at the change between the Irdin Manha and Shara Murun faunas. According to Russell and Zhai (1987), the Irdin Manha fauna displays a significant number of elements in common with the middle Eocene of North America, whereas the Shara Murun fauna has few genera in common and no species shared with North America. As discussed above, this typification of the differences between the Irdinmanhan and Sharamurunian LMAs is somewhat overstated.

Figure 4 presents a visual summary of biostratigraphic evidence regarding genera shared between North America and selected North Asian localities. Data on the ranges of these genera in North America are from Stucky (1992). For primate-bearing localities discussed here and the "classic" Arshanto, Irdin Manha, and Shara Murun faunas, these North American ranges are plotted for each of the shared genera that occur there. The Huadian fauna and the taxa described from Jiangsu are not shown here; no described genera from these localities (with the exception of the long-lived carnivore *Miacis* at Jiangsu) are shared with North America.

Several points are made clear by this summary. A Bridgerian-equivalent age is confirmed for the Arshanto fauna, although several "Uintan" elements are present. An early Uintan age is suggested for Lushi and Irdin Manha, in accordance with previous determinations. However, in contrast to earlier assessments, Heti also appears to be Uintan in age. Correlations of Shara Murun and North American faunas are equivocal; most shared taxa are doubtfully attributed but suggest a Uintan age. The reported *Desmatolagus* sp. is anomalously young for any Eocene interpretation of this site's age. How-

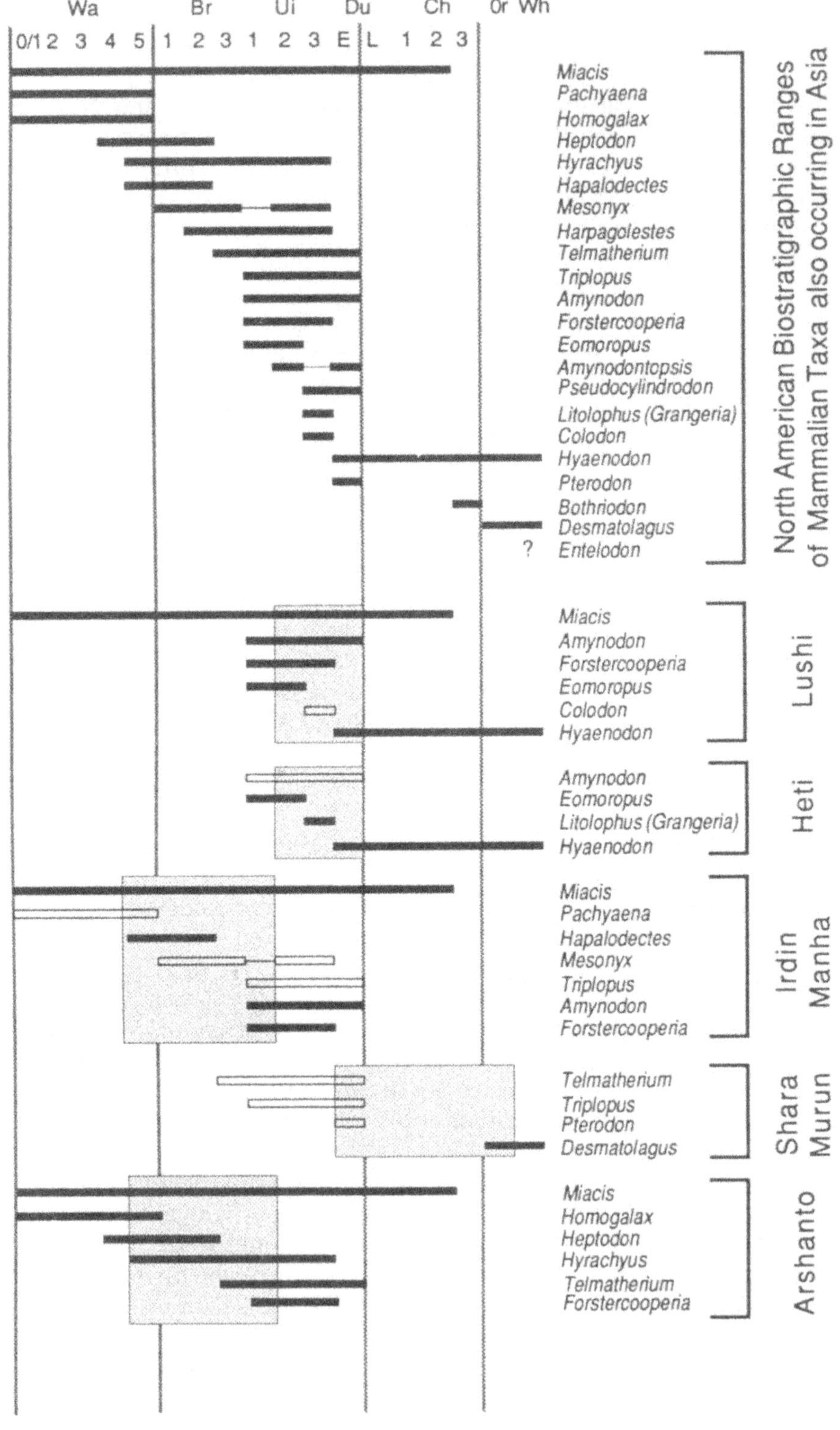

Wa
Br
Ui
Du
Ch
Or Wh
0/1 2 3 4 5 1 2 3 1 2 3 E L 1 2 3
Miacis
Pachyaena
Homogalax
Heptodon
Hyrachyus
Hapalodectes
Mesonyx
Harpagolestes
Telmatherium
Triplopus
Amynodon
Forstercooperia
Eomoropus
Amynodontopsis
Pseudocylindrodon
Litolophus (Grangeria)
Colodon
Hyaenodon
Pterodon
Bothriodon
Desmatolagus
?
Entelodon
North American Biostratigraphic Ranges of Mammalian Taxa also occurring in Asia
Miacis
Amynodon
Forstercooperia
Eomoropus
Colodon
Hyaenodon
Lushi
Amynodon
Eomoropus
Litolophus (Grangeria)
Hyaenodon
Heti
Miacis
Pachyaena
Hapalodectes
Mesonyx
Triplopus
Amynodon
Forstercooperia
Irdin Manha
Telmatherium
Triplopus
Pterodon
Desmatolagus
Shara Murun
Miacis
Homogalax
Heptodon
Hyrachyus
Telmatherium
Forstercooperia
Arshanto

ever, the presence of somewhat younger-occurring North American elements in Asian faunas is not unexpected. Several North American LMAs (i.e., the Bridgerian, Uintan, and Duchesnean) are based on the first occurrences of Asian immigrants to the North American continent. Thus, we should expect to find older representatives of these genera in Asia.

Although good alpha taxonomy and detailed phylogenetic analyses are the most appropriate means to further refine these biochronological assessments (as noted by Stucky, 1992), a new correlational hypothesis can be advanced on current evidence. Lushi, the Jiangsu site, and the Irdin Manhan appear to be temporal correlatives within the Irdinmanhan LMA. Shara Murun and Heti are approximately time equivalent and represent the Sharamurunian LMA, and Member III of the Huadian Formation may also belong there. The Arshantan, Irdinmanhan, and Sharamurunian LMAs are three successive subdivisions of the middle Eocene in China, although the Arshantan and Irdinmanhan appear to be closer in time to each other (based on a greater number of shared species) than either is to the Sharamurunian. Respectively, these three Chinese LMAs correlate with the Bridgerian, early Uintan, and late Uintan–early Duchesnean LMAs in North America. This adjustment in age for the Shara Murun and Heti faunas also best fits with age changes proposed for the Hsanda Gol and Ergilin faunas suggested by Berggren and Prothero (1992), Dashzeveg (1992), and Ducrocq (1993).

Correlation of the Southeast Asian Sites

Although the northern Chinese sites can be regarded as either early or late middle Eocene in age based on their correlation to the revised conception of subepoch boundaries in North American biostratigraphy, the Southeast Asian primates cannot be easily correlated to other biostratigraphies by their associated mammalian faunas. The compositions of the Pondaung and Krabi faunas differ somewhat from one another, and both appear to be endemic when compared with other Asian faunas. No species are found in common, and the only cooccurring genus in China and Southeast Asia is the anthracotheriid artiodactyl *Anthracokeryx*, a form found in the Pondaung, Krabi,

---

Fig. 4. Biostratigraphic ranges of genera shared between North America and selected North Asian localities. For each Asian locality, genera shared with North America are listed, and the North American biostratigraphic range is plotted. The ranges of doubtfully attributed or questionably identified taxa are shown as open bars. The shaded boxes represent approximate temporal limits for each locality based solely on a minimum concurrence of the North American stratigraphic ranges. Vertical lines correspond to epoch and subepoch boundaries. Letter and numeral abbreviations along the top of the chart denote North American Land Mammal Ages and their time-successive subdivisions: Wa, Wasatchian, with five subdivisions; Br, Bridgerian with three subdivisions; Ui, Uintan with three subdivisions; Du, Duchesnean, divided into early and late; Ch, Chadronian; Or, Orellan; Wh, Whitneyan.

and Heti faunas. Amynodontid perissodactyls found in both the Pondaung and northern Asian faunas (as well as North America) differ at the generic level.

The Pondaung, Chindwin–Irrawaddy Basin, Burma

A late Eocene age was long adopted for the Pondaung fauna based on mammalian comparisons and a temporized early 20th century view of the age of the Bartonian Stage. Both Pilgrim (1928, 1941) and Colbert (1938) found the Pondaung anthracotheres to be more primitive than those from the Fayum fauna of Egypt. Because the Fayum mammalian faunas were then considered to be entirely early Oligocene in age, they concluded that the Pondaung forms could not be much older. Hence, a late Eocene age was considered appropriate.

The only comprehensive treatment of mammalian faunal evidence for the age of the Pondaung fauna (Colbert, 1938) also suggested a late Eocene age. Colbert found that the Pondaung mammals were most comparable in stage of morphological evolution to related taxa in North American Uintan and Mongolian Sharamurunian faunas. As discussed above, these faunas, then regarded as late Eocene in age, are now deemed middle Eocene in age.

The only external constraints on the age of the Pondaung beds come from its relationships to overlying and underlying marine sediments. The Eocene sedimentary rocks of this area have been subdivided (from oldest to youngest) into five "formations": the Laungshe, Tilin, Tabyin, Pondaung, and Yaw (see, e.g., Chhibber, 1934; Bender, 1983), shown diagramatically in Fig. 5. It should be noted that the term "formation" as applied to these rocks is not used in the modern sense of the term, i.e., a defined lithostratigraphic unit. There are few clear boundaries between these different "formations" except at the lower boundary of the Yaw shales, where there is a distinct lithological break with the underlying Pondaung sandstones (Stamp, 1922; Chhibber,

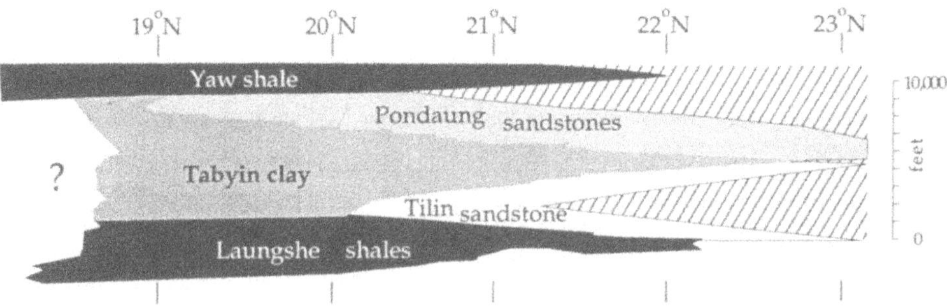

Fig. 5. Diagram showing the stratigraphy of Eocene sediments in the Chindwin–Irrawaddy Basin, Burma (after Stamp, 1922).

1934; Cotter and Clegg, 1938; Bender, 1983). For this reason, the traditional names of "Tabyin clay," "Pondaung sandstones," and "Yaw shale" are used in this discussion in order to reinforce the point that there are no formally defined formations in the Chindwin–Irrawaddy Basin.

The Pondaung sandstones are a complex of largely marine strata that intergrade with the early middle Eocene Tabyin clay (Stamp, 1922; Bender, 1983). The Pondaung sandstones exhibit a gradual change from marine through brackish to freshwater and continental deposition as one moves up-section and northward. Continental beds thicken north of latitude 21°45′ to 23°30′ and include paleosols in which vertebrate remains are found. Neither Stamp nor other geologists (e.g., Cotter and Clegg, 1938) working to the south were able to identify a discrete boundary between the Tabyin clay and the Pondaung sandstones. The lack of a distinctive change or identified unconformity between the Pondaung sandstones and Tabyin clay suggests that there is no marked temporal gap between their respective times of deposition, and it is likely that the two "formations" are facies correlatives at least in part. The upper boundary between the Pondaung sandstones and Yaw shales is sharply defined, and in the south the Yaw shales overlie the Tabyin clays.

The correlation of these marine sediments to the standard marine stages of Europe has not been questioned by recent workers (e.g., Bender, 1983). Traditionally, the Tabyin clay has been assigned to the Lutetian Stage (early middle Eocene) by correlation with the Khirthar Stage in India (see Eames, 1951, and Gingerich and Russell, 1990, for comment on the correlation of the Indo-Pakistan sequences with European standard stages). The Yaw shales have been considered Ludian/Priabonian (upper Eocene) in age (Stamp, 1922; Cotter and Clegg, 1938; Bender, 1983). These assessments suggest a Bartonian age for the Pondaung sandstones based on their intermediate stratigraphic placement.

Faunal correlations to the Javanese marine invertebrate record also support a middle Eocene age for the Pondaung fauna. The Yaw shales have been correlated to the Nanggulan Formation of Java (e.g., Holland, 1926) based on the common occurrence of *Discocylina*. Recent studies of mollusks (Zachello, 1984), calcareous nannoplankton (Okada, 1981), planktonic foraminifera (Purnamaningsih and Harsono, 1981), and magnetostratigraphy (Saito, 1981) indicate that the *Discocylina* beds in the Nanggulan Formation span the latest Lutetian through Priabonian. Thus, the Yaw shales may be older than Priabonian, and the Pondaung sandstones certainly appear to be at least late Bartonian if not older. Reappraisal of the Burmese marine invertebrate faunas and further comparisons to those of Java would further refine the correlations.

A late middle Eocene age for the Pondaung fauna also agrees with the only evidence of shared mammalian genera available for correlation between the northern sites discussed above and the Pondaung, the genus *Anthracokeryx*. As noted above, *A. sinensis* from the Heti fauna is probably latest middle Eocene in age, and Chinese *Anthracokeryx* differs only in minor characters from the two Pondaung species of this genus (R. L. Ciochon and P. A.

Holroyd, *unpublished observations*). Placement of the Pondaung mammal fauna in the late middle Eocene thus best fits with limited faunal evidence.

Krabi, Thailand

The initial description of fossils from Krabi (Suteethorn *et al.*, 1988) suggested a middle Eocene age for the site, based on the more primitive morphology of the anthracotheriid *Siamotherium krabiensis* as compared with that of the Pondaung anthracotheres (presumed to be of late Eocene age). Further description and analysis of the Krabi fauna (Ducrocq *et al.*, 1992) indicate that the Thai fauna is broadly contemporaneous with that of the Pondaung, suggesting a late Eocene age for these mammals (reiterated in Ducrocq, 1993). Ducrocq *et al.* (1992) described a series of new endemic faunal elements for Southeast Asia as well as anthracotheriids congeneric, and possibly conspecific, with those from Pondaung (*Anthracothema* cf. *A. pangan, Anthracokeryx* spp. close to *A. birmanicus* and *A. tenuis*). Metric and morphological comparisons of these Thai anthracotheriids indicate that they are probably conspecific with the Pondaung species (R. L. Ciochon and P. A. Holroyd, *unpublished observations*).

A late Eocene age for the Krabi fauna was based primarily on the similarity of the anthracothere fauna to that of the Pondaung, and there is no reason to doubt this correlation. However, because the Pondaung is better considered late middle Eocene in age, the Krabi fauna, including ?*Hoanghonius,* is probably late middle Eocene in age as well.

Conclusions

Based on correlation of Asian deposits with revised Eocene subepoch boundaries relative to European standard subages (Berggren *et al.*, 1978) and North American LMAs (Swisher and Prothero, 1990; Prothero and Swisher, 1992; Holroyd and Maas, Chapter 11, this volume), it appears that all Eocene primates of Asia (with the exception of *Altanius orlovi*) are middle Eocene in age. These age relationships are shown in Fig. 6.

The effect of this revision is that there are no primates known from late Eocene Asian deposits. This lack may reflect the high latitude distribution of upper Eocene (Ergilian) deposits, as most are concentrated in Mongolia and northern China. Based on limited pollen records from the late Eocene, Leopold *et al.* (1992) concluded that vegetation in the Chinese interior was subtropical woody savanna or open deciuous forest with shrubby areas dominated by arid-adapted genera. The more equable and tropical climes favored by primates are apparently not yet represented in the late Eocene mammalian fossil record in Asia.

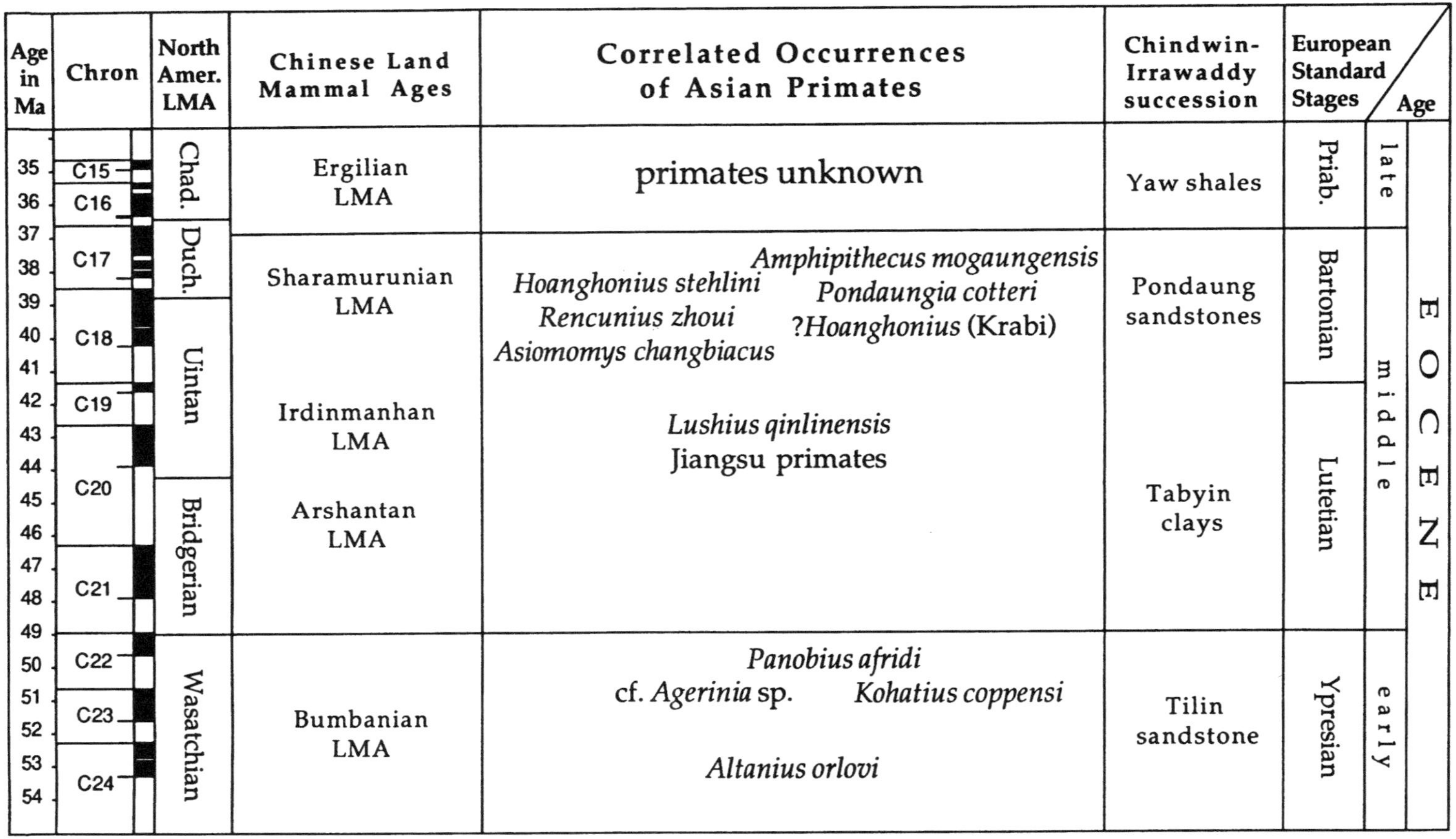

Age in Ma	Chron	North Amer. LMA	Chinese Land Mammal Ages	Correlated Occurrences of Asian Primates	Chindwin-Irrawaddy succession	European Standard Stages	Age
35–36	C15, C16	Chad.	Ergilian LMA	primates unknown	Yaw shales	Priab.	EOCENE: late
37–38	C17	Duch.	Sharamurunian LMA	*Hoanghonius stehlini*, *Rencunius zhoui*, *Asiomomys changbiacus*, *Amphipithecus mogaungensis*, *Pondaungia cotteri*, ?*Hoanghonius* (Krabi)	Pondaung sandstones	Bartonian	middle
39–41	C18	Uintan					
42–43	C19	Uintan	Irdinmanhan LMA	*Lushius qinlinensis*, Jiangsu primates	Tabyin clays	Lutetian	middle
44–48	C20, C21	Bridgerian	Arshantan LMA				
49–54	C22, C23, C24	Wasatchian	Bumbanian LMA	*Panobius afridi*, cf. *Agerinia* sp., *Kohatius coppensi*, *Altanius orlovi*	Tilin sandstone	Ypresian	early

Fig. 6. Summary chart showing the temporal distribution of Asian Eocene primate taxa, incorporating age revisions advocated here.

The revised ages also show that Asia, like other Holarctic continents, hosted a diversity of primates during the middle Eocene. At least two of these (*Asiomomys*, Beard and Wang, 1991; *Pondaungia*, Ciochon and Holroyd, Chapter 6, this volume) are affiliated with those of North America. However, unlike those of North America or Europe, the composition of the middle Eocene Asian primate fauna was not one composed simply of adapids and omomyids as are the primate faunas of Europe and North America. Rather, it may have included groups (represented by some of the undescribed Jiangsu primates and possibly *Amphipithecus*, *Hoanghonius*, and *Rencunius*) divergent from these two main groups of Eocene primates. Whether these are anthropoids, "proto-anthropoids" (as suggested by Rasmussen and Simons, 1992), or new groups only further finds and analysis will demonstrate.

The implications of the revised age of *Hoanghonius*, *Amphipithecus*, and *Pondaungia* for the study of anthropoid origins are primarily one of perspective. In order to understand the unique evolutionary history of the Anthropoidea, it is necessary to place their Eocene origin within a broad and temporally accurate context. Proper temporal correlation permits us to examine the origin and evolution of anthropoids within the greater purview of contemporaneous primate evolution. In the case of the Asian primates, we must recognize that diverse primate assemblages existed on this continent from approximately 45 Ma to 38 Ma, several million years before the well-known Fayum anthropoid radiation.

ACKNOWLEDGMENTS

Our interest in problems of Asian mammalian biochronology grew out of the inspiration of Donald E. Savage. This work has benefited from discussion with M. R. L. Anthony, K. C. Beard, P. D. Gingerich, D. Prothero, Qi Tao, B. A. Williams, T. M. Bown, R. F. Fleming, and three anonymous reviewers provided helpful review and comments on the manuscript.

References

Ba Maw, Ciochon, R. L., and Savage, D. E. 1979. Late Eocene of Burma yields earliest anthropoid primate, *Pondaungia cotteri*. *Nature* **282:**65–67.

Beard, K. C., and Wang Banyue. 1991. Phylogenetic and biogeographic significance of the tarsiiform primate *Asiomomys changbaicus* from the Eocene of Jilin Province, People's Republic of China. *Am. J. Phys. Anthrop.* **85:**159–166.

Bender, F. 1983. *The Geology of Burma. Beiträge zur Regionalen Geologie der Erde*, Vol. 16. Gebrüder Borntraeger, Berlin.

Berggren, W. A., and Prothero, D. R. 1992. Eocene–Oligocene climatic and biotic evolution: An overview. In: D. R. Prothero and W. A. Berggren (eds.), *Eocene–Oligocene Climatic and Biotic Evolution*, pp. 1–28. Princeton University Press, Princeton.

Berggren, W. A., McKenna, M. C., Hardenbol, J., and Obradovich, J. D. 1978. Revised Paleogene polarity time scale. *J. Geol.* **86:**67–81.

Cande, S. C., and Kent, D. V. 1992. A new geomagnetic polarity time scale for the late Cretaceous and Cenozoic. *J. Geophys. Res.* **97**(B10)**:**13917–13951.

Cavalier, C. 1979. La limite Éocène–Oligocene in Europe occidentale. *Univ. L. Pasteur Strasb. Sci. Geol. Mem.* **54:**1–280.

Cavalier, C., and Pomerol, C. 1976. Les rapports entre le Bartonien et le Priabonien. Incidence sur la position de la limite Éocène moyen-Éocène supérieur. *C. R. Somm. Soc. Geol. Fr.* **2:**49–51.

Chhibber, H. L. 1934. *The Geology of Burma.* Macmillan, London.

Ciochon, R. L., and Chiarelli, A. B. 1980. Paleobiogeographic perspectives on the origin of the Platyrrhini. In: R. L. Ciochon and A. B. Chiarelli (eds.), *Evolutionary Biology the New World Monkeys and Continental Drift,* pp. 459–493. Plenum Press, New York.

Ciochon, R. L., and Etler, P. 1994. Reinterpreting past primate diversity. In: R. S. Corruccini and R. L. Ciochon (eds.), *Integrative Paths to the Past: Paleoanthropological Advances in Honor of F. Clark Howell,* pp. 37–68. Prentice Hall, Englewood Cliffs, NJ.

Ciochon, R. L., Savage, D. E., Thaw Tint, and Ba Maw. 1985. Anthropoid origins in Asia? New discovery of *Amphipithecus* from the Eocene of Burma. *Science* **229:**756–759.

Colbert, E. 1937. A new primate from the upper Eocene Pondaung Formation of Burma. *Am. Mus. Novit.* **951.**

Colbert, E. 1938. Fossil mammals from Burma in the American Museum of Natural History. *Bull. Am. Mus. Nat. Hist.* **74:**255–436.

Cotter, G. de P., and Clegg, E. L. G. 1938. Geology of parts of Minbu, Myingyan, Pakkoku, and lower Chindwin Districts, Burma. *Mem. Geol. Surv. India* **72**(1)**:**1–36.

Dashzeveg, D. 1992. Hyracodontids and rhinocerotids (Mammalia, Perissodactyla, Rhinocerotoidea) from the Paleogene of Mongolia. *Palaeovertebrata* **21**(1–2)**:**1–84.

Dashzeveg, D., and McKenna, M. C. 1977. Tarsioid primate from the early Tertiary of the Mongolian People's Republic. *Acta Paleont. Pol.* **22**(2)**:**121–137.

Delson, E. 1977. Vertebrate paleontology, especially of nonhuman primates, in China. In: W. W. Howells and P. Jones Tsuchitani (eds.), *Paleoanthropology in the People's Republic of China. A Trip Report of the American Paleoanthropology Delegation. Committee on Scholarly Communication with the People's Republic of China, Report No. 4,* pp. 40–65. National Academy of Sciences, Washington, DC.

Ding Suyin, Zheng Jianjian, Zhang Yuping, and Tong Yongsheng. 1977. The age and characteristic of the Liuniu and Dongjun faunas, Bose Basin of Guangxi. *Vert. Palas.* **15**(1)**:**35–45.

Ducrocq, S. 1993. Mammals and stratigraphy in Asia: Is the Eocene–Oligocene boundary at the right place? *C. R. Acad. Sci. Paris* [*Ser. II*] **316:**419–426.

Ducrocq, S., Buffetaut, E., Buffetaut-Tong, H., Helmcke-Ingavat, R., Jaeger, J.-J., Jongkanjanasoontorn, Y., and Suteethorn, V. 1992. A lower Tertiary vertebrate fauna from Krabi (South Thailand). *N. Jb. Geol. Palaont. Abh.* **184:**101–122.

Eames, F. E. 1951. A contribution to the study of the Eocene in western Pakistan and western India: D. Discussion of the faunas of certain standard sections, and their bearing on the classification and correlation of the Eocene in western Pakistan and western India. *Q. J. Geol. Soc. Lond.* **107:**173–200.

Evernden, J. F., Savage, D. E., Curtis, G. H., and James, G. T. (1964). Potassium–argon dates and the Cenozoic mammalian chronology of North America. *Am. J. Sci.* **262:**145–198.

Flynn, J. J. 1986. Faunal provinces and the Simpson coefficient. *Contrib. Geol. Univ. Wyoming, Spec. Paper* **3:**317–338.

Gingerich, P. D. 1980. Eocene Adapidae, paleobiogeography, and the origin of South American Platyrrhini. In: R. L. Ciochon and A. B. Chiarelli (eds.), *Evolutionary Biology of the New World Monkeys and Continental Drift,* pp. 123–138. Plenum Press, New York.

Gingerich, P. D., and Russell, D. E. 1990. Dentition of early Eocene *Pakicetus* (Mammalia, Cetacea). *Contrib. Mus. Paleontol. Univ. Michigan* **28:**1–20.

Gingerich, P. D., Dashzeveg, D., and Russell, D. E. 1991. Dentition and systematic relationships of *Altanius orlovi* (Mammalia, Primates) from the early Eocene of Mongolia. *Geobios* **24**(5)**:**637–646.

Holland, T. H. 1926. Classification and correlation of Burma Tertiaries with those of Java and western India. *Mem. Geol. Surv. India* **51**:27.

Holroyd, P. A., and Ciochon, R. L. 1991. A reappraisal of Burmese anthracotheriid artiodactyls. *J. Vert. Paleont.* **11**(3):35A.

Kappelman, J. 1992. The age of the Fayum primates as determined by paleomagnetic reversal stratigraphy. *J. Hum. Evol.* **22**:495–503.

Krause, D. W., and Maas, M. C. 1990. The biogeographic origins of late Paleocene–early Eocene mammalian immigrants to the Western Interior of North America. In: T. M. Bown and K. D. Rose (eds.), *Dawn of the Age of Mammals in the Northern Part of the Rocky Mountain Interior, North America. Geological Society of America, Special Paper* 243, pp. 71–105. Geological Society of America, Boulder, Colorado.

Leopold, E., Gengwu Liu, and Clay-Poole, S. 1992. Low-biomass vegetation in the Oligocene? In: D. R. Prothero and W. A. Berggren (eds.), *Eocene-Oligocene Climatic and Biotic Evolution,* pp. 399–420. Princeton University Press, Princeton.

Okada, H. 1981. Calcareous nannofossils of Cenozoic formations in central Java. In: T. Saito (ed.), *Micropaleontology, Petrology and Lithostratigraphy of Cenozoic Rocks of the Yogyarkarta Region, Central Java,* pp. 25–34. Yamagata University, Yamagata, Japan.

Pilgrim, G. E. 1927. A *Sivapithecus* palate and other primate fossils from India. *Mem. Geol. Surv. India (Paleontol. Indica), N. S.* **14**:1–26.

Pilgrim, G. E. 1928. The Artiodactyla of the Eocene of Burma. *Mem. Geol. Surv. India N. S.* **8**:1–39.

Pilgrim, G. E. 1941. The dispersal of the Artiodactyla. *Biol. Rev. Cambridge Phil. Soc.* **16**:134–163.

Prothero, D. R., and Swisher, C. C. 1992. Magnetostratigraphy and geochronology of the terrestrial Eocene–Oligocene transition in North America. In: D. R. Prothero and W. A. Berggren (eds.), *Eocene–Oligocene Climatic and Biotic Evolution,* pp. 46–73. Princeton University Press, Princeton.

Purnamaningsih, S., and Harsono, P. 1981. Stratigraphy and planktonic foraminifera of the Eocene–Oligocene Nanggulan Formation, central Java. *Publ. Geol. Res. Devel. Centre Paleont. Series* **1**:9–28.

Qi Tao. 1987. The middle Eocene Arshanto fauna (Mammalia) of Inner Mongolia. *Ann. Carnegie Mus.* **56**(1):1–73.

Qi Tao, Zong Guanfu, and Wang Yuanqing. 1991. Discovery of *Lushilagus* and *Miacis* in Jiangsu and its zoogeographical significance. *Vert. Palas.* **29**(1):59–63.

Qiu Zhuding. 1977. Note on the new species of *Anthracokeryx* from Guangxi. *Vert. Palas.* **15**(1):54–65.

Rae, T. C. 1993. A report on the "Anthropoid Origins" conference 6–9 May 1992, Duke University. *J. Hum. Evol.* **24**:239–241.

Rasmussen, D. T., and Simons, E. L. 1992. Paleobiology of the oligopithecines, the earliest known anthropoid primates. *Int. J. Primatol.* **13**(5):477–508.

Rasmussen, D. T., Bown, T. M., and Simons, E. L. 1992. The Eocene–Oligocene transition in continental Africa. In: D. R. Prothero and W. A. Berggren (eds.), *Eocene–Oligocene Climatic and Biotic Evolution,* pp. 548–566. Princeton University Press, Princeton.

Russell, D. E., and Gingerich, P. D. 1980. Un nouveau Primate omomyide dans l'Eocène du Pakistan. *C. R. Acad. Sci. Paris [D]* **291**:621–624.

Russell, D. E., and Gingerich, P. D. 1987. Nouveaux primates de l'Éocène du Pakistan. *C. R. Acad. Sci. Paris [Ser. II]* **304**(5):209–214.

Russell, D. E., and Zhai Ren-jie. 1987. The Paleogene of Asia: Mammals and stratigraphy. *Mem. Mus. Nat. Hist. Nat., Sci. Terre* **52**:1–488.

Saito, T. (ed.). 1981. *Micropaleontology, Petrology and Lithostratigraphy of Cenozoic Rocks of the Yogyarkarta Region, Central Java.* Yamagata University, Yamagata, Japan.

Stamp, L. D. 1922. An outline of the Tertiary geology of Burma. *Geol. Mag.* **59**(11):481–501.

Stucky, R. K. 1992. Mammalian faunas in North America of Bridgerian to early Arikareean "Ages" (Eocene and Oligocene). In: D. R. Prothero and W. A. Berggren (eds.), *Eocene–Oligocene Climatic and Biotic Evolution,* pp. 464–493. Princeton University Press, Princeton.

Suteethorn, V., Buffetaut, E., Helmcke-Ingavat, R., Jaeger, J.-J., and Jongkanjansoontorn, Y. 1988. Oldest known Tertiary mammals from South East Asia: Middle Eocene primate and anthracotheres from Thailand. *N. Jb. Geol. Palaont. Mh.* **9:**563–570.

Swisher, C. C., and Prothero, D. R. 1990. Single-crystal ^{40}Ar–^{39}Ar dating of the Eocene–Oligocene transition. *Science* **249:**760–762.

Tang, Y.-J. 1978. Two new genera of Anthracotheriidae from Guangxi. *Vert. Palas.* **16**(1):13–21.

Tong Yongsheng. 1989. A review of middle and late Eocene mammalian faunas from China. *Vert. Palas.* **28:**663–682.

Wang Banyue. 1992. The Chinese Oligocene: A preliminary review of mammalian localities and local faunas. In: D. R. Prothero and W. A. Berggren (eds.), *Eocene–Oligocene Climatic and Biotic Evolution,* pp. 527–547. Princeton University Press, Princeton.

Wang Banyue, and Li C. 1990. First Paleogene mammalian fauna from northeast China. *Vert. Palas.* **29:**165–205.

Zachello, M. 1984. The Eocene mollusc fauna from Nanggulan (Java) and its palaeogeographic bearing. *Mem. Sci. Geol.* **34:**377–390.

The Asian Origin of Anthropoidea Revisited

6

RUSSELL L. CIOCHON
and PATRICIA A. HOLROYD

Introduction

The concept of an Asian origin for the Anthropoidea has surfaced repeatedly in our attempts to discern the biogeographic and phyletic origins of the suborder (e.g., Pilgrim, 1927; Ba Maw *et al.*, 1979; Gingerich, 1980; Ciochon and Chiarelli, 1980; Ciochon *et al.*, 1985). The primary evidence for such an Asian origin traditionally lies in the poorly known and phylogenetically enigmatic primate genera *Amphipithecus* and *Pondaungia* from the late middle Eocene Pondaung deposits of Burma (now Myanmar).

Amphipithecus and *Pondaungia* have long been considered early anthropoids (Pilgrim, 1927; Colbert, 1937, 1938; Simons, 1971; Kay, 1980; Ba Maw *et al.*, 1979; Ciochon *et al.*, 1985) on the basis of a suite of characters presumed to be shared-derived features of Anthropoidea, e.g., low-crowned, bunodont molars and, in *Amphipithecus,* a fused symphysis in a deep mandibular ramus. Interpretation of these primates as anthropoids has led several workers to infer an Asian origin for Anthropoidea, with the Burmese primates representing either true anthropoids (Pilgrim, 1927; Colbert, 1937, 1938; Simons, 1963, 1965, 1971, 1974; Simons and Pilbeam, 1965; Van Valen, 1969; Szalay

RUSSELL L. CIOCHON • Departments of Anthropology and Pediatric Dentistry, University of Iowa, Iowa City, Iowa 52242. PATRICIA A. HOLROYD • Department of Biological Anthropology and Anatomy, Duke University, Durham, North Carolina 27710, and U.S. Geological Survey, Denver, Colorado 80225.

Anthropoid Origins, edited by John G. Fleagle and Richard F. Kay. Plenum Press, New York, 1994.

and Delson, 1979, *Pondaungia* only; Ciochon and Chiarelli, 1980) or "proto-anthropoids," i.e., part of a paraphyletic stem lineage that lacks some of the definitive characteristics of the group but is more closely related to Anthropoidea than to any other primate group (Rasmussen and Simons, 1992, *Amphipithecus* only; Ciochon and Etler, 1994) or transitional adapid–anthropoids (Gingerich, 1980).

Not all workers have agreed that the Pondaung taxa represent anthropoids or that the fragmentary evidence at hand can even permit us to discern their phylogenetic status. Szalay (1970, 1972; reiterated in Szalay and Delson, 1979) argued for lemuriform status, and von Koenigswald (1965) averred that *Pondaungia* was not a primate at all but a condylarth. Schwartz (1986) considered the molar traits and premolar shape in *Amphipithecus* more similar to lorisids than to any other group. Comparisons of *Pondaungia* with the late Eocene Fayum anthropoid taxon *Oligopithecus* by Rasmussen and Simons (1988) revealed no special similarities between the two genera, and they found the general resemblance between *Amphipithecus* and *Oligopithecus* "not yet strongly compelling" (p. 192) for common ancestry. Fleagle (1988) observed that the broad, low-crowned molars and deep mandibles suggest anthropoid affinities but considered the material insufficient for confirmation. Kay and Williams (Chapter 13, this volume) also found some features shared among *Amphipithecus, Pondaungia,* and anthropoids but considered the sample of Burmese primates too fragmentary to resolve their affinities. Conroy and Bown (1974) and Conroy (1978) reviewed the evidence for an Asian origin of anthropoids and concluded that although Asia could not be precluded on the available evidence, an Asian origin for the group was unlikely.

The equivocal and/or noncommittal viewpoints regarding these Asian genera has stemmed primarily from a lack of adequate fossil material on which to test phylogenetic hypotheses as well as inadequate or differing descriptions of morphology in the few specimens known. The Pondaung primates are known from six gnathic specimens, only one of which preserves relatively unworn or unbroken teeth. No cranial or postcranial material is currently known for either *Amphipithecus* or *Pondaungia.* Consequently, ideas regarding the affinities of these Asian genera have remained unresolved in the absence of new material.

Unfortunately, no additional fossil material has been recovered to resolve this aspect of the problem. However, new, cogent evidence has surfaced in recent years that bears on interpretations of the Pondaung primates. This evidence lies in the morphology of the earliest known anthropoids from the late Eocene to early Oligocene Jebel Qatrani Formation of the Fayum Depression, Egypt. The validity of many of the "anthropoidean features" traditionally used to place the Asian material within or near Anthropoidea has been brought into question by the discovery and description of new primitive anthropoids from Africa: the parapithecids *Qatrania* (Simons and Kay, 1983, 1988) and *Serapia* (Simons, 1992); the oldest propliopithecid, *Catopithecus* (Simons, 1989, 1990); and the primitive anthropoids *Algeripithecus* (Godinot and

Mahboubi, 1992) and *Arsinoea* (Simons, 1992). Each of these lacks the deep mandibular ramus (where known) of many later anthropoids, and *Catopithecus* (along with *Oligopithecus*) lacks the bunodonty elsewhere observed in members of the suborder. The mandibular symphysis is apparently unfused in *Serapia, Arsinoea,* and *Catopithecus* (Simons, 1989, 1992). Thus, although the implications of the morphology of these small, early anthropoids have not been fully explored, the oldest African anthropoids call into question several of the features traditionally used to distinguish the suborder in a paleontological sense and that have been used to ally *Amphipithecus* and *Pondaungia* with it. As noted by Martin (1993), *Amphipithecus* and *Pondaungia* deserve reexamination in light of these finds.

Here we first review the morphology of *Amphipithecus* and *Pondaungia,* specifically addressing published differences in morphological interpretation. Second, we reevaluate the validity of their "anthropoid" characters in order to determine which, if any, represent shared-derived characters that may link them to the definitive anthropoids of the Eocene and Oligocene of Africa. Finally, we attempt, as evidence permits, to determine the proper taxonomic allocation of the Pondaung primates.

Morphological Review of the Pondaung Primates

Pondaungia

Pondaungia cotteri is known from University of California Museum of Paleontology (UCMP) 120377, a right M_2–M_3 (Fig. 1C) described by Ba Maw *et al.* (1979), and the syntype series Geological Survey of India (GSI) D201–203, three jaw fragments (presumed to be from a single individual) consisting of a left maxilla containing M^1 and M^2 (Fig. 1B), a left mandibular ramus with M_2–M_3 (Fig. 1A), and a right mandibular ramus with M_3 only (not shown).

GSI D203 is the only maxillary fragment known for this species and retains M^1–M^2. Both molars possess two distinct labial roots and a single, mesiodistally elongate lingual root. The crowns are subsquare, being longer labially than lingually, and mesiodistally longer than wide. A large protocone and slightly smaller metacone and paracone dominate. No metaconule is apparent on M^1, and only a small one is present on M^2. A paraconule is present and is slightly better developed on M^1. Cresting is generally poorly developed, and the enamel is crenulated. The postprotocrista is mesiodistally expanded to form a pseudohypocone that is undifferentiated from the protocone by a lingual furrow. The postprotocrista and pseudohypocone are confluent with the posterior cingulum. The metacone and paracone are joined by a weak centrocrista. On M^2 a small mesostyle module may have been present; however, this area is poorly preserved, and such an identification is only tentative. An anterior cingulum is well developed on M^2; this area is damaged on M^1.

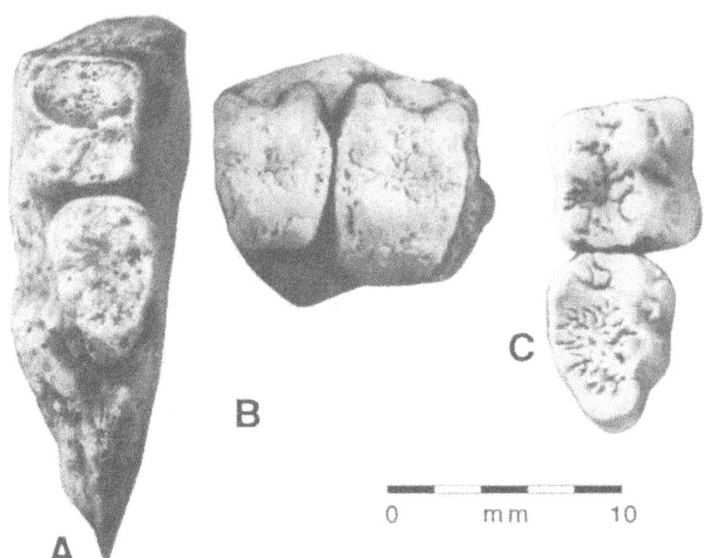

Fig. 1. *Pondaungia cotteri*. (A) GSI D201, left M_2–M_3. (B) GSI D203, left M^1–M^2. (C) UCMP 120377, right M_2–M_3.

Lingual cingula are discontinuously developed on both molars, and a small cuspule occurs on the lingual cingulum of M^1 lingual to the pseudohypocone. This is probably the cusp that other authors have referred to as a hypocone.

The presence or absence of a pseudohypocone and/or a hypocone in *Pondaungia* is critical to understanding its phylogenetic affinities. Pilgrim (1927) and von Koenigswald (1965) noted that the two lingual cusps were linked; or as described here, the postprotocrista is expanded distally to form a fourth major cusp, or pseudohypocone, on the upper molars. Szalay and Delson (1979) did not believe this condition to be the case in *Pondaungia*, pointing out that a pseudohypocone forms by a splitting of the protocone. Clearly, they did not perceive a split between the two lingual cusps, nor do we. However, they did not note that the degree of separation of protocone and hypocone is not the relevant attribute in recognizing the presence of a pseudohypocone. Among primates, pseudohypocones occur uniquely in notharctine adapids and are recognized by their place of origin, independent of their degree of separation from the protocone (see e.g., Beard, 1988 or Covert, 1990). The difference in terminology reflects the fact that the distolingual cusp in notharctines arises from the postprotocrista rather than from the distolingual cingulum (a "true" hypocone). It is clear that in *Pondaungia* the greatest mass of the distolingual corner of the upper molars derives from the postprotocristae and should be recognized as pseudohypocones (see also Pilgrim, 1927; Rasmussen and Simons, 1988). The smaller cingular cusp on the distolingual cingulum may represent a true hypocone in the sense that a true hypocone develops from the distolingual cingulum. However, this cingular cusp is not the dominant feature of this region in the molars and, as discussed below, apparently did not perform a major role in occlusion as the hypocone does in anthropoids.

The mandibular fragments of the type specimen reveal little morphological detail. The teeth are not well preserved, and the enamel cap of the left M_2 trigonid is broken. Consequently, the lower molar morphology of this species is known almost entirely from the better preserved and only slightly worn UCMP 120377. A lengthier description of this specimen is provided in Ba Maw *et al.* (1979); only general configuration and points of morphological contention are discussed here.

The molars are multicuspidate with highly crenulated enamel, and all six main cusps are salient. M_2 is a square tooth on which the trigonid is very slightly wider than, and elevated above, the talonid. The trigonid is formed by a large protoconid joined to the slightly smaller metaconid by a moderately developed crest. A mesially arcing paracristid joins the protoconid to the small paraconid that lies appressed to the anterior surface of the metaconid. A short premetacristid links the metaconid and paraconid to close the trigonid basin lingually. The posterior trigonid wall is low and inclined slightly distally. A small facet* is developed along the distolabial surface of the protoconid, and facet X is present on the distolingual surface of this cusp.

The M_2 talonid basin is shallow. The large hypoconid is anterior to the much smaller entoconid. The cristid obliqua is short and contacts the protoconid near its center. Lingually, the basin is delimited by a low, beaded entocristid. Posteriorly, the talonid is open.

Ba Maw *et al.* (1979) suggested that a centrally placed hypoconulid was present in a worn area on the M_2 of UCMP 120377, and Szalay and Delson (1979) denied the presence of this cusp altogether in the type specimen. Analysis of the wear facets and striae in UCMP 120377, however, leads us to suggest a different interpretation of this area. The wear facet at the posterior portion of the talonid basin possesses weak striae trending roughly parallel to the midline of the tooth. These striae and the position of the facet indicate it is homologous with facet 10. In anthropoids, the centrally positioned hypoconulid typically contributes to the development of two wear facets: (1) facet 6, developed on the mesiolabial surface of the entoconid, which corresponds to the lingual surface of the upper molar postprotocrista and is correlated with the presence of a lingually to centrally placed hypoconulid; and (2) facet 10n, which corresponds to the labial surface of the hypocone, is correlated with the presence of a hypocone developed from the distal cingulum. In *Pondaungia,* this area in M_2 is occupied by a facet continuous with that on the trigonid of M_3 and is homologous to facet 10. Facet 10 corresponds to a surface on the labial side of a lingually deflected postprotocrista and correlates with the development of a pseudohypocone. Thus, if a hypoconulid were present, it

*Facet terminology used here follows that of Kay (1977).

would not have been centrally located but would have been present near the base of the hypoconid, where facets 9 and 10 meet in a small ridge. A very small cusp in this position is present in UCMP 120377 and can also be seen in GSI D201 (contra Szalay and Delson, 1979). This configuration is similar to that in many Eocene primates, particularly notharctines.

M_3 is morphologically similar to M_2 with minor differences in trigonid and talonid construction. The paraconid is smaller and less well differentiated from the metaconid. The hypoconid and entoconid are more anteriorly placed, preventing the development of a facet 1 or facet X. The entoconid is much smaller, becoming virtually subsumed in the entocristid. The heel of the talonid projects distally, and the hypoconulid is slightly labiad of the midline of the tooth.

Amphipithecus

Amphipithecus mogaungensis is known from two specimens: the holotype, American Museum of Natural History (AMNH) 32520 (Figs. 2–4), a left mandibular ramus possessing P_3–M_1 and the roots of C and P_2, described by Colbert (1937, 1938); and Department of Geology, Mandalay University-Primate (DGMU-P) 1 (Fig. 5), a left mandibular fragment possessing worn first and second molars and the anterior root of M_3, described by Ciochon *et al.* (1985).

Both P_3 and P_4 are approximately "teardrop" shaped in occlusal outline with their long axes oblique to the long axis of the molar row ("skewered" in the terminology of Szalay, 1970). Both teeth are mesiodistally longest along their rounded labial border and distolingually attenuated to form a rounded corner at the distal edge of their lingual borders.

This shape is broadly similar to the ovoid premolars of anthropoids; however, the configuration of cusps differs markedly. P_3 is high crowned and dominated by a large protoconid. A small metaconid is located distolingual to the apex of the protoconid and connected to it by a strong crest. A prominent paraconid, subequal to the protoconid in height and joined to it by a strong mesial crest, is present. A short talonid is formed by a moderate hypoconid. The hypoconid is connected to the trigonid by a strong, labially oriented cristid obliqua. No cingula are present, but the metaconid and cristid obliqua delimit a distinct, lingually open distal fovea or lingual depression.

P_4 is similarly constructed, with minor differences. The metaconid is more salient, and the paraconid is smaller. The trigonid is closed lingually by a weak crest extending from the paraconid to the metaconid. The talonid is slightly wider posteriorly, and the distal fovea is limited to the top of the crown, bounded lingually by a slight ridge such that it does not extend down the crown as on P_3.

The first molar is longer than wide with its greatest width across the talonid. The trigonid is smaller than the talonid in length and width and

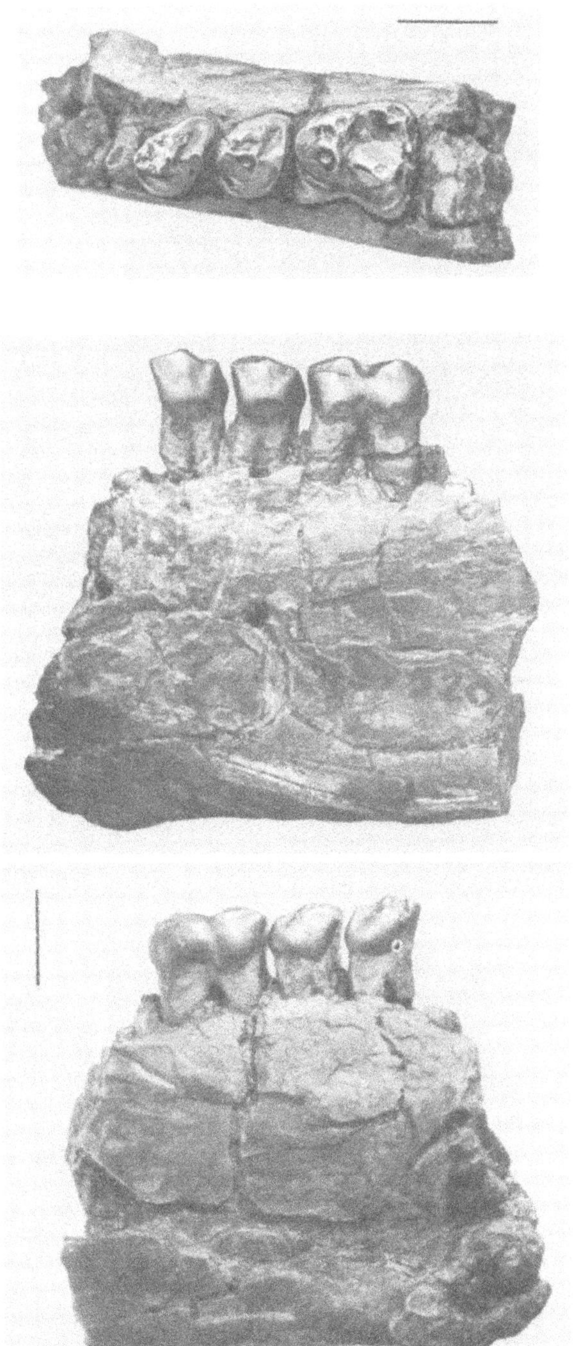

Fig. 2. *Amphipithecus mogaungensis.* AMNH 32520, left P_3–M_1, occlusal, lateral, and medial views. Scale bar, 5 mm.

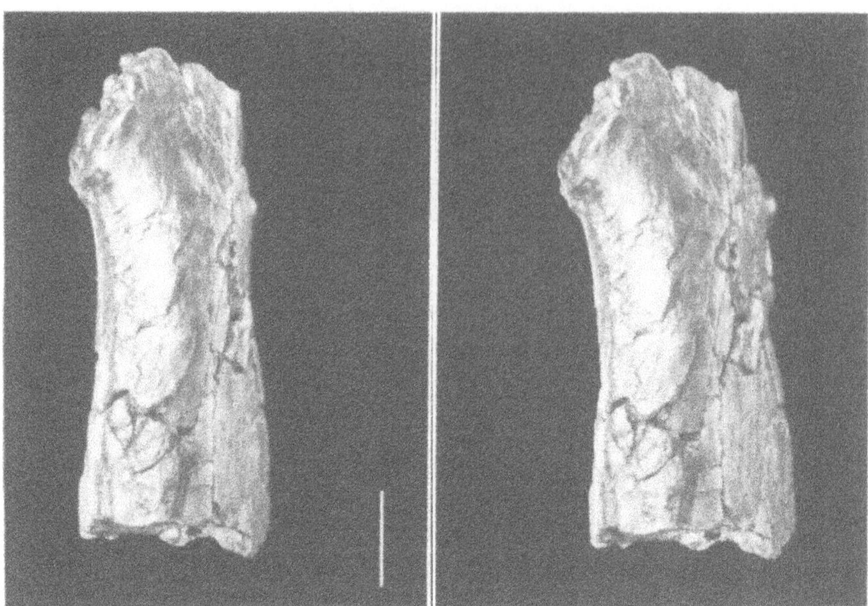

Fig. 3. *Amphipithecus mogaungensis*. Stereo pair, inferior view of mandibular ramus of AMNH 32520. Note scars showing extensive attachment of *M. digastricus* and rugosity of symphyseal region. Scale bar, 5 mm.

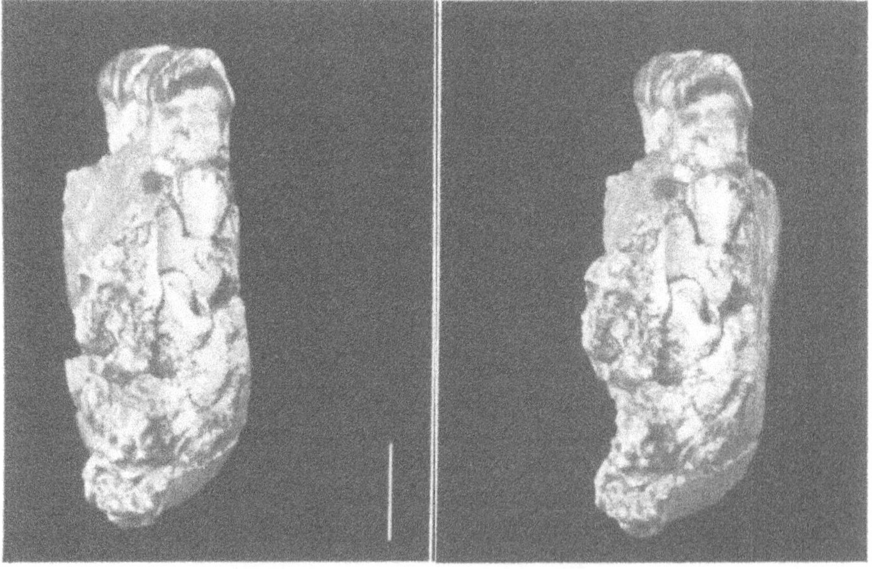

Fig. 4. *Amphipithecus mogaungensis*. Stereo pair, anterior view of AMNH 32520, taken perpendicular to tooth row. Note heavy synostosis of symphyseal region and large canine root and alveolus. Scale bar, 5 mm.

subequal to it in height. The paraconid is prominent and centrally placed, giving the tooth a prow-like appearance. Schwartz (1986) disallowed the presence of a paraconid, instead interpreting it as a massive paracristid. We see no reason to interpret this structure as a crest lacking a cusp. The metaconid is located distolingual to the large protoconid and joined to it by a short crest. The trigonid is closed lingually by a weak crest extending anteriorly from the metaconid. The talonid is wide and bounded labially by a moderately developed cristid obliqua. The entoconid is smaller than, and located slightly posterior to, the hypoconid. A slight crest trending anteriorly from the entoconid to the base of the metaconid closes the basin lingually. A slight posterior cingulum is present, running lingually and upward from the base of the hypoconid to intersect with the low crest joining the entoconid and hypoconid at the position where the enamel is broken.

M_2 is similar in construction to M_1, but the trigonid is wider, and there is no salient paraconid. Ciochon *et al.* (1985) argued that there is no paraconid on the second molar, based on the sole example in DGMU-P1. However, there is a slight centrally located bulge present on the anterior face of the trigonid in this specimen that may represent a reduced paraconid by analogy with the condition observable in the parapithecid *Qatrania wingi* (see Simons and Kay, 1988). However, the specimen is too worn to determine confidently if a reduced or vestigial paraconid was present.

The presence or absence of a molar hypoconulid is also difficult to determine. Szalay (1970, 1972; reiterated in Szalay and Delson, 1979) asserted that AMNH 32530 possessed a small hypoconulid distolingual to the hypoconid at the position where a chip of enamel is missing on M_1. He based this conclusion on the observation that the crest connecting the entoconid and hypoconid appeared continuous except for an interruption by the inferred hypoconulid just distolingual to the hypoconid. Schwartz (1986) miscited Kay (1977, 1980) in suggesting that *Amphipithecus* may have had a small, centrally placed hypoconulid. Ciochon *et al.* (1985) observed no hypoconulid on either M_1 or M_2 of DGMU-P1.

Reexamination of both specimens leads us to believe that Szalay is probably correct, although the damage to the type specimen in this area makes it difficult to determine the development of the hypoconulid. We concur with Szalay on the basis of two additional lines of evidence not previously considered and by drawing morphological analogies with adapids having apparently similar hypoconulids (e.g., *Cantius, Pelycodus*). In these forms, the hypoconulid is small and lacks a dentine core. The hypoconulid typically occurs at the lingual termination of the distal cingulum and, as the molar becomes worn, disappears completely, leaving no trace of a dentine pit or any other evidence of its occurrence. The distal cingulum on AMNH 35230 shows a configuration similar to that in *Cantius* and *Pelycodus,* suggesting a similar positioning for the hypoconulid. Further, the fact that in even slightly worn specimens of these adapids the hypoconulid completely disappears provides an explanation for the apparent lack of this cusp on the moderately worn molars of DGMU-P1.

A

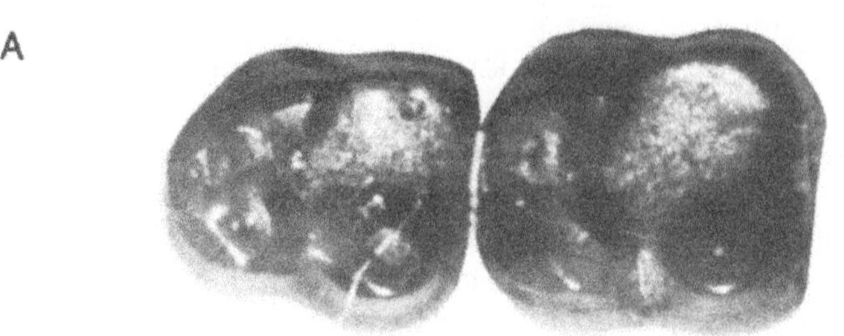

B

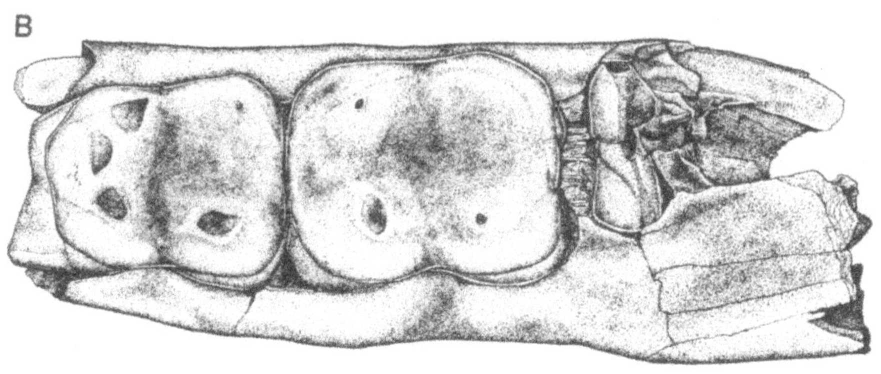

C

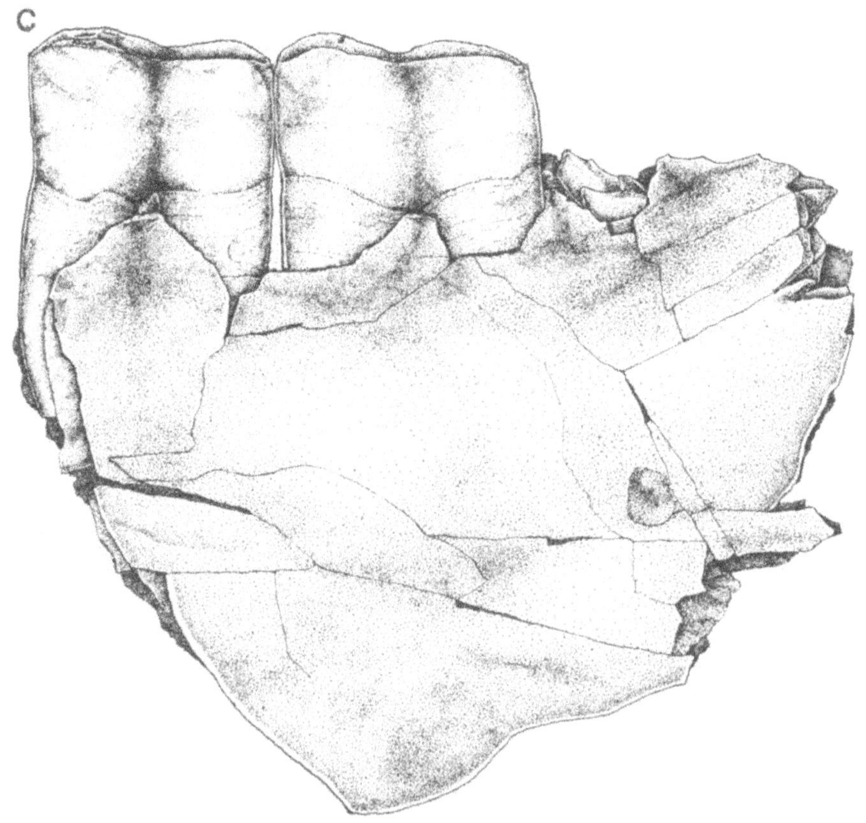

The mandibular corpus of *Amphipithecus* is deep with thick cortical bone exposed at the breaks within the tooth row. Along its inferior surface, the scar for the digastric is extensive and well marked (see Fig. 3). The symphyseal region is incomplete. The posterior portion of a strong inferior transverse torus is present, extending posteriorly to the level of P_2–P_3. An equally well developed superior transverse torus is indicated by a small rugose area separated from the inferior transverse torus by a deep genioglossal pit. The preseved symphyseal surfaces are characterized by deeply convoluted rugosities, and the lower edge of the inferior transverse torus is noticeably "lipped" inferiorly.

Typically, the symphysis of *Amphipithecus* has been interpreted as fused, as in anthropoids (Simons, 1971; Ciochon *et al.*, 1985). Although Szalay (1970) likened it to the unfused symphyses of North American notharctines, he provided no details of these similarities. The rugosities and the inferior lipping of the symphyseal surface indicate an interlocking, unfused symphysis that may have functioned in a manner similar to a fused symphysis (see Ravosa and Hylander, Chapter 14, this volume). The morphology of the symphysis in *Amphipithecus* contrasts strongly with the seamless, fully fused symphyses of anthropoids such as *Apidium, Propliopithecus,* and hominoids with which it has been previously compared. It also contrasts with the unfused symphyses of *Catopithecus, Serapia,* and *Arsinoiea,* in which the surface rugosities are less well developed. Symphyseal morphologies similar to that in *Amphipithecus* have been described in toothcombed prosimians (Beecher, 1979) and in carnivorans (Scapino, 1981). Together, the thick cortical bone, great depth and width of the corpus, strong marking of the digastric muscle, and interlocking but unfused symphysis with well-developed inferior and superior transverse tori indicate a jaw adapted to resist transverse bending stresses. Typically, this combination of features is associated with folivorous dietary habits (Beecher, 1983; Ravosa, 1991) but may also simply be an allometric effect related to increasing body size (Ravosa, 1991).

Assessment of Anthropoid Characters in the Pondaung Primates

Table I lists nine characters that have been used to link *Amphipithecus* and/or *Pondaungia* with Anthropoidea.

Characters 1 and 2, large body size and a deep mandibular corpus, are no longer characters that define Anthropoidea. Most early anthropoids were small in body size and possessed relatively shallow mandibular corpora (*Catopithecus,* Simons, 1989; *Serapia* and *Arsinoea,* Simons, 1992; *Qatrania,* Simons

Fig. 5. *Amphipithecus mogaungensis.* DGMU-P 1, left M_1–M_2 (A,B) Occlusal views. (C) Lateral view.

Table I. Characters of Pondaung Primates That Have Been Used to Substantiate Anthropoid Status

1. Large body size
2. Deep mandibular corpus relative to crown height
3. Fused mandibular ramus possessing superior and inferior transverse tori in *Amphipithecus*
4. Low-crowned, bulbous molars
5. Loss of first lower premolar in *Amphipithecus*
6. Square lower second molar
7. Absence of hypoconulid on M_1 and M_2 in *Amphipithecus* (a similarity to platyrrhines)
8. Mesiobuccally skewed premolars in *Amphipithecus*
9. Presence of facet X in *Pondaungia*

and Kay, 1983, 1988; *Propliopithecus,* Kay *et al.* 1981). Larger body size comparable to that in *Amphipithecus* and *Pondaungia* never developed in parapithecids and was developed independently in propliopithecids and late Paleogene platyrrhines.

Character 3, a fused mandibular ramus with superior and inferior transverse tori, has not been demonstrated for either of the Pondaung primates. This condition was inferred for *Amphipithecus* based on the development of tori and the position of the break opposite the midline of the jaw. However, as discussed above, the symphysis of *Amphipithecus* is unfused. Although the tori are developed in a manner similar to anthropoids such as *Aegyptopithecus* and different than in notharctines (see Beecher 1979, 1983), not all early anthropoids have well-developed tori or even fused symphyses. The parapithecids have relatively weak tori with the inferior transverse torus entirely lacking in smaller individuals. *Catopithecus, Arsinoea,* and *Serapia* possess only moderately developed superior transverse tori, and the degree of mandibular fusion or synostosis is equivocal in all three genera (Simons, 1989, 1992). Although their relatively smooth symphyseal surfaces are covered with irregular grooves and pits (Simons, 1989, 1992), they differ substantially from that of *Amphipithecus* in lacking inferior lipping and being far less rugose. Thus, the development of the tori in *Amphipithecus* is dissimilar from that in any of the Fayum anthropoids and not evidence for anthropoid status.

Although the presence of low crowned, bulbous molars and a square lower second molar (Characters 4 and 6) in the Pondaung primates are similarities to anthropoids, it is also a similarity to a whole range of animals who fed on soft material, e.g., the adapid *Pelycodus,* the omomyid *Rooneyia,* phenacodontid and hyopsodontid condylarths, and dichobunid and helohyid artiodactyls. These are not diagnostic characters of any particular primate group.

Character 5, the loss of the first lower premolar in *Amphipithecus,* is simply a character of most Eocene primates and therefore of little value in discerning its phyletic position.

Character 7, the absence of molar hypoconulids in *Amphipithecus,* has been debated. Ciochon *et al.* (1985) suggested that the absence of a hypoconulid on M_1–M_2 of *Amphipithecus* was a possible synapomorphy between this genus and platyrrhines. As noted above, it is likely that *Amphipithecus* did possess a small hypoconulid. Further, the loss and/or reduction of the hypoconulid occurred convergently in platyrrhines, some adapines, and some notharctines. Although this character may be of value at some levels of phylogenetic analysis, it seems of little utility as a means for bolstering the anthropoid status of the Pondaung primates.

Mesiobuccally skewed premolars (Character 8) in *Amphipithecus* are superficially similar to those of many primates. Reorientation of the premolars along a distolingual–mesiolabial axis occurs convergently in primates several times, e.g., in anthropoids, *Adapis,* and most toothcombed prosimians. The premolars of *Amphipithecus* do not bear special similarity to any of these taxa.

The presence of facet X (Character 9) in *Pondaungia* was thought to be a definitive character linking *Pondaungia* with Old World anthropoids. When Kay (1977) first identified this feature, he suggested that it united Parapithecidae and Propliopithecidae with later Catarrhini and thus was diagnostic of the Old World Anthropoidea. The weight of this character was considered so great, Kay (1980) concluded that the "shared-derived development of facet X definitely links *Pondaungia* with African Oligocene parapithecids and primitive apes" (p. 335). Subsequently, it was determined that facet X arose independently in the Parapithecidae (Simons and Kay, 1983, 1988), the platyrrhine *Cebus* (Kay, 1980), and within the Catarrhini (judging from its absence in *Oligopithecus* and *Catopithecus*). Facet X undoubtedly arose independently in *Pondaungia* as well, in conjunction with the development of a more labiad orientation of the cristid obliqua, a relatively lower and more distally sloping posterior trigonid wall, and a moderately developed postprotocristid. Thus, this character is not diagnostic of catarrhine or even anthropoid status and appears to be of limited phylogenetic value.

Several more robust dental features have been suggested to characterize early anthropoids, and these can be compared with the morphology of the Pondaung primates. These characters are:

1. Uppers molars in which a *Nannopithex* fold is lacking, and the hypocone develops from the distolingual cingulum (e.g., Szalay, 1970; Fleagle and Kay, 1987; Rasmussen and Simons, 1988). The upper molars of *Pondaungia* distinctly differ from the anthropoid plan. A distolingual cingulum is present, as in most Eocene primates, but a pseudohypocone is developed from the protocone, unlike any anthropoid and most prosimian primates.
2. Lower molars in which (a) the paraconid, if present, is centrally placed (Fleagle and Kay, 1987), (b) M_3 is small relative to M_2 (Fleagle and Kay,

1987; Harrison, 1987), and (c) the hypoconulid is distinct and, on M_1–M_2, is lingually (Szalay, 1970; Harrison, 1987; Rasmussen and Simons, 1988) or centrally placed (Fleagle and Kay, 1987). *Pondaungia* clearly differs from anthropoids in the position of both the paraconid and the hypoconulid. *Amphipithecus* is similar to anthropoids in the positioning of the paraconid and is especially similar to *Qatrania* in the prow-like appearance its position gives to the first molar. However, it should also be noted that *Arsinoea* has more lingually placed lower molar paraconids than either *Amphipithecus* or *Qatrania* (Simons, 1992), so even this character may not be diagnostic for Anthropoidea.

Rasmussen and Simons (1988) noted a general resemblance between *Amphipithecus* and *Oligopithecus* but also pointed out that *Amphipithecus* differs from the Fayum genus in having a trigonid that is not raised well above the talonid and in retaining a paraconid on M_2. More importantly, however, the lack of a large, distinct, and cuspate hypoconulid located close to the entoconid, ubiquitously present in early anthropoids, excludes *Amphipithecus* from the Anthropoidea (Szalay, 1970, 1972). Szalay also noted that the small size and placement of this cusp near the base of the hypoconid in *Amphipithecus* is more similar to early adapids.

In sum, the characters previously used to ally the Pondaung primates with Anthropoidea are no longer valid, either being mistaken interpretations of morphology or characters we cannot now sustain as anthropoid synapomorphies. Comparison of *Amphipithecus* and *Pondaungia* with a newer conception of Anthropoidea, based in part on the recently discovered late Eocene primates from Egypt, indicates that *Pondaungia* shows no known shared-derived features with anthropoids. However, *Amphipithecus* remains engimatic.

Allocation of Pondaungia and Amphipithecus

Pondaungia lacks all the features of the Anthropoidea. With its similarities to anthropoids recognized as likely convergences, *Pondaungia*'s greatest resemblances clearly lie with adapids, specifically the Notharctinae, a point previously noted (e.g., Szalay, 1970, 1972; Ba Maw *et al.*, 1979; Gingerich, 1980; Rasmussen and Simons, 1988). It now appears that these resemblances to notharctines represent synapomorphies. *Pondaungia* shares a suite of shared-derived characters with the members of Notharctinae to the exclusion of other adapids (Table II). Although the Notharctinae has not been differentially diagnosed from the Adapinae, Godinot (1992) has recently suggested several derived features (principally of the fourth premolar) that differentiate *Cantius* (and, by homology, all later notharctines) from the more primitive adapid *Donrussellia.* The characters given here as synapomorphies for *Pondaungia* and other notharctines also differentiate them from *Donrussellia* and adapines and demonstrate the monophyly of the notharctine clade.

Table II. Shared-Derived Features of *Pondaungia* and Notharctines

1. Paraconid closely appressed to the metaconid and joined to the protoconid by mesially arcing paracristid
2. Hypoconulid on M_2 small and located at the distolingual base of the hypoconid
3. Postprotocrista of upper molars mesiodistally expanded to form a pseudohypocone
4. M_2 talonid open posteriorly to form a confluent wear facet with trigonid basin of M_3 for occlusion of pseudohypocone
5. Relatively small molar entoconids

Amphipithecus is far more difficult to classify. Features of the mandibular corpus, large body size, and superficial similarities in tooth shape and dental formulae appear to be convergences with unquestioned anthropoids. However, convergences do not refute an anthropoid or protoanthropoid hypothesis. Anthropoid affinities *are* supported by the centrally placed paraconid on M_1 and possibly on M_2. However, the size and position of the hypoconulid appear to refute such a hypothesis.

It is also difficult to place *Amphipithecus* with any certainty in another primate grouping. Although Szalay and Delson (1979) have argued that *Amphipithecus* could be derived from a North American notharctine, there are no shared-derived characters that unequivocally link *Amphipithecus* to the Adapidae. Their cited "adapid" features of *Amphipithecus*—the presence of a paraconid on M_1, the width of the M_1 talonid, and a robust mandibular corpus—are not evidence of shared ancestry. Paraconids are a primitive mammalian feature of no particular phylogenetic relevance in and of themselves, and the other two features are dietary correlates. Notharctines are frugivorous primates of similar body size, and, as noted by Ciochon *et al.* (1985), the shape of the symphysis differs in notharctines and *Amphipithecus*. Kay and Williams (Chapter 13, this volume) also attempted to recognize anthropoid synapomorphies in *Amphipithecus*. They found three shared characters: an enlarged P_4 metaconid, first molar trigonid and talonid of similar height, and weak, rounded molar hypocristids. However, like Kay and Williams (Chapter 13, this volume), we conclude that the current evidence is inadequate to place this genus with confidence.

Revised Placement of Pondaungia

The reappraisal of *Pondaungia* presented here demonstrates that it is not an anthropoid and shows greatest affinities with notharctines but is clearly divergent from the latter. Therefore, a revised diagnosis of this genus is necessary:

Order Primates Linnaeus, 1758
Infraorder Adapiformes Szalay and Delson, 1979

Family Adapidae Trouessart, 1879
Subfamily Notharctinae Trouessart, 1879
Tribe Pondaungini, new

Type Genus: *Pondaungia* Pilgrim, 1927
Diagnosis: Differs from Copelemurini and Notharctini in having crenulate enamel; lower and less distinct molar cusps; smaller and more distally positioned paraconule; less distinct metaconule; postprotocingulae less distally expanded/thickened, especially on M^2; presence of small cingular cusps lingual to pseudohypocone. Differs from copelemurinins and is similar to notharctinins in having distally positioned molar paraconids; shallow M_3 hypoflexid; mesiodistally short, buccolingually wide M_3 hypoconulid; low paracristids; less distinct entoconids; and lingually closed talonids. Differs from notharctinins and is similar to copelemurinins in having mesiodistally extensive postprotocingulae in which the protocone and pseudohypocone are undifferentiated by a lingual furrow. Differs from copelemurinins and other notharctinins but is similar to *Pelycodus* in having poorly developed styles and M^{1-2} postprotocingulae confluent with the postcingulum.

Discussion: Most of the features that differentiate the Pondaungini from other notharctines represent autapomorphies of the Asian taxon. However, the less anteriorly placed paraconule and more poorly developed and less lingually oriented pseudohypocone appear to be more primitive conditions than those found in any North American taxon. Additionally, the similarities to *Pelycodus* noted above may represent primitive character states for the Notharctinae.

Genus *Pondaungia* Pilgrim, 1927

Pondaungia Ba Maw, Ciochon, and Savage, 1979
Type Species: *Pondaungia cotteri* Pilgrim, 1927
Revised Diagnosis: As for tribe.

Pondaungia cotteri Pilgrim, 1927

Pondaungia cotteri Ba Maw, Ciochon, and Savage, 1979
Holotype: GSI-D 201-203.
Type Locality: 1/4 mi west of Pangan village, Myaing townships, Pakkoku district, Burma (see Colbert, 1938, for maps of the area).
Referred Specimens: UCMP 120377 (= DGMU-P4), from UCMP locality V78090, Mogaung Northwest, Burma.
Revised Diagnosis: As for genus and tribe.

Conclusions

The hypothesis of an Asian origin for Anthropoidea is not supported by this analysis of the Pondaung primates, which includes what has long been the only empirical evidence available to bolster such an argument. *Pondaungia* shows clear affinities to notharctines. *Amphipithecus* shows conclusive resemblances to neither the Anthropoidea nor the North American notharctines nor European adapines. To simply classify *Amphipithecus* as either "Anthropoidea, family *incertae sedis*" or as Adapidae ignores the fact that we cannot categorize it with any confidence in either group. The Pondaung primates show unique combinations of primitive and derived characters that are different from those observed elsewhere within the primates but are convergent on those of early Oligocene anthropoids such as *Aegyptopithecus* and *Apidium*.

While diminishing support for a hypothesis of anthropoid origins in southeast Asia, this analysis of the Pondaung primates also demonstrates two additional points relevant to the study of anthropoid origins.

First, the difficulty in allocating *Amphipithecus* and *Pondaungia* suggests that we need rigorously to reassess how we dentally define and recognize each of the Eocene higher taxa. The categorization of the family Adapidae is difficult, and the family has been shown to be paraphyletic with respect to living lemuriformes (Beard *et al.*, 1988). The work of Beard *et al.* (1991) and Kay and Williams (Chapter 13, this volume) shows that the Omomyidae do not necessarily represent a monophyletic group, and this analysis shows we now have very few characters by which to define Anthropoidea beyond their Eocene African locale. Until we can better understand the phylogenetic steps that separate groupings such as Omomyidae, Adapidae, and Anthropoidea, we will continue to be perplexed by a paradox as vexing as that of the "omomyid versus adapid origin" that many of us have spent the last two decades trying to solve.

Second, the origin of anthropoids, whenever and wherever it may have occurred, certainly did not involve a change in grade or a move into an "anthropoid adaptive zone" characterized by a larger-bodied, frugivorous, deep-jawed adaptation as we have long considered it (e.g., Ciochon *et al.*, 1985; Cachel, 1979; Conroy, 1978). Rather, if there is a "monkey grade" that has been independently achieved by the Pondaung primates, by subfossil Malagasy lemurs such as *Hadropithecus*, and by propliopithecids, parapithecids, and later anthropoids, it probably has nothing to do with being an anthropoid in any phylogenetic sense.

Amphipithecus and *Pondaungia* provide an object lesson in how we often use gradistic mental templates in efforts to classify and to discern anthropoid origins. Certainly, the set of largely gradistic characters (many previously championed by one of us: R.L.C.) used to defend anthropoid status for the Pondaung primates points out the need for a more vigorous reexamination of dental and gnathic features of undoubted early anthropoids (such as that by Kay and Williams, Chapter 13, this volume). If we wish to discern the origin of the Anthropoidea, we need to abandon the communally held gradistic concept in which "anthropoid" equals a vague "monkey-like" concept and attempt to delineate the critical derived features that define the group and identify its fossil members.

A shift from gradistic arguments about anthropoid origins to more character-based analyses is necessary at this time. During the course of discussions raised at the conference in Durham, North Carolina from which this volume derives, the suggestion to attempt to list a set of characters delineating Anthropoidea was brushed aside in favor of discussions of the grade achieved or "adaptive rubicon" crossed by Anthropoidea that then differentiated them from the Prosimii. The latter discussions were not particularly fruitful; we cannot attempt to understand how our direct, albeit distant, ancestors differentiated from the antecedents of lemurs, lorises, and tarsiers until we can first recognize which minor osteological and dental variations characterized them. Until we can say what those features were, by whatever method, we cannot begin to grasp the meaning of "anthropoid origins."

The current record of Asian Eocene primates can neither definitively refute nor support an Asian origin of Anthropoidea, although the evidence offered by *Amphipithecus* and *Pondaungia* is considerably weaker than previously believed. A similar conclusion regarding anthropoid origins in Asia was reached by Conroy and Bown (1974) and Conroy (1978). Ironically, more than 15 years later and after this reappraisal of the evidence, we reach a similar conclusion.

Acknowledgments

We are indebted to Donald E. Savage, whose interest in, and enthusiasm for, the study of the Asian Paleogene led us into this project, as well as to the organizers of the conference that led to this volume. Figure 2 and Figures 3 and 4 were prepared, respectively, by C. Tarka and L. Meeker of the American Museum of Natural History. H. H. Covert, R. F. Kay, M. J. Ravosa, and B. A. Williams provided helpful discussion, and B. A. Williams, T. M. Bown, and an anonymous reviewer provided constructive comments on the manuscript. T. M. Bown, P. S. Chatrath, and E. L. Simons provided materials for comparison. We are particularly grateful to E. Delson, who made the type specimen of *Amphipithecus* available for study and made arrangements for the cleaning and rephotographing of AMNH 32530. Finally, we thank our Burmese colleagues, U Thaw Tint and U Ba Maw, who helped recover additional specimens of *Amphipithecus* and *Pondaungia* in 1978 that continued to fuel the ongoing debate on the Asian origin of the Anthropoidea.

References

Ba Maw, Ciochon, R. L., and Savage, D. E. 1979. Late Eocene of Burma yields earliest anthropoid primate, *Pondaungia cotteri*. *Nature* **282**:65–67.

Beard, K. C. 1988. New notharctine primate fossils from the early Eocene of New Mexico and southern Wyoming and the phylogeny of Notharctinae. *Am. J. Phys. Anthropol.* **75:**439–469.

Beard, K. C., Dagosto, M., Gebo, D. L., and Godinot, M. 1988. Interrelationships among primate higher taxa. *Nature* **331:**712–714.

Beard, K. C., Krishtalka, L., and Stucky, R. K. 1991. First skulls of the early Eocene primate *Shoshonius cooperi* and the anthropoid–tarsier dichotomy. *Nature* **349:**64–67.

Beecher, R. M. 1979. Functional significance of the mandibular symphysis. *J. Morphol.* **159:**117–130.

Beecher, R. M. 1983. Evolution of the mandibular symphysis in Notharctinae (Adapidae, Primates). *Int. J. Primatol.* **4:**99–112.

Cachel, S. 1979. A paleoecological model for the origin of higher primates. *J. Hum. Evol.* **8:**351–359.

Ciochon, R. L., and Chiarelli, A. B. 1980. Paleobiogeographic perspectives on the origin of the Platyrrhini. In: R. L. Ciochon and A. B. Chiarelli (eds.), *Evolutionary Biology of the New World Monkeys and Continental Drift,* pp. 459–493. Plenum Press, New York.

Ciochon, R. L., and Etler, P. 1994. Reinterpreting past primate diversity. In: R. S. Corruccini and R. L. Ciochon (eds.), *Integrative Paths to the Past: Paleoanthropological Advances in Honor of F. Clark Howell,* pp. 37–68. Prentice Hall, Englewood Cliffs, NJ.

Ciochon, R. L., Savage, D. E., Thaw Tint, and Ba Maw. 1985. Anthropoid origins in Asia? New discovery of *Amphipithecus* from the Eocene of Burma. *Science* **229:**756–759.

Colbert, E. 1937. A new primate from the upper Eocene Pondaung Formation of Burma. *Am. Mus. Novit.* **951.**

Colbert, E. 1938. Fossil mammals from Burma in the American Museum of Natural History. *Bull. Am. Mus. Nat. Hist.* **74:**255–436.

Conroy, G. C. 1978. Candidates for anthropoid ancestry: Some morphological and palaeozoogeographical considerations. In: D. J. Chivers and A. Joysey (eds.), *Recent Advances in Primatology, Vol. 3, Evolution,* pp. 27–41. Academic Press, London.

Conroy, G. C., and Bown, T. M. 1974. Anthropoid origins and differentiation: the Asian question. *Yearb. Phys. Anthrop.* **18:**1–6.

Covert, H. H. 1990. Phylogenetic relationships among the Notharctinae of North America. *Am. J. Phys. Anthrop.* **81:**381–397.

Fleagle, J. G. 1988. *Primate Adaptation and Evolution.* Academic Press. San Diego.

Fleagle, J. G., and Kay, R. F. 1987. The phyletic position of the Parapithecidae. *J. Hum. Evol.* **16:**483–532.

Gingerich, P. D. 1980. Eocene Adapidae, paleobiogeography, and the origin of South American Platyrrhini. In: R. L. Ciochon and A. B. Chiarelli (eds.), *Evolutionary Biology of the New World Monkeys and Continental Drift,* pp. 123–138. Plenum Press, New York.

Godinot, M. 1992. Apport à la systématique de quatre genres d'Adapiformes (Primates, Eocène). *C. R. Acad. Sci. Paris.* [*Sér. II*] **314:**237–242.

Godinot, M., and Mahboubi, M. 1992. Earliest known simian primate found in Algeria. *Nature* **357:**324–326.

Harrison, T. 1987. The phylogenetic relationships of the early catarrhine primates: A review of the current evidence, *J. Hum. Evol.* **16:**41–80.

Kay, R. F. 1977. The evolution of molar occlusion in the Cercopithecidae and early catarrhines. *Am. J. Phys. Anthrop.* **46:**327–352.

Kay, R. F. 1980. Platyrrhine origins: a reappraisal of the dental evidence. In: R. L. Ciochon and A. B. Chiarelli (eds.), *Evolutionary Biology of the New World Monkeys and Continental Drift,* pp. 309–339. Plenum Press, New York.

Kay, R. F., Fleagle, J. G., and Simons, E. L. 1981. A revision of the Oligocene apes from the Fayum Province, Egypt. *Am. J. Phys. Anthrop.* **55:**293–322.

Koenigswald, G. H. R. von. 1965. Critical observations upon the so-called higher primates from the upper Eocene of Burma. *Proc. Konikl. Nederl. Acad. van Wetenschappen* [*B*] **68:**165–167.

Martin, R. D. 1993. Primate origins: plugging the gaps. *Nature* **363:**223–234.

Pilgrim, G. E. 1927. A *Sivapithecus* palate and other primate fossils from India. *Mem. Geol. Surv. India (Paleontol. Indica), N.S.* **14:**1–26.

Rasmussen, D. T., and Simons, E. L. 1988. New specimens of *Oligopithecus savagei*, early anthropoidean primate from the Fayum, Egypt. *Fol. Primatol.* **51:**182–208.

Rasmussen, D. T., and Simons, E. L. 1992. Paleobiology of the oligopithecines, the earliest known anthropoid primates. *Int. J. Primatol.* **13**(5)**:**477–508.

Ravosa, M. J. 1991. Structural allometry of the prosimian mandibular corpus and symphysis. *J. Hum. Evol.* **20:**3–20.

Scapino, R. 1981. Morphological investigation into functions of the jaw symphysis in carnivorans. *J. Morphol.* **167:**339–375.

Schwartz, J. 1986. Primate systematics and a classification of the order. In: D. R. Swindler and J. Erwin (eds.), *Comparative Primate Biology, Vol. 1: Systematics, Evolution, and Anatomy,* pp. 1–41. Alan R. Liss, New York.

Simons, E. L. 1963. Some fallacies in the study of hominid phylogeny. *Science* **141:**879–889.

Simons, E. L. 1965. New fossil apes from Egypt and the initial differentiation of Hominoidea. *Nature* **205:**135–139.

Simons, E. L. 1971. Relationships of *Amphipithecus* and *Oligopithecus. Nature* **232:**489–491.

Simons, E. L. 1974. Notes on early Tertiary prosimians. In: R. D. Martin, G. A. Doyle, and A. C. Walker (eds.), *Prosimian Biology,* pp. 415–433. Duckworth, London.

Simons, E. L. 1989. Description of two genera and species of Late Eocene Anthropoidea from Egypt. *Proc. Natl. Acad. Sci. USA* **86:**9956–9960.

Simons, E. L. 1990. Discovery of the oldest known anthropoidean skull from the Paleogene of Egypt. *Science* **247:**1567–1569.

Simons, E. L. 1992. Diversity in the early Tertiary anthropoidean radiation in Africa. *Proc. Natl. Acad. Sci. USA* **89:**10743–10747.

Simons, E. L., and Kay, R. F. 1983. *Qatrania,* new basal anthropoid primate from the Fayum, Oligocene of Egypt. *Nature* **304:**624–626.

Simons, E. L., and Kay, R. F. 1988. New material of *Qatrania* from Egypt with comments on the phylogenetic position of the Parapithecidae (Primates, Anthropoidea). *Am. J. Primatol.* **15:**337–347.

Simons, E. L., and Pilbeam, D. R. 1965. Preliminary revision of the Dryopithecinae (Pongidae, Anthropoidea). *Fol. Primatol.* **3:**81–152.

Szalay, F. S. 1970. Late Eocene *Amphipithecus* and the origins of catarrhine primates. *Nature* **227:**355–357.

Szalay, F. S. 1972. *Amphipithecus* revisited. *Nature* **236:**179.

Szalay, F. S., and Delson, E. 1979. *Evolutionary History of the Primates,* Academic Press, New York.

Van Valen, L. 1969. A classification of the Primates. *Am. J. Phys. Anthrop.* **30:**295–296.

Rencunius zhoui, New Primate from the Late Middle Eocene of Henan, China, and a Comparison with Some Early Anthropoidea

7

PHILIP D. GINGERICH,
PATRICIA A. HOLROYD,
and RUSSELL L. CIOCHON

Introduction

Late Eocene primates of Asia are often mentioned in discussions of anthropoid origins. This is in part because of the distinctive morphologies of Asian Eocene primates that document an otherwise hidden diversity of potential anthropoid ancestors. Asia also draws our attention as a large, centrally placed

Publication date: 11/1/94.

PHILIP D. GINGERICH • Museum of Paleontology, University of Michigan, Ann Arbor, Michigan 48109. PATRICIA A. HOLROYD • Department of Biological Anthropology and Anatomy, Duke University, Durham, North Carolina 27710, and U.S. Geological Survey, Denver, Colorado 80225. RUSSELL L. CIOCHON • Departments of Anthropology and Pediatric Dentistry, University of Iowa, Iowa City, Iowa 52242.

Anthropoid Origins, edited by John G. Fleagle and Richard F. Kay. Plenum Press, New York, 1994.

geographic region that is still inadequately known paleontologically. Asia and its Eocene primates are important for understanding both the phylogenetic and biogeographic history of primate and anthropoid diversification.

Amphipithecus and *Pondaungia* from Burma (reviewed by Ciochon and Holroyd, Chapter 6, this volume) have generated the most interest. *Hoanghonius* from the Chaili Member in the upper part of the Heti Formation in Shanxi Province, China, has also received some attention. It has never been labeled an anthropoid, but similarities to the early African anthropoid *Oligopithecus* are interesting. Gingerich (1977) and more recently Rasmussen and Simons (1988) have interpreted *Hoanghonius* as an adapid showing affinities with *Oligopithecus*. Dental similarities between these two taxa are part of the evidence favoring an adapid origin of Anthropoidea. Here we redescribe Chinese specimens previously referred to *Hoanghonius* and place these in a new genus and species, *Rencunius zhoui*.

Hoanghonius stehlini

Hoanghonius stehlini was first described by Zdansky (1930, p. 75) from his Locality 1, the so-called "River Section" locality, on the north bank of the Huanghe [Huang Ho or Yellow River] near Yuanqu city in southern Shanxi Province. This locality is in what is now called the Chaili Member of the upper Heti Formation, which was regarded as late middle Eocene in age by Russell and Zhai (1987, p. 212–215), but is now regarded as late middle Eocene (Holroyd and Ciochon, Chapter 5, this volume). The type specimen is a left dentary with M_{2-3}, and Zdansky also referred an upper molar to this genus and species (for illustrations see Zdansky, 1930, Fig. 7, plate 5, pp. 16–18; Szalay and Delson, 1979, Fig. 136). *Hoanghonius* takes its name from the Yellow River.

The most distinctive characteristics of *Hoanghonius stehlini* noted by Zdansky are the double-cusped trigonid (protoconid and metaconid present, but no paraconid) and the anteriorly sloping "cross-yoke" paracristid connecting these on M_{2-3}. The trigonid is little higher than the talonid. The hypoconid is larger than the entoconid. Anteriorly the hypoconid is connected to the prontoconid. The hypoconulid on M_2 is as large as the entoconid and close to it, forming the posterior corner of the tooth. The talonid is broader than the trigonid, and there is a labial cingulid running along the lateral side of the tooth. M_3 has a strong hypoconulid forming a strong third lobe on the tooth. The labial cingulid on M_3 borders the trigonid and talonid but not the third hypoconulid lobe. The referred upper molar is tritubercular, with the paracone and metacone equally large. The protocone has the appearance of a well-proportioned half-moon with small but distinct conules (paraconule and metaconule) on each arm. The cingulum is complete (as far as preserved), with two inner cusps. Zdansky (1930) referred to these inner cusps as the hypocone and "protostyle"; the latter is now called a pericone.

Zdansky (1930) classified *Hoanghonius* as *incertae sedis* but found that the lower teeth resembled *Smilodectes* most closely and the upper molar resembled *Washakius* or *Hemiacodon,* all now clearly and unquestionably members of the order Primates. *Smilodectes* is a well-known North American middle Eocene notharctine adapid, and *Washakius* and *Hemiacodon* are North American omomyids known since the last century. Uncertainty about the systematic position of *Hoanghonius* within Primates has continued since it was first named. Simpson (1945, p. 64) classified *Hoanghonius* as Prosimii of uncertain infraorder or family. Hill (1955, p. 274) included it "for convenience" in Omomyinae. Romer (1966, p. 382) classified it in Omomyidae. Simons (1972, p. 287) classified it in Omomyidae? *incertae sedis* (see also Conroy and Brown, 1975). Szalay (1974, p. 53) was the first to classify *Hoanghonius* as "probably an adapid," writing "I would not classify it as an omomyid, based upon admittedly scanty material."

In 1975, Gingerich studied the original specimens of *Hoanghonius* in Uppsala and made sharp epoxy casts distributed to colleagues for comparison with other primates. He independently corroborated Szalay's referral of *Hoanghonius* to Adapidae (Gingerich, 1976) and illustrated the close similarity of its M_2 to that of *Oligopithecus* (M_2 was the only comparable tooth known at the time; Gingerich 1977, p. 177). Szalay and Delson (1979, pp. 269–271) confirmed similarity to *Oligopithecus* but reclassified *Hoanghonius* in Omomyidae, *incertae sedis,* writing "Contrary to Gingerich's view, there is no character or combination of characters which would mandate adapid ties for *Hoanghonius.*" A few sentences later they wrote: "Nevertheless, the possible adapid ties of *Hoanghonius,* and also *Oligopithecus,* cannot be dismissed. Were they to be corroborated by better samples of these forms, then *Oligopithecus* should clearly be taken out of the Catarrhini."

Rasmussen and Simons (1988, p. 182), describing new specimens of *Oligopithecus,* favored adapid ties for *Hoanghonius,* writing:

> Among prosimians, the upper teeth of *Oligopithecus* very closely resemble those of *Protoadapis* and allied forms (*Europolemur, Mahgarita, Periconodon, Hoanghonius*) but differ substantially from other prosimian taxa. Most of the dental and osteological resemblances between *Oligopithecus* and the *Protoadapis* group are derived features, thus favoring the hypothesis that *Oligopithecus* and other Anthropoidea are descended from Adapidae.

Thus, *Hoanghonius* is inextricably bound up in the nexus of genera and morphological characteristics connecting the origin of Anthropoidea backward in geological time to Eocene ancestors.

The type specimen of *Hoanghonius stehlini* (unnumbered in Paleontological Museum, University of Uppsala, Sweden) is a left dentary with M_{2-3}. M_2 measures 4.0 mm in anteroposterior length and 3.4 mm in breadth; M_3 measures 4.6 mm in anteroposterior length and 2.8 mm in breadth; and the mandibular ramus is 7.9 mm deep beneath M_1 (measured on original). The referral specimen (also unnumbered in Uppsala) is an isolated right M^2 that measures 3.9 mm in length and 5.9 mm in breadth.

Rencunius zhoui

Woo and Chow (1957) described four new specimens from "Jentsen" [Rencun], a locality on the opposite side of the Huanghe [Yellow river] and 5 km upstream from the *Hoanghonius* type locality. Woo and Chow referred these to *Hoanghonius stehlini* but had no possibility of direct comparison to the original specimens in Uppsala. Study of original specimens of *Hoanghonius* and the referred material and comparison of sharp epoxy casts of these with other Eocene primates indicates that the referred specimens represent a new genus and species: Family Adapidae, Subfamily **Hoanghoniinae,** new subfamily.

Asian *Hoanghonius* and *Rencunius,* described here, differ from other subfamilies of Adapidae in combining strongly developed pericones and hypocones on the lingual cingulum of upper molars with well-developed twinned entoconids and hypoconulids at the posterolingual corners of lower molars. The new *Hoanghonius*-like genus from Thailand (Suteethorn *et al.,* 1988; Ducrocq, 1992) probably belongs in this subfamily as well.

Rencunius zhoui: new genus and species (Figs. 1B, 2C, and 3B)

Holotype: Institute of Vertebrate Paleontology and Paleoanthropology [IVPP] 5312, left dentary with P_4–M_2 (Figs. 1B and 2C). This was also illustrated by Woo and Chow (1957) in their text Figure 2 and plate figures 3a,b, 4a,b, and 5a,b.

Referred specimens: IVPP 5311, right dentary with M_{1-2}, and IVPP 5313, right maxilla with P^4–M^1. Woo and Chow (1957) illustrated a fourth specimen (their Fig. 6, not seen by us) and described this as an isolated left M_1; it is almost certainly a right M_3.

Type locality: Rencun (= Jentsun, or Jentsen of Woo and Chow, 1957; Locality 7 of Zdansky, 1930), Mianchi County, Henan Province, east central China. Latitude and longitude of this locality are approximately 35°03′N, 111°47′E. Chow *et al.* (1973, p. 170, Fig. 2), Li and Ting (1983, p. 32, Fig. 10), and Russell and Zhai (1987, p. 212, Fig. 123) published maps showing this locality.

Age and distribution: Late middle Eocene. The type locality of this genus and species is in the lower part of the Heti Formation, and it is thus older than the type locality of *Hoanghonius stehlini* in the upper part of the Heti Formation. *Rencunius zhoui* is presently known only from the type locality in east central China.

Generic diagnosis: Most similar to *Hoanghonius* but differs in having a higher and more inflated M_2 crown and in retaining a small paraconid on M_2 (minute in the holotype, larger in IVPP 5311); has the twinned entoconid–

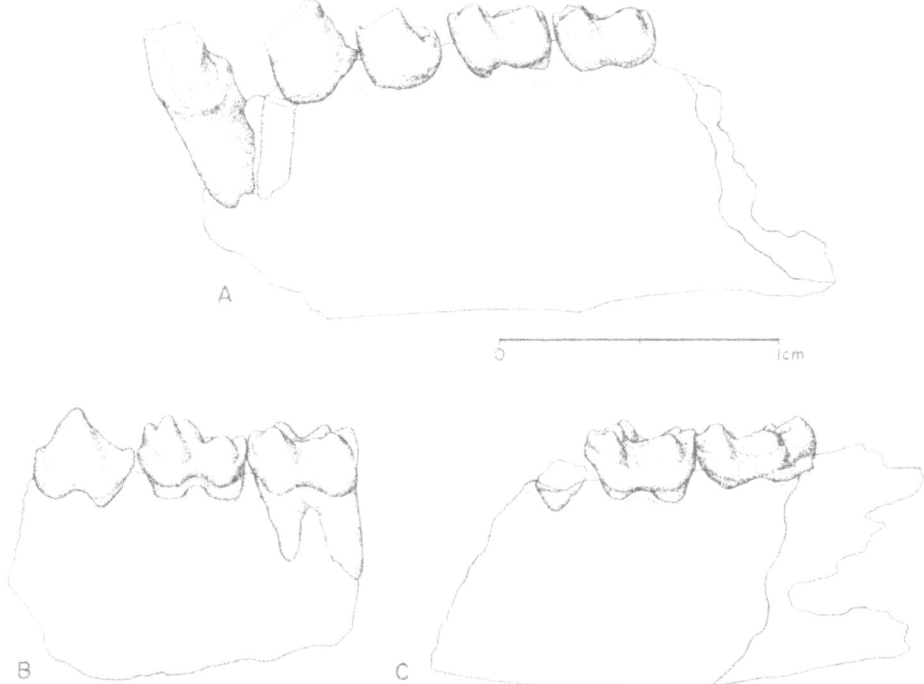

Fig. 1. Comparison of left dentaries of *Oligopithecus*, *Rencunius*, and *Hoanghonius* in lateral view. A: Type specimen of *Oligopithecus savagei*, Cairo Geological Museum 18000, from Quarry E, Fayum Province, Egypt. B: Type specimen of *Rencunius zhoui*, new genus and species, Institute of Vertebrate Paleontology and Paleoanthropology 5312, from Henan Province, China. C: Type of *Hoanghonius stehlini*, University of Uppsala (no number), from Shanxi Province, China. See Fig. 2 for comparison in occlusal view.

hypoconulid seen on M_2 in *Hoanghonius*, but these cusps are larger and more separated in *Rencunius*. The M_3 illustrated by Woo and Chow (1957, Fig. 6) has a trigonid resembling that of *Hoanghonius* in being short anteroposteriorly, narrow in comparison to M_2, and lacking a paraconid, but the talonid is very different in retaining more distinct cusps with a narrower and less rounded hypoconulid lobe. M^1 in *Rencunius* resembles M^2 in *Hoanghonius* in having a large pericone and hypocone on the lingual cingulum but differs in having a much larger and more centrally placed paraconule and metaconule and in having a very distinctive flexure of enamel running along the posterolingual side of the paracone, a small centrocrista within the talon basin.

When *Rencunius* is compared to other primates, the relatively long and narrow P_4 distinguishes it from omomyids. A relatively long and narrow P_4 and M_1, with a small metaconid on P_4 and an open trigonid on M_1, distinguishes it from anthropoids. These are features of Adapidae. The reduced paraconid on M_2 and M_3, twinned entoconid–hypoconulid on M_2, bulbous

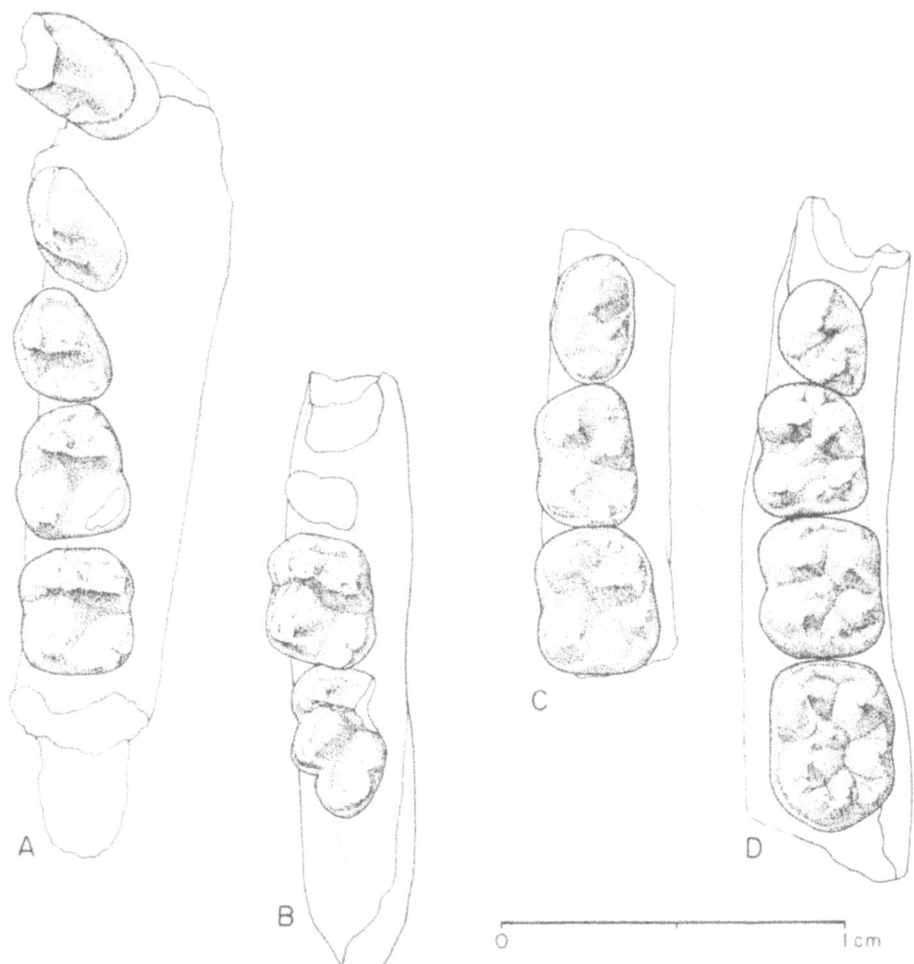

Fig. 2. Comparison of left dentaries of *Oligopithecus, Hoanghonius, Rencunius,* and *Apidium* in occlusal view. A: Type specimen of *Oligopithecus savagei,* Cairo Geological Museum 18000, from Quarry E, Fayum Province, Egypt. B: Type of *Hoanghonius stehlini,* University of Uppsala (no number), from Shanxi Province, China. C: Type specimen of *Rencunius zhoui,* Institute of Vertebrate Paleontology and Paleoanthropology 5312, from Henan Province, China. D: Type specimen of *Apidium phiomense,* American Museum of Natural History 13370, from Fayum Province, Egypt. Note anteroposteriorly elongated P_4 and M_1 and the large paraconid on M_1 distinguishing *Rencunius* from anthropoids *Oligopithecus* and *Apidium.* Note larger size, retention of very small paraconid on M_2, more bulbous cusps, and wider separation of hypoconulid and entoconid distinguishing *Rencunius* from *Hoanghonius.* All four genera share a reduced paraconid and twinned hypoconulid–entoconid on M_2 that gives them a general resemblance in spite of other distinctive specializations.

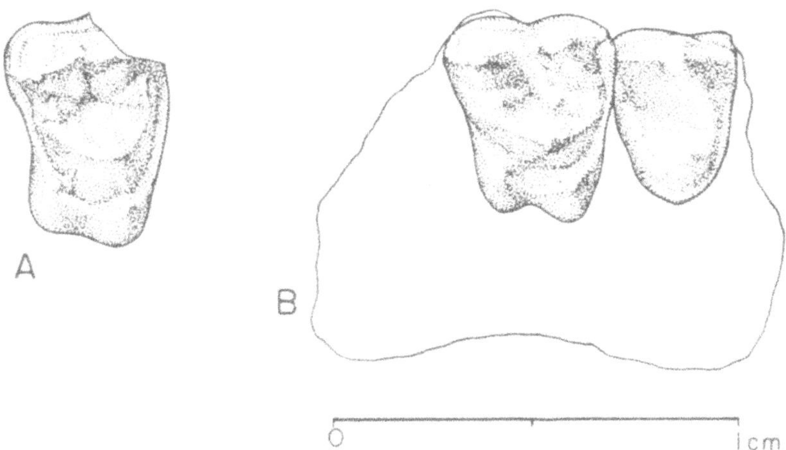

Fig. 3. Comparison of right maxillary cheek teeth in *Hoanghonius* and *Rencunius* in occlusal view. A: Referred M^2 of *Hoanghonius stehlini*, University of Uppsala (no number), from Shanxi Province, China. B: Referred maxilla fragment with P^4–M^1 of *Rencunius zhoui*, Institute of Vertebrate Paleontology and Paleoanthropology 5311, from Henan Province, China. Note larger, more centrally placed conules and flexure of enamel or centrocrista posterolingual to the paracone on M^1. P^4 in *Rencunius* is simple, with a distinct protocone, a single large labial cusp, but no conules or other accessory cusps (if a parastyle was originally present, it has been removed by breakage).

cusps on an inflated crown on M_2 (especially IVPP 5311), distinct talonid cusps on M_3, and large pericone and hypocone on M^1 are resemblances to Anthropoidea that distinguish *Rencunius* from most adapids.

Specific diagnosis: Only species known; diagnosis as for genus.

Etymology: Named for Rencun locality, which is named in turn after Rencun village nearby. Specific name honors Zhou Minchen, Institute of Vertebrate Paleontology and Paleoanthropology, Beijing, who collected these specimens 30 years ago, in recognition of his important contributions to paleomammalogy and paleoprimatology.

Description: The holotype, IVPP 5312 with P_4–M_2, is the most complete specimen of *Rencunius* (Figs. 1B and 2C). P_4 is relatively long and narrow, like *Cantius* and other adapids. Illustrations by Woo and Chow (1957) show that it had a single high central cusp, the protoconid (now broken), with a small but distinct metaconid posterolingual to the protoconid. There is no paraconid on P_4, but there is a small cusp at the base of the crown (hypoconid?). P_4 is bordered labially and lingually by a narrow cingulid. The crown of P_4 in the holotype measures 3.7 mm in anteroposterior length and 2.6 mm in breadth.

M_1 in the holotype is relatively long and narrow by comparison to M_2, with the talonid a little broader than the trigonid. The protoconid, paraconid, and metaconid on the trigonid are all large and well separated. The paraconid is little worn, but enamel at the tips of the protoconid and metaconid is perforated by wear. The hypoconid at the posterolabial corner of the tooth is the largest cusp on the talonid, but the twinned entoconid and hypoconulid at the posterolingual corner of the talonid are also large and subequal in size. The cristid obliqua joins the hypoconid to the metaconid rather than the protoconid or middle of the metacristid (this may be related in some way to development of a distinctive flexure of enamel or centrocrista on M^1). There is a well-developed labial cingulid on M_1 but no lingual cingulid. The crown of M_1 in the holotype measures 3.9 mm in anteroposterior length and 3.0 mm in breadth. M_1 in IVPP 5311 is now broken and adds little to knowledge of *Rencunius*.

M_2 in the holotype is longer and considerably broader than M_1. The trigonid has distinct protoconid and metaconid cusps, but the paraconid is reduced to a minute cusp on the holotype. IVPP 5311 differs in having a larger paraconid. The paraconid is part of a continuous paracristid connecting the protoconid to the metaconid and, with the metacristid, enclosing a distinct anterior fovea. The talonid on M_2 is broader than the trigonid. Enamel at the tips of the protoconid and metaconid is perforated by wear. As in M_1, the hypoconid is large, filling the posterolabial corner of the talonid, and the entoconid and hypoconulid are twinned at the posterolingual corner of the talonid. The latter cusps are both relatively large, and they are more separated than in *Hoanghonius* or *Oligopithecus* (Fig. 2). In *Oligopithecus* and in *Catopithecus* the entoconid is larger than the hypoconulid, and it is more anteriorly positioned. Enamel on the hypoconulid is perforated by wear. The cristid obliqua joins the metacristid a little closer to the protoconid than to the metaconid. There is a well-developed labial cingulid on M_2 and just a hint of a cingulid present on the lingual side of the trigonid (the labial cingulid is stronger, and there is no trace of a lingual cingulid in IVPP 5311). The crown of M_2 in the holotype measures 4.1 mm in anteroposterior length and 3.5 mm in breadth. The crown of M_2 in IVPP 5311 measures 4.0 mm in anteroposterior length and 3.9 mm in breadth. It is broader than M_2 in the type and consequently looks more anthropoid in its proportions.

M_3 is not preserved in the specimens studied, but Woo and Chow (1957) illustrated a tooth (their specimen no. 3, illustrated in their Fig. 6a,b) that appears to be an isolated right M_3. This has the anteroposteriorly short trigonid expected of M_3, with no paraconid and a paracristid connecting the protoconid and metaconid and enclosing a small fovea like that on M_2. The hypoconid, hypoconulid, and entoconid are all relatively large and well separated. Woo and Chow (1957) give measurements (anteroposterior length 4.5 mm, breadth 2.9 mm) indicating that this tooth is much longer than M_1 or M_2 in *Rencunius* and that it is a little shorter and broader than M_3 in *Hoanghonius*.

P^4 is preserved in the right maxillary fragment of referred specimen IVPP 5313 (Fig. 3B). It is simple, with a distinct protocone and a single large lingual cusp (paracone) but no conules. A parastyle may have been present at the anterolabial corner of the crown, but this is broken. A high robust preprotocrista connects the protocone and paracone. The crown is bordered by weak labial and lingual cingula. P^4 measures about 3.1 mm in anteroposterior length (estimate), and 4.2 mm in breadth.

M^1 is also preserved in the referred maxillary fragment, IVPP 5311. M^1 of *Rencunius* is similar in shape to M^2 of *Hoanghonius,* but it is relatively narrower and a little more massively constructed. The protocone, paracone, and metacone enclose a trigon basin, but this is not as simple and open as that of *Hoanghonius:* the paraconule and metaconule are both larger and more centrally positioned, and there is a distinct flexure of enamel or centrocrista located posterolingual to the paracone. The pericone and hypocone are very well developed on the lingual cingulum, and there is a weak labial cingulum as well. M^1 measures 4.2 mm in anteroposterior length and 5.3 mm in breadth.

Discussion: Dashzeveg and McKenna (1977, p. 131) singled out the type specimen of *Rencunius zhoui* (dentary IVPP 5312) from the rest of the Rencun sample and claimed that this specimen is an artiodactyl rather than a primate. No justification was given for separating one specimen from the rest of the sample, and no justification was given for considering this to be an artiodactyl. Delson (1977, p. 49) published a similar conclusion, also without justification. Woo and Chow (1957) considered the IVPP 5312 dentary and the IVPP 5311 maxilla to be parts of the same individual animal, which may or may not be true (see below). Another Eocene mammal from China, *Lantianius,* was originally described as an adapid (Chow, 1964) but later reclassified as an artiodactyl because of its anteroposteriorly elongated premolars, unlike those of any primate (Gingerich, 1976). None of the *Rencunius* specimens has elongated artiodactyl-like premolars. The combination of the moderate length of P_4, the open trigonid and distinct paraconid on M_1, the reduced paraconid and the paracristid enclosing an anterior fovea on the trigonid of M_2, and the basined talonids on M_1 and M_2 indicate that *Rencunius* is a primate. *Rencunius* is an adapid rather than an omomyid because of its (1) relatively large size, (2) longer and narrower P_4 than is typical of omomyids, (3) reduced paraconid on M_2 (and presumably M_3), and (4) twinned hypoconulid–entoconid as found in other Adapidae including *Hoanghonius,* the Thailand adapid, *Indraloris,* and *Sivaladapis.*

Woo and Chow (1957) referred the primate specimens they described from "Jentsen" or Rencun to *Hoanghonius stehlini.* This was conservative and reasonable given the inaccessibility of the *Hoanghonius* type material and an unquestioned general morphological resemblance. It is only by studying originals directly and being able to compare them using sharp epoxy casts that it is evident two genera and species are represented.

Woo and Chow (1957) indicated that the four specimens they described belong to at least three individuals, and they suggested that their specimens no. 2 (IVPP 5312, holotype dentary) and no. 4 (IVPP 5311, maxilla) might belong to a single individual. This appears unlikely because IVPP 5312 is more heavily worn than IVPP 5311; it is the only specimen with perforated enamel on the tips of cusps.

Geology and Mammalian Succession of the Heti Formation

Chow *et al.* (1973, p. 171, Table 1) divided the Heti Formation into two members, an upper Chaili Member and a lower Rencun Member, separated by an unconformity, and each of these members was further subdivided into two stratigraphic intervals. The Heti Formation is about 600 m thick (Li, 1984). *Hoanghonius stehlini* is from the uppermost subinterval (IVPP 5301), whereas *Rencunius zhoui* is from the lowermost subinterval (IVPP 5311-5314; Li and Ting, 1983, pp. 31–32). The most recent review of these faunas (Tong, 1989) indicates that they share a single species, the anthracotheriid artiodactyl *Anthracokeryx sinensis,* and the record from Rencun is recorded as questionable. Thus, it is clear that *Rencunius* is older than *Hoanghonius.* How much older is a difficult problem, but the thickness of the formation, separation of the two genera in distinct faunas at the bottom and top of the formation, and an intervening unconformity, taken together, mean *Rencunius* is probably much older.

Russell and Zhai (1987) and Tong (1989) correlated the Rencun fauna with the Shara Murun fauna in Inner Mongolia, on which the Sharamurunian land-mammal age is based. There are relatively few taxa shared in common between the two faunas, and the correlation is based instead on the high proportion of perissodactyls in both faunas. The Sharamurunian land-mammal age is considered early late Eocene by Russell and Zhai (1987) and late middle Eocene by Tong (1989), which probably reflects recent expansion of the middle Eocene to include the Bartonian marine stage/age. Holroyd and Ciochon (Chapter 5, this volume) consider both Heti Formation faunas, the Chaili Member fauna and the Rencun member fauna, to be late middle Eocene in age.

Rencunius Compared to Hoanghonius and Early Anthropoid Primates

Rencunius is compared to the Chinese adapid *Hoanghonius* and primitive Egyptian Fayum anthropoids *Oligopithecus* and *Apidium* in Figs. 1–3, supplemented by reference to features of *Catopithecus* recently described by Simons (1989). All five genera are similar in size and, to the extent they can be compared, in general primate morphology. Unfortunately, the only tooth identified with certainty that is preserved in common in both *Rencunius* and *Hoanghonius* is M_2. This is generally similar in both. M_2 of *Rencunius* differs from that of *Hoanghonius* in being a little larger, retaining a very small paraconid, having more bulbous cusps, and having a wider separation of the hypoconulid and entoconid. M^1 of *Rencunius* is like M^2 of *Hoanghonius* in that both have conspicuous pericones and hypocones on the lingual cingulum. These teeth differ some in proportion, and M^1 of *Rencunius* has larger conules than M^2 of *Hoanghonius*.

Resemblances of *Hoanghonius* to *Oligopithecus* have been detailed elsewhere (Gingerich, 1977; Rasmussen and Simons, 1988), but here again comparison is limited to M^2. This tooth is a little longer and narrower in *Hoanghonius,* but the two are otherwise very similar in size and shape. Both have the trigonid compressed anteroposteriorly, lacking a paraconid, and both have the twinned hypoconulid and entoconid positioned close together. On the basis of this one comparable tooth, one would expect Asian *Hoanghonius* to have been very similar overall and probably closely related to African *Oligopithecus*. Another slightly older African genus, *Catopithecus,* qualifies these similarities and postulated relationships somewhat. *Catopithecus* and *Oligopithecus* both have a dental formula of 2?-1-2-3, and there is no evidence that *Hoanghonius* or *Rencunius* had reached this advanced state of dental reduction. *Catopithecus* has an M_2 that is similar to that of both *Hoanghonius* and *Oligopithecus,* but it also preserves an M_3 that is conspicuously shorter in length than the M_3 of *Hoanghonius* (relative to M_1 or M_2).

Teeth of *Rencunius* can be compared to those of *Catopithecus* and *Oligopithecus* at four tooth positions. P_4, M_1, and M_2 are all longer anteroposteriorly and narrower buccolingually. P_4 has a buccal cingulid that is lacking in *Oligopithecus.* The metaconid on P_4 is much less developed, and the posterior margin of the tooth is narrower. The paraconid on M_1 is large and well separated from the metaconid. The paracristid on M_2 extends farther forward, and the trigonid is not as compressed anteroposteriorly on either M_1 or M_2. In the upper dentition, M^1 of *Rencunius* differs markedly from that of *Oligopithecus* (Rasmussen and Simons, 1988) and *Catopithecus* (Simons, 1990) in having a large pericone and a large hypocone on the lingual cingulum, with a large paraconule and metaconule, in contrast to the simple *Protoadapis*-like upper M^1 of *Oligopithecus* and *Catopithecus.*

Teeth at five positions in *Rencunius* can be compared to those of *Apidium.* P_4, M_1, and M_2 are all longer anteroposteriorly and narrower buccolingually. P_4 has a buccal cingulid that is lacking in *Apidium,* and, as in *Oligopithecus,* the metaconid is less strongly developed. The paraconid on M_1 is much stronger than that in *Apidium,* in which the paraconid is positioned close to the protoconid and metaconid (and the whole trigonid is much shorter). The position of talonid cusps on M_1 is generally similar. The closest resemblance between *Rencunius* and *Apidium* is in M_2, where the trigonid is a little narrower relative to the talonid, but most cusps and crests are similarly developed and similarly positioned. *Rencunius* lacks the centroconid between the protoconid, metaconid, and hypoconid that is a distinctive characteristic of *Apidium.* P^4 in *Rencunius* differs markedly from P^4 in *Apidium* in lacking any trace of the large centroconule so distinctive of *Apidium.* M^1 is generally similar but differs in having a pericone and hypocone approximately equal in size: the pericone is small in *Apidium,* and the hypocone is greatly enlarged.

Anthropoid Origins

Chinese *Rencunius zhoui* and *Hoanghonius stehlini* are not well enough known, by themselves, to contribute very much to clarification of anthropoid origins. Rather, *Rencunius* and *Hoanghonius* are representative of a general direction and grade of evolution in adapid primates, showing that there was greater morphological and taxonomic diversity in Asia than previously recognized.

The strongest evidence that African *Oligopithecus* is an anthropoid comes from its reduced catarrhine dental formula, with a canine hone on P_3, and from its close dental resemblance to *Catopithecus,* a genus with anthropoid cranial characteristics that include frontal fusion, postorbital closure, and an ectotympanic in the lateral wall of the bulla (Simons, 1990). *Apidium* has the frontal fusion, postorbital closure, and carotid circulation of an anthropoid (Simons, 1959, 1971, 1990; Gingerich, 1973). These cranial characteristics are evolutionarily "derived" or advanced characteristics of early anthropoids that distinguish them from adapids and other prosimian primates. Calling them "derived" or "advanced" means that they are characteristics that appeared later in geological time than adapid and other prosimian cranial characteristics appeared. *Rencunius* and *Hoanghonius* might turn out to be anthropoids when their skulls are known, but this appears unlikely given present evidence. There is no reason to attribute derived character states indicative of anthropoid rank to either *Rencunius* or *Hoanghonius* in the absence of evidence that such diagnostic characteristics had evolved by the late middle Eocene.

There are three distinct superfamilies of prosimian primates living today: Tarsioidea, Lemuroidea, and Lorisoidea (Gingerich, 1984a,b). These are commonly classified in the infra- or suborders Tarsiiformes (Tarsioidea) and Lemuriformes (Lemuroidea + Lorisoidea), which are matched in turn to Eocene Omomyidae and Adapidae, respectively (e.g., Fleagle, 1988). It is not clear whether Lemuroidea and Lorisoidea are grouped together because they belong together or because this grouping is required to make the number of groups of living prosimians match the number of groups of Eocene prosimians. One alternative would be recognition that lorisoids are very different anatomically from lemuroids in spite of their shared tooth comb. Whatever the reason for grouping the living forms, Eocene omomyids were not modern living tarsiers, and Eocene adapids were not modern living lemurs or lorises. Omomyids are sufficiently similar to tarsiers that it is plausible that they are broadly ancestral to tarsiers, and, likewise, adapids are sufficiently similar to lemurs and to lorises that it is plausible they are broadly ancestral to both.

However, no adapid has a tooth comb, and this must have evolved from spatulate incisors if lemurs and lorises are descendants of adapids. Alternatively, the spatulate incisors of adapids may have evolved as a derived characteristic from the procumbent pointed incisors of an earlier primate that independently gave rise to lemurs and lorises. The presence of pointed incisors in Paleocene proprimates, Eocene omomyids, and later colugos, tree shrews, tarsiers, lemurs, and lorises suggests that pointed incisors are primitive in primates. Whether omomyids and adapids are ancestral to living prosimians or not, omomyids and adapids were generally similar to each other in evolutionary grade, and both were generally similar in grade to prosimians living today. In spite of their similarities, adapids consistently differ from omomyids, where these characteristics are known, in having a primitively free ectotympanic in the middle ear and in having a derived anterior dentition with overlapping canines and spatulate upper and lower incisors.

Anthropoids are different from both adapids and omomyids in many characteristics. Living anthropoids are clearly a third group distinct from living tarsiiform and lemuriform primates, but this does not mean that they were so clearly distinct from Omomyidae and Adapidae in the Eocene. Anthropoids are widely recognized to be more advanced in grade than living prosimians, with different behaviors and life history strategies and, as a consequence, different morphological specializations such as enlarged brains, frontal fusion, and postorbital closure. These anthropoid characteristics are unknown in undisputed Eocene primates, which means that anthropoids are more advanced than most, and possibly all, Eocene primates. Enlarged brains, frontal fusion, and postorbital closure are unknown in putative anthropoids *Biretia* (Bonis *et al.*, 1988) and *Algeripithecus* (Godinot and Mahboubi, 1992); the age of *Biretia* is poorly constrained, and the Eocene age of the early anthropoid *Catopithecus* is also debatable (Simons, 1989; Gingerich, 1993).

Conservatively, it is not possible to trace the origin of anthropoids farther and farther back in time as a "third group" separate from adapids and omomyids without some clear evidence that such a third group existed in the Eocene. Anthropoids may have been part of a third group of primates in the Eocene, but, if so, such a third group will not be very interesting until it is known from fossils unequivocally different from omomyids or adapids. *Rencunius* joins *Hoanghonius, Amphipithecus,* and *Pondaungia* in Asia (Szalay, 1970; Gingerich, 1981; Ciochon *et al.*, 1985), *Oligopithecus* and other Fayum taxa, *Algeripithecus, Djebelemur,* and possibly *Biretia* from North Africa (Rasmussen and Simons, 1988; Simons, 1971, 1989, 1992; Godinot and Mahboubi, 1992; Hartenberger and Marandat, 1992; see also Bonis *et al.*, 1988), *Cercamonius* in Europe (Gingerich, 1975), and *Mahgarita* in North America (Rasmussen, 1990) as intermediate fossils of uncertain systematic position. Some appear to link anthropoids to a broadly adapoid origin, but this may reflect convergence in dental and gnathic morphology related to similar diets and behavioral strategies. Knowing which are anthropoids, which are adapoids, and which might represent a new third group of Eocene primates different from omomyids and adapids will require better Eocene fossils that preserve anatomic evidence from the cranium as well as the dentition.

ACKNOWLEDGMENTS

We thank Drs. Richard Reyment and Niall Mateer (Uppsala) and Drs. Zhou Minchen, Zhai Renjie, and Qi Tao (Beijing) for access to specimens described here (P.D.G. in 1975 and 1979, respectively). Dennis Etler translated important papers on Chinese faunas, and Karen Klitz drew the figures.

References

Bonis, L. de, Jaeger, J.-J., Coiffait, B., and Coiffait, P.-E. 1988. Découverte du plus ancien primate Catarrhinien connu dans l'Éocene supérieur d'Afrique du Nord. *C. R. Acad. Sci.* **306:**929–934.

Chow, M.-C. 1964. A lemuroid primate from the Eocene of Lantian, Shensi. *Verteb. PalAsiat.* **8:**260–262.

Chow, M.-C., Li, C.-K., and Chang, Y.-P. 1973. Late Eocene mammalian faunas of Honan and Shansi with notes on some vertebrate fossils collected therefrom. *Verteb. PalAsiat.* **11:**165–181.

Ciochon, R. L., Savage, D. E., Thaw Tint, and Ba Maw. 1985. Anthropoid origins in Asia? New discovery of *Amphipithecus* from the Eocene of Burma. *Science* **229:**756–759.

Conroy, G. C., and Bown, T. M. 1975. Anthropoid origins and differentiation: The Asian question. *Yearb. Phys. Anthrop.* **18:**1–6.

Dashzeveg, D., and McKenna, M. C. 1977. Tarsioid primate from the early Tertiary of the Mongolian People's Republic. *Acta Palaeontol. Pol.* **22:**119–137.

Delson, E. 1977. Vertebrate paleontology, especially of nonhuman primates, in China. In: W. W. Howells and P. Jones Tsuchitani (eds.), *Paleoanthropology in the People's Republic of China, Report no. 4,* pp. 40–65. National Academy of Science, Washington, D.C.

Ducrocq, S. 1992. Etude biochronologique des bassins continentaux tertiaires du Sud-Est asiatique: Contribution des mammifères. Thesis, Université Montpellier II.

Fleagle, J. G. 1988. *Primate Adaptation and Evolution.* Academic Press, San Diego.

Gingerich, P. D. 1973. Anatomy of the temporal bone in the Oligocene anthropoid *Apidium* and the origin of Anthropoidea. *Fol. Primatol.* **19;**329–337.

Gingerich, P. D. 1975. A new genus of Adapidae (Mammalia, Primates) from the late Eocene of southern France, and its significance for the origin of higher primates. *Contrib. Mus. Paleontol. Univ. Michigan* **24:**163–170.

Gingerich, P. D. 1976. Systematic position of the alleged primate *Lantianius xiehuensis* Chow, 1964, from the Eocene of China. *J. Mammal.* **57:**1994–198.

Gingerich, P. D. 1977. Radiation of Eocene Adapidae in Europe. *Geobios, [Mem. Spec.]* **1:**165–182.

Gingerich, P. D. 1981. Eocene Adapidae, paleobiogeography, and the origin of South American Platyrrhini. In: R. L. Ciochon and A. B. Chiarelli (eds.), *Evolutionary Biology of the New World Monkeys and Continental Drift,* pp. 123–138. Plenum Press, New York.

Gingerich, P. D. 1984a. Primate evolution: Evidence from the fossil record, comparative morphology, and molecular biology. *Yearb. Phys. Anthropol.* **27:**57–72.

Gingerich, P. D. 1984b. Primate evolution. In: P. D. Gingerich and C. E. Badgley (eds.), *Mammals:*

Notes for a Short Course, pp. 167–181. Paleontological Society Short Course, University of Tennessee Department of Geological Sciences (Studies in Geology no. 8), Knoxville.

Gingerich, P. D. 1993. Oligocene age of the Gebel Qatrani Formation, Fayum, Egypt. *J. Hum. Evol.* **24:**207–218.

Godinot, M., and Mahboubi, M. 1992. Earliest known simian primate found in Algeria. *Nature* **357:**324–326.

Hartenberger, J.-L., and Marandat, B. 1992. A new genus and species of an early Eocene primate from North Africa. *Hum. Evol.* **7:**9–16.

Hill, W. C. O. 1955. *Primates: Comparative Anatomy and Taxonomy. II. Haplorhini: Tarsioidea.* Edinburgh University Press, Edinburgh.

Li, C.-K., and Ting, S.-Y. 1983. The Paleogene mammals of China. *Bull. Carnegie Mus. Nat. Hist.* **21:**1–93.

Li, Y.-T. (ed.). 1984. *The Tertiary System of China. Stratigraphy of China,* Vol. 13. Geological Publishing House, Beijing. [In Chinese, translated by D. Etler]

Rasmussen, D. T. 1990. The phylogenetic position of *Mahgarita stevensi:* Protoanthropoid or lemuroid? *Int. J. Primatol.* **11:**439–469.

Rasmussen, D. T., and Simons, E. L. 1988. New specimens of *Oligopithecus savagei,* early Oligocene primate from the Fayum, Egypt. *Fol. Primatol.* **51:**182–208.

Romer, A. S. 1966. *Vertebrate Paleontology.* University of Chicago Press, Chicago.

Russell, D. E., and Zhai, R.-J. 1987. The Paleogene of Asia: mammals and stratigraphy. *Mem. Mus. Nat. Hist. Nat. Sci. Terre* **52:**1–488.

Simons, E. L. 1959. An anthropoid frontal bone from the Fayum Oligocene of Egypt: The oldest skull fragment of a higher primate. *Am. Mus. Novit.* **1976:**1–16.

Simons, E. L. 1971. Relationships of *Amphipithecus* and *Oligopithecus. Nature* **232:**489–491.

Simons, E. L. 1972. *Primate Evolution: An Introduction to Man's Place in Nature.* Macmillan, New York.

Simons, E. L. 1989. Description of two genera and species of late Eocene Anthropoidea from Egypt. *Proc. Natl. Acad. Sci. USA* **86:**9956–9960.

Simons, E. L. 1990. Discovery of the oldest known anthropoidean skull from the Paleogene of Egypt. *Science* **247:**1567–1569.

Simons, E. L. 1992. Diversity in the early Tertiary anthropoidean radiation in Africa. *Proc. Natl. Acad. Sci. USA* **89:**10743–10747.

Simpson, G. G. 1945. The principles of classification and a classification of mammals. *Bull. Am. Mus. Nat. Hist.* **85:**1–350.

Suteethorn, V., Buffetaut, E., Helmcke-Ingavat, R., Jaeger, J.-J., and Jongkanjanasoontorn, Y. 1988. Oldest known Tertiary mammals from South East Asia: Middle Eocene primate and anthracotheres from Thailand. *Neues Jahrb. Geol. Palaontol. Mittheilungen* **1988:**563–570.

Szalay, F. S. 1970. Late Eocene *Amphipithecus* and the origins of catarrhine primates. *Nature* **227:**355–357.

Szalay, F. S. 1974. A review of some recent advances in paleoprimatology. *Yearb. Phys. Anthrop.* **17:**39–64.

Szalay, F. S., and Delson, E. 1979. *Evolutionary History of the Primates.* Academic Press, New York.

Tong, Y.-S. 1989. A review of middle and late Eocene mammalian faunas from China. *Acta Palaeontol. Sin.* **28:**663–682.

Woo, J.-K., and Chow, M.-C. 1957. New materials of the earliest primate known in China, *Hoanghonius stehlini. Vert. Palas.* **1:**267–272.

Zdansky, O. 1930. Die alttertiären Säugetiere Chinas nebst stratigraphischen Bemerkungen. *Palaeontol. Sin. [C]* **6**(2)**:**1–87.

The Eocene Origin of Anthropoid Primates

Adaptation, Evolution, and Diversity

8

ELWYN L. SIMONS, D. TAB RASMUSSEN, THOMAS M. BOWN, and PRITHIJIT S. CHATRATH

Introduction

One of the few previous attempts to model anthropoid origins emphasized apparent global climatic changes at the Eocene–Oligocene boundary (now placed at 34 MA) that may have served as the driving force for changes in primate geographic distribution southward and for the evolutionary origin of key new dietary and foraging adaptations (Cachel, 1979, 1981). The most recent proponent of the idea that anthropoid origin was a geologically sudden event associated with profound environmental change at the Eocene–Oligocene transition has been Gingerich (1993). In recent years, however, the view of the Eocene–Oligocene boundary as an important threshold for anthropoid ori-

ELWYN L. SIMONS • Duke Primate Center, Duke University, Durham, North Carolina 27705. D. TAB RASMUSSEN • Department of Anthropology, Washington University, St. Louis, Missouri 63130. THOMAS M. BOWN • Paleontology and Stratigraphy Branch, U.S. Geological Survey, Denver, Colorado 80225. PRITHIJIT S. CHATRATH • Duke University Primate Center, Durham, North Carolina 27706.

Anthropoid Origins, edited by John G. Fleagle and Richard F. Kay. Plenum Press, New York, 1994.

gins has been supplanted by new interpretations based on research in geology and dating (Bown and Kraus, 1988; Van Couvering and Harris, 1991; Kappelman *et al.*, 1992; Rasmussen *et al.*, 1992; Gingerich, 1993), Afro-Arabian paleontology (de Bonis *et al.*, 1988; Thomas *et al.*, 1988, 1989, 1991; Simons, 1989, 1990, 1992; Godinot, Chapter 10, this volume; Godinot and Mahboubi, 1992; Hartenberger and Marandat, 1992), and functional anatomy (Rasmussen and Simons, 1992). Evidence now demonstrates that the anthropoid clade underwent a significant radiation in the late Eocene of Africa, where prosimians were apparently scarce, and that these early anthropoids did not differ appreciably from their prosimian contemporaries on other continents in basic dietary and sensory adaptations or in size. The origin of anthropoids thus appears to have been a relatively subtle phenomenon rather than a dramatic and revolutionary adaptive radiation temporally associated with the extinction of Laurasian prosimian faunas.

Until recently, the earliest undoubted anthropoids came from the upper sequence of the Jebel Qatrani Formation, Fayum Province, Egypt (Simons, 1965, 1967, 1974, 1983, 1987; Conroy, 1976; Fleagle, 1980; Fleagle and Simons, 1982a,b; Fleagle *et al.*, 1980; Kay *et al.*, 1981; Gebo and Simons, 1987). The entire 340-m-deep formation was considered to be Oligocene in age (Simons, 1968; Fleagle *et al.*, 1986). The anthropoids from the upper sequence (top 190 m) include the well-known taxa *Aegyptopithecus, Propliopithecus, Apidium,* and *Parapithecus,* which share several specialized features with extant anthropoids that are absent in most prosimians, including (1) complete postorbital closure, (2) fused metopic suture, (3) annular ectotympanic bones attached to the lateral bullar margin, (4) flat-crowned bunodont molars, (5) deep mandibles with broadly curved angular regions, steep ascending rami, and fused symphyses; (6) small capitular tails of the distal humerus, and (7) reduced or absent third trochanters of the femur (Kay and Simons, 1980; Fleagle and Kay, 1985, 1987). For many years, the earliest primate specimens from the lower sequence of the Jebel Qatrani Formation were restricted to a single fragmentary jaw and several isolated teeth of *Oligopithecus* (Simons, 1962), the primitive molar morphology of which led to disagreement over whether it was an anthropoid or a prosimian (Szalay, 1970, 1976; Simons, 1971; Gingerich, 1977; Szalay and Li, 1986; Delson and Rosenberger, 1980; Rasmussen and Simons, 1988). During the 1980s, several partial jaws of a diminutive anthropoid, *Qatrania wingi,* were also discovered at quarry E in the lower sequence, the same quarry that has produced all specimens of *Oligopithecus* (Simons and Kay, 1983, 1988).

New finds of still older primate species have come from near the base of the Jebel Qatrani Formation (Simons, 1986, 1989, 1992). Some of these resemble *Oligopithecus,* and others are extremely divergent from anyone's preconceptions of what a primitive anthropoid should look like. Additional anthropoids that apparently predate the Fayum's upper-sequence primates have also been found in other countries, including a collection of jaws and teeth from near the coast of Oman (Thomas *et al.*, 1988, 1989, 1991), several teeth

Table I. Afro-Arabian Primates of Eocene or Early Oligocene Age Found outside the Fayum Depression of Egypt, Arranged in Order of Publication

Locality and species	Described fossil material
Malembe, Angola[a]	
Propliopithecus sp.	One lower canine (Pickford, 1986)
Taqah and Thaytiniti, Oman[a]	
cf. *Oligopithecus savagei*	One lower molar (Thomas *et al.*, 1988, 1989)
Propliopithecus markgrafi	Several upper and lower jaws and teeth (Thomas *et al.*, 1988, 1991)
Bir el Ater, Algeria	
Biretia piveteaui	One lower molar (de Bonis *et al.*, 1988)
Adrar Mgorn 1, Morocco[b]	
Altiatlasius koulchii	Ten isolated teeth (Sigé *et al.*, 1990)
Glib Zegdou, Algeria	
Algeripithecus minutus	Isolated upper and lower teeth (Godinot and Mahboubi, 1992)
Tabelia hammadae	Isolated upper and lower teeth
cf. *Tabelia* sp.	Isolated teeth (Godinot, Chapter 10, this volume)
Chambi, Tunisia[c]	
Djebelemur martinezi	Lower jaw; upper teeth (Hartenberger and Marandat, 1992)
Unnamed taxon, ?primate	One lower molar (Court, 1993)

[a]Probably early Oligocene in age.
[b]None of the Paleocene mammal teeth referred to this taxon are of anthropoids.
[c]Neither of the Chambi primates is likely to be an anthropoid. *Djebelemur* is a cercamoniine adapoid (Hartenberger and Marandat, 1992), but Godinot (chapter 10, this volume) considers the upper teeth to be those of an anthropoid because of their greater bunodonty while at the same time dismissing the possibility that the lower teeth have anything to do with anthropoid origins. We concur with Court's (1993) suggestion that the Chambi tooth may be of a sivaladapine primate; the pecular entoconid/hypoconulid pairing is also found to a lesser extent in some Eocene anthropoids and cercamoniines (e.g., Fig. 2).

from Algeria (de Bonis *et al.*, 1988; Godinot and Mahboubi, 1992; Godinot, Chapter 10, this volume), and what are sure to be controversial teeth and a lower jaw from Tunisia (Hartenberger and Marandat, 1992; Court, 1993). The new finds are summarized in Table I.

Age of the Early Afro-Arabian Primates

Determining the age of early African primates relies on the following major lines of evidence: (1) paleomagnetic sections that have been compiled for Thaytiniti and Taqah, Oman (Thomas *et al.*, 1989) and for the Fayum (Kappelman *et al.*, 1992); (2) radiometric dates from basalts that cap the Fayum section (Fleagle *et al.*, 1986; Kappelman *et al.*, 1992); (3) correlation of fossil charophytes (algal cysts) found at Algerian, Tunisian, and Omani sites (Gevin *et al.*, 1974; Coiffait *et al.*, 1984; Sassi *et al.*, 1984; Thomas *et al.*, 1989; Godinot and Mahboubi, 1992); (4) analysis of erosional unconformities and marine regressive events that may correlate with the Eocene–Oligocene boundary (Bown and Kraus, 1988; Van Couvering and Harris, 1991; Rasmussen *et al.*, 1992; Gingerich, 1993); and (5) mammalian faunal correlations among all sites (Rasmussen *et al.*, 1992; Thomas *et al.*, 1991; Court and Hartenberger, 1992). There is no universal agreement on all details pertaining to the age of these early anthropoids, but a generally consistent picture is beginning to emerge.

Kappelman *et al.* (1992) presented an analysis of the paleomagnetic section of the Fayum. Their preferred correlation of polarity events (among several possible correlations that were considered in detail) suggests that all Fayum primates occur between chrons C15r and C13, inclusive (Table II). The Eocene–Oligocene boundary falls within the upper part of chron C13r. In the Fayum, the boundary probably coincides with the marine regression associated with a major unconformity separating the lower and upper sequences of the Jebel Qatrani Formation (Bown and Kraus, 1988; the old "Lower and Upper Fossil Wood Zones"). The preferred polarity correlation of Kappelman *et al.* (1992) corresponds precisely with conclusions based primarily on erosional unconformities and mammal correlations (Rasmussen *et al.*, 1992). These results indicate an early Oligocene age for the upper-sequence primate fauna (dominated by species of *Propliopithecus, Aegyptopithecus, Apidium,* and *Parapithecus*), and a late Eocene age is indicated for the lower-sequence primates (Table I) of quarries L-41 (chron C15r) and E (lower part of chron C13r).

Table II. Ages of Fayum Primates Based on the Preferred Paleomagnetic Correlation of Kappelman *et al.* (1992)[a]

Quarry	Species	Chron	Age (MA)
Quarries I, M	*Aegyptopithecus zeuxis*	C13	33.1–33.4
	Propliopithecus chirobates		
	Apidium phiomense		
	Parapithecus grangeri		
	Qatrania fleaglei		
Quarries V, G	*Propliopithecus haeckeli*	C13r	33.8–34.0
	Propliopithecus ankeli		
	Apidium moustafai		
Unconformity containing Eocene/Oligocene boundary			34.0
Quarry E	*Oligopithecus savagei*	C13r	34.0–35.1
	Qatrania wingi		
Quarry L-41	*Catopithecus browni*	C15r	35.6–35.9
	Proteopithecus sylviae		
	Serapia eocaena		
	Arsinoea kallimos		
	Pleisopithecus teras		

[a]Absolute dates are based on revised polarity times scales (Odin and Montanari, 1989; Swisher and Prothero, 1990; Berggren *et al.*, 1992); the Eocene–Oligocene boundary is younger than previously believed.

The age of the Omani primates apparently falls between the ages of Fayum quarries E and V, based on paleomagnetic correlations and faunal comparisons. The Omani fossils came from the lower part of a major magnetic reversal period, which was briefly interrupted at a level just above the fossils by a brief period of normal polarity (Thomas *et al.*, 1989). Such a period of reversal containing within it a brief interruption occurs in the late Eocene (C15r) and in the earliest Oligocene (C13r). One difference between chrons C13r and C15r is that C13r is preceded by a very long period of normal polarity, whereas C15r has a briefer period of normal polarity between it and the reversed period of C16r. Because Thomas *et al.* (1991) found a long period of normal polarity in the sediments below their observed reversal, they logically concluded that it was the younger C13r (Thomas *et al.*, 1989). This conclusion is consistent with fossil comparisons to the Fayum primates. The Omani anthropoids include: (1) a lower molar tentatively referred to *Oligopithecus* (Thomas *et al.*, 1988), a genus otherwise known only from the Fayum's quarry E; (2) teeth and jaws that closely resemble primitive species of *Propliopithecus* (Thomas *et al.*, 1991), a genus that first appears in the Fayum at quarry V (Table II). Therefore, on the basis of the primate comparisons, the Omani site may be between quarry E and V in age. Kappelman *et al.* (1992) placed quarries V and E in chron C13r, which matches perfectly with the correlation of Thomas *et al.* (1991).

Algeripithecus minutus from Glib Zegdou, Algeria, is said to be early or middle Eocene in age based on correlation of charophytes, although documentation of such an early age is currently lacking (work in progress on charophytes has been cited by Godinot and Mahboubi, 1992). The fossil hyracoids from Glib Zegdou and the nearby correlative Gour Lazib (Sudre, 1979) compare well with those from the Jebel Qatrani Formation above quarry L-41 (Rasmussen, 1989; Rasmussen *et al.*, 1992). In particular, the genus *Titanohyrax* undergoes substantial evolutionary change up through the Fayum section, making it useful for faunal correlations; species of *Titanohyrax* described by Sudre (1979) are not as primitive as an undescribed species from quarry L-41. Dental features present in the only known tooth of *Biretia piveteaui* of the Nementcha region (de Bonis *et al.*, 1988) suggest an age for that species younger than the Fayum's lower-sequence primates (Rasmussen *et al.*, 1992). Only the fossils of Chambi, Tunisia (Hartenberger *et al.*, 1985; Sigé, 1985; Crochet, 1986; Hartenberger, 1986; Hartenberger and Marandat, 1992; Court, 1993) appear on the basis of faunal correlations to be substantially older than the material from Algeria and Egypt, including a primitive species of *Titanohyrax* (Court and Hartenberger, 1992). Chambi charophytes suggest a Bartonian age (late middle Eocene; Sassi *et al.*, 1984). A hypothetical overall view of the age sequence of Paleogene African primates is presented in Table III.

Not all recent researchers have agreed with the above overview. Van Couvering and Harris (1991) suggested a late Eocene age for the entire Jebel Qatrani Formation based on their interpretation that there are no major

Table III. Hypothetical Relative Age Sequence of Afro-Arabian Primate-Bearing Sites[a]

Site	Primates
Early Oligocene (chron C13; Stampian)	
Malembe, Angola	*Propliopithecus*
Fayum quarries I, M	*Propliopithecus, Aegyptopithecus, Apidium, Parapithecus, Qatrania, Afrotarsius*
Latest Eocene and earliest Oligocene (chron C13r)	
Fayum quarries V, G	*Propliopithecus, Apidium*
Thaytiniti and Taqah, Oman	*Propliopithecus, Oligopithecus,* omomyid
Possible Eocene–Oligocene boundary	
Fayum quarry E	*Oligopithecus, Qatrania,* omomyid
?Bir el Ater, Algeria	*Biretia*
Late Eocene (near chron C15r; Priabonian)	
?Glib Zegdou, Algeria	*Algeripithecus, Tabelia*
Fayum quarry L-41	*Catopithecus, Proteopithecus, Arsinoea, Serapia, Plesiopithecus*
Early to middle Eocene (substantially older than chron C15r; Bartonian)	
Chambi, Tunisia	*Djebelemur,* unnamed ?sivaladapine
Paleocene (Thanetian)	
Adrar Mgorn, Morocco	*Altiatlasius*

[a]For European reference levels and correlation tables, see Schmidt-Kittler (1987) and Sudre *et al.* (1992).

unconformities within the combined section of the Eocene Qasr el Sagha Formation and the overlying Jebel Qatrani Formation that could span the Eocene–Oligocene regression event. In contrast, Gingerich (1993) concluded that the entire Jebel Qatrani Formation was Oligocene on the basis of his interpretation of sedimentation rates and his belief that only the erosional unconformity at the Qasr el Sagha/Jebel Qatrani contact was of a magnitude sufficient to span the Eocene–Oligocene regression. Researchers working in Algeria have consistently claimed an early to middle Eocene age for fossil-bearing deposits there; we eagerly await detailed documentation of the charophyte correlations from the key Algerian localities.

Adaptations of Eocene Anthropoids

Oligopithecines

Structural adaptations of the extinct oligopithecines have been described in other publications (Rasmussen and Simons, 1992; Gebo *et al.*, Chapter 9, this volume) and will be reviewed briefly here, with a few new observations based on unpublished material. *Proteopithecus sylviae* was included in the discussion of oligopithecines by Rasmussen and Simons (1992), but new, unpublished specimens of this species show retention of three premolars in the lower jaw, leaving its taxonomic situation uncertain. For now, the certain oligopithecines include only *Catopithecus* and *Oligopithecus*. Allometric regressions relating molar size to body size give the following estimated body weight ranges for the oligopithecines: *O. savagei*, 700–1000 g; *C. browni*, 600–900 g. The M_2 from Oman referred to as ?*Oligopithecus savagei* by Thomas *et al.* (1988, 1989) is on the large end of the range for *O. savagei* of quarry E. These weight estimates, which are compatible with known mandibular and cranial dimensions, are based in part on the range of overlap of confidence intervals derived from several anthropoid regression curves (Gingerich *et al.*, 1982; Conroy, 1987) and in part on a lower estimated body size derived from the distinct molar–body scaling relationships that occur among extant prosimians (Conroy, 1987). Thus, the oligopithecines appear to have ranged in size from tamarins (*Leontopithecus, Saguinus*) to small cebids such as squirrel monkeys (*Saimiri*).

Dietary adaptations of the oligopithecines reconstructed by comparative study of molar occlusal structures suggest that both cutting and grinding features were important. Cutting edges present in oligopithecines include a continuous crescentic cristid obliqua and posthypocristid on the lower molars, a corresponding crescent formed of the pre- and postprotocristae on the upper molars, and an abruptly raised trigonid lacking wear facet X on its distal slope (Figs. 1 and 2). Among extant species, cutting edges are associated with both folivory and insectivory (Kay, 1977, 1984; Maier, 1984). That the oligopithecines were not folivorous is indicated by the absence of specific buccal shearing crests and styles that occur among folivores and by small body size. Grinding structures, such as closely paired, bunodont entoconid and hypoconulid, are present but not as strongly developed as in several of the other non-oligopithecine small genera such as *Qatrania, Algeripithecus,* and *Arsinoea.* Thus, the dietary adaptations of the oligopithecines are inferred to be insectivorous/frugivorous. This is consistent with known dietary preferences of living primates within the size range of the oligopithecines (cebids, lorisoids).

Unpublished specimens of *Catopithecus browni* preserve upper and lower incisors, which are spatulate in shape like those of later anthropoids. The upper central incisors are larger than the upper laterals, and the lower laterals are larger than the lower centrals, which is the same relationship seen in typical anthropoids.

The reconstructed orbit diameter (11–12 mm) and skull length (45–55 mm) of *Catopithecus* may be applied to allometric regressions of orbit size as a function of skull length in living primates (Fig. 3; Kay and Cartmill, 1977; Rasmussen and Simons, 1992). These indicate that *Catopithecus* falls easily within the range of variation of living diurnal primates but well outside the range of nocturnal ones.

Catopithecus has a relatively broad estimated interorbital breadth (IOB), falling near the high end of the range of living anthropoids on the basis of an

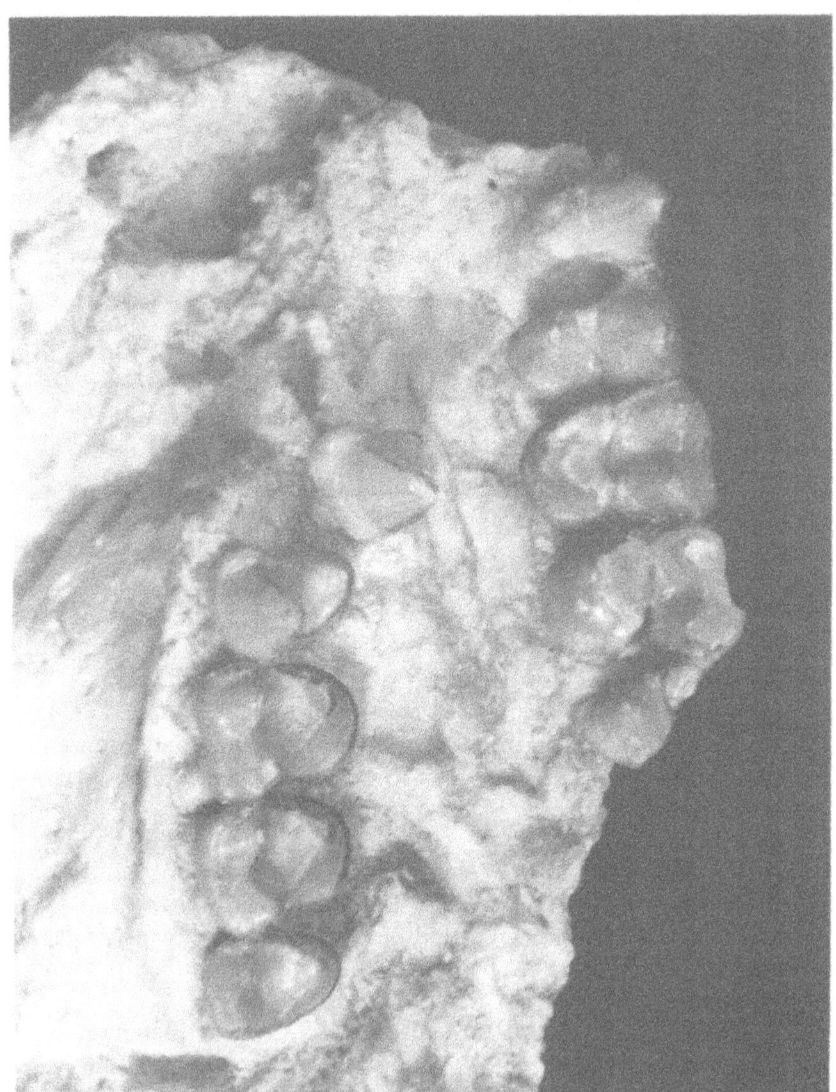

Fig. 1. Upper dentition of *Catopithecus browni* (DPC 8772) showing right and left P^2–M^3 (M^3 on the left side is fragmentary). Note the crescent-shaped lingual shearing crests (pre- and postprotocristae), the small molar hypocones, and the simple buccal cusps. The alveolus for the right canine is visible in front of P^2.

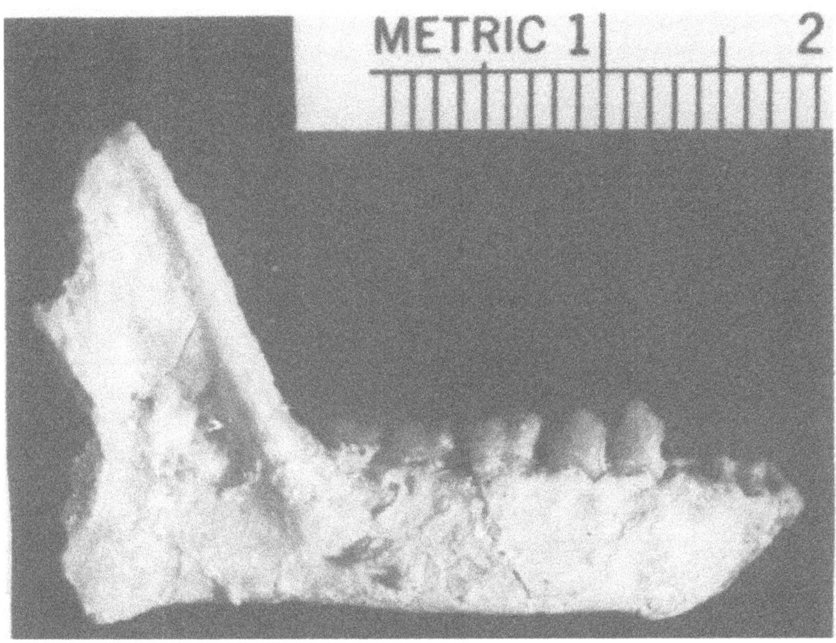

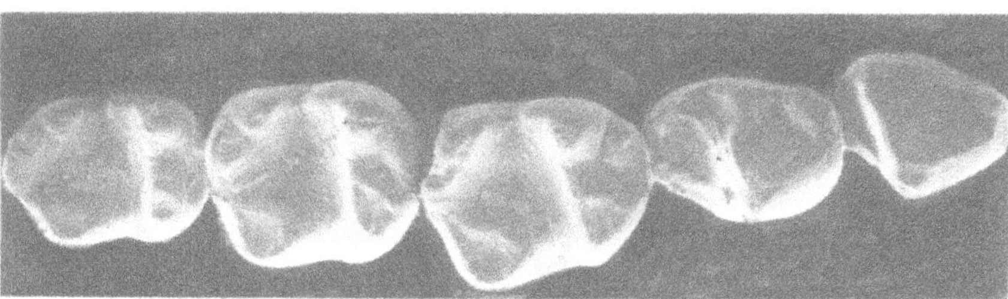

Fig. 2. Lower dentition of *Catopithecus browni* (type specimen, CGM 41885) viewed in buccal aspect (top) and occlusal aspect (bottom). In the occlusal view, note the development of lingual grinding structures, including the robust metaconid and the bulbous, twinned entoconid and hypoconulid. In the buccal view, the alveoli for I_1, I_2, and C can be discerned, demonstrating a catarrhine 2-1-2-3 dental formula.

allometric regression of IOB to skull length (Rasmussen and Simons, 1992). The IOB of *Catopithecus* is outside the range of living ceboids and small catarrhines. It falls within the range of callitrichids and near the low end of the range of extant prosimians, which rely more heavily than anthropoids do on olfaction. The relatively greatest IOBs among extant anthropoids occur in callitrichids, which more than any other extant anthropoids utilize olfaction in territorial marking and social communication (Epple, 1985, 1986; Bartecki and Heymann, 1990). The shared broad IOB of oligopithecines, callitrichids, and prosimians is interpreted to be a primitive feature reflecting a relatively

Fig. 3. Cranium of *Catopithecus browni* (DPC 8772). Although badly crushed, and with the right maxilla and zygomatic bones displaced backward and upward, the structure and relative size of several features can be assessed. Note the relatively small orbits, correlated with diurnal habits, the relatively broad interorbital region, comparable to those of callitrichids, and the relatively small braincase with truncated occipital region. The numbers identify the following structures: (1) the point on the frontal bone where the temporal lines approach the skull midline; (2) the position of the missing right canine as determined by the empty alveolus above it; (3) fragment of the right zygomatic bone lying beneath the zygomatic process of the frontal bone; (4) external acoustic meatus, ringed by a platyrrhine-like tympanic bone.

heavy reliance on olfaction; alternative biomechanical or architectural hypotheses should also be evaluated.

Brain size of *Catopithecus* cannot be measured directly because of damage to the specimen. Unlike the ballooned-out occipital regions of callitrichids and small cebids, *Catopithecus* appears to have a truncated occipital region (Fig. 3). Temporal lines of *Catopithecus* rise steeply above the orbital region and converge at or near the midline to form a low sagittal crest or ridge, a feature not present in modern anthropoids of similar size. These attributes reflect the small brain size of *Catopithecus*, which, like the brain of the younger Fayum anthropoid *Aegyptopithecus,* was similar in size to the brain in modern prosimians and comparatively smaller than the brains of most living anthropoids (Radinsky, 1973, 1974; Conroy, 1987).

Postcranial elements of primates recovered from quarry L-41 now include a distal humerus and a proximal femur of *Catopithecus* (Gebo *et al.*, Chapter 9, this volume). Like the cranial and dental remains, these postcranial bones also show several prosimian-like features, such as the retention of very large third trochanters on the femur and long capitular tails of the humerus. Comparison of these fossils to the homologous elements in living anthropoids and prosimians suggests that *Catopithecus* probably utilized arboreal quadrupedalism and leaping as its primary mode of locomotion. The range of movement would probably correspond in general terms to the locomotor patterns observed among extant callitrichids, small cebids, and cheirogaleids.

Thus, as might be expected, the oligopithecines represent a distinctive mosaic of prosimian and anthropoid features. Their tribosphenic molar structure, mandibles with hooked angular processes and sloping ascending rami, and broad IOB are prosimian-like, but their postorbital closure, fused metopic suture, and reduced premolar formula are anthropoid traits.

Parapithecids and Diminutive Anthropoids, Incertae Sedis

Most of the small, bunodont Eocene anthropoids currently known pose taxonomic problems as they may represent new families. Some of these, such as *Qatrania wingi, Algeripithecus minutus,* and *Serapia eocaena,* present dental evidence that favors a taxonomic placement with younger parapithecids (Simons and Kay, 1983, 1988; Simons, 1992; Godinot, Chapter 10, this volume). The others are adaptively so divergent (*Arsinoea, Plesiopithecus*), so generalized (*Proteopithecus*), or so poorly known (*Biretia, Tabelia*) as to defy precise taxonomic placement. The bizarre *Plesiopithecus teras* will be discussed under a separate heading.

These varied Eocene anthropoids are small, being comparable to the extant callitrichids in body size range. *Algeripithecus minutus* is smaller than living marmosets (Godinot and Mahboubi, 1992; Godinot, Chapter 10, this volume). *Proteopithecus, Qatrania,* and *Algeripithecus* are remarkable for their extreme upper molar bunodonty in conjunction with a lack of conules, enlarged lingual grinding structures (pericone, hypocone), and very small body size (Fig. 4; this assemblage of traits is also evident in the European genus *Periconodon:* Stehlin, 1916; Tattersall and Schwartz, 1983; Godinot, 1988, Chapter 10, this volume; Rasmussen, Chapter 12, this volume). These dental attributes suggest a frugivorous diet; presumably the protein component of the diet came from insects rather than foliage, judging from the primates' small size (Kay, 1984). The molar surfaces of *Arsinoea* are unusually flat, suggesting that grinding of resistant items (seeds?) was perhaps important. The

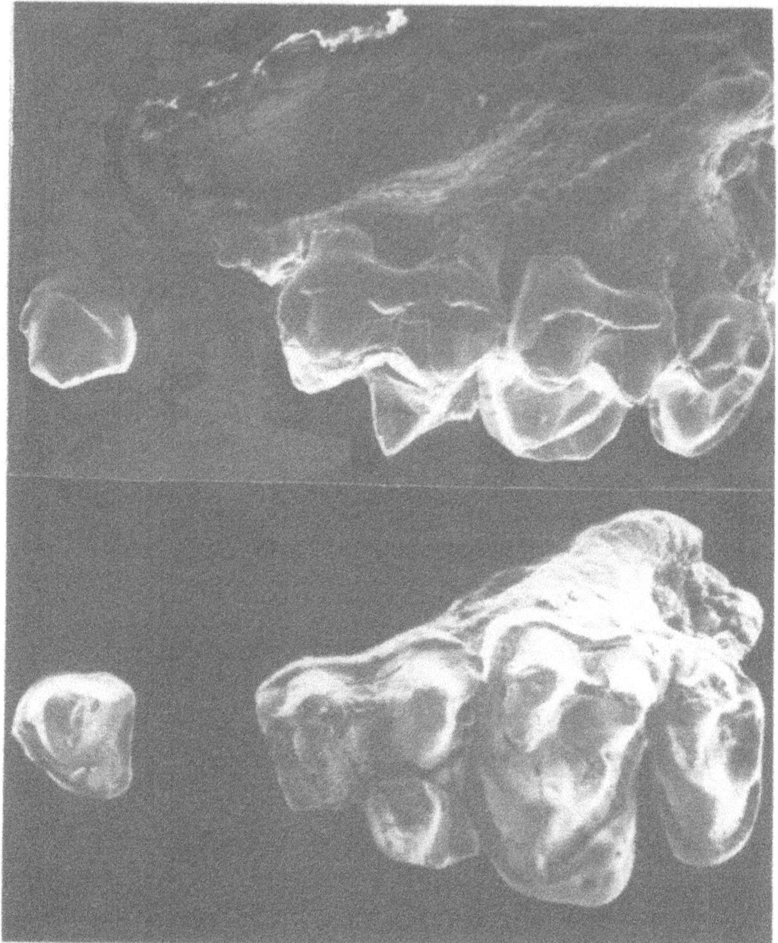

Fig. 4. Upper dentition of *Proteopithecus sylviae* (CGM 41886) viewed in oblique buccal–occlusal aspect (top) and directly perpendicular to the occlusal plane (bottom), showing its more bulbous cusp structure and larger hypocone than in *C. browni* (compare Fig. 1).

preserved lower incisors of *Arsinoea* are comparatively small, spatulate, vertically implanted teeth similar to those of later anthropoids (Simons, 1992); it offers no evidence of exudate-feeding specializations or tarsier-like insect capture. *Serapia* may have lacked adult incisors (Fig. 5), as in the parapithecid *Parapithecus grangeri* (Kay and Simons, 1983; Simons, 1986), but better specimens will be required to confirm this.

A proximal femur and an innominate bone from quarry L-41 are of very small size, and they presumably belong to either *Proteopithecus* or *Arsinoea;* based on abundance, *Proteopithecus* is the most likely candidate (Gebo *et al.*,

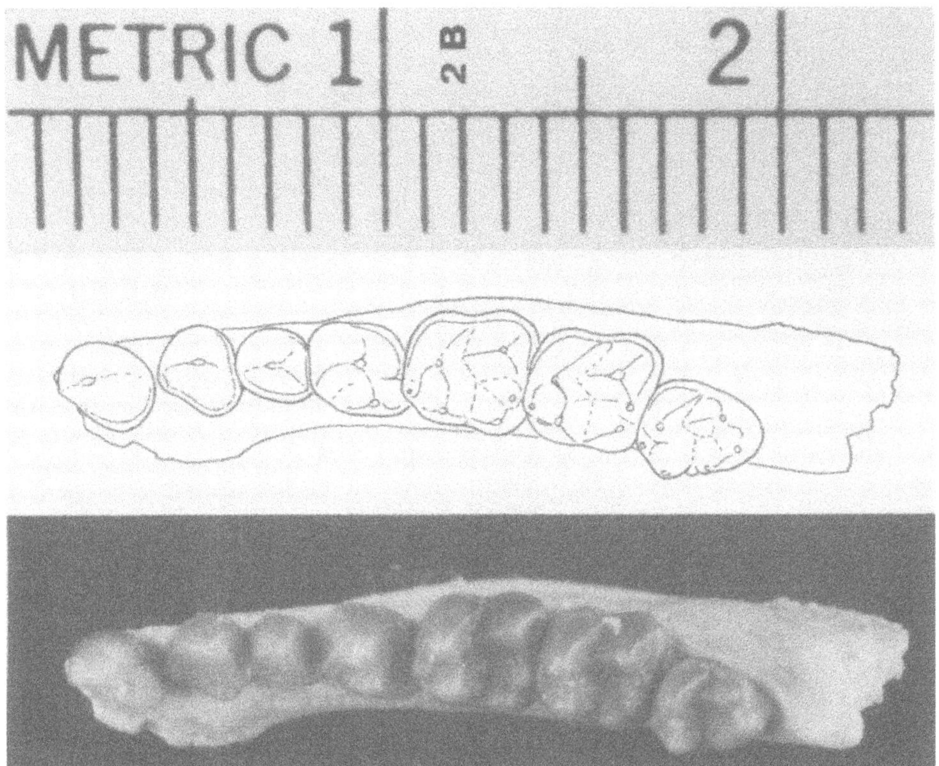

Fig. 5. Lower dentition of the parapithecid *Serapia eocaena* (CGM 42286) in occlusal view: left, photograph; center, simplified line drawing; right, metric scale. The jaw is broken at or near the symphysis; no incisor sockets are evident, but small or shallow ones could have been lost through damage (*P. grangeri* and possibly *P. fraasi* lacked adult incisors). The P_2 is distinctly larger than P_3; among other parapithecids, *P. fraasi* is the one with the relatively largest P_2. The presence of a raised trigonid with paraconids still intact in a primitive parapithecid indicates that these molar attributes were lost in parallel among later members of Parapithecidae and Propliopithecidae.

Chapter 9, this volume). These postcranial bones indicate a flexible quadrupedal and leaping animal similar in locomotor repertoire to small platyrrhines and cheirogaleids.

Plesiopithecus teras

The most surprising of all the early African discoveries is the enigmatic jaw of *P. teras* (Fig. 6). This animal, about the size of *Catopithecus browni,* has several obvious but superficial similarities to plesiadapiforms and squirrels

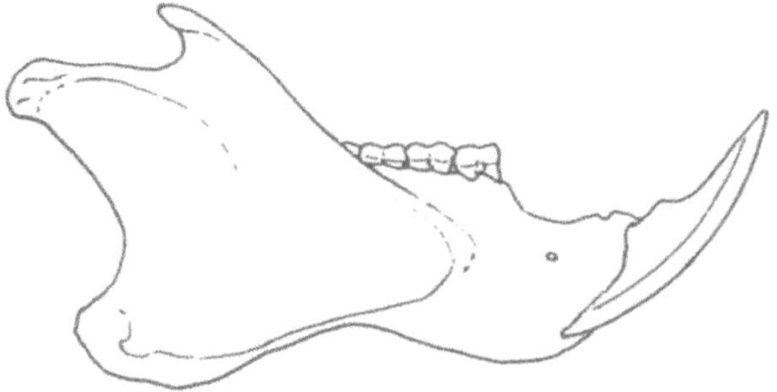

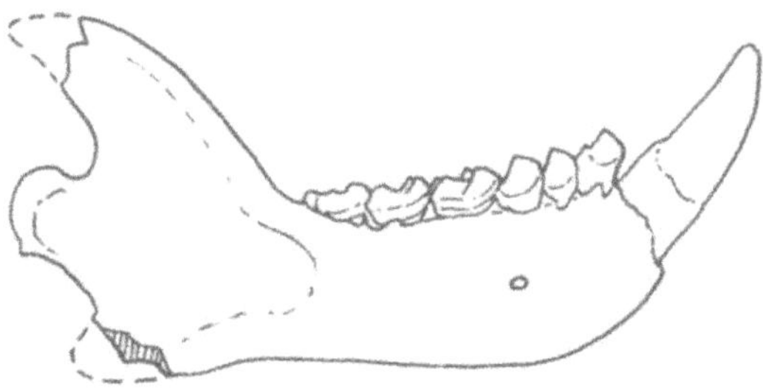

Fig. 6. Comparison of the mandible of *Plesiopithecus teras* (bottom) to that of a tree squirrel (*Sciurus carolinensis;* top) in buccal view. Drawn to same scale. We interpret the lower dental formula of *P. teras* to be 0-1-3-3, as in *Parapithecus* and perhaps *Serapia;* the alternative would be 1-0-3-3 (the squirrel is 1-0-1-3).

(Simons, 1992). The lower molars, particularly M_1, resemble those of anthropoids, but it is possible that *P. teras* might belong to a prosimian group otherwise unknown. The most anterior tooth in the jaw is elongated and curved; whether this tooth is a canine or an incisor is not obvious (Fig. 7). No incisors could have been present mesial to this tooth, as it directly abuts against the unfused symphysis. Arguments in favor of this enlarged tooth being the canine would rely on the following observations: (1) it is the largest and stoutest of the anterior teeth in the mandible, which in all other known anthropoids is the canine; (2) there is no other reasonable candidate for the canine behind the front tooth (if the front tooth is an incisor, the canine has been lost); (3) the structure of the tooth itself bears some similarities to primate canines that do not occur in incisors (the base of the crown is inflated, with weakly developed basal cingulum or shoulder and with a well-developed distal cristid descending to the distal heel); and (4) among other Fayum primates, at least one (*Parapithecus grangeri*) and possibly three (*Parapithecus fraasi, Serapia eocaena*) have undergone incisor loss with the result that the large canine is the most mesial tooth in the adult jaw (Kay and Simons, 1983; Simons, 1986; Simons and Rasmussen, 1991). The lack of a diastema behind the front tooth is not a reliable indicator that lateral incisors and canines have not been lost in the tooth-row. None of the above points prove that the tooth is the canine, but the arguments available to support its identification as an incisor are even more flimsy.

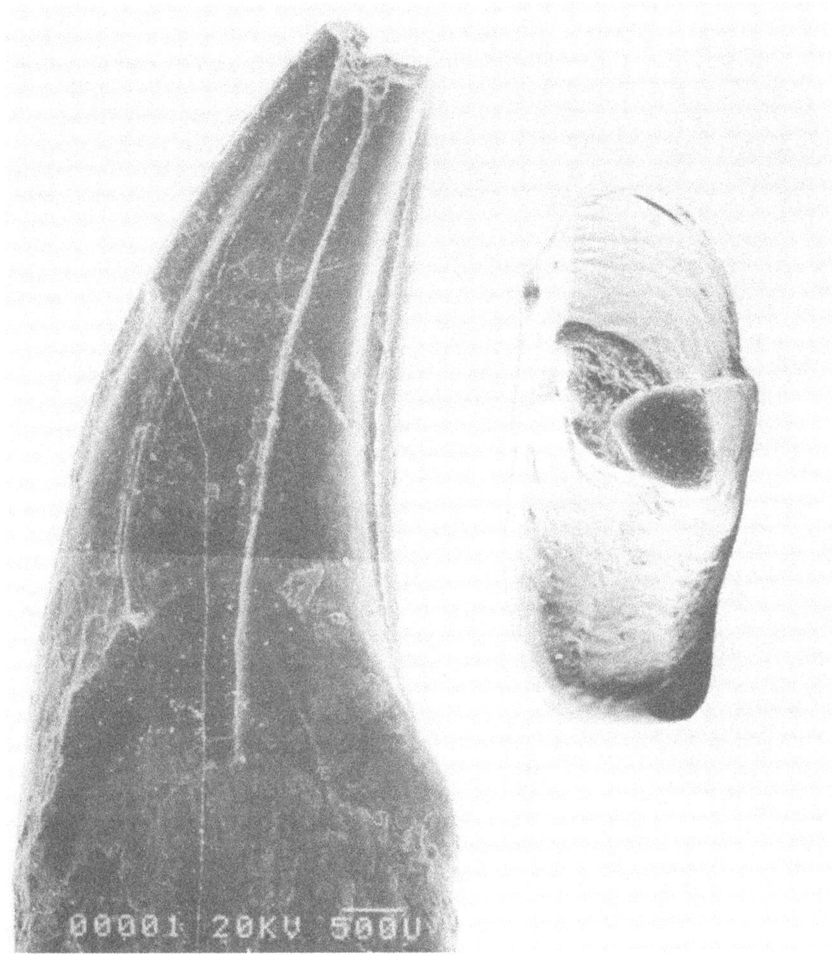

Fig. 7. Anterior tooth of *Plesiopithecus teras* (CGM 42291) in mesiolingual view (left) and in direct apical view (right). The tip of the tooth is broken off; the smooth dark area at the tip visible in the apical view is an air bubble in the cast. The sharp edge on the buccal side is a postmortem crack. The apical view reveals no obvious striations caused by wear; the concentric horizontal bands are perikymata. SEM micrographs by G. Michael Veith, Washington University.

Whether it proves to be an incisor or a canine, the enlarged anterior tooth and the shape of the jaw suggest adaptations for gnawing on resistant food items, as in squirrels (*e.g.*, Emry and Thorington, 1984; Atchley *et al.*, 1992). Unfortunately, the very tip of the enlarged tooth of *Plesiopithecus* is broken off, so patterns of wear at the tip cannot be evaluated (Fig. 7). The molars, especially M_2, also show some similarities to squirrel teeth in that they have mesiodistally compressed trigonids, a high anterior transverse crest (from

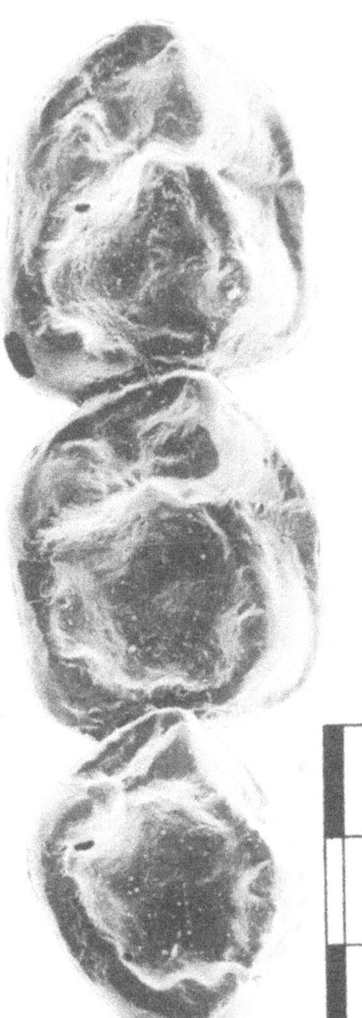

Fig. 8. Lower right molar series of *Plesiopithecus teras* (CGM 42291) in occlusal view. Of all the teeth, the first molar most closely resembles those of other primates. Unlike molars of other Eocene anthropoids, the paraconid is completely reduced, and the talonid cusps are less bulbous; compare to the oligopithecine dentition in Fig. 2. Bar scale, 3 mm. SEM micrograph by G. Michael Veith, Washington University.

protoconid to metaconid), and a shallow, open talonid basin with the bordering rims pushed to the edge of the tooth in occlusal view (Figs. 8–10). The Fayum fauna is unusual among Eocene and Oligocene mammalian faunas in lacking an arboreal, sciuromorph, squirrel-sized rodent (the only rodents present in the Fayum are heavily lophed phiomyids; Wood, 1968). We hypothesize that *Plesiopithecus* filled the niche in the food chain of the African Paleogene occupied elsewhere by plesiadapiforms and arboreal squirrels (Emry and Thorington, 1984; Maas *et al.*, 1988).

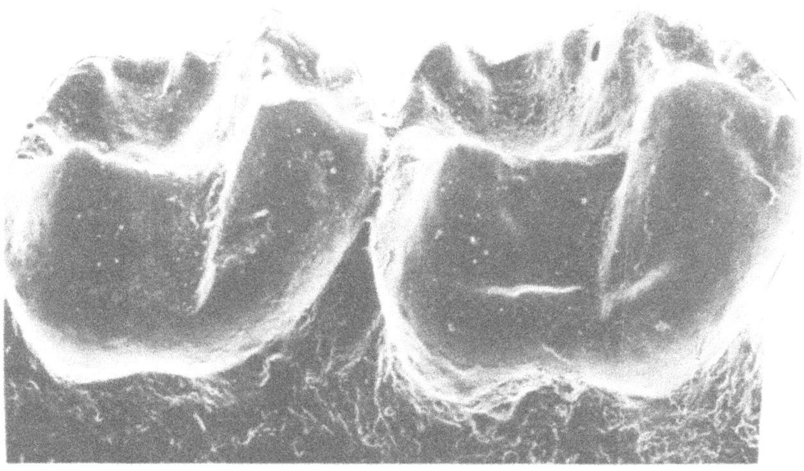

Fig. 9. Lower first and second molar of *Plesiopithecus teras* (CGM 42291) in oblique buccal–occlusal view, showing the relatively high transverse crest formed between the protoconid and metaconid, the loss of a paraconid, and the low cusps and nubs circling the talonid rim. SEM micrograph by G. Michael Veith, Washington University.

The Origin of Anthropoids

The paleobiological evidence suggests that the anthropoid clade was established by late Eocene times, with divergence among platyrrhines, parapithecids, and catarrhines already accomplished. However, despite cladistic differentiation, many of the Eocene anthropoids were still somewhat prosimian-like. In body size, diet, and locomotion the earliest anthropoids were probably similar to Eocene prosimian species found among the omomyines of North America and the cercamoniines of Eurasia (Covert, 1986; Franzen, Chapter 4, this volume, 1994). Taken together, these new analyses suggest that the origin of the anthropoid clade did not involve fundamental changes in diet, sensory adaptations, or body size. Like many of the Eocene prosimians on other continents, the Eocene anthropoids were probably small-bodied, small-brained, diurnal, quadrupedal, and leaping, fruit- and insect-eating primates, possibly depending heavily on olfaction, that lived in warm, moist tropical forests (the Fayum's environment as inferred by Bown *et al.*, 1982; Olson and Rasmussen, 1986). Note that many of these paleobiological inferences are based only on analysis of oligopithecines; greater adaptive diversity surely awaits discovery among the parapithecoids.

The origin of the anthropoid clade cannot be viewed as a major, pervasive evolutionary revolution associated with climatic and floral changes taking place about synchronously with an Eocene–Oligocene boundary climatic

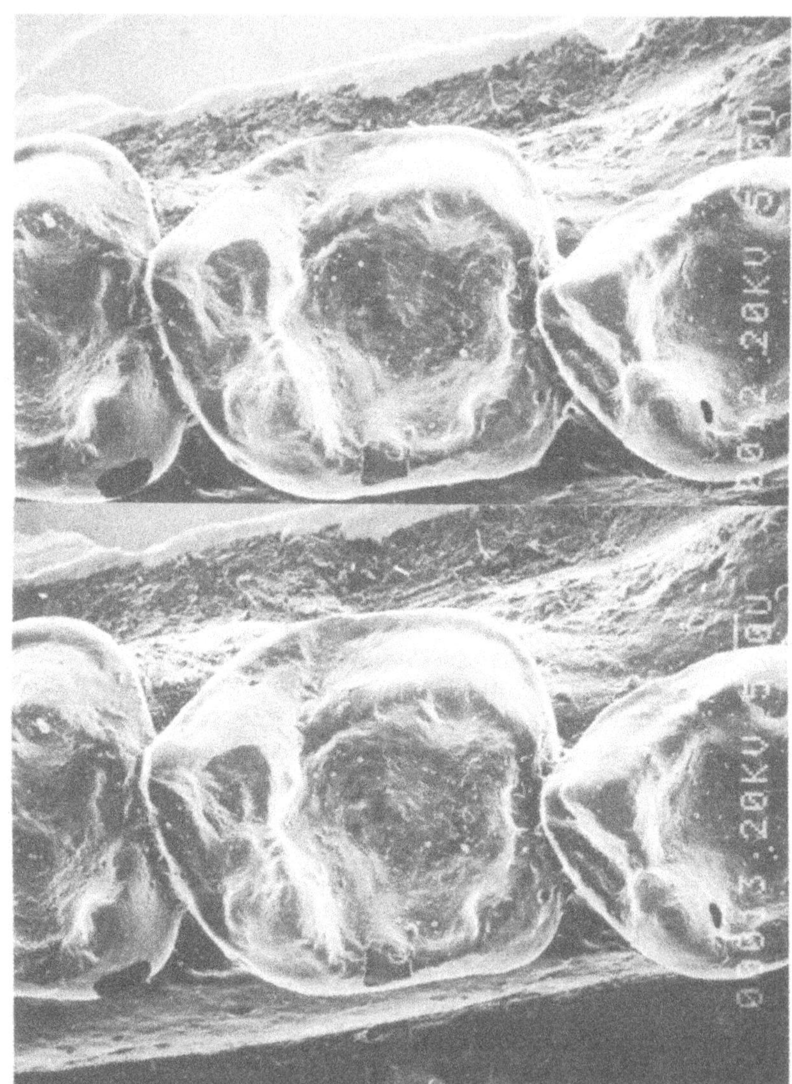

Fig. 10. Stereo pair of the lower second molar of *Plesiopithecus teras* (CGM 42291) in direct occlusal view, showing the broad, shallow, bowl-like talonid and the dominant protoconid and metaconid. SEM micrograph by G. Michael Veith, Washington University.

event (Prothero and Berggren, 1992). The assemblage of anthropoid attributes observed among extant species, including relatively flat-crowned molar relief, large brains, and reduced olfaction, did not characterize the earliest representatives of the suborder and did not evolve together as part of a single adaptive shift or evolutionary jump. Rather, anthropoid characteristics accumulated piecemeal over millions of years. The evolutionary transition between primate suborders could easily have resulted from known microevolutionary processes rather than a single or a few dubious macroevolutionary jumps. The very recognition and easy definition of Anthropoidea as a taxon were consequences of the extinction of the lineages now being discovered in the African Eocene.

A corollary of the above observations is that early anthropoids found in the fossil record will be increasingly difficult to identify with certainty in the absence of excellent cranial material. This will be especially true if any Eocene anthropoids occurred in Eurasia as well as Africa (Rasmussen, Chapter 12, this volume), where the geographic position of any dental specimens found would tend to bias investigators to interpret them as cercamoniines. Although the potential difficulties in distinguishing early anthropoids from Eocene prosimians may be disconcerting, it is a necessary consequence of the immense improvements that have been made in the primate fossil record during the last decades. The confidence that systematists have in delimiting higher taxa is inversely proportional to the density and resolution of the fossil record. Because it is clear that all anthropoid synapomorphies did not evolve simultaneously, systematists will have to be satisfied with arbitrarily picking one or a few traits of shared derivation to define the suborder. Ultimately, the evolutionary transition between primate suborders will be narrowed down to a transition between genera.

It would not be possible to infer the details of this early anthropoid radiation directly from the comparative study of extant or late Tertiary anthropoids (Andrews, 1985; Harrison, 1987). Although neontological studies are important for many kinds of evolutionary questions, unraveling further details of the prosimian–anthropoid transition will depend more than ever on paleontological work in Africa and Eurasia, particularly as it leads to finding new well-preserved crania of intermediate taxa.

Acknowledgments

We thank Marc Godinot and two anonymous reviewers for their many useful comments on the manuscript, and Ken Rose for insights about enlarged anterior teeth in Paleogene primates. The specimens described here were prepared by E.L.S. and P.S.C. The research was supported by grants from the National Science Foundation (BNS 86-07392, BNS-88-09776, and BNS 91-08445). We thank the staff of the Egyptian Geological Survey and

Mining Authority and the Director and staff of the Cairo Geological Museum, Maadi, for support and assistance in Egypt. This is Duke Primate Center publication no. 567.

References

Andrews, P. 1985. Family group systematics and evolution among catarrhine primates. In: E. Delson (ed.), *Ancestors, the Hard Evidence,* pp. 14–22. Alan R. Liss, New York.

Atchley, W. R., Cowley, D. E., Vogl, C., and McLellan, T. 1992. Evolutionary divergence, shape change, and genetic correlation structure in the rodent mandible. *Syst. Biol.* **41:**196–221.

Bartecki, U., and Heymann, E. W. 1990. Field observations on scent-marking behaviour in saddle-backed tamarins, *Saguinus fuscicollis. J. Zool.* **220:**87–99.

Berggren, W. A., Kent, D. V., Obradovich, J. D., and Swisher, C. C. III. 1992. Toward a revised paleogene geochronology. In: W. A. Berggren and D. R. Prothero (eds.), *Eocene–Oligocene Climatic and Biotic Evolution,* pp. 29–45. Princeton University Press, Princeton.

Bonis, L. de, Jaeger, J.-J., Coiffait, B., and Coiffait, P.-E. 1988. Découverte du plus ancien primate Catarrhinien connu dans l'Éocène supérieur d'Afrique du Nord. *C.R. Acad. Sci. Paris [II]* **306:**929–934.

Bown, T. M., and Kraus, M. J. 1988. Geology and paleoenvironment of the Oligocene Jebel Qatrani Formation and adjacent rocks, Fayum Depression, Egypt. *U.S. Geol. Surv. Prof. Paper* **1452:**1–64.

Bown, T. M., Kraus, M. J., Wing, S. L., Fleagle, J. G., Tiffney, B. H., Simons, E. L., and Vondra, C. F. 1982. The Fayum primate forest revisited. *J. Hum. Evol.* **1:**603–632.

Cachel, S. M. 1979. A functional analysis of the primate mastication system and the origin of the anthropoid postorbital septum. *Am. J. Phys. Anthropol.* **50:**1–18.

Cachel, S. M. 1981. Plate tectonics and the problem of anthropoid origins. *Yearb. Phys. Anthropol.* **24:**139–172.

Coiffait, P.-E., Coiffait, B., Jaeger, J.-J., and Mahboubi, M. 1984. Un nouveau gisement à mammifères fossiles d'âge Éocène superiéur sur le versant sud des Nementcha (Algérie orientale): découverte des plus anciens rongeurs d'Afrique. *C.R. Acad. Sci. Paris [II]* **299:**893–898.

Conroy, G. 1976. Primate postcranial remains from the Oligocene of Egypt. *Contrib. Primatol.* **8:** 1–134.

Conroy, G. 1987. Problems of body-weight estimation in fossil primates. *Int. J. Primatol.* **8:**115–137.

Court, N. 1993. An enigmatic new mammal from the Eocene of North Africa. *J. Vert. Paleontol.* **13:**267–269.

Court, N., and Hartenberger, J.-L. 1992. A new species of the hyracoid mammal *Titanohyrax* from the Eocene of Tunisia. *Palaeontology* **35:**309–317.

Covert, H. H. 1986. Biology of early Cenozoic primates. In: D. R. Swindler and J. Erwin (eds.), *Comparative Primate Biology, Vol. 1, Systematics, Evolution, and Anatomy,* pp. 335–359. Alan R. Liss, New York.

Crochet, J.-Y. 1986. *Kasserinotherium tunisiense* nov. gen., nov. sp., troisième marsupial découverte en Afrique. *C.R. Acad. Sci. Paris [II]* **302:**923–926.

Delson, E., and Rosenberger, A. L. 1980. Phyletic perspectives on platyrrhine origins and anthropoid relationships. In: R. L. Ciochon and A. B. Chiarelli (eds.), *Evolutionary Biology of New World Monkeys and Continental Drift,* pp. 445–458. Plenum Press, New York.

Emry, R. J., and Thorington, R. W., Jr. 1984. The tree squirrel *Sciurus* (Sciuridae, Rodentia) as a living fossil. In: N. Eldredge and S. M. Stanley (eds.), *Living Fossils,* pp. 23–31. Springer-Verlag, Berlin.

Epple, G. 1985. The primates, I: Order Anthropoidea. In: R. E. Brown and D. W. MacDonald (eds.), *Social Odours in Mammals, Vol. 2,* pp. 739–769. Clarendon Press, Oxford.

Epple, G. 1986. Communication by chemical signals. In: G. Mitchell and J. Erwin (eds.), *Comparative Primate Biology: Behavior, Conservation and Ecology*, pp. 531–580. Alan R. Liss, New York.

Fleagle, J. G. 1980. Locomotor behavior of the earliest anthropoids: A review of the current evidence. *Z. Morphol. Anthropol.* **71:**149–156.

Fleagle, J. G., and Kay, R. F. 1985. The paleobiology of catarrhines. In: E. Delson (ed.). *Ancestors: The Hard Evidence*, pp. 23–36. Alan R. Liss, New York.

Fleagle, J. G., and Kay, R. F. 1987. The phyletic position of the Parapithecidae. *J. Hum. Evol.* **16:**483–532.

Fleagle, J. G., and Simons, E. L. 1982a. The humerus of *Aegyptopithecus zeuxis:* A primitive anthropoid. *Am. J. Phys. Anthropol.* **59:**175–193.

Fleagle, J. G., and Simons, E. L. 1982b. Skeletal remains of *Propliopithecus chirobates* from the Egyptian Oligocene. *Fol. Primatol.* **39:**161–177.

Fleagle, J. G., Kay, R. F., and Simons, E. L. 1980. Sexual dimorphism in early anthropoids. *Nature* **287:**328–330.

Fleagle, J. G., Bown, T. M., Obradovich, J., and Simons, E. L. 1986. Age of the earliest African anthropoids. *Science* **234:**1247–1249.

Gebo, D. L., and Simons, E. L. 1987. Morphology and locomotor adaptations of the foot in early Oligocene anthropoids. *Am. J. Phys. Anthropol.* **74:**83–101.

Gevin, P., Feist, M., and Mongereau, N. 1974. Découverte de charophytes au Gliv Zegdou (frontiere algero-marocaine). *Bull. Soc. Hist. Nat. Afr. Nord. Alger.* **65:**371–375.

Gingerich, P. D. 1977. New species of Eocene primates and the phylogeny of European Adapidae. *Fol. Primatol.* **28:**60–80.

Gingerich, P. D. 1993. Oligocene age of the Gebel Qatrani Formation, Fayum, Egypt. *J. Hum. Evol.* **24:**207–218.

Gingerich, P. D., Smith, H. S., and Rosenberg, K. 1982. Allometric scaling in the dentition of primates and prediction of body weight from tooth size in fossils. *Am. J. Phys. Anthropol.* **58:**81–100.

Godinot, M. 1988. Les primates adapidés de Bouxwiller (Eocène Moyen, Alsace) et leur apport à la compréhension de la faune de Messel et à l'évolution des Anchomomyini. *Cour. Forsch. Inst. Senckenberg* **107:**383–407.

Godinot, M., and Mahboubi, M. 1992. Earliest known simian primate found in Algeria. *Nature* **357:**324–326.

Harrison, T. 1987. The phylogenetic relationships of the early catarrhine primates: A review of the current evidence. *J. Hum. Evol.* **16:**41–80.

Hartenberger, J.-L. 1986. Hypothèse paléontologique sur l'origine des Macroscelidea (Mammalia). *C.R. Acad. Sci., Paris [II]* **302:**247–249.

Hartenberger, J.-L., and Marandat, B. 1992. A new genus and species of an early Eocene primate from North Africa. *Hum. Evol.* **7:**9–16.

Hartenberger, J.-L., Martinez, C., and Ben Said, A. 1985. Découverte de mammifères d'âge Éocène inférieur en Tunisie centrale. *C.R. Acad. Sci. Paris [II]* **301:**649–652.

Kappelman, J., Simons, E. L., and Swisher, C. C. III. 1992. New age determinations for the Eocene–Oligocene boundary sediments in the Fayum depression, northern Egypt. *J. Geol.* **100:**647–667.

Kay, R. F. 1977. The evolution of molar occlusion in the Cercopithecidae and early catarrhines. *Am. J. Phys. Anthropol* **46:**327–352.

Kay, R. F. 1984. On the use of anatomical features to infer foraging behavior in extinct primates. In: P. S. Rodman and J. G. H. Cant, (eds.), *Adaptations for Foraging in Nonhuman Primates*, pp. 21–53. Columbia University Press, New York.

Kay, R. F., and Cartmill, M. 1977. Cranial morphology and adaptations of *Palaechthon nacimienti* and other Paromomyidae (Plesiadapoidea, ?Primates), with a description of a new genus and species. *J. Hum. Evol.* **6:**19–35.

Kay, R. F., and Simons, E. L. 1980. The ecology of Oligocene African Anthropoidea. *Int. J. Primatol.* **1:**21–37.

Kay, R. F., and Simons, E. L. 1983. Dental formulae and dental eruption patterns in Parapithecidae. *Am. J. Phys. Anthropol.* **63:**353–375.

Kay, R. F., Fleagle, J. G., and Simons, E. L. 1981. A revision of the Oligocene apes from the Fayum Province, Egypt. *Am. J. Phys. Anthropol.* **55:**293–322.

Maas, M. C., Kraus, D. W., and Strait, S. G. 1988. Decline and extinction of Plesiadapiformes (?Primates: Mammalia) in North America: Displacement or replacement? *Paleobiology* **14:** 23–29.

Maier, W. 1984. Tooth morphology and dietary specialization. In: D. J. Chivers, B. A. Wood, and A. Bilsborough (eds.), *Food Acquisition and Processing in Primates,* pp. 303–330. Plenum Press, London.

Odin, G. S., and Montanari, A. 1989. Age radiometrique et stratotype de la limite Eocene–Oligocene. *C.R. Acad. Sci. Paris [II]* **309:**1939–1945.

Olson, S. L., and Rasmussen, D. T. 1986. Paleoenvironment of the earliest hominoids: New evidence from the Oligocene avifauna of Egypt. *Science* **233:**1202–1204.

Pickford, M. 1986. Première découverte d'une faune mammalienne terrestre paléogène d'Afrique sub-saharienne. *C.R. Acad. Sci Paris [II]* **302:**1205–1210.

Prothero, D. R., and Berggren, W. A. (eds.). 1992. *Eocene–Oligocene Climatic and Biotic Evolution.* Princeton University Press, Princeton.

Radinsky, L. 1973. *Aegyptopithecus* endocasts: Oldest record of a pongid brain. *Am. J. Phys. Anthropol.* **39:**239–248.

Radinsky, L. 1974. The fossil evidence of anthropoid brain evolution. *Am. J. Phys. Anthropol.* **41:**15–27.

Rasmussen, D. T. 1989. The evolution of the Hyracoidea: A review of the fossil evidence. In: D. R. Prothero and R. M. Schoch (eds.), *The Evolution of Perissodactyls,* pp. 57–78. Oxford University Press, New York.

Rasmussen, D. T., and Simons, E. L. 1988. New specimens of *Oligopithecus savagei,* early Oligocene primate from Egypt. *Fol. Primatol.* **51:**182–208.

Rasmussen, D. T., and Simons, E. L. 1992. Paleobiology of the oligopithecines, the world's earliest known anthropoid primates. *Int. J. Primatol.* **13:**477–508.

Rasmussen, D. T., Bown, T. M., and Simons, E. L. 1992. The Eocene–Oligocene transition in continental Africa. In: D. R. Prothero and W. A. Berggren (eds.), *Eocene–Oligocene Climatic and Biotic Evolution,* pp. 548–566. Princeton University Press, Princeton.

Sassi, S., Trait, G., Truc, G., and Millot, G. 1984. Découverte de l'Éocène continental en Tunisie centrale: La formation du Jebel Chambi et ses encroûtements carbonates. *C.R. Acad. Sci. Paris [II]* **299:**357–364.

Schmidt-Kittler, N. (ed.). 19987. European reference levels and correlation tables. *Munch. Geowiss. Abh. [A]* **10:**15–31.

Sigé, B. 1985. Chiroptères de l'Éocène inférieur de Tunisie. In: *110ᵉ Congrès de la Société des Savantes, Montpellier,* p. 307.

Sigé, B., Jaeger, J.-J., Sudre, J., and Vianey-Liaud, M. 1990. *Altiatlasius koulchii* n.gen.et sp., primate omomyidé du Paléocène supérieur du Maroc, et les origines des Euprimates. *Palaeontographica [A]* **214:**31–56.

Simons, E. L. 1962. Two new primate species from the African Oligocene. *Postilla* **64:**1–12.

Simons, E. L. 1965. New fossil apes from Egypt and the initial differentiation of Hominoidea. *Nature* **205:**135–139.

Simons, E. L. 1967. The earliest apes. *Sci. Am.* **217:**28–35.

Simons, E. L. 1968. Early Cenozoic mammalian faunas, Fayum Province, Egypt. Part I. African Oligocene mammals: Introduction, history of study, and faunal succession. *Bull. Peabody Mus. Nat. Hist. Yale Univ.* **28:**1–21.

Simons, E. L. 1971. Relationships of *Amphipithecus* and *Oligopithecus. Nature* **232:**489–491.

Simons, E. L. 1974. *Parapithecus grangeri* (Parapithecidae, Old World Higher Primates): New species from the Oligocene of Egypt and the initial differentiation of Cercopithecoidea. *Postilla* **166:**1–12.

Simons, E. L. 1983. Recent advances in knowledge of the earliest catarrhines of the Egyptian Oligocene (including the most ancient known presumed ancestors of man). *Pont. Acad. Sci. Scripta Varia* **50:**11–27.

Simons, E. L. 1986. *Parapithecus grangeri* of the African Oligocene: An archaic catarrhine without lower incisors. *J. Hum. Evol.* **15:**205–213.

Simons, E. L. 1987. New faces of *Aegyptopithecus* from the Oligocene of Egypt. *J. Hum. Evol.* **16:**273–289.

Simons, E. L. 1989. Description of two genera and species of late Eocene Anthropoidea from Egypt. *Proc. Natl. Acad. Sci. USA* **86:**9956–9960.

Simons, E. L. 1990. Discovery of the oldest known anthropoidean skull from the Paleogene of Egypt. *Science* **247:**1507–1509.

Simons, E. L. 1992. Diversity in the early Tertiary anthropoidean radiation in Africa. *Proc. Natl. Acad. Sci. USA* **89:**10743–10747.

Simons, E. L., and Kay, R. F. 1983. *Qatrania,* new basal anthropoid primate from the Fayum, Oligocene of Egypt. *Nature* **304:**624–626.

Simons, E. L., and Kay, R. F. 1988. New material of *Qatrania* from Egypt with comments on the phylogenetic position of the Parapithecidae (Primates, Anthropoidea). *Am. J. Primatol.* **15:**337–347.

Simons, E. L., and Rasmussen, D. T. 1991. The generic classification of Fayum Anthropoidea. *Int. J. Primatol.* **12:**163–178.

Stehlin, H. G. 1916. Die Säugetiere des schweizerischen Eozäns. Critischer Catalog der Materialien, part 7, second half. *Abh. Schweiz. Pal. Ges.* **41:**1299–1552.

Sudre, J. 1979. Nouveaux mammifères éocènes du Sahara occidental. *Palaeovertebrata* **9:**83–115.

Sudre, J., de Bonis, L., Brunet, M., Crochet, J.-Y., Duranthon, F., Godinot, M., Hartenberger, J.-L., Jehenne, Y., Legendre, S., Marandat, B., Remy, J. A., Ringeade, M., Sigé, B., and Vianey-Liaud, M. 1992. La biochronologie mammalienne du Paléogène au Nord et au Sud des Pyrénées: État de la question. *C.R. Acad. Sci. Paris [II]* **314:**631–636.

Swisher, C. C. III, and Prothero, D. R. 1990. Single-crystal $^{40}Ar/^{39}Ar$ dating of the Eocene–Oligocene transition in North America. *Science* **249:**760–762.

Szalay, F. S. 1970. Late Eocene *Amphipithecus* and the origins of catarrhine primates. *Nature* **227:**355–357.

Szalay, F. S. 1976. Systematics of the Omomyidae (Tarsiiformes, Primates): Taxonomy, phylogeny and adaptations. *Bull. Am. Mus. Nat. Hist.* **156:**157–450.

Szalay, F. S., and Li, C. K. 1986. Middle Paleocene euprimate from southern China and the distribution of primates in the Paleogene. *J. Hum. Evol.* **15:**387–397.

Thomas, H., Roger, J., Sen, S., and Al-Sulaimani, Z. 1988. Découverte des plus anciens "Anthropoïdes" du continent arabo-africain et d'un primate tarsiiforme dans l'Oligocène du Sultanat d'Oman. *C.R. Acad. Sci. Paris [II]* **306:**823–829.

Thomas, H., Roger, J., Sen, S., Bourdillon-de-Grissac, C., and Al-Sulaimani, Z. 1989. Découverte de vertébrés fossiles dans l'Oligocène inférieur du Dhofar (Sultanat d'Oman). *Geobios* **22:** 101–120.

Thomas, H., Sen, S., Roger, J., and Al-Sulaimani, Z. 1991. The discovery of *Moeripithecus markgrafi* Schlosser (Propliopithecidae, Anthropoidea, Primates), in the Ashawq Formation (early Oligocene of Dhofar Province, Sultanate of Oman). *J. Hum. Evol.* **20:**33–49.

Van Couvering, J. A., and Harris, J. A. 1991. Late Eocene age of Fayum mammal faunas. *J. Hum. Evol.* **21:**241–260.

Wood, A. E. 1968. Early Cenozoic mammalian faunas, Fayum Province, Egypt, Part II: The African Oligocene Rodentia. *Bull. Peabody Mus. Nat. Hist.* **28:**23–105.

Eocene Anthropoid Postcrania from the Fayum, Egypt

9

DANIEL L. GEBO, ELWYN L. SIMONS, D. TAB RASMUSSEN, and MARIAN DAGOSTO

Introduction

The anthropoids from the Fayum Province, Egypt, represent the best-preserved and most diverse assemblage of early anthropoids found anywhere in the world. However, the best known of the Fayum primates—*Aegyptopithecus zeuxis, Propliopithecus chirobates, Apidium phiomense,* and *Parapithecus grangeri*—occur near the top of the 340m-deep Jebel Qatrani Formation (Simons, 1965, 1967, 1974, 1987; Kay, 1977; Fleagle and Simons, 1978, 1982a,b; Gebo and Simons, 1987; Kay *et al.,* 1981; Fleagle and Kay, 1983, 1985, 1987). Fossil primates millions of years older have been found at lower stratigraphic levels, and these older species differ appreciably from the primates of the uppermost quarries, reflecting significant evolutionary changes that took place dur-

DANIEL L. GEBO • Department of Anthropology, Northern Illinois University, DeKalb, Illinois 60115. ELWYN L. SIMONS • Duke Primate Center, Duke University, Durham, North Carolina 27705. D. TAB RASMUSSEN • Department of Anthropology, Washington University, St. Louis, Missouri 63130. MARIAN DAGOSTO • Department of Cell and Molecular Biology, Northwestern University Medical School, Chicago, Illinois 60611, and Department of Mammalogy, American Museum of Natural History, New York, New York 10024.

Anthropoid Origins, edited by John G. Fleagle and Richard F. Kay. Plenum Press, New York, 1994.

ing the time represented by the deposition of the Jebel Qatrani Formation (Simons, 1962; Simons and Kay, 1983; Rasmussen and Simons, 1988). The oldest primates yet found in the Fayum have been described recently on the basis of dental and cranial specimens found at quarry L-41, which lies near the base of the Jebel Qatrani Formation (Simons, 1989, 1990, 1992; Rasmussen and Simons, 1992). These L-41 primates are among the most ancient and structurally most primitive anthropoids found anywhere in the world.

Quarry L-41 consists of fine-grained green to greenish-gray mudstones with abundant evaporites (Simons, 1989; Rasmussen and Simons, 1991). Fossils occur in incredible concentrations here, with fossil hyracoids and fish making up the bulk of the material. Unfortunately, primates are relatively rare in the deposit, and the specimens are often badly crushed and distorted as a consequence of diagenetic processes (Rasmussen and Simons, 1991). Quarry L-41 lies just 47 m above the base of the Jebel Qatrani Formation, and just below a major unconformity that separates it from all younger primate-bearing levels (Bown and Kraus, 1988). The next oldest Fayum primates are *Oligopithecus savagei, Qatrania wingi,* and an unnamed omomyid that occur at quarry E, 50 m and two erosional unconformities above quarry L-41 (Simons and Kay, 1983); Rasmussen and Simons, 1988; Simons *et al.,* 1987). The oldest primate postcrania that have been described come from the Fayum's uppermost levels, 200 m, five unconformities, and probably 2–3 million years above quarry L-41 (Preuschoft, 1974; Fleagle *et al.,* 1975; Conroy, 1976a,b; Fleagle and Simons, 1978, 1979, 1982a,b, 1983, 1985; Schön Ybarra and Conroy, 1978; Anapol, 1983; Gebo and Simons, 1987).

The age of quarry L-41 is late Eocene, based on paleomagnetic correlations (Kappelman *et al.,* 1992), on faunal correlations with other Afro-Arabian sites (Rasmussen *et al.,* 1992), and on geological interpretations of unconformities that may correspond to the marine regressive event coincident with the Eocene–Oligocene boundary (Plaziat, 1981; Bown and Kraus, 1988; Rasmussen *et al.,* 1992). One recent paper resurrected the idea held by researchers early in the century that the entire Jebel Qatrani Formation is Eocene (Van Couvering and Harris, 1991), but this conclusion is based on the premise that there are no unconformities below the top of the section, although, in fact, there are at least 12 (Bown and Kraus, 1988; Rasmussen *et al.,* 1992). In contrast, Gingerich (1993) concluded that the entire formation is Oligocene; Gingerich emphasized a single unconformity at the base of the formation at the expense of other major unconformities higher in the section, and he rejected what Kappelman *et al.* (1992) preferred as correlations of the paleomagnetic reversal scale. Kappelman *et al.* (1992) preferred to place Chron C13r (containing the Eocene–Oligocene boundary, 34 MA) near the middle of the Jebel Qatrani Formation, in agreement with the results of Rasmussen *et al.* (1992), although Kappelman *et al.* also discussed three alternative correlations. Kappelman *et al.* (1992) further concluded that quarry L-41 lies within Chron C15r (35.6–35.9 MA), making it Priabonian or late Eocene in age.

The anthropoid species from quarry L-41 are of great interest for helping to establish primitive anthropoid morphology, for reconstructing the behavioral adaptations of early anthropoids, and for contributing to our understanding of the many unsolved phylogenetic issues pertaining to anthropoid origins and the early anthropoid radiations. Despite the growing number of putative Eocene and early Oligocene anthropoids now known by teeth from sites in Burma (Ciochon *et al.*, 1985), Oman (Thomas *et al.*, 1988, 1989), and Algeria (de Bonis *et al.*, 1988; Godinot and Mahboubi, 1991, 1992) in addition to those from the Fayum, the limb elements described here are the first Eocene anthropoid postcrania yet described.

Material

Five anthropoid postcranial specimens have been found at quarry L-41, but the allocation of these to specific taxa is not easy. *C. browni* is the most common of the primates from L-41; the size of its dentition resembles that of *Saimiri* or large *Callimico* (Rasmussen and Simons, 1992). The remaining four species fit into two size categories. *Plesiopithecus teras* and *Serapia eocaena* are comparable in size to *C. browni* (the holotype of *Serapia* is slightly smaller than known specimens of *Catopithecus* and *Plesiopithecus*). *Proteopithecus sylviae* and *Arsinoea kallimos* resemble each other in size but are distinctly smaller than the other three L-41 taxa, being within the size range of tamarins (*Saguinus* and *Leontopithecus*). Apart from size, taxonomic affiliation may provide some clues for allocation of postcrania. Among the group of three larger species, *Catopithecus* is an oligopithecine, and *Serapia* is a parapithecid. *Plesiopithecus* is extremely unusual in its dental structure (Simons, 1992), and what its postcranium looked like is anyone's guess; it may be "parapithecoid" if the absence of adult lower incisors is homologous to the condition in *Parapithecus grangeri*. Among the two smaller taxa, both *Proteopithecus* and *Arsinoea* may be "parapithecoid"; they retain three premolars but defy simple placement in either Parapithecidae or Propliopithecidae because *Proteopithecus* has very generalized molar structure and *Arsinoea* is specialized. *Proteopithecus* is the second most common species at L-41, represented by several upper and lower jaws; therefore, it is probable that tamarin-sized elements recovered from L-41 would belong to it rather than to *Arsinoea*.

Given the above information on L-41 primates, the largest, squirrel-monkey-sized postcranial fragments are here allocated to *C. browni* based on size and on apparent propliopithecid structural affinities that will be described and discussed below. These specimens are: DPC 7328, a distal fragment minus the medial part of a right humerus; DPC 8204, a distal, medial fragment of a left humerus; and DPC 8256, the proximal end of a right femur. *Serapia* is tentatively ruled out because it is rare and a parapithecid; *Plesiopithecus* is tentatively ruled out because it is rare and because there is no reason to expect propliopithecid features in this odd primate. Smaller specimens comparable in size to the corresponding elements of tamarins most likely belong to *Proteopithecus,* but it is possible that they belong to rare *Arsinoea.* The small postcranial specimens are DPC 7529, the proximal end of a right femur, and DPC 9278, a para-acetabular fragment of a right innominate. The size of the femoral head of DPC 7529 fits well in the acetabulum of DPC 9278, suggesting that they represent a single species. All allocations must remain tentative, however, until the samples of dentitions and postcrania from L-41 are much larger than they are currently; for now, the larger specimens will be referred to *Catopithecus browni,* the smaller specimens to *Proteopithecus sylviae.*

Table I lists measurements taken on the L-41 postcranial specimens following Schultz (1969), Szalay and Dagosto (1980), and Dagosto and Schmid (1994).

Descriptions and Comparisons

Distolateral Fragment of Humerus, DPC 8204

This fragmentary specimen allocated to *C. browni* consists of the distal part of the shaft, the capitulum, the radial fossa, and the lateral epicondylar region; parts that are missing are the trochlea and medial epicondyle (Fig. 1). The full diameter of the shaft is preserved for a distance of about 1 cm proximal to the entepicondylar foramen. DPC 8204 preserves a large brachial flange, which is similar in relative size to those of *Apidium* and the upper-sequence propliopithecids (Fig. 2) and to living and extinct prosimians. As in *Apidium, Aegyptopithecus,* and *Propliopithecus,* an entepicondylar foramen is retained (Conroy, 1976a; Fleagle and Simons, 1982a,b; Fleagle and Kay, 1987; Senut, 1989). This foramen has been lost independently in several different lineages of platyrrhines (Ford, 1986; Meldrum *et al.,* 1990; Gebo, 1993) and in living catarrhines (Fleagle, 1983; Fleagle and Simons, 1983; Harrison, 1987).

The capitulum is very round and possesses a long and prominent capitular tail (the length is 46–57% of the estimated width of the capitulum, Table II). Long capitular tails are often found in prosimian primates but are only moderately developed in New and Old World monkeys and in the higher-level Fayum anthropoids (Table II, Fig. 3). There is a sharp demarcation defined by a groove between the start of the capitular tail and the lateral edge of the capitulum. The sharply demarcated capitular tail and the very rounded capitulum in DPC 8204 are morphologically similar to humeri attributed to *Microchoerus* and *Necrolemur.* This sharp demarcation differs from the more shallow separation seen in *Apidium* and upper-level propliopithecids, which also differ from the L-41 specimen in possessing a more mediolaterally elongated capitulum and significantly shorter capitular tails (Fig. 2).

Table I. Measurements of L-41 Postcranial Elements[a]

	DPC 8256	DPC 7529
Femora		
1. Femoral head width	4.6	3.8
2. Femoral head height	5.7	4.3
3. Length of the lesser trochanter	8.2	5.9
4. Length of femoral head and neck	8.2	6.2
5. Angle of femoral head and neck	58°	56°
6. Angle of lesser trochanter	48°	40°
Humeri	DPC 8204	DPC 7328
7. Width of capitulum (mediolat.)	2.8[b]	2.7[b]
8. Height of capitulum (proximodist.)	3.2	3.4
9. Width of anterior trochlea	—	4.3
10. Maximum height of anterior trochlea	—	4.4
11. Width of posterior trochlea	—	6.1
12. Posterior height midtrochlea	—	2.9
13. Length of medial epicondyle	—	3.6
14. Capitular tail length	1.6	—
15. Capitular width with zona conoidea	—	3.5[b]
16. Articular width	—	7.6[b]
17. Maximum trochlear height	—	4.4
18. Trochlear breadth	—	2.8
19. Capitular breadth	—	3.2[b]
Innominate	DPC 9278	
20. Diameter of the acetabulum	6.2	
21. Depth of the acetabulum	3.6	
22. Width of the incisura acetabuli	2.1	
23. Dorsal breadth of the acetabulum	6.3	
24. Medial breadth of the acetabulum	7.4	
25. Ventral breadth of the acetabulum	4.9	
26. Length of the ilium	23.1[b]	
27. Maximum width of gluteal plane	8.8	
28. Maximum width of distal iliac plane	4.7	
29. Maximum width of proximal iliac plane	2.3	

[a]Measurements 15 to 19 are from Szalay and Dagosto (1980), and measurements 20 to 26 are from Schultz (1969); all lengths are in millimeters.
[b]Estimated values.

Above the capitulum, the radial fossa is moderately deep, as it is in *Apidium*, propliopithecids, and many other quadrupedal primates (Conroy, 1976a; Szalay and Dagosto, 1980; Meldrum *et al.*, 1990). The lateral epicondylar region, in terms of curvature and width, appears to be very similar in shape to this same region in humeri attributed to *Apidium phiomense* and upper-level propliopithecids (Fig. 2).

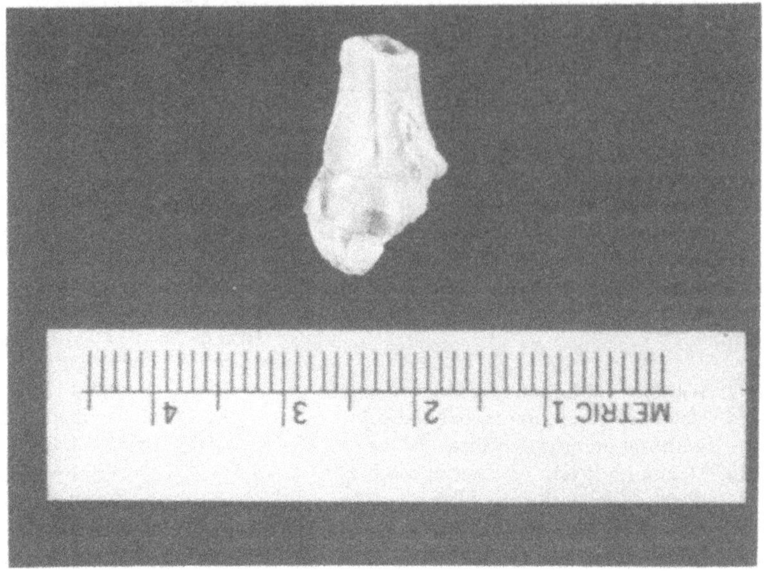

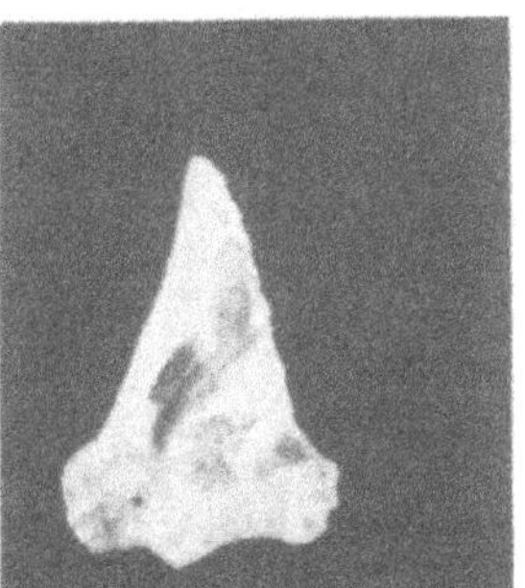

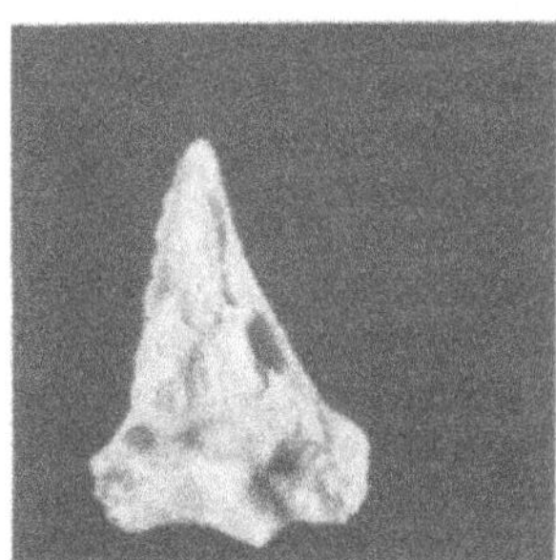

Fig. 1. Anthropoid distal humeri from quarry L-41: DPC 8204 in anterior view (top) and DPC 7328 (below) in anterior (left) and posterior (right) views.

Overall, DPC 8204 is similar to a wide variety of primates. There is no assemblage of derived features that DPC 8204 shares with any one particular primate taxon. It differs from the humeri of *Apidium* and the upper-level propliopithecids in its rounder shape of the capitulum, its demarcation between the capitulum and the capitular tail, and in the greater length of the capitular tail.

Distomedial Fragment of Humerus, DPC 7328

This specimen is very fragmentary, retaining only the trochlea, a portion of the capitulum, the medial epicondyle, the entepicondylar foramen, and a short segment of the medial shaft (Fig. 1). It is approximately the same size as

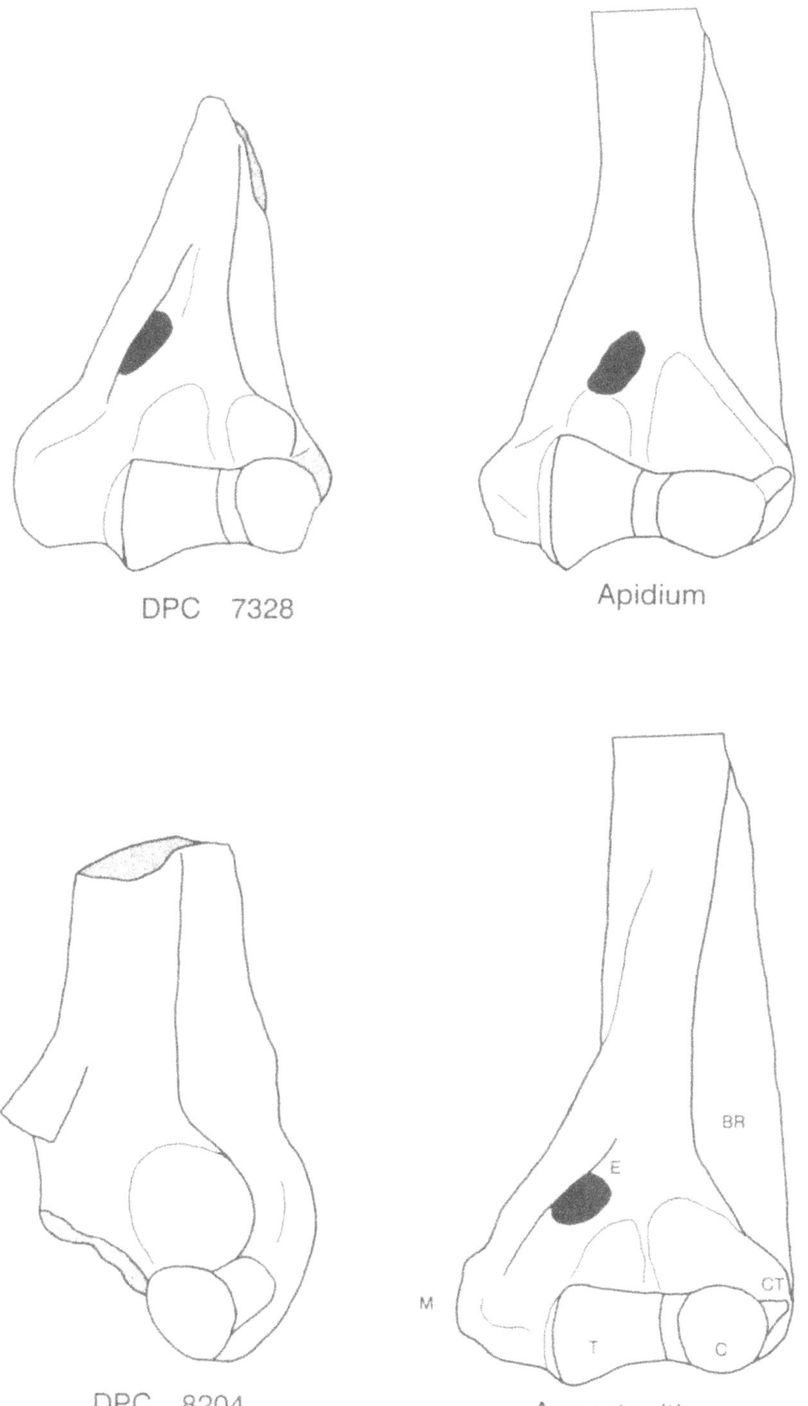

Fig. 2. Anterior views of distal humeri: DPC 7328 and DPC 8204 (reversed), *Catopithecus browni; Apidium phiomense* (DPC 1008, reversed); and *Aegyptopithecus zeuxis* (DPC 1275). M, medial epicondyle; T, trochlea; C, capitulum; CT, capitular tail; E, entepicondylar foramen; BR, brachial flange.

Table II. Relative Size of the Capitular Tail: Capitular Tail Length/Mediolateral Width of the Capitulum

Taxon	*n*	Ratio
Eocene primates		
Cantius trigonodus	1	44%
Smilodectes gracilis	1	44%
Notharctus tenebrosus	1	53%
Adapis parisiensis	1	36%
Microchoerus sp.	1	47%
Necrolemur antiquus	1	43%
Extant prosimians		
Tarsius syrichta	1	50%
Tarsius bancanus	2	44%
Cheirogaleus medius	1	49%
Cheirogaleus major	2	43%
Galago crassicaudatus	1	49%
Galago senegalensis	5	39%
Nycticebus coucang	4	37%
Perodicticus potto	2	39%
Lemur catta	4	40%
Lemur macaco	3	37%
Lemur rubriventer	1	28%
Varecia variegata	2	26%
Hapalemur griseus	1	33%
Avahi laniger	1	37%
Propithecus verreauxi	1	34%
Fayum anthropoids		
DPC 8204	1	46%[a]
Apidium phiomense	3	33%
Aegyptopithecus zeuxis	1	31%
Platyrrhines		
Cebupithecia sarmientoi	1	25%
Callimico goeldii	2	29%
Cebuella pygmaea	4	24%
Callithrix jacchus	5	34%
Saguinus leucopus	3	33%
Saguinus fuscus	2	29%
Leontopithecus rosalia	4	26%
Saimiri sciureus	5	35%
Cebus apella	5	33%
Aotus trivirgatus	5	34%
Callicebus torquatus	4	33%
Chiropotes satanas	5	25%
Pithecia pithecia	5	26%
Alouatta seniculus	5	28%
Lagothrix lagothricha	4	26%
Ateles fusciceps	2	23%
Extant catarrhines[b]		
Cercopithecus neglectus	4	31%

(*continued*)

Table II. (*Continued*)

Taxon	*n*	Ratio
Cercopithecus talapoin	2	29%
Cercopithecus aethiops	1	35%
Cercocebus torquatus	4	37%
Macaca sylvanus	2	38%
Macaca mulatta	2	36%
Macaca fascicularis	5	38%
Papio papio	5	35%
Colobus guereza	5	27%
Presbytis entellus	4	28%
Nasalis larvatus	5	21%

[a]The range for DPC 8204 is based on the estimated width of the capitulum (absolute length is 2.8 mm; estimated length is 3.5 mm).
[b]Extant hominoids lack capitular tails.

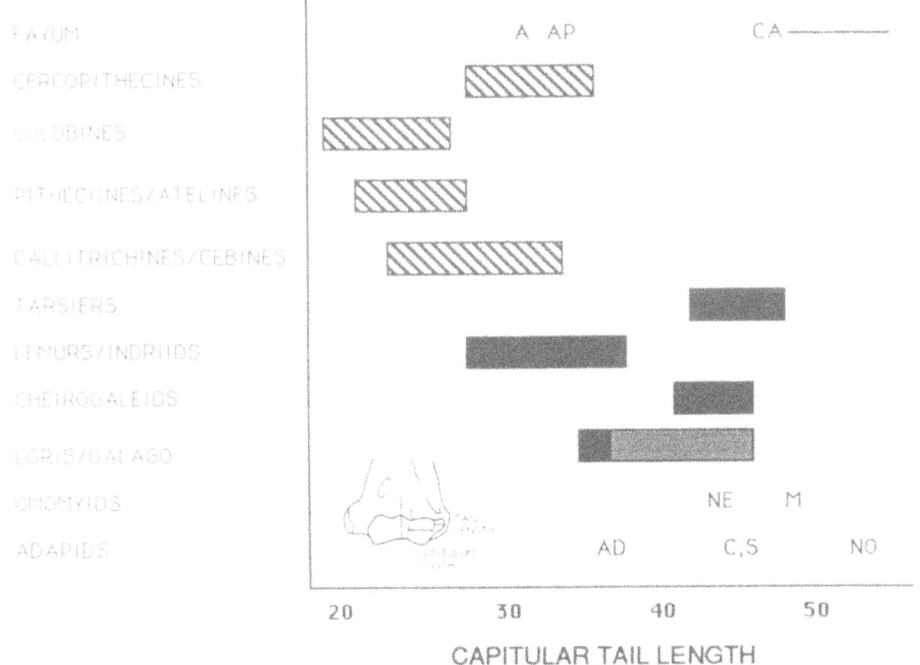

Fig. 3. Capitular tail length as a percentage of capitulum diameter. A, *Aegyptopithecus;* AP, *Apidium;* CA, *Catopithecus,* DPC 8204; NE, *Necrolemur;* M, *Microchoerus;* AD, *Adapis;* C, *Cantius;* S, *Smilodectes;* NO, *Notharctus.*

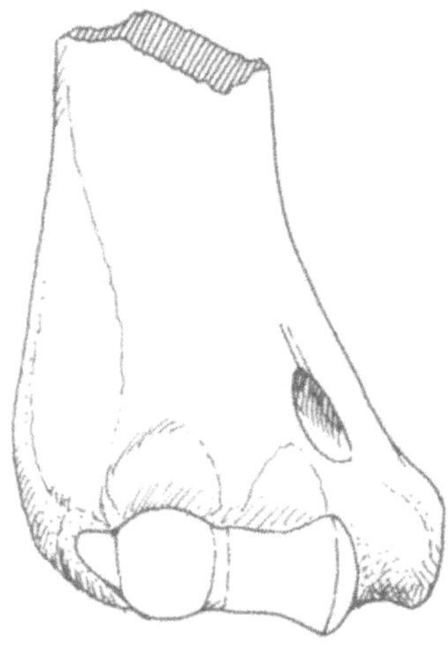

Fig. 4. Composite drawing of DPC 8204 and DPC 7328, prepared from micrographs of the two specimens, with the image of DPC 7328 reversed, and with the image of DPC 8204 (from a larger individual) reduced to 87% of original so that the distance between the proximal rim of the entepicondylar foramen and the medial border of the capitulum match in the two specimens.

DPC 8204, and both humeri are tentatively allocated to *Catopithecus browni*. A composite drawing made from DPC 8204 and DPC 7328 appears in Fig. 4.

The entepicondylar foramen present in this humerus is small compared to the size of this foramen in *Apidium* (Fig. 2). The entepicondylar foramen is placed far medially, with only a narrow span closing the foramen, unlike the more massive strut enclosing a more centrally placed foramen found in most living prosimians. The large medial epicondyle projects medially rather than curving posteriorly as in humeri of Fayum primates or platyrrhines (Conroy, 1976a; Szalay and Dagosto, 1980; Meldrum *et al.*, 1990).

DPC 7328 resembles the humerus of *Apidium* in the shape of the trochlea, and it resembles that of *Aegyptopithecus* in the relative sizes of the trochlea and capitulum (Fig. 2). In DPC 7328, as in most primates, the proximodistal height of the trochlea's medial edge is greater than the height of its lateral edge. However, the difference between the heights of the medial and lateral edges is greater than that found in most primates; the ratio of lateral to medial proximodistal height is 61% in DPC 7328, 60% in *Apidium,* and 72% in *Aegyptopithecus*. Thus, *Apidium* and DPC 7328 share a distinctly flared shape of the trochlea.

DPC 7328 differs from *Apidium* in the relative sizes of the trochlea and the capitulum. The ratio of capitulum size to trochlear size (capitulum mediolateral breadth/trochlea mediolateral breadth) is 63% in DPC 7328 (partially estimated), 74% in *Aegyptopithecus,* and 125% in *Apidium.* In the latter genus, the trochlea is very short, and the capitulum and the zona conoidea are both very wide (Fig. 2). The relatively large articular surface of the trochlea (especially anteriorly) provides for a large surface area for articulation with the ulna, a characteristic of anthropoids (Szalay and Dagosto, 1980).

DPC 7328 has a very slightly curved distal projection of the medial border of the trochlea, a feature partially related to the medial flare of the trochlea described above. This is a similarity to the Fayum's upper-sequence anthropoids and platyrrhines. In callitrichids (except *Leontopithecus*) the medial border of the trochlea does not extend distally beyond the rest of the trochlea. *Leontopithecus, Saimiri, Callicebus,* and *Aotus* have a slight distal curvature of the trochlea like DPC 7328, while *Cebus* has a much more prominent distal curve to its medial trochlea.

The posterior trochlear articular surface of DPC 7328 is fairly high proximodistally, as in anthropoids (Szalay and Dagosto, 1980). The oblique angle of the posterior medial trochlear rim is similar to the condition observed in *Apidium* and the upper-sequence propliopithecids. The depth of the olecranon fossa is difficult to determine because of *in situ* flattening in this area. Posteriorly, a dorsoepitrochlear fossa is present as in *Apidium* and propliopithecines, but this feature has a variable distribution among primate taxa (Conroy, 1976a; Fleagle and Simons, 1982a; Harrison, 1987; Fleagle and Kay, 1987; Ford, 1988; Dagosto, 1990).

DPC 7328 appears to represent a nontranslatory type of elbow (Rose, 1988), or it is only slightly translatory relative to living and extinct platyrrhines (Fig. 5). In catarrhines, especially terrestrial cercopithecids and living apes, there is little or no angle of translation for the ulna. In some cercopithecids, the angle of translation is slight (<10°), and this is the situation found in DPC 7328, *Apidium,* and *Aegyptopithecus* (Fig. 5). This slightly translatory or nontranslatory type of trochlea in catarrhine primates contrasts with the translatory trochlea of platyrrhines (Rose, 1988), which generally possess a much greater angle of translation (25–30°). Some individuals in the genera of *Callicebus, Aotus, Ateles,* and *Leontopithecus* were found to display a lower angle of translation, but no measured individual in any platyrrhine species displayed an angle of translation less than 15°. Accordingly, in terms of ulnar translation, DPC 7328, *Apidium,* and *Aegyptopithecus* all share a catarrhine-like trochlea.

Proximal End of Large Femur, DPC 8256

The larger of the two femoral fragments from quarry L-41 (DPC 8256) is broken at about the midpoint of the shaft (Fig. 6). It is of a size compatible with *Catopithecus browni.* Unfortunately, the proximal portion has suffered some crushing and consequent distortion of shape. The femoral head and neck are obliquely oriented (58° relative to the shaft; see Dagosto and Schmid, 1994), which is similar to the angle in *Apidium* (57°) (Fig. 7) and other anthropoids. DPC 8256 shows a shallow notch between the greater trochanter and the femoral head, as does the smaller L-41 femur described below (DPC 7529), but this feature is wider and flatter in *Apidium* (Fig. 7). Femoral head and neck length is relatively short (Fig. 8) compared to those of other anthropoids including *Apidium* (head and neck length/shaft width of DPC 8256, 1.5–1.7; the diameter of the shaft can only be partially estimated in DPC 8256;

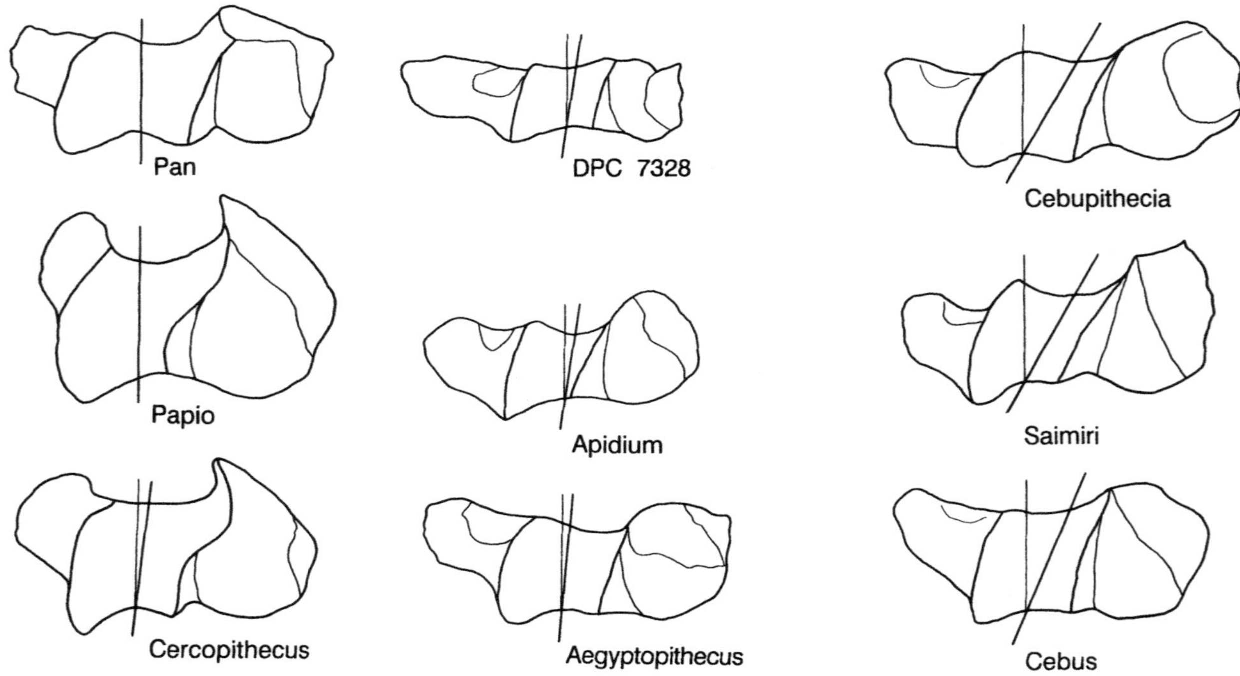

Fig. 5. Distal view of anthropoid humeri. The angle of ulnar translation is drawn for a variety of living and extinct anthropoids. Note that the angle of translation is low to zero for living catarrhines and for the Fayum anthropoids, whereas platyrrhines exhibit greater angles of ulnar translation.

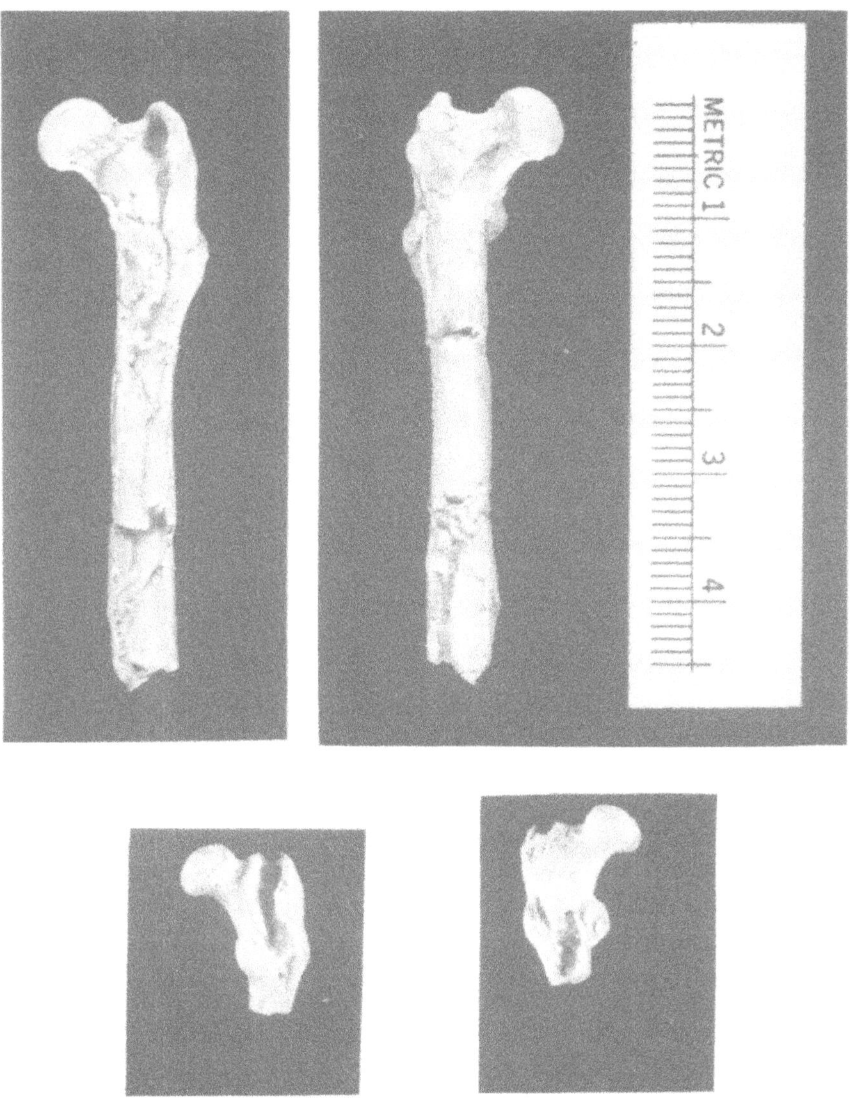

Fig. 6. Posterior and anterior views of femora from quarry L-41: DPC 8256 (top) and DPC 7529 (below).

Apidium, 1.9), whereas the femoral neck is long as in most anthropoids. The shape of the articular surface of the femoral head in DPC 8256 differs from that of *Apidium* and most closely resembles that of several platyrrhines, especially *Callicebus.* In *Apidium,* femoral head shape is more oval dorsally with anterior and posterior articular edges obliquely oriented. DPC 8256 also shows less anteroposterior bending of the proximal femur relative to *Apidium,*

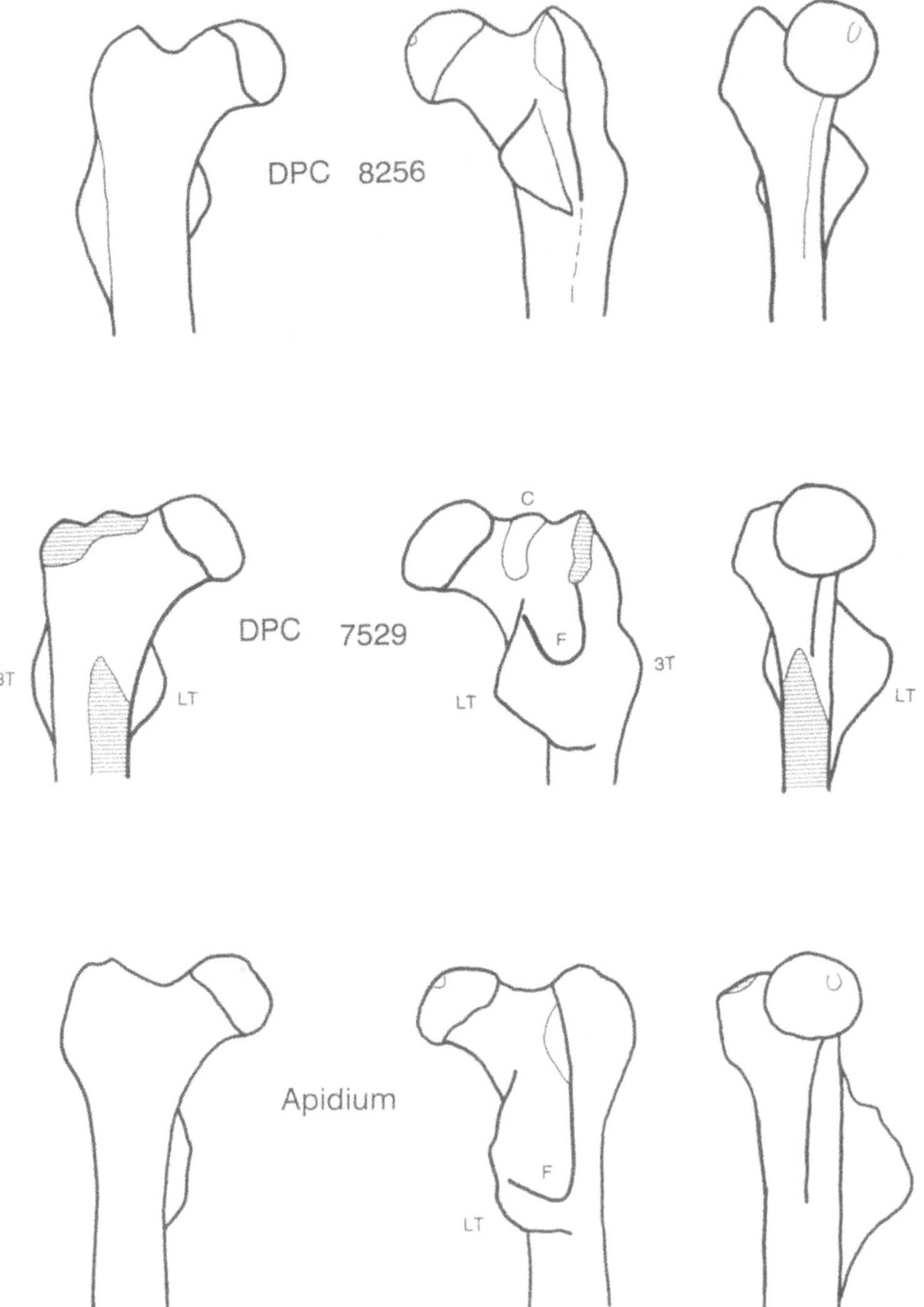

Fig. 7. Anterior, posterior, and oblique views of Fayum proximal femora. Note the loss of the third trochanter in *Apidium* (DPC 1074, reversed). 3T, third trochanter; LT, lesser trochanter; F, closed trochanteric fossa; C, crista paratrochanterica.

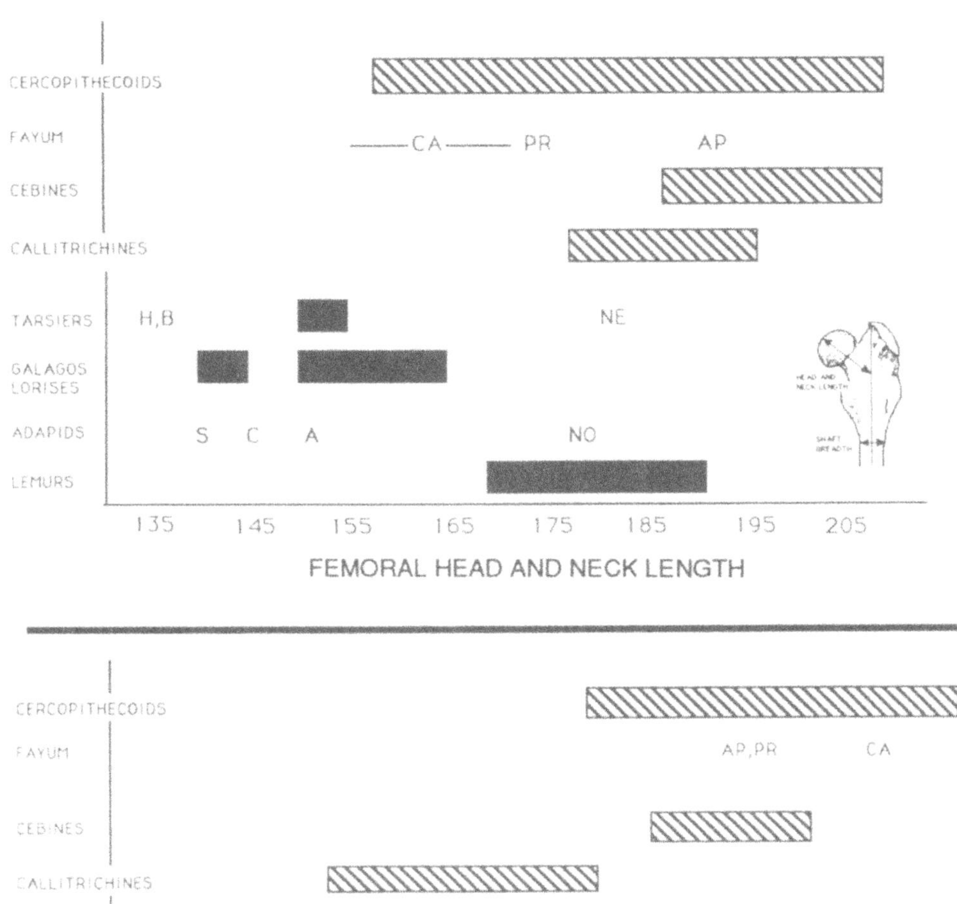

Fig. 8. Ratio of femoral head and neck length relative to shaft breadth (top) and angle of lesser trochanter to head angle (bottom) in a selection of fossil and extant primates. A, *Adapis;* AP, *Apidium;* B, Bridger omomyid; C, *Cantius;* CA, *Catopithecus,* DPC 8204; H, *Hemiacodon;* M, *Microchoerus;* NE, *Necrolemur;* NO, *Notharctus;* PR, *Proteopithecus,* DPC 7529; S, *Smilodectes.*

but it is similar in this respect to the femora of *Callicebus* and *Saimiri.* On the anterior surface of the proximal femur there is slight anterior cresting along the greater trochanter, with the greater trochanter overhanging the shaft and a triangular-shaped fossa below for the origin of vastus lateralis.

A prominent third trochanter is present in DPC 8256 (Figs. 6 and 7). Prominent third trochanters are found in a variety of living and extinct prosimians (Gregory, 1920; Simpson, 1940; Rose and Walker, 1985; Ford, 1980, 1988). The position of the third trochanter in DPC 8256 is similar to that in the few specimens of *Saimiri* and *Aotus* that retain a definable third trochanter, but the third trochanter is generally reduced to a crest along the femoral shaft in living platyrrhines and is even more diminished in living catarrhines (Ford, 1980, 1988; Harrison, 1987; Hershkovitz, 1988; Dagosto, 1990). A partial femoral fragment of a shaft from quarry I (best attributable to *Aegyptopithecus zeuxis,* DPC 8709) has only slight development of a third trochanter, actually little more than a roughened, low rise for attachment of the gluteal superficialis muscle, but other quarry I and V specimens attributed to propliopithecids possess prominent third trochanters. *Apidium* has no identifiable third trochanter.

The lesser trochanter is very large and is oriented posteriorly (48°), as it also is in *Apidium* (39°) and in cebines (Fig. 8). The large lesser trochanter in DPC 8256 has a small lateral extension, but it does not appear to contact the crest of the trochanteric fossa (Fib. 7). Thus, the intertrochanteric area is not "walled off" as it is in *Apidium* (Fleagle and Kay, 1987).

The posterior surface of the greater trochanter is not as wide as that in *Apidium.* The trochanteric ridge runs straight distally, not obliquely as in living catarrhines. This ridge or crest is very well developed, and it is similar to that of *Apidium* and the smaller L-41 femur (DPC 7529). Prominent development of this ridge is unlike that of most platyrrhines or living and extinct prosimians. The long trochanteric fossa in DPC 8256 is similar to that of *Apidium* and other platyrrhines.

DPC 8256 differs from specimens of *Apidium phiomense* in the following ways: (1) a distinct third trochanter is present; (2) the greater trochanter is less robust; (3) the greater trochanter appears to lack the distinctive "ventral curvature" of *Apidium;* (4) the intertrochanteric area is not walled off as it is in *Apidium;* (5) the femoral head is round rather than oval; and (6) the lesser trochanter is large but not plate-like.

Proximal End of Small Femur, DPC 7529

This specimen is much smaller than DPC 8256 (Fig. 6), being comparable in size to the femora of tamarins or medium-sized galagos. Thus, it does not belong to *C. browni.* The most probable allocation based on abundance would be *Proteopithecus sylviae,* but the femur could represent *Arsinoea.* DPC 7529 is in three separate pieces including the proximal third of the femur, a portion of the midshaft, and a piece of the distal shaft that continues to the proximal edge of the patellar rims. All three pieces together measure 54.3 mm in length. Like the previous specimen, DPC 7529 retains much of the morphology of the proximal end, and therefore it can be compared directly to the larger specimen. Several significant structural differences can be observed when the two L-41 femora are compared.

The femoral head and neck of the small femur (DPC 7529) are obliquely oriented at an angle of 56°, similar to those of other anthropoids. The relative length of the femoral head and neck is slightly greater in DPC 7529 than in the large L-41 femur (DPC 8256) and notably shorter than in *Apidium* (head and neck length/shaft width: DPC 7529, 1.7; DPC 8256, 1.5–1.7; *Apidium,* 1.9) (Fig. 8). The lesser trochanter of DPC 7529 is very large and flange-like (Fig. 7) and similar to that of prosimians and platyrrhines rather than being shaped like a compressed tubercle as in catarrhines. Although the lesser trochanter is large and expanded in both of the L-41 femora, it is not as expanded nor as rectangular in shape as it is in *Apidium* (Fig. 7). The posterior orientation of the lesser trochanter in DPC 7529 (40°) and *Apidium* (39°) is less than that in DPC 8256 (48°) but similar to that in cebines (35° to 44°) and living catarrhines (33° to 55°) (Fig. 8). The third trochanter is positioned at the distal end of the greater trochanter, and it is more prominent than that in any living anthropoid. The size of this trochanter is most similar to those of prosimians with the exception of *Necrolemur,* where it is smaller. In DPC 8256, the third trochanter is far more proximally located in its termination relative to this trochanter's position in DPC 7529 (Fig. 7).

The small L-41 femur has a large crista paratrochanterica (Fig. 7; Ford, 1986; Ford and Morgan, 1986; Hershkovitz, 1988). The posterior crest of the greater trochanter is pronounced and meets the lateral crest of the lesser trochanter. These crests close the trochanteric fossa, as is also the case in *Apidium* (Fig. 7). The relatively greater size of the crests in *Apidium* makes the intertrochanteric region appear even more closed. Thus, the DPC 7529 femur shows a less extreme and a more primitive version of the posterior femoral fossa than *Apidium phiomense.* The large L-41 femur (DPC 8256) is slightly damaged, but it almost certainly has a more open intertrochanteric region. Some platyrrhines on occasion (e.g., *Aotus,* one out of five specimens) possess a slight ridge that connects the lesser trochanter with the posterior trochanteric crest, but in no platyrrhine is this feature as well developed as in *Apidium.* A large oblique trochanteric crest is observed in living and extinct catarrhines and occasionally in *Cebus,* although it is differently formed in the latter genus (Hershkovitz, 1988). The width of the posterior surface of the greater trochanter is large in DPC 7329, like that of *Apidium,* while the trochanteric fossa is short compared to that in *Apidium.* The DPC 7529 femur is also similar in size and shape to an undescribed parapithecid femur (DPC 8708) from quarry V, although this specimen lacks a third trochanter.

DPC 7529 is structurally similar to DPC 8256 except: (1) it has the region of the trochanteric fossa sealed off in the manner of *Apidium,* with a distinct ridge connecting the lesser and greater trochanters to each other; (2) the third trochanter is more proximally positioned, being located directly across from the lesser trochanter; (3) the greater trochanter does have a slight "ventral curve" to its proximal end, like the condition in *Apidium;* (4) the femoral head is more rounded in DPC 8256; and (5) the trochanteric fossa is shorter. Both of the L-41 femora possess prominent third trochanters along with large, posteriorly oriented lesser trochanters. In sum, the small femur from L-41 (DPC 7529) differs from the large femur (DPC 8256) in the same ways that distinguish DPC 8256 from *Apidium.*

Partial Innominate Bone, DPC 9278

The innominate from quarry L-41 (DPC 9278) preserves only the region around the acetabulum and the root of the ilium (Fig. 9). The cranial portion of the ilium, the caudal portion of the ischium, and the ventral portion of the pubis have been lost. The caudal portion of the auricular surface is present. This innominate is about the size of innominates in *Saguinus,* and it articulates well with the small L-41 femur (DPC 7529).

The innominate fragment has a very sharply defined ridge separating the iliac and gluteal planes (Fig. 10). It appears that the definition of the innominate into three distinct planes is probably even more fully developed than the tripartite arrangement noted in *Apidium* (Fleagle and Simons, 1979). The gluteal plane is wide (gluteal width/ilium length: DPC 9278, 38%; *Apidium,* 44% Table III), and in this respect it is most similar to several New and Old World monkeys as well as the larger Malagasy lemurs (Fig. 11). Fleagle and Kay (1987) hypothesized that a wide gluteal plane is an anthropoid synapomorphy. Table III shows that relative to ilium length, the gluteal plane is narrow (below 30%) in lorisoids, tarsiers, and Eocene primates, but the larger Malagasy lemurs overlap in their gluteal expansion with anthropoids. DPC 9278 and specimens of *Apidium* both fall within this wider gluteal range (Fig. 11).

The crest that separates the gluteal and iliac planes is larger in DPC 9278 than it is in *Apidium* (Fig. 10). In *Aotus, Callicebus,* and all callitrichid genera, the superior edge of margo acetabuli, the ridge separating the iliac and the gluteal planes, is long but begins a short distance away from the acetabulum. Thus, in profile, the edge of the acetabulum gives way to a short spatial gap, followed by a ridge (margo acetabuli) that slopes up to the highest point of the crest along the ilium. In *Saimiri* and *Cebus* the ridge begins at the acetabulum and continues up along the ilium, as in DPC 9278. However, the ridge in these two platyrrhines differs from DPC 9278 in being much lower.

In DPC 9278, the caudal width of the iliac plane relative to the diameter of the acetabulum is extremely wide, with a ratio of 76% compared to 58% in *Apidium* (Table IV). As the iliac plane extends cranially away from the acetabular region, it becomes smaller in living and extinct prosimians, most non-

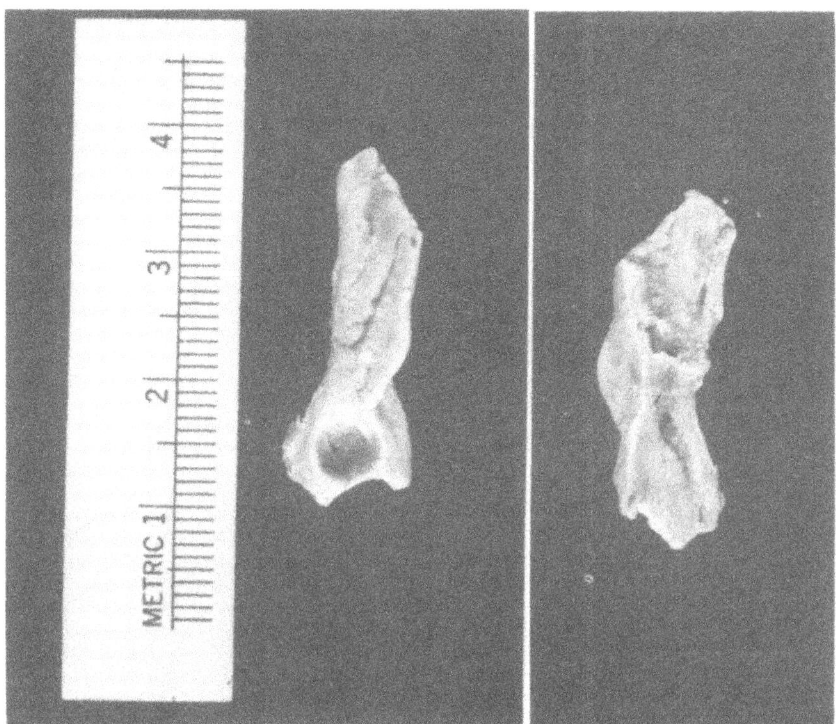

Fig. 9. Innominate bone from quarry L-41 (DPC 9278).

ateline platyrrhines, DPC 9278, and *Apidium* (Table IV). In contrast, living catarrhines and atelines show this plane to increase in width relative to the acetabular region (Table IV). A caudal-to-cranial ratio for the widths of the iliac plane measured in these two locations supports this observation (Table IV, Fig. 11).

The following is based on Schultz's (1969) measurements and ratios (Table I). The ratio for maximum depth of the acetabulum relative to its diameter is 58% in DPC 9278, which is closest to *Gorilla, Pan,* and males of *Macaca, Cercopithecus,* and *Presbytis.* The acetabular depth for DPC 9278 is deeper than that in platyrrhines, which range between 52% and 55% for this ratio (Schultz, 1969). The relative width of the acetabular notch is 34% for DPC 9278 and best matches that of *Pan* and then *Papio.* The relative thickness of the ventral-to-dorsal acetabular walls ratio is 78% for DPC 9278, and this value is similar to that of *Cebus* and males of the genus *Pan.* The value of this ratio is low, like those of all anthropoids, which have greater dorsal acetabular breadths than do prosimians (Schultz, 1969). Last, the diameter of the acetabulum relative to

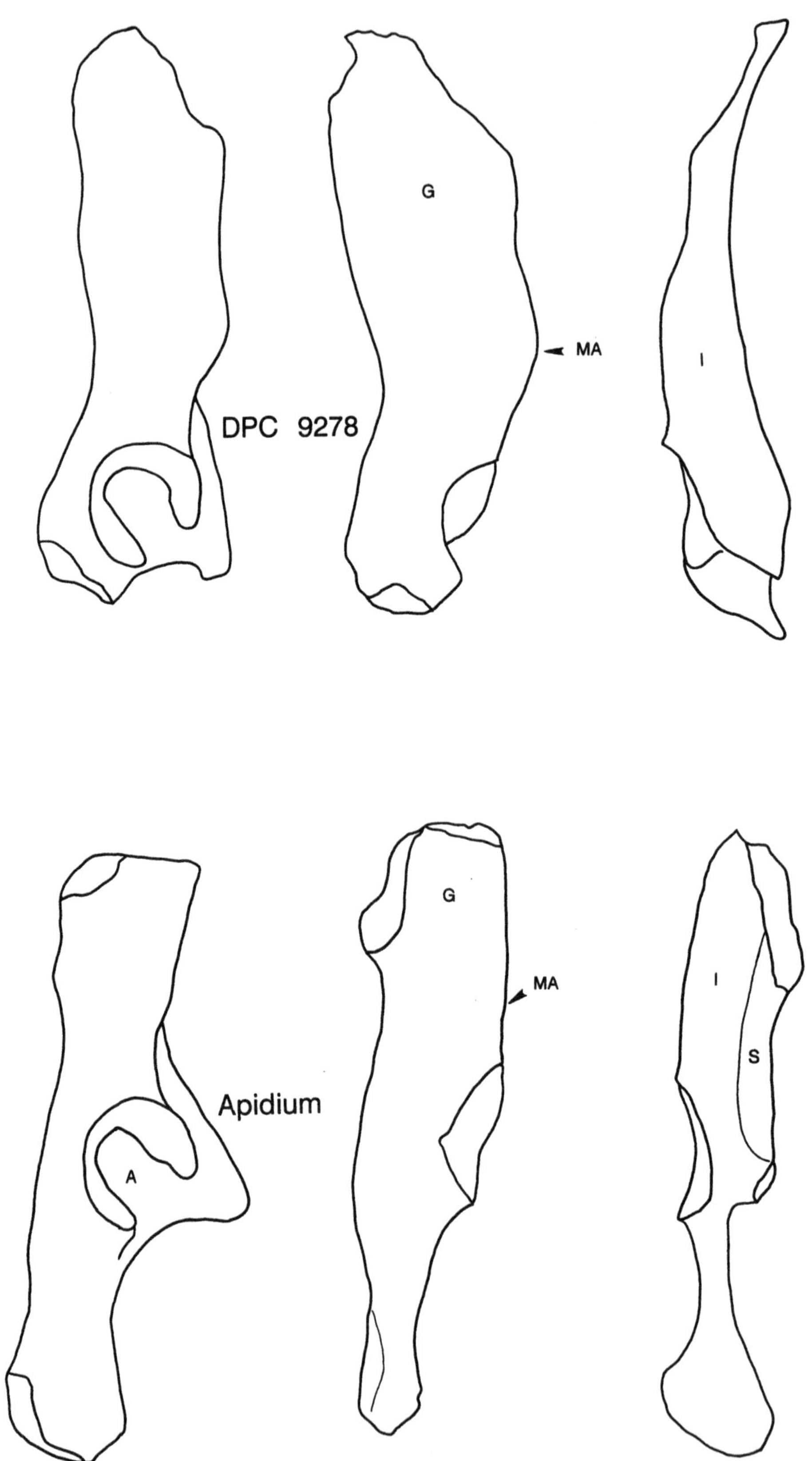
G
MA
I
DPC 9278
G
MA
I
S
Apidium
A

Table III. Relative Shape[a] of the Innominate in Primates

Taxon	Ratio range
Cheirogaleids	18–23
Lorisids	23–29
Lemurids	27–40
Indriids	35–42
Tarsiids	19–24
Callitrichines	32–39
Cebines	29–36
Pitheciines	33–34
Atelines	35–41
Cercopithecines	35–46
Colobines	41–44
Hylobatids	40–45
Pongids	61–89
Cantius	20
Smilodectes	23
Hemiacodon	31
Cebupithecia	39
Apidium	44
DPC 9278	38

[a]Ratio for gluteal width = 100 × (gluteal width/ilium length).

all three acetabular breadth measurements (Table I) is 34% and similar to those of *Macaca, Papio, Presbytis,* female *Cercopithecus,* male *Nasalis,* female *Lemur,* and *Nycticebus.*

When the four pelvic ratios used by Schultz (1969) are compared, it is interesting that the best match with DPC 9278 is found among the Old World monkeys and apes rather than with platyrrhines or prosimians. Schultz's ratios do show strong sexual differences within some species, even within a few species lacking notable size dimorphism, like *Hylobates* or *Ateles.* The ratios may in some cases differ from 5% to 13% between the sexes. In most catarrhine species however, the sexes differ by only 1% or 2%.

Overall, the DPC 9278 innominate is extremely similar in most respects to the innominate of *Apidium phiomense* and, to a lesser extent, to those of *Saimiri*

Fig. 10. Fayum innominates. Lateral, dorsal, and ventral views of DPC 9278 and *Apidium phiomense* (DPC 1036, reversed). A, acetabulum; MA, margo acetabuli; G, gluteal plane; I, iliac plane; S, sacral plane. Note the well-elevated ridge (margo acetabuli, arrowheads) separating the iliac and gluteal planes, especially in DPC 9278.

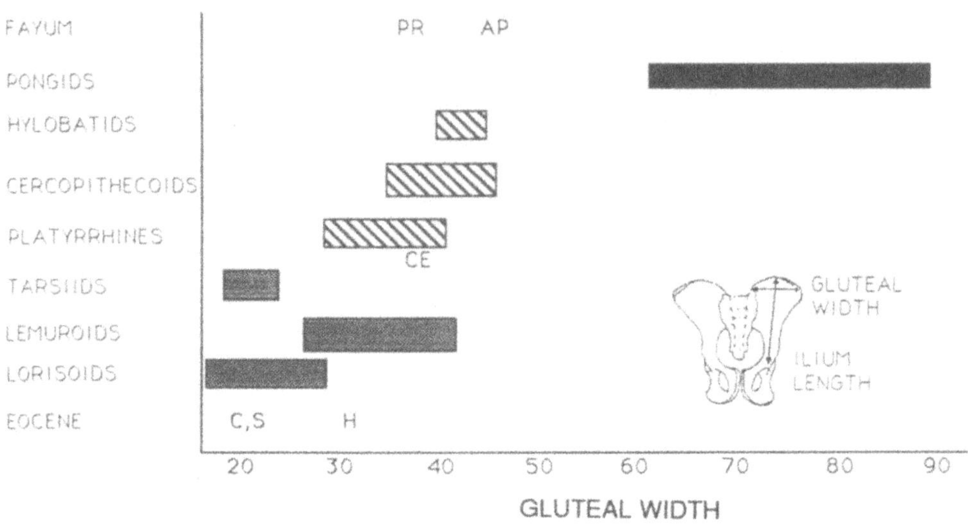

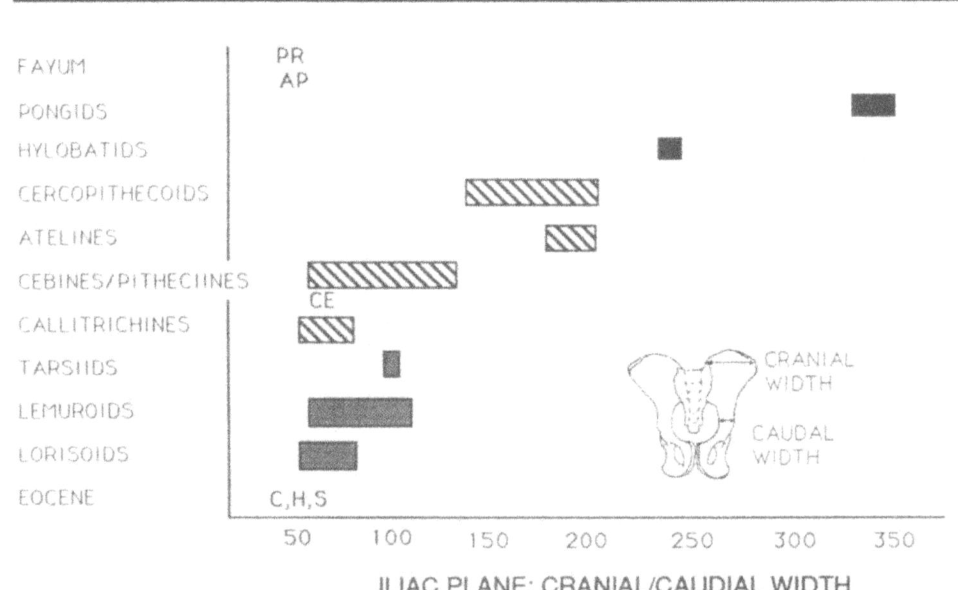

Fig. 11. Ratios of gluteal width relative to ilium length (top) and of the cranial width relative to caudal width of the iliac plane (bottom). AP, *Apidium;* C, *Cantius;* CE, *Cebupithecia;* H, *Hemiacodon;* PR, *Proteopithecus,* DPC 9278.

Table IV. Relative Proportions of the Iliac Plane in Primates

Taxon	Ratio range[a] Index 1	2	3
Cheirogaleids	29–37	44–46	64–83
Lorisids	19–44	36–52	52–83
Lemurids	31–59	50–82	54–91
Indriids	39–65	67–70	57–104
Tarsiids	35–36	36	97–98
Callitrichines	29–45	47–63	50–78
Cebines	21–66	40–53	55–128
Pitheciines	49–67	50–56	98–120
Atelines	99–125	54–65	174–199
Cercopithecines	75–109	48–63	137–201
Colobines	80–104	50–68	150–184
Hylobatids	139–160	60–69	233–239
Pongids	224–306	67–90	329–351
Cantius	28	74	39
Smilodectes	29	55	53
Hemiacodon	26	66	39
Cebupithecia	38	56	67
Apidium	26	58	45
DPC 9278	37	76	49

[a]Index 1 = cranial iliac plane width/acetabular diameter. Index 2 = caudal iliac plane width/acetabular diameter. Index 3 = cranial/caudal iliac plane width.

and *Cebus*. The well-developed crest (margo acetabuli) separating the gluteal and iliac planes in both *Apidium* and DPC 9278 is especially distinctive.

Discussion

Locomotor Adaptation

The two smaller limb elements show that the hip region of at least one L-41 genus (*Proteopithecus*, *Arsinoea?*) is similar anatomically and functionally to that of *Apidium phiomense*. The hip suggests a preference for fore-and-aft movements associated with rapid arboreal quadrupedalism or perhaps with leaping. The shape of the femoral head implies more frequent fore-and-aft movements at the hip relative to abduction, adduction, or rotations of the thigh during complex movements such as climbing. The size and position of the third trochanter suggest significant femoral extension capabilities. Likewise, the large gluteal plane suggests powerful femoral extensors. The large lesser trochanter for the attachment of iliopsoas implies significant femoral flexion and rotation during climbing (Taylor, 1976) although its more posterior position implies greater rotation. The dorsal breadth of the acetabulum suggests greater vertical forces being applied in a horizontal or quadrupedal posture. Given this admittedly meager evidence, it appears that the best locomotor interpretion for the smaller L-41 primate is that of an arboreal quadruped most similar to small cebids such as *Cebus, Aotus,* and *Callicebus.* More hindlimb elements will certainly be needed to verify and refine this interpretation.

For *Catopithecus,* locomotor habits can be reconstructed from the distal humerus and the proximal femur. The distolateral humerus fragment (DPC 8204) suggests that rotation (round capitulum) as well as flexion (brachial flange, capitular tail, and radial fossa depth) at the elbow was well developed. These two features are best observed in frequently climbing and quadrupedal species of living primates, particularly prosimians. The distomedial humerus fragment (DPC 7328) shows medial expansion of the trochlea with slight distal projection of the medial edge, a large trochlear surface area, a deep fossa, a round capitulum, and a wide medial epicondyle. This morphology is characteristically found in arboreal quadrupeds (Conroy, 1976a; Szalay and Dagosto, 1980; Fleagle and Kay, 1987). Even if the two humeral fragments do not actually belong to the same taxon, they are nevertheless consistent with a common locomotor pattern.

The femur (DPC 8256) suggests that *Catopithecus* was not a specialized leaper. The round femoral head and the angle of the head and neck are features that are more likely associated with relatively complex hip movements, normally associated with climbing and quadrupedal primates. Again, the large lesser and third trochanters imply significant femoral flexion, extension, abduction, and medial rotation of the thigh, as they did in the smaller L-41 specimens. *Catopithecus* was perhaps a less rapid arboreal quadruped than the species represented by the smaller fossils, and it possibly climbed more frequently. On the basis of this sparse evidence, *Catopithecus* appears to have been more oriented toward quadrupedalism and climbing than was the smaller taxon from L-41. This is similar to the locomotor interpretations previous authors have made concerning the upper sequence propliopithecines (Conroy, 1976a; Fleagle and Simons, 1982a,b; Gebo and Simons, 1987).

Anthropoid Status

The L-41 postcrania are allocated to Anthropoidea on the basis of three factors. First, the only known primates from L-41 are anthropoids (Simons, 1989, 1990, 1992). Prosimian remains from other Fayum quarries are exceedingly rare (Simons *et al.,* 1986; Simons and Bown, 1985). Second, the L-41 elements show strong phenetic similarities to *Apidium phiomense* and other upper-sequence anthropoids. Likewise, although the L-41 elements display many primitive euprimate-like features, they do not show any strong phenetic similarity with any single living or extinct prosimian taxon. Third, the L-41 postcranial elements do share some attributes with other anthropoids that possibly represent anthropoid synapomorphies, as discussed below.

In the postcranium as in the teeth and skull, the L-41 primates lack several features often believed to be anthropoid synapomorphies. For example, a reduced or absent third trochanter and a reduced lesser trochanter are characteristic of anthropoids, but are decidedly unlike the condition seen in the L-41 specimens (see Ford, 1980, 1986, 1988; Harrison, 1987). On the other hand, the dorsoepitrochlear fossa of the humerus is present on the L-41 specimen and also in most anthropoids, but this character has a variable distribution among primates and is difficult to characterize as an anthropoid synapomorphy (Dagosto, 1990). Like those of other anthropoids, the L-41 femora do show a moderate notch between femoral head and the greater trochanter (Ford, 1988), and the trochlear surface on DPC 7328 is well developed (Szalay and Dagosto, 1980). The L-41 femora also display a posteriorly oriented lesser trochanter, and this may turn out to be an anthropoid characteristic as well (see Dagosto and Schmid, 1994). The only feature that the distomedial humeral fragment (DPC 7328) shares exclusively with living catarrhines is a nontranslatory or only slightly translatory type of elbow (Rose, 1988). Thus, besides general, phenetic similarities to anthropoids, there are only a few features that could stand as shared-derived features linking the L-41 elements with Anthropoidea.

Comparison with Eocene Prosimians

We have compared the limb bones from L-41 with those of known Eocene prosimians—adapids and omomyids. With the exception of the few general similarities already discussed, we have been unable to identify convincing derived features shared exclusively between the L-41 taxa and any Eocene prosimian. They share few similarities other than primitive euprimate resemblances. Thus, the new material from L-41 does not suggest any special phylogenetic connections with either Adapidae or Omomyidae, nor does the new evidence compellingly exclude either Eocene family from consideration as a potential ancestral group.

Phylogenetic Inferences about Early Anthropoids

The postcranial elements that are available from quarry L-41 show a mix of parapithecid and propliopithecine attributes. In the case of the proximal femora, the smaller specimen allocated to *Proteopithecus* looks decidedly more parapithecid-like than the larger specimen allocated to *Catopithecus* (for example, in the intertrochanteric region). In the distal humerus, parapithecid-like features (the medially flared trochlea) and propliopithecine features (the trochlea/capitulum breadth ratio) are evident in two specimens that are possibly from the same primate species. If DPC 8204 and DPC 7328 do in fact belong to *C. browni,* this would suggest a primitive mosaic of parapithecid and propliopithecid characteristics.

The L-41 postcranial elements are certainly more primitive in some respects than any that have been found in the upper sequence of the Fayum. As older dental and postcranial remains are found from the lower localities at the Fayum, the differences between the families Parapithecidae and Propliopithecidae become increasingly slight. However, even in the upper sequence the two families share quite a number of postcranial features in common, particularly above the knee. The older taxa from L-41 certainly lie closer to the common ancestor of the two families than do the upper-sequence species. The relatively indistinct family divisions at the Fayum's lower levels may indicate that Parapithecidae and Propliopithecidae diverged from each other in Africa not long before the deposition of quarry L-41.

The primitive structure of the L-41 primates raises important questions about evolutionary convergence and reversal. Several hypotheses have been presented concerning the phylogenetic relationships of Fayum primates. With the allocation of L-41 postcrania currently tentative, it is impossible to definitively resolve the issues, but it is noteworthy that whichever phylogenetic hypothesis proves to be the correct one, significant amount of evolutionary convergence and reversal are required (Fig. 12).

Our own conclusions on the phylogenetic issues are the following. Early anthropoids are much more primitive than would be predicted by morphotype reconstructions or parsimony analyses based only on examination of later anthropoid taxa. The stratigraphic record in the Fayum suggests that some dental and gnathic features evolved in parallel up through the section. For example, the lowering of the trigonid, the acquisition of wear facet X, the enlargement of the hypocone, the deepening of the mandible, and the rounding of the gonial angle occurred in parallel between the *Catopithecus–Oligopithecus –Propliopithecus* lineage and the *Qatrania–Apidium* lineage (Fleagle and Kay, 1987; Rasmussen and Simons, 1988). We conclude that the same evolutionary pattern occurred for some postcranial characters, such as the loss of the third trochanter, the reduction of the capitular tail, and the reduction of the margo acetabuli.

The alternative to this hypothesis of parallel evolution is that all of the upper-sequence anthropoids (including the parapithecids, *Apidium* and *Parapithecus,* together with the propliopithecines, *Propliopithecus* and *Aegyptopithecus*) form a clade with respect to the lower-sequence species. This possibility has been considered by previous authors (Rasmussen and Simons, 1988; Fleagle and Kay, 1987). These studies concluded that *Oligopithecus* does belong with propliopithecines, and *Qatrania* does belong with upper-sequence parapithecids. Fitting a clade exclusively to the upper-sequence taxa does not eliminate the problem of convergent evolution; it simply "passes the buck" of

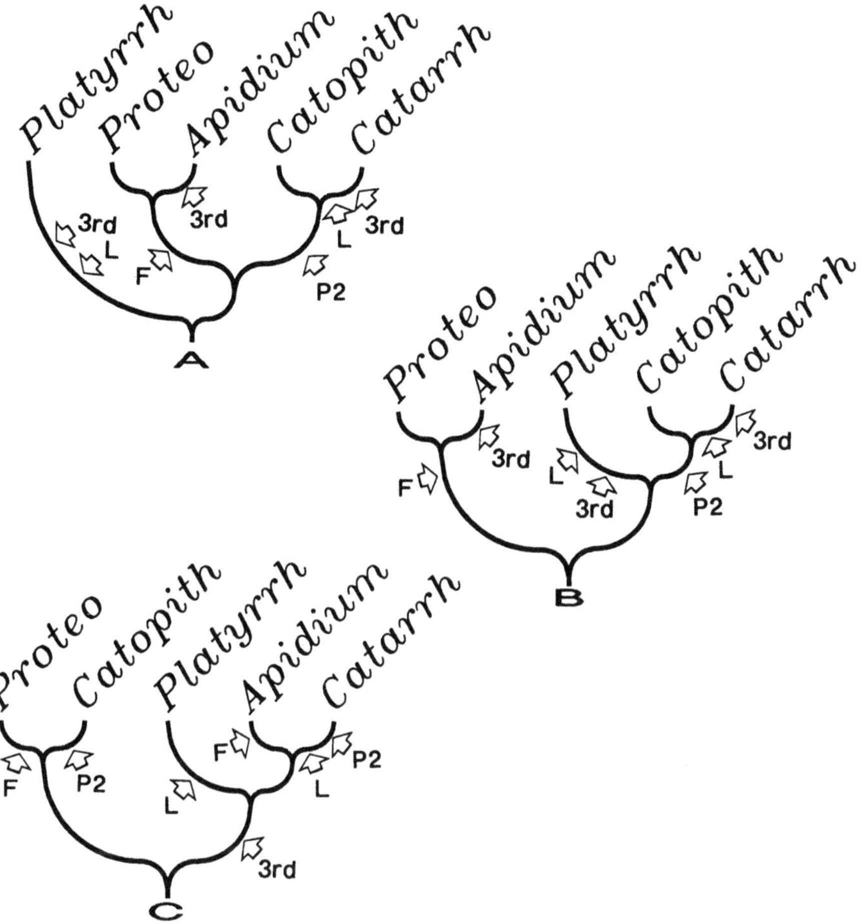

Fig. 12. Three hypothetical phylogenies of anthropoid primates with information about character convergence and evolution. (A) A traditional phylogeny showing oligopithecines aligned with Catarrhini, and with all known Old World anthropoids united with each other relative to Platyrrhini. (B) The "preplatyrrhine" hypothesis of Fleagle and Kay (1987) showing parapithecids diverging before the platyrrhine–catarrhine split. (C) What we will call the "Eocene protoanthropoid" hypothesis discussed by Kay and Williams (Chapter 1, this volume), in which the oligopithecines are excluded from a clade containing Oligocene parapithecids and propliopithecines. Several evolutionary changes in character states of the femur are mapped onto these hypothetical phylogenies: 3rd, loss of the third trochanter; F, walling off of the intertrochanteric fossa by crests; L, shift of orientation of the lesser trochanter from medial to posterior. The evolutionary loss of the second premolar is designated by P2.

evolutionary convergence onto other features (e.g., loss of second premolar and acquisition of a nontranslatory type of elbow joint, found in *Catopithecus, Aegyptopithecus,* and later catarrhines; Rose, 1988). Furthermore, such an arrangement ignores the Fayum's stratigraphic context and the meaning of morphologically intermediate taxa such as *Propliopithecus markgrafi.* We conclude that the parallel evolution hypothesis is still the most reasonable one and that the same family-level split seen in the Fayum's uppermost levels can be extended downsection to quarry L-41 as well.

The small femur from quarry L-41 exhibits much greater development of parapithecid-like features than does the larger femur of the oligopithecine *Catopithecus browni.* Whether the small femur belongs to *Proteopithecus* or to *Arsinoea,* it does share several similarities with parapithecids, such as the walling off of the intertrochanteric fossa. This suggests the evolutionary divergence between parapithecids and the two-premolared propliopithecids, such as *Catopithecus,* had already taken place. This hypothesis certainly requires a significant amount of parallel evolution of dental and postcranial features in the three major anthropoid lineages: parapithecids, platyrrhines, and catarrhines. However, at least for the dental features, which are better sampled, this conclusion can be justified on the basis of change preserved in stratigraphic sequence. *Catopithecus* of the lowest levels is more primitive in dental structure than *Oligopithecus* from quarry E, which in turn shares specific resemblances with Fayum and Omani specimens of *Propliopithecus markgrafi* (hypoconulid placed near entoconid, lower molar crests sharply defined), which in turn resemble *Propliopithecus chirobates* of the uppermost quarries. The only alternative would be to assume that the similarities between *Oligopithecus* and *P. markgrafi* were obtained convergently or, conversely, that the similarities between *P. markgrafi* and other species of *Propliopithecus* are convergent.

An analogous case of independent lineages evolving specialized dental features in parallel up through the Fayum section is documented for fossil hyracoids (Rasmussen and Simons, 1991). Such cases of parallel evolution among recently diverged taxa are probably not uncommon and presumably result from similar underlying genetic systems operating within a shared ecological context. We propose that the very primitive postcrania from quarry L-41, with their large third trochanters and long capitular tails, represent the ancestral anthropoid condition, which was then modified at least three times in parallel among parapithecids, platyrrhines, and catarrhines.

If our interpretation of anthropoid evolution in the Fayum is correct, this casts some doubt on the logic used in formulating the "preplatyrrhine" hypothesis (Harrison, 1987; Fleagle and Kay, 1987; Fig. 12B). This hypothesis was based on several very primitive prosimian-like attributes evident in parapithecids that were absent in catarrhines and platyrrhines. However, if all three anthropoid lineages (Parapithecidae, Platyrrhini, Catarrhini) were ancestrally very primitive, there is no special evidence in favor of an early parapithecid divergence.

Acknowledgments

We thank Pat Holroyd and Prithijit Chatrath for providing expert preparatory and curatorial work on the fossils; Mr. Chatrath also serves as field manager of the Fayum operations. Field work at quarry L-41 is done in cooperation with the Egyptian Geological Survey and with the help of the Cairo Geological Museum staff; funding is provided by the National Science Foundation. We also thank the staff of the Field Museum of Natural History (Chicago, IL) for access to their collections. This is Duke Primate Center publication 587.

References

Anapol, F. 1983. The scapula of *Apidium phiomense:* A small anthropoid from the Oligocene of Egypt. *Fol. Primatol.* **40:**11–31.

Bonis, L. de, Jaeger, J.-J., Coiffat, B., and Coiffat, P. E. 1988. Découverte du plus ancien primate catarrhinien connu dans l'Éocène supérieur d'Afrique du Nord. *C. R. Acad. Sci. Paris [II]* **306:**929–938.

Bown, T. M., and Kraus, M. D. 1988. Geology and Paleoenvironment of the Oligocene Jebel Qatrani Formation and Adjacent Rocks, Fayum Depression, Egypt. U.S. Geological Survey Professional Paper 1452.

Ciochon, R. L., Savage, D. E., Tint, T., and Ba Maw. 1985. Anthropoid origins in Asia? New discovery of *Amphipithecus* from the Eocene of Burma. *Science* **229:**756–759.

Conroy, G. C. 1976a. Primate postcranial remains from the Oligocene of Egypt. In: F. S. Szalay (ed.), *Contributions to Primatology,* pp. 1–134. S. Karger, New York.

Conroy, G. C. 1976b. Hallucal metatarsal joint in an Oligocene anthropoid *Aegyptopithecus zeuxis. Nature* **262:**684–686.

Dagosto, M. 1990. Models for the origin of the anthropoid postcranium. *J. Hum. Evol.* **19:**121–139.

Dagosto, M., and Schmid, P. 1994. Proximal femoral anatomy of omomyiform primates. *J. Hum. Evol.* (in press).

Fleagle, J. G. 1983. Locomotor adaptations of Oligocene and Miocene hominoids and their phyletic implications. In: R. L. Ciochon and A. B. Chiarelli (eds.), *New Interpretations of Ape and Human Ancestry,* pp. 301–324. Plenum Press, New York.

Fleagle, J. G., and Kay, R. F. 1983. New interpretations of the phyletic position of Oligocene hominoids. In: R. L. Ciochon and A. B. Chiarelli (eds.), *New Interpretations of Ape and Human Ancestry,* pp. 181–210. Plenum Press, New York.

Fleagle, J. G., and Kay, R. F. 1985. The paleobiology of catarrhines. In: E. Delson (ed.), *Ancestors: the Hard Evidence,* pp. 23–36. Alan R. Liss, New York.

Fleagle, J. G., and Kay, R. F. 1987. The phyletic position of the Parapithecidae. *J. Hum. Evol.* **16:**483–532.

Fleagle, J. G., and Simons, E. L. 1978. Humeral morphology of the earliest apes. *Nature* **276:** 705–707.

Fleagle, J. G., and Simons, E. L. 1979. Anatomy of the bony pelvis in parapithecid primates. *Fol. Primatol.* **31:**176–186.

Fleagle, J. G. and Simons, E. L. 1982a. The humerus of *Aegyptopithecus zeuxis:* A primitive anthropoid. *Am. J. Phys. Anthrop.* **59:**175–193.

Fleagle, J. G., and Simons, E. L. 1982b. Skeletal remains of *Propliopithecus chirobates* from the Egyptian Oligocene. *Fol. Primatol.* **39:**161–177.

Fleagle, J. G., and Simons, E. L. 1983. The tibio-fibular articulation in *Apidium phiomense,* an Oligocene anthropoid. *Nature* **301:**238–239.

Fleagle, J. G., and Simons, E. L. 1985. Skeletal anatomy of *Apidium phiomense,* an Oligocene anthropoid. *Am. J. Phys. Anthrop.* **66:**168.

Fleagle, J. G., Simons, E. L., and Conroy, G. C. 1975. An ape limb bone from the Oligocene of Egypt. *Science* **189:**135–137.

Ford, S. M. 1980. Phylogenetic relationships of the Platyrrhini—the evidence of the femur. In: R. L. Ciochon and A. B. Chiarelli (eds.), *Evolutionary Biology of the New World Monkeys and Continental Drift,* pp. 317–329. Plenum Press, New York.

Ford, S. M. 1986. Systematics of the New World monkeys. In: D. R. Swindler and J. Erwin (eds.), *Comparative Primate Biology, Vol. 1, Systematics, Evolution and Anatomy,* pp. 73–135. Alan R. Liss, New York.

Ford, S. M. 1988. Postcranial adaptations of the earliest platyrrhine. *J. Hum. Evol.* **17:**155–192.

Ford, S. M., and Morgan, G. S. 1986. A new ceboid femur from the Late Pleistocene of Jamaica. *J. Vert. Paleontol.* **6:**281–289.

Gebo, D. L. 1993. Locomotor adaptation and limb morphology in early African anthropoids. In: D. L. Gebo (ed.), *Postcranial Adaptation in Non-Human Primates,* pp. 220–234. Northern Illinois University Press, DeKalb.

Gebo, D. L., and Simons, E. L. 1987. Morphology and locomotor adaptations of the foot in early Oligocene anthropoids. *Am. J. Phys. Anthrop.* **74:**83–101.

Gingerich, P. D. 1993. Oligocene age of the Gebel Qatrani Formation, Fayum, Egypt. *J. Hum. Evol.* **24:**207–218.

Godinot, M., and Mahboubi, M. 1991. First comparisons between the primates from Glib Zegdou (?Middle Eocene, Algeria) and European primates. In: *Monument Grube Messel—Perspectives and Relationships.* Hessisches Landesmuseum, Darmstadt.

Godinot, M., and Mahboubi, M. 1992. Earliest known simian primate found in Algeria. Nature **357:**324–326.

Gregory, W. K. 1920. On the structure and relations of the Eocene lemur *Notharctus* to the Adapidae and to other primates. II. On the classification and phylogeny of the Lemuroidea. *Bull. Geol. Soc. Am.* **26:**419–446.

Harrison, T. 1987. The phylogenetic relationships of the early catarrhine primates: A review of the current evidence. *J. Hum. Evol.* **16:**41–79.

Hershkovitz, P. 1988. The subfossil monkey femur and subfossil monkey tibia of the Antilles: A review. *Int. J. Primatol.* **9:**365–384.

Kappelman, J., Simons, E. L., and Swisher, C. C. III. 1992. New age determinations for the Eocene–Oligocene boundary sediments in the Fayum depression, northern Egypt. *J. Geol.* **100:**647–667.

Kay, R. F. 1977. The evolution of molar occlusion in the Cercopithecidae and early catarrhines. *Am. J. Phys. Anthrop.* **46:**327–352.

Kay, R. F., Fleagle, J. G., and Simons, E. L. 1981. A revision of the Oligocene apes from the Fayum Province, Egypt. *Am. J. Phys. Anthrop.* **55:**293–322.

Meldrum, D. J., Fleagle, J. G., and Kay, R. F. 1990. Partial humeri of two Miocene Colombian primates. *Am. J. Phys. Anthropol.* **81:**413–422.

Plaziat, J.-C. 1981. Late Cretaceous to late Eocene palaeogeographic evolution of southwest Europe. *Palaeogeogr. Palaeoclimatol. Palaeoecol.* **36:**263–320.

Preuschoft, H. 1974. Body posture and mode of locomotion in fossil primates: Method and example—*Aegyptopithecus zeuxis.* In: *Proceedings from the Symposia of the Fifth Congress of the International Primatological Society 1974,* pp. 345–359. Japan Science Press, Tokyo.

Rasmussen, D. T. and Simons, E. L. 1988. New specimens of *Oligopithecus savagei,* early Oligocene primate from the Fayum, Egypt. *Fol. Primatol.* **51:**182–208.

Rasmussen, D. T., and Simons, E. L. 1991. The oldest Egyptian hyracoids (Mammalia: Pliohyracidae): New species of *Saghatherium* and *Thyrohyrax* from the Fayum. *N. Jb. Geol. Palaont. Abh.* **182:** 187–209.

Rasmussen, D. T., and Simons, E. L. 1992. Paleobiology of the oligopithecines, the world's earliest known anthropoid primates. *Int. J. Primatol.* **13:**477–508.

Rasmussen, D. T., Bown, T. M. and Simons, E. L. 1992. The Eocene–Oligocene transition in continental Africa. In: D. R. Prothero and W. A. Berggren (eds.), *Eocene–Oligocene Climatic and Biotic Evolution,* pp. 548–566. Princeton University Press, Princeton.

Rose, K. D., and Walker, A. 1985. The skeleton of early Eocene *Cantius*, oldest lemuriform primate. *Am. J. Phys. Anthrop.* **66:**73–89.

Rose, M. D. 1988. Another look of the anthropoid elbow. *J. Hum. Evol.* **17:**193–224.

Schön Ybarra, M., and Conroy, G. C. 1978. Nonmetric features in the ulna of *Aegyptopithecus, Alouatta, Ateles,* and *Lagothrix. Fol. Primatol.* **29:**178–195.

Schultz, A. H. 1969. Observations on the acetabulum of primates. *Fol. Primatol.* **11:**181–199.

Senut, B. 1989. *Le Conde des Primates Hominoïdes—Anatomie, Fonction, Taxonomie, Évolution.* CNRS, Paris.

Simons, E. L. 1962. Two new species from the African Oligocene. *Postilla* **64:**1–12.

Simons, E. L. 1965. New fossil apes from Egypt and the initial differentiation of the Hominoidea. *Nature* **205:**135–139.

Simons, E. L. 1967. The earliest apes. *Sci. Am.* **217:**28–35.

Simons, E. L. 1974. *Parapithecus grangeri* (Parapithecidae, Old World Higher Primates): New species from the Oligocene of Egypt and the initial differentiation of the Cercopithecoidea. *Postilla* **166:**1–12.

Simons, E. L. 1987. New faces of *Aegyptopithecus* from the Oligocene of Egypt. *J. Hum. Evol.* **16:**273–289.

Simons, E. L. 1989. Description of two genera and species of late Eocene Anthropoidea from Egypt. *Proc. Natl. Acad. Sci. USA* **86:**9956–9960.

Simons, E. L. 1990. Discovery of the oldest known anthropoidean skull from the Paleogene of Egypt. *Science* **247;**1567–1569.

Simons, E. L. 1992. Diversity in the early Tertiary anthropoidean radiation in Africa. *Proc. Natl. Acad. Sci. USA* **89:**10743–10747.

Simons, E. L., and Kay, R. F. 1983. *Qatrania,* a new basal anthropoid from the Fayum, Oligocene of Egypt. *Nature* **304:**624–626.

Simons, E. L., Bown, T. M., and Rasmussen, D. T. 1987. Discovery of two additional prosimian families (Omomyidae, Lorisidae) in the African Oligocene. *J. Hum. Evol.* **15:**431–437.

Simpson, G. G. 1940. Studies on the earliest primates. *Bull. Am. Mus. Nat. Hist.* **77:**185–212.

Szalay, F. S., and Dagosto, M. 1980. Locomotor adaptations as reflected in the humerus of Paleogene primates. *Fol. Primatol.* **34:**1–45.

Taylor, M. E. 1976. The functional anatomy of the hindlimb of some African Viverridae (Carnivora). *J. Morphol.* **148:**227–254.

Thomas, H., Roger, J., Sen, S., and Al-Sulaimani, Z. 1988. Découverte des plus anciens ⟨⟨Anthropoïdes⟩⟩ du continent arabo-africain et d'un Primate tarsiiforme dans l'Oligocène du Sultanat d'Oman. *C. R. Acad. Sci. Paris [II]* **306:**823–829.

Thomas, H., Roger, J., Sen, S., Bourdillon-de-Grissac, C., and Al-Sulaimani, Z. 1989. Découverte de Vertébrés fossiles dans l'Oligocène inférieur de Dhofar (Sultanat d'Oman). *Geobios* **22:** 101–120.

Van Couvering, J. A., and Harris, J. A. 1991. Late Eocene Age of Fayum mammal faunas. *J. Hum. Evol.* **21:**241–260.

Early North African Primates and Their Significance for the Origin of Simiiformes (= Anthropoidea)

10

MARC GODINOT

Introduction

Twelve years ago, just after the publication of the volume dedicated to platyrrhine origins (Ciochon and Chiarelli, 1980), Hoffstetter made a remarkable attempt to synthesize information and debates about simiiform (= anthropoid) evolution. What he wrote then holds true today:"We are still far away from a consensus about phylogeny and biogeographical history, that is to say about the evolution, of these animals [primates]" (Hoffstetter, 1982, p. 242, translated). In this paper, Hoffstetter drew attention to the inconveniences of the vernacular term "anthropoid," ambiguous in many languages. In all these languages, anthropoid means apes, often great apes only, which is quite different from "Anthropoidea." Furthermore, the Code of Zoological Nomenclature recommends that the suffix *-oidea* be reserved for super-

MARC GODINOT • Laboratoire de Paléontologie, Institut des Sciences de l'Evolution, Université Montpellier II, F-34095 Montpellier Cédex 5, France.
Anthropoid Origins, edited by John G. Fleagle and Richard F. Kay. Plenum Press, New York, 1994.

families, an almost universal use in mammalogy. Hoffstetter added: "As Szalay and Delson (1979) revealed the existence of a genus *Anthropus,* synonym of *Homo,* Anthropoidea is a synonym of Hominoidea and must be definitely rejected" (Hoffstetter, 1982, translated). I follow the authors who, since this time, use Simiiformes Hoffstetter, 1974 instead of Anthropoidea Mivart, 1864 (vernacular: simiiform or simian).

Until recently, there was no known primate earlier than those from the Fayum from anywhere in Africa. This situation allowed debates and speculations about the question of simiiform origins. The different options considered a decade ago can be summarized briefly. Simons (1976) doubted the close affinity of *Tarsius* and simiiforms and instead saw the early euprimate radiation as an "unresolved trichotomy" dating back to at least the early Paleocene, with one branch forming the basal "Anthropoidea." This early group was supposed to have evolved on the northern continents, and platyrrhines and catarrhines to have later entered their respective southern continents in middle Eocene time (around 45 MA). Simons later abandoned this view. In their very useful book, Szalay and Delson (1979) proposed an evolutionary scheme in which platyrrhines and cartarrhines branched off from a tarsiiform omomyid ancestor during middle Eocene times (around 40–45 MA). Cartmill and Kay (1978) had advocated a related but different phylogenetic relationship in considering living Tarsiidae as the sister group of simiiforms. Because I share this point of view and am skeptical about the relationship of *Tarsius* and know Eocene fossils, I use instead Omomyiformes (Schmid, 1982) for omomyids and microchoerids (and not tarsiids). Gingerich (1980) reaffirmed his view that a transition between adapiforms and simiiforms took place during the late Eocene in Asia, with *Pondaungia, Amphipithecus,* and the African *Oligopithecus* being transitional forms. Hoffstetter (1982) predicted that simiiforms probably differentiated in Africa and that their separation from their sister group, the tarsiiforms (including omomyids), had to be very ancient, Paleocene (around 53–60 MA), and he designated the unknown intermediates between the common ancestor with tarsiiforms and simiiforms as protosimiiforms. (Hoffstetter had a precise diagnosis of simiiforms including many shared derived characters; hence, his definition of protosimiiforms as a paraphyletic stem group was clear.)

Many new studies have been published during the past decade, e.g., about the basicranial evidence (MacPhee, 1981; MacPhee and Cartmill, 1986) and about postcranials (Gebo, 1986, 1988, 1989; Szalay and Dagosto, 1980, 1988; Ford, 1988; Dagosto, 1988, 1990). Nevertheless, the debate continues. Many authors still favor an omomyid, or tarsiid, hypothesis for simiiform origins; however, the new Fayum oligopithecines *Catopithecus* and *Proteopithecus,* as well as new fossils of *Oligopithecus,* were interpreted to confirm an adapiform hypothesis (Rasmussen and Simons, 1988; Simons, 1989, 1990).

In recent years, several older African mammalian faunas were discovered that have yielded fragmentary but interesting material. From the late Eocene locality of Bir el Ater in the Nementcha mountains (Algeria), one isolated lower molar was described as *Biretia piveteaui* and interpreted as an early catarrhine (de Bonis *et al.*, 1988). From the late Paleocene (Thanetian) of Morocco, Sigé *et al.* (1990) described *Altiatlasius koulchii,* based on isolated teeth. Detailed comparisons with plesiadapiforms and euprimates led these authors to interpret this genus as an omomyid closely related to simiiform origins. This interpretation, publicized by Gingerich (1990), has not been discussed until now. The unknown dental formula of *Altiatlasius* renders a precise statement difficult to make (Rose *et al.*, Chapter 1, this volume), but see below. From the Algerian Glib Zegdou locality, probably late early Eocene, three isolated teeth were described as *Algeripithecus minutus* (Godinot and Mahboubi, 1992). Following a brief character analysis, we interpreted this genus as a simiiform, more closely related to propliopithecines than to oligopithecines. We concluded that this discovery definitely ruled out a connection between simiiforms and middle or late Eocene European adapiforms. From the Tunisian Chambi locality, which is possibly early Eocene, the new primate *Djebelemur martinezi* was recently described (Hartenberger and Marandat, 1992). These authors allocate this genus to the adapiforms ("Adapidae"). They also mention similarities with simiiforms, which would confirm rooting of the latter in the adapiforms. We subsequently described the additional material recovered from Glib Zegdou and named the new taxon *Tabelia hammadae* (Godinot and Mahboubi, 1994). We considered *Algeripithecus, Tabelia,* another species designated cf. *Tabelia,* and the upper molars referred to *Djebelemur* as small simiiforms, which will be informally referred to here as "small bunodont simians." Because doubt remains about the association of the referred upper molars with the type mandible of *Djebelemur,* that name will be used for the mandible only, and the upper molars from Chambi will be designated "cf. *Djebelemur*" (see Discussion). Small bunodont simians plus *Djebelemur* will be informally called "early African simians." New small primates, possibly adapiforms, were found in the new Oman localities, which have yielded primates close to the Fayum ones (Thomas *et al.*, 1988, 1991; Gheerbrant and Thomas, 1992). These small species will not be taken into account until their complete description is published.

The age of these North African faunas is critical for the interpretation of their contained primates. Recently Rasmussen *et al.* (1992) questioned the ages previously proposed for the faunas of Nementcha, Glib Zegdou, and Chambi and suggested much later dates. About Nementcha, three of the taxa used by Rasmussen *et al.* (1992) are families, i.e., taxa too broad to allow precise comparisons. Among these is the family Phiomyidae, which is indeed "widespread through the Fayum section" (Rasmussen *et al.*, 1992, p. 557). However, the only described Nementcha phiomyid is the genus *Protophiomys,* interpreted by Jaeger *et al.* (1985) as more primitive than the Fayum genera, implying that Nementcha is older than Fayum sediments. The primate *Biretia piveteaui* from Nementcha is bunodont and has a distinct wear facet X and a distal fovea (de Bonis *et al.*, 1988). Because primates having these attributes are "notably absent below the level of Quarry V in the Fayum," Rasmussen *et al.* (1992) suggest that Nementcha deposits may be similar in age to the upper Fayum sequence. However, such reasoning would hold true only if the ancestors of the later Fayum primates were precisely identified below the level of Quarry V, thereby giving a time interval for the development of the stated characters. Such is not the case. Few specialists would propose that propliopithecines descended directly from the known oligopithecine genera (see other contributions to this volume). The characters of *Biretia* simply reflect that it is not an oligopithecine and in no way imply an age equivalent to the upper Fayum sequence. On the contrary, its small size and morphology fit well with the evidence from geology and rodents that Nementcha has a late Eocene age (Coiffait *et al.*, 1984; Jaeger *et al.*, 1985; de Bonis *et al.*, 1988). The Bir el Ater locality is a beach deposit that also yielded ostracods of latest Lutetian or earliest Bartonian age, suggesting for this locality an early late or even late middle Eocene age (J.-J. Jaeger, *personal communication*).

I disagree with statements made by Rasmussen *et al.* (1992) about Glib Zegdou hyracoids. They write about *Microhyrax* that it "does generally resemble Fayum taxa." However, this genus was never found in the very abundant Fayum fauna, and its validity was not questioned by Rasmussen (1989), who made comparisons with other Fayum genera and meanwhile agreed with the primitive nature of the primitive characters listed by Sudre (1979). *Microhyrax* remains a primitive hyracoid genus never found until now in the Fayum. About *Titanohyrax,* Rasmussen *et al.* (1992) write that "a large species of *Titanohyrax* from Glib Zegdou is similar in size and structure to *T. ultimus* of FFZ4" (Fayum upper sequence). However, Sudre (1979) had described several characters, including brachyodonty, showing that *T. mongereaui* from Glib Zegdou was more primitive than *T. ultimus* from the Fayum. Rasmussen *et al.* (1992) propose that "the L-41 species [related to *Titanohyrax*] may also be more primitive than the teeth attributed to this genus from Glib Zegdou," but they do not give any character supporting this statement (they mention that the L-41 species have no metastylid, but only upper teeth are known from the Glib Zegdou *T. mongereaui*). The two new Fayum species "related to *Titanohyrax*" would be meaningful if they were shown to be ancestral to this genus, but this appears less likely in view of the discovery of a new Chambi *Titanohyrax* (see below). Rasmussen *et al.* (1992) also did not mention that Sudre (1979) had described characters of *Megalohyrax gevini* from the near and stratigraphically correlated Gour Lazib localities, which he interpreted as more primitive than in *M. eocaenus* from the Fayum. Thus, published information on Glib Zegdou and Gour Lazib hyracoids shows that they are more primitive than their congeneric species from the Fayum (but detailed comparisons with the L-41 hyracoids remain to be done).

Likewise, Rasmussen *et al.*'s interpretation of Chambi hyracoids is difficult to accept. The first mention of Chambi hyracoids was: "cf. *Pachyhyrax* de petite taille; cf. *Sagatherium*" (Hartenberger *et al.*, 1985). From these tentative attributions Rasmussen *et al.* (1992) speculated that the Chambi fauna had to be geologically younger, because *Pachyhyrax* occurs only at the uppermost levels of the Fayum sequence. However, the actual study of these specimens by Court and Hartenberger (1992) concludes that this material belongs to a new species of *Titanohyrax*, more primitive by its small size and low crown height than any previously known species of the genus. They also suggest a close affinity with *T. mongereaui* from Glib Zegdou, adding evidence for the probable older age of both localities. Hyracoids were abundant in Africa and diversified early (Rasmussen, 1989). They should be an important source of biostratigraphic information once their lineages are better studied. On the whole, the published evidence concerning hyracoids does not support the proposition of Rasmussen *et al.* (1992) that forces the age of Glib Zegdou and Chambi into that of the Fayum sequence. On the contrary, this evidence agrees with the previously older ages proposed by their discoverers.

Studies of the Glib Zegdou and Chambi faunas are continuing and give other clues to their age. The primates include three genera unknown from the Fayum; each is smaller and more primitive in many ways than its closest Fayum relative (see below). Their general degree of evolution suggests, in my opinion, a late early to early late Eocene age. The rodents from both localities represent three new genera of a new family, which is rooted among the most primitive rodents, the ischyromyids; some of these genera are considered ancestral to the Nementcha anomalurids (Vianey-Liaud *et al.*, 1994). These last authors consider this rodent radiation as preceding a later immigration of phiomyids into Africa. In any case, the recognizable derivation of these rodents from ischyromyids confirms an age between late early and middle Eocene (Vianey-Liaud *et al.*, 1994). Sigé (1985) studied the bats from the Fayum and described the genus *Philisis*. Also, he has studied the bats from Chambi and listed a number of characters of *Dizzya* from Chambi, which he interpreted as more generalized in comparison with *Philisis*. Sigé concluded that *Dizzya* was ancestral to or close to the ancestry of *Philisis* (Sigé, 1991, p. 362). Hence, the evidence from bats corroborates for Chambi an age older than that of the Fayum. Other isolated teeth from Glib Zegdou have condylarthran affinities, which is also the case for *Chambius* from Chambi, described as a macroscelidid (Hartenberger, 1986); these taxa give a more primitive stamp to these faunas in comparison with the Fayum.

In summary of the above, the mammalian evidence on the whole clearly suggests a late early or middle Eocene age for Glib Zegdou and Chambi. However, it is difficult to be more precise in the absence of similar and better-dated African faunas and in the absence of close affinities with mammalian faunas from neighboring areas. The nonmammalian evidence comes from the study of charophytes. In fact, these algae allowed the first recognition of Eocene beds in the Dra region (Grambast and Lavocat, 1959). Later they allowed the suggestion of an early Eocene age for Glib Zegdou (Gevin *et al.*, 1974). The evidence from charophytes should not be underestimated, as it relies on several species in common with the European early Eocene (M. Feist and F. Mebrouk, *personal communication;* Mebrouk's thesis on this subject should soon be published). However, the rodents and primates are too advanced to be earliest Eocene, and Sigé also found surprising the presence of rhinolophoid and vespertilionoid bats in the "early Eocene" of Chambi. More data and better correlations will be needed to define more precisely the age of Glib Zegdou and Chambi, but this age in any case must lie between the later early Eocene and the later middle Eocene. This age is then much older than the Fayum sequence, especially if the Gebel Qatrani Formation is mostly Oligocene (Fleagle *et al.*, 1986) or even entirely Oligocene (Gingerich, 1993). These Maghreb faunas progressively disclose the earlier mammalian history of Africa, and their primates add critical information on early simian history.

In order to reassess these fossil primates, their most important characters will be analyzed, and polarities and homologies will be searched for.

Character Analyses

Upper Molar Characters

Certainly the most remarkable characters found in *Algeripithecus* are those of the upper molars, especially M^2. These characters include labial expansion of the paracone and to a lesser degree the metacone, both having low labial slopes, lingual expansion of protocone wall, bunodonty, presence of a large hypocone, and the distinctive characters of the lingual cingulum. In our first inquiry we had not realized that some late Wasatchian and Bridgerian anaptomorphines showed a number of these characters (Bown, 1979; Bown and Rose, 1987). Hence, these genera, *Absarokius, Strigorhysis, and Gazinius,* will be taken into account with the relevant adapiforms and simiiforms. Also, we had not considered the new material of *Moeripithecus* (Thomas *et al.*, 1991). This genus has upper molars in several respects more primitive than those of *Aegyptopithecus* and *Propliopithecus* and is thus very helpful in reconstructing a propliopithecine morphotype.

Labial Expansion of the Paracone

In *Absarokius,* the labial extension of the paracone is as developed, and its labial slope is almost as low, as in *Algeripithecus* (Figs. 1 and 2). This character is just as marked in *Strigorhysis* but less so in *Gazinius* (Figs. 1 and 4). Such a paracone extension is incipient in some *Teilhardina* (e.g., *T. crassidens,* Bown and Rose, 1987) and some *Tetonius* but is not found in other anaptomorphines. Labial extension of the paracone is very rare in omomyines (*Rooneyia*), and I have not detected it in any microchoerine. Hence, it is not frequent in omomyiforms as a whole. A labial expansion of the paracone is exceptional in adapiforms. It is sometimes incipient in the most bunodont forms, such as some larger *Cantius* and *Pelycodus*. Only one of the three species of the European cercamoniine *Periconodon* shows such a development: only *P. huerzeleri*

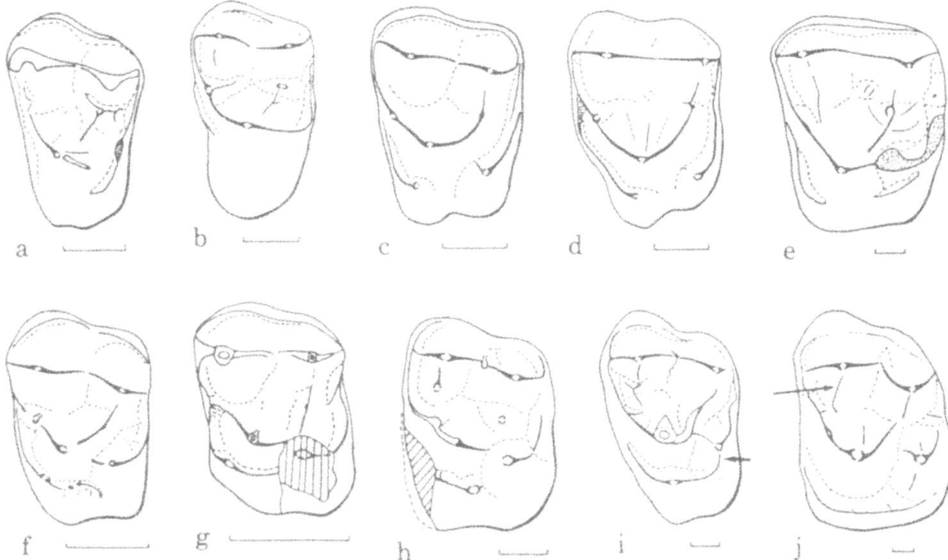

Fig. 1. Second upper molars of a series of fossil primates, all in occlusal view and drawn to the same transverse width. All scale bars are 1 mm. On Figs. 1, 2, 4, and 5, some drawings are reversed right to left to make all teeth appear as left upper molars, with their posterior side on the right. a, *Absarokius abbotti;* b, *Strigorhysis bridgerensis;* c, *Periconodon helveticus;* d, *P. huerzeleri;* e, *Pelycodus* sp.; f, *Algeripithecus minutus;* g, cf. *Djebelemur;* h, *Apidium moustafai;* i, *Moeripithecus markgrafi;* j, *Aegyptopithecus zeuxis.* The arrow on *Moeripithecus* (i) points to the newly defined hypocrista (see text), and that on *Aegyptopithecus* (j) points to the hypoparacrista. Note the narrow lingual part and primitive absence of hypocone in the two omomyids (a and b) and the nonexpanded paracone in the most bunodont adapiform *Pelycodus* (e).

has a limited labial expansion of its paracone, but the cusp is not expanded in either *P. helveticus* or *P. jaegeri* (Figs. 1 and 2; Stehlin, 1912; Godinot, 1988a).

Labial expansion of the paracone seems to typify propliopithecines. It is present and well marked in *Aegyptopithecus* and in *Moeripithecus,* underlined in occlusal view by the lingual shift of the metacone (Fig. 1). It seems also to be present in *Propliopithecus chirobates* (Kay *et al.*, 1981). The situation in parapithecids is more difficult to establish. An examination of the photographs published by Szalay and Delson (1979) had led us to conclude that parapithecids did not possess that character (Godinot and Mahboubi, 1992). However, we had not realized that a drawing of *Parapithecus* clearly showed paracone and metacone labial expansion (Szalay and Delson, 1979, Fig. 150). Confirmation was given by a close examination of some casts of *Apidium moustafai* (YPM 20913), *A. phiomense* (the type maxilla YPM 23995), and *Simonsius grangeri* (DPC 2372). It is clear that a paracone labial expansion and low slope are present in both genera (Figs. 1 and 2). The labial protrusion is less marked than in *Algeripithecus,* but it appears still to be quite strong if the labial cingulum attenuation is taken into account. The character is not present on the

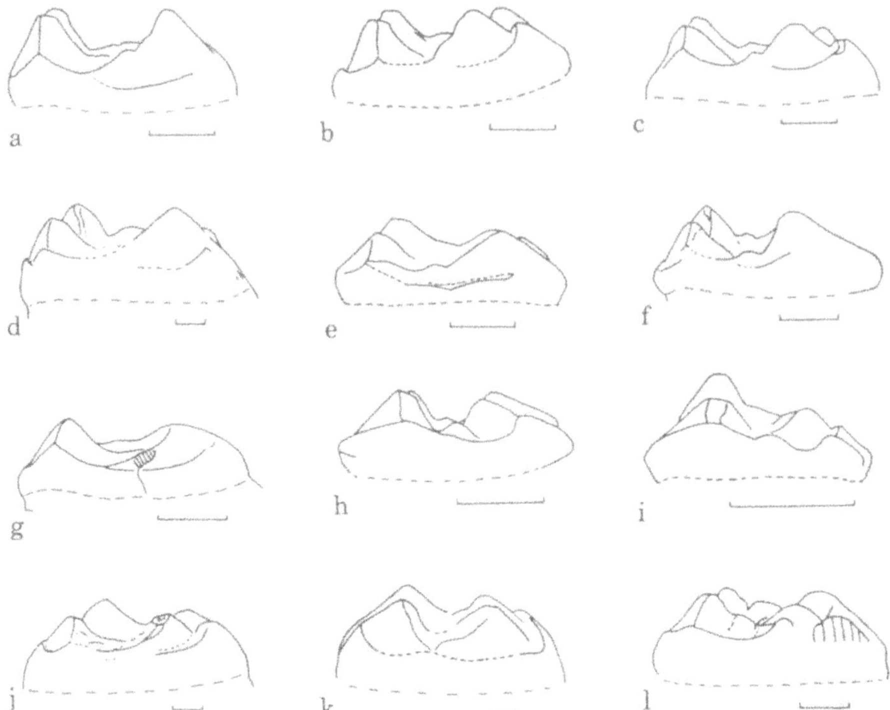

Fig. 2. Anterior profile views of the second upper molars in a series of fossil primates, all drawn to the same width. Scale bars are 1 mm. The lingual side is toward the right. a, *Donrussellia provincialis;* b, *Periconodon helveticus;* c, *P. huerzeleri;* d, *Pelycodus* sp.; e, *Absarokius abbotti;* f, *Strigorhysis bridgerensis;* g, *Tabelia hammadae;* h, *Algeripithecus minutus;* i, cf. *Djebelemur;* j, *Moeripithecus markgrafi;* k, *Aegyptopithecus zeuxis;* l, *Apidium moustafai.* Note the high and pointed cusps in the primitive adapiform *Donrussellia* (a) and the different stages of cusp lowering and rounding in the other taxa.

partial upper molar of *Qatrania* (Simons and Kay, 1988), but this tooth probably is an M^1. From these data, I conclude that lingual expansion of the paracone, despite its less salient nature in occlusal view than in propliopithecines, is present in *Apidium* and *Parapithecus* and probably is typical of parapithecids. In fact, this character presents some variation in the small bunodont simians too. Labial protrusion is much less marked in cf. *Djebelemur* than in *Algeripithecus* and *Tabelia* and represents either a secondary reduction or a misinterpretation of an M^1 as an M^2 (Figs. 1 and 2).

Some flaring of the paracone is present in *Proteopithecus,* but it does not look similar to the character studied in the other genera, as it occurs on teeth having very high cusps and crests (Simons, 1989). Furthermore, this character is clearly absent in *Catopithecus* and *Oligopithecus* (Rasmussen and Simons, 1988). Thus, paracone labial expansion seems typically absent in oligopithecines. I consider *Saimiri* to be the closest living approximation of the platyrrhine M^2 morphotype, because this genus retains the most transversely elongated upper molars (see also Kinzey, 1973; Rosenberger, 1977; Kay, 1980). *Saimiri* has no paracone labial expansion (there is a labial paracone expansion in several callitrichines, however, on M^1; e.g., *Callimico*, Kay, 1980; Fig. 3). There is no strong paracone labial expansion in *Branisella* and *Szalatavus* (Hoffstetter, 1969; Rosenberger *et al.*, 1991). I will provisionally assume that the character was absent in these genera, but specimens or casts should be studied to confirm this (a moderately low labial slope might be present in *Branisella*).

In a labial view, the M^2 of many of these taxa has a peculiar shape because the notch between paracone and metacone is shallow, its two slopes are low, the cingulum is broadly curved, the parastyle has been lost, and the base of the crown is high (Fig. 3). There are variations. The condition with a very high cingulum is achieved in *Algeripithecus*, cf. *Djebelemur*, *Absarokius*, and *Gazinius;* in *Apidium* with further reduction of the cingulum; in *Moeripithecus* and *Aegyptopithecus* (in the last genus, it is further modified; the cingulum is very thin and seems to be duplicated on CGM 40237; there are probably variations of the details). These labial characters are usually less marked on the M^1, including those attributed to *Algeripithecus minutus.* Other less bunodont primates have on their M^2 a deeper notch between paracone and metacone and a lower cingulum (Fig. 3). Even in one of the most bunodont adapiforms, *Pelycodus jarrovii,* there are slight modifications of the cingulum, possibly some dorsal shift, but the morphology remains less specialized, and this in a much larger species. The aspect of *Altanius* is very strange, with a relatively deep notch, a very high labial wall, and a low cingulum (Fig. 6). This appears derived and very unlike other euprimates, adding evidence for the plesiadapiform affinities of this genus (see Rose *et al.*, Chapter 1, this volume).

On the presumed M^2 of *Tabelia,* the notch is low, and the cingulum seems less high than in *A. minutus,* but the exact crown height is obscured by wear of the crown base. The M^2 of cf. *Djebelemur* seems to show in labial view a real increase in crown height, but this should be measured (a difficult task because the ventral limit of the crown often is not clear). The appearance of basal high-crownedness in the most bunodont teeth is partly a result of the occlusal displacement of the cingulum but also possibly represents in part a real increase (Figs. 3 and 6).

Paracone labial expansion and associated cingular modifications are interpreted by me to be derived characters. The fact that the mentioned anaptomorphines have no hypocone, added to the absence of these characters in the earliest omomyids, show that these labial characters resulted from a convergent development in the two groups (a direct origin of these North African genera from one of the North American anaptomorphines seems unlikely for many other reasons; see below). A limited paracone labial expansion also must be convergent between one species of *Periconodon* and the North African groups studied here (see below). Because it is generally uncommon in primates, paracone labial expansion is potentially a derived homology shared by the small bunodont simians and propliopithecines and parapithecids.

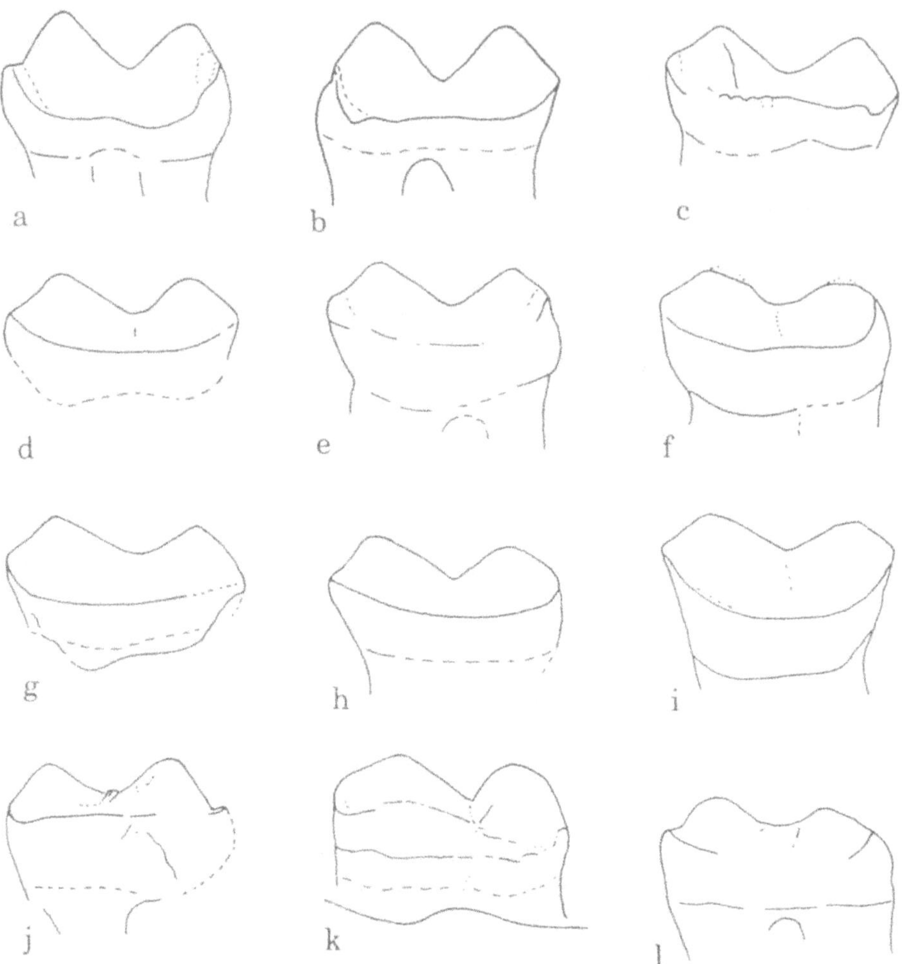

Fig. 3. Labial views of the second upper molars in a series of fossil primates, all drawn to the same anteroposterior length. On Figs. 3 and 6, some teeth are reversed right to left to make them appear as right upper molars, with their paracone and parastyle on the left side. a, *Teilhardina belgica;* b, *Donrussellia provincialis;* c, *Pelycodus* sp.; d, *Absarokius abbotti;* e, *Strigorhysis bridgerensis;* f, *Gazinius amplus;* g, *Tabelia hammadae;* h, *Algeripithecus minutus;* i, cf. *Djebelemur;* j, *Moeripithecus markgrafi;* k, *Aegyptopithecus zeuxis;* l, *Apidium moustafai.* Note high pointed cusps and narrow and low-situated cingulum in the two most primitive taxa (a and b) and different degrees of cusp and notch lowering and of cingulum thickening in the other taxa.

Lingual Expansion of the Protocone

Another of the noticeable characters of *Algeripithecus* is the lingual expansion and associated shallow slope of the protocone wall. This feature may be partly obscured when the lingual cingulum is very thick, as in propliopithecines, or when a large pericone modifies the profile in anterior view. However, it can

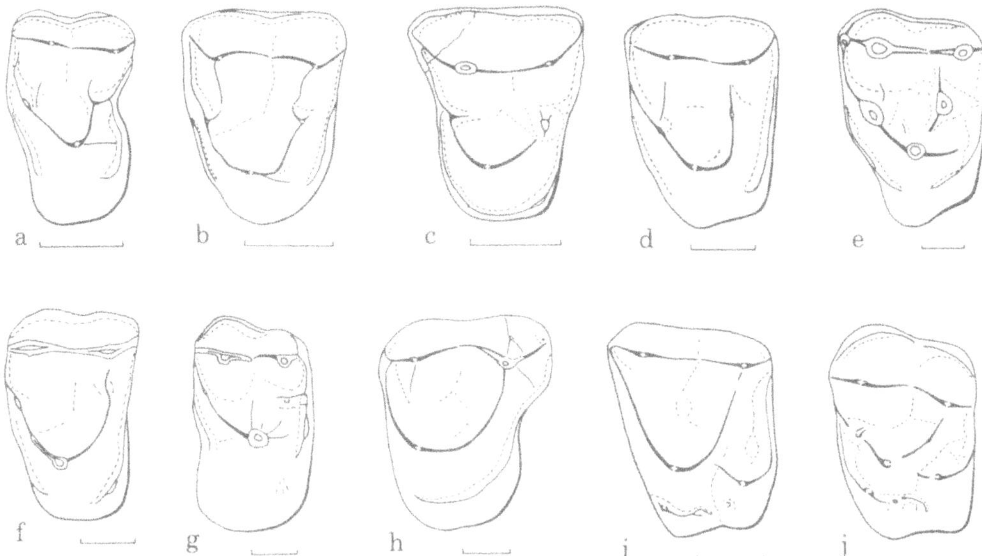

Fig. 4. Second upper molars of a series of fossil and living primates, all in occlusal view and drawn at the same transverse width. All scale bars are 1 mm. a, *Teilhardina belgica;* b, *Nannopithex zuccolae;* c, *Altiatlasius koulchii;* d, *Donrussellia provincialis;* e, *Cantius eppsi;* f, *Tarsius* sp.; g, *Gazinius amplus;* h, *Oligopithecus savagei;* i, *Saimiri sciureus;* j, *Algeripithecus minutus.* Note the long lingual slope of the protocone in *Teilhardina* (a), *Cantius* (e), and *Gazinius* (g). It is modified by the cingular structures in *Oligopithecus* (h) and *Algeripithecus* (j).

still be studied separately in a number of genera. Lingual expansion of the protocone wall is common in Eocene primates and in plesiadapiforms, which renders its polarity difficult to assess (Fig. 4). Because it is present in *Teilhardina belgica,* in early *Cantius,* in *Altanius orlovi* (Gingerich *et al.,* 1991), and in the small bunodont simians, it could be a primitive euprimate character. In support of this, Szalay (1976) and Rosenberger and Szalay (1980) considered the overall shape of the cheek teeth as primitive in *Teilhardina.* However, lingual expansion is not marked in *Donrussellia provincialis,* the most primitive adapiform (Godinot, 1981). Nor is it marked in *Omomys* (Szalay, 1976) and presumably other primitive omomyines, nor in the oldest microchoerine (Godinot *et al.,* 1992), suggesting that nonexpansion was probably the primitive omomyiform condition. Lingual expansion is also small in the Paleocene *Altiatlasius* (Sigé *et al.,* 1990). It seems to me that lingual expansion of the protocone is more likely a derived condition frequently occurring in early groups than a primitive primate trait (see also Rose *et al.,* Chapter 1, this volume).

A marked lingual expansion can be observed in *Strigorhysis,* which possesses a broader and more extended lingual part than *Absarokius.* Lingual extension is even more exaggerated in the larger genus *Gazinius* (Fig. 4). In these anaptomorphines, other characters and stratigraphic position show the lingual protocone wall to increase in the most derived of these genera (Bown, 1979; Bown and Rose, 1987). Here the polarity also is toward an increase in this character. A long lingual protocone wall is present in early *Cantius* and remains a very frequent character in notharctines, becoming secondarily reduced in the largest *Notharctus* species (Gregory, 1920). This character is less frequent in European cercamoniines. It developed in some *Periconodon,* with a maximum in *P. huerzeleri* (Figs. 1 and 2), but without reaching the extreme low slope of *Algeripithecus.*

In the small bunodont simians, protocone lingual expansion and shallow slope are less marked in cf. *Djebelemur* than in the two other genera. This may well be secondary in this genus, linked to an overall reduced transverse length (if the upper molar in question really is an M^2; it might still be an M^1). Lingual protocone expansion and a shallow protocone lingual slope occur in parapithecids, being a part of their bunodonty. When a large lingual cingulum partly covers it, the slope of the remaining protocone is low (*Apidium moustafai* in Fig. 2), and when the lingual cingulum is absent or reduced, the long protocone wall is evident, as in *Simonsius grangeri* (Szalay and Delson, 1979, Fig. 150). The case of propliopithecines is more difficult to assess: they have a marked lingual development, but mainly of the cingulum. The lingual slope of the protocone in *Aegyptopithecus* is not as low as in *Algeripithecus* (Godinot and Mahboubi, 1992). Surprisingly, this slope is quite steep on the M^2 of *Moeripithecus* (Fig. 2), being lower on the M^1 of the same specimen. I do not see a good reason to consider a possible secondary reincrease of this slope in *Moeripithecus,* which is otherwise bunodont, has transversely elongated upper molars, and has strong cingular characters. Hence, I now consider that the primitive steep-slope configuration must be retained in *Moeripithecus* and indicates the state of the character in the propliopithecine morphotype. The moderately shallow slope present in *Aegyptopithecus* is then considered convergent with that of the other bunodont forms.

In oligopithecines, the state of this character is difficult to evaluate. They appear to have less exaggerated morphological differences between M^1 and M^2 than in the more bunodont simians. From the published accounts emphasizing similarities with European cercamoniines (Rasmussen and Simons, 1988; Simons, 1989) I had supposed that *Oligopithecus* had a relatively steep lingual slope. However, examination of a cast revealed a long and moderately low lingual slope (Fig. 5). This seems to be the case in *Proteopithecus.* I was unable to evaluate this character in *Catopithecus* because published photographs show the tooth rows with inclination (Simons, 1989, 1990). Profile views of the upper teeth of these two genera were not published. I will provisionally assume for oligopithecines a moderate protocone lingual expansion and low slope as in *Oligopithecus.*

For platyrrhines, the lingual slope of the protocone seems to be steep in the Deseadan *Branisella. Saimiri* has a very steep lingual protocone slope (Fig. 5). Early platyrrhines are provisionally assumed to exhibit no significant protocone lingual expansion but marked lingual cingular development.

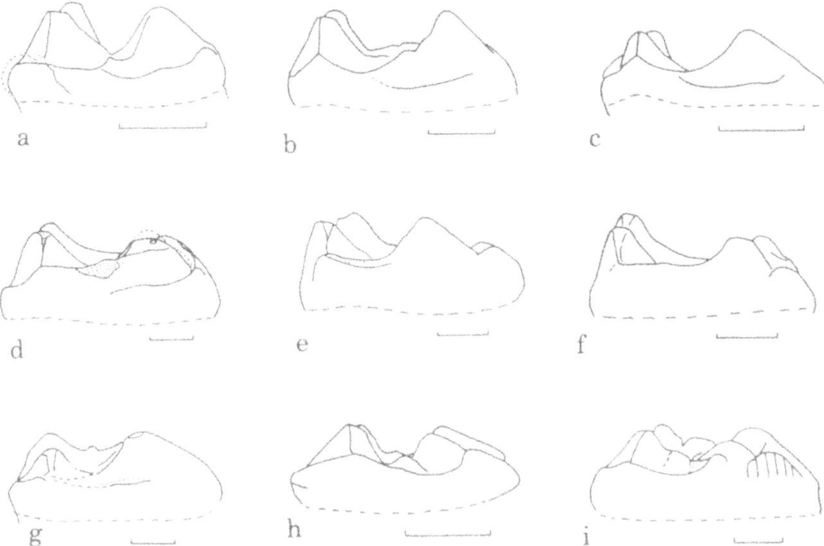

Fig. 5. Anterior views of the second upper molars in a series of fossil and living primates, showing the increase in bunodonty from primitive forms (top row) to derived, very bunodont forms (bottom row). All are drawn to the same transverse width, and all scale bars are 1 mm. a, *Altiatlasius koulchii;* b, *Donrussellia provincialis;* c, *Teilhardina belgica;* d, *Europolemur dunaifi;* e, *Oligopithecus savagei;* f, *Saimiri sciureus;* g, *Gazinius amplus;* h, *Algeripithecus minutus;* i, *Apidium moustafai.* The relatively large and bunodont adapiform *Europolemur* does not show the protocone expansion present in *Oligopithecus.* Note also that *Teilhardina* (c) is advanced over *Donrussellia* (b) in cusp lowering and that *Altiatlasius* (a) further differs from them by higher, probably primitive labial cusps.

In view of this distribution, a moderate protocone lingual expansion is potentially a homologous derived character in all the African forms considered. (It would shift as a simian character if it were shown to be present in early platyrrhines too.) A similar state in some *Periconodon* species is probably convergent (see other characters below). A marked protocone lingual expansion and shallow slope is a synapomorphy of the *Algeripithecus* group and parapithecids, which is probably homologous. For the cited anaptomorphines, which have a different trigon construction, convergence is clear.

Bunodonty

Bunodonty is a term used to characterize teeth having a low relief, rounded cusps, and poorly developed shearing crests. It is often correlated with several of the preceding characters, e.g., the peculiar labial shape of some M^2 or the low lingual slopes of protocone and hypocone (Figs. 5 and 6). However, many of the characters discussed above do occur independently of each other; e.g., a long lingual slope of the protocone can develop in conjunc-

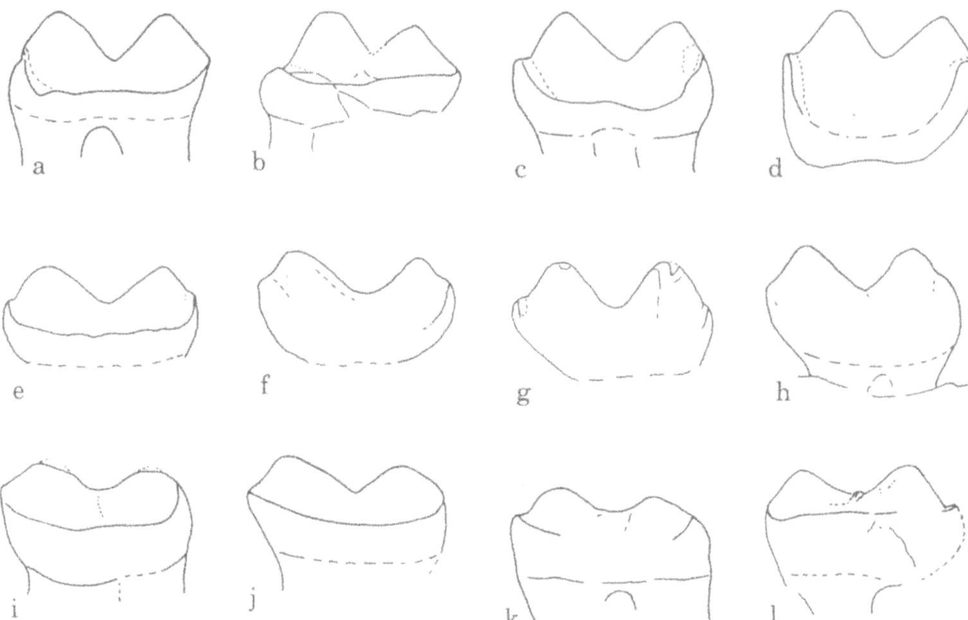

Fig. 6. Labial views of the second upper molars in a series of fossil primates, showing the increase in bunodonty from primitive forms (top row) to derived, very bunodont forms (bottom row). All are drawn to the same anteroposterior length. a, *Donrussellia provincialis;* b, *Altiatlasius koulchii;* c, *Teilhardina belgica;* d, *Altanius orlovi;* e, *Europolemur dunaifi;* f, *Periconodon* sp.; g, *Oligopithecus savagei;* h, *Saimiri sciureus.;* i, *Gazinius amplus;* j, *Algeripithecus minutus;* k, *Apidium moustafai;* l, *Moeripithecus markgrafi.* Note also the secondary cingulum reduction in *Periconodon* (f), *Oligopithecus* (g), and *Saimiri* (h) and the very peculiar aspect of *Altanius* (d), with the basal expansion of its crown suggesting plesiadapiform affinities.

tion with the retention of a high protocone (*Teilhardina,* two *Periconodon* species), labial expansion of the paracone can occur without lingual expansion of the protocone (*Perodicticus*), and bunodonty may exist without much paracone lingual expansion (e.g., modern hominoids). Hence, the characters mentioned above must be considered separately from bunodonty. Bunodonty is common in large living primates (hominoids, a number of cercopithecoids, and some platyrrhines) but not in living strepsirhines. It occurs in several omomyiforms but is very rare in adapiforms. Furthermore, what is remarkable in *Algeripithecus* is the extreme degree of bunodonty achieved by such a small species (Godinot and Mahboubi, 1992). In other groups, similar-sized species usually have much higher cusps and crests. Parapithecids also are extremely bunodont, as shown by the large cusps of *Qatrania* (Simons and Kay, 1988). Propliopithecines have a less extreme but still high degree of bunodonty, as can be seen on profile views (Fig. 2). Oligopithecines are not bunodont but have cusps and shearing crests broadly similar to those of adapiforms such as *Protoadapis* and *Europolemur* (Fig. 5; Rasmussen and Simons, 1988; Simons,

1989). Assessment of the platyrrhine morphotype is difficult: *Saimiri* is not bunodont, nor does it possess a very high relief. *Branisella* seems to have possessed a moderate degree of bunodonty (Hoffstetter, 1969). The platyrrhine morphotype might have possessed a low to moderate degree of bunodonty. The result of this distribution is a synapomorphy among small bunodont simians, propliopithecines, and parapithecids for a high degree of bunodonty and a further shared derived state in the extreme degree of bunodonty achieved in the small bunodont simians and parapithecids.

Hypocones

A molar hypocone is common in primates. In adapiforms, *Donrussellia* and primitive *Cantius* have no hypocone. This cusp later develops in larger species of *Cantius, Protoadapis,* and *Europolemur.* It is well known that North American notharctines developed a hypocone located on the protocone fold, not homologous to the more common cingular hypocones (Stehlin, 1912), called a pseudhypocone. There is no hypocone in the anaptomorphine *Teilhardina,* in the primitive omomyine *Chumashius* (Szalay, 1976), or in the earliest microchoerine *Nannopithex zuccolae* (Figs. 4 and 7). Hence, in omomyiforms too, the hypocone developed independently a number of times in different lineages. Not surprisingly, hypocone shape varies in the different groups in which it developed. One common type of hypocone is a cusp growing at the lingual end of the posterior cingulum, as in *Periconodon* (Fig. 1). In this case, the crest issuing from the cingulum is called a posthypocone crista, and if a smaller anterolabial crest is present, it is called a prehypocone crista (Szalay and Delson, 1979, Fig. 6). I propose to call this type a semicrestiform hypocone (with or without a prehypocone crista). Another common type is the crestiform hypocone, issued from a continuous cingulum and then maintaining two crests linking its summit to these cingula. This type also may have a prehypocone crista in the same position as in the other type, rendering it necessary to use another name for the crest running anterolingually or lingually from the summit. This might be called either a hypohypocone crista, by analogy with the two other hypocone crests, or hypohypocrista by analogy with the lingual crests issued from paracone and metacone (hypoparacrista and hypometacrista). Because "hypohypo" is awkward, I propose simply to call this crest a hypocrista (e.g., *Moeripithecus,* Fig. 1). When growing, the hypocone can retain crests linking it to the cingulum from which it developed, but these links may also be lost or obscured when the hypocone becomes large and bulbous.

The first remarkable aspect of the hypocone of *Algeripithecus* is its large size relative to tooth size (Figs. 1, 4, and 7). In many Eocene primates, hypocone development seems to be correlated with increase in size, probably for functional reasons, and possibly also because of developmental constraints. However, they develop in different sizes in different groups. In early Eocene adapiforms, the hypocone develops in relatively large species such as *Cantius*

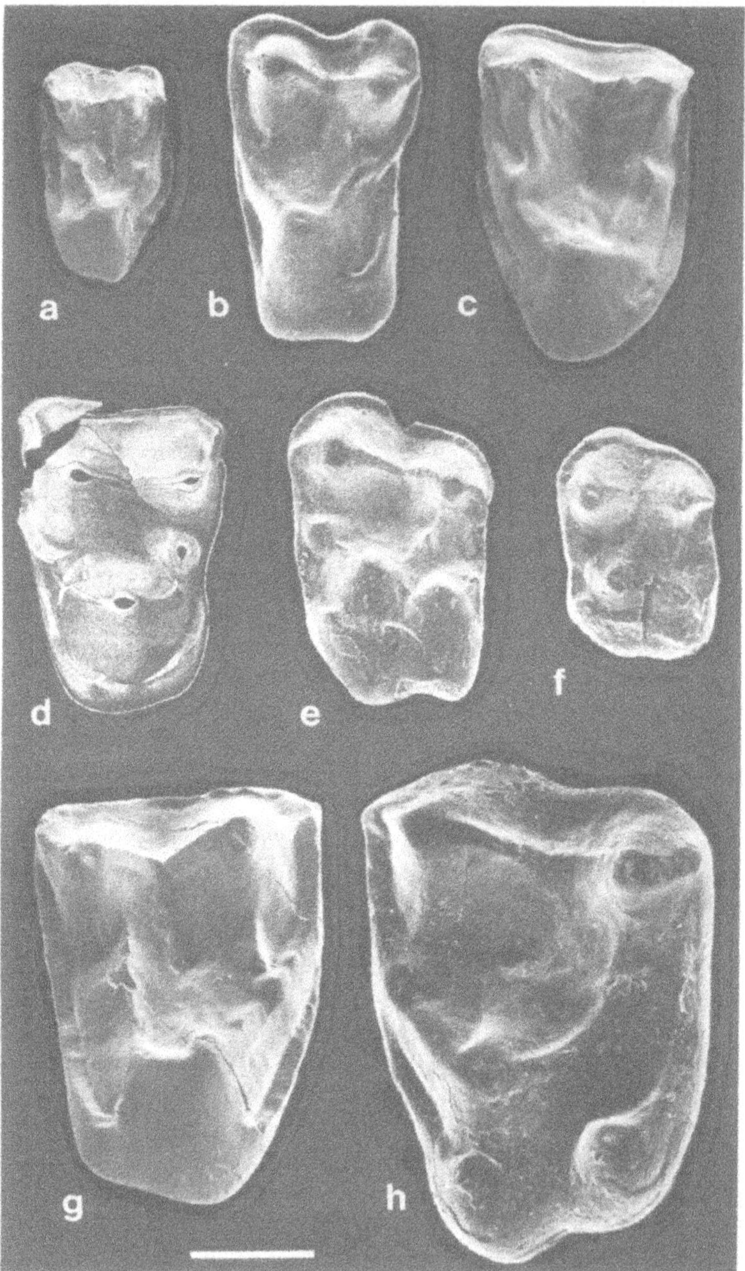

Fig. 7. Second upper molars of a series of fossil primates, all in occlusal views and at the same scale to allow a comparison of their relative sizes (scale bar is 1 mm). Top row, an Asiatic plesiadapiform, *Altanius orlovi* (a) and two European omomyiforms, *Teilhardina belgica* (b) and *Nannopithex zuccolae* (c). Middle row, three early North African primates, *Altiatlasius koulchii* (d), *Algeripithecus minutus* (e), and cf. *Djebelemur* (f). Bottom row, two European adapiforms, *Don-*

and *Protoadapis* (it may later by secondarily reduced, e.g., in the tiny *Anchomomys gaillardi*, probably correlated with size reduction, and also in some very large species, e.g., late *Leptadapis* sp.). Only adapines might have had a morphotype possessing a relatively large hypocone, as suggested by *Microadapis* and *L. ruetimeyeri*. However, both are already much larger than *Algeripithecus*. In omomyiforms, a hypocone starts to develop in species having a size very close to that of *A. minutus*, but it develops slowly and remains relatively small (e.g., the late middle Eocene *Nannopithex filholi*). It becomes middle-sized in some later and larger omomyiforms (e.g., *Necrolemur antiquus*, *Rooneyia*, *Microchoerus*). Hence, it seems probable that in *Algeripithecus minutus* the very large size of the hypocone in relation to body size typifies a group quite distinct from both adapiforms and omomyiforms. Because parapithecids also possess extremely large hypocones, e.g., larger than the protocone in some *Apidium*, and they also do not have a very large size, this character is a good potential homology between them and two of the small bunodont simians (*Algeripithecus* and *Djebelemur; Tabelia* has a smaller hypocone and is probably primitive for that).

The large cuspidate hypocone of *Aegyptopithecus* was considered as having lost its major link with the lingual cingulum and as probably shared-derived with that of *Algeripithecus* (Godinot and Mahboubi, 1992). However, the smaller size of this cusp and its well-developed hypocrista in *Moeripithecus* (Thomas *et al.*, 1991) probably more closely reflect the primitive condition for propliopithecines. Convergence in hypocone growth of *Aegyptopithecus* is not surprising in view of the very common size increase of this cusp in primates. Convergence in reduction of the hypocrista is also not surprising, as it is a consequence of a similar inflation of the cuspidate hypocone. In fact, this crest can still be discerned on M^2 of *Aegyptopithecus*, as shown in Fig. 1 and on the M^1 on Fig. 10 of Hiiemäe and Kay (1973). This new morphotypic condition for propliopithecines does not allow us to separate them from parapithecids on the basis of hypocone shape. In parapithecids, in spite of the large size and rounded shape of the hypocone, the hypocrista often can be observed (e.g., YMP 23995, type maxilla of *Apidium phiomense*, YPM 20913, *Apidium moustafai*, and *Simonsius grangeri* in Fig. 150 of Szalay and Delson, 1979). Both parapithecids and propliopithecines clearly derive from forms having a crestiform hypocone with a prehypocone crista. The fact that *Tabelia* has a smaller crestiform hypocone suggests that the small bunodont simians also probably are derived from a form with a crestiform hypocone. *Algeripithecus* has probably lost its hypocrista in relation to its marked hypocone size increase and swelling and is thus apomorphous for this loss.

---

russellia provincialis (g), and *Periconodon huerzeleri* (h). All are scanning electron micrographs of casts except for d, which is reproduced from a drawing by Sigé *et al.* (1990). Note the small size of the two early African bunodont forms, e and f, in comparison with the earliest Eocene *Teilhardina* (b) and *Donrussellia* (g). The species of *Periconodon* (h) showing the most convergent characters with simians is larger and in fact much less bunodont.

In oligopithecines, *Catopithecus* and *Oligopithecus* have a small crestiform hypocone. *Proteopithecus* has a larger hypocone, which remains crestiform and without a prehypocone crista. Hence, it is a type of hypocone quite distinct from that of the preceding groups (or that at least developed in a distinct way). In platyrrhines, the M^2 of *Saimiri* shows a hypocone sometimes without a clear hypocrista, sometimes with a not very salient prehypocone crista, but also often with these structures well developed (e.g., Fig. 4; Fig. 6a of Kinzey, 1973; Fig. 7c of Hiiemäe and Kay, 1973). *Branisella* shows a crestiform hypocone without a prehypocone crista. *Szalatavus* has a smaller hypocone, probably of the same crestiform type (Rosenberger *et al.,* 1991). The platyrrhine morphotype probably had an M^2 with a crestiform hypocone and without a prehypocone crista. This in turn provides a shared derived character linking platyrrhines to oligopithecines.

Lingual Cingulum and Pericone

A continuous lingual cingulum on M^2 is irregularly present in Eocene taxa. It is possibly always absent in anaptomorphines (rare on M^1; two individuals illustrated by Bown and Rose, 1987) and is rare in microchoerines. A continuous lingual cingulum on M^2 is more common in omomyines and was considered a shared-derived character of the two genera *Omomys* and *Chumashius* by Szalay (1976). I agree with this interpretation. The resulting view is that the continuous lingual cingulum was absent in early omomyiforms and developed early in only one of the three subfamilies, the Omomyinae. A similar mosaic picture is found in the adapiforms, where this cingulum is absent in primitive forms such as *Donrussellia* and in some later forms such as *Anchomomys, Buxella,* and *Pronycticebus* but is present in some later and larger cercamoniines (*Europolemur klatti, Mahgarita*). It is present in the early *Cantius eppsi* and remains very frequent in later notharctines (Gregory, 1920; Gingerich and Simons, 1977), but there are some exceptions (e.g., one *Cantius frugivorus* out of four, and *C. angulatus* and *Copelemur tutus* illustrated in Beard, 1988). In adapines, a continuous lingual cingulum is frequent; however, it is not present in one of the earliest, *Leptadapis (Paradapis) ruetimeyeri.* Hence, it was probably not present in the adapine morphotype and developed later in several lineages. In adapiforms, too, the picture is that a continuous lingual cingulum was not present in the earliest forms. Only in one subfamily, the notharctines, did it develop early enough (and still at a rather large size) to typify this group.

In *Algeripithecus minutus* an incomplete lingual cingulum is present; irregularities on the low slope give the impression of a large structure, but the discontinuous crests are not thick (Fig. 7), and the other M^2 shows a large interruption of the lingual cingulum. A continuous lingual cingulum clearly is present in cf. *Djebelemur.* It is weak and interrupted in *Tabelia.* However, *Tabelia* shows a hypocrista, which suggests that the small bunodont simians derive from a form with a crestiform hypocone and continuous cingulum. This last structure appears in the process of reduction in *A. minutus* and *Tabelia,* probably because of their extreme lingual sloping.

A continuous lingual cingulum is present on the upper molars of propliopithecines, on which it is very thick, and in oligopithecines. It is present in some parapithecids (e.g., *Apidium moustafai,* Fig. 1) and absent in others, e.g., *Simonsius grangeri* (Szalay and Delson, 1979, Fig. 150). However, in this latter case, the crestiform hypocone points toward its derivation from a continuous cingulum, probably lost secondarily from the extremely shallow sloping of the protocone wall. A continuous lingual cingulum is present in *Branisella* and present but attenuated in *Szalatavus.* It is variable but often present in *Saimiri.* It was probably present in the platyrrhine ancestor (Kinzey, 1973; Kay, 1980). Because it is present in all these groups, the lingual cingulum is probably a character typical of simian primates. It must have been acquired by them very early and is a potential homology of small bunodont simians and undoubted later simiiforms. Its presence in *Altiatlasius* might be a shared derived feature with simiiforms (Godinot and Mahboubi, 1992). Here, too, acquisition of this character at a very small body size helps to show probable convergence in other groups that acquired it at larger sizes (Fig. 7).

In some taxa, instead of a continuous lingual cingulum, an irregular bumpy chain of cuspules links the hypocone and the lingual extremity of the anterior cingulum, which also can be enlarged into a cusp, the pericone. Such a chain of cuspules exists in *Nannopithex filholi* but not in earlier *Nannopithex.* A bumpy link, and sometimes a crest, is found as a variation on the upper molars of *Periconodon jaegeri,* which has a large pericone. *P. huerzeleri* also has a variably present lingual cingulum, and this unusual variability in lingual cingular characters may be typical of the genus (Godinot, 1988a). A small pericone is present on *A. minutus* M^2 and in cf. *Djebelemur* but not in *Tabelia.* A large pericone is present in *Moeripithecus* but not in *Aegyptopithecus* (?lost). It is also large and crestiform in some parapithecids, e.g., *Apidium,* but absent in some others (*Simonsius grangeri*). A pericone is variably present in *Saimiri* and some other platyrrhines (Kinzey, 1973). The pericone seems absent in oligopithecines. The swelling on *Proteopithecus* M^2, considered a possible primitive retention by Simons (1989), is not a crestiform pericone but the base of a swelling running down from the protocone summit, as developed in *Necrolemur.* Because a pericone is still rare in adapiforms and not very common in omomyiforms, its frequent occurrence in North African groups creates a suspicion that a pericone was present at some early stage of simian evolution. It also would have been frequently lost, or alternatively, it must have been frequently acquired. One view of parsimony is to suppose its early acquisition (there is already a very small one in *Altiatlasius*), development in many groups, and later secondary losses. Larger forms often reduced lingual cingular structures, as well illustrated in platyrrhines (Kinzey, 1973). However, several losses in early forms would also be necessary, and pericone development may also be subject to convergence, especially in a group having early acquired a continuous lingual cingulum. The preceding pericone scenario is weakly based, and it will not be given weight in the phylogenetic analysis until more material allows a stronger statement. However, the early acquisition of a continuous lingual cingulum, and the general importance of lingual cingular structures, including the hypocone, very probably typify early simiiforms.

Abbreviation of the Trigon

Some additional characters of the M^2 trigon must be mentioned. The trigon basin is transversely short in *Algeripithecus*. However, it will not be scored separately, because it might be the result of paracone and protocone modifications. Paracone labial expansion and protocone lingual expansion were treated as if they were real expansions resulting in lower external slopes. This can be the case, for example, when protocone lingual expansion occurs without much trigon basin narrowing, as in *Teilhardina* or *Cantius eppsi* (Fig. 4). Likewise, paracone labial expansion occurred without much trigon basin narrowing in *Absarokius* (Bown and Rose, 1987). However, one wonders whether the lowering of the external slopes in *Algeripithecus* pushed the cusp summits toward the center of the tooth, resulting in trigon basin narrowing. This might also apply to *Periconodon,* which has a narrow trigon basin. However, trigon basin shortening might also be independent, correlated with the transverse shortening trend (see below). One or the other may have occurred in later anaptomorphines such as *Strigorhysis bridgerensis* or *Gazinius,* which also have short trigon basins. Trigon basin shortness is derived; it is more advanced in *Algeripithecus* than in propliopithecines, but it should be scrutinized to understand if it is a separate character of its own.

Metacone Position

A lingual shift of the metacone relative to the paracone can be seen on *Algeripithecus* M^2, which is also well marked on propliopithecine M^2. However, it has not occurred in cf. *Djebelemur* (if its upper molar is not an M^1) or on the M^2 of *Tabelia.* Such a metacone shift seems to be present in *Proteopithecus* and *Catopithecus* but less marked than in *Algeripithecus.* A well-marked metacone shift is clear on *Apidium* M^2, but it seems less strong in *Simonsius grangeri.* It occurs in *Saimiri* and several living platyrrhines but is absent in *Branisella* and *Szalatavus.* The preceding distribution suggests that this character developed or increased in several groups (it is incipient in *Altiatlasius,* Sigé *et al.,* 1990). Hence, the marked metacone lingual shift present in propliopithecines and *Algeripithecus* is a synapomorphy of weak significance between them. Convergence in this character might be linked to some M^3 reduction, but such a link is not necessary. Many anaptomorphines have a small M^3 without such a shift, and some other taxa have this metacone lingual shift without a reduced M^3, e.g., some *Leptadapis* or *Macrotarsius.* In fact, this character is not common in adapiforms and omomyiforms, in contrast to simiiforms.

Paraconule and Metaconule

The paraconule and metaconule are generally considered present in early euprimates, as in *Teilhardina* and primitive omomyiforms, and in *Donrussellia,* which also illustrates the early decrease in size of the metaconule in adapiforms. A secondary enlargement and sometimes a multiplication of cuspules can occur in some larger bunodont forms (*Microchoerus, Rooneyia,* parapithecids). Both conules are lost in *Moeripithecus,* and this is also the case in my opinion for *Aegyptopithecus* (Fig. 1; Hiiemäe and Kay, 1973, Fig. 10) despite irregularities in the crests or the junction of a small crest to the preprotocrista, which were interpreted as small conules by some authors (e.g., Szalay and Delson, 1979, Fig. 150; Kay, 1980, Fig. 13). Hence, propliopithecines have lost their conules, as is the case for oligopithecines; this represents a potentially homologous synapomorphy among them (Simons, 1989). However, conule reduction is a common evolutionary trend in primates, which diminishes the value of this shared derived character. The case of parapithecids is obscured by their secondary increase and multiplication of cuspules. It can be assumed that their morphotype had conules.

Because a decrease of the metaconule and realization of an almost continuous crest from protocone to metacone (crista obliqua) is very general in simiiforms, the question is whether this might have been the case in the parapithecid ancestor? Because remnants of a hypometacrista can be seen on some parapithecids (Szalay and Delson, 1979, Fig. 150), I conclude that this probably was the case. This small crest must be, in such bunodont forms, a heritage. Among platyrrhines, *Branisella* has no paraconule and only a small remnant of metaconule (Hoffstetter, 1969). However, because conules are found on the upper molars of *Callimico,* they must have been present in the platyrrhine morphotype (Kay, 1980). There may well have also been a small metaconule on a postprotocrista reaching the metacone, as such a crest is present in *Saimiri.* Hence, a conule situation as in *Algeripithecus* might closely approach the situation in a simiiform morphotype, which is not very far from the morphology of *Donrussellia provincialis* (Fig. 7), but retains a larger paraconule.

Hypoparacrista

The hypoparacrista is a crest running lingually from the paracone. A long hypoparacrista can be seen in *Aegyptopithecus* (Fig. 1), in which it comes close to the preprotocrista. A similar morphology occurs in *Propliopithecus chirobates* (Kay *et al.*, 1981). This crest is shorter, between a lower point on the anterior side of the paracone and better isolated form the preprotocrista, in *Moeripithecus.* A hypoparacrista, isolated from the preprotocrista by a groove and bearing a small cusp, can be seen in *Apidium* and *Parapithecus* and seems typical of parapithecids. Simons (1989, p. 9959) described how, in *Proteopithecus,* "a crest running mesiolingually from the paracone joins the crest between the basal cingulum and the apex of the protocone." However, the figure shows the usual mesial preparacrista and a lingual prolongation of the paracone (probably crested, but I cannot see for sure) that is isolated from the preprotocrista by a deep groove. In *Oligopithecus,* too, a hypoparacrista is limited to the tip of the paracone (Rasmussen and Simons, 1988). Although I cannot discern it on the *Catopithecus* illustrations, I will assume that a hypoparacrista limited to the paracone wall is an oligopithecine morphotypic character. This crest does not exist in any of the small bunodont simians, a likely primitive state.

The case of platyrrhines is difficult to analyze. A ridge on the paracone of *Saimiri* possibly represents it. Because this crest is absent in *Branisella* and *Szalatavus,* I will assume that it was absent in the platyrrhine morphotype too. A hypoparacrista limited to the paracone wall is a potential synapomorphy among oligopithecines, parapithecids, and propliopithecines. In this case, a prolongation of this crest could be shared by propliopithecines and parapithecids, and this crest would have grown further and nearly reached the preprotocrista in *Aegyptopithecus* and *Propliopithecus.* It is also possible that the morphology of oligopithecines is not homologous with that of the others. A hypoparacrista is not unique to these African groups, as it can be seen on some omomyiforms (e.g., *Teilhardina tenuicula,* Bown and Rose, 1987). It is frequent in microchoerines, including the early *Nannopithex.* In omomyiforms, it seems to be mainly a prolongation of the postparaconule crista. This crest is rarely found in adapiforms; however, its presence in *Donrussellia* (Fig. 7) might indicate a common heritage with early omomyiforms. Hence, an alternative view could be proposed, in which the hypoparacrista would be a primitive euprimate character, early lost in adapiforms, entirely retained in propliopithecines, isolated by a small groove in parapithecids, and more reduced in oligopithecines. Its loss in *Algeripithecus* and *Altiatlasius* would be autapomorphous and exclude them from later simiiform ancestry. This scenario appears less likely in view of all the other evidence. However, the first alternative implies convergence between propliopithecines and some primitive primates, and also among simiiform groups if the small bunodont simians have a close relationship to any one of them.

Evaluating Upper Molar Characters

In order to help evaluate these upper molar characters, I will mention two evolutionary trends. The notion of an evolutionary trend is sometimes criticized, but it is very helpful when used in a descriptive sense, without any finalist connotation. It often becomes critical to evaluate the significance of numerous dental traits. Several of the characters that were discussed, such as the lingual cingulum and the hypocone, have the result, when they develop, of squaring and/or reinforcing the lingual part of these teeth. Such a trend seems quite general in primates. The group started with triangular and lingually narrow upper molars inherited from their primitive mammalian ancestors and evolved toward the more rectangular, lingually squared, molars of most later groups. Such a trend may of course suffer some exceptions. Callitrichids are one; they probably secondarily lost the hypocone (Rosenberger, 1977), and their upper molars remain lingually broad. Another exception can happen when a strong protocone lingual expansion occurs in a species without a hypocone and affects only the anterior part of the tooth, thus creating a posterolingual concavity in occlusal view (primitive *Cantius eppsi*). Otherwise, the trend seems very general.

Squaring and reinforcing the lingual part of the molars probably has a functional significance. It may be linked to transverse movements increasing with more vegetarian diets, whereas primitive insectivorous species had predominantly vertical movements. This trend would also be correlated with an increase in body size, because this in turn is often linked to dietary shifts toward more plant food (Kay and Simons, 1980; Kay and Covert, 1984). There might be other reasons, such as an increase of crushing relative to vertical shearing, allometric effects, etc. From this perspective, the lingual characters of the cingulum and associated cusps of the upper molars appear to follow the same trend: they are derived when they develop. However, it changes our view of one of these characters, the protocone fold or "*Nannopithex* fold," which was often assumed to be primitive in omomyids (Szalay, 1976; Szalay and Delson, 1979) and primitive and subsequently lost in simiiforms (Kay, 1977, 1980). I think that the protocone fold is one of many ways to increase or broaden the lingual part of the upper molars. It developed as a derived trait in several lineages in euprimates (e.g., *Cantius, Nannopithex*). Hence absence of a protocone fold is probably primitive for euprimates. Simiiforms and their ancestors probably never had a protocone fold. Likewise, the earliest adapiforms did not possess this character, which developed in notharctines only. Rarely, this protocone fold evolved into a crest continuous with the posterior cingulum, a morphology convergent with the postprotocingulum of plesiadapiforms (*Pelycodus, Strigorhysis,* Fig. 1; Bown, 1979; see also Rose *et al.*, Chapter 1, this volume).

The second evolutionary trend is the general tendency to decrease upper molar transversness (a trend also called "lengthening," e.g., Simons, 1989). Again, primitive mammalian ancestors had transversely elongated molars; many advanced groups have shorter, almost square upper molars; and it is common to consider that relatively squared upper molars are derived relative to more transverse ones, e.g., in the omomyids *Ourayia* and *Macrotarsius* (Szalay and Delson, 1979, p. 238), in *Adapis* and late *Leptadapis* relative to earlier adapiforms, and in atelines and other platyrrhines relative to *Saimiri.* Exceptions to this broad trend are the labial paracone and lingual protocone expansions when they really increase tooth transverseness, which then becomes a secondary effect of bunodonty. The small bunodont simians studied here have relatively transverse M^2, which may have secondarily shortened,

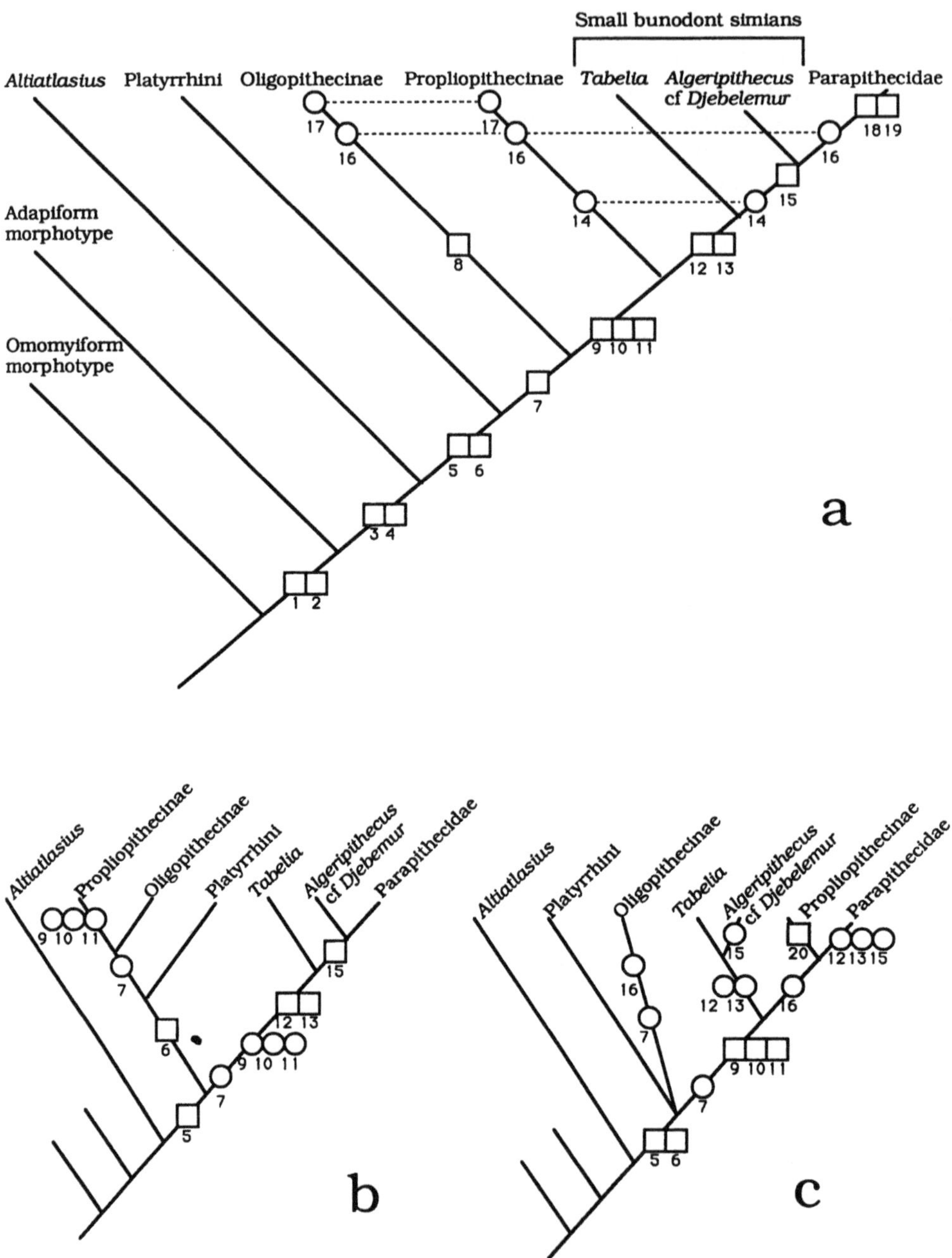

Fig. 8. Cladograms of early simiiform relationships based solely on derived characters of the second upper molars. Derived characters assumed to be homologous are indicated by squares, and derived characters assumed to be convergent are indicated by circles. 1, reduction of metaconule; 2, loss of postmetaconule crista; 3, continuous lingual cingulum; 4, low degree of bunodonty; 5, small crestiform hypocone; 6, M^2 lingual expansion with cingular structures, thick cingulum, hypocone, and eventual pericone; 7, moderate protocone lingual expansion; 8, large

already slightly in cf. *Djebelemur* relative to the two others, and in parapithecids relative to them. This probable secondary labial retraction preserves the peculiarities of the cingulum and labial aspect.

Oligopithecines are interesting because *Catopithecus* has a transversely shorter M^2 than *Proteopithecus* and propliopithecines. The morphology of *Catopithecus* recalls the situation in some late Eocene *Leptadapis*, which reduced their hypocone. The M^1 of *Oligopithecus* also has a small hypocone, and both *Oligopithecus* and *Catopithecus* seem derived over propliopithecines by their upper molar transverse shortness and probable secondary hypocone reduction. These advanced characters render more dubious for oligopithecines their role of possible "intermediates" between propliopithecines and an ancestral group.

Upper molar transverse shortening might also partly explain the frequent tendency for conule reduction. Such common trends show that derived characters such as transverse shortness or reduced conules may have value within a group, e.g., within a genus or within a tribe, but are of little help in deciphering more distant phylogenetic relationships. (Would they not be misleading if scored and introduced in a computer analysis?)

A first discussion of the phylogenetic relationships of *Algeripithecus* and its relatives can be done solely on the basis of the M^2 characters. At the present stage of knowledge, I prefer not to score and use all the characters, because they have very unequal values. Cingular characters are much more subject to reversals than others. Characters that evolved convergently a number of times were encountered, e.g., lingual shift of the metacone, conule reduction; they are of little help. The following hypotheses are based on a limited number of characters and apparently can make sense morphologically and functionally.

My favored hypothesis is shown on the cladogram of Fig. 8a. In contrast with our earlier views (Godinot and Mahboubi, 1992), the closest affinity of the small bunodont simians has shifted from propliopithecines to parapithecids. This is mainly because I realized that parapithecids probably share

---

crestiform hypocone without a prehypocone crista; 9, high degree of bunodonty; 10, marked paracone labial expansion; 11, marked difference between M^1 and M^2, the latter being more transversely elongated; 12, extreme bunodonty; 13, marked protocone lingual expansion; 14, large cuspidate hypocone with prehypocone crista; 15, very large hypocone relative to body size; 16, hypoparacrista; 17, loss of paraconule; 18, extra cusps added; 19, secondary transverse shortening; 20, reincrease of protocone lingual slope, and some diminution of bunodonty. Cladogram a places taxa along a morphocline from low to extreme bunodonty. It places the small bunodont simians close to parapithecids, and this group closer to propliopithecines than to oligopithecines. It is favored here because it is the most parsimonious; however, the true story will very probably show some convergence in the development of bunodonty. Cladogram b emphasizes the significance of strongly developed lingual cingular structures (6), which could link platyrrhines with oligopithecines and propliopithecines. It implies convergence in bunodonty in propliopithecines and other early bunodont simians. In cladogram c, the hypoparacrista (16) is emphasized as homologous in Propliopithecinae and Parapithecidae, implying convergence in extreme bunodonty and very large hypocone relative to body size in parapithecids and small bunodont simians.

their peculiar labial characters, which are attenuated in occlusal view, and because the inclusion of *Moeripithecus* modified the propliopithecine morphotype (small crestiform hypocone), showing convergence in hypocone shape between *Algeripithecus* and *Aegyptopithecus* instead of homologous similarity. Using family-group morphotypes for other primates simplifies the analysis by suppressing all convergences between individual genera. This holds true for *Absarokius* and *Periconodon,* which could be thrown in the middle, considering, e.g., that *Absarokius* derives from a small bunodont simian by losing the hypocone. However, considering an anaptomorphine (or omomyiform) and an anchomomyin (or adapiform) morphotype places these as outgroups and avoids the consideration of a number of useless hypotheses (implying highly improbable scenarios). Further justification of this choice will arise from the final discussion.

Other hypotheses can be considered, for example, uniting together platyrrhines, oligopithecines, and propliopithecines by their possession of strongly developed lingual cingular structures without protocone lingual expansion (Fig. 8b). A moderate lowering of the lingual protocone slope would unite oligopithecines and propliopithecines (or alternatively, there might have been an initially moderately low protocone lingual slope and a secondary decrease in platyrrhines). This would lead us to consider the high degree of bunodonty and paracone lingual expansion as convergently evolved in parapithecids and propliopithecines. This seems less likely but is not impossible. Another possibility is to consider the small bunodont simians as a stem group to the later bunodont forms (Fig. 8c). This view would be supported by the common possession by propliopithecines and parapithecids of a hypoparacrista (nonhomologous with that of oligopithecines) but would necessitate some reduction of bunodonty in propliopithecine ancestry, something probably less likely. Other possibilities are numerous. Choosing with more certainty among various hypotheses would require a much better knowledge of morphological trends and function in small bunodont teeth. This will become possible when more complete fossils are found.

Lower Molar Characters

Because only M_3 is known in *Tabelia* and *Algeripithecus,* this tooth will be discussed first. Comments on M_{1-2} of *Djebelemur* will follow. In M_3 trigonid construction, the three small African simians show differences that in good part justify their generic separation, but they also show some similarities. All three have lost the paraconid. *Djebelemur* and *Tabelia* have a continuous crest uniting the protoconid and the metaconid anteriorly to form a complete paralophid (Figs. 9 and 10). *Algeripithecus* has a discontinuity as the paracristid, after its lingual turn, continues with a ventral inclination and tapers, whereas the thick premetacristid curves labially in a more dorsal position.

Paraconid reduction is a rule in primate evolution. It usually proceeded

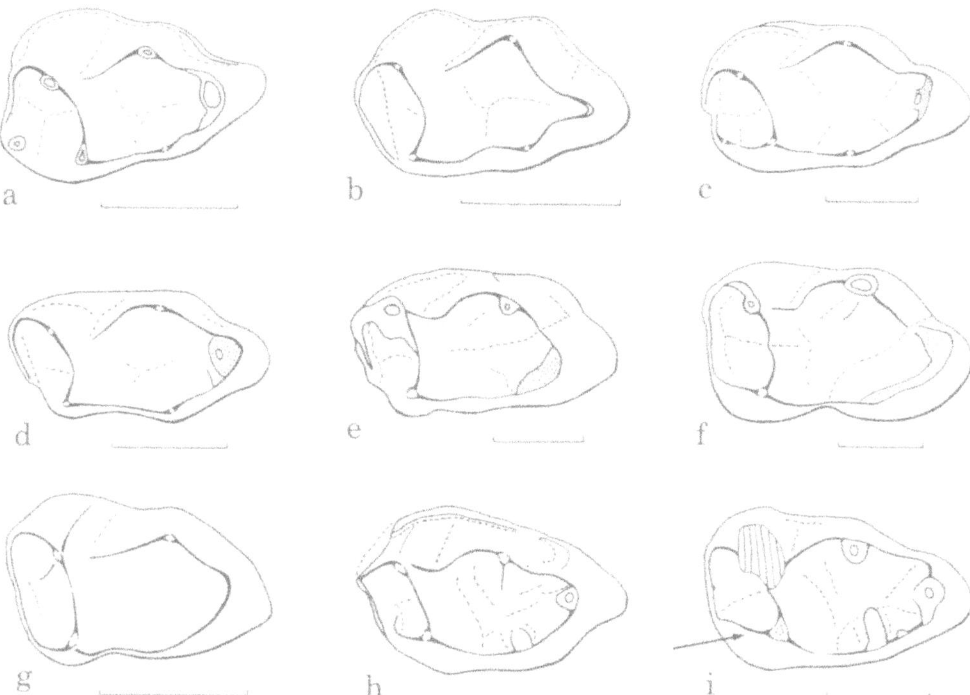

Fig. 9. Lower third molars of a series of fossil primates in occlusal views, all drawn to the same anteroposterior length (scale bars are 1 mm). Some are reversed right to left to make all appear as right M_3. a, *Teilhardina belgica;* b, *Pseudoloris parvulus;* c, *Donrussellia provincialis;* d, *Anchomomys gaillardi;* e, *Periconodon huerzeleri;* f, *Agerinia roselli;* g, *Djebelemur martinezi;* h, *Algeripithecus minutus;* i, *Tabelia hammadae.* The primitive *Teilhardina* (a) and *Donrussellia* (c) have a paraconid. The three simians (g to i) and *Agerinia* (f) have a premetacristid (black arrow). Note the important broadening and posterior elongation of the talonid in the adapiforms d to f, incipient in the earliest of them, *Donrussellia* (c).

with a gradient from M_3, the most affected, to M_1 which is less or later affected. Hence, parallelisms and convergences are commonplace in trigonid structure. Hartenberger and Marandat (1992) compared the continuous paralophid of *Djebelemur* with the morphology in *Agerinia* and in the isolated teeth found in Pakistan and attributed to an adapiform (Russell and Gingerich, 1987). They considered that it was an adapiform character. However, a continuous paralophid is common in simiiforms and is also realized in other groups. In omomyiforms, *Chumashius* and *Utahia* come close, and it is a frequent variation in microchoerines (some *Nannopithex filholi, Necrolemur,* and *Microchooerus*). In adapiforms, a complete paralophid is found in some notharctines (their morphology often suggests a fusion of the paraconid into the metaconid summit). In cercamoniines, it is much more rare. Usually the paracristid remains isolated from the metaconid wall, even in the rare cases in which a premetacristid is variably present (e.g., *Periconodon*). However, some

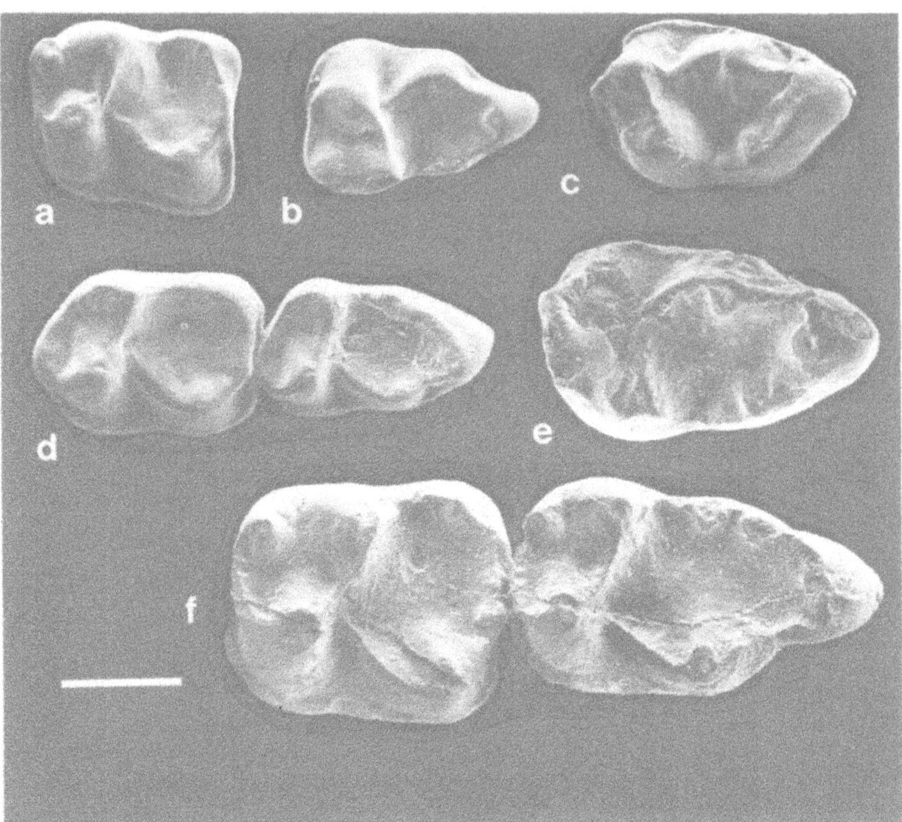

Fig. 10. Lower molars of a series of fossil primates in occlusal views, all at the same scale to allow comparison of their relative sizes (scale bar is 1 mm). All are scanning electron micrographs of casts. a and b, *Teilhardina belgica,* left M_2 (a) and right M_3 (b); c, *Algeripithecus minutus,* right M_3; d, *Djebelemur martinezi,* left M_2 and M_3; e, *Tabelia hammadae,* right M_3; f, *Donrussellia provincialis,* left M_2 and M_3. Note that the M_3 of *Algeripithecus* (c) is more bunodont than that of *Djebelemur* (d) and that both are smaller than the earliest Eocene adapiform, *Donrussellia* (f).

individuals attain a continuous paralophid, e.g., one M_3 of *Europolemur dunaifi* from Bouxwiller and some *Periconodon. Agerinia* is the only European adapiform possessing a constant complete paralophid (Fig. 9; and its subfamilial assignment is problematic). Even in *Agerinia,* the paralophid makes a deep ventral concavity in anterior view, whereas *Djebelemur* and *Tabelia* have a more horizontal one, more similar in this to many later African simiiforms. Here the difference is also the lower metaconid in *Djebelemur* as in simiiforms, making the metaconid summit closer to the paracristid. Continuous paralophid and low metaconid are also frequent in living lemuriforms, but the anterior part of *Djebelemur*'s mandible shows that it cannot be a lemuriform (see below).

The continuous paralophid is also realized on M_1 in *Djebelemur,* which is derived compared with the small paraconid retained on M_1 in several oligopithecines, parapithecids, and propliopithecines. For this character, *Djebelemur* is autapomorphous and excluded from direct ancestry of later Fayum simians (this is not known, and cannot be guessed from their M_3, in *Algeripithecus* and *Tabelia*).

The paralophid discontinuity in *Algeripithecus* is interesting because it shows the presence of a premetacristid distinct from the paracristid. This character was considered by Kay (1980) to be present in the ancestral platyrrhine and known to have evolved convergently in adapiforms and omomyiforms, and I agree with this. In these latter groups, it is quite rare (e.g., some microchoerines and some *Periconodon*). Although it is not unique to them, the presence of a premetacristid may well be one of the characters typical of simiiforms (one element of their unique combination of characters). On the whole, anterior trigonid characters of the small African simians do not show significant similarities with either adapiforms or omomyiforms. On the contrary, the presence of a premetacristid and the frequent horizontal paralophid are similarities with later simians.

In the three small African simians, the M_3 talonid is transversely narrower than the trigonid, which gives these teeth a contour with a general posterior narrowing (Figs. 9 and 10). They also have a moderate length. This stands in contrast with the morphology of adapiforms, in which the M_3 are long, and the talonids have the same breadth or are broader than the trigonids (an exception is in some large notharctines such as *Pelycodus,* which have a broader anterior part and then a convergent triangular M_3, which, however, remains long). The hypoconid in adapiforms is more peripheral, being both more posterior and closer to the labial border. This usually results in a transversely broad talonid basin (very broad in *Anchomomys* and *Agerinia,* Fig. 9, which were cited for similarities with *Djebelemur*). This character seems already to separate *Donrussellia provincialis,* a primitive adapiform with a broad talonid basin, and *Teilhardina belgica,* which has a narrower basin and an overall contour much closer to that of the small African simians. This is probably one more derived adapiform aspect of *Donrussellia* when contrasted with *Teilhardina* (see Godinot, 1992a; Rose *et al.*, Chapter 1, this volume; Fig. 9 also shows that paraconid reduction seems more advanced in *Donrussellia* than in *Teilhardina*).

The third lobe of M_3 is transversely broad in *Tabelia* and *Djebelemur* and slightly narrower in *Algeripithecus.* A slightly narrower hypoconulid, situated in a transversely median position as in *Teilhardina* and *Donrussellia,* might approximate the euprimate morphotype (however see *Altiatlasius* below). Broadening of this lobe happened a number of times. In cercamoniines, *Anchomomys gaillardi* has a very elongated third lobe, which remains relatively narrow (or became secondarily narrow, as happened in the small insectivorous *Pseudoloris*). Others have a long third lobe usually broader and shifted lingually (*Periconodon, Agerinia;* Fig. 9). Broadening can become extreme in omomyiforms as well, e.g., *Nannopithex* and later *Microchoerus.* Third-lobe broadening and posterior extension are certainly derived when they occur. Small African simians seem to have some M_3 third-lobe broadening without posterior extension. This contrasts with adapiforms and, on the contrary, represents a likely step toward the short and posteriorly round M_3 of later simiiforms. Similar convergent evolution is found in those omomyiforms that do not lengthen their M_3, e.g., several anaptomorphines and *Necrolemur.*

On the whole, for these posterior M_3 characters, the small African simians stand in sharp contrast with adapiforms but have some similarities with some omomyiforms. Their overall posterior narrowing, lack of talonid basin broadening, and third lobe broadening without posterior extension make them structural intermediates between early euprimates and later simians with posteriorly round M_3.

A distinctive feature of the M_3 of *Algeripithecus* is that it has a moderately high crown (Godinot and Mahboubi, 1992). Comparison of the lateral view of the M_3 of *A. minutus* with that of the similar-sized *Teilhardina belgica* shows that its crown is higher. This is not because the metaconid is higher; on the contrary, it is slightly lower. The basal part of the crown is generally higher, giving it a more robust aspect. This appears to be the same high-crowned character that is often used to describe catarrhine teeth, but in an incipient state. This character is lacking in *Djebelemur* and apparently also in *Tabelia,* although wear makes it more difficult to judge. This incipient high-crownedness of *Algeripithecus,* remarkable in view of its small size, is a synapomorphy with later catarrhines.

A cingulid linking hypoconid and hypoconulid on M_3 was cited as a possible synapomorphy between *Algeripithecus* and propliopithecines (Godinot and Mahboubi, 1992). This character is absent in *Djebelemur* and *Tabelia.* It is a trait that is very rare in adapiforms, but there is a posterior prolongation of the cingulid in *Teilhardina* and *Pseudoloris,* in which it does not curve dorsally as in *Algeripithecus.* A cingulid is absent in *Catopithecus* and parapithecids. This cingulid, which is found in *Aegyptopithecus,* obscured in *Moeripithecus,* and present in *P. chirobates* (Simons *et al.*, 1987), might be a character of the propliopithecine morphotype. It may be a shared-derived similarity of propliopithecines with *A. minutus* but might well be convergent.

Concerning the M_{1-2}, known only in *Djebelemur,* two aspects will be mentioned. First, there are very low metaconids, especially on M_1. This is striking in lingual view and stands in sharp contrast with adapiforms (Fig. 12). Some omomyiforms also exhibit such a low metaconid, e.g., early members of the *Pseudoloris* group and probably some North American omomyids. The similarity between *Djebelemur* and early *Pseudoloris* is even more striking if the still elongated M_1 trigonid and almost continuous paralophid are taken into account. A low metaconid is frequent in living lemuriforms, and metaconid lowering is probably a common evolutionary trend in primate dental evolution. The low M_{1-2} metaconid of *Djebelemur* is potentially homologous with the similar morphology found in later simiiforms.

The second aspect of *Djebelemur* M_{1-2} is the absence of a hypoconulid (Fig. 10). The situation in *Algeripithecus* and *Tabelia* is unknown. However, when a large hypoconulid is present on M_{1-2}, the entoconid is usually well formed on M_3. Because this cusp is little differentiated in *A. minutus* and in *Tabelia,* I would guess that the hypoconulid was absent on their M_{1-2}, as in *Djebelemur.* The presence of a well-formed hypoconulid was often cited as linking parapithecids with catarrhines (Kay, 1980; Delson and Rosenberger, 1980). Alternatively, Fleagle and Kay (1987) proposed that the M_{1-2} hypoconulid, also present in *Oligopithecus,* might be a primitive simiiform character subsequently lost in platyrrhines. However, the morphology of the hypoconulid is so different in oligopithecines and the others that a parallel development of this cusp in the two groups seems likely. A large hypoconulid is developed in *Microchoerus* and some other omomyiforms and is linked to cusp proliferation and/or bunodonty. Also, its loss in platyrrhines would appear surprising unless it were very small. Hence, I would suspect that primitive simiiforms had no hypoconulid, and the absence of this cusp in *Djebelemur* would not preclude this genus from being one of them. A discussion of this character in small bunodont simians is delayed until definitive evidence is found for them.

Another noticeable character of *Djebelemur* lower molars is their slight waisting, clearly seen on M_2 and M_3, which is a character frequent in simiiforms.

Lower Premolar Characters and Anterior Tooth Reduction

Perhaps the most surprising characters of *Djebelemur* are those of its P_4 and P_3 (Figs. 11 and 13). P_4 is very simple, without paraconid or metaconid, and scarcely enlarged posteriorly. P_3 is unusually elongated and narrow. By comparison, the two known premolars of *Algeripithecus* must also be P_4 and P_3 (Godinot and Mahboubi, 1994). Its P_3 is higher but shows an anterior elevation of its cingulum and a probably correlated anterior projection of the protoconid. The premolars of *Algeripithecus* and *Djebelemur* look remarkably different from those of *Teilhardina* and *Donrussellia.* On the P_4 of these two taxa, the postprotocristid has a posterolingual orientation, going toward a small metaconid and in this way delineating an incipient trigonid (more easily recognized in occlusal view; Fig. 11). In *Djebelemur* the postprotocristid remains more median. It has a small talonid cusp from which two cingulids slope away ventrally. The lingual cingulum remains low and does not curve dorsally toward a metaconid as in *Donrussellia* and *Teilhardina,* in which this delineates an incipient talonid basin. The P_3 is also simple and narrow, with a postprotocristid oriented posterolabially and not posterolingually as in *Donrussellia* and *Teilhardina.* For these P_{3-4} characters, *Djebelemur* could be more primitive than early omomyiforms and adapiforms, which would imply very

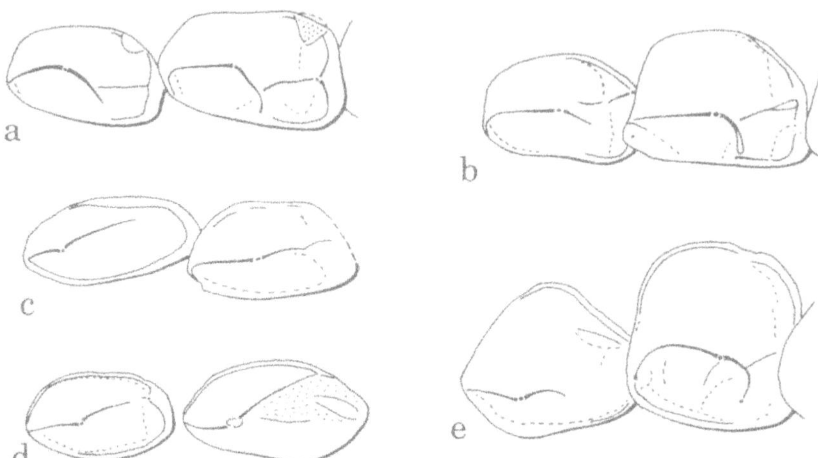

Fig. 11. Lower third and fourth premolars of selected fossil primates in occlusal views, all drawn to the same anteroposterior length. Some of them are reversed top to bottom to make all appear as right teeth. a, *Donrussellia provincialis;* b, *Teilhardina belgica;* c, *Djebelemur martinezi;* d, *Algeripithecus minutus;* e, *Nannopithex filholi.* Two isolated teeth having markedly different wear stages are shown in d. All other are in place in jaws. However, displacement of P_3 in the jaw of *Djebelemur* (c) exaggerates its length and consequently also diminishes the size of the two teeth relative to the other species. Note the extreme broadening of the premolars in *Nannopithex* (e), a character that is incipient in *Teilhardina* (b).

simple P_{3-4} in the euprimate (= primate) morphotype, or it could be derived, with P_{3-4} secondarily simplified from a morphology closer to *Donrussellia* (see below).

The P_{3-4} of *Djebelemur* are very simple but not primitive in all aspects. In lateral view their main cusp appears very low, much lower than that in *Teilhardina* and *Donrussellia,* a character that is probably derived relative to these latter taxa (Fig. 12). In lingual view, these premolars show an anterior elevation of their crown, especially of their basal cingulum (relatively to the dentary or to their own roots). This makes the protoconid appear to project anteriorly, especially that of P_3 (Fig. 12h). M_1 and P_4 overlap each other, and this probably also was the case between P_4 and P_3 (P_3 is displaced on the *Djebelemur* mandible by a fissure of the dentary). The basal cingulum of these teeth is also quite thick. All these characters are derived compared to the more horizontal and nonoverlaping premolars of *Teilhardina* and *Donrussellia* (Fig. 12). These characters are also similarities with some microchoerines, e.g., some *Nannopithex* (Fig. 12e). This is a convergently developed similarity: P_{3-4} of *Algeripithecus* and *Djebelemur* do not have the broadness found in microchoerines and many other omomyiforms, including *Teilhardina* (Fig. 11). Also, anterior premolar reduction is advanced in microchoerines over *Djebelemur,* and their incisor enlargement is a specialization excluding microchoerines from being possibly ancestral to the small African simians. The posterolingual narrowness

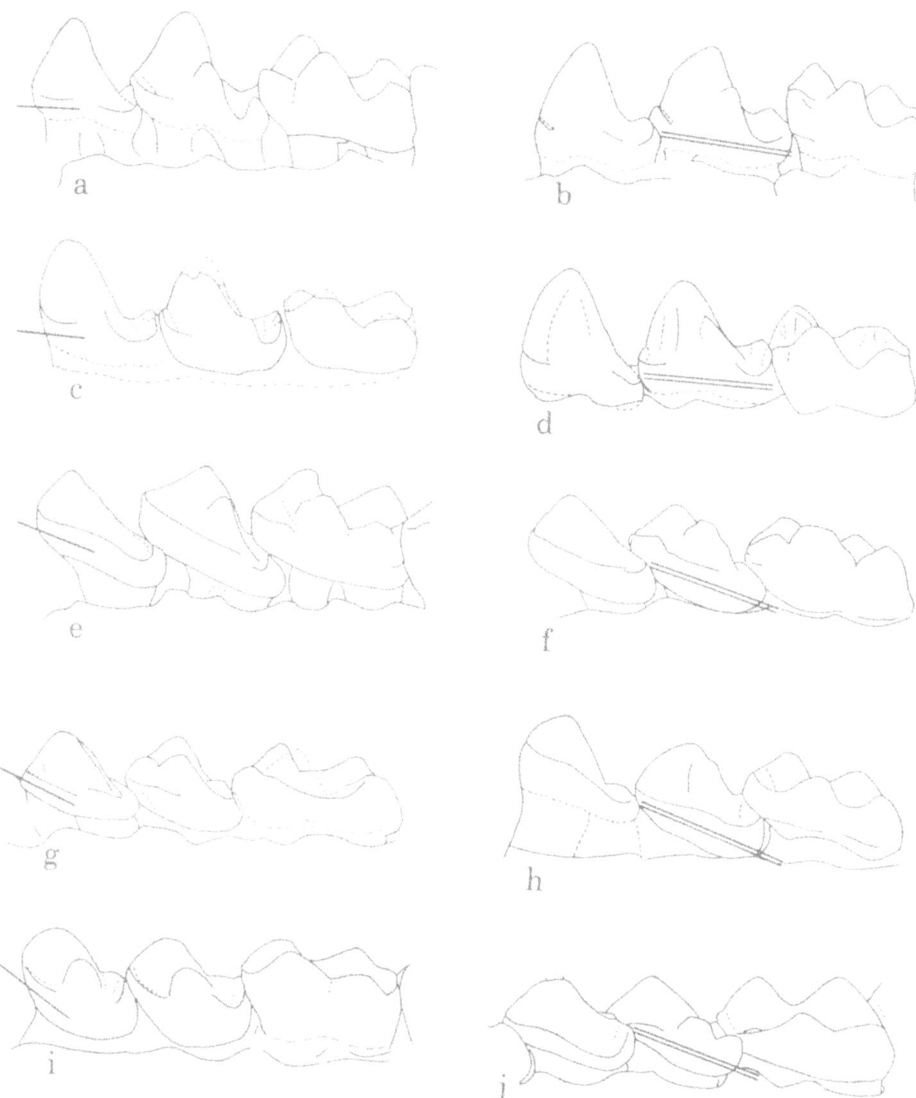

Fig. 12. Lower P_3 to M_1 on jaws in lingual profile views, all drawn to the same anteroposterior length. All show definitive premolars with the exception of j, which shows a juvenile with dP_2 to dP_4. Some are reversed right to left to make all appear as right jaws. a, *Teilhardina belgica;* b, *Donrussellia provincialis;* c, *Anchomomys* aff. *quercyi;* d, *Europolemur klatti;* e, *Nannopithex filholi;* f, *Microchoerus erinaceus;* g, *M. edwardsi;* h, *Djebelemur martinezi;* i, *Parapithecus fraasi;* j, *Saimiri sciureus,* juvenile. The bar on the specimens of the left column underlines the subhorizontal setting of the anterior part of P_3 in a primitive omomyiform (a) and an adapiform (c), and its anterior inclination in microchoerids (e and g) and a simian (i). The double lines on the specimens of the right column underline the subhorizontal setting of the base of the P_4 crown in adapiforms (b, d) and its anterior dorsal sloping in a microchoerid (f), *Djebelemur* (h), and the dP_4 of a living simian (j). The resemblance between *Djebelemur* (h) and *Nannopithex* (e) suggests an analogous process of premolar crowding preceding premolar molarization in simians and microchoerids, contrasting with elongation of a subhorizontal P_4 in adapiforms (c, d).

of P_{3-4} in *Algeripithecus* and *Djebelemur* might be a correlated response to the peculiar crowding of these teeth. In this case, the posterolingually narrow P_4 would be derived relative to the more squared morphologies of horizontal premolars as in *Donrussellia.*

The contrast is striking between the P_{3-4} of *Djebelemur* and those of adapiforms. Primitive adapiforms have P_4 and P_3 in an anteroposterior line with little or no overlap, and the bases of their crowns are parallel to the dorsal border of the dentary. They have rather high protoconids. Many have a metaconid (small in *Donrussellia*), but some lack this cusp (some *Europolemur,* anchomomyins). Anchomomyins are noteworthy for retaining relatively narrow P_{3-4}, more blade-like than in other tribes (this is one of the reasons I followed Szalay and Delson, 1979, in treating them as a tribe; Godinot, 1988a,b). The latest known *Anchomomys, H. (Huerzeleris) quercyi,* has the most molarized premolars in the tribe. Its P_{3-4} are more anteroposteriorly elongated than in earlier species, with P_3 having a well-isolated talonid cusp and P_4 having a low lingual cusp, possibly a metaconid, close to the posterolingual base of the main cusp (Fig. 12c; Godinot, 1988b). This latest anchomomyin shows no shortening and broadening of its P_{3-4} but, in contrast, elongation and growth of the talonid, i.e., a type of molarization recalling Adapidae *sensu stricto* and not later simians. The only known P_4 of *Periconodon* is that of the type of *P. huerzeleri* (Gingerich, 1977a). This tooth is slightly elongated anteroposteriorly and, without apical wear, would be high. It bears a well-formed postprotocristid running down and joining the summit of the talonid cusp. A swelling runs down the posterolingual border of the main cusp; it joins at a very low point the lingual cingulum issued from the talonid cusp. Hence, an incipient talonid basin may be detected in occlusal view. This tooth is more elongated and blade-like than that in *Donrussellia provincialis, Algeripithecus,* and *Djebelemur.*

The fact that anchomomyin P_{3-4}s are narrow and relatively simple makes them superficially similar to those of the small African simians. However, they remain high and more blade-like and lie horizontally and behind each other. P_4 has a median postprotocristid; the two teeth have neither the lingual cingular characters nor the anterior main cusp projection of the small African simians. Advanced anchomomyins show a slow molarization by elongation and development of a talonid, as occurred in other adapiforms. In sum, P_{3-4} of anchomomyins, in their morphology and in their known evolution, do not show any significant morphological change toward simian characters (only the limited P_4 lowering of the latest *Anchomomys* is convergent with simians). *Europolemur* shows some similarities with *Djebelemur,* with no metaconid on P_4 and a very limited lowering of the protoconid on P_4 and of the metaconid on M_1. However, this occurred at a very large size compared with *Djebelemur,* its premolars remain high (P_3 very high; Fig. 12d), horizontal, and with adapiform lingual cingular characters. Likewise, the larger *Protoadapis (Cercamonius) brachyrhynchus* has a thicker but otherwise cercamoniin-like P_4 (Stehlin, 1912; Tattersall and Schwartz, 1983), and the alveoli for P_3 do not clearly indicate an obliquely oriented P_3 (*contra* Gingerich, 1975). On the whole, despite the limited reduction in anterior premolars seen in some cercamoniines and emphasized by some authors, this group, including anchomomyins, does not show any significant morphological change in the shape of its P_{3-4} that would go toward the distinctive simian condition. P_4 molarization, on the contrary, follows a divergent pathway. This is the strongest—and in my opinion the most definitive—argument against a special phylogenetic relationship between simians and cercamoniins.

Algeripithecus and *Djebelemur* have P_{3-4} that show an evolution of their own, unlike that of any adapiform, with distant analogy with microchoerines (and probably some other omomyiforms), but with a unique combination of primitive (narrowness, simplicity, absence of metaconid) and derived characters (protoconid low, thick and continuous lingual cingulum, incipient crowding by anteriorly sloping protoconid). They clearly belong to a third group, whose distant sister-group relationship with the others mentioned would be difficult to hypothesize; they must be rooted in a very primitive euprimate state.

On the other hand, is it possible to link the P_{3-4} of *Algeripithecus* and *Djebelemur* to those of later simians, many of which have much more molarized premolars? First, it must be recalled that the P_4 of *Qatrania* is quite primitive and simple. Its small metaconid suggests that early simiiforms had at least as primitive a P_4 (Simons and Kay, 1988). *Qatrania* is early Oligocene or late Eocene in age. Primitive morphology preserved so late suggests that earlier simiiforms probably had simple premolars. Another interesting aspect of later simian premolars is that, even though larger and transformed, their premolars have an anterodorsal inclination and some posterior inclination of premolar roots. It is slightly visible on the P_3 of *Propliopithecus chirobates,* clearer on P_3 and P_4 of *Oligopithecus savagei,* and most visible on the P_{2-4} of *Parapithecus fraasi* (Fig. 12i).

The same premolar characters are still visible in some living platyrrhines, e.g., in *Callicebus, Saimiri,* and *Cebuella* (Hershkovitz, 1977, Fig. V22) and in juvenile series of *Callithrix, Callimico,* and *Saimiri* (Fig. 12j; Hershkovitz, 1977, Fig. V28). Hence, premolar crowding occurred in simians in a way reminiscent of microchoerines but very different from adapiforms (there are a few convergent characters in some large and late adapiforms, e.g., some crowding and lingual expansion of premolar crowns in *Leptadapis,* but this represents only a distant similarity in a cranially lemur-like form). Not surprisingly, the closest analogues to early simians in the P_{3-4} morphology of Eocene prosimians is found in some late Eocene microchoerines. There are similarities in the overall shape of P_4, especially in lingual view, between *Oligopithecus* and *Microchoerus erinaceus,* which has a trigonid less closed, and *M. edwardsi,* which has a trigonid slightly more molarized (the paraconid is well formed; Fig. 12f,g). The longer talonid in *Oligopithecus* and different outline in occlusal view show that they pertain to different groups. However, the shape similarities that are there can be interpreted as parallel processes of premolar molarization. This shows that the similarities that have been observed between *Djebelemur* and

Nannopithex, anterodorsal inclination and peculiar crowding of P_{3-4}, are characters well suited to be those of ancestral simians.

In *Qatrania* and *Parapithecus* the P/4 is remarkable for its relatively narrow posterior part, which contrasts with the broadening of the same part in many adapiforms and omomyiforms. Such a morphology seems a good step toward later simian premolar evolution, in which the P_4 becomes shorter, shifted, in a way often preserving some posterolingual narrowness. Such a stage is still visible on the least molarized platyrrhines, for example, on a juvenile specimen of *Saimiri,* in which dP_3 has a metaconid lower than the protoconid, a posterior part lower than the trigonid, and a narrow posterolingual part. This is quite close to *Parapithecus* morphology and is a likely primitive simian morphology. Likewise, the P_3 of *Parapithecus* is relatively elongated and posteriorly narrow, an unusual condition in omomyiforms and adapiforms but a condition for which *Djebelemur* and *Algeripithecus* offer possible ancestral stages. Hence, if parapithecids are considered as reflecting a step in simian premolar evolution, *Algeripithecus* and *Djebelemur* have a suitable ancestral morphology for them, as they possess P_{3-4} with an unusually narrow posterolingual part.

The P_3 of *Algeripithecus* has an anterior root smaller than the posterior one. After examination of the jaw of *Djebelemur,* I think that the small alveolus described by Hartenberger and Marandat (1992), considered by them anterior to P_3 but covered by it in occlusal view, is more probably the anterior P_3 alveolus than the alveolus of a P_2 (because of the fissure of the specimen, the P_3 is slightly detached from its original position). Hence, instead of a two-rooted P_2 or single-rooted P_2 and P_1, I think that there is only one alveolus for a single-rooted P_2 and the larger anterior alveolus. Because the anterior part of the jaw is broken, covered with glue, and was coated for scanning electron microscopy, precise observation is difficult. However, a small piece of transverse septum can be found near the anterior border of the larger alveolus. This small piece of septum shows that the anterior jaw did not contain one large tooth but a single-rooted medium-sized tooth, presumably the canine, and something smaller anteriorly, presumably one or two small incisors (Hartenberger and Marandat, 1992).

Comparing the shape of the jaw of *Djebelemur* with those of other Eocene primates close in size shows that its mandible is much shorter than that of *Donrussellia,* slightly longer than that of *Pseudoloris* (Fig. 13), and probably very close to that of *Teilhardina belgica.* Its shape in lateral view shows that it is relatively deep under M_3 and becomes more shallow anteriorly. This gives it an outline recalling the much larger *Donrussellia magna* (Godinot *et al.,* 1987). Such an outline differs from *Pseudoloris,* other microchoerids, and many omomyids, in which an anterior reinforcement of the jaw, linked to the enlarged anterior incisor, leads to an anterior part as deep as or deeper than the posterior part. A similar anterior depth increase is present in a tooth-combed primate of comparable size such as *Microcebus.* The clear absence of anterior depth increase in *Djebelemur* is important to show that it cannot be a lemuri-

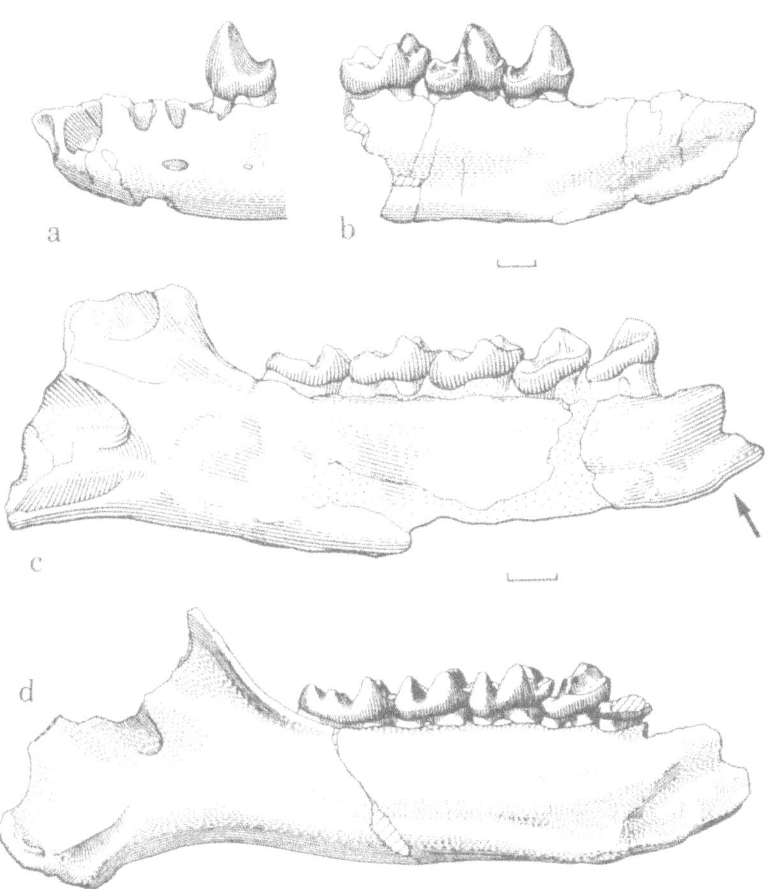

Fig. 13. Three small fossil primate jaws in lateral views and not to scale (scale bars are 1 mm). a and b, *Donrussellia provincialis;* c, *Djebelemur martinezi;* d, *Pseudoloris parvulus.* Three are lingual views (b, c, d), and one is labial (a). The jaw of *Pseudoloris* is almost complete anteriorly. That of *Donrussellia* is less well preserved, but the labial view (a) shows that the anterior lower border is only slightly deformed and that little is missing anteriorly. The jaw of *Djebelemur* is somewhat less complete, but it is intact in the anterior portion (arrow), revealing a posterior increase in height under M_3 and, more importantly, a low jaw under P_3 with a tapering of the anterior portion and a low ventral slope (primitive condition). The low anterior portion of the jaw in *Djebelemur* gives clear evidence that it could not have supported an anterior tooth comb and thus that *Djebelemur* can not be a lemuriform. The low height of P_3 and P_4 and the low metaconid of M_1 are considered by the author to be simian characters. a and b are from the unpublished *Thèse d'Etat* of the author (1983); c is a courtesy of J.-L. Hartenberger, from Hartenberger and Marandat (1992); and d is from Godinot (1988b).

form. On the whole, *Djebelemur* presents an anterior jaw shorter than that of *Donrussellia,* advanced over it, with associated reduced premolars, but preserving a primitive thin anterior extremity. This jaw does not show any trace of symphyseal fusion or modification from appression. Such a situation is linked to its very small size, as jaw fusion is correlated to size increase (Beecher, 1983; Ravosa, 1991; Ravosa and Hylander, Chapter 14, this volume). The small size of the early African simians implies that they had separate jaws. Jaw fusion happened later, probably in parallel in different groups, and can no longer be considered a defining simiiform character, as was done by most authors until now (Delson and Rosenberger, 1980; Kay, 1980; Hoffstetter, 1982; Harrison, 1987; Andrews, 1988; Fleagle, 1988; Martin, 1990).

Discussion

Affinities of the New Early North African Primates

Algeripithecus, Tabelia, cf. *Tabelia,* and the upper molars designated cf. *Djebelemur* appear very similar to each other by their peculiar very bunodont upper molars. A close association of *Djebelemur* with them depends on the association of its type mandible with the bunodont upper molars from Chambi, and this association is problematic (Drs. E. Gheerbrant and C. Beard both called my attention to this point). I cannot make a definitive choice on this point at the moment. The M^2 from Chambi seems slightly too small to pertain to the same individual as the mandible, but the difference is not great enough to exclude intraspecific variation. The lower molars of *Djebelemur* are crested and seem less bunodont than the M^2. However, these lower molars have a low relief, and their labial outlines do not seem incompatible. A more detailed study of occlusion is difficult because the M^2 is worn and incomplete, and there are not enough teeth of *Algeripithecus* to understand the occlusion in these forms (they are difficult to compare with bunodont anaptomorphines because the lower molars of the latter are so different, with much higher trigonids). Also, the M^2 of cf. *Djebelemur* is less bunodont than that of *Algeripithecus.* The labial expansion of this M^2 is less exaggerated, and the lingual slope of its protocone is less low. Its trigon basin is also slightly more transversely elongated. The upper molars of some taxa, e.g., *Proteopithecus,* appear much more bunodont than the lower molars do (D. T. Rasmussen, *personal communication*). Hence, could this M^2 still be associated with the mandible? One clear difference remains, which is that the M_3 attributed to *Algeripithecus* is more bunodont and has more rounded cusps than the M_3 of *Djebelemur.* Without more material, a risk exists that the M^2 and M^3 pertain to a more bunodont form than *Djebelemur* (this suspicion is enhanced by the presence of a badly preserved but larger and less bunodont M^3 in the Chambi material, which might fit better with *Djebelemur* lower teeth). This uncertainty led me to discuss the upper molars from Chambi separately as cf. *Djebelemur* and the *Djebelemur* mandible and to avoid using for all these taxa a formal taxonomic category derived from the "Djebelemurinae" briefly alluded to by Hartenberger and Marandat (1992). In any case, I am impressed by the similarities in M_3 and in P_{3-4} construction, and I think that *Algeripithecus* and *Djebelemur* are probably members of the same broad category of primates. This justifies a common character analysis and a common discussion for them (Godinot and Mahboubi, 1994). However, if *Djebelemur* pertained to a less bunodont group than *Algeripithecus,* their phylogenetic relationships within this broad taxon would be quite different, and a common familial name for them would be misleading.

From the preceding character analyses, the main similarities and differences of the new small North African primates with known groups of Eocene prosimians can be summarized: their very small size makes them closer to early omomyiforms; their bunodonty and their M_3 not being elongated and without broadening of the talonid basin recall characters found in some anaptomorphines; however, their large upper molar hypocone and their lack of P_{3-4} broadening show that they cannot be referred to anaptomorphines. Likewise, there are a few similarities with microchoerids, such as the low metaconid of *Djebelemur* M_1 and that of early *Pseudoloris,* a distant analogy in premolar appression, anterior sloping, and molarization. However, the large hypocone and general pattern of the upper molars, the different M_3, and the much less broadened premolars show that they are a very different group. These small African primates do not fit in omomyiforms. In spite of a few similarities of the upper molars with those of *Periconodon,* these small bunodont primates have little in common with adapiforms. Cercamoniines all have an elongated M_3 with a broad talonid basin, P_{3-4} horizontal, not recovering, and without a continuous lingual cingulum. Anchomomyins, including *Periconodon,* have high and blade-like P_{3-4} that become molarized by posterior elongation and talonid development. All this is markedly unlike the small African forms. Bunodonty is rare in adapiforms. Large *Cantius* and *Pelycodus* differ entirely from the small African forms. In cercamoniines, only one species of *Periconodon,* noticeably larger, shows an incipient bunodonty on its upper molars. However, *Periconodon* upper molars remain far from the small bunodont African forms (Figs. 1 and 7), and all its other characters show that there is nothing more than a partial convergence in upper molar morphology between them. On the whole, the new North African primates can not be referred to adapiforms and thus do not fit in any known group of Eocene prosimians.

In our analysis of the first three isolated teeth of *Algeripithecus* and in our description of further teeth, we emphasized their similarities with later simiiforms (Godinot and Mahboubi, 1992, 1994). The preceding analysis, which is much more detailed than previous ones, confirms this conclusion. The characters of the P_3 and P_4 especially add evidence that the peculiar mode of molarization of these teeth is that suitable to reach the primitive simiiform premolars as found in parapithecids and some living platyrrhines. This holds true also for the M_3 characters, which were found to reflect likely ancestral states for later simians. Hence, there is no doubt for me that the three new North African primates are primitive simiiforms. This view also fits with the well-known African endemism and with the importance of the later simian radiation in Africa, recently augmented by new Fayum discoveries (Simons, 1992).

Until now it has been difficult to find derived dental characters defining simiiforms (Kay, 1980). A list of such traits was proposed by Harrison (1987), but it contained a number of primitive euprimate characters. The new early North African simians allow a better delineation of their differences with similar-sized and similarly primitive prosimians. The series of characters that are interpreted as shared-derived simian characters include moderate bunodonty on M^2; a continuous lingual cingulum and well-developed lingual cingular structures, including a crestiform hypocone, present in small forms (the continuous lingual cingulum alone could also be a haplorhine character; see below); a postprotocrista running toward the metacone summit (incomplete crista obliqua) and a reduced metaconule; a premetacristid on the lower molars and tendency to realize a continuous horizontal paralophid (except in *Algeripithecus*); the anterodorsal inclination of the crown on P_{3-4}, which has a low summit, a continuous lingual cingulum, and a relatively narrow posterolingual aspect of P_4. The continuous lingual cingulum of the upper molars and their crista obliqua were listed by Harrison (1987). Several of these characters appear in some omomyiforms or adapiforms (Kay, 1980; see above), but their occurrence together is assumed here to typify simiiforms.

Comments on the Adapiform Hypothesis

In our first description of *Algeripithecus,* we emphasized that these early simiiforms ruled out the adapiform hypothesis of simiiform origins based on middle or late Eocene cercamoniines (Godinot and Mahboubi, 1992). This claim deserves further comments, because the adapiform hypothesis received increased attention with the discovery of additional oligopithecines (Rasmussen and Simons, 1988; Simons, 1989, 1990; Rasmussen, 1990). In the initial formulation of the adapiform hypothesis, Gingerich (1975) emphasized resemblances between *Protoadapis (Cercamonius) brachyrhynchus* and simians, especially in the P_3 dental hone and large interlocking canines. The P_3 hone was not actually observed and would need a confirmation, as early *Protoadapis* had a P_3 taller than P_4 (Russell *et al.,* 1967) and without a honing facet. In any case, such a hone and large interlocking canines are subject to convergence, as shown in the same paper (Gingerich, 1975): the P_2 hone evolved in *Leptadapis* is exactly the same as those in many platyrrhines, without any phylogenetic meaning. This character and others like the quadrate form of the molars are the result of morphological trends and adaptative convergence. They have been uncritically cited (Gingerich, 1980; Andrews, 1988).

Other similarities were added, between the M^2 of *Periconodon* and that of *Hoanghonius*, between the M_2 of *Hoanghonius* and that of *Oligopithecus* (Gingerich, 1977b), and between an upper molar of *Oligopithecus* and one of *Protoadapis* (Rasmussen and Simons, 1988). However, such isolated similarities have little significance if they are not reconciled with the surrounding evidence. Incidentally, stratophenetics is a very powerful and interesting approach, but when applied without precise stratigraphic correlations on insufficient phenetic evidence (Gingerich, 1977b, 1984), it loses its strength. The major difficulty with the adapiform hypothesis is that its proponents never achieved a coherent scheme of relationships that takes into account all the available information. For example, resemblances between *Oligopithecus* and the Asian *Hoanghonius* cannot be extended to European cercamoniines, which never have twinned entoconid and hypoconulid. Should *Oligopithecus* be rooted in European or in Asian forms? Or, if a graded series *Protoadapis–Oligopithecus–Aegyptopithecus* is considered (Rasmussen and Simons, 1988, Fig. 7 and text), where would the smaller and highly derived parapithecids then fit? This scheme leaves many unanswered, and in my opinion unsolvable, difficulties.

The resemblances between the upper molars of *Oligopithecus* and *Protoadapis* are real, but there are also differences between the two groups: in the European forms, the continuous lingual cingulum occurs only in later forms such as *Europolemur,* where it is accompanied by a large hypocone; a paraconule is retained; and a lingual spur running down from the summit of the paracone is not present (Russell *et al.*, 1967; Godinot, 1988a). The anterior shifting of the protocone and resulting asymmetry of the trigon are more accentuated in *Oligopithecus* (Rasmussen and Simons, 1988). The P^{3-4} of *Oligopithecus* described by these authors is more transversely elongated than in the *Protoadapis–Europolemur* group. Its narrowness is, in my opinion, a primitive character shared by most other Eocene simiiforms, showing that they probably never went through a stage of P^{3-4} broadness and squareness as in the cercamoniines (even *Donrussellia*). European adapiforms never show indications of twinned entoconid–hypoconulid on their lower molars. Furthermore, *Oligopithecus* M_1 has a low metaconid. This character was found in *Djebelemur* and is especially unlike the morphology of adapiforms. In lingual view, the base of the crown of M_1 also appears relatively high, suggesting that an incipient high-crownedness might be present in oligopithecines. The unworn M_2 of *Oligopithecus* figured by Rasmussen and Simons (1988, Fig. 4) also shows a premetacristid and a labial shift of the metaconid summit, two characters known in simiiforms but rare in adapiforms (the second is unknown in small species but present in some much larger and bunodont forms as *Pelycodus*). The characters of the upper and lower molars of *Oligopithecus* taken together suggest that occlusion in this genus was very different from that in cercamoniines. I emphasized above how premolar evolution distinguishes simians and *Oligopithecus* from cercamoniines. On the whole, based on dental morphology, despite an amazing superficial similarity in the upper molars of two genera, not only is there no evidence of a close affinity between oligopithecines and cercamoniines, but there is ample evidence for the contrary.

Rasmussen (1990) tried to reconcile the adapiform hypothesis with the cranial evidence. However, his argument is weakened by the omission of critical taxa. Adding *Adapis,* which is well known cranially and also an omomyiform, would have considerably changed the conclusions. On the contrary, the discovery that oligopithecines are definitive simians, with postorbital closure and loss of an interfrontal suture (Simons, 1990), increases the difficulty of rooting them within cercamoniines, in which these characters are not present. The adapiform hypothesis also was hard to reconcile with the postcranial evidence linking adapiforms and lemuriforms within Strepsirhini (MacPhee and Cartmill, 1986; Gebo, 1986; Dagosto, 1988; Godinot and Beard, 1991). It is true that cercamoniines are more primitive and less well known than other adapiform subfamilies, but there is little positive evidence to question their place close to them (however, see Franzen, 1987; Franzen, Chapter 4, this volume). Recently, humerus remains from Prajoux probably pertaining to *P. (C.) brachyrhynchus* have been found (Fig. 14). Preliminary study reveals that they are strikingly notharctine-like, with an indriid-like distal extremity (Godinot, 1992b). *Cercamonius* had a kind of vertical clinging and leaping adaptation commonly found in strepsirhines. Its morphology is very different from that of the most leaping simians, the parapithecids (Fleagle and Simons, 1983; Fleagle and Kay, 1987). Thus, the cranial and postcranial evidence, like the teeth, does not support the adapiform hypothesis.

Instead of pointing toward a possible "link" between propliopithecines and adapiforms, oligopithecines show that the simian radiation in Africa was a

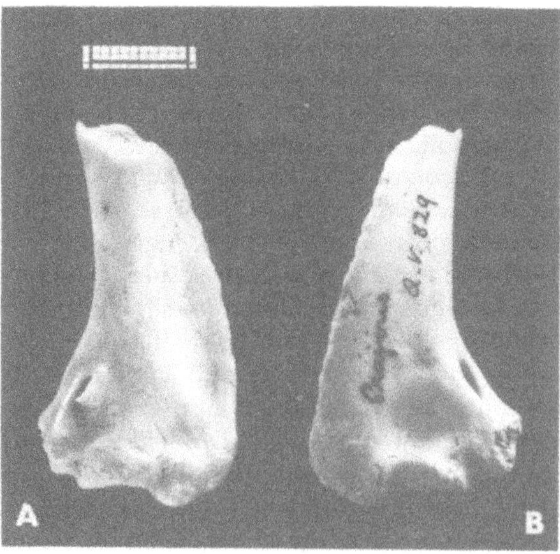

Fig. 14. Distal humeral extremity from the Prajous collection (Basel Museum) attributed to *Protoadapis (Cercamonius) brachyrhyncus* from the same locality. A is anterior and B is posterior view. Scale bar is 1 cm. The articular extremity shows a high and narrow trochlea well separated from the capitulum, very similar to those of *Smilodectes* and living indriids. Together with the extended brachioradialis crest, it suggests a locomotor repertoire including a high proportion of vertical clinging and leaping. This morphology and the implied locomotor mode are unlike those known in undoubted simiiforms.

broad and ancient one, with widely divergent groups in the late Eocene. This view is dramatically confirmed by the recent discovery of three new simian genera from the Fayum (Simons, 1992). In this context, the hypothesis of *Algeripithecus* and its relatives as early simians fits very well. They confirm the antiquity of the simian radiation in Africa. For those who believed in an ancient presence of simians in Africa (e.g. Simons, 1976; Hoffstetter, 1977, 1982; Martin, 1990), the new North African simians give the first positive evidence confirming the prediction. This in turn corroborates that the middle Eocene adapiform hypothesis is no longer tenable.

The Early Simiiform Radiation

The study of the M^2 characters led to a cladogram linking *Algeripithecus* and its relatives to parapithecids (Fig. 8). The evidence from lower tooth characters does not change very much the likely relationships, because their premolars and M_3 are essentially primitive in comparison with later simiiforms. Nevertheless, a few characters can be added to the cladograms: presence of three premolars in these early North African simians, seen in *Djebelemur,* assumed in the small bunodonts; absence of hypoconulid, seen in *Djebelemur,* possible but not sure in the others; and incipient M_3 high-crownedness in *Algeripithecus.* Different hypotheses of relationships will be briefly mentioned, again based on a limited number of characters having some morphological and, one hopes, functional coherence.

Cladogram a (Fig. 15) is the same as that in Fig. 8a. It is obtained if bunodonty is emphasized as important and not subject to much parallelism. The many shared-derived similarities between *Algeripithecus* and parapithecids, including extreme bunodonty and very large hypocone, make it attractive. In this case, the centrally placed hypoconulid of parapithecids and propliopithecines has to be convergent. Conversely, if the hypoconulid is considered a shared-derived trait in parapithecids and propliopithecines, these can be linked together as on cladogram b, a hypothesis that still preserves a coherent view about bunodonty: only the extreme level reached in the small bunodont simians and parapithecids has to be convergent, something not unlikely in forms already well engaged in this specialization. Hence, cladogram b is for me a second likely, but less attractive, hypothesis.

On both cladograms a and b (Fig. 15), the small bunodont simians are closely related to parapithecids and propliopithecines. These taxa should be considered together as catarrhines, typified by their precocious high bunodonty. It implies that early catarrhines had three premolars, something that was already accepted by authors ranking parapithecids within catarrhines (Simons, 1970, 1972; Kay, 1977; Szalay and Delson, 1979). It also implies that these early catarrhines had no symphyseal fusion. As discussed above, convergent symphyseal fusion in catarrhines and platyrrhines is implied by size increases in these groups. These hypotheses also suggest that some oligo-

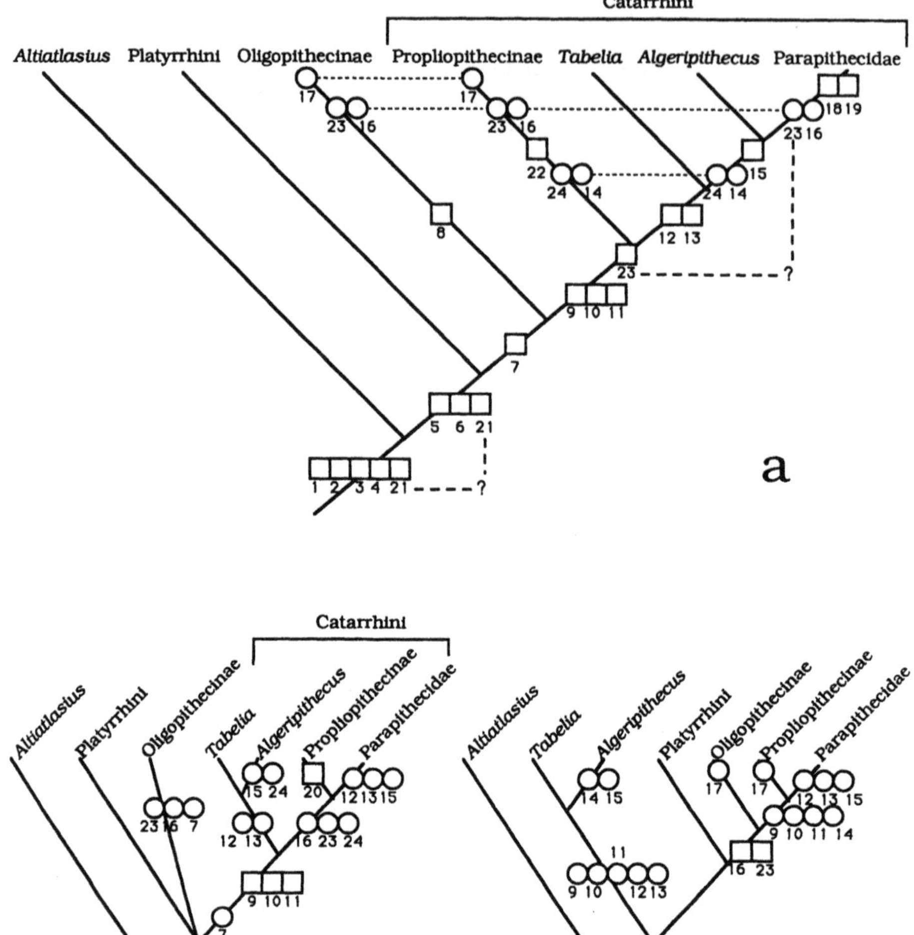

Fig. 15. Cladograms of early simian relationships based on a limited number of characters from the available dentition. They are similar to those of Fig. 8, because the lower teeth of early simians show predominantly primitive characters and do not notably alter the preceding schemes based on M^2 characters. Derived dental characters are assumed to be homologous (squares) or convergent (circles). Characters 1 to 20 are the same as on Fig. 8; 21, three premolars; 22, two premolars; 23, cuspidate hypoconulid on M_{1-2}; 24, some lower molar high-crownedness. Note that these cladograms do not include the *Djebelemur* mandible, because its association with the upper molars is not secure.

Cladogram a is basically a morphocline of increasing bunodonty. It places the early bunodont simians close to parapithecids. Cladogram b is a variation of the same general scheme in which the small bunodont simians appear as the stem group of other bunodont simians, implying convergence only in the extreme states of bunodonty reached by these early bunodonts and parapithecids. Both hypotheses suggest a possible definition of catarrhines essentially based on bunodonty and excluding oligopithecines. Cladogram c shows *Algeripithecus* and *Tabelia* as the stem group of all other simians. It shows many more convergences than the others and appears less likely.

pithecines convergently lost P_2, which is not surprising in view of the common primate trend of reducing the anterior dentition, and in fact recently confirmed by the mention of three premolars in *Proteopithecus* (Simons, 1992). Also, the astonishing resemblance between P_{3-4} in *Oligopithecus* and *Aegyptopithecus* is probably a result of convergence and can be explained by the broadly similar process of premolar molarization in early simians (in any case, the rest of their dentition implies a long separate history for these two genera). These hypotheses viewing the early bunodont simians as catarrhines also imply an earlier split of platyrrhines and oligopithecines, which would go back to the early Eocene or before.

Bunodonty is exceptional in adapiforms. It is a rare character in Eocene primates, although it developed in anaptomorphines. If it developed very early in simiiforms, its presence would be the rule in this group, and its increase in many lineages would become probable. A different schema could be proposed with a moderate bunodonty shared by the small North African simians and the ancestor of platyrrhines and oligopithecines. Other characters, such as the hypoparacrista, could be emphasized as uniting oligopithecines, parapithecids, and propliopithecines (Fig. 15, cladogram c). In such a view, the small bunodont simians would be precociously specialized forms having striking convergences with parapithecids. This is not impossible within one group, simiiforms (similar heritage and "potentialities"), but is less parsimonious and appears to me less likely.

If further material showed that the *Djebelemur* mandible is not associated with the small bunodont upper molars from Chambi, this genus would represent a less bunodont group. It then might well be near the ancestry of oligopithecines, a place that would not dramatically change the main conclusions of this study.

Further progress in deciphering simiiform relationships should come from further analysis of all groups. The late Eocene *Biretia* would give interesting clues if more material were found (de Bonis *et al.*, 1988). The dental morphology of oligopithecines will have to be studied, and their morphotype evaluated. Was this morphotype a genus with a large hypocone such as *Proteopithecus*? Can its morphology be derived from a moderately bunodont form? Do the lower molars have a low metaconid and a degree of high-crownedness as do *Oligopithecus*? These questions are critical for an evaluation of the oligopithecine morphotype and its potential relationship with platyrrhines. The third lobe of M_3 appears very short in *Catopithecus* (Simons, 1989); is it comparable to that of platyrrhines? Is it typical of oligopithecines? Why could not oligopithecines be platyrrhines? Concerning platyrrhines, I find it very interesting that *Branisella* seems to have some degree of bunodonty and accentuated lingual cingular structures (Hoffstetter, 1969; Wolff, 1984). This and the fact that its M^2 retains some lingual narrowness are potentially critical to root platyrrhines within early African simians.

An increased understanding of early simiiforms and catarrhines leads to another look at *Amphipithecus* and *Pondaungia*, the two controversial late Eocene Burmese primates (Pilgrim, 1927; Colbert, 1937; Maw *et al.*, 1979). These genera often were treated separately in the literature (e.g., Szalay and Delson, 1979). For a long time, the anterior mandible of *Amphipithecus* had no teeth in common with the two fragments of *Pondaungia,* rendering direct comparison impossible between them. The recovery of a fragmentary mandible of *Amphipithecus* with an M_2 (Ciochon *et al.*, 1985) allows such a comparison. Both share many distinctive characters: high crowns, bunodonty, interrupted protocristid, and oblique orientation of the crests running from the summit of the metaconid. These underline in my opinion their probably close relationship. Taking them together as Amphipithecinae allows us to add the information derived from each genus to look for possible relationships.

Following the preceding character analysis, I consider as very significant the fact that the M_2 talonid basin in *Pondaungia* is narrower than that in *Pelycodus,* the most bunodont adapiform. The proportion of the third lobe of M_3 and the relative narrowness of the talonid basin (compared with bunodont notharctines) are similar in *Pondaungia* and *Tabelia* despite their enormous size difference. On all molars and premolars, amphipithecines have low cusps and a labial basal inflation of the crown, which is high. This entails an extreme degree of bunodonty recalling African catarrhines and very unlike anything known in adapiforms. The transverseness of *Pondaungia*'s M^2 is similar to that in *Algeripithecus* and propliopithecines, whereas *Pelycodus* already has a reduced M^2 transverseness. The P_{3-4} of *Amphipithecus* are very low and have a triangular outline in occlusal view with a posterolingual expansion, narrow on P_3. This type of premolar morphology is much closer to premolar molarization as analyzed in simiiforms and opposite to what is known in adapiforms. Hence, after having counted these genera as late Eocene Asian adapiforms (Godinot, 1990, Fig. 1), I now think that they must be simiiforms and probably early catarrhines (an opinion defended by Simons, 1971, 1972; shared by Szalay and Delson, 1979, for *Pondaungia;* for a different view, see Holroyd and Ciochon, Chapter 5, this volume). *Algeripithecus* and its relatives show the early presence of very bunodont catarrhines having three premolars in Africa. They might have found their way to Asia, or the reverse.

On the whole, our understanding of the early simian radiation remains poor because the material at hand is fragmentary and because the gaps in the fossil record are still enormous. However, the new North African primates add critical information. Their most surprising character is the high degree of bunodonty achieved in such tiny primates. By comparison with other Eocene primates, I tend to give a high weight to this character. It leads me to place them with catarrhines, along with hominoids, cercopithecoids, propliopithecines, parapithecids, and probably amphipithecines. This concept is quite different from that of many recent authors but is similar to the view of Kay (1977). It is close to the view of those who included parapithecids in catarrhines, e.g., Simons (1972) and Szalay and Delson (1979), but these authors also included *Oligopithecus.* A high degree of bunodonty is not simply one character but is a complex of dental traits linked to a modified occlusion (Kay, 1977), probably also linked with other jaw and skull muscular characters, and reflecting a change in diet. Compared with similar-sized insectivorous primates, it reveals a major adaptive shift that, if realized only once, would provide a good definition for a higher taxon like Catarrhini.

Simiiform Origins

Algeripithecus and its relatives are the oldest known simians. The character analyses showed that they cannot be closely related to any known adapiform or omomyiform. Their combination of characters shows that they cannot be rooted in *Teilhardina* or *Donrussellia,* the earliest and most primitive omomyiform and adapiform, respectively. In short, they look more primitive than *Donrussellia* by M_3 characters, and their P_{3-4} show a line of evolution without the broadening already under way in *Teilhardina.* In a way, an omomyiform more primitive than *Teilhardina,* with premolars closer to those of *Steinius* (Rose and Bown, 1991), would appear the closest approximation to known prosimians. Furthermore, the small bunodont simians have later retained sizes close to that of *Teilhardina,* whereas bunodonty in adapiforms developed at much larger sizes. Hence, on the whole these fossils strongly suggest that earlier simiiforms were differentiated as a third group of small species in the earliest Eocene, in Africa or elsewhere. Would there by any other taxon that could be near the origin of simiiforms?

The first important fossil to consider in this context is the late Paleocene *Altiatlasius* from Morocco. This genus was assigned to the Omomyidae, subfamily Anaptomorphinae, and was interpreted as having a sister-group relationship with simiiforms, mainly because of its relative bunodonty (Sigé *et al.,* 1990). Another possible shared derived character with simians, the continuous lingual cingulum with incipient hypocone and pericone, was added (Godinot and Mahboubi, 1992). Because this genus is still incompletely known, it may seem difficult to be certain that it is a euprimate, much less the oldest (see Rose *et al.,* Chapter 1, this volume). Hence, it is important to reappraise its characters.

Some of the teeth of *Altiatlasius* look very strange for a primate. If it had been found isolated, my first guess for the M^1 THR144 would be that it is not a primate: the parastyle is labially extended; the metacone is smaller than the paracone; the postprotocrista runs toward the metastyle; and in posterior view the protocone and metacone have a slightly convex outline, which gives an "insectivore" impression. Likewise, the large lower molar THR146 is striking by its enormous protoconid and comparatively quite small and rounded talonid basin (a marked difference from both *Donrussellia* and *Teilhardina*). It looks really distinct from well-known early euprimates. However, the upper M^1 fits well with the M^2-type THR141 (Figs. 4 and 7); it has a broad protocone compared to insectivores. The M^2 looks primate-like and advanced (see below). The large inferior molar is close to THR135, which has a less predominant protoconid, a comparatively larger talonid, and looks definitely primate-like. Also, the characters of the two "extreme" specimens are likely primitive characters consistent with a Paleocene age. On the whole, the differences between individual teeth are not large enough to dismember the assemblage into several distant taxa. In the teeth other than the two extremes mentioned, there are a lot of characters that fit with an early primate (Sigé *et al.*, 1990).

Because the primate identification is critical, and Sigé *et al.* (1990) reported differences with insectivores *sensu lato* and hyopsodontid condylarthrans, I also made comparisons with small arctocyonids. Members of this group sometimes show similarities with primates. Some early arctocyonids have large talonids on their lower molars, and some have bunodont, transversely elongated M^2 with a lingual cingulum, which recall bunodont simians (examples in Van Valen, 1978). Also, one "Arctocyonidae indet" is mentioned in a North African Thanetian locality (Gheerbrant *et al.*, 1993). The smallest arctocyonids that I was able to compare, *Protungulatum, Prothryptacodon,* and *Oxyprimus,* are all much larger than *Altiatlasius.* All of them differ from *Altiatlasius* by upper molars that are much broader lingually, have a postprotocrista always running toward the metastyle, have conules closer to the protocone summit, and have a lingual cingulum higher on the protocone wall and sloping labially, whereas the cingulum is more basal and horizontal in *Altiatlasius.* On the lower molars, these arctocyonids have an anteroposteriorly shorter trigonid with a more reduced paraconid advanced over *Altiatlasius* morphology, and at the same time they have a much higher trigonid, more primitive in this character. The trigonid is lower on larger arctocyonids, which, however, are more specialized by their lingually open talonid basin. The comparisons show that *Altiatlasius* is certainly not an arctocyonid. For the characters mentioned this genus is more primate-like, and I agree with Sigé *et al.* (1990) that it is very probably a primate, hence the oldest one. Close comparisons with the earliest undoubted euprimates (= primates) are justified.

The continuous lingual cingulum with incipient cuspules of *Altiatlasius* M^2 does not exist in *Teilhardina, Donrussellia,* and their close relatives; it only developed in some later prosimians. Hence, it could really be shared-derived with later simiiforms and fits with the high development of these structures in the new small simians. On *Altiatlasius* M^2, the loss of the postmetaconule crista is advanced over *Teilhardina* and many other omomyiforms. Its postprotocrista curves toward the metacone summit, as also does its prolongation after the metaconule (probably the premetaconule crista). A direct link between protocone and metacone as in simians is not far from being achieved. It would also recall *Donrussellia* if the protocone were not so broad, another condition likely advanced over *Teilhardina* and *Donrussellia.*

The lower premolar of *Altiatlasius* is also interesting. By comparison with *Algeripithecus* and *Djebelemur,* it is probably a P_4 (it is lower than the P_3 of *Algeripithecus*). It is simple, without a metaconid and without any closing of an incipient talonid basin as was described in *Teilhardina* and *Donrussellia.* This P_4 has a lingual base with an incipient continuous rim as for the cingulum of

Djebelemur (this nearly achieved horizontal cingulum is not apparent enough on the drawing published by Sigé *et al.*, 1990; this character contributes significantly to the primate appearance of this tooth). It also shows the posterolingual narrowness found in many simians. The low summit of this P_4 is advanced over that of *Teilhardina* and *Donrussellia* in the direction of *Algeripithecus* and *Djebelemur.*

Other peculiar characters of *Altiatlasius* are those of the lower molars. Their talonids are narrower than in the earliest prosimians, and they have a more rounded outline in occlusal view, especially posterolingually. In *Teilhardina* and *Donrussellia,* the talonid basin is broader, the hypoconid is higher, and its crests are more angulated; the crests of the entoconid also are more angulated. The interesting point is that *Djebelemur* too has a relatively narrow talonid basin and rounded apical crests (Fig. 10; Fig. 3 of Hartenberger and Marandat, 1992). These characters might be typical of simiiforms and might have been primitively retained in them. Broadening and angular cresting of the talonid would have been more advanced in early prosimians. Another noticeable character is the relatively low metaconid of *Altiatlasius* in lingual view. By comparison, both *Teilhardina* and *Donrussellia* retain a much higher metaconid, especially on M_1 (probably primitive). Because *Djebelemur* is remarkable for its low metaconid, accentuated on M_1, this might be another simiiform character derived over the earliest prosimians.

Another intriguing tooth of *Altiatlasius* is THR136, interpreted as an M_2 by Sigé *et al.* (1990). This tooth has a paraconid more labially situated and somewhat smaller than those on the others. Its talonid is slightly more posteriorly extended, and it would appear more so if the base of the crown were intact. Its talonid appears very round in occlusal outline, with a talonid basin projecting more posteriorly than on the others. Furthermore, the posterior border presents a duplicated crestiform hypoconulid that is higher than the entoconid (there is some deformation of the crown, however, mainly on the labial side, where the hypoconid is too close to the protoconid). Hence, I suspect that this tooth might be an M_3. I would also identify THR135 as an M_2 instead of an M_1, because it has a talonid basin broader than the M_1 THR146. If THR136 really is the M_3 of *Altiatlasius,* its third lobe is especially small for a primate. This would be a strong difference from early prosimians, a less dramatic difference from known early simiiforms, but a condition that would give a possible ancestral state for the platyrrhine M_3, which remain so strange within primates by their lack of an M_3 third lobe (Kay, 1980). Hence, the polarity of the M_3 third lobe in euprimates may have to be reconsidered. However, because this tooth is deformed, it will not be included as an M_3 in the character analysis until better material is found.

On the whole, *Altiatlasius* has a few characters more primitive than those in any known euprimate, fitting well with its Paleocene age, and a larger number of characters, either primitive or derived, that suggest a link closer to early simians than to the earliest known prosimians, *Teilhardina* and *Donrussellia* (Fig. 16). Hence, *Altiatlasius* should not be classified as an Omomyidae

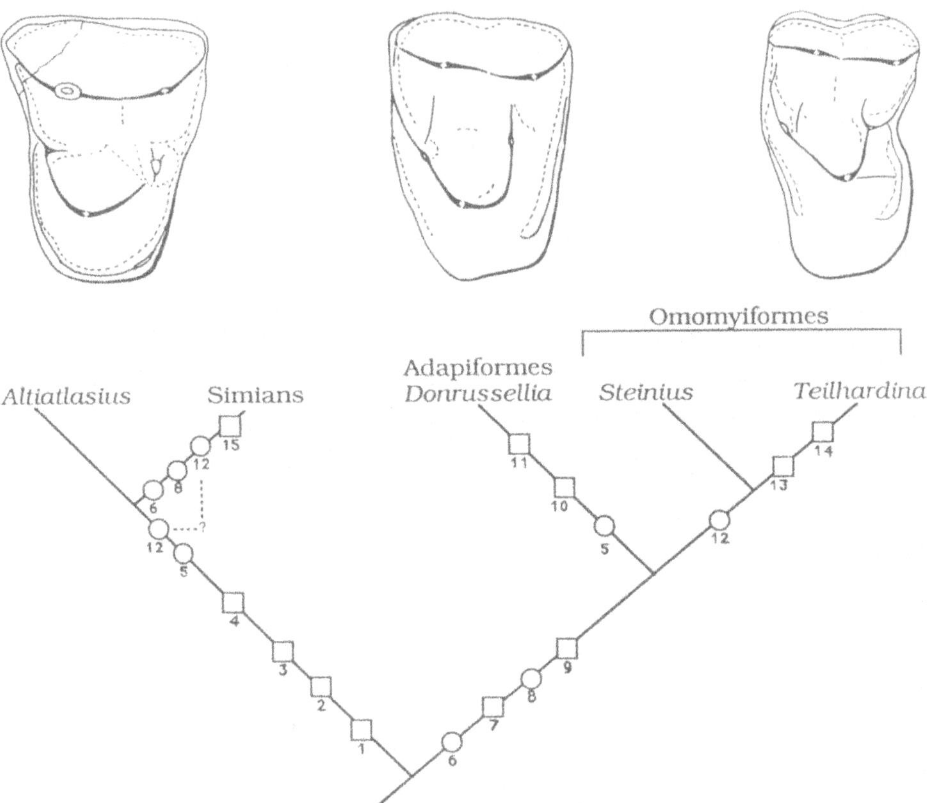

Fig. 16. Cladogram of possible phylogenetic relationships of *Altiatlasius* and the earliest undoubted primate (= euprimate) genera. The three drawings of M^2 in occlusal view are those of *Altiatlasius koulchii, Donrussellia provincialis,* and *Teilhardina belgica.* Derived characters are assumed to be homologous (squares) or convergent (circles). 1, modest bunodonty; 2, M^2 with broad protocone and continuous lingual cingulum; 3, metaconid relatively lower; 4, P_4 low, simple, with incipient continuous lingual cingulum (other premolars presumably also low); 5, reduced metaconule, loss of postmetaconule crista, incomplete crista obliqua (not reaching the metacone summit); 6, reduced stylar shelf; 7, broader talonid basin and higher hypoconid; 8, some paraconid reduction; 9, small metaconid on P_4; 10, straighter postcristid with incipient notch isolating the entoconid; 11, broader talonid basin on M_3; 12, some anterior jaw and premolar shortening; 13, P_{3-4} broader and shorter; 14, M^3 reduced; 15, further P_{3-4} crowding, with anterodorsal inclination; small crestiform hypocone, and other typical simian characters. Note that the morphological differences between the earliest adapiform and the earliest omomyiform imply a certain amount of evolution between them. If *Altiatlasius* is a primate, a view adopted here, its primitive characters imply derived states in the others (6 to 9), and some of its characters shared with simians (1 to 5) make it a member of the stem group of simiiforms, that is, a protosimiiform. This hypothesis implies a Paleocene split between adapiforms and omomyiforms (Strepsirhini if lemuriforms fit here), and the suborder containing the simiiforms (Haplorhini if *Tarsius* could be included there).

as was done by Sigé *et al.* (1990). It is not known if *Altiatlasius* would have the characters to be called a real simiiform, but it certainly points toward very early primates closely related to them, a concept recalling the Protosimiiformes of Hoffstetter (1977). *Altiatlasius* could provisionally be maintained without a familial attribution, as a protosimiiform. The number of derived characters of *Altiatlasius* is in good accord with the discovery that early simians were already quite specialized in the lower Eocene. Such marked specializations take time to be realized and increase the likelihood that they would be started in the late Paleocene. In turn, accepting *Altiatlasius* as an early euprimate modifies the concept of the euprimate morphotype, which includes characters as primitive as a small stylar shelf and large parastyle on the upper molars, and high protoconid, very large paraconid, and small talonid basin on the lower molars, especially M_1. Late Paleocene euprimates must have included two groups, one the protosimiiforms, the other the common stem of adapiforms and omomyiforms.

Altiatlasius must also be compared with *Altanius,* which is considered by some to be the most primitive euprimate (Gingerich *et al.,* 1991). However, *Altanius* has on its upper molars a postprotocingulum, larger conules, a very bizarre shape of the labial tubercles (Figs. 6 and 7), and an extreme narrowing of the trigonid on M_1. All these derived characters link it to the plesiadapiforms (see also Rose *et al.,* Chapter 1, this volume). *Altiatlasius* looks more distinct from omomyids than *Altanius* does, but its derived characters link it with early simians, whereas those of *Altanius* link it with early plesiadapiforms.

Another step toward deciphering early primate relationships would be to find a link between the earliest simians and tarsiids, as I consider the haplorhine concept probably to have some content. For this question, *Afrotarsius* would be critical if it were a real African tarsiid (Simons and Bown, 1985). However, the affinities of *Afrotarsius,* which is much younger than the primates discussed above, are doubtful for me. This fossil has some characters very rare in primates: the paraconid has the same shape in M_{1-3} and is nearly crestiform, M_3 has almost no third lobe, and the metaconid is very high [in lingual view, the metaconid is higher than in *Tarsius* and *Pseudoloris* (Fig. 17); I have found a similar height only in some galagos]. A crestiform paraconid centrally located is rare in euprimates but common in some insectivores such as adapisoricids, which also have a very high metaconid. Simons and Bown (1985) emphasized similarities of *Afrotarsius* with the late Eocene *Pseudoloris* and the living *Tarsius.* However, these genera differ in important characters: *Afrotarsius* M_3 has a much higher protoconid than the two others, it has a higher metaconid with a more vertical posterior slope (Fig. 17), and it has no third lobe, whereas the two others have a well-developed one.

These are numerous differences and also suggest possible similarities with other insectivorous mammals. M_1 larger than M_2 is not common in primates but exists in some *Necrolemur,* in *Catopithecus,* and certainly other primates (but in neither the *Tarsius* I have seen nor *Pseudoloris*). Hence, if *Afrotarsius* were a primate, it would be a very strange one. Much more discus-

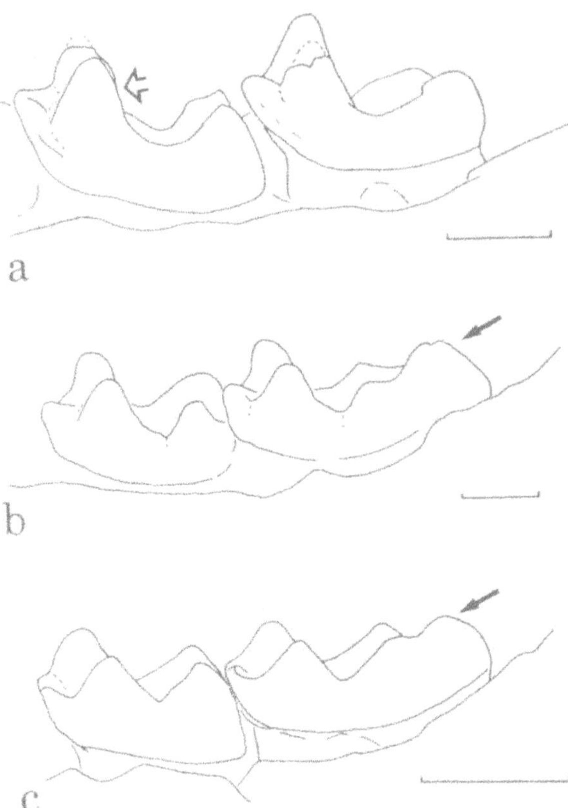

Fig. 17. Right mandibles with M_2 and M_3 of *Afrotarsius chatrathi* (a), *Tarsius* sp. (b), and *Pseudoloris parvulus* (c). All are lingual views drawn to the same anteroposterior length. All scale bars are 1 mm. The white arrow points to the very steep and high metaconid in *Afrotarsius*. The two black arrows point to the hypoconulid, posteriorly extended into a third lobe, in *Tarsius* and *Pseudoloris*.

sion is needed to show why it is not some kind of insectivore *sensu lato*. To include *Afrotarsius* in the discussion of early primates would be very far-reaching, leading to inclusion of more insectivore-like dental characters in the euprimate morphotype and changing the biogeographical perspective. This appears to me premature in view of the uncertainties about the affinities of this genus. Other mentions of African and Arabian omomyids or "tarsiiforms" are based on very fragmentary evidence (Simons *et al.*, 1986; Thomas *et al.*, 1988). This evidence seems to me insufficient for a familial identification or to deduce that omomyiforms were ever present in Africa. Likewise, the presence of a lorisid in the Fayum (Simons *et al.*, 1986) needs further confirmation. In view of the well-developed M^3 cingulum in oligopithecines (Simons, 1989) and the extended M^3 talon described in the primitive simian *Branisella* (Wolff, 1984), the large size of this M^3 and its differences with living lorisids enumerated by Simons *et al.* (1986) might have a simpler interpretation with an oligopithecine M^3.

The living *Tarsius* is the main source of the haplorhine concept. Its dental morphology is that of a small insectivorous primate, but it also presents several characters more primitive than in other living prosimians: there is a paraconid on M_{1-3}. *Tarsius* upper molars have no hypocone, whereas this cusp is usually well developed in small insectivorous primates, and they are transversely as elongated as in *Donrussellia* and more than in *Altiatlasius* (Fig. 4). The specimen that I have has no continuous lingual cingulum (Fig. 4). However, such a cingulum with small irregularities sometimes suggesting a hypocone is described by Maier (1980). Instead of being a simiiform character, the continuous lingual cingulum might be a haplorhine one. *Tarsius* has molars probably retaining some primitive characters, which point toward a very ancient branching with other groups. If it were certain that transversness of simple pointed molars cannot reincrease, this would imply a branching at a stage more primitive than *Altiatlasius*. Ancestral tarsiids and protosimiiforms must have split very early, in the Paleocene according to Hoffstetter (1977) and Beard *et al.* (1991). *Tarsius* dental characters may well reflect an antiquity of this event greater than the late Thanetian, but they do not exclude a slightly later split.

Conclusion

The two genera *Algeripithecus* and *Tabelia* and the upper molars referred to as cf. *Djebelemur* from the late lower Eocene or early middle Eocene of Algeria and Tunisia are small primates typified by very bunodont upper molars. They are certainly early simiiforms. The characters of these bunodont genera suggest a closer relationship with parapithecids than with propliopithecines. If the small bunodont teeth from Chambi are associated with the mandible of *Djebelemur*, this genus also pertains to the same group. If *Djebelemur* had non-bunodont upper molars, it could represent an early member of a less bunodont group, oligopithecines (Fig. 18), platyrrhines, or another unrecognized group. These small bunodont simians together with parapithecids and propliopithecines may be united in catarrhines *sensu lato*, defined by their high degree of bunodonty. Their early members had three premolars. The Asiatic Amphipithecinae probably pertain to the same group.

Oligopithecines and platyrrhines probably branched off earlier than these fossils, i.e., in the early Eocene or Paleocene. Large gaps remain in the record, and it is still difficult to elaborate hypotheses about a more precise branching within simiiforms (the place of oligopithecines and platyrrhines) and a schema relating the tarsier ancestors to the preceding groups (Fig. 18). The dichotomies between primate infraorders took place during the Paleocene. Tiny species such as the early bunodont simians change the view of simiiforms as inferred from their later representatives. *Djebelemur* had an unfused mandible, and a similar morphology for other early simiiforms is implied by their small size. Very simple premolars as in *Algeripithecus* and *Djebelemur* fit with their early age. A number of three premolars as in *Djebelemur* probably was common in these forms.

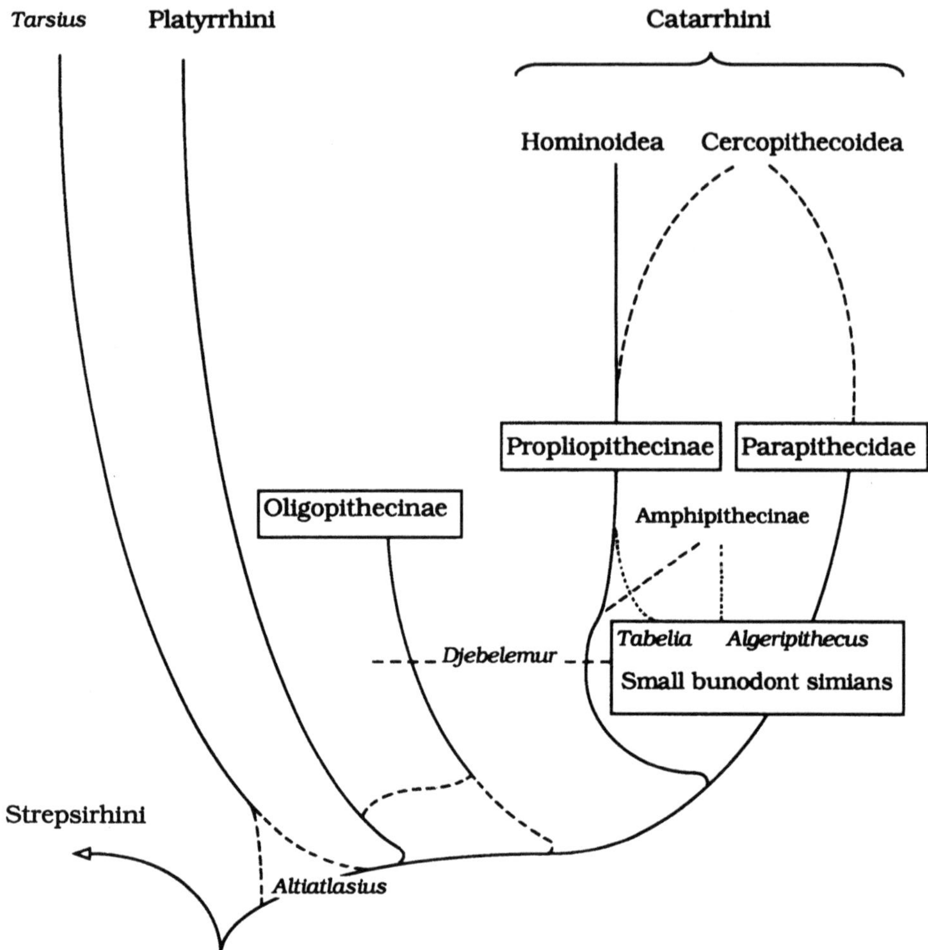

Fig. 18. A view of the evolutionary position of the small early African simians as proposed in this study. The new small bunodont forms from North Africa are members of a group that probably contained the ancestors of parapithecids. It is possible but less likely that some of them might have had a close relationship with propliopithecines or amphipithecines. In any case, all these highly derived, very bunodont taxa suggest a grouping of all of them as Catarrhini, thus primarily defined on their bunodonty (and related skull characters) and reflecting an early dietary specialization. The place of the *Djebelemur* mandible is not settled but could well be with a more crested group such as oligopithecines. No clear hypothesis was found relating Platyrrhini and Oligopithecinae. Likewise, if *Altiatlasius* is suggested as a possible protosimiiform, the branching of *Tarsius* could be proposed as either before or after this genus.

Comparisons of the small bunodont simians with adapiforms show no close relationship between them. Analysis of the evolutionary trends in European cercamoniines seems definitely to rule out the possibility of rooting early simians in them. Omomyiforms seem closer, but the early North African simians are outside their known radiation. Comparisons with the late Paleocene *Altiatlasius* from Morocco suggest a close relationship, that is, a Paleocene differentiation of early simiiforms (or protosimiiforms), which are distinct from a group including known adapiforms and omomyiforms.

Until now, no adapiform, omomyiform, tarsiid, or lorisid has been shown by sufficiently convincing evidence to be present in the Paleogene of Africa. This could change in the future. However, the new finds, instead of confirming Africa as the cradle of all primate infraorders, confirm and amplify an early simiiform radiation in Africa, coherent with the marked faunal endemism of this continent.

Addendum

Two important papers have appeared since this chapter was completed. They deserve some comment. Gheerbrant *et al.* (1993) described *Omanodon* and *Shizarodon* as possible adapiforms from the lower Oligocene from Oman. However, I do not agree with their interpretation. These authors interpret the reduced metaconid and the anteriorly extended paracristid on M_1 as having evolved from an *Anchomomys*-like morphology. However, an anteriorly extended paracristid can hardly be derived from the shorter anchomomyine M_1 paracristid. Reduction of the metaconid is not found in adapiforms to my knowledge. These structurally important characters suggest *Omanodon* and *Shizarodon* belong to another group. A low and reduced metaconid is found in some omomyiforms, lemuriforms, and simiiforms. Both characters together are found in the simiiform *Djebelemur.* Gheerbrant *et al.* (1993) recognize that the P_4 attributed to *Omanodon* has nothing to do with adapiforms. I agree with their discussion of the attribution of this tooth, and consequently I think that *Omanodon* can not be an adapiform. This P_4 shows similarities with *Microcebus* and with some simiiforms (Gheerbrant *et al.*, 1993). The upper molars of *Omanodon* have characters similar to oligopithecines: anteriorly placed protocone, and wide notch between paracone and metacone.

This new Oman material is difficult to interpret. It is hard to definitely rule out lemuriform affinities for these two genera. However, based on all the present evidence, and particularly similarities to *Djebelemur* and to oligopithecines, I favor simiiform affinities for these two genera. This fits with the P^3 and P_4 characters analyzed by Gheerbrant *et al.* (1993), and with the upper molar the size of *Shizarodon* that is oligopithecine-like. Also, the abundance of oligopithecines in the Fayum faunas renders possible the contemporaneous existence of other small, highly crested simiiforms nearby. In this interpretation, there is still no definitive adapiform found until now on the Afro-Arabian continent.

The recent description of *Eosimias* from Shanghuang, China, as a middle Eocene simiiform (Beard *et al.*, 1994) is controversial, and I do not share the conclusions of these authors. I will comment only on some critical points.

These authors emphasize that the metaconid widely separated from an enlarged protoconid is a synapomorphy of *Eosimias* with simians. I am not sure of what the "enlarged" protoconid means, but a tall protoconid is likely to be primitive in early primates. The M_2 cusps are higher in *Eosimias* than in *Cantius* and *Steinius:* if this height is not primitive, it is derived in a direction opposite to simian structure. A widening between protoconid and metaconid occurred in nonsimians, e.g., in *Microchoerus* compared with *Nannopithex,* in large adapines compared with primitive adapiforms. A character resulting from such a trend should be used with caution, in relation to size, functional adaptation, and possible morphological trends. Furthermore, transverse widening of the trigonid seems already more accentuated on the M_1 of *Eosimias* than on the M_1 of the later Fayum *Catopithecus,* and it is clearly more accentuated on the M_2 of *Eosimias* than on the M_2 of the Miocene *Neosaimiri* illustrated by Beard *et al.* (1994). There was at least some convergence along the polarity assumed by these authors, and simple homology between these taxa for this character is not at all evident. The most striking character of *Eosimias* molars is the anteroposterior narrowness of its trigonid, in marked contrast with euprimates bearing a similar-sized paraconid. Such an anteroposteriorly short trigonid with narrow protoconid on M_1 contrasts especially with *Djebelemur* and several undoubted Fayum simiiforms, and suggests nonsimian affinities. Beard *et al.* (1994) mention the "specifically simian-like mode of dental reduction and premolar compaction in *Eosimias,*" but its subhorizontal P_4 (which is not obliquely oriented) does not fit with the model I inferred from *Algeripithecus* and *Djebelemur.* On the whole, I doubt that the synapomorphies proposed by Beard *et al.* (1994) would be real homologies between *Eosimias* and simians, and I do not accept this genus as a primitive simiiform. For these reasons, I see no need to modify my hypotheses about the early North African simiiforms or about simiiform origins. More material and more detailed descriptions and analyses are needed to know what *Eosimias* is. If it really is a euprimate, tarsiiform affinities would be less unlikely, but the documentation of a new group of primates would also be possible.

Acknowledgments

I am very grateful to the organizers for inviting me and supporting my participation at the Duke Conference. M. Mahboubi discovered the specimens that we described together and that led to this work. B. Sigé and J.-L. Hartenberger allowed the study of specimens in their care. This work was facilitated by casts given or exchanged by T. M. Bown, P. D. Gingerich, R. F. Kay, K. D. Rose, D. E. Russell, D. E. Savage, E. L. Simons, and H. Thomas. Helpful comments on the manuscript were provided by K. C. Beard, D. T. Rasmussen, and R. F. Kay. E. L. Simons commented the fossils. J.-J. Jaeger discussed the age of Nementcha. Help in English editing was provided by W. P. Luckett, K. C. Beard, and R. F. Kay. The scanning electron micrographs were made by A. Rossi and printed by M. Pons, and D. Crès made the diagrams.

References

Andrews, P. 1988. A phylogenetic analysis of the Primates. In: M. J. Benton (ed.), *The Phylogeny and Classification of the Tetrapods, Vol. 2: Mammals,* pp. 143–175. Clarendon Press, Oxford.

Beard, K. C. 1988. New notharctine primate fossils from the early Eocene of New Mexico and southern Wyoming and the phylogeny of Notharctinae. *Am. J. Phys. Anthropol.* **75:**439–469.

Beard, K. C., Krishtalka, L., and Stucky, R. K. 1991. First skulls of the Eocene primate *Shoshonius cooperi* and the anthropoid–tarsier dichotomy. *Nature* **349:**64–67.

Beard, K. C., Qi, T., Dawson, M. R., Wang, B., and Li, C. 1994. A diverse new primate fauna from middle Eocene fissure-fillings in southeastern China. *Nature* **368:**604–609.

Beecher, R. M. 1983. Evolution of the mandibular symphysis in Notharctinae (Adapidae, Primates). *Int. J. Primatol.* **4:**99–112.

Bonis, L. de, Jaeger, J.-J., Coiffait, B., and Coiffait, P.-E. 1988. Découverte du plus ancien primate catarrhinien connu dans l'Eocène supérieur d'Afrique du Nord. *C. R. Acad. Sci. Paris [Ser. II]* **306:**929–934.

Bown, T. M. 1979. New omomyid primates (Haplorhini, Tarsiiformes) from middle Eocene rocks of west-central Hot Springs County, Wyoming. *Fol. Primatol.* **31:**48–73.

Bown, T. M., and Rose, K. D. 1987. Patterns of dental evolution in early Eocene anaptomorphine primates (Omomyidae) from the Big Horn Basin, Wyoming. *Paleont. Soc. Mem.* **23:**1–162.

Cartmill, M., and Kay, R. F. 1978. Craniodental morphology, *Tarsius* affinities, and primate suborders. In: D. J. Chivers and K. A. Joysey (eds.), *Recent Advances in Primatology, Vol. 3, Evolution,* pp. 205–214. Academic Press, London.

Ciochon, R. L., and Chiarelli, A. B. (eds.). 1980. *Evolutionary Biology of the New World Monkeys and Continental Drift.* Plenum Press, New York.

Ciochon, R. L., Savage, D. E., Tint, T., and Maw, B. 1985. Anthropoid origins in Asia? New discovery of *Amphipithecus* from the Eocene of Burma. *Science* **229:**756–759.

Coiffait, P.-E., Coiffait, B., Jaeger, J.-J., and Mahboubi, M. 1984. Un nouveau gisement à Mammifères fossiles d'âge Eocène supérieur sur le versant sud des Nementcha (Algérie orientale): découverte des plus anciens rongeurs d'Afrique. *C. R. Acad. Sci. Paris [Ser. II]* **299:**893–898.

Colbert, E. H. 1937. A new primate from the upper Eocene Pondaung Formation of Burma. *Am. Mus. Novit.* **951:**1–18.

Court, N., and Hartenberger, J.-L. 1992. A new species of the hyracoid mammal *Titanohyrax* from the Eocene of Tunisia. *Paleontology* **35:**309–317.

Dagosto, M. 1988. Implications of postcranial evidence for the origin of Euprimates. *J. Hum. Evol.* **17:**35–56.

Dagosto, M. 1990. Models for the origin of the anthropoid postcranium. *J. Hum. Evol.* **19:**121–139.

Delson, E., and Rosenberger, A. L. 1980. Phyletic perspectives on platyrrhine origins and anthropoid relationships. In: R. L. Ciochon and A. B. Chiarelli (eds.), *Evolutionary Biology of the New World Monkeys and Continental Drift,* pp. 445–458. Plenum Press, New York.

Fleagle, J. G. 1988. *Primate Adaptation and Evolution.* Academic Press, San Diego.

Fleagle, J. G., and Kay, R. F. 1987. The phyletic position of the Parapithecidae. *J. Hum. Evol.* **16:**483–532.

Fleagle, J. G., and Simons, E. L. 1983. The tibio–fibular articulation in *Apidium phiomense,* an Oligocene anthropoid. *Nature* **301:**238–239.

Fleagle, J. G., Bown, T. M., Obradovich, J. D., and Simons, E. L. 1986. How old are the Fayum primates? In: J. G. Else and P. C. Lee (eds.), *Primate Evolution*, pp. 3–17. Cambridge University Press, Cambridge.

Ford, S. M. 1988. Postcranial adaptations of the earliest platyrrhine. *J. Hum. Evol.* **17:**155–192.

Franzen, J. L. 1987. Ein neuer Primate aus dem Mitteleozän der Grube Messel (Deutschland, S-Hessen). *Cour. Forsch. Inst. Senckenberg* **91:**151–187.

Gebo, D. L. 1986. Anthropoid origins—the foot evidence. *J. Hum. Evol.* **15:**421–430.

Gebo, D. L. 1988. Foot morphology and locomotor adaptation in Eocene primates. *Fol. Primatol.* **50:**3–41.

Gebo, D. L. 1989. Locomotor and phylogenetic considerations in anthropoid evolution. *J. Hum. Evol.* **18:**201–233.

Gevin, P., Feist, M., and Mongereau, N. 1974. Découverte de charophytes d'âge éocène au Glib Zegdou (Sahara algérien). *Bull. Soc. Hist. Nat. Afr. Nord* **65:**371–376.

Gheerbrant, E., and Thomas, H. 1992. The two first possible new adapids from the Arabian Peninsula (Taqah, early Oligocene of Sultanate of Oman). In: *Abstracts of the 14th Congress of the International Primatological Society*, pp. 349–350. Sicop Press, Strasbourg.

Gheerbrant, E., Cappetta, H., Feist, M., Jaeger, J.-J., Sudre, J., Vianey-Liaud, M., and Sigé, B. 1993. La succession des faunes de vertébrés d'âge paleéocène supérieur et éocène inférieur dans le bassin d'Ouarzazate, Maroc. Contexte géologique, portée biostratigraphique et paléogéographique. *Newsl. Stratigr.* **28:**33–58.

Gheerbrant, E., Thomas, H., Roger, J., Sen, S., and Al-Sulaimani, Z. 1993. Deux nouveaux primates dans l'Oligocène inférieur de Taqah (Sultanat d'Oman): premiers adapiformes (?Anchomomyini) de la péninsule arabique. *Palaeovertebrata* **22:**141–196.

Gingerich, P. D. 1975. A new genus of Adapidae (Mammalia, Primates) from the late Eocene of southern France, and its significance for the origin of higher primates. *Contrib. Mus. Paleontol. Univ. Michigan* **24:**163–170.

Gingerich, P. D. 1977a. New species of Eocene primates and the phylogeny of European Adapidae. *Fol. Primatol.* **28:**60–80.

Gingerich, P. D. 1977b. Radiation of Eocene Adapidae in Europe. *Geobios Mém. Spéc.* **1:**165–182.

Gingerich, P. D. 1980. Eocene Adapidae, paleobiogeography, and the origin of South American Platyrrhini. In: R. L. Ciochon and A. B. Chiarelli (eds.), *Evolutionary Biology of the New World Monkeys and Continental drift*, pp. 123–138. Plenum Press, New York.

Gingerich, P. D. 1984. Primate evolution: Evidence from the fossil record, comparative morphology, and molecular biology. *Yearb. Phys. Anthropol.* **27:**57–72.

Gingerich, P. D. 1990. African dawn for primates. *Nature* **346:**411.

Gingerich, P. D. 1993. Oligocene age of the Gebel Qatrani Formation, Fayum, Egypt. *J. Hum. Evol.* **24:**207–218.

Gingerich, P. D., and Simons, E. L. 1977. Systematics, phylogeny, and evolution of early Eocene Adapidae (Mammalia, Primates) in North America. *Contrib. Mus. Paleontol. Univ. Michigan* **24:**245–279.

Gingerich, P. D., Dashzeveg, D., and Russell, D. E. 1991. Dentition and systematic relationships of *Altanius orlovi* (Mammalia, Primates) from the early Eocene of Mongolia. *Geobios* **24:**637–646.

Godinot, M. 1981. Les mammifères de Rians (Eocène Inférieur, Provence). *Palaeovertebrata* **10:**43–126.

Godinot, M. 1988a. Les primates adapidés de Bouxwiller (Eocène Moyen, Alsace) et leur apport à la compréhension de la faune de Messel et à l'évolution des Anchomomyini. *Cour. Forsch. Inst. Senckenberg* **107:**383–407.

Godinot, M. 1988b. Le gisement du Bretou (Phosphorites du Quercy, Tarn-et-Garonne, France) et sa faune de vertébrés de l'Eocène supérieur, VII, Primates. *Palaeontographica* **205:**113–127.

Godinot, M. 1990. An introduction to the history of primate locomotion. In: F. K. Jouffroy, M. H. Stack, and C. Niemitz (eds.), *Gravity, Posture and Locomotion in Primates*, pp. 45–60. Il Sedicesimo, Florence.

Godinot, M. 1992a. Apport à la systématique de quatre genres d'Adapiformes (Primates, Eocène). *C. R. Acad. Sci. Paris [Ser. II]* **314**:237–242.
Godinot, M. 1992b. New Eocene adapiform humeri from Europe. In: *Abstracts of the 14th Congress of the International Primatological Society,* pp. 285–286. Sicop Press, Strasbourg.
Godinot, M., and Beard, K. C. 1991. Fossil primate hands: A review and an evolutionary inquiry emphasizing early forms *Hum. Evol.* **6**:307–354.
Godinot, M., and Mahboubi, M. 1992. Earliest known simian primate found in Algeria. *Nature* **357**:324–326.
Godinot, M., and Mahboubi, M. 1994. Les petits primates simiiformes de Glib Zegdou (Eocène, Algérie). *C. R. Acad. Sci. Paris [Sér. II]* **319**:357–364.
Godinot, M., Crochet, J.-Y., Hartenberger, J.-L., Lange-Badré, B., Russell, D. E., and Sigé, B. 1987. Nouvelles données sur les mammifères de Palette (Eocène inférieur, Provence). *Münch. Geowiss. Abh.* **10**:273–288.
Godinot, M., Russell, D. E., and Louis, P. 1992. Oldest known *Nannopithex* (Primates, Omomyiformes) from the early Eocene of France. *Fol. Primatol.* **58**:32–40.
Grambast, L., and Lavocat, R. 1959. Sur la présence dans la région du Dra (Sahara nord-occidental) de couches éocènes datées par les charophytes. *C. R. Som. Soc. Géol. Fr.* **6**:153–154.
Gregory, W. K. 1920. On the structure and relations of *Notharctus,* an American Eocene primate. *Mem. Am. Mus. Nat. Hist.* **3**: 49–243.
Harrison, T. 1987. The phylogenetic relationships of the early catarrhine primates: A review of the current evidence. *J. Hum. Evol.* **16**:41–80.
Hartenberger, J.-L. 1986. Hypothèse paléontologique sur l'origine des Macroscelidea (Mammalia). *C. R. Acad. Sci. Paris [Ser. II]* **302**:247–249.
Hartenberger, J.-L., and Marandat, B. 1992. A new genus and species of an early Eocene primate from North Africa. *Hum. Evol.* **7**:9–16.
Hartenberger, J.-L., Martinez, C., and Ben Saïd, N. 1985. Découverte d'un gisement de Mammifères de l'Eocène inférieur en Tunisie. *C. R. Acad. Sci. Paris [Ser. II]* **301**:649–652.
Hershkovitz, P. 1977. *Living New World Monkeys (Platyrrhini), Vol. 1.* University of Chicago Press, Chicago.
Hiiemäe, K. M., and Kay, R. F. 1973. Evolutionary trends in the dynamics of primate mastication. In: M. R. Zingeser (ed.), *Craniofacial Biology of Primates,* pp. 28–64. Karger, Basel.
Hoffstetter, R. 1969. Un primate de l'Oligocène inférieur sud-américain: *Branisella boliviana* gen. et sp. nov. *C. R. Acad. Sci. Paris [D]* **269**:434–437.
Hoffstetter, R. 1977. Phylogénie des primates; Confrontation des résultats obtenus par les diverses voies d'approche du problème. *Bull. Mém. Soc. Anthropol. Paris [Ser. XIII]* **4**:327–346.
Hoffstetter, R. 1982. Les Primates Simiiformes (= Anthropoidea) (Compréhension, phylogénie, histoire biogéographique). *Ann. Paléont.* **68**:241–290.
Jaeger, J.-J., Denys, C., and Coiffait, B. 1985. New Phiomorpha and Anomaluridae from the late Eocene of North-West Africa: Phylogenetic implications. In: W. P. Luckett and J.-L. Hartenberger (eds.), *Evolutionary Relationships among Rodents - A Multidisciplinary Analysis,* pp. 567–588. Plenum Press, New York.
Kay, R. F. 1977. The evolution of molar occlusion in the Cercopithecidae and early catarrhines. *Am. J. Phys. Anthrop.* **46**:327–352.
Kay, R. F. 1980. Platyrrhine origins—a reappraisal of the dental evidence. In: R. L. Ciochon and A. B. Chiarelli (eds.), *Evolutionary Biology of the New World Monkeys and Continental Drift,* pp. 159–188. Plenum Press, New York.
Kay, R. F., and Covert, H. H. 1984. Anatomy and behavior of extinct primates. In: D. J. Chivers, B. A. Wood, and A. Bilsborough (eds.), *Food Acquisition and Processing in Primates,* pp. 467–508. Plenum Press, New York.
Kay, R. F., and Simons, E. L. 1980. The ecology of Oligocene African Anthropoidea. *Int. J. Primatol.* **1**:21–37.
Kay, R. F., Fleagle, J. G., and Simons, E. L. 1981. A revision of the Oligocene apes of the Fayum Province, Egypt. *Am. J. Phys. Anthrop.* **55**:293–322.

Kinzey, W. G. 1973. Reduction of the cingulum in Ceboidea. In: M. R. Zingeser (ed.), *Craniofacial Biology of Primates,* pp. 101–127. Karger, Basel.

MacPhee, R. D. E. 1981. Auditory regions of primates and eutherian insectivores. *Contrib. Primatol.* **18:**1–282.

MacPhee, R. D. E., and Cartmill, M. 1986. Basicranial structures and primate systematics. In: D. R. Swindler and J. Erwin (eds.), *Comparative Primate Biology, 1, Evolution and Anatomy,* pp. 219–275. Alan R. Liss, New York.

Maier, W. 1980. Konstruktionsmorphologische Untersuchungen am Gebiss der rezenten Prosimiae (Primates). *Abh. Senckenb. Naturforsch. Ges.* **538:**1–158.

Martin, R. D. 1990. *Primate Origins and Evolution. A Phylogenetic Reconstruction.* Chapman and Hall, London.

Maw, B., Ciochon, R. L., and Savage, D. E. 1979. Late Eocene of Burma yields earliest anthropoid primate, *Pondaungia cotteri. Nature* **282:**65–67.

Pilgrim, G. E. 1927. A *Sivapithecus* palate and other primate fossils from India. *Mem. Geol. Surv. India (Paleont. Indica)* **14:**1–26.

Rasmussen, D. T. 1989. The evolution of the Hyracoidea: A review of the fossil evidence. In: D. R. Prothero and R. M. Schoch (eds.), *The Evolution of Perissodactyls,* pp. 57–78. Oxford University Press, Oxford.

Rasmussen, D. T. 1990. The phylogenetic position of *Mahgarita stevensi:* Protoanthropoid or lemuroid? *Int. J. Primatol.* **11:**439–468.

Rasmussen, D. T., and Simons, E. L. 1988. New specimens of *Oligopithecus savagei,* early Oligocene primate from the Fayum, Egypt. *Folia Primatol.* **51:**182–208.

Rasmussen, D. T., Bown, T. M., and Simons, E. L. 1992. The Eocene–Oligocene transition in continental Africa. In: D. R. Prothero and W. A. Berggren (eds.), *Eocene–Oligocene Climatic and Biotic Evolution,* pp. 548–566. Princeton University Press, Princeton.

Ravosa, M. J. 1991. Structural allometry of the prosimian mandibular corpus and symphysis. *J. Hum. Evol.* **20:**3–20.

Rose, K. D., and Bown, T. M. 1991. Additional fossil evidence on the differentiation of the earliest euprimates. *Proc. Natl. Acad. Sci. USA* **88:**98–101.

Rosenberger, A. L. 1977. *Xenothrix* and ceboid phylogeny. *J. Hum. Evol.* **6:**461–481.

Rosenberger, A. L., and Szalay, F. S. 1980. On the tarsiiform origin of Anthropoidea. In: R. L. Ciochon and A. B. Chiarelli (eds.), *Evolutionary Biology of the New World Monkeys and Continental Drift,* pp. 139–157. Plenum Press, New York.

Rosenberger, A. L., Hartwig, W. C., and Wolff, R. G. 1991. *Szalatavus attricuspis,* an early platyrrhine primate. *Fol. Primatol.* **56:**225–233.

Russell, D. E., and Gingerich, P. D. 1987. Nouveaux primates de l'Eocène du Pakistan. *C. R. Acad. Sci. Paris [Ser. 2]* **304:**209–214.

Russell, D. E., Louis, P., and Savage, D. E. 1967. Primates of the French early Eocene. *Univ. Calif. Publ. Geol. Sci.* **73:**1–46.

Schmid, P. 1982. *Die systematische Revision der europäischen Microchoeridae Lydekker, 1887 (Omomyiformes, Primates).* Juris, Zurich.

Sigé, B. 1985. Les chiroptères oligocènes du Fayum, Egypte. *Geol. Palaeontol.* **19:**161–189.

Sigé, B. 1991. Rhinolophoidea et Vespertilionoidea (Chiroptera) du Chambi (Eocène inférieur de Tunisie)—aspects biostratigraphique, biogéographique et paléoécologique de l'origine des chiroptères modernes. *N. Jb. Geol. Paläont. Abh.* **182:**355–376.

Sigé, B., Jaeger, J.-J., Sudre, J., and Vianey-Liaud, M. 1990. *Altiatlasius koulchii* n. gen. n. sp., primate omomyidé du Paléocène du Maroc, et les origines des euprimates. *Palaeontographica* **214:**31–56.

Simons, E. L. 1970. Review of the phyletic relationships of Oligocene and Miocene Catarrhini. In: A. A. Dahlberg (ed.), *Dental Morphology and Evolution,* pp. 193–208. University of Chicago Press, Chicago.

Simons, E. L. 1971. Relationships of *Amphipithecus* and *Oligopithecus. Nature* **232:**489–491.

Simons, E. L. 1972. *Primate Evolution: An Introduction to Man's Place in Nature.* Macmillan, New York.

Simons, E. L., 1976. The fossil record of primate phylogeny. In: M. Goodman, R. E. Tashian, and J. E. Tashian (eds.), *Molecular Anthropology*, pp. 35–62. Plenum Press, New York.

Simons, E. L. 1989. Description of two genera and species of late Eocene Anthropoidea. *Proc. Natl. Acad. Sci. USA* **86:**9956–9960.

Simons, E. L. 1990. Discovery of the oldest known anthropoidean skull from the Paleogene of Egypt. *Science* **247:**1567–1569.

Simons, E. L. 1992. Diversity in the early Tertiary anthropoidean radiation in Africa. *Proc. Natl. Acad. Sci. USA* **89:**10743–10747.

Simons, E. L., and Bown, T. M. 1985. *Afrotarsius chatrathi,* first tarsiiform primate (? Tarsiidae) from Africa. *Nature* **313:**475–477.

Simons, E. L., and Kay, R. F. 1988. New material of *Qatrania* from Egypt with comments on the phylogenetic position of the Parapithecidae (Primates, Anthropoidea). *Am. J. Primatol.* **15:**337–347.

Simons, E. L., Bown, T. M., and Rasmussen, D. T. 1986. Discovery of two additional prosimian primate families (Omomyidae, Lorisidae) in the African Oligocene. *J. Hum. Evol.* **15:**431–437.

Simons, E. L., Rasmussen, D. T., and Gebo, D. L. 1987. A new species of *Propliopithecus* from the Fayum, Egypt. *Am. J. Phys. Anthrop.* **73:**139–147.

Stehlin, H. G. 1912. Die Säugetiere des schweizerischen Eocaens. Siebenter Teil, erste Hälfte. *Abh. Schweiz. Paläont. Ges.* **38:**1163–1298.

Sudre, J. 1979. Nouveaux mammifères du Sahara occidental. *Palaeovertebrata* **9:**83–115.

Szalay, F. S. 1976. Systematics of the Omomyidae (Tarsiiformes, Primates). Taxonomy, phylogeny, and adaptations. *Bull. Am. Mus. Nat. Hist.* **156:**157–450.

Szalay, F. S., and Dagosto, M. 1980. Locomotor adaptations as reflected on the humerus of Paleogene primates. *Fol. Primatol.* **34:**1–45.

Szalay, F. S., and Dagosto, M. 1988. Evolution of hallucial grasping in the primates. *J. Hum. Evol.* **17:**1–33.

Szalay, F. S., and Delson, E. 1979. *Evolutionary History of the Primates.* Academic Press, New York.

Tattersall, I., and Schwartz, J. H. 1983. A revision of the European Eocene primate genus *Protoadapis* and some allied forms. *Am. Mus. Novit.* **2762:**1–16.

Thomas, H., Roger, J., Sen, S., and Al-Sulaimani, Z. 1988. Découverte des plus anciens "Anthropoïdes" du continent arabo-africain et d'un primate tarsiiforme dans l'Oligocène du Sultanat d'Oman. *C. R. Acad. Sci. Paris [Ser. II]* **306:**823–829.

Thomas, H., Sen, S., Roger, J., and Al-Sulaimani, Z. 1991. The discovery of *Moeripithecus markgrafi* Schlosser (Propliopithecidae, Anthropoidea, Primates), in the Ashawq Formation (early Oligocene of Dhofar Province, Sultanate of Oman). *J. Hum. Evol.* **20:**33–49.

Van Valen, L. 1978. The beginning of the age of mammals. *Evol. Theory* **4:**45–80.

Vianey-Liaud, M., Jaeger, J.-J., Hartenberger, J.-L., and Mahboubi, M. 1994. Les rongeurs de l'Eocène d'Afrique nord-occidentale [Glib Zegdou (Algérie) et Chambi (Tunisie)] et l'origine des Anomaluridae. *Palaeovertebrata* **23:**93–118.

Wolff, R. G. 1984. New specimens of the primate *Branisella boliviana* from the early Oligocene of Salla, Bolivia. *J. Vert. Paleontol.* **4:**570–574.

Paleogeography, Paleobiogeography, and Anthropoid Origins

11

PATRICIA A. HOLROYD and
MARY C. MAAS

Introduction

The study of anthropoid origins has long been tied to hypotheses postulating Eocene dispersal of early anthropoids or their precursors into Africa from Europe (e.g., Gingerich, 1975; Rasmussen and Simons, 1988), from Asia (e.g., Gingerich, 1980; Ciochon and Chiarelli, 1980; Ciochon *et al.*, 1985; Rosenberger, 1986), or from South America (e.g., Szalay, 1976). Similarly, investigations of platyrrhine origins have focused on either a North American (e.g., Gingerich, 1980; Hoffstetter, 1972; Wood, 1980; Rosenberger, 1986) or an African source (e.g., Lavocat, 1974, 1980; Ciochon and Chiarelli, 1980; Fleagle, 1986). These different scenarios have been based in large part on putative ancestor-descendant relationships and the identification of early anthropoids or protoanthropoids in the presumed source areas but also have relied on current understanding of temporal relationships between faunas and reconstructions of Eocene paleogeography and paleobiogeography to determine the probable timing, mode, and route of dispersal.

PATRICIA A. HOLROYD • Department of Biological Anthropology and Anatomy, Duke University, Durham, North Carolina 27710, and U.S. Geological Survey, Denver, Colorado 80225. MARY C. MAAS • Department of Biological Anthropology and Anatomy, Duke University, Durham, North Carolina 27710.

Anthropoid Origins, edited by John G. Fleagle and Richard F. Kay. Plenum Press, New York, 1994.

For example, some postulated anthropoid origin scenarios have been tied to reported large-scale influxes of mammals into Africa, either from Europe or from Asia (e.g., Coryndon and Savage, 1973; Rasmussen and Simons, 1988). Similarly, joint migration of platyrrhine and caviomorph rodent ancestors from either North America or Africa has been proposed to explain the appearance of anthropoids in the New World (e.g., Wood, 1980; Lavocat, 1980; Hoffstetter, 1972). Still a third scenario features "*en masse*" intercontinental migration of a more diverse assemblage of taxa between Africa and the northern continents (Rosenberger, 1986).

In other cases, choice of a particular paleogeographic reconstruction has influenced the interpretation of the timing and mode of anthropoid dispersal. For example, Gingerich (1980) argued for a North American origin for platyrrhines, in part on the basis of the paleogeographic reconstruction by Smith and Briden (1977), which showed that South America was closer to North America than to Africa in the late Eocene. In contrast, Ciochon and Chiarelli (1980) argued that platyrrhine dispersal from Africa was more probable, based in part on a different paleogeographic reconstruction for the same time period. Their reconstruction showed Africa and South America closer to one another and connected by island arcs in the late Eocene (Sclater *et al.*, 1977; Sibuet and Mascle, 1978; Van Andel *et al.*, 1977; Tarling, 1980).

Anthropoid and protoanthropoid dispersal scenarios have focused almost exclusively on the late Eocene and early Oligocene (e.g., Cooke, 1968; Coryndon and Savage, 1973; Ciochon and Chiarelli, 1980; Gingerich, 1980; Rosenberger, 1986; Rasmussen and Simons, 1988), largely because of the long-accepted early Oligocene age for Fayum primates and the late Eocene age for the putative anthropoids from the Pondaung of Burma. The recent redating of all or part of the Fayum primate fauna to the late Eocene (see Simons, 1989, 1990; Kappelman, 1992; Van Couvering and Harris, 1991; Rasmussen *et al.*, 1992) and reports of early Eocene anthropoids from Algeria (Godinot and Mahboubi, 1992) suggest that anthropoid dispersal may have occurred as early as the late Paleocene. Thus, possible paleogeographic and paleobiogeographic settings for anthropoid dispersal must consider the late Paleocene through middle Eocene as well as the late Eocene.

Krause and Maas (1990) recently reviewed global paleogeography and paleobiogeography for the earliest part of this period. They argued that faunal exchange between Europe and Africa across the Tethys was possible during the early Paleogene and more likely than dispersal between Africa and Asia. These assessments are supported by the recent description of the omomyid *Altiatlasius* from the upper Paleocene of Adrar Mgorn, Morocco (Sigé *et al.*, 1990) and the early Eocene anthropoid *Algeripithecus* from Algeria (Godinot and Mahboubi, 1992). Krause and Maas (1990) concluded that the late Paleocene–early Eocene African faunas, though poorly known, showed general Euramerican affinities but were largely isolated by the early Eocene.

Together, the paleogeography and paleobiogeography of both the Old

World and New World from at least the late Paleocene to the late Eocene provides the framework for hypotheses and scenarios of anthropoid origins and dispersal. Krause and Maas (1990) have previously presented data relevant to mammalian dispersals for late Paleocene time. Here we present a synthesis of current views on early through late Eocene paleogeography for the three regions most relevant to the study of anthropoid origins and dispersal: (1) the Mediterranean Tethys; (2) the Atlantic Ocean; and (3) the Caribbean and Middle America in the New World. Based on these paleogeographic reconstructions, we evaluate the potential for dispersal along these routes and assess the likelihood of anthropoid dispersal by discussion of known phylogenetic relationships of possible mammalian dispersers and through the analysis of overall similarity among Eocene continental mammalian faunas. These data are used to examine five biogeographic hypotheses relevant to anthropoid origins: (1) anthropoid (or protoanthropoid) dispersal to Africa from Asia; (2) anthropoid (or protoanthropoid) dispersal to Africa from Europe; (3) platyrrhine (or protoplatyrrhine) dispersal from North America to South America; (4) platyrrhine (or protoplatyrrhine) dispersal from Africa to South America; and (5) endemic African origin for Anthropoidea.

Eocene Paleogeography

Mechanisms of Faunal Interchange

Dispersal of mammals from one land mass to another can occur by way of land corridors, filter bridges, sweepstakes routes, or "Noah's arks" (Simpson, 1940, 1965; McKenna, 1973). Dispersal across land corridors involves large-scale faunal interchange between areas, is largely independent of environmental constraints, and can take place in either direction. Filter bridges include dispersal routes that cross a partial barrier and therefore are more selective and potentially more limited in scale. Factors limiting faunal exchange across filter bridges may include latitude, local environment, and geographic position of the corridor, as in the case of an intercontinental isthmus. The third mechanism, sweepstakes dispersal (sometimes called waif dispersal), involves chance dispersal across a barrier, typically by way of island-hopping across a body of water. Sweepstakes dispersal is therefore the most unpredictable and, for any given taxa, the least likely of dispersal mechanisms. The final mechanism, dubbed "Noah's arks" (McKenna, 1973), involves tectonic movement of land masses via continental drift. The relative importance of these different mechanisms in the dispersal history of anthropoids and their ancestors depends on global geography during the latest Paleocene and Eocene.

Old World Paleogeography

At the end of the Cretaceous, several areas of emergent land served as connections between the Afro-Arabian continent and Europe (Gheerbrant, 1990). At the beginning of the Tertiary, however, many of these areas were submerged in the Tethys, and no land connections of Paleogene age are known with certainty between the Afro-Arabian continent and any other area. Thus, the base of the Tertiary has traditionally marked the beginning of the continent's "period of isolation."

Figure 1 presents a palinspastic reconstruction of the Mediterranean Tethys showing areas of emergent land approximately at the Eocene–Oligocene boundary [correlated to Anomaly 13 of the Geomagnetic Reversal Time Scale (GRTS)]. The map is based primarily on the work of Dercourt *et al.* (1986, plate 7). Modifications to their reconstruction consist of the addition of more emergent areas in three locations: (1) the western Mediterranean area, based on Plazait (1981); (2) Great Britain and in the area of the Dinaro Hellenic Chain, based on Hooker (1986); and (3) eastern north Africa, based on Bown and Kraus (1988) and Bown (*personal communication,* 1992). The reconstruction of the eastern African coastline in the late Eocene follows the correlation of rock units in Egypt proposed by Rasmussen *et al.* (1992) and Kappelman (1992).

Most of the emergent areas shown in Fig. 1 were present since the beginning of the Eocene. Major exceptions are (1) the greater development of land in the areas of the Alpine and Dinaro Hellenic Chains in the early to middle Eocene; (2) the infilling of much of the North African coast and Arabian peninsula during middle to late Eocene time as a result of dropping sea levels; (3) the expansion of emergent land west of the Hindu Kush as a result of middle and late Eocene vulcanism; and (4) the significant northward movement of both the Indian subcontinent and the eastern part of the Afro-Arabian continent. Therefore, the amount of emergent land was at its maximum at the end of the late Eocene, but all the potential dispersal routes were present throughout the Eocene. If the probability of dispersal of land mammals increases with the amount of emergent land, dispersal would have been most likely in the late Eocene.

The geological evidence synthesized in the Fig. 1 suggests that the Afro-Arabian continent was relatively isolated from Europe, Asia, and the Indian subcontinent throughout the Paleocene and Eocene (e.g., Dercourt *et al.,* 1986, and references contained therein). Certainly, no land corridors were present. A number of workers, however, have postulated the presence of filter bridges or sweepstakes routes between the Afro-Arabian continent and other land masses during this period of isolation. These routes may have been used by early anthropoids or their precursors. Four main routes have been proposed:

1. *Southwestern Europe to northwestern Africa.* This route appears to be the most likely for faunal interchange between the Afro-Arabian conti-

Fig. 1. Palinspastic reconstruction of the Mediterranean Tethys, correlated to Chron 13 of the GRTS (latest Eocene–earliest Oligocene). Solid lines delimit emergent areas; bodies of water are shown as gray; and dashed lines indicate some current coastlines. Modified from Dercourt *et al.* (1986, plate 7).

nent and Europe because of the close proximity of the two areas. The presence of a thryonomyoid rodent in the early Oligocene of Spain (Hugueney and Adrover, 1991) indicates that at least one group probably used this route for sweepstakes dispersal.

2. *Europe and the Arabian Peninsula.* Gheerbrant (1990) suggested this sweepstakes route between Africa and Europe via a chain of barrier islands as the probable path for the interchange of mammals between the two continents during the early Eocene, although he considered northward dispersal more likely than southward movement. An island-hopping transtethyan route between Europe, Africa, and Asia also has been proposed by Thewissen (1990a).
3. *Asia and the Arabian Peninsula.* Dispersal of small mammals from western Asia directly into Africa also has been postulated, although no specific routes were proposed (e.g., Wood, 1985; Flynn *et al.*, 1986). Prasad *et al.* (1986) suggested that this area served as a dispersal corridor for Asiatic faunas crossing to the Indian subcontinent, and Gheerbrant (1990) included it as a possible route between Africa and Asia in the early Eocene (although Gheerbrant considered dispersal *from* Africa *to* Asia to be more likely). In either case, an Asia–Arabian Peninsula route presumably would involve sweepstakes dispersal via Tethyan island chains.
4. *The Indian subcontinent and the Arabian Peninsula.* Briggs (1989) proposed that the Indian subcontinent could have served as part of a land connection between Asia and the Arabian Peninsula during the Cretaceous and the Paleocene (but see Thewissen, 1990a; Thewissen and McKenna, 1992). Other proposals for potential routes between Africa and India include a Late Cretaceous filter corridor (Sahni, 1984), a simple transtethyan sweepstakes route between the Indian subcontinent and the Arabian Peninsula (Hussain *et al.*, 1978), and early Eocene sweepstakes dispersal among the Indian subcontinent, Africa, and Europe via east–west-trending Tethyan island arcs (Thewissen *et al.*, 1987; Thewissen, 1990a).

In summary, there are no known land corridors between the Afro-Arabian land mass and any other continental area during the Eocene. The only types of transtethyan dispersal that could have been accomplished during this time would have been by sweepstakes routes, either directly across the Tethys or by way of island chains.

Sea level was at a minimum, and emergent areas at a maximum, during the late Eocene, and therefore dispersal of land mammals onto the Afro-Arabian continent would have been most likely during this time. Estimates of open water distances in the late Eocene Mediterranean Tethys range from less than 100 km, between western Europe and Africa to more than 1400 km between the Indian subcontinent and the Arabian Peninsula (see Dercourt *et al.*, 1986). Therefore, the most likely late Eocene dispersal route for land mammals (i.e., that which minimizes the distance traveled and the amount of water traversed) was between western Europe and western Africa. Dispersal from either Europe or Asia to the Arabian platform also was possible, given the likelihood of additional emergent land in the eastern part of the Mediterranean Tethys. Dispersal directly from India was the least probable but could most easily have been accomplished by crossing to Asia and island hopping to the Arabian plate.

In addition to relative distances between land masses, however, marine circulation patterns also must be considered in evaluating the relative potential for dispersal along different sweepstakes routes. Previous reconstructions of Tethyan paleocurrents, inferred from the distribution of invertebrates, featured a strong east–west-trending surface current (Luyendyk *et al.*, 1972; Berggren and Hollister, 1974). This current would have facilitated sweepstakes dispersal preferentially from Asia to Africa. However, more recent computer modeling of surface currents based on a much revised continental paleogeography suggests a radically different surface current pattern for the Mediterranean Tethys, although one not inconsistent with the invertebrate evidence (Barron and Peterson, 1989, 1991). Barron and Peterson's (1989) model for the mid-Cretaceous shows a gyre-like clockwise circulation in the Mediterranean Tethys producing a generally easterly flow along the southern border of Europe and a generally northern flow off the coast of the Afro-Arabian continent. This current pattern apparently continued through the Paleogene in the Mediterranean Tethys (Barron and Peterson, 1991). These surface current patterns suggest that current-aided sweepstakes dispersal into Africa would be unlikely in the Paleogene; rather, dispersal out of Africa would have a higher probability of success.

The Spreading Atlantic

Africa and South America last had direct terrestrial contact in late Albian or early Cenomanian time (*ca.* 100 Ma) (Popoff *et al.*, 1989). Subsequently, these continents have separated by successive cycles of sea floor spreading along the mid-Atlantic ridge. This process recently has been correlated to the geomagnetic time scale (Nürnberg and Muller, 1991), which permits revised estimates of the relative motion and distances between South America and Africa during the Paleogene. Figure 2 shows the relative distances between the two continents in the middle Paleocene (Chron 27) and in the late Eocene (Chron 16). During the Paleocene, the minimum distance between South America and Africa was approximately 850 km. In the late Eocene, the distance was approximately 1700 km for a transatlantic route between western Africa and eastern Brazil. Because there currently are no detailed reconstructions of South Atlantic coastal paleogeography, actual distances may have been somewhat greater or less than these estimates.

Two volcanic island arcs may have been intermittently present across the

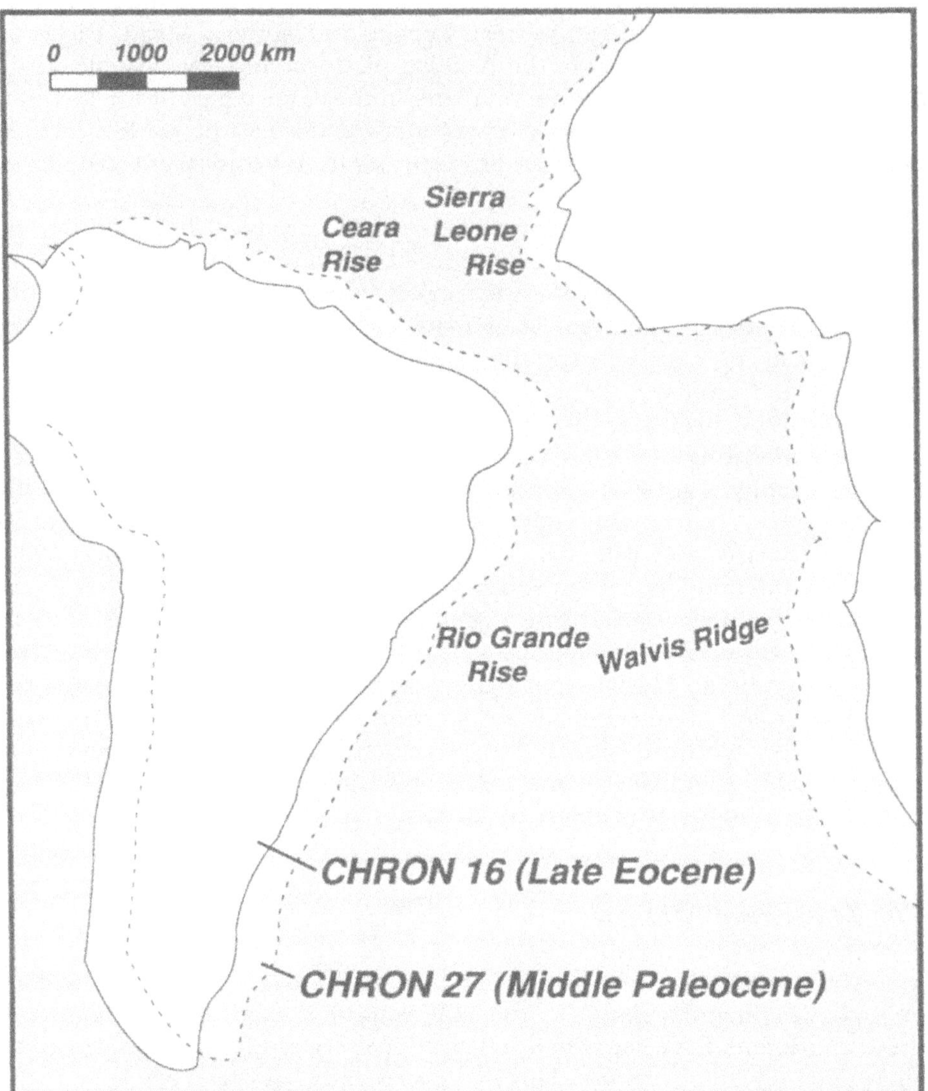

Fig. 2. Schematic diagram showing relative positions of Africa and South America in the middle Paleocene (Chron 27) (dashed line) and late Eocene (Chron 16) (solid line). Continental margins are based on continental shelf margins and do not reflect paleocoastlines. Based on Nürnberg and Muller (1991, Figs. 15 and 16).

spreading South Atlantic (e.g., Tarling, 1980). The more prominent, a chain along the Ceara and Sierra Leone Rises and extending between western Africa and northern South America, formed approximately 80 Ma. It is thought to have been subaeriel throughout much of the early Tertiary, forming islands hundreds of square kilometers in size (Tarling, 1980). A second chain may have existed further south during the Cretaceous and Eocene, along the Rio Grande Rise and Walvis Ridge between southwest Africa and southern Brazil but would not have provided as substantial terrestrial environments as the more northern chain (Van Andel *et al.*, 1977; Tarling, 1980). In any event, transatlantic sweepstakes dispersal of early anthropoids would involve the crossing of considerable distances (> 200 km), even with the benefit of large islands linking the two continents. Because of these large distances, paleocurrent patterns are critical in assessing the likelihood of transatlantic sweepstakes dispersal.

All reconstructions of ocean surface circulation for the South Atlantic Eocene share two common elements: (1) a counterclockwise trending current, which would have flowed north and east across the Walvis Ridge and Rio Grande Rise; and (2) a southwesterly current running from the southern border of the bulge of Africa toward the southern coast of modern Brazil (Barron and Peterson, 1991; Berggren and Hollister, 1974). The northeastward flow of water across the Walvis Ridge and Rio Grand Rise would appear to preclude westward movement across this area, but the southwesterly current in the northern part of the South Atlantic could possibly have aided westward dispersal of African mammals to South America.

Interpretations of the exact positions of currents through the narrowest point between the continents vary, however. Berggren and Hollister (1974) posited a northwestward-flowing current that fed South Atlantic waters to the North Atlantic and a southwesterly current flowing out of the Mediterranean Tethys along the northern coast of Africa and across the Atlantic to the northern coast of South America. In contrast, Barron and Petterson (1991) show strong southwesterly current vectors bridging the continents further south, approximately at the position of the Sierra Leone Rise. In either case, currents could have aided westward migrants, but the more southerly positioned current reconstruction of Barron and Peterson (1991) would have facilitated island hopping in the vicinity of island chains along the Sierra Leone Rise.

New World Paleogeography

Stephan *et al.* (1990) have presented the most recent synthesis of geological data on the paleogeography of the Caribbean region. Figure 3 presents their reconstructions of the region for the middle and late Eocene. As in the Tethyan region, no continuous land corridors between continents were present in the Caribbean at any time during the Eocene. Also like the Tethyan region, emergent land was at a maximum at the end of the Eocene. The western portion of Mexico and Central America as far south as present-day Honduras was emergent throughout the Eocene. During the early and middle Eocene an island arc emerged along Cuba and Hispaniola, curving eastward from the vicinity of the Yucatan to eastern Venezuela. This island arc ex-

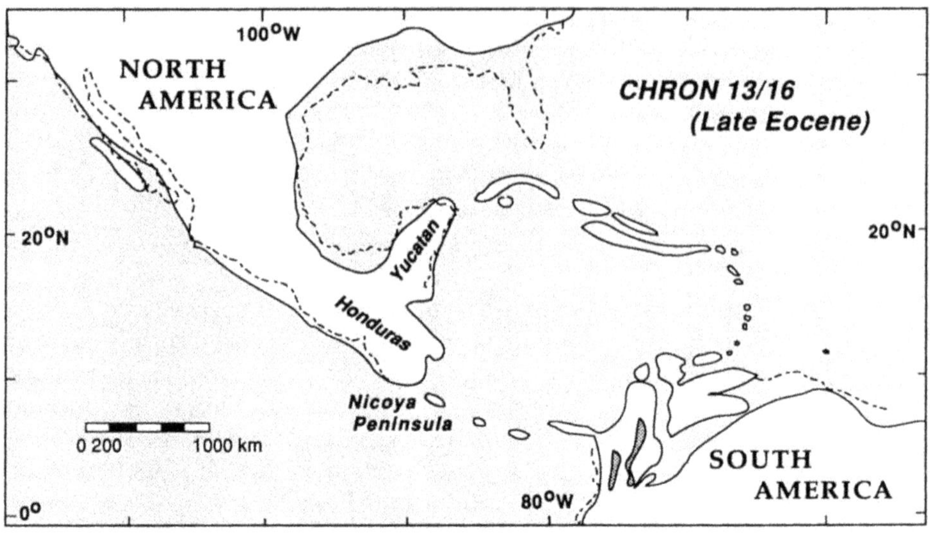

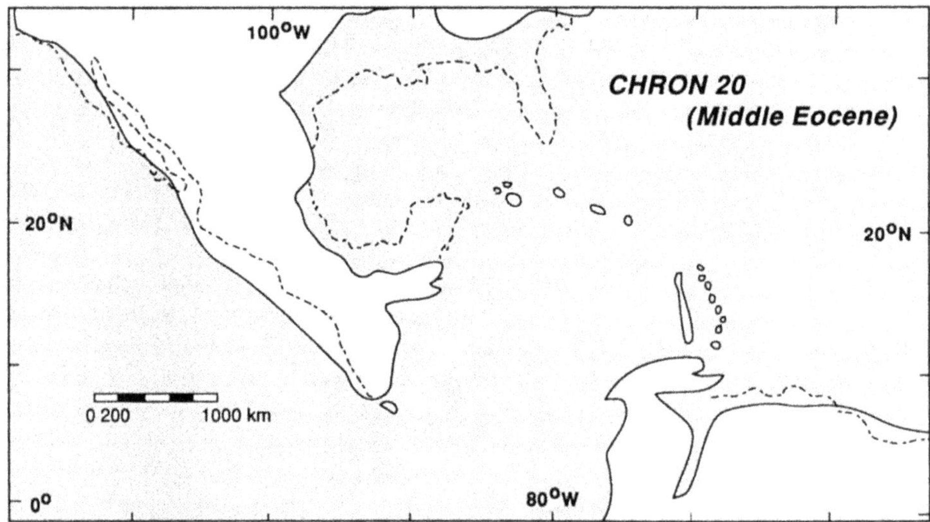

Fig. 3. Reconstructions of the Caribbean region correlated to Chrons 13/16 (late Eocene) and Chron 21 (middle Eocene) showing emergent land and relative positions of the continents. Redrawn from Stephan *et al.* (1990, plates 10 and 11) and based on a Mercator projection.

panded during the late Eocene, the Yucatan Peninsula became terrestrial, and a second, smaller island arc developed between Honduras and western Colombia.

Based on this paleogeographic reconstruction, sweepstakes routes were the only potential routes for land mammal dispersal between North and South America. The most probable route (i.e., that with the greatest amount of emergent land) would have been one from eastern Mexico through the Cuba–Hispaniola island arc to Venezuela, at some time from the middle Eocene through late Eocene. The less likely route, from southern Mexico–Honduras to western Colombia via the western Central American island arc, would only have been possible at the end of the Eocene. The distances of open water to be traversed along either route exceeded 800 km until the latest Eocene, when minimum distances along an island-hopping route would have been approximately 200 km.

Paleocurrent reconstructions for the Caribbean and Pacific coast of southern North America all indicate the presence of increasingly stronger northward currents between South America and North America during the Paleogene (Luyendyk *et al.*, 1972; Berggren and Hollister, 1974; Barron and Peterson, 1989, 1991). This current pattern presumably would have hindered sweepstakes dispersal from North America to South America. Thus, although island-hopping routes similar in distance to those postulated between Africa and South America existed between the two continents of the New World, surface currents would *not* have aided north-to-south dispersal in the Caribbean region.

Paleobiogeography

Phylogenetic Evidence for Mammalian Dispersal

As noted above, several authors have constructed anthropoid dispersal scenarios around general mammalian dispersal hypotheses. For the origin of African anthropoids, this evidence has traditionally been based on the putative phylogenetic and time–stratigraphic relationships of Fayum artiodactyls, creodonts, rodents, and primates (and to a lesser extent that of marsupials) to related taxa in Asia and Europe. For the origin of South American anthropoids, the evidence has been drawn from the reported relationships between South American caviomorph rodents and primates with North American and African taxa. These arguments are reviewed below.

Dispersal into Africa

Artiodactyl evidence for faunal exchange is primarily based on the presence of the anthracotheriid *Bothriogenys* in the Egyptian Fayum fauna and of anthracotheres in the Asian Eocene and in the latest Eocene and early Oligocene of Europe. The majority of workers have considered the anthracotheriid *Bothriogenys* to have its origin in Asia (e.g., Pilgrim 1928, 1941; Colbert, 1938), probably from the Chinese and Burmese genus *Anthracokeryx*. An origin of anthracotheres in Asia is bolstered by the evidence of an even more primitive genus, *Siamotherium*, in the Krabi fauna of Thailand (Suteethorn *et al.*, 1988) and the fact that the probable sister taxon of anthracotheres, the Helohyidae (Coombs and Coombs, 1977; Gentry and Hooker, 1988), are exclusively Asian and North American in distribution.

Black (1978), however, suggested a European origin, based on the view that anthracotheres first appear in the middle Eocene of Europe and on resemblances between *Bothriogenys* and European *Bothriodon* and *Anthracotherium*. However, the genera from the middle Eocene of Germany previously described as anthracotheriids have been reassigned to the Haplobunodontidae (Franzen and Haubold, 1989) and are probably only distantly related to anthracotheriids (Gentry and Hooker, 1988). Certainly, haplobunodontids are already too derived in their premolar morphology to be ancestral to *Bothriogenys*. The only European Eocene anthracotheriid, *Elomeryx crispus*, first appears at a time approximately contemporaneous with the appearance of *Bothriogenys* in the Fayum fauna, and this species continued into the European early Oligocene (Hellmund, 1991). However, the oldest specimens of *Elomeryx crispus* are insufficient to determine whether this taxon is more or less derived than early *Bothriogenys*. Thus, it is not possible to determine whether *Elomeryx* and *Bothriogenys* represent separate offshoots of Asian stock or are more closely related to one another than to any Asian taxon. Even if the latter relationship is correct, the direction of migration is not determinable.

If the biogeographic origins of *Bothriogenys* lie in Asia, the probable timing of an African anthracothere immigration is conjectural. If helohyids are the sister taxon of anthracotheres, then the family Anthracotheriidae must have diverged from other artiodactyls by at least the early middle Eocene, when the genus *Helohyus* first appears in the Bridgerian of North America. *Bothriogenys*' first known occurrence is in the late Eocene Qasr el Sagha Formation of Egypt (Simons, 1968); thus, in the absence of more precise phylogenetic evidence for the relationships of *Bothriogenys* to the Asian Anthracotheriidae, a migration from Asia some time between the early Eocene and late Eocene is the best resolution currently possible.

Only two creodont genera are shared between Africa and the northern continents, the hyaenodontids *Pterodon* and *Apterodon*. *Pterodon* first appears in the late middle Eocene of Asia (*P. dakhoensis* at Heti, China: Chow, 1957) and then in the late Eocene of Europe (*P. dasyuroides* at Quercy: Lange-Badré, 1979) and of Africa (*P. africanus* and *P. phiomensis* in the lower sequence of the Jebel Qatrani Formation: Osborn, 1909). The Asian species is demonstrably the most primitive, retaining a molar metaconid that is lost in all other species of the genus. The European species is the most derived, having reduced premolars and incisor number (Lange-Badré, 1979). Thus, an Asian, or even an African, origin of this genus best fits the known distribution of taxa and characters.

Apterodon's earliest occurrence is known from the upper Eocene Qasr el Sagha Formation, Fayum Province, Egypt. In Europe, this genus appears subsequent to the "Grand Coupure" in lower Oligocene deposits (Lange,

1967). Tilden *et al.* (1990) examined the phylogenetic relationships of *Apterodon* spp. and found that the European species was derived by contrast with late Eocene species from the Fayum fauna, and Simons and Gingerich (1976) considered *Apterodon* to be of African origin.

The relationships of other African creodonts to those of Europe and Asia are far more conjectural. Only four other genera have been described from Africa: *Koholia* from the early Eocene of Algeria and *Metasinopa, "Sinopa,"* and *Masrasector* from the late Eocene to early Oligocene of Egypt and also Oman for *Masrasector.* All are very poorly known. *Koholia* has been allocated to a separate subfamily and does not appear closely related to non-African Eocene forms (Crochet, 1988). Generalized ancestor–descendent or sister-group relationships between the Egyptian genera and either North American or European forms have been suggested (e.g., Schlosser, 1911; Simons and Gingerich, 1974; Savage, 1978; Barry, 1988), but none strongly support any particular biogeographic origin (see, e.g., Barry, 1988). The African proviverrines may be African endemics or of Eurasian ancestry. However, to determine which of these biogeographic alternatives may be correct, additional material and more precise phylogenetic hypotheses based on such material will be necessary.

A European origin of the Fayum Oligocene peradictid marsupial *Peratherium africanus* has been suggested based on similarities to European late Eocene *P. cuvieri* and early Oligocene *P. perrierense,* with a late Eocene dispersal event considered most likely (Bown and Simons, 1984; Simons and Bown, 1984). Crochet *et al.* (1992), however, do not consider this argument convincing and allocate *P. africanus* to a new genus, *Qatranitherium.* They suggest that the differences between the African peradictids (*Q. africanus* and the early Eocene African *Kasserinotherium*) and Holarctic peradictids indicate a degree of African endemism for the peradictids. Further, they suggest that the trans-Tethyan dispersal that resulted in the presence of the peradictids in Africa and Holarctica is more likely to have occurred near the Paleocene–Eocene boundary, when peradictids first appear in Holarctica.

Arguments concerning the biogeographic origins of Eocene–Oligocene African hystricognath rodents (thryonomyoids or phiomyids) focus on their putative origin out of one of four groups: (1) the European Theridomyidae, having an early middle Eocene to Oligocene distribution (e.g., Lavocat, 1969, 1980); (2) the Eocene group "Franimorpha," largely based on middle to late Eocene North American taxa (e.g., Wood, 1972); (3) the early Eocene Indo-Pakistan family Chappatimyidae (e.g., Hussain *et al.*, 1978; Jaeger *et al.*, 1985); or (4) Asian early to late Eocene ctenodactyloids (e.g., Flynn *et al.*, 1986), principally known from Chinese localities.

Currently, both the theridomyid and franimorph hypotheses are largely rejected. Dawson (1977) and Korth (1984) demonstrated that Wood's (1974) Franimorpha was baseless on morphological grounds and allocated these taxa to the North American Ischyromyidae. Jaeger *et al.* (1985) presented evidence based on analysis of new hystricognaths from Algeria that theridomyids were too derived to give rise to the African forms, and theridomyids have been linked with anomalurids rather than thryonomyoids (Hartenberger, 1990). The focus of study has now shifted to the role of the Indo-Pakistan chapptimyids and Chinese ctenodactyloids in the origin of the Thryonomyoidea.

An ancestor–descendent relationship or sister-group relationship between the Chappatimyidae and Thryonomyoidea has been proposed, with a probable common origin from ctenodactyloids (e.g., Hussein *et al.*, 1978; Jaeger *et al.*, 1985; Jaeger, 1988). Flynn *et al.* (1986) also favored a close relationship between chappatimyids and thryonomyoids, although they also admitted the possibility that thryonomyoids might be a sister taxon to both ctenodactyloids and chappatimyids. Each of these hypotheses suggests an Asian origin for the African Thryonomyoidea, although the possible timing of such an origin and migration into Africa could range from the earliest Eocene to the late Eocene, based on our current understanding of the phylogenetic relationships of ctenodactyloid, chappatimyid, and thryonomyoid rodents.

The identification of a non-African origin for anthropoid primates has focused largely on the possible ancestral relationship between European protoadapines and Fayum anthropoids (e.g., Gingerich, 1975; Rasmussen and Simons, 1988; Franzen, 1987) or the identification of the Asian taxa *Pondaungia* and *Amphipithecus* as anthropoids (e.g., Ba Maw *et al.*, 1979; Ciochon and Chiarelli, 1980; Gingerich, 1980; Ciochon *et al.*, 1985). These hypotheses are reviewed elsewhere in this volume (Rasmussen, Chapter 12, this volume, for the protoadapine hypothesis; Ciochon and Holroyd, Chapter 6, this volume, on *Pondaungia* and *Amphipithecus*). Currently, there is little strong evidence for an Asian origin of Anthropoidea based on the Burmese primates, although the possibility cannot be eliminated that *Amphipithecus* may represent an early anthropoid (Ciochon and Holroyd, Chapter 6; Kay and Williams, Chapter 13, this volume). The case for a close phyletic relationship between protoadapines and anthropoids is still a matter of considerable debate (see Rasmussen, Chapter 12; Ross, Chapter 15; Kay and Williams, Chapter 13, this volume). In either the Asian origin or European origin case, however, an immigration into Africa appears to have taken place between the middle and late Eocene. Either hypothesis would be refuted by the presence of earlier anthropoids in Africa, evidence that has been presented by Godinot and Mahboubi (1992) in the form of *Algeripithecus* from the early Eocene of Algeria.

In summary, current phylogenetic evidence for mammalian migration into Africa points to probable African endemicity for apterodonine creodonts, marsupials, and anthropoid primates. Asian origins are indicated for rodents, anthracotheriid artiodactyls, and possibly some hyaenodontine creodonts. Evidence for migration from Europe into Africa is possibly indicated by relationships among proviverrine creodonts but excluded for other groups. Present knowledge of phyletic relationships does not permit a more precise approximation of timing of possible migration than between early and late Eocene, nor does it elucidate the mode of these presumed dispersal events.

Dispersal into South America

Platyrrhine primates first appear in the fossil record of South America in the late Oligocene; this appearance coincides with the first appearance of the hystricognathous caviomorph rodents. Both are unknown in the Eocene fossil record of the continent. For this reason, researchers have attempted to examine their shared biogeographic history as a clue to the possible source area for the South American anthropoids.

Arguments regarding the biogeographic origin of the Platyrrhini have focused on the possible phyletic relationship of this group to North American adapids (e.g., Simons, 1976; Gingerich, 1980; Rasmussen, 1990) or omomyids (e.g., Szalay, 1976; Rosenberger and Szalay, 1980) and to the African Anthropoidea (e.g., Hoffstetter, 1972). Recent investigations have upheld the monophyly of the Anthropoidea relative to both Omomyidae and Adapidae (e.g., Kay and Williams, Chapter 13, this volume). As discussed above, the relationship between the North American protoadapine *Mahgarita* and anthropoids has not been resolved.

The origin of the Caviomorpha has been placed within either North American franimorphs or African thryonomyoids. These divergent views have most recently been summarized in Wood (1980, 1985) for North American origin and Lavocat (1980) for African origin. As discussed above, a North American franimorph origin for hystricognaths has been refuted on the grounds that the Franimorpha is morphologically baseless. An African origin has been strongly supported by several lines of evidence. Lavocat (1969, 1971, 1972, 1974, 1976, 1980; Lavocat and Parent, 1985) has pointed out in detail the resemblances between thryonomyoids and caviomorphs. The unity of the Hystricognathi relative to other rodent groups has also been upheld based on a diverse array of anatomic systems (Bugge 1971, 1985, cranial foramina; Woods and Hermanson, 1985, myology; Luckett, 1980, fetal membranes; Luckett, 1985, dentition and placentation; Durette-Desset, 1971, nematode parasites) and based on parsimony studies of character distributions in living and fossil rodents (e.g., Hartenberger, 1985; Flynn *et al.*, 1986; Jaeger, 1988).

Thus, an African origin for both the Platyrrhini and Caviomorpha appears to be best supported by current phylogenetic evidence. However, the phylogenetic evidence for either group does not provide a means for establishing the timing for a transatlantic dispersal. Most dispersal hypotheses have suggested a late Eocene to Oligocene dispersal between South American and Africa. However, workers have not yet been able to demonstrate caviomorph monophyly (see, e.g., Luckett, 1980, 1985; Woods and Hermanson, 1985), and no attempt has been made to conduct a study of character distributions incorporating both the Eocene–Oligocene African thryonomyoids and early caviomorphs. The possible relationships between early African primates (*Altiatlasius* and *Algeripithecus*) and platyrrhines are unknown; even the relationships between platyrrhines and the Fayum anthropoids have not been clarified (see, e.g., Fleagle and Kay, 1987; Harrison, 1987; Ford, 1988; Kay and Williams, Chapter 13, this volume). Until a better understanding of these relationships is achieved or more conclusive fossil evidence is recovered, any attempt to determine the timing of a transatlantic migration event from phylogenetic evidence would be premature.

Intercontinental Faunal Resemblances

Because phylogenetic evidence is largely inconclusive, the probability of intercontinental faunal exchange having occurred is also examined through the analysis of overall faunal resemblances. In order to determine the likelihood of faunal interchange during the Eocene, mammalian faunal lists were compiled for the early, middle, and late Eocene of Africa, Asia (including the Indian subcontinent), Europe, North America, and South America. For Africa, the lists were derived principally from Sudre (1979), Bown *et al.* (1982), Coiffait *et al.* (1984), Jaeger *et al.* (1985), de Bonis *et al.* (1988), and Thomas *et al.* (1989, 1991), with additions from Mahboubi *et al.* (1983, 1984, 1986), Simons and Kay (1983, 1988), Crochet (1984, 1986, 1988), Hartenberger *et al.* (1985), Hartenberger (1986), Simons *et al.* (1986, 1991), Bown and Simons (1987), Simons (1989), Crochet *et al.* (1990), Godinot and Mahboubi (1992), and unpublished work by T. M. Bown, P. A. Holroyd, D. T. Rasmussen, and E. L. Simons on Fayum faunas. The major source for Asian faunas was Russell and Zhai (1987), with additions from Dawson *et al.* (1984), Qi (1987), Suteethorn *et al.* (1988), Beard and Wang (1990), Dashzeveg (1990), Ducrocq *et al.* (1992), Thewissen *et al.* (1987), and Wang and Li (1990). Marshall *et al.* (1983) was the major source for South American faunal lists. Major sources for North America include the continental faunal lists compiled by Savage and Russell (1983), Stucky (1984, 1992), and Prothero (1985). The North American lists were updated from the recent primary literature and taxonomic reviews, including Lillegraven (1980), Emry (1981, 1990), Bown and Schankler (1982), Korth (1984), Storer (1984, 1988, 1990), Maas (1985), Novacek *et al.* (1985), Wilson (1986), Bown and Rose (1987), Gingerich (1987, 1989, 1991), Beard and Houde (1989), Gingerich and Deutsch (1989), Gunnell (1989), Prothero and Schoch (1989), Thewissen and Gingerich (1989), Kelly (1990), Mason (1990), McKenna (1990), Thewissen (1990b), Gunnell and Gingerich (1991), and Beard *et al.* (1992). European Eocene faunal lists were drawn largely from Russell *et al.* (1982) and Savage and Russell (1983), with additions from Radulesco *et al.* (1976), Lange-Badré (1979), Sen and Heintz

(1979), Godinot (1981, 1987, 1988), Lange-Badré and Godinot (1982), Godinot *et al.* (1987), Russell and Godinot (1988), Haubold (1989), Sudre *et al.* (1990), Crochet (1991), and Hooker (1991).

Correlation

Recent revision of the position of the Eocene–Oligocene boundary in North America and reappraisal of this boundary worldwide has resulted in new age correlations for mammalian faunas in North America, Africa, and, in part, Asia (see Swisher and Prothero, 1990; Kappelman, 1992; Berggren and Prothero, 1992; Prothero and Swisher, 1992; Rasmussen *et al.*, 1992; Holroyd and Ciochon, Chapter 5, this volume). Our assignments of biostratigraphic units (land mammal ages, mammalian reference levels, etc.) and/or localities from different continents to the early, middle, or late Eocene are summarized in Figure 4. Note that no middle Eocene sites are known from Africa; the only middle Eocene site previously reported from Africa (Gour Lazib, Sudre 1979) is now regarded as early Eocene in age on the basis of its correlation with Glib Zegdou, recently redated to the early Eocene (Godinot and Mahboubi, 1992). Also, no South American faunas of late Eocene age are known; Divisaderan and Deseadan faunas, previously considered late Eocene, are now thought to be latest Oligocene in age (MacFadden *et al.*, 1985; Berggren and Prothero, 1992).

Several problems were identified in intercontinental correlations of Eocene and Oligocene localities. These include correlation of Asian sites, the age of the Duchesnean North American land mammal age (NALMA), the age of the Indo-Pakistan mammalian faunas, and the biostratigraphic placement of certain African sites. The problems associated with correlating Asian sites to NALMAs and to subepoch boundaries are also discussed by Holroyd and Ciochon (Chapter 5, this volume).

The placement of the Duchesnean NALMA in relation to subepoch boundaries is currently uncertain (e.g., Wilson, 1986; Lucas, 1992; Prothero and Swisher, 1992). It generally is considered late Eocene in age (e.g., Krishtalka *et al.*, 1987; Berggren and Prothero, 1992; Prothero and Swisher, 1992). However, comparisons with the late middle Eocene Bartonian European standard stage suggests that it may be older. Figure 5 represents a preliminary correlation of the European standard stages (on which subepoch boundaries are based) with the GRTS and with North American magnetostratigraphy and land mammal ages. No radiometric dates are given because such dates for the middle portion of the Eocene are currently suspect (see Prothero and Swisher, 1992, for discussion), and revisions to the entire Paleogene time scale are in progress. The Duchesnean extends from within Chron 18 of the GRTS (Flynn, 1986a) to a time just younger than the beginning of Chron 16 (Prothero and Swisher, 1992). The Bartonian is aligned with the GRTS via correlations with marine zones and extends from approximately Chron 18 to the midpoint of the youngest episode of normal polarity in Chron 17 (Aubry *et al.*, 1988). Thus, most of the Duchesnean correlates with the middle Eocene, although a small part [approximately 0.2 to 0.3 million years, based on Prothero and Swisher's (1992) recalculations for the radiometric ages of Chrons 16 and 17] may lie within the late Eocene.

	EARLY EOCENE	MIDDLE EOCENE	LATE EOCENE
AFRICA	Brezina, El Kohol, Gour Lazib & Glib Zegdou, Algeria; Chambi, Tunisia		Nementcha, Algeria; Dor el Talha, Libya; Dhofar, Oman; Qasr el Sagha Fm. & lower sequence of Jebel Qatrani Fm., Fayum, Egypt
ASIA	Bumbanian LMA and Eocene fossil mammal localities of the Indian subcontinent	Arshanto, Irdin Manhan and Sharamurunian LMA; Pondaung of Burma; Krabi, Thailand	Ergilin and Ulangochulan LMA
NORTH AMERICA	Wasatchian LMA	Bridgerian & Uintan LMAs and early Duchesnean faunas	Late Duchesnean faunas and Chadronian LMA
EUROPE	Ypresian Standard Stage--MP7-10	Lutetian and Bartonian Standard Stages-- MP 11-16	Priabonian Standard Stage-- MP 17-20
SOUTH AMERICA	Casamayoran LMA	Mustersan LMA	

Fig. 4. Summary chart showing the partitioning of mammalian faunal zonations along Eocene subepoch boundaries used in this analysis.

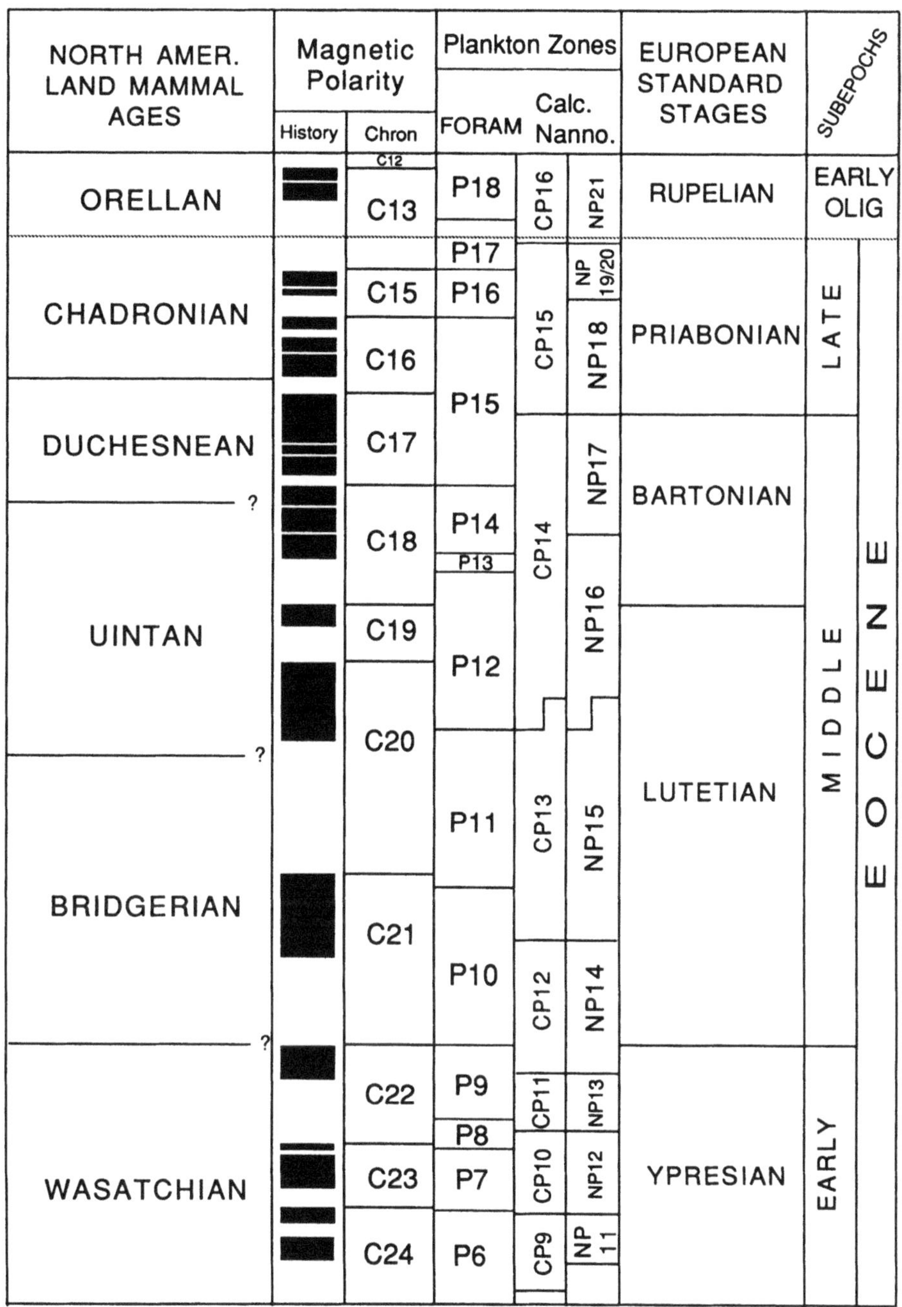

Fig. 5. A preliminary correlation of European Standard Stages, North American Land Mammal Ages, and planktonic zones to the Geomagnetic Reversal Time Scale (GRTS). Correlation of European Standard Stages and planktonic zones to the polarity history is from Aubry *et al.* (1988). Correlation of North American Land Mammal Ages to the GRTS is from Prothero and Swisher (1992) for the Duchesnean and Chadronian and Flynn (1986a) for the Bridgerian/Uintan. The boundary between the Uintan and Duchesnean is placed on the basis of discussion in Prothero and Swisher (1992) and Lucas (1992).

This correlation of the Duchesnean is consistent with the biostratigraphic conclusions of Wilson (1986). Wilson recognized an early and late Duchesnean and considered the former a subage of the middle Eocene Uintan NALMA and the latter a subage of the late Eocene Chadronian NALMA. The late Duchesnean includes the Porvenir and Lapoint faunas. All other Duchesnean age localities are included in the early Duchesnean. We have followed that convention here and include the early Duchesnean, the Bridgerian, and Uintan in the middle Eocene, and the late Duchesnean, along with the Chadronian, in the late Eocene.

The age of faunas from the Indian subcontinent also is somewhat problematic. Traditionally a middle Eocene age was assigned to most vertebrate fossil localities in India and Pakistan. A latest early Eocene age is now indicated for some or all of these sites (see Gingerich and Russell, 1990). The Indian subcontinent faunas are here considered part of the composite Asian fauna. The Indian subcontinent had collided with Asia by the early Eocene (e.g., Patriat and Achache, 1984; Garzanti *et al.,* 1987), and although the Indian faunas are generally considered to be endemic in most respects, they were Asian faunas in a strict geographic sense throughout the Eocene.

The age of certain African localities is the final correlation question to be considered. Rasmussen *et al.* (1992) have presented new interpretations for the age of several sites, three of which, Glib Zegdou, Algeria, and Taqah and Thaytiniti, Oman, contain anthropoid primates. Glib Zegdou (correlative with the nearby Gour Lazib) was traditionally considered early or middle Eocene in age (Sudre, 1979), but based on reports of Fayum hyracoid genera at Gour Lazib, Rasmussen *et al.* (1992) suggested that both Gour Lazib and Glib Zegdou are probably late Eocene to early Oligocene in age. However, more recently Godinot and Mahboubi (1992), in their description of the anthropoid *Algeripithecus,* reported that new data from charophytes indicate an early Eocene age for Glib Zegdou, and this interpretation is adopted here.

A late Eocene age was suggested by Rasmussen *et al.* (1992) for the localities of Thaytiniti and Taqah (Dhofar, Oman) in contrast to the early Oligocene age given by Thomas *et al.* (1989). Rasmussen *et al.*'s conclusions were based on comparison with the fauna of Fayum Faunal Zone 2, in the lower sequence of the Jebel Qatrani Formation, as well as a numerical age of 35.8 Ma given for Thaytiniti by Thomas *et al.* (1989). However, this figure is not a radiometric age but is based on chronostratigraphic correlation, specificially, the identification of the lower normal polarity event in the Thaytiniti magnetostratigraphic section as Chron 13 (earliest Oligocene). Consequently, since Chron 13, using newly revised dates for the Chron boundaries (Swisher and Prothero, 1990), is no older than 33.7 Ma, there is a discrepancy between the age based on correlation with the Fayum fauna (late Eocene) and the age based on magnetostratigraphy (earliest Oligocene). A likely explanation of this apparent discrepancy lies in the possibility that the normal-polarity event identified by Thomas *et al.* (1989) as Chron 13 actually represents Chron 15 (late Eocene), also represented in the lower sequence of the Jebel Qatrani Formation (Kappelman, 1992). We suspect that this may be the case and therefore consider Thaytiniti and Taqah to be late Eocene in age. A late Eocene age makes these localities the earliest African occurrences of propliopithecine primates (*Moeripithecus markgrafi,* Thomas *et al.*, 1991) and of a possible tarsiiform (Thomas *et al.*, 1988, 1989).

Methods

Measures of intercontinental faunal similarity allow us to assess the degree of faunal interchange that actually took place across the potential dispersal routes described above. These measures do not address the question of direction of dispersal; they simply document whether faunal exchange took place and, to some extent, its magnitude. Geographic barriers and/or ecological differences between continents limit faunal exchange. A complete barrier between continents presumably would result in significantly endemic faunas. Opening of ecological/geographic barriers, as in the case of faunal exchange across land corridors, could drive wholesale faunal exchange, resulting in cosmopolitan faunas. Alternatively, the scale of faunal exchange might be more limited, as in the case of filter bridges or, if dispersal is very restricted, sweepstakes routes. Such limited faunal exchange would be expected to reduce, but not eliminate, continental endemism.

The similarity measure most commonly applied to the vertebrate fossil record is Simpson's Faunal Resemblance Index (FRI) (e.g., Simpson, 1943, 1960; Van Couvering and Van Couvering, 1976; Flynn, 1986b; Krause and Maas, 1990; Anthony and Maas, 1990). It is most commonly used when taxonomic lists differ markedly in size (e.g., Van Couvering and Van Couvering, 1976).

Simpson's FRI is calculated as:

$$\text{FRI} = (N_c \,/\, N_1) \times 100$$

where N_c is the number of taxa shared between two faunas, and N_1 is the number of taxa in the smaller of the two faunas (Simpson, 1960). Very low FRIs (e.g., <20) occur if few taxa are shared; high FRIs (e.g., >80) occur if a large proportion of the taxa in the smaller of the two faunas are also found in the larger fauna. Generic FRIs over 50 are characteristic of areas separated by corridors thousands of kilometers long (Simpson, 1965; McKenna, 1973). The index is likely to be anomolously low if both samples are very small (i.e., <15 taxa) because of sampling error (Simpson, 1947; Van Couvering and Van Couvering, 1976).

Although Simpson's FRI is widely used, it also has been criticized (e.g., Cheetham and Hazel, 1969; Henderson and Heron, 1977; Raup and Crick, 1979; Alroy, 1992). Alroy (1992) contends that all similarity and dissimilarity measures are constrained by incomplete sampling, although different indices respond differently to sampling differences. He argues that since the nature of sampling differences cannot be known, the choice of similarity measures necessarily must be arbitrary. He suggests an alternative to similarity or dissimilarity indices, one that uses only the observed conjunctions among faunal lists (i.e., those taxa that occur on more than one list).

A serious limitation to Simpson's FRI is that it does not allow statistical evaluation of the significance of faunal similarity or dissimilarity (e.g., Simberloff, 1978). In order to address this problem, Raup and Crick (1979) developed a probability measure designed to test the significance of shared taxa in relation to a null hypothesis of random distribution of taxa among faunas. Their method compares the observed number of taxa in common between two faunas (e.g., the late Eocene of Africa and Europe) with a randomly generated distribution of the number of taxa expected to be shared between the two faunas by chance alone. Alroy (1992) has criticized the logic of the Raup–Crick approach, because its use both as a test statistic and as a similarity measure implies simultaneously that there are only stochastic differences between faunas (when used as a test statistic) and that there are important nonstochastic differences (when used as a similarity index). In this study we consider the Raup–Crick statistic only as a probability measure, to test the null hypothesis of random distribution of taxa among continents, and not as a similarity measure.

The use of both the Simpson similarity index and the Raup–Crick test statistic allows a more thorough evaluation of the biological significance of the observed patterns of faunal similarity and dissimilarity. Whereas the Simpson index is a straightforward and intuitive comparison of two faunas, it also is subject to influences of different sampling regimes. The Raup–Crick method determines the probability that the degree of similarity between faunas could have occurred simply by chance but cannot be used as a measure of similarity.

Derivation of the Raup–Crick test distribution for each pair of faunas entails three steps:

1. For each comparison (e.g., Africa–Europe), a pair of simulated faunal lists are constructed by randomly drawing taxa from a pool consisting of all taxa known for a given time period. The probability of occurrence of a taxon in the pool is equal to the number of faunas in which the taxon actually appears. The number of taxa drawn from the pool for each simulated faunal list of the pair is equal to the observed number of taxa in each.
2. The number of taxa in common between the two simulated lists (k_{exp}) is counted; k_{exp} represents one point in the expected frequency distribution for a given comparison.
3. Steps 1 and 2 are repeated N times. The higher the number of simulations (N) the more precise is the expected distribution. For each of the comparisons made here, the expected distribution is based on 3000 simulations.

The observed number of shared taxa (k_{obs}) is compared to the expected distribution. The percentile in the expected frequency distribution in which k_{obs} fails is the Raup–Crick probability. The process is repeated for each faunal pair comparison. In this case the test is one-tailed because, given the paleogeographic isolation of Africa in the Eocene, we consider it unlikely that the observed frequency would fail in the upper tail of the distribution. In the case of a one-tailed test, and with $p \leq 0.05$, the faunas are considered significantly endemic if the observed number of common taxa falls in the lower 5% of the frequency distribution (i.e., if the Raup–Crick test statistic is 5 or less). A nonsignificant Raup–Crick probability (i.e., a Raup–Crick test statistic greater than 5) indicates that the number of shared taxa is not less than the number expected in common if taxa were randomly distributed among faunas. Our assumption is that a nonsignificant outcome could be the result of limited faunal exchange, as in the case of sweepstakes dispersal, whereas a significant outcome implies that the probability of even limited faunal exchange is very low.

Comparisons and Results

Table I summarizes Simpson's FRI comparing early and late Eocene African faunas with those for the early, middle, and late Eocene of Europe, Asia, North America, and South America. The Raup–Crick test statistic probabilities are reported in Table II. In addition to comparing penecontemporaneous faunas, we compare both the early Eocene African fauna and the late African fauna with middle Eocene faunas from other continents. Comparison of early Eocene African faunas with middle Eocene faunas allows us to evaluate the potential for dispersal from Africa to other continents between the early and middle Eocene. Comparison of late Eocene African faunas with middle Eocene faunas allows us to assess more completely the likelihood of dispersal into Africa between the middle and late Eocene.

For the early Eocene, the low generic FRIs indicate that the African mammalian fauna clearly was strongly endemic with respect to other continental faunas. Familial FRIs also are low, and even comparisons with North America and with Europe are below the range expected for faunas separated by long land corridors (e.g., McKenna, 1973). On paleogeographic grounds, it is likely that the familial similarity reflects earlier Paleogene, or even Late Cretaceous, dispersal patterns (e.g., Gheerbrant, 1990; Krause and Maas, 1990).

The early Eocene African fauna is known from only 17 genera, six of which are classified as family *incertae sedis,* and the low FRIs may in part reflect this incomplete sampling. The Raup–Crick statistic allows us to evaluate the probability that the low similarity, especially at the generic level, reflects the effects of chance. The Raup–Crick statistic supports the hypothesis of significant endemicity between Africa and other continents at the generic level. However, at a family level the early Eocene African fauna does not appear to

Table I. Simpson's Faunal Resemblance Index[a]

	Europe	India/Asia	North America	South America
Early Eocene	$n = 60/35$	$n = 68/31$	$n = 139/47$	$n = 63/22$
Africa (early Eocene) $n = 17/8$	0/25	0/13	0/25	0/0
Middle Eocene	$n = 103/42$	$n = 151/55$	$n = 207/67$	$n = 35/14$
Africa (early Eocene) $n = 17/8$	0/25	0/13	0/25	0/0
Middle Eocene	$n = 103/42$	$n = 151/55$	$n = 207/67$	$n = 35/14$
Africa (late Eocene) $n = 35/17$	6/24	0/18	6/24	0/0
Late Eocene	$n = 96/39$	$n = 60/28$	$n = 129/53$	—
Africa (late Eocene) $n = 35/17$	6/24	0/18	6/24	—

[a]Numerator = number of genera; denominator = number of families.

be significantly endemic with respect to Europe or Asia, and we therefore cannot reject the null hypothesis of random distribution of families among continents. It also should be pointed out that the two families whose distributions extend beyond Africa (Omomyidae? and Hyaenodontidae) are virtually cosmopolitan in distribution (except South America) and therefore are not

Table II. Raup–Crick Test Statistic[a]

	Europe	India/Asia	North America	South America
Early Eocene	$n = 60/35$	$n = 68/31$	$n = 139/47$	$n = 63/22$
Africa (early Eocene) $n = 17/8$	1/17	0/7	0/5	1/4
Middle Eocene	$n = 103/42$	$n = 151/55$	$n = 207/67$	$n = 35/14$
Africa (early Eocene) $n = 17/8$	1/29	0/3	0/5	13/17
Middle Eocene	$n = 103/42$	$n = 151/55$	$n = 207/67$	$n = 35/14$
Africa (late Eocene) $n = 35/17$	0/15	0/1	0/0	3/5
Late Eocene	$n = 96/39$	$n = 60/28$	$n = 129/53$	—
Africa (late Eocene) $n = 35/17$	0/6	0/11	0/0	—

[a]Numerator = number of genera; denominator = number of families.

very useful in evaluating relative dispersal potential between Africa and the various Holarctic continents.

An increased level of faunal similarity in comparisons of the early Eocene of Africa to the middle Eocene of other continents would support the notion that African taxa had migrated to other continents between the early and middle Eocene. However, there is little difference in FRIs or Raup–Crick probabilities when the African early Eocene is compared to either the early or middle Eocene of Asia, Europe, or North America. Interestingly, in contrast to the early Eocene comparison, the Raup–Crick statistic indicates that the faunas of the African early Eocene and South American middle Eocene were not significantly endemic with respect to one another. Therefore, we must consider the possibility of limited dispersal between Africa and South America between the early and middle Eocene.

The better-known late Eocene African fauna shows a pattern of generic endemicity relative to penecontemporaneous faunas that is similar to the early Eocene pattern. Europe and North America are equally similar to Africa, and both are slightly more similar to Africa than the Asian fauna. No late Eocene South American faunas are known. Generic FRIs are slightly higher than for the early Eocene, as expected, given the greater number of taxa known from the African late Eocene fauna. As for the early Eocene, the Raup–Crick statistics support an interpretation of significant generic endemism, but at a family level the only significantly endemic comparison is between the late Eocene of Africa and that of Asia. Interestingly, the probabilities of a random distribution of families are much lower than for the early Eocene.

Comparison of the late Eocene African fauna with middle Eocene faunas demonstrates a pattern of generic endemism almost identical to that between the late Eocene faunas. At the family level, the late Eocene African fauna is significantly endemic with respect to the middle Eocene faunas of Asia, North America, and South America, though not with respect to Europe. The comparison therefore does not support the idea that the middle to late Eocene was a period of increased dispersal into Africa from other continents.

In part, the low level of faunal resemblance between Africa and other continents is likely to be a function of the small number of taxa known from the African Eocene, particularly the early Eocene. In this sense, the results underscore the danger of overestimating the importance of shared taxa in evaluating biogeographic hypotheses when sample sizes are small. Alroy's (1992) alternative to traditional similarity matrices, the conjunction matrix, may resolve this problem. However, using the Raup–Crick test statistic we can evaluate the statistical significance of the observed frequency of shared taxa and can conclude with some confidence that it is unlikely that even limited dispersal occurred between Africa and other continents during the early Eocene. The maintenance of this high level of endemicity in the African fauna during the late Eocene supports the biogeographic conclusion that Africa was virtually isolated during the approximately 11 million years of the middle and late Eocene.

Discussion

By combining the paleogeographic data for probable dispersal routes with analysis of intercontinental faunal similarity, the probability of the different scenarios for anthropoid dispersal in the Eocene can be evaluated.

Anthropoid or Protoanthropoid Dispersal from Asia to Africa

If anthropoids arose in Asia, dispersal into Africa could only have been accomplished by sweepstakes dispersal along an island-hopping route or, potentially, a terrestrial filter route. Africa shows the least similarity with the Asian fauna of all the Holarctic continental faunas. The significant generic endemism of Africa with respect to Asia suggests that it is unlikely that extensive faunal interchange occurred between Africa and Asia during the Eocene. The only probable Eocene dispersers are rodents and anthracotheres. Given the low (though not significantly low) level of familial similarity between the two continents throughout the Eocene, it seems more likely that most faunal interchange between Africa and Asia took place in the Paleocene rather than the Eocene (e.g., Krause and Maas, 1990). In sum, in the absence of any strong paleogeographic or paleobiogeographic evidence to support dispersal of primates from Asia to Africa, we consider an Asian origin for anthropoids to be an unlikely scenario.

Anthropoid or Protoanthropoid Dispersal from Europe to Africa

On paleogeographic grounds alone, Eocene dispersal of primates is most probable between Europe and Africa by way of the short (<100 km) transtethyan sweepstakes route from southwestern Europe to northwestern Africa. The paleocurrent data, however, suggest that sweepstakes dispersal between the two continents would have been more likely to occur in the opposite direction, from Africa to Europe. In fact, there is little indication that extensive faunal exchange did occur along a Euro-African transtethyan route. Faunal resemblance between Africa and Europe, particularly at a generic level, is low throughout the Eocene, and the known phylogenetic relationships between European and African Eocene mammals do not strongly support dispersal from Europe to Africa. Interestingly, although at a family level the

African fauna is never significantly endemic with respect to Europe, the probability that the low resemblance is a result of chance is lowest in the late Eocene, the time when emergent land was maximized and thus intercontinental dispersal would seem most likely.

There is no indication of wholesale faunal exchange as asserted in some anthropoid dispersal scenarios (e.g., Schlosser, 1911; Rasmussen and Simons, 1988). Although limited faunal exchange (via sweepstakes dispersal) between Europe and Africa probably did take place, the timing and, importantly, the direction of exchange are uncertain. It may have taken place during the late Paleocene and early Eocene (Gheerbrant, 1990; Krause and Maas, 1990), or even between the early and middle Eocene, or middle and late Eocene (based on nonsignificant family-level Raup–Crick values for comparisons between African early Eocene and European middle Eocene, and between European middle Eocene and African late Eocene faunas). In any event, whether or not these possible exchanges involved anthropoids or their ancestors has yet to be determined.

Endemic African Origin of Anthropoids

On the basis of paleogeography, phylogenetic evidence, and overall faunal resemblance, an endemic African origin of anthropoids is the most likely of the scenarios evaluated here. From any source area, an extra-African origin of anthropoids would require dispersal across extensive water barriers, and recent ocean current models indicate that sweepstakes dispersal to northern Africa would have been hindered by the offshore direction of prevailing currents. Moreover, although on paleogeographic grounds the most probable timing of such a dispersal would have been during the late Eocene, Africa shows the greatest faunal endemism at that time. In fact, the African mammalian fauna was significantly endemic at a generic level throughout the Eocene, and, though generally not significantly endemic at a family level, the slight faunal similarities with other continents are likely to reflect earlier instances of faunal exchange, possibly during the Paleocene. In the simplest of all scenarios, anthropoids or their precursors would have been present in Africa during the Eocene and possibly the Paleocene, and no dispersal would be necessary to explain their presence in the early Eocene of Algeria or the late Eocene Fayum fauna.

Platyrrhine Dispersal from North America to South America

Paleogeographic data, phylogenetic data, and faunal comparisons indicate that a North American origin for anthropoids is unlikely. North America shows significant faunal endemicity compared to South America (Raup–Crick probabilities for early and for middle Eocene North America/South America comparisons are zero at both generic and familial levels: Anthony *et al.*, 1992). The distances to be traversed between South America and North America in the Eocene were as great as those between South America and Africa, and the northerly direction of prevailing currents during the Eocene would have reduced the likelihood of successful southward sweepstakes dispersal even more.

Platyrrhine Dispersal from Africa to South America

Limited dispersal between Africa and South America during the Eocene cannot be ruled out on the basis of faunal resemblance analysis. Although Africa and South America share no genera or families, Raup–Crick tests are significant, indicating that some limited dispersal may have occurred during the Eocene. In particular, the fact that the middle Eocene South American fauna was not significantly endemic with respect to the early Eocene African fauna raises the possibility that some dispersal may have taken place between the early and middle Eocene, presumably from Africa to South America. Anthropoid holophyly also supports this hypothesis. However, the distance to be covered along a transatlantic sweepstakes route is considerable (probably greater than 1000 km at any point in the Eocene). Even with the aid of an island arc along the Sierra Leone Rise, dispersal would require crossing more than 200 km of open ocean. However, in contrast to the North America–South America route, ocean currents would have been favorable for sweepstakes dispersal from Africa to South America. In sum, because South America was significantly endemic with respect to North America but not Africa during the Eocene, and because prevailing current direction would favor sweepstakes dispersal from Africa to South America but hinder dispersal from North America to South America, an African origin for platyrrhines is the more likely of the two scenarios.

Conclusions

Analysis of paleogeographic and paleobioeographic data indicates that a European or Asian origin for anthropoids, with subsequent dispersal to Africa, cannot be rejected. However, neither scenario is strongly supported. Limited faunal exchange by way of sweepstakes dispersal may have occurred between Africa and other continents from the late Paleocene throughout the Eocene, but there is no reason to expect that anthropoids or protanthropoids were among the waif taxa. Indeed, paleocurrent reconstruction suggests that sweepstakes dispersal in a direction northward from Africa and across the Tethys would have been the most likely. The simplest explanation for the presence of anthropoids in the late Eocene of the Fayum and the early Eocene of Algeria, in terms of paleogeographic and paleobiogeographic evidence, is one of anthropoid endemism in Africa throughout the Eocene. Africa also appears to be a more likely source than North America for the South American anthropoids. With respect to the ultimate origin of Anthropoidea, we conclude that attention can most productively be focussed on the Paleogene of Africa itself.

Acknowledgments

We are grateful to the conference organizers—Drs. John Fleagle, Richard Kay, and Elwyn Simons—for inviting us to participate in the symposium on anthropoid origins on which this book is based. This chapter has benefited from discussions with Mr. J. Alroy and Mr. M. R. L. Anthony, Drs. T. M. Bown, J. G. Fleagle, P. D. Gingerich, D. W. Krause, D. R. Prothero, K. D. Rose, and J. G. M. Thewissen. M. R. L. Anthony wrote the programs used for computing the biogeographic indices.

References

Alroy, J. 1992. Conjunction among taxonomic distributions and the Miocene mammalian biochronology of the Great Plains. *Paleobiology* **18**:326–343.

Andrews, C. W. 1906. *Catalogue of the Tertiary Vertebrata of the Fayum, Egypt.* British Museum (Natural History), London.

Anthony, M. R. L., and Maas, M. C. 1990. Biogeographic provinciality in North American Paleocene mammalian faunas. *J. Vertebr. Paleontol.* **9**:12A.

Anthony, M. R. L., Holroyd, P., and Maas, M. C. 1992. Statistical measure of faunal similarity: An example from Eocene continental mammal biotas. *J. Vertebr. Paleontol.* **12**(3):16A.

Aubry, M.-P., Berggren, W. A., Kent, D. V., Flynn, J. J., Klitgord, K. D., Obradovich, J. D., and Prothero, D. R. 1988. Paleogene geochronology: An integrated approach. *Paleoceanography* **3**(6):707–742.

Ba Maw, Ciochon, R. L., and Savage, D. E. 1979. Late Eocene of Burma yields earliest anthropoid primate, *Pondaungia cotteri. Nature* **282**:65–67.

Barron, E. J., and Peterson, W. H. 1989. Model simulation of the Cretaceous ocean circulation. *Science* **244**:684–686.

Barron, E. J., and Peterson, W. H. 1991. The Cenozoic ocean circulation based on ocean General Circulation Model results. *Palaeogeogr. Palaeoclim. Palaeoecol.* **83**:1–28.

Barry, J. C. 1988. *Dissopsalis,* a middle and late Miocene proviverrine creodont (Mammalia) from Pakistan and Kenya. *J. Vertebr. Paleontol.* **8**:25–45.

Beard, K. C., and Houde, P. 1989. An unusual assemblage of diminutive plesiadapiforms (Mammalia, ?Primates) from the early Eocene of the Clark's Fork Basin, Wyoming. *J. Vertebr. Paleontol.* **9**:388–399.

Beard, K. C., and Wang Banyue. 1990. Phylogenetic and biogeographic significance of the tarsiiform primate *Asiomomys changbaicus* from the Eocene of Jilin Province, People's Republic of China. *Am. J. Phys. Anthrop.* **85:**159–166.

Beard, K. C., Sigé, B., and Krishtalka, L. 1992. A primitive vespertilionoid bat from the early Eocene of central Wyoming. *C. R. Acad. Sci. Paris [Sér. II]* **314:**735–741.

Berggren, W. A., and Hollister, C. 1974. Paleogeography, paleobiogeography and the history of the circulation in the Atlantic Ocean. In: W. W. Hay (ed.), *Studies in Paleo-Oceanography,* pp. 126–186. Special Publication 20, Society of Economic Paleontologists and Mineralogists, Tulsa, Oklahoma.

Berggren, W. A., and Prothero, D. R. 1992. Eocene–Oligocene climatic and biotic evolution: An overview. In: D. R. Prothero and W. A. Berggren (eds.), *Eocene–Oligocene Climatic and Biotic Evolution,* pp. 1–28. Princeton University Press, Princeton.

Black, C. C. 1978. Anthracotheriidae. In: V. J. Maglio and H. B. S. Cooke (eds.), *Evolution of African Mammals,* pp. 423–434. Harvard University Press, Cambridge.

Bonis, L. de, Jaeger, J.-J., Coiffait, B., and Coiffait, P.-É. 1988. Découverte du plus ancien primate Catarrhinien connu dans l'Éocène supérieur d'Afrique du Nord. *C. R. Acad. Sci. Paris [Sér. II]* **306:**929–934.

Bown, T. M., and Kraus, M. J. 1988. Geology and paleoenvironment of the Oligocene Jebel Qatrani Formation and adjacent rocks, Fayum Depression, Egypt. U.S. Geological Survey Professional Paper 1452. U. S. Government Printing Office, Washington.

Bown, T. M., and Rose, K. D. 1987. Patterns of dental evolution in early Eocene anaptomorphine primates (Omomyidae) from the Bighorn Basin, Wyoming. Paleontological Society Memoir 23. *J. Paleont.* **61:**1–162.

Bown, T. M., and Schankler, D. 1982. A review of the Proteutheria and Insectivora of the Willwood Formation (lower Eocene), Bighorn Basin, Wyoming. U.S. Geological Survey Bulletin 1523. U.S. Government Printing Office, Washington, D.C.

Bown, T. M., and Simons, E. L. 1984. First record of marsupials (Metatheria: Polyprotodonta) from the Oligocene of Africa. *Nature* **308:**447–449.

Bown, T. M., and Simons, E. L. 1987. New Oligocene Ptolemaiidae (Mammalia: ?Pantolesta) from the Jebel Qatrani Formation, Fayum Depression, Egypt. *J. Vert. Paleontol.* **7:**311–324.

Bown, T. M., Kraus, M. J., Wing, S. L., Fleagle, J. G., Tiffney, B. H., Simons, E. L., Vondra, C. F. 1982. The Fayum primate forest revisited. *J. Hum. Evol.* **11:**603–632.

Briggs, J. C. 1989. The historic biogeography of India: isolation or contact? *Syst. Zool.* **38:**322–332.

Bugge, J. 1971. The cephalic arterial system in New and Old World hystricomorphs, and in bathyergoids, with special reference to the systematic classification of rodents. *Acta Anat.* **80:**516–536.

Bugge, J. 1985. Systematic value of the carotid arterial pattern in rodents. In: W. P. Luckett and J.-L. Hartenberger (eds.), *Evolutionary Relationships Among Rodents: A Multidisciplinary Analysis,* pp. 355–380. NATO Scientific Affairs Division, Plenum Press, New York.

Cheetham, A. H., and Hazel, J. E. 1969. Binary (presence–absence) similarity coefficients. *J. Paleontol.* **43:**1130–1136.

Chow Minchen. 1957. On some Eocene and Oligocene mammals from Kwangsi and Yunnan. *Vert. Palas.* **1**(3)**:**201–214.

Ciochon, R. L., and Chiarelli, A. B. 1980. Paleobiogeographic perspectives on the origin of the Platyrrhini. In: R. L. Ciochon and A. B. Chiarelli (eds.), *Evolutionary Biology of the New World Monkeys and Continental Drift* pp. 459–463. Plenum Press, New York.

Ciochon, R. L., Savage, D. E., Thaw Tint, and Ba Maw. 1985. Anthropoid origins in Asia? New discovery of *Amphipithecus* from the Eocene of Burma. *Science* **229:**756–759.

Coiffait, P.-É., Coiffait, B., Jaeger, J.-J., and Mahboubi, M. 1984. Un nouveau gisement à Mammifères fossiles d'âge Éocène supérieur sur le versant sud des Nementcha (Algérie orientale): Découverte des plus anciens Rongeurs d'Afrique, *C. R. Acad. Sci. Paris [Sér. II]* **299:**893–898.

Colbert, E. 1938. Fossil mammals from Burma in the American Museum of Natural History. *Bull. Am. Mus. Nat. Hist.* **74:**255–436.

Cooke, H. B. S. 1968. Evolution of mammals on southern continents. II. The fossil mammal fauna of Africa. *Q. Rev. Biol.* **43:**234–264.

Coombs, W. P., and Coombs, M. C. 1977. The origin of anthracotheres. *N. Jb. Geol. Paläont. Mh.* **1977**(10):584–599.

Coryndon, S. C., and Savage, R. J. G. 1973. The origin and affinities of African mammal faunas. In: *Organisms and Continents Through Time, Special Papers in Palaeontology 12,* pp. 121–135. Palaeontological Association, London.

Crochet, J.-Y. 1984. *Garatherium mahboubii* nov. gen., nov. sp., marsupial de l'Eocène inférieur d'El Kohol (Sud-Oranais, Algérie). *Ann. Paleontol.* **70:**275–294.

Crochet, J.-Y. 1986. *Kasserinotherium tunisiense* nov. gen., nov. sp., troisième marsupial découvert en Afrique (Eocène inférieur de Tunisie). *C. R. Acad. Sci. Paris [Sér. II]* **302:**923–926.

Crochet, J.-Y. 1988. Le plus ancien créodonte africain: *Koholia atlasense* nov. gen., nov. sp. (Eocène inférieur d'El Kohol, Atlas saharien, Algérie) *C. R. Acad. Sci. Paris [Sér. II]* **307:**1795–1798.

Crochet, J.-Y. 1991. A propos de quelques Créodontes proviverrinés de l'Eocène supérieur du sud de la France. *N. Jb. Geol. Paläont. Abh.* **182:**99–115.

Crochet, J.-Y., Thomas, H., Roger, J., Sen, S., and Al-Sulaimani, Z. 1990. Premiere découverte d'un créodonte dans la peninsule Arabique: *Masrasector ligabeui* nov. sp. (Oligocene inférieur de Taqah, Formation d'Ashawq, Sultanat d'Oman). *C. R. Acad. Sci. Paris [Sér. II]* **311:**1455–1460.

Crochet, J.-Y., Thomas, H., Sen, S., Roger, J., Gheerbrant, E., and Al-Sulaimani, Z. 1992. Découverte d'un Péradectidé (Marsupialia) dans l'Oligocene inférieur du Sultanat d'Oman: Nouvelles données sur la paléobiogéographie des Marsupiaux de la plaque arabo-africaine. *C. R. Acad. Sci. Paris [Sér. II]* **314:**539–545.

Dashzeveg, D. 1990. The earliest rodents (Rodentia, Ctenodactyloidea) of central Asia. *Acta. Zool. Cracov.* **33**(2)**:**11–35.

Dawson, M. R. 1977. Late Eocene rodent radiations: North America, Europe and Asia. *Géobios Mém. Spéc.* **1:**195–209.

Dawson, M. R., Li, C., and Qi, T. 1984. Eocene ctenodactyloid rodents (Mammalia) of eastern and central Asia. In: Mengel, R. M. (ed.), *Papers in Vertebrate Paleontology Honoring Robert Warren Wilson, Special Publication of Carnegie Museum of Natural History No. 9,* pp. 138–150. Carnegie Museum, Pittsburgh.

Dercourt, J., Zohenshain, L. P., Ricou, L.-E., Kazmin, V. G., Le Pichon, X., Knipper, A. L., Grandjacquet, C., Sbortshikov, I. M., Geyssant, J., Lepvrier, C., Perchersky, D. H., Boulin, J., Sibuet, J.-C., Savostin, L. A., Sorokhtin, O., Westphal, M., Bazhenov, M. L., Lauer, J. P., and Biju-Duval, B. 1986. Geological evolution of the Tethys belt from the Atlantic to the Pamirs since the Lias. *Tectonophysics* **123:**241–315.

Ducrocq, S., Buffetaut, E., Buffetaut-Tong, H., Helmcke-Ingavat, R., Jaeger, J.-J., Jongkanjanasoontorn, Y., and Suteethorn, V. 1992. A lower Tertiary vertebrate fauna from Krabi (South Thailand). *N. Jb. Geol. Paläont. Abh.* **184:**101–122.

Durette-Desset, M. C. 1971. Essai de classification des nematodes heligmosomes. Correlations avec la paleobiogeographie des hotes. *Mém. Mus. Nat. Hist. Nat. N.S. [Sér A] Zool.* **49:**1–126.

Emry, R. J. 1981. Additions to the mammalian fauna of the type Duchesnean, with comments on the status of the Duchesnean "Age." *J. Paleontol.* **55:**563–570.

Emry, R. J. 1990. Mammals of the Bridgerian (middle Eocene) Elderberry Canyon Local Fauna of eastern Nevada. In: T. M. Bown and K. D. Rose (Eds.), *Dawn of the Age of Mammals in the Northern Part of the Rocky Mountain Interior, North America. Geological Society of America Special Paper 243,* pp. 187–210. Geological Society of America, Boulder, Colorado.

Fleagle, J. G. 1986. Early anthropoid evolution in Africa and South America. In: J. G. Else and P. C. Lee (eds.), *Primate Evolution,* pp. 133–142. Cambridge University Press, Cambridge.

Fleagle, J. G., and Kay, R. F. 1987. The phyletic position of the Parapithecidae, *J. Hum. Evol.* **16:**483–532.

Flynn, J. J. 1986a. Correlation and geochronology of middle Eocene strata from the western United States. *Palaeogeogr. Palaeoclimatol. Palaeoecol.* **55:**335–406.

Flynn, J. J. 1986b. Faunal provinces and the Simpson Coefficient. *Contrib. Geol. Univ. Wyo. Spec. Paper* **3**:317–338.

Flynn, L. J., Jacobs, L. L., and Cheema, I. U. 1986. Baluchimyinae, a new ctenodactyloid rodent subfamily from the Miocene of Baluchistan. *Am. Mus. Novit.* **2841**:1–58.

Ford, S. M. 1988. Postcranial adaptations of the earliest platyrrhine. *J. Hum. Evol.* **17**:155–192.

Franzen, J. L. 1987. Ein neuer Primate aus dem Mitteleozän der Grube Messel (Deutschland, S-Hessen). *Cour. Forsch. Inst. Senckenberg* **91**:151–187.

Franzen, J. L., and Haubold, H. 1989. Artiodactyla aus den Eozänen braunkohlen des Geiseltales bei Halle (DDR). *Palaeovertebrata* **19**(3):131–160.

Garzanti, E., Baud, A., and Mascle, G. 1987. Sedimentary record of the northward flight of India and its collision with Eurasia (Ladakh Himalaya, India). *Geodinam. Acta (Paris)* **1**:297–312.

Gentry, A. W., and Hooker, J. J. 1988. The phylogeny of the Artiodactyla. In: M. J. Benton (ed.), *The Phylogeny and Classification of the Tetrapods, Vol. 2: Mammals,* pp. 235–272. Clarendon Press, Oxford.

Gheerbrant, E. 1990. On the early biogeographical history of the African placentals. *Hist. Biol.* **4**:107–116.

Gingerich, P. D. 1975. A new genus of Adapidae (Mammalia, Primates) from the late Eocene of southern France, and its significance for the origin of higher primates. *Contrib. Mus. Paleontol. Univ. Michigan* **24**:163–170.

Gingerich, P. D. 1980. Eocene Adapidae, paleobiogeography, and the origin of South American Platyrrhini. In: R. L. Ciochon and A. B. Chiarelli (eds.), *Evolutionary Biology of the New World Monkeys and Continental Drift,* pp. 123–138. Plenum Press, New York.

Gingerich, P. D. 1987. Early Eocene bats (Mammalia, Chiroptera) and other vertebrates in freshwater limestones of the Willwood Formation, Clarks Fork Basin, Wyoming. *Contrib. Mus. Paleontol. Univ. Michigan* **7**:270–320.

Gingerich, P. D. 1989. New earliest Wasatchian mammalian fauna from the Eocene of northwestern Wyoming: composition and diversity in a rarely sampled high-floodplain assemblage. University of Michigan Papers on Paleontology No. 28, Ann Arbor.

Gingerich, P. D. 1991. Systematics and evolution of early Eocene Perissodactyla (Mammalia) in the Clarks Fork Basin, Wyoming. *Contrib. Mus. Paleontol. Univ. Michigan* **28**:181–213.

Gingerich, P. D., and Deutsch, H. A. 1989. Systematics and evolution of early Eocene Hyaenodontidae (Mammalia, Creodonta) in the Clarks Fork Basin, Wyoming. *Contrib. Mus. Paleontol. Univ. Michigan* **24**:245–279.

Gingerich, P. D., and Russell, D. E. 1990. Dentition of early Eocene *Pakicetus* (Mammalia, Cetacea). *Contrib. Mus. Paleontol. Univ. Michigan* **28**:1–20.

Godinot, M. 1981. Les Mammiféres de Rians (Éocène inferieur, Provence). *Palaeovertebrata* **10**:43–126.

Godinot, M. 1987. Mammalian Reference Levels MP 1–10. *Münch. Geowiss. Abh. [A]* **10**:22–23.

Godinot, M. 1988. Les primates adapidés de Bouxwiller (Éocène moyen, Alsace) et leur apport à la compréhension de la faune de Messel et à l'évolution des Anchomomyini. *Cour. Forsch. Inst. Senckenberg* **107**:383–407.

Godinot, M., and Mahboubi, M. 1992. Earliest known simian primate found in Algeria. *Nature* **357**:324–326.

Godinot, M., Crochet, J.-Y., Hartenberger, J.-L., Lange-Badré, B., Russell, D. E., and Sigé, B. 1987. Nouvelles données sur les mammiferes de Palette (Écoéne inférieur, Provence). *Münch. Geowiss. Abh. [A]* **10**:273–288.

Gunnell, G. F. 1989. Evolutionary history of Microsyopoidea (Mammalia, ?Primates) and the relationship between plesiadapiformes and primates. University of Michigan Papers on Paleontology No. 27, Ann Arbor.

Gunnell, G. F., and Gingerich, P. D. 1991. Systematics and evolution of late Paleocene and early Eocene Oxyaenidae (Mammalia, Creodonta) in the Clarks Fork Basin, Wyoming. *Contrib. Mus. Paleontol. Univ. Michigan* **29**:141–180.

Harrison, T. 1987. The phylogenetic relationships of the early catarrhine primates: a review of the current evidence. *J. Hum. Evol.* **16**:41–80.

Hartenberger, J.-L. 1985. The order Rodentia: major questions of their evolutionary origin, relationships and suprafamilial systematics. In: W. P. Luckett and J.-L. Hartenberger (eds.), *Evolutionary Relationships among Rodents: A Multidisciplinary Analysis,* pp. 1–32. Plenum Press, New York.

Hartenberger, J.-L. 1986. Hypothèse paléontologique sur l'origine des Macroscelidea (Mammalia). *C. R. Acad. Sci. Paris [Sér. II]* **302:**247–249.

Hartenberger, J.-L. 1990. L'origine des Theridomyoidea (Mammalia, Rodentia): données nouvelles et hypothèses. *C. R. Acad. Sci. Paris [Sér. II]* **311:**1017–1023.

Hartenberger, J.-L., Martinez, C., and Ben Saïd, A. 1985. Découverte de mammifères d'âge Éocène inférieur en Tunisie Centrale, *C. R. Acad. Sci. Paris [Sér. II]* **301:**649–652.

Haubold, H. 1989. Die referenzfauna des Geiseltalium, MP Levels 11 bis 13 (Mitteleozän, Lutetium). *Palaeovertebrata* **19:**81–93.

Hellmund, M. 1991. Revision der Europäischen species der gattung *Elomeryx* Marsh 1894 (Anthracotheriidae, Artiodactyla, Mammalia)—odontologische untersuchungen. *Palaeontographica [A]* **220:**1–101.

Henderson, R. A., and Heron, M. L. 1977. A probabilistic method of paleobiogeographic analysis. *Lethaia* **10:**1–15.

Hoffstetter, R. 1972. Relationships, origins and history of the ceboid monkeys and caviomorph rodents: a modern reinterpretation. In: Dobzhansky, T., Hecht, M. K., and Sterre, W. C. (eds.), *Evolutionary Biology, Vol. 6,* pp. 323–347. New York: Appleton-Century-Crofts.

Hooker, J. J. 1986. Mammals from the Bartonian (middle/late Eocene) of the Hampshire Basin, southern England. *Bull. Br. Mus. Nat. Hist. (Geol.)* **39:**191–478.

Hooker, J. J. 1991. Two new pseudosciurids (Rodentia, Mammalia) from the English late Eocene, and their implications for phylogeny and speciation. *Bull. Br. Mus. Nat. Hist. (Geol.)* **47:** 35–50.

Hugueney, M., and Adrover, R. 1991. *Sacaresia moyaeponsi* nov. gen. nov. sp., rongeur thryonomyide (Mammalia) dans le Paleogene de Majorque (Baleares, Espagne). *Geobios* **24:**207–214.

Hussain, S. T., de Bruijn, H., and Leinders, J. M. 1978. Middle Eocene rodents from the Kala Chitta Range (Punjab, Pakistan), *Proc. Kon. Ned. Akad. Wetensch. [B]* **81:**74–112.

Jaeger, J.-J. 1988. Rodent phylogeny: New data and old problems. In: M. J. Benton (ed.), *The Phylogeny and Classification of the Tetrapods, Vol. 2: Mammals,* pp. 177–199. Clarendon Press, Oxford.

Jaeger, J.-J., Denys, C., and Coiffait, B. 1985. New Phiomorpha and Anomaluridae from the late Eocene of North West Africa. In: W. P. Luckett and J. L. Hartenberger (eds.), *Evolutionary Relationships among Rodents, a Multidisciplinary Analysis,* pp. 567–589. Plenum Press, New York.

Kappelman, J. 1992. The age of the Fayum primates as determined by paleomagnetic reversal stratigraphy. *J. Hum. Evol.* **22:**495–503.

Kelly, T. S. 1990. Biostratigraphy of Uintan and Duchesnean land mammal assemblages from the middle member of the Sespe Formation, Simi Valley, California. *Contrib. Sci. Nat. Hist. Mus. Los Angeles County* **149:**1–42.

Korth, W. W. 1984. Earliest Tertiary evolution and radiation of rodents in North America. *Bull. Carnegie Mus. Nat. Hist.* **24:**1–71.

Krause, D. W., and Maas, M. C. 1990. The biogeographic origins of late Paleocene—early Eocene mammalian immigrants to the Western Interior of North America. In: T. M. Bown and K. D. Rose (eds.), *Dawn of the Age of Mammals in the Northern Part of the Rocky Mountain Interior, North America,* pp. 71–106. Geological Society of America Special Paper 243. Geological Society of America, Boulder, Colorado.

Krishtalka, L., Stucky, R. K., West, R. M., McKenna, M. C., Black, C. C., Bown, T. M., Dawson, M. R., Golz, D. J., Lillegraven, J. A., and Turnbull, W. D. 1987. Eocene (Wasatchian through Duchesnean) biochronology of North America. In: M. O. Woodburne (ed.), *Cenozoic Mammals of North America, Geochronology and Biostratigraphy,* pp. 77–117. University of California Press, Berkeley.

Lange, B. 1967. Créodontes des phosphorites du Quercy: *Apterodon gaudryi*. *Ann. Pal. Vert.* **53**(2):79–90.

Lange-Badré, B. 1979. Les créodontes (Mammalia) d'Europe occidentale de l'Éocène supérieur a l'Oligocéne supérieur. *Mém. Mus. Nat. Hist. Nat. [C]* **42:**1–249.

Lange-Badré, B., and Godinot, M. 1982. Sur la presénce du genre nord-américain *Arfia* Van Valen (Creodonta, Mammalia) dans la faune de Dormaal (Eocène inférieur de Belgique). *C. R. Acad. Sci. Paris [Sér. II]* **294:**471–476.

Lavocat, R. 1969. La systématique des rongeurs hystricomorphes et la dérive des continents. *C. R. Acad. Sci. [D]* **269:**1496–1497.

Lavocat, R. 1971. Affinités systématiques des Caviomorphes et des Phiomorphes et origine Africaine des Caviomorphes. *An. Acad. Bras. Cienc.* **43:**515–522.

Lavocat, R. 1972. Miocene rodents of East Africa and Oligocene rodents of Bolivia. In: *20th Symposium on Vertebrate Palaeontology and Comparative Anatomy.*

Lavocat, R. 1974. The interrelationships between the African and South American rodents and their bearing on the problem of the origins of South American monkeys. *J. Hum. Evol.* **3:**323–326.

Lavocat, R. 1976. Rongeurs caviomorphes de l'Oligocène de Bolivie: II. Rongeurs du bassin Déséadien de Salla-Luribay. *Palaeovertebrata* **7:**15–90.

Lavocat, R. 1977. Sur l'origine des faunes sud-amèricaines de Mammiféres du Mésozoique terminal et du Cénozoique ancien. *C. R. Acad. Sci. [D]* **285:**1423–1426.

Lavocat, R. 1980. The implications of rodent paleontology and biogeography to the geographical sources and origin of the platyrrhine primates. In: R. L. Ciochon and A. B. Chiarelli (eds.), *Evolutionary Biology of the New World Monkeys and Continental Drift,* pp. 93–102. Plenum Press, New York.

Lavocat, R., and Parent, J.-P. 1985. Phylogenetic analysis of middle ear features in fossil and living rodents. In: W. P. Luckett and J.-L. Hartenberger (eds.), *Evolutionary Relationships among Rodents: A Multidisciplinary Analysis,* pp. 333–354. NATO Scientific Affairs Division, Plenum Press, New York.

Lillegraven, J. A. 1980. Primates from later Eocene rocks of southern California. *J. Mamm.* **61:**181–204.

Lucas, S. G. 1992. Redefinition of the Duchesnean land mammal "age," late Eocene of western North America. In: D. R. Prothero and W. A. Berggren (eds.), *Eocene–Oligocene Climatic and Biotic Evolution,* pp. 85–105. Princeton University Press, Princeton.

Luckett, W. P. 1980. Monophyletic or diphyletic origins of Anthropoidea and Hystricognathi: Evidence of the fetal membranes. In: R. L. Ciochon and A. B. Chiarelli (eds.), pp. 347–368. *Evolutionary Biology of the New World Monkeys and Continental Drift,* pp. 347–368. Plenum Press, New York.

Luckett, W. P. 1985. Superordinal and intraordinal affinities of rodents: developmental evidence from the dentition and placentation. In: W. P. Luckett and J.-L. Hartenberger (eds.), *Evolutionary Relationships among Rodents: A Multidisciplinary Analysis,* pp. 227–275. NATO Scientific Affairs Division, Plenum Press, New York.

Luyendyk, B., Forsyth, D., and Phillips, J. 1972. An experimental approach to the paleocirculation of oceanic surface waters. *Geol. Soc. Am. Bull.* **83:**2649–2664.

Maas, M. C. 1985. Taphonomy of a late Eocene microvertebrate locality, Wind River Basin, Wyoming (U.S.A.). *Palaeogeogr. Palaeoclimatol. Palaeoecol.* **52:**123–142.

MacFadden, B. J., Campbell, E. E., Cifelli, R. L., Siles, O., Johnson, N. M., Naeser, C. W., and Zeitler, P. K. 1985. Magnetic polarity, stratigraphy and mammalian faunas of the Deseadean (late Oligocene–early Miocene) Salla Beds of northern Bolivia. *J. Geol.* **93:**23–250.

Mahboubi, M., Ameur, R., Crochet, J. Y., and Jaeger, J.-J. 1983. Première découverte d'un marsupial en Afrique. *C. R. Acad. Sci. Paris [D]* **297:**691–694.

Mahboubi, M., Ameur, R., Crochet, J. Y., and Jaeger, J.-J. 1984. Earliest known proboscidean from early Eocene of north-west Africa, *Nature* **308:**543–544.

Mahboubi, M., Ameur, R., Crochet, J. Y., and Jaeger, J.-J. 1986. El Kohol (Saharan Atlas, Algeria): A new Eocene mammal locality in Northwestern Africa. *Palaeontographica [A]* **192:**15–49.

Marshall, L. G., Hoffstetter, R., and Pascual, R. 1983. Mammals and biostratigraphy: Geochronology of the continental mammal-bearing Tertiary of South America. *Palaeovert. Mem. Extra.* **1983:**1–93.

Mason, M. A. 1990. New fossil primates from the Uintan (Eocene) of southern California. *Paleobios* **13**(49):1–7.

McKenna, M. C. 1973. Sweepstakes, filters, corridors, Noah's arks, and beached Viking funeral ships in palaeogeography. In: D. H. Tarling and S. K. Runcorn (eds.), *Implications of Continental Drift to the Earth Sciences,* pp. 295–308. Academic Press, London.

McKenna, M. C. 1990. Plagiomenids (Mammalia: ?Dermoptera) from the Oligocene of Oregon, Montana, and South Dakota, and middle Eocene of northwestern Wyoming. In: T. M. Bown and K. D. Rose (eds), *Dawn of the Age of Mammals in the Northern Part of the Rocky Mountain Interior, North America,* pp. 211–234. Geological Society of America Special Paper 243. Geological Society of America, Boulder, Colorado.

Novacek, M. J., Bown, T. M., and Schankler, D. 1985. On the classification of the early Tertiary Erinaceomorpha (Insectivora, Mammalia). *Am. Mus. Novit.* **2813:**1–22.

Nürnberg, D., and Muller, R. D. 1991. The tectonic evolution of the South Atlantic from late Jurassic to present. *Tectonophysics* **191:**27–53.

Osborn, H. F. 1909. New carnivorous mammals from the Fayum Oligocene, Egypt. *Bull. Am. Mus. Nat. Hist.* **26:**415–424.

Patriat, P., and Achache, J. 1984. India–Eurasia collision chronology has implications for crustal shortening and driving mechanism of plates. *Nature* **311:**615–621.

Pilgrim, G. E. 1928. The Artiodactyla of the Eocene of Burma. *Mem. Geol. Surv. India N. S.* **8:** 1–39.

Pilgrim, G. E. 1941. The dispersal of the Artiodactyla. *Biol. Rev. Camb. Phil. Soc.* **16:**134–163.

Plazait, J.-C. 1981. Late Cretaceous to late Eocene palaeogeographic evolution of southwest Europe. *Palaeogeogr. Palaeoclimatol. Palaeoecol.* **36:**263–320.

Popoff, M., Ojah, H. A., and Baudin, P. 1989. Opening of the South Atlantic and Mesozoic rifting in the Benue Trough. *Geochronique* **30:**64–65.

Prasad, G. V. R., Sahni, A., and Gupta, V. J. 1986. Fossil assemblages from infra- and intertrappan beds of Asifabad, Andraha Pradesh and their geological implications. *Geosci. J.* **7:**163–180.

Prothero, D. R. 1985. North American mammalian diversity and Eocene–Oligocene extinctions. *Paleobiology* **11:**389–405.

Prothero, D. R., and Schoch, R. 1989. Origin and evolution of the Perissodactyla: Summary and synthesis. In: D. R. Prothero and R. M. Schoch (eds.), *The Evolution of Perissodactyls,* pp. 504–537. Oxford University Press, New York.

Prothero, D. R., and Swisher, C. C. 1992. Magnetostratigraphy and geochronology of the terrestrial Eocene–Oligocene transition in North America. In: D. R. Prothero and W. A. Berggren (eds.), *Eocene–Oligocene Climatic and Biotic Evolution,* pp. 46–73. Princeton University Press, Princeton.

Qi Tao. 1987. The middle Eocene Arshanto fauna (Mammalia) of Inner Mongolia. *Ann. Carnegie Mus.* **56**(1):1–73.

Radulesco, C., Iliesco, G., and Iliesco, M. 1976. Un Embrithopode nouveau (Mammalia) dans le Paléogène de la dépression de Hateg (Roumanie) et la géologie de la région. *N. Jb. Geol. Paläont. Mh.* **11:**690–698.

Rasmussen, D. T. 1990. The phylogenetic position of *Mahgarita stevensi:* protoanthropoid or lemuroid? *Int. J. Primatol.* **11:**437–467.

Rasmussen, D. T., and Simons, E. L. 1988. New specimens of *Oligopithecus savagei,* early anthropoidean primate from the Fayum, Egypt. *Fol. Primatol.* **51:**182–208.

Rasmussen, D. T., Bown, T. M., and Simons, E. L. 1992. The Eocene–Oligocene transition in North Africa. In: D. R. Prothero and W. A. Berggren (eds.), *Eocene–Oligocene Climatic and Biotic Evolution,* pp. 548–566. Princeton University Press, Princeton.

Raup, D. M., and Crick, R. E. 1979. Measurement of faunal similarity in paleontology. *J. Paleontol.* **53:**1213–1227.

Rosenberger, A. L. 1986. Platyrrhines, catarrhines and the anthropoid transition. In: B. Wood,

L. Martin, and P. Andrews (eds.), *Major Topics in Primate and Human Evolution,* pp. 66–89. Cambridge University Press, Cambridge.

Rosenberger, A. L., and Szalay, F. S. 1980. On the tarsiiform origins of Anthropoidea. In: R. L. Ciochon and A. B. Chiarelli (eds.), *Evolutionary Biology of the New World Monkeys and Continental Drift,* pp. 139–157. Plenum Press, New York.

Russell, D. E., and Godinot, M. 1988. The Paroxyclaenidae (Mammalia) and a new form from the early Eocene of Palette, France. *Paläont. Z.* **62:**319–331.

Russell, D. E., and Zhai Ren-jie 1987. The Paleogene of Asia: Mammals and stratigraphy. *Mém. Mus. Nat. Hist. Nat. Sci. Terre* **52:**1–488.

Russell, D. E., Hartenberger, J.-L., Pomerol, C., Sen, S., Schmidt-Kittler, N., and Vianey-Liaud, M. 1982. Mammals and stratigraphy: the Paleogene of Europe. *Palaeovert. Mem. Extra.* **1982:**1–77.

Sahni, A. 1984. Cretaceous–Paleocene terrestrial faunas of India: Lack of endemism during drifting of the Indian plate. *Science* **226:**441–443.

Savage, D. E., and Russell, D. E. 1983. *Mammalian Paleofaunas of the World.* Addison-Wesley, Reading, MA.

Savage, R. J. G. 1978. Carnivora. In: V. J. Maglio and H. B. S. Cooke (eds.), *Evolution of African Mammals,* pp. 249–267. Harvard University Press, Cambridge.

Schlosser, M. 1911. Beiträge zur Kenntnis der Oligozänen Landsäugetiere dem Fayum (Ägypten). *Beitr. Palaont. Geol. Öst.-Ungarns Orients* **24:**51–167.

Schlater, J. G., Hellinger, S., and Tapscott, C. 1977. The paleobathymetry of the Atlantic Ocean from the Jurassic to the present. *J. Geol.* **85:**509–522.

Sen, S., and Heintz, E. 1979. *Palaeoamasia kansui* Ozansoy 1966, Embrithopode (Mammalia) de l'Éocène d'Anatolie. *Ann. Paléont. (Vertébr.).* **65**(I)**:**73–91.

Sibuet, J. C., and Mascle, J. 1978. Plate kinematic implications of Atlantic equatorial fracture zone trends. *J. Geophys. Res.* **83:**3401–3421.

Sigé, B., Jaeger, J.-J., Sudre, J., and Vianey-Liaud, M. 1990. *Altiatlasius koulchii* n. gen. et sp., primate omomyidé du Paléocene supérieur du Maroc, et les origins des euprimates. *Palaeontographica [A]* **214:**31–56.

Simberloff, D. 1978. Using island biogeographic distributions to determine if colonization is stochastic. *Am. Nat.* **112:**713–726.

Simons, E. L. 1968. Early Cenozoic mammalian faunas, Fayum Province, Egypt, Part I: African Oligocene Mammals: Introduction, history of study, and faunal succession. *Peabody Mus. Bull.* **28:**1–22.

Simons, E. L. 1976. The fossil record of primate phylogeny. In: M. Goodman, E. Tashian, and J. H. Tashian (eds.), *Molecular Anthropology,* pp. 35–62. Plenum Press, New York.

Simons, E. L. 1989. Description of two genera and species of late Eocene Anthropoidea from Egypt. *Proc. Natl. Acad. Sci. USA* **86:**9956–9960.

Simons, E. L. 1990. Discovery of the oldest known anthropoidean skull from the Paleogene of Egypt. *Science* **247:**1567–1569.

Simons, E. L., and Bown, T. M. 1984. A new species of *Peratherium* (Didelphidae: Polyprotodonta): the first African marsupial. *J. Mamm.* **65:**539–548.

Simons, E. L., and Gingerich, P. D. 1974. New carnivorous mammals from the Oligocene of Egypt. *Ann. Geol. Surv. Egypt* **4:**157–166.

Simons, E. L., and Gingerich, P. D. 1976. A new species of *Apterodon* (Mammalia, Creodonta) from the upper Eocene Qasr el-Sagha Formation of Egypt. *Postilla* **168:**1–9.

Simons, E. L., and Kay, R. F. 1983. *Qatrania,* new basal anthropoid primate from the Fayum, Oligocene of Egypt, *Nature* **304:**624–626.

Simons, E. L., and Kay, R. F. 1988. New material of *Qatrania* from Egypt with comments on the phylogenetic position of the Parapithecidae (Primates, Anthropoidea). *Am. J. Primatol.* **15:**337–347.

Simons, E. L., Bown, T. M., and Rasmussen, D. T. 1986. Discovery of two additional prosimian primate families (Omomyidae, Lorisidae) in the African Oligocene. *J. Hum. Evol.* **15:**431–437.

Simons, E. L., Holroyd, P. A., and Bown, T. M. 1991. Early Tertiary elephant shrews from Egypt and the origin of the Macroscelidea. *Proc. Natl. Acad. Sci. USA* **88:**9734–9737.

Simpson, G. G. 1940. Mammals and land bridges. *J. Wash. Acad. Sci.* **30:**137–163.

Simpson, G. G. 1943. Mammals and the nature of continents. *Am. J. Sci.* **241:**1–31.

Simpson, G. G. 1947. Holarctic mammalian faunas and continental relationships during the Cenozoic. *Bull. Geol. Soc. Am.* **58:**613–688.

Simpson, G. G. 1960. Notes on the measurement of faunal resemblance. *Am. J. Sci.* **258A:**300–311.

Simpson, G. G. 1965. *The Geography of Evolution.* Capricorn Books, New York.

Smith, A. G., and Briden, J. C. 1977. *Mesozoic and Cenozoic Paleocontinental Maps.* Cambridge University Press, Cambridge.

Stephan, J. F., Mercier de Lepinay, B., Calais, E., Tardy, M., Beck, C., Carfantan, J.-C., Olivet, J.-L., Vila, J.-M., Bouysse, P., Mauffret, A., Bourgois, J., Thery, J.-M., Tournon, J., Blanchet, R., and Dercourt, J. 1990. Paleogeodynamic maps of the Caribbean: 14 steps from Lias to Present. *Bull. Soc. Geol. Fr. Paris Suppl. 8* **7**(6)**:**915–919.

Storer, J. E. 1984. Mammals of the Swift Current Creek Local Fauna (Eocene: Uintan), Saskatchewan, *Natural History Contributions No. 7.* Saskatchewan Museum of Natural History, Regina.

Storer, J. E. 1988. The rodents of the Lac Pelletier Lower Fauna, late Eocene (Duchesnean) of Saskatchewan. *J. Vertebr. Paleontol.* **8:**84–101.

Storer, J. E. 1990. Primates of the Lac Pelletier Lower Fauna (Eocene: Duchesnean), Saskatchewan. *Can. J. Earth. Sci.* **27:**520–524.

Stucky, R. K. 1984. The Wasatchian–Bridgerian Land Mammal Age boundary (early to middle Eocene) in western North America. *Ann. Carnegie Mus.* **53:**347–382.

Stucky, R. K. 1992. Mammalian faunas in North America of Bridgerian to early Arikareean "Ages" (Eocene and Oligocene). In: D. R. Prothero and W. A. Berggren (eds.), *Eocene-Oligocene Climatic and Biotic Evolution,* pp. 464–493. Princeton University Press, Princeton.

Sudre, J. 1979. Nouveaux mammifères Eocènes du Sahara occidental. *Palaeovertebrata* **9**(3)**:**83–115.

Sudre, J., Sigé, B., Remy, J. A., Marandat, B., Hartenberger, J.-L., Godinot, M., and Crochet J.-Y. 1990. Une faune du niveau d'Egerkingen (MP 14; Bartonien inferieur) dans les phosphorites du Quercy (sud de la France). *Palaeovertebrata* **20**(1)**:**1–32.

Suteethorn, V., Buffetaut, E., Helmcke-Ingavat, R., Jaeger, J.-J., and Jongkanjansoontorn, Y. 1988. Oldest known Tertiary mammals from South East Asia: Middle Eocene primate and anthracotheres from Thailand. *N. Jb. Geol. Paläont. Mh.* **9:**563–570.

Swisher, C. C., and Prothero, D. R. 1990. Single-crystal ^{40}Ar–^{39}Ar dating of the Eocene–Oligocene transition. *Science* **249:**760–762.

Szalay, F. S. 1976. Systematics of the Omomyidae (Tarsiiformes, Primates): Taxonomy, phylogeny, and adaptations. *Bull. Am. Mus. Nat. Hist.* **156:**157–450.

Tarling, D. H. 1980. Continental drift and the positioning of the circum-Atlantic continents throughout the last 100 million years of Earth history. In: R. L. Ciochon and A. B. Chiarelli (eds.), pp. 1–41. *Evolutionary Biology of the New World Monkeys and Continental Drift,* Plenum Press, New York.

Thewissen, J. G. M. 1990a. Comment on "Paleontological view of the ages of the Deccan Traps, the Cretaceous/Tertiary boundary, and the India–Asia collision." *Geology* **18:**185,187–188.

Thewissen, J. G. M. 1990b. Evolution of Paleocene and Eocene Phenacodontidae (Mammalia, Condylarthra). University of Michigan Papers on Paleontology No. 29, Ann Arbor.

Thewissen, J. G. M., and Gingerich, P. D. 1989. Skull and endocranial cast of *Eoryctes melanus,* a new palaeoryctid (Mammalia: Insectivora) from the early Eocene of western North America. *J. Vertebr. Paleontol.* **9**(4)**:**459–470.

Thewissen, J. G. M., and McKenna, M. C. 1992. Paleobiogeography of Indo-Pakistan: A response to Briggs, Patterson, and Owen. *Syst. Biol.* **41:**248–251.

Thewissen, J. G. M., Gingerich, P. D., and Russell, D. E. 1987. Artiodactyla and Perissodactyla (Mammalia) from the early-middle Eocene Kuldana Formation of Kohat (Pakistan). *Contrib. Mus. Palentol. Univ. Michigan* **27:**247–274.

Thomas, H., Roger, J., Sen, S., and Al-Sulaimani, Z. 1988. Découverte des plus anciens ≪Anthropoides≫ du continent arabo-africain et d'un Primate tarsiiforme dans l'Oligocène du Sultanat d'Oman, *C. R. Acad. Sci. Paris [Sér. II]* **306:**823–829.

Thomas, H., Roger, J., Sen, S., Bourdillon-de-Grissac, C., and Al-Sulaimani, Z. 1989. Découverte de vertébrés fossiles dans l'Oligocène inférieur du Dhofar (Sultanat d'Oman) *Geobios* **22**(1)**:**101–120.

Thomas, H., Sen, S., Roger, J., and Al-Sulaimani, Z. 1991. The discovery of *Moeripithecus markgrafi* Schlosser (Propliopithecidae, Anthropoidea, Primates), in the Ashawq Formation (Early Oligocene of Dhofar Province, Sultanate of Oman). *J. Hum. Evol.* **20:**33–49.

Tilden, C. D., Holroyd, P. A., and Simons, E. L. 1990. Phyletic affinities of *Apterodon* (Hyaenodontidae, Creodonta). *J. Vertebr. Paleontol.* **9**(3)**:**46A.

Van Andel, T. H., Thiede, J., Schlater, J. G., and Hay, W. W. 1977. Depositional history of the South Atlantic Ocean during the last 125 million years. *J. Geol.* **85:**651–698.

Van Couvering, J. A., and Harris, J. A. 1991. Late Eocene age of Fayum mammal faunas. *J. Hum. Evol.* **21:**241–260.

Van Couvering, J. A. H., and Van Couvering, J. A. 1976. Early Miocene mammal fossils from East Africa: Aspects of geology, faunistics, and paleoecology. In: G. L. Isaac and E. R. McCown (eds.), *Human Origins,* pp. 155–206. W. A. Benjamin, Menlo Park, CA.

Wang Banyue, and Li C. 1990. First Paleogene mammalian fauna from northeast China. *Vert. Palas.* **28:**165–205.

Wilson, J. A. 1986. Stratigraphic occurrence and correlation of early Tertiary vertebrate faunas, Trans-Pecos Texas: Agua Fria–Green Valley areas. *J. Vertebr. Paleontol.* **6:**350–373.

Wood, A. E. 1972. An Eocene hystricognathous rodent from Texas: its significance in interpretations of continental drift. *Science* **175:**1250–1251.

Wood, A. E. 1974. The evolution of the Old World and New World hystricomorphs. In I. W. Rowlands and B. J. Weir (eds.) *The Biology of Hystricomorph Rodents. Symp. Zool. Soc. Lond.* **34:**21–60.

Wood, A. E. 1980. The origin of the caviomorph rodents from a source in Middle America: A clue to the area of origin of the platyrrhine primates. In: R. L. Ciochon and A. B. Chiarelli (eds.), *Evolutionary Biology of the New World Monkeys and Continental Drift,* pp. 79–91. Plenum Press, New York.

Wood, A. E. 1985. The relationships, origin and dispersal of the hystricognathous rodents. In: W. P. Luckett and J.-L. Hartenberger (eds.), *Evolutionary Relationships among Rodents: A Multidisciplinary Analysis,* pp. 475–513. Plenum Press, New York.

Woods, C. A., and Hermanson, J. W. 1985. Myology of hystricognath rodents: An analysis of form, function and phylogeny. In: W. P. Luckett and J.-L. Hartenberger (eds.), *Evolutionary Relationships among Rodents: A Multidisciplinary Analysis,* pp. 515–548. Plenum Press, New York.

The Different Meanings of a Tarsioid–Anthropoid Clade and a New Model of Anthropoid Origin

12

D. TAB RASMUSSEN

Introduction

Since the early part of this century there has been widespread but not universal acceptance of the idea that among the living prosimians the species most closely related to anthropoids are the extant tarsiers. This idea of a tarsier–anthropoid clade among the living primates was first clearly formulated by Hubrecht (1896, 1902, 1908, 1909), Wortman (1903, 1904a,b), and Pocock (1918) on the basis of placentation, intrabullar carotid circulation, and the structure of the upper lip and rhinarium, among other features. Since then, additional morphological, biochemical, and molecular resemblances between tarsiers and anthropoids have been identified and promoted as shared-derived features indicating common ancestry (e.g., Luckett, 1974; Noback, 1975; Hershkovitz, 1977; Cartmill, 1980; Baba *et al.*, 1982; MacPhee and Cartmill, 1986; Pollock and Mullin, 1987). Contradictory and ambiguous evidence regularly appears (Starck, 1956; Barnicot and Hewett-Emmett, 1974; Sarich and Cronin, 1976; Cronin and Sarich, 1980; Bonner *et al.*, 1980; Mai,

D. TAB RASMUSSEN • Department of Anthropology, Washington University, St. Louis, Missouri 63130.

Anthropoid Origins, edited by John G. Fleagle and Richard F. Kay. Plenum Press, New York, 1994.

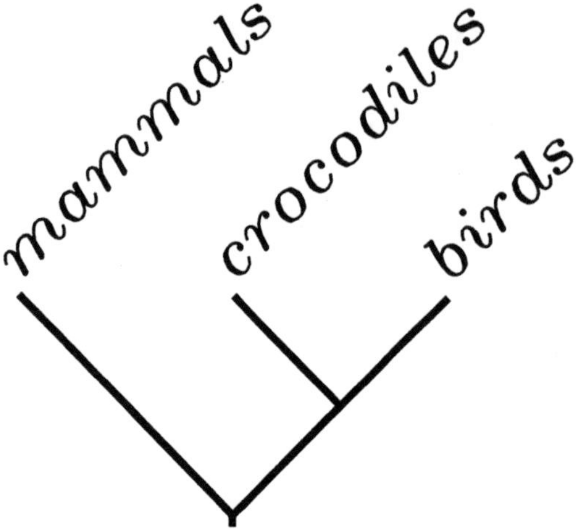

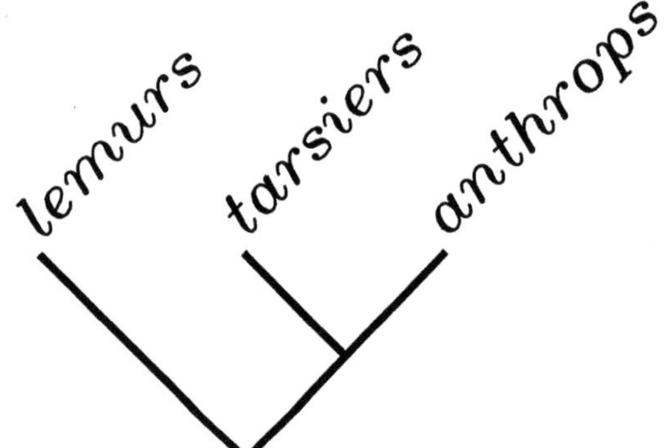

Fig. 1. Clade membership does not necessarily provide much specific information about ancestry. The top illustration shows the cladistic relationships among the classes Mammalia, Aves, and Crocodilia; no zoologist would seriously consider that the actual historical evolutionary sequence was from mammal to crocodile to bird. The cladistic relationship among the Lemuriformes, Tarsiiformes, and Anthropoidea is shown in the bottom illustration; it has often been interpreted as a sequence of evolutionary grades, from lemur to tarsier to anthropoid.

1985; Poorman *et al.,* 1985), but usually these results have been ignored or are considered to have resulted from methodological errors or convergent evolution.

The acceptance of a tarsier–anthropoid clade among the living primates has exerted a powerful influence on how primatologists have viewed the evolutionary origin of anthropoids. To many researchers, a tarsier–anthropoid clade suggested that anthropoids arose directly from a tarsier-like prosimian (e.g., Wood Jones, 1929, and many subsequent authors). This hypothetical tarsioid ancestral stock, in turn, was believed to be anciently derived from a very early and primitive "lemur-like" prosimian. The three major branches on a cladogram of the primates—lemur, tarsier, and anthropoid—thus became converted into a natural scale, which has often been presented in the primary literature and in textbooks as sequential "grades" of primate evolution (Fig. 1).

This simplistic conversion of a cladogram into an evolutionary sequence is not valid. The reasons can easily be seen if one considers the cladistic relationships among selected other vertebrate higher taxa. For example, the fact that crocodiles and birds are more closely related to each other than either group is to mammals reveals very little about the structure and adaptations of their common ancestor—certainly, the ancestor was neither bird nor crocodile (Fig. 1). Similarly, the fact that mammals form an outgroup of the bird–crocodile clade does not mean that birds or crocodiles evolved from a mammal, or that mammals are primitive for the bird–crocodile clade.

There are several reasons why evolutionary sequences cannot be read from cladograms. Animal lineages usually contain a mosaic of primitive and specialized features; no single branch of a cladogram will be all "primitive" while another is all "derived." Another reason is that the extant species on which cladograms are based comprise only an impoverished sample of the structural and taxonomic diversity that has actually evolved within higher-level taxa. The basal radiations of many vertebrate groups were much more "bushy" than can be appreciated from the species that happened to survive to the present. Basal bushes probably often contained animals that were radically different from any of the living descendents in terms of their structure, adaptations, and ecological relationships (Fig. 2). Who could have predicted a theropod dinosaur as a bird ancestor (Ostrom, 1975, 1990) in the absence of a fossil record?

The Poverty of Neontological Data

The title of this section sounds like an oxymoron; after all, it is the paleontological record that is clearly poverty stricken. The study of fossil vertebrates relies on fragmentary osteological evidence from a very few taxa scattered through enormous tracts of time. Anything that is discovered about past organisms must usually be a bare fraction of what can be learned about any living species. Neontological information about living organisms, in the form of morphological, behavioral, ecological, and molecular data, far outweighs the corresponding information available from extinct species.

In acknowledging the superiority of neontological data for many kinds of questions, it is important not to fall into the trap of therefore concluding that all inferences about past life based on neontological data are superior to inferences based on paleontological data. For example, molecular data may provide a hypothetical phylogeny for a group of living mammals under study, and perhaps even divergence dates may be estimated from molecules, but molecular data cannot reveal the structural adaptations and the ecological relationships of extinct organisms. Genealogies and dates do not provide direct information about the adaptations of extinct taxa, their taxonomic diversity, the evolutionary processes that shaped them, the biogeographic patterns of dispersal, climatic and environmental correlates, and other factors that contribute to an understanding of the full richness of the evolutionary process. Ultimately, the fossil record, despite its obvious drawbacks, repre-

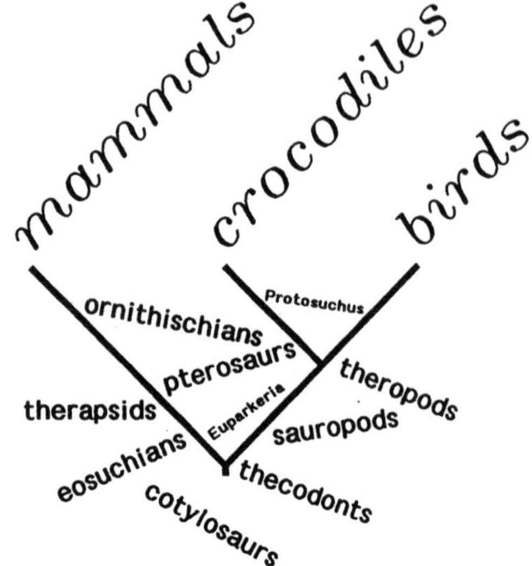

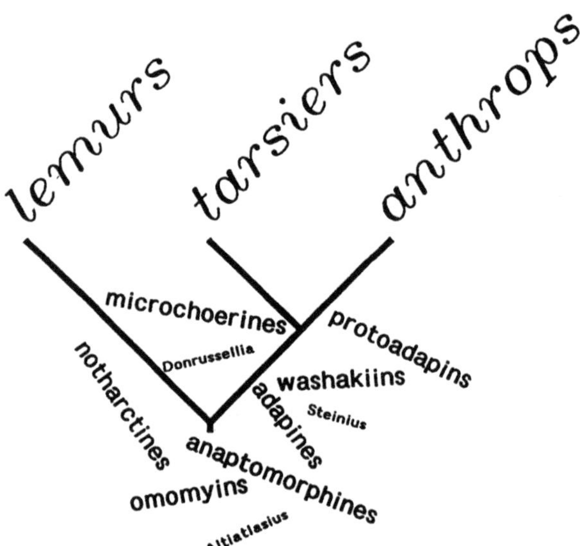

Fig. 2. The living species that are available for molecular and other neontological studies are but a small fraction of the much greater diversity that occurred in the past. The phylogenetic positions of extinct species cannot be inferred directly from the extant ones but must be determined by analysis of the fossil record. The top illustration shows several Mesozoic vertebrate taxa whose existence or phylogenetic significance could not be reconstructed solely on the basis of extant forms. Similarly, the bottom illustration shows Paleogene primates whose position relative to the cladogram can only be learned from fossil evidence. (The positions of the extinct taxa shown here are random or arbitrary and are intended for illustrative purposes only.)

sents the only available avenue for answering many important evolutionary questions.

Even when research questions can be addressed with neontological data, these data are not infallible. Returning to the bird–crocodile–mammal example, several recent neontological studies relying on physiological, morphological, and molecular data have actually revived the old idea that the homeothermic birds and mammals are closer to each other than either is to crocodilians (Gardiner, 1982; Løvtrup, 1985; Hedges *et al.*, 1990). In response, Gauthier *et al.* (1988), Donaghue *et al.* (1989), and Marshall (1992) have shown that the neontological studies in this case are misleading. Donaghue *et al.* (1989) convincingly demonstrated that despite the many drawbacks of the fossil record, fossil forms are pivotal for accurate reconstructions of phylogenies. Fossils simply cannot be hung like ornaments on the trees generated by neontological data.

The modest conclusion that tarsiers and anthropoids shared a common ancestor with each other more recently than with lorisoids or Malagasy prosimians tells us practically nothing about the actual evolutionary origin of anthropoids. When and where did anthropoids originate? What was the ancestor like from an adaptive point of view? What was the sequence in which anthropoid traits were acquired? Did the origin occur relatively suddenly, or was "anthropoidization" a slow accumulation of minor changes over a very long period of time? Was the prosimian precursor of anthropoids like a tarsier, a lemur, a mixture of both, or something radically different from either? The fascination among many primatologists with merely delimiting cladistic relationships as an end in itself stops far short of all these crucial evolutionary questions.

The Use of Morphotypes

One way in which neontological data are used in an explicit attempt to reconstruct the morphology and adaptations of extinct primates is through the construction of morphotypes. Several criteria can be used to choose which character states among living taxa might have been present in hypothetical ancestors: commonality and outgroup comparisons are two of the most common criteria, but these methods and others are all fairly arbitrary. There are many philosophical and logical reasons to expect that morphotype reconstruction based on living taxa will not be particularly accurate. The organisms that survive in the present and that are available to neontologists are such a tiny fraction of the biotic diversity of the past that there is simply no reasonable way to put together whole extinct animals using a patchwork of pieces assembled from living species. Entire radiations have become extinct, the problem of distinguishing specializations from ancestral conditions remains fairly intractable, and by all indications, convergence and parallelism run rampant.

More important than these theoretical concerns, however, is the fact that empirical comparisons often show that the anticipated morphotypes differ spectacularly from observed basal members once these are found as fossils (e.g., Marshall and Schultze, 1992). For an illustration of the inadequacy of morphotypes, one need look no further than the recently discovered Eocene anthropoids from quarry L-41 of the Fayum. The best known of these, *Catopithecus browni,* is more primitive ("prosimian-like") in its teeth, jaws, and postcranium than could ever have been predicted from living taxa on the basis of morphotype reconstruction (Simons, 1989, 1990; Rasmussen and Simons,

1992; Gebo *et al.*, Chapter 9, this volume). *Catopithecus* may even lie within the catarrhine clade, based on its specialized dental formula and other dental and postcranial resemblances to *Oligopithecus, Propliopithecus,* and *Aegyptopithecus* (Fleagle and Kay, 1987; Rasmussen and Simons, 1988; Gebo *et al.*, Chapter 9, this volume). Thus, many of the "anthropoid" characters shared by platyrrhines, catarrhines, and parapithecids must have evolved convergently from a common ancestor that lacked them.

These problems in morphotype reconstruction highlight the fact that acknowledgment of a tarsier–anthropoid clade among living primates places few restrictions on what the ancestral group actually looked like. There is no rational basis for concluding that the ancestor was a blend of tarsier and anthropoid features or that it was primarily tarsiiform in structure. Tarsier features and anthropoid features may have evolved in their respective lineages well after the divergence from a common ancestor. The only way to develop a fairly complete understanding of what the ancestral group was like is to find evidence of it in the fossil record.

Dangers in Reading the Fossil Record

It is easy to criticize hypothetical morphotypes and other neontological inferences, but of course, reading the fossil record is also fraught with imposing difficulties. Paleontologists are confronted by the persistent problems of convergence, parallelism, evolutionary reversal, and identification of ancestral and specialized character states. Evidence drawn from different anatomic systems may suggest substantially different phylogenetic arrangements. In addition to these standard anatomic problems, the fossil record is laden with additional difficulties. Knowledge of most taxa is restricted to teeth and jaws. Researchers may inadvertently mistake temporal sequences of fossils as evolutionary sequences when they are not; they must remember that old things are not necessarily ancestral or primitive things.

The bottom line is that paleontological and neontological data both have their drawbacks, and both have their strengths. However, the strengths and weaknesses are not general properties that apply equally to all evolutionary questions or phylogenetic problems. The ongoing debate about which kind of data is best misses the point entirely; the important question is, which kind of data is applicable to the particular question being asked? Molecules and fossils are both tools to answer evolutionary questions, but all tools in science always have a specific application and specific limitations.

The Different Meanings of a Tarsier–Anthropoid Clade

It should now be evident why acknowledgment of a tarsier–anthropoid clade among living taxa may take on very different meanings when applied to the problem of anthropoid origins. Depending on which osteological features are considered to reflect the tarsier–anthropoid clade (as defined by soft tissues and molecules), different researchers have variably included or excluded a bewildering array of extinct primate groups from the clade containing tarsiers and anthropoids. Much of the disagreement among researchers grows out of fair, objective analyses that nevertheless lead to different interpretations and conclusions as a simple consequence of the many normal difficulties that encumber any phylogenetic study (see Cartmill, Chapter 16, this volume). However, much of the debate, in my opinion, is unnecessarily complicated by the subconscious or sometimes explicit belief on the part of some researchers that the tarsier–anthropoid clade *among extant taxa* dictates that the anthropoid ancestor was tarsiiform. The search for anthropoid "synapomorphies" among prosimians becomes tautological if one selects (and defines as apomorphous) only distinctly tarsiiform traits among the hundreds of osteological characters available for comparison because it is already believed that the ancestor must be tarsiiform.

In this chapter, I will focus on four possible phylogenies that bear on the question of anthropoid origins, each of which potentially comprises innumerable subtle permutations. Each of these hypothetical phylogenies has acquired proponents among modern researchers and has undergone intensive scrutiny in the literature. A major thesis of this review is that each of these competing hypotheses has waved the flag of a tarsier–anthropoid clade, and yet this is irrelevant to testing them against each other. Only paleontological evidence can possibly provide an empirical test. After reviewing the four most prominent versions of a tarsioid–anthropoid clade and how recent fossil evidence impacts each one, I will present a new model of the phylogenetic origin of anthropoids.

A Strict Tarsius–Anthropoid Clade

This hypothesis holds that extant *Tarsius* and anthropoids are more closely related to each other than they are to any of the Eocene prosimians (Fig. 3). The greatest strength of this hypothesis is the shared occurrence of zygomatic–alisphenoid contact (Cartmill and Kay, 1978; Schmid, 1981, 1982; Cartmill, 1980; Cartmill *et al.*, 1981; MacPhee and Cartmill, 1986).

Although zygomatic–alisphenoid contact and the associated variable degree of postorbital closure are unique to tarsiers and anthropoids among mammals, many researchers have concluded that the contact between the two bones evolved convergently. Researchers have doubted that postorbital closure is synapomorphic for two main reasons: Eocene tarsioids lack closure; and there are differences between tarsiers and anthropoids in the structure and development of the postorbital region. In anthropoids, postorbital closure probably evolved as a consequence of temporal muscles developing a

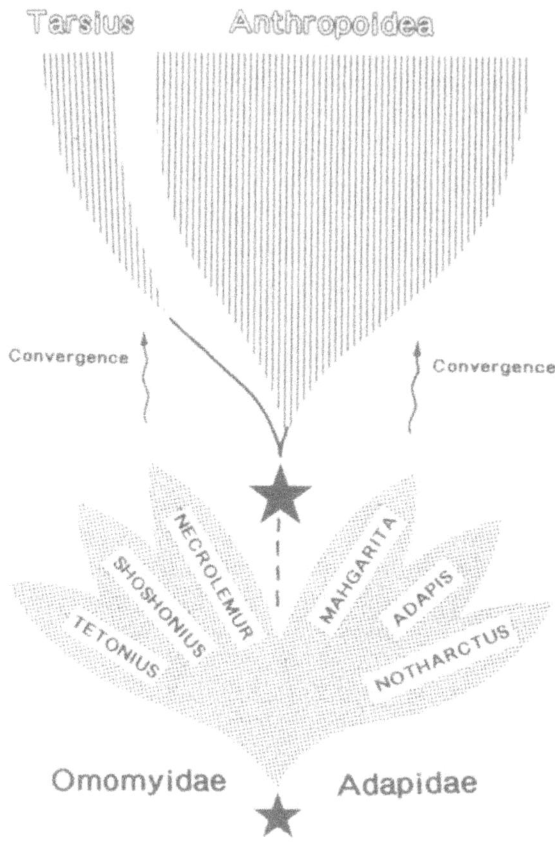

Fig. 3. A strict *Tarsius*–Anthropoidea clade (TAC). This hypothesis suggests that the extant taxa *Tarsius* and Anthropoidea (vertical hatching) are more closely related to each other than to any of the known Eocene primate radiations (stippled). The hypothetical TAC ancestor (top star) possessed sutural contact between the zygomatic and alisphenoid bones at the back of the orbit. This hypothesis implies that similarities between omomyids on the one hand and *Tarsius* on the other (Table I) arose by convergent evolution (wavy arrow) or were primitive for the TAC; similarly, resemblances between *Mahgarita* and early anthropoids either arose convergently (wavy arrow) or were primitive for the TAC. For this hypothesis, the relationship of the families Omomyidae and Adapidae to each other is not of explicit concern.

more anterior position that wrapped around the posterior and superior part of the eye, whereas in tarsiers it resulted from the enlarged orbit being forced back into the temporal muscle. The conclusion that tarsier postorbital structure is not synapomorphic with that of anthropoids has been expressed in several publications that explicitly evaluated postorbital structure (Le Gros Clark, 1959; Simons and Russell, 1960; Tattersall, 1973; Cachel, 1979; Simons and Rasmussen, 1989), and it has also been implied in other works on cranial structure or tarsiiform phylogeny (e.g., Gazin, 1958; Spatz, 1968; Szalay, 1976; Rosenberger, 1985). Several features of the middle ear have also been interpreted as synapomorphies of a tarsier–anthropoid clade (MacPhee and Cartmill, 1986), but there are disagreements among researchers about the interpretation of intrabullar homologies. For example, in my interpretation, the "transverse septum" of tarsiers is entirely different from the intrabullar septum observed in anthropoids, but others believe they are homologous (MacPhee and Cartmill, 1986; Simons and Rasmussen, 1989; Rasmussen, 1990).

The hypothesis of a strict tarsier–anthropoid clade (TAC) must also be evaluated in light of the structural features shared by omomyids and *Tarsius* that are not present in anthropoids. The basicrania of *Necrolemur, Shoshonius,* and *Tetonius* have been shown to share an impressive array of derived similarities with the skull of *Tarsius* (Table I; Wortman, 1903, 1904b; Stehlin, 1912; Simons and Russell, 1960; Simons, 1961, 1974; Rosenberger, 1985; Beard *et al.,* 1991). If the hypothesis of a strict tarsier–anthropoid clade is correct, then these features have either been lost secondarily in the anthropoid lineage after its split with the true tarsier lineage, or the similarities between omomyids and *Tarsius* are convergent.

An interesting consequence arising from the homoplasy required by a

Table I. Cranial Specializations of *Tarsius* Shared with *Necrolemur, Shoshonius,* and *Tetonius*[a]

1. Extensive lateral pterygoid plates overlap lateral bullar wall
2. Basioccipital flange overlaps medial bullar wall[b]
3. Gutter-like mandibular fossa and corresponding shape of condyle
4. Intrabullar ectotympanic attached to inside of lateral bullar wall[c]
5. Acoustic meatus including a tubular extension lateral to bulla[c]
6. Carotid enters bulla medially or anteriorly[d,e]
7. Maxillary palate V-shaped and converging anteriorly in ventral view
8. Presence of a distinct posterior palatine torus
9. Great inflation of auditory bullae
10. Pterygoid plates originate far medially
11. Peaked and narrow posterior nares
12. Presence of a suprameatal foramen[f]
13. Distinct groove at intraorbital junction of rostrum and braincase[d]
14. Sharply pointed upper and lower incisors[g]
15. Relatively large orbits

[a]List of traits compiled from Stehlin (1916), Simons and Russell (1960), Simons (1961), Rosenberger (1985), and Beard *et al.* (1991). The traits are listed here without consideration of functional integration or relative weights for phylogenetic studies.

[b]Reported only for *Tarsius* and *Shoshonius.*

[c]The ectotympanic element of *Necrolemur* and *Tarsius* probably comprises only the most medial, anular part of the tubular apparatus, which in tarsiiforms is intrabullar and appressed by varying degrees to the intrabullar wall; the "extrabullar" tube may be bulla itself (cf. Simons, 1961; Conroy, 1980). Extrabullar tubular extension is not reported in *Shoshonius.*

[d]Reported only for *Tarsius* and *Necrolemur.*

[e]The intrabullar septum of *Tarsius* stretches from the anteriorly placed carotid canal to the medial bullar wall, precisely in the position where the posterior carotid foramen occurs in *Necrolemur.* In my interpretation, this strongly suggests that the condition in *Tarsius* evolved from one like that seen in *Necrolemur* (Simons and Russell, 1960). The intrabullar arrangment of septa in *Shoshonius* is reported to be distinctly different.

[f]Condition not yet reported in *Tetonius.*

[g]Anterior teeth of *Tetonius* and *Shoshonius* not represented with certainty in the fossil record. Isolated incisors possibly of *Tetonius* (Szalay, 1976) are sharply pointed, as are those of *Dyseolemur,* a close relative of *Shoshonius* (Rasmussen *et al.,* 1994).

strict tarsier–anthropoid clade that often escapes notice in the literature is that it leaves wide open the question of whether the early Eocene ancestor of the TAC was an omomyid, an adapoid, or something else. The assumption has apparently been that the ancestor is an omomyid (Schmid, 1981; MacPhee and Cartmill, 1986), but notice that such a conclusion does not logically follow if the cranial resemblances between the omomyids on the one hand and *Tarsius* on the other arose convergently (as opposed to having them lost secondarily in anthropoids). If the shared specializations (Table I) are dismissed in order to tie *Tarsius* exclusively to anthropoids, why should these same characters then be judged acceptable for tying omomyids to the combined clade of *Tarsius* and Anthropoidea?

Of course, this problem is negated if the ancestor of the TAC was extremely tarsier-like, and in the lineage to anthropoids these tarsioid specializations were secondarily lost. However, supporters of the TAC have not favored this interpretation. Cartmill (1980) hypothesized that the common ancestor was "monkeylike," thereby implying that at least several of the shared similarities between omomyids and *Tarsius* were indeed attained convergently. Schmid (1981) concluded that the specializations of *Tarsius* were obtained after the split from the anthropoid stem line. In my opinion, this line of reasoning strains an already tenuous hypothesis. I suggest that supporters of the TAC should more explicitly address the problem of evolutionary reversal of basicranial, dental, and postcranial features in the lineage leading to Anthropoidea, and they should seriously question the idea that the TAC ancestor was very "monkey-like." It seems to me, given the known fossil record of omomyids, that the hypothesis of a tarsier–anthropoid clade is now very dependent on the premise that the TAC common ancestor was extremely tarsier-like, and that wholesale evolutionary reversal of tarsioid specializations at the base of the anthropoid stem is a key requirement of the theory.

The Omomyid–Anthropoid Hypothesis

The omomyid theory of anthropoid origins (Fig. 4) has a long history, and it has been one of the most prominent hypotheses of anthropoid origins in recent years, especially following Gazin (1958; see also Szalay, 1976; Szalay *et al.*, 1987; Rosenberger and Szalay, 1980; Rosenberger *et al.*, 1985; Beard *et al.*, 1988; Covert and Williams, 1992). Like the other three hypotheses, there is some osteological evidence in its favor, and on the other hand, it too requires an uncomfortably high degree of convergence or reversal. In my interpretation, the omomyid hypothesis has risen to a prominent position because of the incorrect assumption that a tarsier–anthropoid clade among extant primates constrains the ancestor to be a tarsiiform species, and, as a consequence of this assumption, researchers have sought an unspecialized tarsiiform that is suitable as an anthropoid ancestor. Such a hypothetical animal

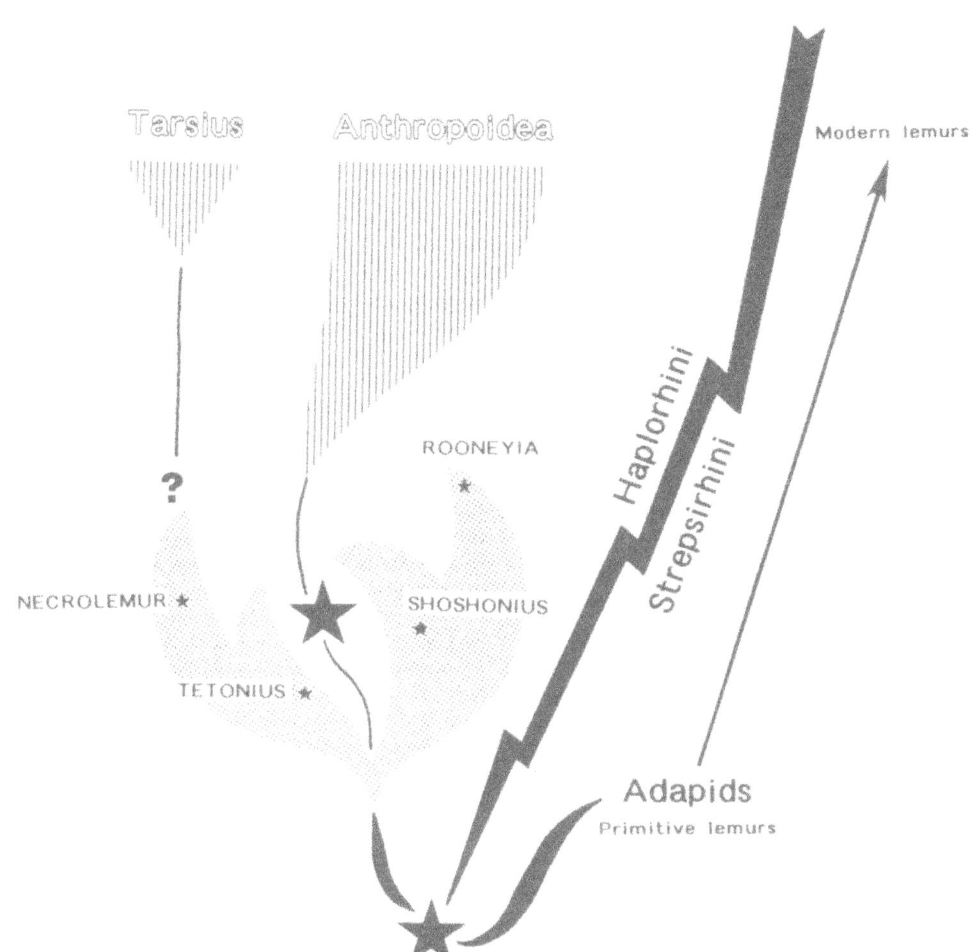

Fig. 4. An omomyid–anthropoid clade. This hypothesis suggests that both *Tarsius* and Anthropoidea emerged from Eocene tarsiiform primates belonging to the family Omomyidae. Several subfamilies have been considered as ancestors (Microchoerinae, containing *Necrolemur;* Anaptomorphinae containing *Tetonius;* Omomyinae containing *Shoshonius* and *Rooneyia*), but all well-known taxa are too specialized to serve as anthropoid ancestors, and so support for this hypothesis lies in the reconstruction of a hypothetical morphotype (top star). *Tarsius* is shown here coming from near the microchoerine *Necrolemur,* but other omomyid sources of origin are also possible. The nontarsiiform primate family Adapidae is excluded from consideration because it is believed to belong with extant lemuroid and lorisoid prosimians in Strepsirhini. The haplorhine–strepsirhine dichotomy is considered to be traceable back to the common ancestor of omomyids and adapoids (bottom star).

—a tarsiiform primate that is not *too* tarsiiform—can presumably be found only among omomyids, especially among the larger-bodied, dentally generalized North American omomyines.

The arguments for and against an omomyid ancestry for anthropoids have been debated in detail (Szalay, 1976; Szalay *et al.*, 1987; Archibald, 1977; Cartmill and Kay, 1978; Gingerich and Schoeninger, 1977; Gingerich, 1981a,b, 1984; Schmid, 1981, 1982; Kay, 1980; Rosenberger *et al.*, 1985; Rasmussen, 1986, 1990; Beard *et al.*, 1988; Covert and Williams, 1992). Tautological thinking resulting from neontological expectations has often entered the analysis: character states common to both omomyids and adapoids, such as "large promontory canal" or "short face," are alleged to be acceptable "synapomorphies" only of an omomyid–anthropoid clade because of the neontological evidence that is inaccurately interpreted as supporting only a tarsiiform origin of anthropoids.

In my opinion, the strongest osteological feature specifically linking omomyids and anthropoids to the exclusion of adapoids is the configuration of the tibiotalar joint (Gebo, 1986, 1988; Beard *et al.*, 1988). However, even for this character, polarity is uncertain, justification for a binary coding of this character is unavailable, quantitative studies of interspecific variation in this trait have not been undertaken, and the distribution of character states is unknown among the cercamoniine adapoids that form the basis of the modern adapoid–anthropoid hypothesis. Cranial traits that have been presented in support of an omomyid–anthropoid clade are entirely inadequate, as they are either shared by adapoids as well or are absent in early anthropoids (Rasmussen, 1990).

All omomyid taxa for which cranial anatomy is well known (*Tetonius, Shoshonius, Rooneyia, Necrolemur*) are, in my opinion, too specialized to have given rise to anthropoids (Le Gros Clark, 1959; Conroy, 1980; Beard *et al.*, 1991). The shared similarities among omomyids imply that their common ancestor was too specialized as well. Proponents of an omomyid ancestry of anthropoids have intimated that cutting and pasting the right parts together, for example, the incisor and canine roots of *Washakius* (Covert and Williams, 1992), the molars of *Chumashius* (Kay, 1980), the lower leg of *Hemiacodon* (Dagosto, 1985), and so forth, can produce a suitably primitive anthropoid ancestor in the end. This may be true to an extent, but it is impossible to "morphotype away" all the omomyid specializations, especially since it is the shared specializations, not the oddly distributed adapoid-like features, that should appear in a hypothetical morphotype.

The use of hypothetical morphotypes that only loosely resemble actual primate fossils or that rely on arbitrarily selected character states is usually a circular procedure (Szalay, 1976; Szalay *et al.*, 1987; Rosenberger *et al.*, 1985; Fleagle and Kay, 1987). The criteria used in the construction of a morphotype are intimately linked with preconceptions about the phylogenetic problem being investigated. Choosing, coding, and estimating the polarity of characters is already a dangerously tautological procedure; relying too heavily on morphotypes makes the problem even worse. For example, Rosenberger *et al.* (1985) reconstructed an upper incisor morphotype for omomyids that quite simply is not known to exist in omomyids, and they then used their morphotype in support of their arguments for an omomyid ancestry of Anthropoidea (Franzen, Chapter 4, this volume; Rasmussen et al., 1994).

The Pretarsioid Hypothesis

Over the years, many researchers have noted that all of the well-known omomyids are too specialized to give rise to anthropoids. This observation has received renewed attention because of the discovery of nearly complete crania of *Shoshonius* that show that *Shoshonius,* by the early Eocene, was already notably specialized in the direction of modern *Tarsius* (Beard *et al.,* 1991). The identical conclusion was first published, to my knowledge, by Wortman (1903, 1904b) with respect to the skull of *Tetonius,* and it has subsequently been repeated with respect to the skull of *Necrolemur* as well. The occurrence of distinctly tarsier-like skulls in each of the three major recognized divisions of the family Omomyidae (*Tetonius* in Anaptomorphinae, *Shoshonius* in Omomyinae, and *Necrolemur* in Microchoerinae) provides compelling evidence that the ancestral omomyid and possibly all omomyids are too specialized to have given rise to Anthropoidea. If one believes that adapoids are strepsirhines, and that the ancestral anthropoid must be "haplorhine," the solution to this dilemma is to postulate the existence of an as yet undiscovered, early, pretarsioid member of Tarsiiformes that gave rise to the ancestral anthropoid stock (Fig. 5). A convenient place to hide this ancestor is in the poorly known Paleocene and Eocene of Africa (Conroy, 1980; Sigé *et al.,* 1990), although Asia is also a candidate (Conroy and Bown, 1974).

The pretarsioid hypothesis has received much support through the years. For example, Le Gros Clark (1959, pp. 331–332) wrote after describing the anatomic specializations of omomyids:

> It is thus only possible to maintain the thesis of a tarsioid ancestry for the higher Primates if it is supposed that the latter took their origin (presumably in Palaeocene times) from a basal pro-tarsioid stock before the characteristic tarsioid specializations had become established. But, on the morphological criteria by which the infraorder Tarsiiformes is usually defined, it is doubtful whether such a basal stock could be properly termed tarsioid.

The strength of the pretarsioid hypothesis is that it takes into account the fact that none of the known omomyids can be reasonably viewed as an ancestral anthropoid. The weaknesses of the hypothesis are that it is not easily subjected to scientific refutation and that it hangs on to the hope of a "tarsiiform" ancestor in an illogical fashion. It essentially grows out of the belief that a "haplorhine" non-omomyid prosimian will be recognizable when it is finally found and that it will not be an adapoid.

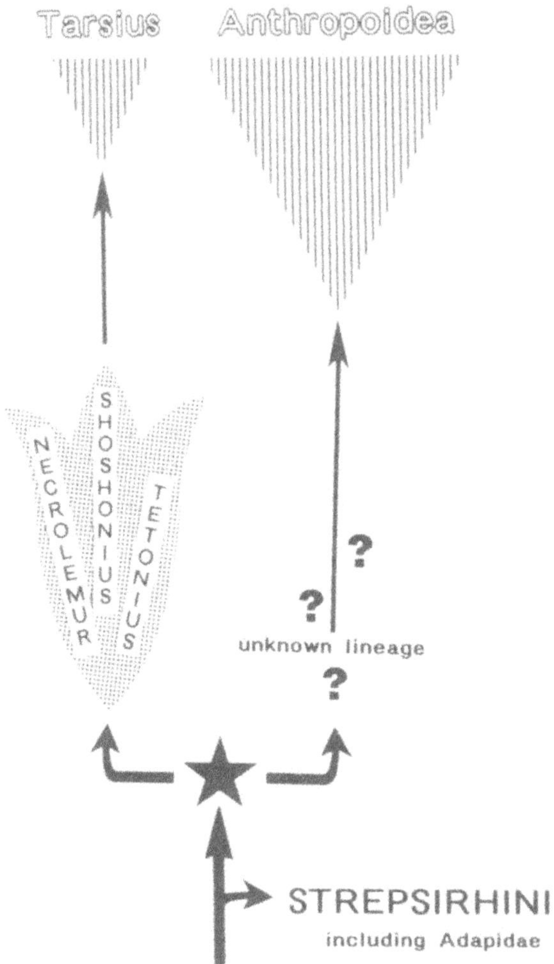

Fig. 5. A pretarsioid clade. This hypothesis suggests that the tarsiiform primates are the sister taxon of Anthropoidea, but they did not give rise directly to Anthropoidea. Rather, anthropoids shared a hypothetical "haplorhine" ancestor (star) that postdated the branching off of "strepsirhine" primates, including adapoids. The anthropoid stem lineage is hypothesized to be still unknown in the fossil record.

The Adapoid–Anthropoid Hypothesis

As most often construed today, the adapoid–anthropoid hypothesis is similar to the pretarsioid one but with a single important exception: the generalized anthropoid stock is not unknown; it is the adapoid subfamily Cercamoniinae* (Fig. 6).

Despite the protests of a few researchers opposed to an adapoid origin (e.g., Szalay *et al.*, 1987), the fact is that certain adapoids do bear many resem-

*In this chapter I break with my earlier use of the tribe Protoadapini (Szalay and Delson, 1979) and recognize the adapoid subfamily Cercamoniinae (Gingerich, 1975) as used by Franzen (Chapter 4, this volume). The generic content of old Protoadapini and new Cercamoniinae is about the same.

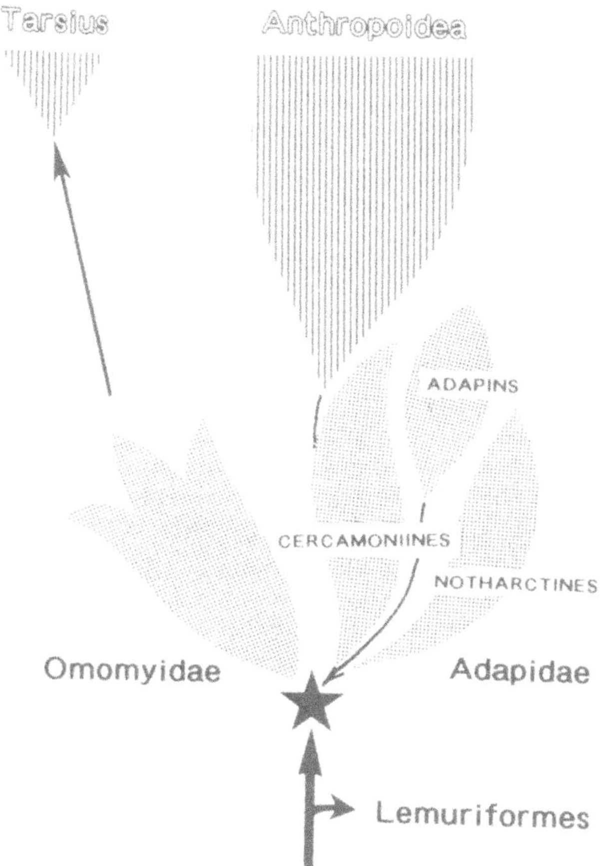

Fig. 6. An adapoid–anthropoid clade. This hypothesis suggests that tarsiiform primates form a distinct, specialized monophyletic clade (left) and that a group of adapoids, probably the subfamily Cercamoniinae, gave rise to anthropoids (right). This hypothesis suggests that omomyids and adapoids are the Eocene members of a clade that also includes *Tarsius* and Anthropoidea to the exclusion of Lemuriformes.

blances to primitive anthropoids. What this means in terms of phylogeny is, unfortunately, far from clear. An argument that the adapoid–anthropoid similarities are taxonomically significant has sometimes forcefully, if inadvertently, come from researchers who otherwise favor a "haplorhine" origin of anthropoids. For example, Szalay (1970) insisted that the Fayum anthropoid *Oligopithecus* was an adapid. Similarly, Le Gros Clark (1959), while espousing a pretarsioid ancestry for anthropoids because he believed that haplorhinism excluded adapoids from consideration, nevertheless concluded that the two omomyids that were the best potential candidates for anthropoid ancestry were *Periconodon* and especially *Caenopithecus,* now both believed to be cercamoniines (but see below).

The first clear formulation of the idea that adapoids belonged within a clade containing tarsiers, omomyids, and anthropoids, and that adapoids were closer than tarsiiforms to the basal stock of anthropoids, was by Wortman (1903, 1904a,b). In a series of papers, Wortman criticized Cope, Gregory, and others for their gradistic classifications of the primates and for their views that most Eocene taxa should be dumped together into a primitive "Mesodonta," which Wortman derisively referred to as a Condylarthra of the primates. Wortman's work—which, by the way, is one of many classic examples of precise "Hennigian" thinking long before Hennig came along—emphasized that natural groupings of taxa in which all members were descended from a common ancestor should be identified and utilized whenever possible in paleontological work. Primarily on the basis of basicranial and facial characters, Wortman recognized that tarsiers were closer to anthropoids than were any other living prosimians. In recognition of this, he classified tarsiers in "Anthropoidea" (equivalent in logic to some current authors' use of Haplorhini). Other Eocene taxa that shared specializations with tarsiers, such as *Tetonius* and *Necrolemur,* were placed with *Tarsius* in "Paleopithecini" (equivalent in logic but not in content, because of poor knowledge of many fossil taxa, to our current use of Tarsiiformes).

Eocene taxa that were more closely allied to catarrhines and platyrrhines than they were to lemuriforms, but that were not specialized toward tarsiers, were classified with catarrhines and platyrrhines in Neopithecini. The five families contained therein were Adapidae, Cebidae, Cercopithecidae, Simiidae, and Hominidae. Neopithecini contained Eocene adapoids known by fossil crania (*Adapis* and *Notharctus*) plus a few taxa we now recognize as omomyines, such as *Omomys* and *Washakius,* that Wortman hesitantly allocated here on the basis of dental structure.* Wortman (1903) went on to write of the Eocene members of Neopithecini:

> They occupy a position intermediate in many respects between the remaining Anthropoidea and the Paleopithecini. . . . It is in just such a group as that which includes *Adapis, Notharctus,* and *Limnotherium* [= *Notharctus*], that we must seek for the beginnings of the higher monkeys and apes which follow.

Fifteen years later, Pocock (1918) erected the taxon Haplorhini on the basis of his studies of the soft tissues of extant primates, and after that anything in the fossil record that was vaguely lemur-like was usually uncritically dumped into the taxon Strepsirrhini purely for phenetic reasons or on the basis of shared primitive characters. A gradistic view of primate evolution (lemur to tarsier to monkey to ape) became popular and persists today, even though it is now often cloaked in the guise of cladistic jargon. Interpretations

*Wortman is occasionally credited with founding the omomyid theory of anthropoid origins (e.g., Franzen, 1987). Wortman did indeed note dental similarities between *Omomys* and cebids, which were later played up by other authors (e.g., Gazin, 1958), but Wortman developed his concept of Neopithecini on the basis of adapid cranial morphology. He certainly would have transferred *Omomys* and *Washakius* to Paleopithecini given the fossil evidence now available to us, especially if he had been familiar with omomyine skulls such as those of *Rooneyia* and *Shoshonius.*

of early Tertiary forms have been unnecessarily and inaccurately biased against the possibility of nontarsiiform "haplorhines" ever since. The many researchers still intrigued by similarities between adapoids and anthropoids generally deferred to the superior neontological data, which, as is now clear, was erroneously applied and interpreted, thus giving rise to the adapid–omomyid "paradox" as outlined by Gingerich (1981a).

Nowhere has the misuse of the strepsirhine–haplorhine dichotomy been more damaging than in the study of early Tertiary primates. To drive this neontological wedge back into the depths of the Eocene defies logic and empiricism (while satisfying the thirst for a superficially tidy monophyly). The misuse of the dichotomy has been facilitated by the frequent division of Eocene euprimates into two families, Adapidae and Omomyidae, which provides an all too convenient crack for the wedge. What the new paleontological evidence demonstrates unambiguously is that the early representatives of Adapoidea and Omomyidae were very closely related to each other.

Euprimate species known near the base of the Eocene in North America, Europe, and Asia now include *Donrussellia provincialis, Donrussellia magna, Cantius torresi, Cantius eppsi, Teilhardina americana, Teilhardina belgica, Steinius vespertinus, Altanius orlovi,* and *Panobius afridi.* These defy simple separation as haplorhines and strepsirhines, or even as adapoids and omomyids. They represent taxa in or near the root of the common radiation of adapoids and omomyids. The dental differences between the species tagged as tarsiiform and those labeled strepsirhine are extremely slight (see especially Szalay, 1976; Godinot *et al.,* 1987; Rose and Bown, 1991; Gingerich *et al.,* 1991). All of the early Eocene species were probably more closely related to each other from a genetic point of view (as would be revealed, for example, if DNA sequencing could be applied) than any of them are to modern tarsiers, lemurs, or anthropoids. How is it that the strepsirhine–haplorhine dichotomy seduces us into believing that one subset of these nearly identical early Eocene primates is acceptable as an anthropoid stem lineage while another subset is not?

After Wortman, the first explicit mention I have found that adapoids might be "persistently primitive haplorhines" rather than strepsirhines was by Cartmill and Kay (1978). Subsequent researchers revived Wortman's logic in full detail (Gingerich and Schoeninger, 1977; Greenfield, 1983; Rasmussen, 1986); these researchers and others have identified adapoid features that they believe indicate a special relationship to Anthropoidea, although not all have espoused separating adapoids from extant strepsirhines (Gingerich, 1975, 1977; Gingerich and Schoeninger, 1977; Ciochon *et al.,* 1985; Franzen, 1987; Rasmussen and Simons, 1988; Simons, 1989, 1990; Rasmussen, 1990; Hartenberger and Marandat, 1992).

One of the strengths of the adapoid–anthropoid hypothesis is that specific fossil species—rather than vague, preselected "morphotypes"—have been identified as potential anthropoid relatives. The adapoid–anthropoid hypothesis is therefore relatively testable, at least to the extent that any phylogenetic hypothesis can truly be called "testable." The anthropoid-like adapoids come from a relatively narrow taxonomic grouping of early to middle Eocene species including *Cercamonius* (Gingerich, 1975), *Periconodon* (Gingerich, 1980), *Protoadapis* and *Hoanghonius* (Rasmussen & Simons, 1988), *Mahgarita* (Wilson & Szalay, 1976; Rasmussen, 1990), and *Djebelemur* (Hartenberger and Marandat, 1992). A corollary of the specificity of the adapoid–anthropoid hypothesis is that the well-known notharctines and late Eocene adapins—which form most primatologists' image of "adapids"—are only indirectly relevant to testing the hypothesis. The only recent case I know of where a particular real species was used to support the omomyid theory was when *Chumashius balchi* was identified as the possible sister taxon of anthropoids (Cartmill and Kay, 1978; Kay, 1980; see Rasmussen and Simons, 1988).

One apparent weakness of the adapoid–anthropoid hypothesis is that similarities shared by adapoids and anthropoids to the exclusion of omomyids have been labeled as primitive for euprimates and therefore dismissed as containing no useful phylogenetic information. There are several problems with this criticism of the adapoid–anthropoid hypothesis: (1) the taxa that do share similarities to anthropoids, such as *Mahgarita,* are significantly different from the early euprimates and therefore cannot be viewed as uniformly primitive; (2) primitiveness is often defined by circular "morphotype" reasoning; indeed, similarities shared by adapoids and anthropoids are designated primitive by some researchers precisely because the two groups are considered unrelated; (3) anything nontarsiiform in the Eocene is often uncritically considered primitive; (4) even if some adapoid character states are primitive for euprimates, but the corresponding character states in omomyids are too specialized, then the primitive condition is left as the most reasonable ancestral condition for anthropoids; (5) there are theoretical and mathematical reasons to believe that the cladistic value of primitive traits is not completely insignificant (Sober, 1988); and, (6) most ironically, the omomyid morphotypes that are reconstructed by proponents of an omomyid ancestry are probably at least as primitive as the actual adapoid species under consideration (Szalay *et al.*, 1987; Rosenberger *et al.*, 1985; see Rasmussen, 1990). Dismissing adapoids as potential anthropoid ancestors because they are "primitive" will be successful only if a better ancestral fit can be found among real members of the anaptomorphines, omomyines, or microchoerines.

A New Model of Anthropoid Origin: The Para-Tethyan Theory

Several recent paleontological discoveries have changed our impressions of early anthropoid primates from Africa: they have become older, smaller, structurally more primitive, and geographically more widespread (Simons, 1989, 1990; de Bonis *et al.*, 1988; Thomas *et al.*, 1988, 1989, 1991; Godinot and Mahboubi, 1992; Pickford, 1986a; Rasmussen *et al.*, 1992; Rasmussen and Simons, 1992; Godinot, Chapter 10, this volume). The Fayum primates from quarry L-41 are late Eocene as judged from paleomagnetic data, analysis of marine regressive events reflected in erosional unconformities, and faunal correlations with other Afro-Arabian sites (Rasmussen *et al.*, 1992; Kappelman *et al.*, 1992; Simons *et al.*, Chapter 8, this volume). The Eocene anthropoids are small, ranging in size from callitrichids to squirrel monkeys (Rasmussen and Simons, 1992). Structurally, they are extremely primitive, retaining numerous prosimian-like features in the teeth, jaws, and postcranium (Rasmussen and Simons, 1992; Gebo *et al.*, Chapter 9, this volume). If clues derived from these primitive anthropoids are reliable, they suggest that the anthropoid stem group lived during the middle Eocene and that its members were small prosimian-like animals.

Recent research has also led to a growing appreciation of a significant amount of mammalian interchange between the Afro-Arabian continent and the northern continents during the Paleocene, Eocene, and Oligocene. Shared taxa include the paleoryctids *Paleoryctes* and *Cimolestes* (Gheerbrant, 1990), the didelphid marsupial *Peratherium* (Bown and Simons, 1984), early pangolins (Gebo and Rasmussen, 1985), anaptomorphine omomyids (Simons *et al.*, 1987), hyaenodontine creodonts such as *Apterodon* and *Pterodon* (Savage, 1978; Tilden *et al.*, 1990), and anthracotheres (Black, 1978), and a little later in the Oligocene, even such classic African endemics as arsinoitheres, proboscideans, and giant hyracoids appear in Asia (Pickford, 1986b). It would be highly unlikely that such a broad-scale interchange took place without adapoids or anthropoids being involved, especially before the Grande Coupure, when climatic conditions were still tolerable for primates in the north. Primates are mobile mammals unusually adept at dispersing onto islands or isolated continental masses: a cercamoniine inexplicably appears in Texas, another in Tunisia, a tarsiid turns up in Egypt, platyrrhines get to South America, notharctines may well have reached Burma in the form of *Pondaungia,* and lemurs thrive out on Madagascar.

Biogeographers usually think of the basic units of their science as being continents, plates, and islands. The shores of the Tethyan region (and later the Mediterranean) may have always displayed a regional ecological unity even though different shores are biogeographically catalogued as belonging to as many as three continents: Africa, Europe, and Asia. The pivotal position of the Tethyan as a junction among continents also casts it in a key role for animal dispersal and gene flow. Arguments about the continent of origin of anthropoids may miss the target. Could it be that the complex ring of tropical lowland forest of the para-Tethyan region—whether northern Africa, Arabia, or southern Europe—was the center of diversification of early anthropoids?

Recently, Eocene primates of Eurasian heritage have been reported in Africa (Hartenberger and Marandat, 1992; Court, 1993*). Given the much better fossil record from Europe than Africa, it should be easier—in fact, it

*From the published description, illustrations, and comparisons, I concur wholeheartedly that the Chambi fossil looks like a sivaladapine primate rather than a hyracoid or hyopsodontid condylarth.

should be expected!—to find early anthropoids of "African extraction" in Eurasia. The pretarsioid hypothesis revived by Beard *et al.* (1991) holds that the ancestral anthropoid stem is unknown or is hopelessly isolated in Africa, where Eocene fossil deposits are so frustratingly rare. It is a much more likely proposition that anthropoids are already known in the Eurasian Eocene but have not yet been recognized on the basis of dental characters. It is possible that some of the small Eocene primates of Europe and Asia known only by their teeth and jaws are actually early anthropoids closely related to those from Egypt, Oman, and Algeria. The idea that early anthropoids could exist unrecognized in the known fossil record is not far fetched; after all, *Oligopithecus* was misidentified as an adapid (Szalay, 1970; Gingerich, 1977), and why could similar errors not be made on other continents for the more primitive ancestors of *Oligopithecus?*

If not for the different continent of origin, I do not believe that the Fayum's Eocene anthropoids could be reliably distinguished by dentitions from certain Eurasian primates, putatively adapoids. In particular, I suggest that fossils allocated to Lutetian and Bartonian species of Cercamoniinae should be reexamined very closely, including *Periconodon* and *Hoanghonius* (Gingerich, 1977; Szalay and Delson, 1979; Godinot, 1988; Franzen, Chapter 4, this volume). When considering just the known dental morphology and not the continent of origin, I see no reason to exclude *Periconodon* from Anthropoidea. Like the known Eocene anthropoids of Africa, *Periconodon* is a small-bodied, low-crowned, markedly bunodont species with remarkable development of upper molar lingual structures including hypocone and pericone. The major difference between the upper teeth of *Periconodon* and the Fayum's *Proteopithecus* is that the pericone of the latter is a slight, noncuspate fold of the lingual enamel, whereas that of the former is a distinct inflated cusp, as in *Apidium.*

In order to emphasize the point that known Eurasian taxa could be anthropoids, I will present a precise prediction that will be testable with future fossil finds: *Periconodon* is an early anthropoid, possibly even a primitive parapithecid. When the cranium is found, I predict it will have postorbital closure formed by a broadened zygomatic bone extending medially to the wall of the temporal fossa.

The new phylogenetic model of anthropoid origin can be summarized as follows. The neontological evidence cited in the first paragraph of this chapter suggests, if anything, a very ancient common ancestry of tarsiiforms and anthropoids. The fossil record of tarsiiform primates pushes the point of the split back to the late Paleocene. At that time, the tarsiiform lineage diverged from a primitive adapoid lineage, the latter representing the stem group of Anthropoidea (Wortman's Neopithecini). From this early, primitive adapoid resembling *Donrussellia magna* (Godinot, 1988), a radiation of cercamoniines produced several divergent lineages in Europe and Asia (Russell and Gingerich, 1987), including at least one that migrated to North America (*Mah-*

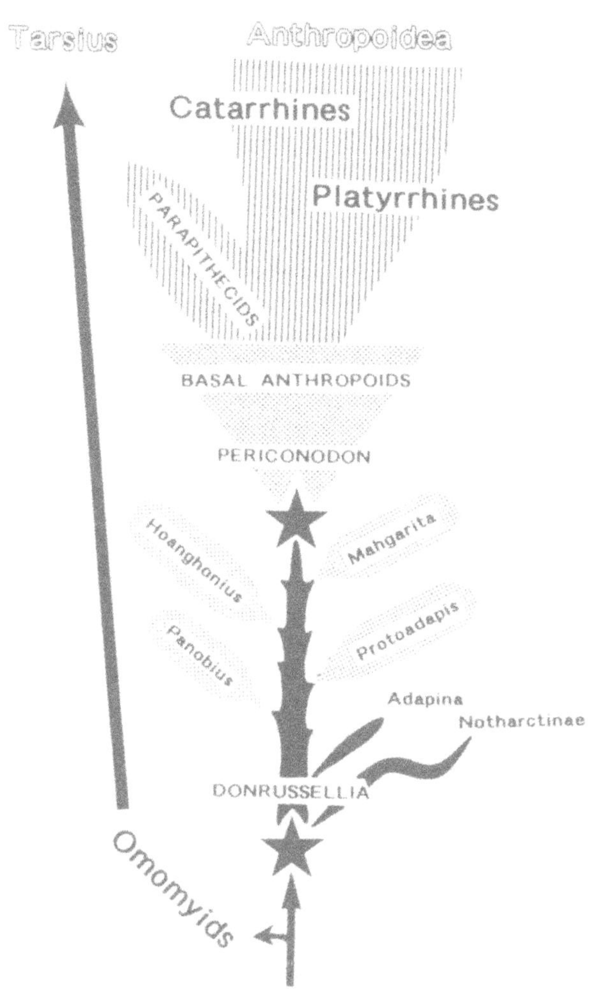

Fig. 7. A new model of anthropoid origin. The late Paleocene ancestor of the adapoid clade (bottom star) probably resembled *Donrussellia magna,* from the earliest Ypersian of France (Godinot *et al.,* 1987). North American northarctines and European adapines diverged from the branch leading to anthropoids during the late Paleocene or early Eocene (Gingerich, 1986; Godinot, 1988). The lineage leading to anthropoids is postulated to have diversified in Eurasia and possibly Africa, giving rise to a nested sequence of increasingly anthropoid-like taxa, including *Mahgarita,* which is known by complete cranial material. The common ancestor of Anthropoidea (top star), as arbitrarily defined by postorbital closure, was not dentally or cranially much different from contemporaneous cercamoniines; *Periconodon* is illustrated in a hypothetical position as a basal anthropoid (top star). There was a late Eocene radiation of basal anthropoids containing species such as *Catopithecus* that cannot be easily separated from cercamoniines without cranial evidence (stippled). Finally, in the early Oligocene, anthropoids of modern aspect appear (vertical hatching). The relative branching sequence of Adapina and Notharctinae should probably be reversed (Franzen, Chapter 4, this volume).

garita stevensi). One radiation centered in the para-Tethyan region began to accumulate several features that, in hindsight, we judge to be anthropoid characters. A radiation of small-bodied species belonging to this clade, resembling and possibly including *Periconodon,* contained the earliest true anthropoids with postorbital closure. These diverse cercamoniines and early anthropoids retained a para-Tethyan distribution as the end of the Eocene approached. At that time, climatic changes shrunk the acceptable primate habitat southward. Finally, the larger-bodied anthropoids of modern aspect appeared in the early Oligocene of the Fayum, south of the Tethys Sea.

ACKNOWLEDGMENTS

For careful review of earlier versions of the manuscript I thank Glenn Conroy, Ellen Miller, Sam Senturia, Myron Shekelle, and several anonymous reviewers. I also wish to acknowledge with gratitude the never-ending education provided to me by Ken Rose, Tom Bown, Dan Gebo, Marian Dagosto, Elwyn Simons, and Friderun Ankel-Simons, all of whom have served through the years as stimulating influences and critical checks on the development of ideas expressed in this paper. I also thank Prithijit Chatrath for his support and help in the lab and in the field.

References

Archibald, J. D. 1977. Ectotympanic bone and internal carotid circulation of eutherians in reference to anthropoid origins. *J. Hum. Evol.* **6:**609–622.

Baba, M. L., Weiss, M. L., Goodman, M., and Czelusniak, J. 1982. The case of the tarsier hemoglobin. *Syst. Zool.* **31:**156–165.

Barnicott, N. A., and Hewett-Emmett, D. 1974. Electrophoretic studies on prosimian blood proteins. In: R. D. Martin, G. A. Doyle, and A. C. Walker (eds.), *Prosimian Biology,* pp. 891–902. Duckworth, London.

Beard, K. C., Dagosto, M., Gebo, D. L., and Godinot, M. 1988. Interrelationships among primate higher taxa. *Nature* **331:**712–714.

Beard, K. C., Krishtalka, L., Stucky, R. K. 1991. First skulls of the early Eocene primate *Shoshonius cooperi* and the anthropoid–tarsier dichotomy. *Nature* **349:**64–67.

Black, C. C. 1978. Anthracotheriidae. In: V. J. Maglio and H. B. S. Cooke (eds.), *Evolution of African Mammals,* pp. 423–434. Harvard University Press, Cambridge.

Bonis, L. de, Jaeger, J.-J., Coiffait, B., and Coiffait, P.-E. 1988. Découverte du plus ancien primate catarrhinien connu dans l'Éocène supérieur d'Afrique du Nord. *C. R. Acad. Sci. Paris [II]* **306:**929–934.

Bonner, T. I., Heinemann, R., and Tadaro, G. J. 1980. Evolution of DNA sequences has been retarded in Malagasy primates. *Nature* **286:**420–423.

Bown, T. M., and Simons, E. L. 1984. First record of marsupials (Metatheria: Polyprotodonta) from the Oligocene in Africa. *Nature* **308:**447–449.

Cachel, S. M. 1979. A functional analysis of the primate masticatory system and the origin of the anthropoid post-orbital septum. *Am. J. Phys. Anthropol.* **50:**1–17.

Cartmill, M. 1980. Morphology, function, and evolution of the anthropoid postorbital septum. In: R. L. Ciochon and A. B. Chiarelli (eds.), *Evolutionary Biology of the New World Monkeys and Continental Drift,* pp. 243–274. Plenum Press, New York.

Cartmill, M., and Kay, R. F. 1978. Cranio-dental morphology, tarsier affinities, and primate suborders. In: D. G. Chivers and K. A. Joyssey (eds.), *Recent Advances in Primatology, Vol. 3, Evolution,* pp. 205–213. Academic Press, London.

Cartmill, M., MacPhee, R. D. E., and Simons, E. L. 1981. Anatomy of the temporal bone in early anthropoids, with remarks on the problem of anthropoid origins. *Am. J. Phys. Anthropol.* **56:**3–21.

Ciochon, R. L., Savage, D. E., Thaw Tint, and Ba Maw. 1985. Anthropoid origins in Asia? New discovery of *Amphipithecus* from the Eocene of Burma. *Science* **229:**756–759.

Conroy, G. C. 1980. Ontogeny, auditory structures, and primate evolution. *Am. J. Phys. Anthropol.* **52:**443–451.

Conroy, G. C., and Bown, T. M. 1974. Anthropoid origins and differentiation: The Asian question. *Yearkb. Phys. Anthropol.* **18:**1–6.

Court, N. 1993. An enigmatic new mammal from the Eocene of North Africa. *J. Vert. Paleontol.* **13:**267–269.

Covert, H. H., and Williams, B. A. 1992. The anterior lower dentition of *Washakius insignis* and adapid–anthropoidean affinities. *J. Hum. Evol.* 21:463–467.

Cronin, J. E., and Sarich, V. M. 1980. Tupaiid and archonta phylogeny: The macromolecular evidence. In: W. P. Luckett (ed.), *Comparative Biology and Evolutionary Relationships of Tree Shrews,* pp. 293–312. Plenum Press, New York.

Dagosto, M. 1985. The distal tibia of primates with special reference to the Omomyidae. *Int. J. Primatol.* **6:**45–75.

Donaghue, M. J., Doyle, J. A., Gauthier, J., Kluge, A. G., and Rowe, T. 1989. The importance of fossils in phylogeny reconstruction. *Annu. Rev. Ecol. Syst.* **20:**431–460.

Fleagle, J. G., and Kay, R. F. 1987. The phyletic position of the Parapithecidae. *J. Hum. Evol.* **16:**383–532.

Franzen, J. L. 1987. Ein neuer Primate aus dem Mitteleozän der Grube Messel (Deutschland, S-Hessen). *Cour. Forsch. Inst. Senckenberg* **91:**151–187.

Gardiner, B. G. 1982. Tetrapod classification. *Zool. J. Linn. Soc.* **74:**207–232.

Gauthier, J., Kluge, A. G., and Rowe, T. 1988. Amniote phylogeny and the importance of fossils. *Cladistics* **4:**105–209.

Gazin, C. L. 1958. A review of the middle and upper Eocene primates of North America. *Smithson. Misc. Coll.* **136:**1–112.

Gebo, D. L. 1986. Anthropoid origins—the foot evidence. *J. Hum. Evol.* **15:**421–430.

Gebo, D. L. 1988. Foot morphology and locomotor adaptation in Eocene primates. *Fol. Primatol.* **50:**3–41.

Gebo, D. L., and Rasmussen, D. T. 1985. The earliest fossil pangolin (Pholidota: Manidae) from Africa. *J. Mammal.* **66:**538–541.

Gheerbrant, E. 1990. On the early biogeographical history of the African placentals. *Hist. Biol.* **4:**107–116.

Gingerich, P. D. 1975. A new genus of Adapidae (Mammalia: Primates) from the late Eocene of southern France, and its significance for the origin of higher primates. *Contrib. Mus. Paleontol. Univ. Michigan* **24:**163–170.

Gingerich, P. D. 1977. Radiation of Eocene Adapidae in Europe. *Geobios. Mem. Spec.* **1:**165–182.

Gingerich, P. D. 1980. Dental and cranial adaptations in Eocene Adapidae. *Z. Morphol. Anthropol.* **71:**135–142.

Gingerich, P. D. 1981a. Early Cenozoic Omomyidae and the evolutionary history of the tarsiiform primates. *J. Hum. Evol.* **10:**345–374.

Gingerich, P. D. 1981b. Eocene Adapidae, paleobiogeography, and the origin of South American Platyrrhini. In: R. L. Ciochon and A. B. Chiarelli (eds.), *Evolutionary Biology of the New World Monkeys and Continental Drift,* pp. 123–138. Plenum Press, New York.

Gingerich, P. D. 1984. Primate evolution: Evidence from the fossil record, comparative morphology, and molecular biology. *Yearb. Phys. Anthropol.* **27:**57–72.

Gingerich, P. D. 1986. Early Eocene *Cantius torresi*—oldest primate of modern aspect from North America. *Nature* **320:**319–321.

Gingerich, P. D., and Schoeninger, M. 1977. The fossil record and primate phylogeny. *J. Hum. Evol.* **6:**483–505.

Gingerich, P. D., Dashzeveg, D., and Russell, D. E. 1991. Dentition and systematic relationships of *Altanius orlovi* (Mammalia, Primates) from the early Eocene of Mongolia. *Geobios* **24**(5): 637–646.

Godinot, M. 1988. Les primates adapidés de Bouxwiller (Eocène Moyen, Alsace) et leur apport à la compréhension de la faune de Messel et à l'évolution des Anchomomyini. *Cour. Forsch. Inst. Senckenberg* **107:**383–407.

Godinot, M., and Mahboubi, M. 1992. Earliest known simian primate found in Algeria. *Nature* **357:**324–326.

Godinot, M., Crochet, J.-Y., Hartenberger, J.-L., Lange-Badré, B., Russell, D. E., and Sigé, B. 1987. Nouvelles dominées sur les mammifères de Palette (Eocène inférieur, Provence). *Munch. Geowiss. Abh. [A]* **10:**273–288.

Greenfield, L. O. 1983. Recent advances and suggestions for expansion of the field of human origins. *Pontif. Akad. Sci. Scr. Var.* **50:**29–40.

Hartenberger, J.-L., and Marandat, B. 1992. A new genus and species of an early Eocene primate from North Africa. *Hum. Evol.* **7:**9–16.

Hedges, S. B., Moberg, K. D., and Maxson, L. R. 1990. Tetrapod phylogeny inferred from 18S and 28S ribosomal RNA sequences and a review of the evidence for amniote relationships. *Mol. Biol. Evol.* **7:**606–633.

Hershkovitz, P. 1977. *Living New World Monkeys (Platyrrhini), Vol. 1, With an Introduction to the Primates.* University of Chicago Press, Chicago.

Hubrecht, A. A. W. 1896. Die Keimblase von *Tarsius,* ein Hülfsmittel zur schärferen Definition gewisser Säugetierordnungen. In: *Festschrift für C. Gegenbauer II,* pp. 147–178. Engelmann, Leipzig.

Hubrecht, A. A. W. 1902. Furchung und Keimblattbildung bei *Tarsius spectrum. Verh. Kon. Akad. Wetensch. Amsterdam* **8**(6)**:**1–114.

Hubrecht, A. A. W. 1908. Early ontogenetic phenomena in mammals and their bearing on our interpretation of the phylogeny of the vertebrates. *Q. Rev. Microsc. Sci.* **53:**1–181.

Hubrecht, A. A. W. 1909. *Die Säugetierontogenese in ihrer Bedeutung für die Phylogenie der Wirbeltiere.* Gustav Fischer, Jena.

Kappelman, J., Simons, E. L., and Swisher, C. C. III. 1992. New age determinations for the Eocene–Oligocene boundary sediments in the Fayum depression, northern Egypt. *J. Geol.* **100:**647–667.

Kay, R. F. 1980. Platyrrhine origins: A reappraisal of the dental evidence. In: R. L. Ciochon, and A. B. Chiarelli (eds.), *Evolutionary Biology of the New World Monkeys and Continental Drift,* pp. 159–188. Plenum Press, New York.

Le Gros Clark, W. E. 1959. *The Antecedants of Man: An Introduction to the Evolution of the Primates.* Edinburgh University Press, Edinburgh.

Løvtrup, S. 1985. On the classification of the taxon Tetrapoda. *Syst. Zool.* **34:**463–470.

Luckett, W. P. 1974. Comparative development and evolution of the placenta in primates. *Contrib. Primatol.* **3:**142–234.

MacPhee, R. D. E., and Cartmill, M. 1986. Basicranial structures and primate systematics. In: D. R. Swindler and J. Erwin (eds.), *Comparative Primate Biology, Vol. 1, Systematics, Evolution, and Anatomy,* pp. 219–276. Alan R. Liss, New York.

Mai, L. L. 1985. Chromosomes and the taxonomic status of the genus *Tarsius:* Preliminary results. *J. Hum. Evol.* **14:**229–240.

Marshall, C. R. 1992. Substitution bias, weighted parsimony, and amniote phylogeny as inferred from 18S rRNA sequences. *Mol. Biol. Evol.* **9:**370–373.

Marshall, C. R., and Schultze, H.-P. 1992. Relative importance of molecular, neontological and paleontological data in understanding the biology of the vertebrate invasion of land. *J. Mol. Evol.* **35:**93–101.

Noback, C. R. 1975. The visual system of primates in phylogenetic studies. In: W. P. Luckett and F. S. Szalay (eds.), *Phylogeny of the Primates,* pp. 199–218. Plenum Press, New York.

Ostrom, J. H. 1975. On the origin of *Archaeopteryx* and the ancestry of birds. *Coll. Int. C.N.R.S.* **218:**519–532.

Ostrom, J. H. 1990. Dromaeosauridae. In: D. B. Weishampel, P. Dodson, and H. Osmólska, (eds.), *The Dinosaurs,* pp. 269–279. University of California Press, Berkeley.

Pickford, M. 1986a. Premiere découverte d'une faune mammalienne terrestre paleogène d'Afrique sub-saharienne. *C. R. Acad. Sci. Paris [II]* **302:**1205–1210.

Pickford, M. 1986b. Premiere découverte d'une hyracoide paleogène en Eurasie. *C.R. Acad. Sci. Paris [II]* **303:**1251–1254.

Pocock, R. I. 1918. On the external characters of the lemurs and of *Tarsius. Proc. Zool. Soc. Lond.* **1918:**19–53.

Pollock, J. I., and Mullin, R. J. 1987. Vitamin C biosynthesis in prosimians: Evidence for the anthropoid affinities of *Tarsius*. *Am. J. Phys. Anthropol.* **73:**65–70.

Poorman, P. A., Cartmill, M., MacPhee, R.D.E., and Moses, M. J. 1985. The G-banded karyotype of *Tarsius bancanus* and its implications for primate phylogeny. *Am. J. Phys. Anthropol.* **66:**215.

Rasmussen, D. T. 1986. Anthropoid origins: A possible solution to the Adapidae–Omomyidae paradox. *J. Human. Evol.* **15:**1–12.

Rasmussen, D. T. 1990. The phylogenetic position of *Mahgarita stevensi:* Protoanthropoid or lemuroid? *Int. J. Primatol.* **11:**439–469.

Rasmussen, D. T., and Simons, E. L. 1988. New specimens of *Oligopithecus savagei*, early Oligocene primate from the Fayum, Egypt. *Fol. Primatol.* **51:**182–208.

Rasmussen, D. T., and Simons, E. L. 1992. Paleobiology of the oligopithecines, the world's earliest known anthropoid primates. *Int. J. Primatol.* **13:**477–508.

Rasmussen, D. T., Bown, T. M., and Simons, E. L. 1992. The Eocene–Oligocene transition in continental Africa. In: D. R. Prothero and W. A. Berggren (eds.), *Eocene–Oligocene Climatic and Biotic Evolution*, pp. 548–566. Princeton University Press, Princeton.

Rasmussen, D. T., Shekelle, M., Walsh, S. L., and Riney, B. O. 1994. The dentition of *Dyseolemur* and comments on the use of the anterior teeth in primate systematics. *J. Hum. Evol.* (in press).

Rose, K. D., and Bown, T. M. 1991. Additional fossil evidence on the differentiation of the earliest euprimates. *Proc. Natl. Acad. Sci. USA* **88:**98–101.

Rosenberger, A. L. 1985. In favor of the necrolemur–tarsier hypothesis. *Fol. Primatol.* **45:**179–194.

Rosenberger, A. L., and Szalay, F. S. 1980. On the tarsiiform origins of Anthropoidea. In: R. L. Ciochon, and A. B. Chiarelli (eds.), *Evolutionary Biology of the New World Monkeys and Continental Drift*, pp. 139–157. Plenum Press, New York.

Rosenberger, A. L., Strasser, E., and Delson, E. 1985. Anterior dentition of *Notharctus* and the adapid–anthropoid hypothesis. *Fol. Primatol.* **44:**15–39.

Russell, D. E., and Gingerich, P. D. 1987. Nouveaux primates de l'Éocène du Pakistan. *C.R. Acad. Sci. Paris [II]* **304:**209–214.

Sarich, V. M., and Cronin, J. E. 1976. Molecular systematics of the primates. In: M. Goodman and R. E. Tashian (eds.), *Molecular Anthropology*, pp. 141–170. Plenum Press, New York.

Savage, R. J. G. 1978. Carnivora. In: V. J. Maglio and H. B. S. Cooke (eds.), *Evolution of African Mammals*, pp. 249–267. Harvard University Press, Cambridge.

Schmid, P. 1981. Comparison of Eocene nonadapids and *Tarsius*. In: A. B. Chiarelli and R. S. Corruccini (eds.), *Primate Evolutionary Biology*, pp. 6–13. Springer-Verlag, Berlin.

Schmid, P. 1982. *Die systematische Revision der europäischen Microchoeridae Lydekker, 1887 (Omomyiformes, Primates).* Juris Druck + Verlag, Zurich.

Sigé, B, Jaeger, J.-J., Sudre, J., and Vianey-Liaud, M. 1990. *Altiatlasius koulchii* n.gen.et sp., primate omomyidé du Paléocène supérieur du Maroc, et les origines des Euprimates. *Palaeontographica [A]* **214:**31–56.

Simons, E. L. 1961. Notes on Eocene tarsioids and a revision of some Necrolemurinae. *Bull. Br. Mus. Nat. Hist. Geol.* **5:**45–69.

Simons, E. L. 1974. Notes on early Tertiary prosimians. In: R. D. Martin, G. A. Doyle, and A. C. Walker (eds.), *Prosimian Biology*, pp. 415–433. Duckworth, London.

Simons, E. L. 1989. Description of two new genera and species of late Eocene Anthropoidea from Egypt. *Proc. Natl. Acad. Sci. USA* **86:**9956–9960.

Simons, E. L. 1990. Discovery of the oldest known anthropoidean skull from the Paleogene of Egypt. *Science* **247:**1507–1509.

Simons, E. L., and Rasmussen, D. T. 1989. Cranial morphology of *Aegyptopithecus* and *Tarsius* and the question of the tarsier–anthropoidean clade. *Am. J. Phys. Anthropol.* **79:**1–23.

Simons, E. L., and Russell, D. E. 1960. Notes on the cranial anatomy of *Necrolemur*. *Breviora. Mus. Comp. Zoology* **127:**1–14.

Simons, E. L., Bown, T. M., and Rasmussen, D. T. 1987. Discovery of two additional prosimian primate families (Omomyidae, Lorisidae) in the African Oliogocene. *J. Human Evol.* **15:**431–437.

Sober, E. 1988. *Reconstructing the Past: Parsimony, Evolution, and Inference.* MIT Press, Cambridge.

Spatz, W. B. 1968. Die Bedeutung der Augen für die sagittale Gestaltung des Schädels von *Tarsius* (Prosimiae, Tarsiiformes). *Fol. Primatol.* **9:**22–40.

Starck, D. 1956. Primitiventwicklung und Plazentation der Primaten. In: H. Hofer, A. H. Schultz, and D. Starck (eds.), *Primatologia, Vol. 1,* pp. 723–886. Basel, Karger.

Stehlin, H. G. 1912. Die Säugetiere des schweizerischen Eozäns. Critischer Catalog der Materialen, part 7, first half. *Abh. Schweiz. Pal. Ges.* **38:**1165–1298.

Stehlin, H. G. 1916. Die Säugetiere des schweizerischen Eozäns. Critischer Catalog der Materialen, part 7, second half. *Abh. Schweiz. Pal. Ges.* **41:**1299–1552.

Szalay, F. S. 1970. Late Eocene *Amphipithecus* and the origins of catarrhine primates. *Nature* **227:**355–357.

Szalay, F. S. 1976. Systematics of the Omomyidae (Tarsiiformes, Primates): Taxonomy, phylogeny and adaptations. *Bull. Am. Mus. Nat. Hist.* **156:**157–450.

Szalay, F. S., and Delson, E. 1979. *Evolutionary History of the Primates.* Academic Press, New York.

Szalay, F. S., Rosenberger, A. L., and Dagosto, M. 1987. Diagnosis and differentiation of the Order Primates. *Yearb. Phys. Anthropol.* **30:**75–105.

Tattersall, I. 1973. Cranial anatomy of the Archaeolemurinae (Lemuroidea, Primates). *Anthropol. Pap. Am. Mus. Nat. Hist.* **52:**1–110.

Thomas, H., Roger, J., Sen, S., and Al-Sulaimani, Z. 1988. Découverte des plus anciens "Anthropoïdes" du continent arabo-africain et d'un primate tarsiiforme dans l'Oligocène du Sultanat d'Oman. *C.R. Acad. Sci. Paris [II]* **306:**823–829.

Thomas, H., Roger, J., Sen, S., Bourdillon-de-Grissac, C., and Al-Sulaimani, Z. 1989. Découverte de vertébrés fossiles dans l'Oligocène inférieur du Dhofar (Sultanat d'Oman). *Geobios* **22:**101–120.

Thomas, H., Sen, S., Roger, J., and Al-Sulaimani, Z. 1991. The discovery of *Moeripithecus markgrafi* Schlosser (Propliopithecidae, Anthrpooidea, Primates), in the Ashawq Formation (early Oligocene of Dhofar Province, Sultanate of Oman). *J. Hum. Evol.* **20:**33–49.

Tilden, C. D., Holroyd, P. A., and Simons, E. L. 1990. Phyletic affinities of *Apterodon* (Hyaenodontidae, Creodonta). *J. Vert. Paleontol.* **10**(3)[Suppl]:46A.

Wilson, J. A., and Szalay, F. S. 1976. New adapid primate of European affinities from Texas. *Fol. Primatol.* **25:**294–312.

Wood Jones, F. 1929. *Man's Place among the Mammals.* Edward Arnold, London.

Wortman, J. L. 1903. Studies of Eocene Mammalia in the Marsh collection, Peabody Museum, Part II, Primates, Art. XLI, Classification of the primates. *Am. J. Sci. [Ser 4]* **15:**399–414.

Wortman, J. L. 1904a. Studies of Eocene Mammalia in the Marsh collection, Peabody Museum, Part II, Primates, Art. II, Suborder Anthropoidea. *Am. J. Sci. [Ser. 4]* **17:**23–33.

Wortman, J. L. 1904b. Studies of Eocene Mammalia in the Marsh collection, Peabody Museum, Part II, Primates, Art. XVIII, On the affinities of Omomyinae. *Am. J. Sci. [Ser. 4]* **17:**203–214.

Dental Evidence for Anthropoid Origins

13

RICHARD F. KAY and
BLYTHE A. WILLIAMS

Introduction

Over the past decade, many new finds of African Eocene and Oligocene monkeys and Holarctic Eocene primates have rekindled long-standing debates concerning the origins and early diversification of the Anthropoidea. These debates have centered around several related questions:

1. Is Anthropoidea a monophyletic group, and, if so, what did the last common ancestor of the anthropoids look like?
2. From which known group of Holarctic primates (if any) did anthropoids evolve?
3. How do known Eocene and Oligocene anthropoids of Africa [Parapithecidae, Propliopithecidae (*Propliopithecus* and *Aegyptopithecus*), Oligopithecidae] relate to the Miocene–Recent Platyrrhini and Catarrhini of the New and Old World, respectively?

Many alternative answers to these questions have been proposed. With respect to anthropoid monophyly, older views such as those of W. K. Gregory (1922) that catarrhines and platyrrhines evolved from separate and not very

RICHARD F. KAY • Department of Biological Anthropology and Anatomy, Duke University, Durham, North Carolina 27710. BLYTHE A. WILLIAMS • Department of Anthropology, University of Colorado, Boulder, Colorado 80309. *Present address:* Department of Biological Anthropology and Anatomy, Duke University, Durham, North Carolina 27710.
Anthropoid Origins, edited by John G. Fleagle and Richard F. Kay. Plenum Press, New York, 1994.

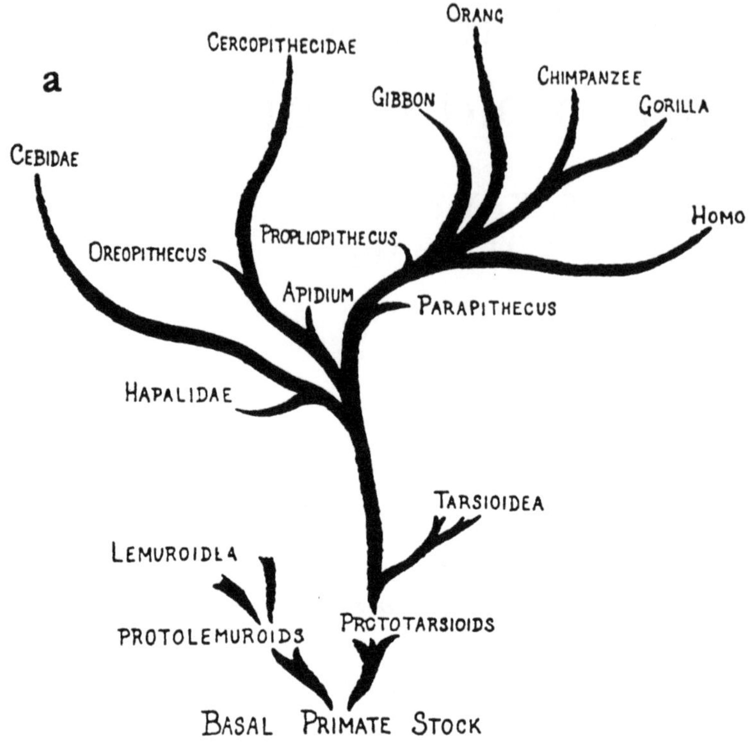
a
Cercopithecidae
Orang
Chimpanzee
Gibbon
Gorilla
Cebidae
Homo
Oreopithecus
Propliopithecus
Apidium
Parapithecus
Hapalidae
Tarsioidea
Lemuroidea
Protolemuroids
Prototarsioids
Basal Primate Stock

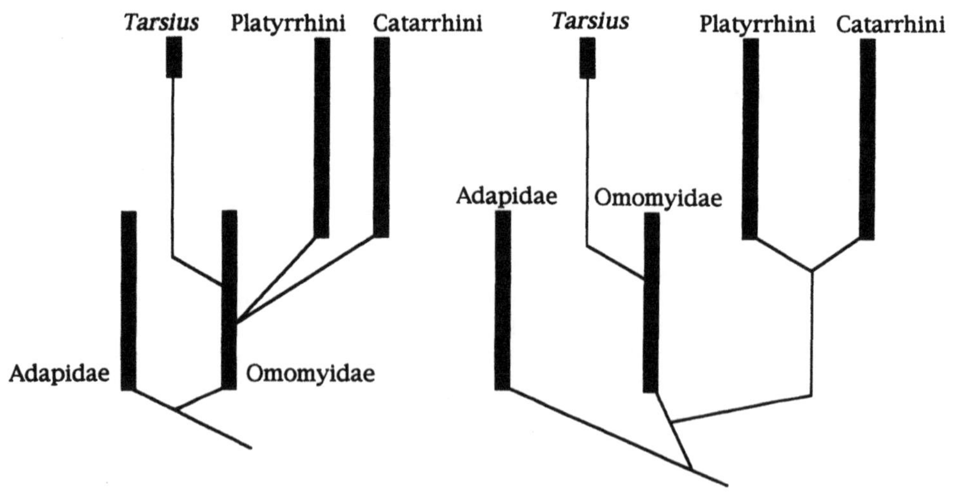
b
c
Tarsius
Platyrrhini
Catarrhini
Adapidae
Omomyidae
Tarsius
Platyrrhini
Catarrhini
Adapidae
Omomyidae

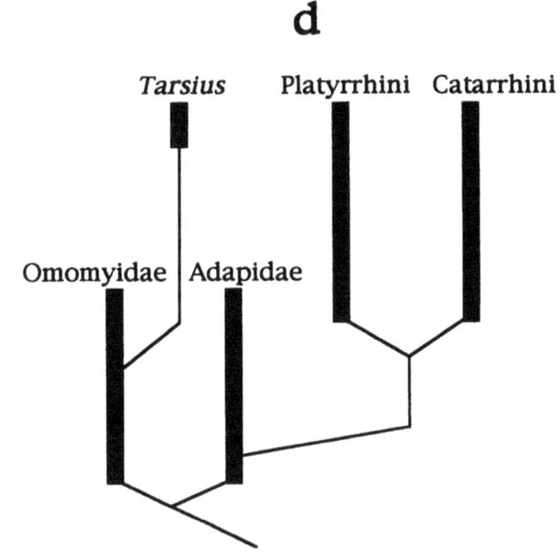

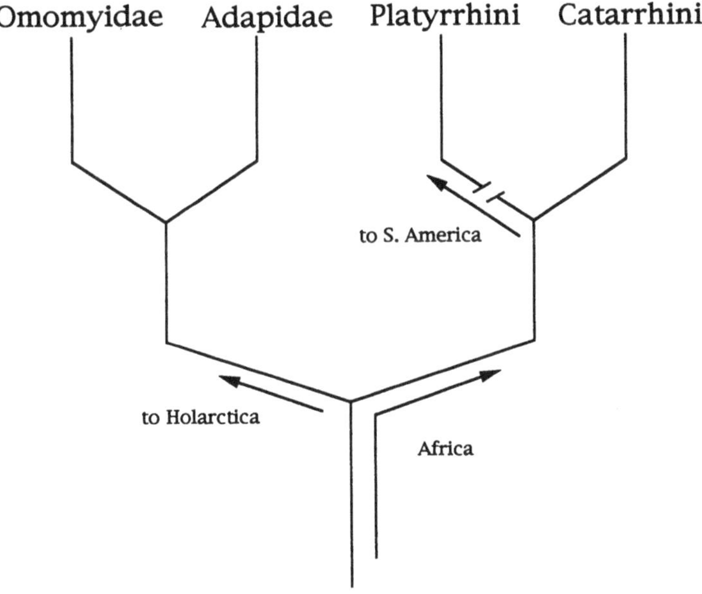

Fig. 1. Contrasting views of anthropoid ancestry. (a) A proposed omomyid ancestor for Anthropoidea, from Le Gros Clark (1934). "Prototarsioid" is taken to mean omomyid. (b) Anthropoid phylogeny, redrawn from Szalay (1976, Fig. 176, Page 420). (c) anthropoid phylogeny redrawn from Hoffstetter (1980a, Figs 1,2). (d) An adapid stem group for Anthropoidea, redrawn from Gingerich (1976). (e) A possible alternative view of anthropoid origins in which Anthropoidea is the sister group of *both* Tarsiiformes and Strepsirhini.

closely related Eocene "tarsioids" have now largely been abandoned. However, there remains a debate as to whether some or all monkey-like traits evolved independently in the two hemispheres or just once in a stem anthropoid (see Rosenberger, 1986, for a thorough review). More recent discussions of anthropoid origins recognize shared–derived features (postorbital closure, peculiarities of the ear region, and a fused mandibular symphysis) that support anthropoid monophyly (Delson and Rosenberger, 1980; Hoffstetter, 1980a,b; Rosenberger, 1986; Szalay, 1976; Szalay and Delson, 1979).

The second question, "Which Holarctic group is the sister group to anthropoids?" has proponents advocating one of two positions, each with deep historical roots (Fig. 1). Some researchers have contended that anthropoids evolved from the Omomyidae (e.g., Hoffstetter, 1980a,b; Szalay, 1976; Szalay and Delson, 1979) and have even suggested that omomyids must be paraphyletic with respect to both *Tarsius* and Anthropoidea (i.e., the ancestry of these modern haplorhines is to be found in different omomyid species). Others (e.g., Franzen, 1987; Gingerich, 1980; Rasmussen, 1990) have argued for an ancestry or sister-group relationship between some or all Adapidae and Anthropoidea. These two points of view are foreshadowed in the works of Le Gros Clark (1934), Patterson (1954), and others. With the recent recovery of older Eocene fossil anthropoids from Africa (Godinot, Chapter 10, this volume; Godinot and Mahboubi, 1992; Simons, 1992), it may now be just as plausible to propose that both Omomyidae and Adapidae are the sister group of Anthropoidea, that is, that a fundamental dichotomy exists between omomyids and adapids of the northern continents and Anthropoidea of Africa and South America.

The third area of disagreement and debate concerns cladogenesis of the Anthropoidea and centers on questions regarding the phyletic position of the Oligopithecidae and Parapithecidae in relation to platyrrhines and various groups of catarrhines. Oligopithecids (Fig. 2a–d) (*Oligopithecus* and *Catopithecus*, of the Fayum late Eocene/early Oligocene) are assigned by Rasmussen and Simons (1988; Simons, 1989, 1990) to Propliopithecidae and recognized by them as in or near the ancestry of Hominoidea. At the opposite extreme, Gingerich (1977, 1980) argued that *Oligopithecus* is very similar dentally to some Adapidae and may even be an adapid. Others have argued that *Oligopithecus* is a stem catarrhine (Kay, 1977; Szalay and Delson, 1979) or stem anthropoid (Hoffstetter, 1980a,b).

The phyletic position of the Parapithecidae similarly has been debated (Fig. 3). Simons (1974) regarded them as near the ancestry of Old World monkeys (Cercopithecidae), whereas Fleagle and Kay (1987) followed Delson (1975), Hoffstetter (1977, 1980a,b), and Harrison (1987) in placing parapithecids as the sister group of all other living and fossil anthropoids.

As a contribution to these continuing phylogenetic debates we present here a cladistic assessment of the dental evidence for early primate cladogenesis and early anthropoid diversification. We recognize that dental evidence taken alone cannot be expected to supply a definitive answer. Other

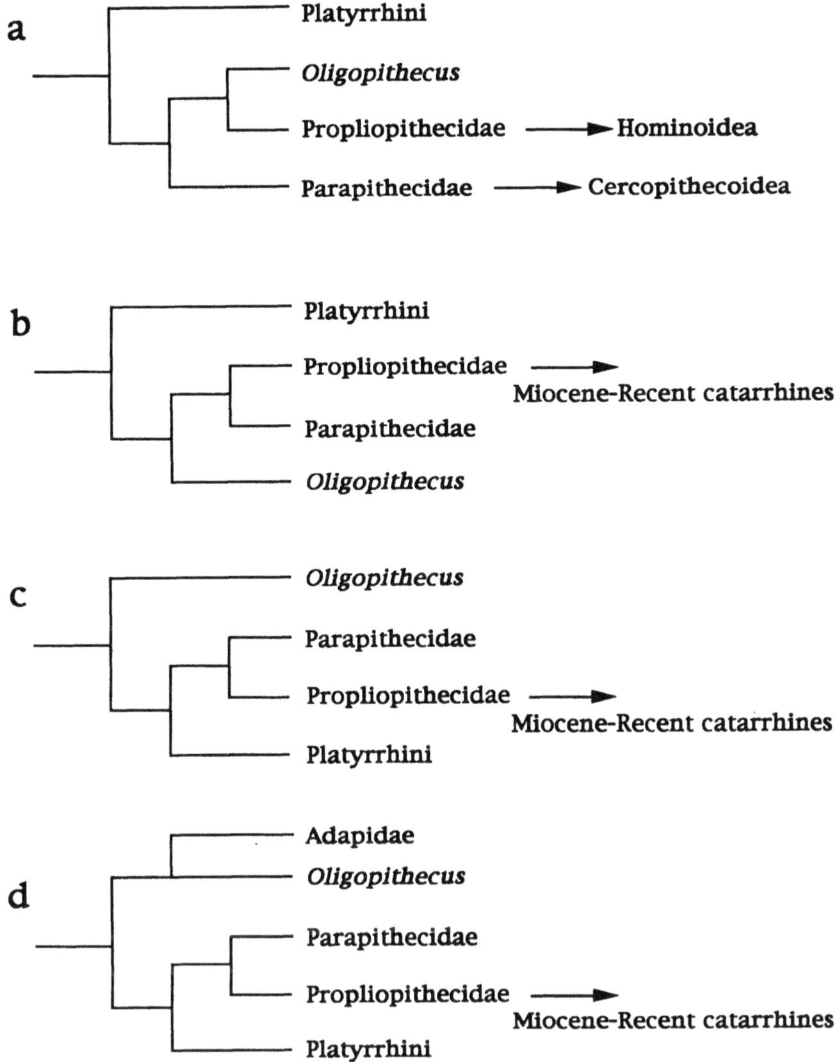

Fig. 2. Hypotheses about the phyletic position of Oligopithecidae: (a) as the sister group to Hominoidea; (b) as a sister group to Catarrhini; (c) as a sister group to all Anthropoidea; (d) as an adapid. See text for references.

anatomic data from the cranium and postcranium are available for study (see Covert and Williams, Chapter 2, Dagosto and Gebo, Chapter 17, Ford, Chapter 18, Gebo *et al.*, Chapter 9, Ross, Chapter 15, this volume) and may present contrasting phylogenetic evidence to that of the dentition. We add, however, that most of the fossil evidence about primate cladogenesis comes from the dentition and that this will continue to be the case for the foreseeable future.

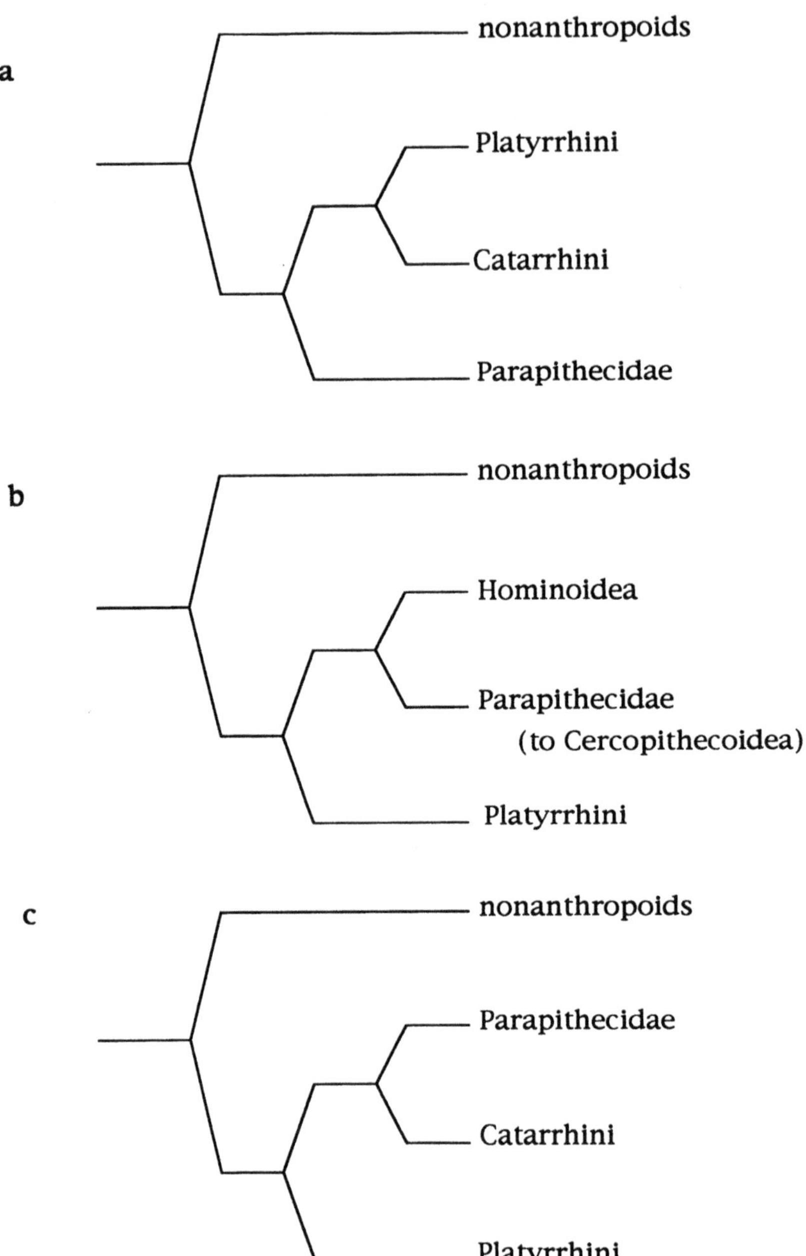

Fig. 3. Hypotheses about the phyletic position of Parapithecidae: (a) as sister group of both platyrrhines and catarrhines; (b) as sister group of Cercopithecidea; (c) as sister group of catarrhines. See text for references.

Materials and Methods

For this study, 157 dental characters were analyzed, of which 30 were characters of the incisors, 9 of the canines, 48 of the premolars, and 70 of the molars. A list of these characters and their states is given in the Appendix. The character–taxon matrix is presented in Table I. Originally formulated for the study of platyrrhine and parapithecid phylogeny (Fleagle and Kay, 1987; Kay, 1990), this character list was modified and augmented to add many characters and states that document the additional range of variation encompassed within the Eocene primate groups Adapidae and Omomyidae. Two-, three- and four-state characters were used to document the range of morphology represented within the taxa studied. Many subtle differences were not coded here. For instance, we coded P_4 metaconid size as a three-state character. This leaves most of the Omomyidae and Adapidae with the same scoring, namely, "P_4 metaconid small." We realize that this description leaves unreported a substantial amount of variation in metaconid size that may contain phylogenetically significant data for the study of the cladogenesis of those Eocene groups. Work in progress by Williams addresses this deficiency in our study.

Twenty-eight fossil and extant taxa were examined in this analysis: 6 extant platyrrhines, 9 Eocene/Oligocene anthropoids, *Tarsius* and 12 Holarctic Eocene primates (6 omomyids and 6 adapids). A partial list of specimens examined is given in Table II. We also consulted the extensive literature for information about the character states of taxa not examined personally by us. In addition, a number of less-well-known taxa were examined and will be discussed where pertinent (Table I). To avoid confusion with taxonomic terminology, we include in Table III a hierarchic listing of all the ordinal and family-level taxa referred to in this paper. The term "prosimian" is used informally in the text to refer to a possibly paraphyletic group of non-anthropoid primates. This, and any other informal term, is enclosed in quotation marks.

Selection Criteria

The taxa were selected on the basis of the following criteria:

1. *Importance in the debate on anthropoid origins.* Eocene taxa were chosen that have been suggested to share unique similarities with anthropoids (e.g., *Mahgarita, Omomys*) and to sample the range of dental variation within omomyids and adapids generally.
2. *Relevance to early anthropoid cladogenesis.* The Fayum taxa *Serapia, Arsinoea, Apidium, Simonsius, Parapithecus, Qatrania, Oligopithecus, Aegyptopithecus,* and *Catopithecus* were chosen because they represent the earliest well-known anthropoids and thus are of utmost importance in

Table I. Character–Taxon Matrix of Dental Characters Described in Appendix

	i1	i2	i3	i4	i5	i6	i7	i8	i9	i10	i11	i12
Cantius sp.	1	0	0	1	0	1	0	0	0	0	0	0
Pronycticebus gaudryi	1	9	9	9	0	1	0	9	0	9	0	0
Adapis parisiensis	1	0	0	1	0	1	0	0	0	0	0	0
Leptadapis magnus	1	9	9	0	9	0	9	9	0	0	0	0
Protoadapis curvicuspidens	1	9	9	9	0	1	9	9	9	9	9	9
Mahgarita stevensi	1	9	0	1	0	0	9	9	9	9	1	9
Teihardina americana	1	1	9	9	1	3	9	9	9	9	9	9
Steinius vespertinus	1	1	9	9	2	3	9	9	9	9	1	9
Nannopithex sp.	1	1	0	1	2	3	9	9	2	9	1	2
Ourayia uintensis	1	1	9	9	1	3	9	9	1	0	1	1
Omomys carteri	1	1	9	9	2	3	9	0	9	9	1	1
Tarsius sp.	2	9	9	1	9	9	9	1	1	0	1	1
Washakius insignis	1	0	0	1	0	2	9	0	9	9	0	9
Parapithecus fraasi	9	0	0	1	0	9	0	0	0	0	0	0
Qatrania wingi	9	9	9	9	9	9	9	9	9	9	9	9
Serapia eocaena	9	9	9	9	9	9	9	9	9	9	9	9
Arsinoea kallamos	1	0	0	1	1	0	9	1	1	0	0	0
Apidium phiomense	1	0	0	1	0	1	0	0	0	0	0	0
Simonsius grangeri	3	9	9	9	9	9	9	9	9	9	9	9
Oligopithecus savagei	9	9	9	9	9	9	9	9	9	9	9	9
Catopithecus browni	9	9	9	9	9	9	9	9	9	9	9	9
Aegyptopithecus zeuxis	1	0	0	0	0	1	0	0	1	0	0	0
Saguinus oedipus	1	0	0	1	1	1	0	0	0	0	0	0
Callimico goeldii	1	0	0	1	1	2	0	0	0	0	0	0
Aotus trivirgatus	1	0	0	1	1	2	0	0	0	0	0	0
Callicebus moloch	1	0	0	1	0	1	1	1	1	0	0	0
Saimiri sciureus	1	0	0	1	2	1	0	0	0	0	0	0
Dolichocebus gaimanensis	1	9	9	9	9	9	0	0	0	9	9	0
Purgatorius unio	0	9	9	9	9	9	9	9	9	9	9	9
Afrotarsius chatrathi	9	9	9	9	9	9	9	9	9	9	9	9
Pondaungia cotteri	9	9	9	9	9	9	9	9	9	9	9	9
Amphipithecus mogaungensis	9	9	9	9	9	9	9	9	9	9	9	9
Proteopithecus silviae	9	9	9	9	9	9	9	9	9	9	9	9

i13	i14	i15	i16	i17	c1	c2	c3	c4	c5	p1	p2	p3	p4	p5
0	0	1	0	1	9	9	9	9	9	0	0	9	2	0
0	0	1	0	9	2	9	9	9	9	0	0	1	2	0
0	0	1	0	2	2	9	0	1	1	0	0	1	2	0
0	0	1	9	2	2	9	1	1	0	0	0	1	2	0
9	9	9	9	9	2	9	0	0	9	0	0	1	2	0
0	0	9	9	9	2	9	0	0	0	1	0	0	2	0
0	9	9	9	9	1	9	0	0	9	0	0	0	2	0
0	9	9	9	9	1	9	9	9	9	0	0	0	2	0
0	1	1	0	2	1	9	0	0	0	1	1	9	1	2
0	0	9	0	2	1	9	9	9	9	1	0	0	2	0
0	9	0	0	2	1	9	0	0	0	1	0	0	2	0
0	2	9	0	0	2	0	0	1	0	1	0	0	1	2
0	1	1	0	2	1	9	0	0	9	1	0	0	2	1
0	0	1	0	2	1	9	0	0	0	1	0	0	2	0
9	9	9	9	9	9	9	9	9	9	9	0	0	2	0
9	9	9	9	9	2	9	1	0	0	1	0	0	2	2
0	0	1	0	2	1	9	1	0	0	1	0	9	9	2
0	0	1	0	2	1	2	0	0	0	1	0	0	2	0
9	9	9	9	9	2	9	0	0	0	1	0	0	2	0
9	9	9	9	9	1	9	0	0	0	9	1	9	2	0
9	9	9	9	9	9	9	9	9	9	1	1	9	2	0
0	0	1	0	1	2	2	0	0	0	1	1	9	2	0
0	0	1	0	9	1	0	0	0	0	1	0	0	0	0
0	0	1	0	9	1	0	0	0	0	1	0	0	0	0
0	0	1	0	9	1	0	0	0	0	1	0	0	0	0
0	0	1	0	9	1	0	0	0	0	1	0	0	0	0
0	0	1	0	9	1	2	0	0	0	1	0	0	0	0
0	0	1	9	9	1	1	0	0	0	1	0	0	0	0
9	9	9	9	9	1	9	9	9	9	0	0	1	2	0
9	9	9	9	9	9	9	9	9	9	9	9	9	9	9
9	9	9	9	9	9	9	9	9	9	9	9	9	9	9
9	9	9	9	9	9	9	9	9	9	9	0	0	1	0
9	9	9	9	9	9	9	9	9	9	9	0	9	9	9

(*continued*)

Table I. (*Continued*)

	p6	p7	p8	p9	p10	p11	p12	p13	p14	p15	p16	p17
Cantius sp.	2	1	0	1	0	1	0	2	0	0	0	1
Pronycticebus gaudryi	2	1	0	1	9	1	0	9	0	9	0	1
Adapis parisiensis	2	9	9	9	9	2	1	2	1	0	0	2
Leptadapis magnus	2	1	0	0	2	2	1	2	0	0	0	2
Protoadapis curvicuspidens	2	2	0	9	2	2	1	2	0	0	0	1
Mahgarita stevensi	2	2	9	9	2	1	0	2	9	0	0	0
Teihardina americana	2	1	9	1	1	1	0	2	0	0	0	1
Steinius vespertinus	2	2	9	9	1	1	0	9	0	9	0	1
Nannopithex sp.	9	2	9	9	0	1	0	2	0	9	0	1
Ourayia uintensis	2	0	0	1	1	1	0	2	0	0	0	1
Omomys carteri	2	2	9	9	2	1	0	2	0	0	0	1
Tarsius sp.	2	2	9	9	0	1	0	2	0	0	0	1
Washakius insignis	1	1	0	2	1	0	0	2	0	0	1	1
Parapithecus fraasi	2	2	9	9	1	0	0	2	0	0	1	1
Qatrania wingi	9	1	0	1	9	1	0	9	0	9	9	1
Serapia eocaena	2	1	0	1	2	1	0	0	0	0	0	1
Arsinoea kallamos	2	2	9	9	1	1	0	2	0	0	1	1
Apidium phiomense	2	2	9	9	1	0	0	0	0	0	1	1
Simonsius grangeri	1	1	9	9	1	0	0	2	0	0	1	1
Oligopithecus savagei	2	1	0	1	2	0	1	9	1	9	0	2
Catopithecus browni	1	1	0	2	2	1	1	9	1	9	0	2
Aegyptopithecus zeuxis	2	2	9	9	1	1	1	9	1	9	0	2
Saguinus oedipus	2	2	9	9	1	1	0	0	0	0	0	2
Callimico goeldii	2	2	9	9	1	1	1	0	1	0	2	2
Aotus trivirgatus	2	2	9	9	1	1	1	2	1	0	2	2
Callicebus moloch	2	2	9	9	1	0	1	2	1	0	2	2
Saimiri sciureus	2	2	9	9	1	0	0	2	1	0	2	2
Dolichocebus gaimanensis	2	2	9	9	1	1	1	2	1	0	1	2
Purgatorius unio	2	0	0	1	0	2	9	2	0	0	0	0
Afrotarsius chatrathi	9	9	9	9	9	9	9	9	9	9	9	9
Pondaungia cotteri	9	9	9	9	9	9	9	9	9	9	9	9
Amphipithecus mogaungensis	0	0	0	3	1	2	0	9	0	9	1	2
Proteopithecus silviae	9	9	9	9	9	9	9	9	9	9	9	9

p18	p19	p20	p21	022	p23	p24	p25	p26	p27	p28	p29	m1	m2
1	0	0	0	1	1	0	1	1	1	9	0	0	1
1	0	2	0	1	1	0	2	9	0	9	0	0	1
1	0	2	0	1	1	0	0	0	2	2	9	0	1
1	0	1	0	1	1	0	0	0	2	2	0	0	1
1	1	1	0	1	1	0	0	1	0	9	0	0	1
1	0	1	9	1	1	0	1	0	0	9	0	0	1
1	1	1	0	1	1	0	2	9	0	0	0	0	1
1	2	2	0	1	1	0	2	9	0	9	0	0	1
1	0	0	0	1	1	0	2	9	0	9	2	0	1
1	1	2	1	1	0	0	1	0	0	0	0	0	1
1	2	2	0	1	0	0	1	0	0	0	0	0	1
1	1	1	0	1	1	0	1	0	0	2	1	0	1
1	0	0	0	1	1	0	1	0	0	0	0	0	1
1	0	0	1	1	1	1	0	1	0	1	2	0	1
1	9	0	1	9	1	1	1	1	0	9	0	0	1
1	0	1	0	1	0	0	1	1	0	1	1	0	1
0	1	1	0	1	1	0	1	1	0	0	1	0	1
1	0	0	1	1	1	1	0	1	0	1	2	0	1
1	0	0	1	1	0	1	2	0	0	0	1	0	1
1	0	2	0	1	0	0	1	0	0	9	0	0	1
1	0	1	0	1	0	0	2	0	0	9	0	0	1
0	0	0	0	1	0	0	1	0	0	9	1	0	1
0	1	1	0	0	0	0	2	9	0	0	0	1	1
0	1	1	0	0	0	0	2	9	0	0	0	0	1
0	1	1	0	0	0	0	2	9	0	0	0	0	1
0	1	2	0	0	0	0	2	9	0	0	0	0	1
0	2	2	0	0	0	0	2	9	0	0	1	0	1
1	0	2	0	1	0	0	2	0	0	0	1	0	9
1	1	1	9	1	1	0	1	0	0	0	0	0	1
9	9	9	9	9	9	9	9	9	9	9	9	0	1
9	9	9	9	9	9	9	9	9	9	9	9	0	9
0	1	1	1	1	1	0	1	1	1	9	2	0	1
9	9	9	9	9	9	9	9	9	9	9	9	9	9

(*continued*)

Table I. (*Continued*)

	m3	m4	m5	m6	m7	m8	m9	m10	m11	m12	m13	m14
Cantius sp.	1	1	2	1	1	1	2	3	0	0	0	1
Pronycticebus gaudryi	1	1	2	2	1	2	0	0	1	0	0	0
Adapis parisiensis	1	1	2	1	1	9	9	9	9	2	9	9
Leptadapis magnus	1	1	2	1	1	9	9	9	0	2	0	1
Protoadapis curvicuspidens	1	1	2	2	1	1	1	9	0	0	0	1
Mahgarita stevensi	1	1	2	1	1	9	9	9	0	9	0	1
Teihardina americana	1	1	2	1	0	2	3	3	0	0	0	1
Steinius vespertinus	1	1	2	2	1	2	3	3	0	0	0	1
Nannopithex sp.	1	1	2	1	1	1	1	1	0	0	0	1
Ourayia uintensis	1	1	1	2	1	2	1	1	0	0	0	1
Omomys carteri	1	1	2	1	1	2	1	1	0	0	0	1
Tarsius sp.	1	1	2	1	1	1	1	1	0	0	0	1
Washakius insignis	1	1	2	1	1	1	1	1	0	2	1	2
Parapithecus fraasi	1	1	0	0	0	9	9	9	0	0	1	1
Qatrania wingi	1	1	0	0	0	1	9	9	0	0	0	1
Serapia eocaena	1	1	0	1	0	1	1	1	0	0	0	1
Arsinoea kallamos	1	1	0	1	0	2	3	9	0	0	0	1
Apidium phiomense	1	1	2	1	1	9	9	9	0	0	0	2
Simonsius grangeri	1	1	0	0	0	9	9	9	0	0	0	1
Oligopithecus savagei	1	9	9	1	9	1	9	9	0	0	9	9
Catopithecus browni	1	1	0	1	1	1	9	9	0	0	0	1
Aegyptopithecus zeuxis	1	1	2	1	1	9	9	9	0	0	0	1
Saguinus oedipus	0	9	9	1	9	9	9	9	0	0	9	9
Callimico goeldii	1	0	0	1	0	9	9	9	0	0	9	0
Aotus trivirgatus	1	0	1	1	0	9	9	9	0	0	9	0
Callicebus moloch	1	0	1	1	0	9	9	9	0	0	9	0
Saimiri sciureus	1	0	0	1	0	9	9	9	0	0	9	0
Dolichocebus gaimanensis	1	9	9	9	9	9	9	9	9	9	9	9
Purgatorius unio	1	1	2	1	1	2	2	2	0	0	0	1
Afrotarsius chatrathi	1	1	0	1	0	1	1	1	0	0	0	1
Pondaungia cotteri	1	1	2	1	1	9	3	3	9	9	9	1
Amphipithecus mogaungensis	1	1	9	1	9	1	9	9	0	0	9	9
Proteopithecus silviae	9	9	9	9	9	9	9	9	9	9	9	9

m15	m16	m17	m18	m19	m20	m21	m22	m23	m24	m25	m26	m27	m28
0	1	0	0	1	2	0	0	0	1	3	2	2	2
0	0	0	0	1	1	0	0	0	1	3	2	3	3
0	9	0	0	1	0	0	0	0	1	3	1	3	2
0	2	0	0	1	0	0	0	0	1	2	1	3	2
0	0	0	0	1	1	0	0	0	1	3	2	3	3
0	9	0	9	1	9	0	0	0	1	3	2	9	9
0	0	0	0	1	2	0	0	0	1	3	2	3	3
0	1	0	0	1	2	0	0	0	1	3	2	3	3
0	1	0	0	1	1	0	0	0	1	3	2	3	3
1	2	0	0	1	2	0	2	1	1	3	2	2	2
0	1	0	0	1	2	0	0	0	1	3	2	3	3
0	0	0	0	1	2	0	0	0	1	3	2	3	3
1	1	0	0	1	2	0	0	0	1	3	2	2	2
0	2	1	1	1	0	1	1	1	0	3	0	0	0
0	1	1	0	1	1	0	0	0	1	3	0	0	0
0	2	1	0	1	1	0	0	0	1	3	0	0	0
0	2	1	0	1	1	0	0	0	1	3	1	1	1
0	2	1	1	1	0	1	1	1	0	3	0	0	0
0	2	1	1	1	0	1	1	1	0	3	0	0	0
0	2	0	0	0	1	0	0	0	1	2	0	1	1
0	2	0	0	0	1	0	0	0	1	3	1	1	1
0	2	1	1	0	0	1	0	0	0	3	0	0	0
0	2	0	1	0	1	0	0	0	1	1	2	2	3
0	2	0	1	0	0	0	0	0	1	1	2	3	3
0	2	0	1	0	0	0	0	0	1	3	2	3	3
0	2	1	1	0	0	0	0	0	1	3	1	1	1
0	2	0	1	0	0	0	0	0	1	3	2	2	2
0	2	0	9	9	9	9	9	9	1	9	9	9	9
0	0	0	0	1	2	0	0	0	1	3	2	1	1
0	1	0	0	0	2	0	0	0	1	3	2	2	2
1	9	1	9	9	2	9	9	9	0	3	2	9	2
0	2	1	9	1	1	0	0	9	0	3	2	2	9
0	9	9	9	9	9	9	9	9	9	9	9	9	9

(*continued*)

Table I. (*Continued*)

	m29	m30	m31	m32	m33	m34	m35	m36	m37	m38	m39	m40
Cantius sp.	0	2	1	1	0	2	1	1	2	2	2	1
Pronycticebus gaudryi	0	9	1	1	0	2	0	9	2	2	2	1
Adapis parisiensis	0	2	1	1	1	1	1	0	2	2	2	0
Leptadapis magnus	0	2	1	1	1	2	1	0	2	2	2	0
Protoadapis curvicuspidens	2	9	1	2	1	2	1	1	2	2	2	2
Mahgarita stevensi	1	9	1	9	1	9	2	2	2	2	2	2
Teihardina americana	1	9	2	1	0	1	0	9	2	2	2	1
Steinius vespertinus	1	9	1	1	0	1	0	1	2	2	2	1
Nannopithex sp.	1	9	1	1	1	2	0	9	2	2	2	1
Ourayia uintensis	0	1	2	1	1	1	1	1	2	2	2	1
Omomys carteri	1	9	1	0	0	1	0	0	2	2	2	1
Tarsius sp.	1	9	2	0	0	0	0	0	2	2	2	2
Washakius insignis	0	1	2	1	1	2	0	1	0	2	2	0
Parapithecus fraasi	1	1	0	1	1	1	0	0	2	1	1	1
Qatrania wingi	1	1	0	0	0	0	0	0	2	1	1	1
Serapia eocaena	2	1	0	9	0	0	0	0	2	2	1	0
Arsinoea kallamos	1	1	0	1	0	9	9	1	2	1	1	2
Apidium phiomense	0	1	0	1	0	0	0	0	0	1	1	1
Simonsius grangeri	2	1	1	0	0	0	0	0	2	1	1	1
Oligopithecus savagei	9	0	1	0	0	0	0	9	2	2	9	1
Catopithecus browni	1	0	1	0	0	0	0	0	2	1	1	0
Aegyptopithecus zeuxis	0	1	0	0	0	0	0	0	2	1	1	1
Saguinus oedipus	9	9	1	0	0	1	1	9	2	2	9	1
Callimico goeldii	3	9	2	0	0	0	0	0	2	2	0	1
Aotus trivirgatus	3	9	1	0	0	2	2	2	2	1	0	1
Callicebus moloch	3	1	1	0	0	2	2	2	2	1	0	1
Saimiri sciureus	3	1	1	1	0	0	0	1	2	2	0	1
Dolichocebus gaimanensis	9	9	9	9	9	9	9	9	2	9	9	9
Purgatorius unio	1	1	2	2	2	2	1	1	2	2	2	1
Afrotarsius chatrathi	2	1	2	1	1	0	0	0	2	2	2	1
Pondaungia cotteri	0	1	0	9	2	9	9	9	9	9	9	2
Amphipithecus mogaungensis	9	2	0	1	2	1	9	9	2	1	9	2
Proteopithecus silviae	9	9	9	9	9	9	9	9	9	9	9	9

m41	m42	m43	m44	m45	m46	m47	I1	I2	I3	I4	I5	I6	I7
0	0	1	1	3	2	2	9	9	9	9	9	9	9
0	0	1	0	1	2	2	9	9	9	9	9	9	9
0	0	1	9	1	2	2	1	0	9	0	9	2	2
0	0	1	0	2	2	2	9	9	9	9	9	9	9
0	0	1	1	1	2	2	9	9	9	9	9	9	9
0	0	1	0	1	2	2	9	9	9	9	9	9	9
0	0	1	1	1	2	2	9	9	9	9	9	9	9
0	0	1	0	3	2	1	9	9	9	9	9	9	9
0	0	1	1	2	2	2	9	9	9	9	9	9	9
0	0	1	1	1	1	1	1	9	9	2	0	2	2
0	0	1	1	1	1	1	9	9	9	9	9	9	9
0	0	1	1	2	1	1	2	0	0	2	0	0	1
0	0	1	1	1	2	2	9	9	9	9	9	9	9
1	0	1	2	1	2	1	9	9	9	9	9	9	9
1	0	1	0	1	2	1	9	9	9	9	9	9	9
1	0	1	1	1	1	1	9	9	9	9	9	9	9
0	0	1	2	0	1	1	9	9	9	9	9	9	9
1	0	1	2	1	2	1	9	9	9	9	9	9	9
1	0	1	1	1	1	1	9	9	9	9	9	9	9
0	0	1	0	1	1	1	9	9	9	9	9	9	9
0	0	1	0	0	1	1	9	9	9	9	9	9	9
1	0	1	1	1	1	1	9	9	9	9	9	9	9
0	0	1	0	1	1	0	1	0	0	2	1	1	1
0	0	1	0	0	1	1	1	0	0	1	0	2	1
0	0	1	0	0	1	1	1	0	0	2	2	2	0
1	0	0	0	0	1	1	1	0	0	1	1	1	0
0	0	1	1	1	2	1	1	0	0	2	2	1	9
9	9	1	9	0	9	1	1	0	0	0	9	1	0
0	0	1	0	1	2	2	9	9	9	9	9	9	9
0	0	1	0	3	2	2	9	9	9	9	9	9	9
9	9	9	1	0	9	0	9	9	9	9	9	9	9
0	0	1	1	0	2	1	9	9	9	9	9	9	9
9	9	9	9	9	9	9	9	9	9	9	9	9	9

(*continued*)

Table I. (*Continued*)

	I8	I9	I10	I11	I12	I13	C1	C2	C3	C4	P1	P2
Cantius sp.	9	9	9	9	9	9	9	9	9	9	9	2
Pronycticebus gaudryi	9	9	9	9	9	9	0	9	9	9	2	2
Adapis parisiensis	0	9	2	9	9	1	0	0	9	9	1	2
Leptadapis magnus	9	9	9	9	9	9	9	0	9	9	1	2
Protoadapis curvicuspidens	9	9	9	9	9	9	9	9	9	9	9	9
Mahgarita stevensi	9	9	9	9	9	9	0	2	1	1	0	2
Teihardina americana	9	9	9	9	9	9	9	9	9	9	9	9
Steinius vespertinus	9	9	9	9	9	9	9	9	9	9	9	2
Nannopithex sp.	9	9	9	9	9	9	9	9	9	9	1	2
Ourayia uintensis	0	0	1	0	0	0	0	9	0	0	9	2
Omomys carteri	9	9	9	9	9	9	9	9	9	9	0	2
Tarsius sp.	2	0	9	2	0	1	1	1	1	1	0	2
Washakius insignis	9	9	9	9	9	9	9	1	9	9	0	2
Parapithecus fraasi	9	9	9	9	9	9	9	1	9	9	9	9
Qatrania wingi	9	9	9	9	9	9	9	9	9	9	9	9
Serapia eocaena	9	9	9	9	9	9	9	9	9	9	9	9
Arsinoea kallamos	9	9	9	9	9	9	9	9	9	9	9	9
Apidium phiomense	0	9	9	9	9	9	9	1	9	9	2	2
Simonsius grangeri	9	9	9	9	9	9	9	1	9	9	1	2
Oligopithecus savagei	9	9	9	9	9	9	9	3	9	9	9	2
Catopithecus browni	9	9	9	9	9	9	0	9	1	0	9	1
Aegyptopithecus zeuxis	9	9	9	9	9	9	0	3	1	1	9	2
Saguinus oedipus	0	0	0	1	0	0	0	1	1	0	0	0
Callimico goeldii	0	0	0	0	0	0	0	1	9	9	0	1
Aotus trivirgatus	0	0	0	2	0	0	1	1	0	1	1	1
Callicebus moloch	0	0	0	2	1	0	1	1	0	2	1	1
Saimiri sciureus	0	0	0	2	0	0	1	1	1	2	0	0
Dolichocebus gaimanensis	0	0	0	1	0	0	0	1	1	1	1	1
Purgatorius unio	9	9	9	9	9	9	9	9	9	9	9	9
Afrotarsius chatrathi	9	9	9	9	9	9	9	9	9	9	9	9
Pondaungia cotteri	9	9	9	9	9	9	9	9	9	9	9	9
Amphipithecus mogaungensis	9	9	9	9	9	9	9	9	9	9	9	9
Proteopithecus silviae	9	9	9	9	9	9	9	9	9	9	0	9

P3	P4	P5	P6	P7	P8	P9	P10	P11	P12	P13	P14	P15	P16
2	9	1	9	1	0	0	0	1	9	0	2	0	0
2	9	0	9	1	0	0	0	1	9	0	2	0	0
2	1	1	2	2	0	1	1	1	1	0	2	1	0
2	0	2	2	2	1	0	1	1	1	0	2	0	0
9	9	9	9	1	0	9	0	1	9	3	2	1	0
2	0	1	3	1	0	0	0	1	1	0	1	1	1
2	9	0	9	1	0	0	0	0	9	0	2	1	0
2	9	2	9	1	0	0	0	0	9	0	2	0	1
2	0	1	1	1	0	0	0	0	1	0	2	0	1
2	0	0	2	1	0	0	0	1	1	2	2	0	0
2	0	0	2	1	0	0	0	9	1	0	2	9	9
2	0	0	3	1	0	1	0	0	1	0	2	1	0
2	0	1	1	1	0	0	0	0	1	2	2	0	1
9	9	9	9	9	9	9	9	9	9	9	9	9	9
9	9	9	9	9	9	9	9	9	9	9	9	9	9
9	9	9	9	9	9	9	9	9	9	9	9	9	9
9	9	9	9	9	9	9	9	9	9	9	9	9	9
2	1	1	1	1	0	0	0	0	0	3	0	0	0
2	0	1	1	1	0	0	0	0	0	0	0	1	0
2	9	9	9	0	0	9	0	0	9	1	2	0	1
2	9	1	9	1	0	0	0	0	9	0	2	0	0
2	9	1	9	1	0	0	0	0	9	2	2	1	0
0	0	0	0	1	0	9	0	0	1	0	2	1	0
0	2	0	0	1	0	0	0	0	0	0	2	1	0
1	2	1	0	1	0	0	0	0	0	0	2	1	0
1	1	0	0	1	0	0	0	0	0	3	2	1	0
1	1	1	0	1	0	0	0	0	1	1	2	1	0
1	9	0	0	1	0	0	0	0	9	2	2	0	1
2	9	1	9	0	0	9	1	9	9	0	2	0	0
9	9	9	9	9	9	9	9	9	9	9	9	9	9
9	9	9	9	9	9	9	9	9	9	9	9	9	9
9	9	9	9	9	9	9	9	9	9	9	9	9	9
9	9	9	0	9	9	0	9	9	1	9	9	9	9

(*continued*)

Table I. (*Continued*)

	P17	P18	P19	M1	M2	M3	M4	M5	M6	M7	M8	M9
Cantius sp.	0	1	0	0	0	0	2	1	0	0	1	0
Pronycticebus gaudryi	1	1	0	0	0	1	2	0	0	0	1	0
Adapis parisiensis	1	0	1	0	0	2	2	0	0	0	1	0
Leptadapis magnus	1	0	1	0	0	2	2	1	0	0	1	0
Protoadapis curvicuspidens	9	0	0	0	9	2	9	0	0	9	1	0
Mahgarita stevensi	0	1	1	0	0	2	2	0	0	0	2	0
Teihardina americana	0	9	0	0	0	0	2	2	0	0	1	0
Steinius vespertinus	0	1	0	0	0	0	2	2	0	0	1	0
Nannopithex sp.	0	0	0	0	0	1	1	2	0	0	1	0
Ourayia uintensis	1	1	0	0	0	1	2	0	0	0	1	0
Omomys carteri	0	1	0	0	0	1	2	0	0	0	1	0
Tarsius sp.	0	9	0	0	0	1	1	0	0	0	1	0
Washakius insignis	1	1	0	0	0	0	2	1	0	1	1	0
Parapithecus fraasi	9	9	9	9	9	9	9	9	9	9	9	9
Qatrania wingi	9	9	9	0	9	1	9	0	0	0	2	0
Serapia eocaena	9	9	9	9	9	9	9	9	9	9	9	9
Arsinoea kallamos	9	9	9	9	9	9	9	9	9	9	9	9
Apidium phiomense	0	0	0	0	0	1	2	0	0	0	2	2
Simonsius grangeri	0	0	0	0	0	2	1	0	0	0	2	0
Oligopithecus savagei	0	9	1	0	9	1	9	0	0	9	0	0
Catopithecus browni	0	1	1	0	0	1	2	0	0	9	0	0
Aegyptopithecus zeuxis	0	0	1	0	0	1	2	0	0	9	1	1
Saguinus oedipus	0	0	0	0	9	0	0	0	0	9	0	0
Callimico goeldii	0	1	0	0	1	1	0	0	0	0	1	0
Aotus trivirgatus	0	0	0	1	2	2	1	0	0	9	0	0
Callicebus moloch	0	0	1	0	1	1	1	0	0	0	0	0
Saimiri sciureus	0	0	1	1	2	0	1	0	0	9	0	0
Dolichocebus gaimanensis	1	0	1	0	1	0	2	0	0	9	1	0
Purgatorius unio	0	9	0	0	9	0	2	1	0	0	1	0
Afrotarsius chatrathi	9	9	9	9	9	9	9	9	9	9	9	9
Pondaungia cotteri	9	9	9	9	9	9	2	9	9	9	9	9
Amphipithecus mogaungensis	9	9	9	9	9	9	9	9	9	9	9	9
Proteopithecus silviae	9	0	9	0	1	0	1	0	0	9	0	0

M10	M11	M12	M13	M14	M15	M16	M17	M18	M19	M20	M21	M22	M23
2	2	9	0	0	0	2	0	2	1	0	2	1	1
1	1	1	0	0	0	1	0	0	1	0	2	9	1
0	0	0	0	0	0	0	0	0	1	0	2	2	1
0	1	0	0	0	0	0	0	1	1	0	2	2	1
1	1	1	0	9	0	0	0	0	0	0	2	1	1
0	0	0	2	0	0	1	0	0	1	0	2	2	1
2	2	9	0	0	0	1	0	0	1	0	2	2	0
2	2	9	0	0	0	1	0	0	1	0	2	0	0
1	1	1	0	0	0	1	2	0	1	0	2	0	0
1	1	1	0	0	0	1	2	0	1	1	2	2	1
1	1	1	0	0	0	1	0	0	1	0	2	2	0
2	2	9	0	0	0	1	0	0	0	0	2	2	0
1	1	1	0	0	0	2	2	9	1	1	2	1	0
9	9	9	9	9	9	9	9	9	9	9	9	9	0
0	9	0	0	9	0	2	0	1	9	0	9	2	0
9	9	9	9	9	9	9	9	9	9	9	9	9	0
9	9	9	9	9	9	9	9	9	9	9	9	9	0
0	0	1	1	0	1	2	0	2	1	2	1	2	1
0	0	0	0	0	1	2	0	2	1	0	2	2	1
1	9	0	0	9	9	0	0	0	1	0	1	2	1
1	1	0	0	0	9	0	0	0	1	0	2	2	1
0	0	0	0	0	0	0	0	0	0	0	1	2	1
2	2	9	9	9	9	0	2	0	0	0	1	1	0
1	2	0	0	0	0	1	2	0	0	0	0	1	0
0	0	0	2	9	9	0	0	0	0	0	0	0	1
0	0	0	2	2	9	2	0	1	1	0	0	1	1
0	0	0	1	9	9	0	0	0	0	1	0	1	1
0	0	0	1	9	0	0	2	0	0	0	1	1	1
1	1	0	0	0	0	1	0	0	1	0	2	1	0
9	9	9	9	9	9	9	9	9	9	9	9	9	9
0	0	0	9	9	9	9	9	9	9	0	9	9	2
9	9	9	9	9	9	9	9	9	9	9	9	9	9
9	1	1	0	0	9	0	2	0	1	0	2	2	0

Table II. A Partial List of Specimens Examined in This Study[a]

Species, provenance, and age	Specimens[b]	Description
Purgatorius unio	UCMP 107406	P_2–M_3
Paleocene, North America	LACM 28128	P^4–M^2
Teilhardina americana	UW 8961	P^4–M^3
early Eocene, North America	UW 6896	C–M_3
Steinius vespertinus	USGS 502	P^3–M^3
early Eocene, North America	USGS 5995	P^3–M^3
	USGS 25026	P_4–M_3, alveoli of I_1–P_3
	USGS 28326	P_3–M_1
	AMNH 16835	M_{1-3}
Omomys carteri	YPM 11854	M^{1-3}
middle Eocene, North America	USNM 13289	P_3–M_3
	AMNH 19036	P_4–M_3
	USNM 363843	P_4–M_2
	USNM 17788	P_4–M_3
	UW 2995	P_3–M_1
Ourayia uintensis	PU 16431	I^1–M^3
middle Eocene, North America	PU 11236	I_1, P_2–M_3
Washakius insignis	UW 10275	P^2–M^3, canine alveolus
middle Eocene, North America		
	UW 12580	I_1–M_2
	UW 13426	P_3–M_2
	UW 1319	P_4–M_3
	AMNH 12039	P_2–M_1
	UW 13535	M_3
Nannopithex sp.	no #	I_1–M_3
middle Eocene, Europe	Halle 4254	P_2–M_3, canine alveolus
	Halle 4255	I_1–P_4, M_{2-3}
Nannopithex pollicaris (= *filholi*)	EM 6	P^2–M^3
middle Eocene, Europe		
Tarsius sp.	Duke Primate Center specimen	Full dentition
Recent		
Cantius eppsi	BMNH M 15145	P^3–M^3
early Eocene, Europe	BMNH M 13773	P_3–M_3
Mahgarita stevensi	TMM 41578-9	C–M^3
late Eocene, North America	TMM 41578-8	C, P_3–M_3
Protoadapis curvicuspidens	MHN AL = 517	M_{2-3}
late Eocene, Europe	MHN Louis 15 Ma	C_1, P_3–M_3
Adapis parisiensis	MNHN-Paris	I^1–P^3, alveol. P^4
late Eocene, Europe		
	MNHN-Paris	C–M^3
	MCZ 8887	P^4–M^3
	MNHN-Paris	I_1–P_3
	MCZ 8891	P_2–M_3
	MCZ 8892	P_1–M_3
	YPM 30440	P_1–M_3

(*continued*)

Table II. (*Continued*)

Species, provenance, and age	Specimens[b]	Description
Leptadapis magnus	MCZ 8884	P^1–M^3
late Eocene, Europe	BMNH 10265	C
	Basel QD 34	I_{1-2}, P_2–M_3
	MNHN QU no #	C–M_3
	MNHN QU 10973	I_2–C, P_2–M_3
Pronycticebus gaudryi	MNHN-QU 11056	P^3–M^3, alveol. C–P^2
late Eocene, Europe		
	MNHN-QU 11057	P_3–M_3, alveol. P_{1-2}
Apidium phiomense	DPC 1048	P^2–M^3
early Oligocene, Egypt	DPC 3080	I_1–M_3
Parapithecus fraasi	TYPE, Stuttgart	I_2–M_3, I_2–C, P_3–M_3
early Oligocene, Egypt		
Simonsius grangeri	DPC 2373	P^3–M^3
early Oligocene, Egypt	DPC 1123	P^3–M^2
	DPC 1090	P^2–M^1
	DPC 2807	C–M_3
	DPC 1009	P_3–M_2
	DPC 3110	M_{1-3}
Serapia eocaena	CGM 42286	C–M_3
late Eocene, Egypt		
Oatrania wingi	DPC 6125	P_4–M_3
late Eocene or early Oligocene, Egypt	DPC 2804	M^1
Arsinoea kallimos	CGM 42310	I^1–M_3
late Eocene or early Oligocene, Egypt		
Oligopithecus savagei	DPC 5407	P^4
early Oligocene, Egypt	CGM 29627	C–M_2
	DPC 6062	M_1
	DPC 6020	M_1
	DPC 3170	Half of an upper premolar
Catopithecus browni	DPC 8772	P^3–M^3
late Eocene or early Oligocene, Egypt	DPC 6134	C
	CGM 41885	P_3–M_3, alveol. I_{1-2}, C
	DPC fn90-971	P_4–M_3
Aegyptopithecus zeuxis	CGM 40237	C–M^3
early Oligocene, Egypt	DPC 1028	C–M_3
Dolichocebus gaimanensis	Isolated teeth at Buenos Aires	
early Miocene, Argentina		
Saguinus oedipus	USNM (VZ) 301646, 301653	Full dentitions
Recent		
Callimico goeldii	USNM (VZ) 303323, 395455, 399073	Full dentitions
Recent		
Aotus trivirgatus	USNM (VZ) 396715, 396716, 396720, 396722, 396724	Full dentitions
Recent		

(*continued*)

Table II. (*Continued*)

Species, provenance, and age	Specimens[b]	Description
Callicebus moloch Recent	USNM (VZ) 397977, 220293	Full dentitions
Saimiri sciureus Recent	USNM (VZ) 398676, 398691, 398704, 398708, 398709	Full dentitions
Proteopithecus sylviae late Eocene or early Oligocene, Egypt	CGM 41886	P^2, M^1–M^3

[a]We also consulted the extensive literature for information about the character states of taxa not examined personally by us.

[b]Abbreviations: AMNH, American Museum of Natural History; USNM, National Museum of Natural History, Smithsonian Institution; USNM (VZ), National Museum of Natural History, Smithsonian Institution, Vertebrate Zoology collections; YPM, Yale Peabody Museum; MCZ, Museum of Comparative Zoology (Harvard University); LACM, Los Angeles County Museum; TMM, Texas Memorial Museum, Austin TX; UW, University of Wyoming Geological Museum; PU, Princeton University; USGS, United States Geological Survey; MNHN, Museum National d'Histoire Naturelle (Paris or QU, Quercy); Halle, Geological and Paleontological Institute (Univ. of Halle); Basel, Naturhistorisches Museum (Basel); DPC, Duke Primate Center; CGM, Cairo Geological Museum.

Table III. Scheme of Classification of Primates Followed in the Text[a]

Order Primates Linneaus, 1758
- Semiorder Strepsirhini Geoffroy, 1812 [new rank]
  - Suborder Lemuriformes Gregory, 1915
  - Suborder Adapiformes Szalay and Delson, 1979
    - Family Adapidae Trouessart, 1879
      - *Adapis*
      - *Leptadapis*
      - *Mahgarita*
      - *Protoadapis*
      - *Cantius*
      - *Pronycticebus*
      - *Pondaungia cotteri*
- Semiorder Haplorhini Pocock, 1918 [new rank]
  - Suborder Tarsiiformes Gregory, 1915
    - Family Omomyidae Trouessart, 1879
      - *Omomys*
      - *Chumashius*
      - *Ourayia*
      - *Teilhardina*
      - *Steinius*
      - *Nannopithex*
      - *Washakius*
    - Family Tarsiidae Gray, 1825
      - *Tarsius*

(*continued*)

Table III. (*Continued*)

Suborder Anthropoidea Mivart, 1864
  Infraorder Platyrrhini Geoffroy, 1812
    Family Cebidae Bonaparte, 1831
      Dolichocebus gaimanensis
      Saimiri sciureus
      Saguinus oedipus
      Callimico goeldii
      Aotus trivirgatus
      Callicebus moloch
  Infraorder Catarrhini Geoffroy, 1812
    Superfamily Pliopithecoidea Zapfe, 1960
    Family Pliopithecidae Zapfe, 1960
    Family Propliopithecidae Straus, 1961
      Propliopithecus hackeli
      Propliopithecus chirobates
      Propliopithecus ankeli
      Moeripithecus markgrafi
      Aegyptopithecus zeuxis
    (also including superfamilies Cercopithecoidea, Hominoidea, and Proconsuloidea)
  Infraorder Parapithecoidea Schlosser, 1911 [new rank]
    Family Parapithecidae Schlosser, 1911
      Subfamily Parapithecinae Schlosser, 1911
        Apidium phiomense
        Apidium moustafai
        Parapithecus fraasi
        Simonsius grangeri
      Subfamily Qatraniinae [new]
        Qatrania wingi
        Qatrania fleaglei
        Serapia eocaena
        Arsinoea kallimos
  Infraorder *incertae sedis*
    Family Oligopithecidae Simons, 1989 [new rank]
      Catopithecus browni
      Oligopithecus savagei
    Family Afrotarsiidae Ginsburg and Mein, 1987
      Afrotarsius chatrathi
    Family *incertae sedis*
      Proteopithecus sylviae
Semiorder *Incertae sedis*
      Amphipithecus mogaungensis
      Plesiopithecus teras
      Algeripithecus minutus

[a]Nonanthropoid taxa and Miocene–Recent anthropoid taxa used in our cladistic analysis or mentioned in the text are classified to the level of Family. Eocene and early Oligocene anthropoid taxa are classified to the genus level.

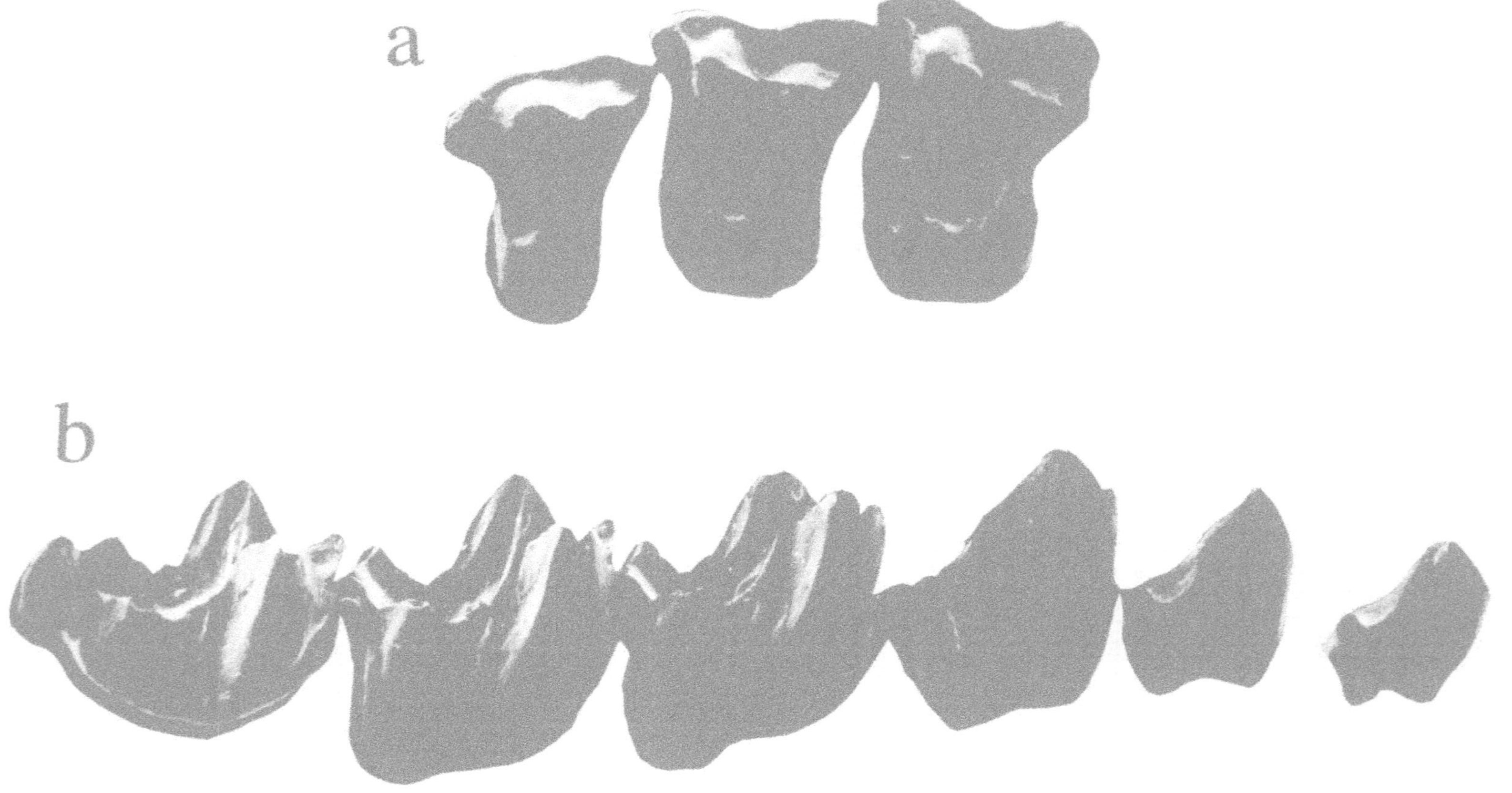

Fig. 4. *Purgatorius unio.* (a) LACM 28128, occlusal view of left P^4–M^2, M^1 length 2.0 mm. (b) UCMP 107406, lateral view of right P_2–M_3, M_1 length 2.1 mm.

understanding anthropoid origins and cladogenesis. We accept that *Aegyptopithecus* is phylogenetically a catarrhine, e.g., the sister taxon of Miocene–Recent catarrhines (Fleagle and Kay, 1983; Harrison, 1987; Szalay and Delson, 1979).

3. *Relative completeness.* Many fossil taxa are known only from fragmentary material. Missing data can render problematic any analysis of phylogenetic relationships among groups; thus, we chose taxa for which at least 65% of the data were available. In the data sets (Table I) no distinction was made between data that are unavailable because of evolutionary change and data that are missing because material is unavailable. In both cases the character is scored as "9." In the first case, for example, when a tooth is missing in one species that was present in its ancestor, morphological traits of the tooth must be scored as missing; in the second case traits cannot be scored because the available material does not preserve the dental structures being scored.
4. *Relative primitiveness.* The Eocene "prosimians" and extant platyrrhine taxa also were selected on the basis of their relative primitiveness. Highly derived taxa such as advanced notharctines, some large-bodied platyrrhines, or some gum-eating callitrichines were not included in the analysis because we wished to understand early cladogenesis in the major lineages rather than cladogenesis in terminal branches of these clades.
5. *Outgroups for rooting the analysis.* Ideally, outgroups should be morphologically conservative taxa, as close as possible in time and phylogenetic position to the last common ancestor of the primates. Paleocene Plesiadapiformes come close to fulfilling these criteria; frequently they are included as the most primitive members of primates (e.g., Szalay and Delson, 1979), although a relationship of some or all of them to Dermoptera or other archontans has recently been proposed (Beard, 1990; Kay *et al.*, 1990, 1992). Also, the most primitive plesiadapiform, *Purgatorius unio* (Fig. 4), predates the first appearance of primates.

However, the dental structure of Plesiadapiformes may be specialized in several ways; this would alter the way primates are rooted by this outgroup and consequently affect the interpretation of primate cladogenesis. For example, plesiadapiforms have relatively enlarged, procumbent lower central incisors; lower lateral incisors are much smaller or absent. This is a similarity to many omomyid primates, but we suspect the resemblence results from convergence and that the ancestral primate had smaller, vertically implanted lower incisors with the first smaller than the second (Cartmill and Kay, 1978; Covert and Williams, 1991; Rose and Bown, 1991; Rose *et al.*, Chapter 1, this volume; Szalay *et al.*, 1987). However, the alternative, that the ancestral primate resembled plesiadapiforms in having a large procumbent central incisor and a smaller lateral one, also has proponents (Gingerich and Schoeninger, 1977;

Rasmussen, 1990). Because there is disagreement as to which of several incisor character states is plesiomorphic, the incisor characters of *Purgatorius* are coded as missing in the analysis.

Another group frequently discussed in connection with primate origins is Scandentia, the tree shrews. However, we have not used any living tree shrews to root our analyses because living tree shrews would have been separated from a last common ancestor (LCA) with primates by at least 60 million years, and their dental anatomy may not closely resemble that of the LCA (Butler, 1980).

Taxa Known Only from Fragments

A number of taxa have figured prominently in discussions about anthropoid origins but are too fragmentary for inclusion in the formal analysis. These will be discussed separately but not formally analyzed. One useful procedure with respect to these taxa is to place them in all possible positions of the maximum-parsimony tree to see which phylogenetic placements are most plausible. This approach utilizes best the capabilities of the phylogenetic analysis program MacClade, Version 3.0 (Maddison and Maddison, 1992).

Dental Anatomy and Systematics

In this study we have restricted ourselves to dental characters alone. In so doing we have chosen to analyze one anatomic system as fully as possible to see what it can tell us about phylogeny. This is not to say that we believe the key to unlocking the "true" phylogeny is this one anatomic region. To the contrary, the best approach would be to integrate all available information from other anatomic regions as well as behavioral and genetic data from living taxa. We see the results described below only as a contribution to the reconstruction of the phylogeny of Anthropoidea. Our findings may be substantiated or weakened by other anatomic information. Having said this, it is important to remember that much of what we know about primate phylogeny today is based, of necessity, on the study of the dentition alone because the teeth of primates are the most often preserved anatomic structures in the primate fossil record. In many cases they are the sole evidence documenting the existence of taxa. In contrast, cranial and postcranial anatomy is much less commonly preserved. The exclusive use of data from the latter anatomic system presents the danger of substantial errors in interpreting phylogeny. For example, we have information about the distribution of characters of otic and orbital anatomy in just a handful of the many adapid, omomyid, and early anthropoid genera documented by teeth (Ross, Chapter 15, this volume). Postcranial anatomy is even less well known for most taxa; no complete skeleton is available for any early anthropoid or omomyid, and isolated skeletal parts are often difficult to refer to a specific taxon (Covert and Williams, Chapter 2, Dagosto and Gebo, Chapter 17, Ford, Chapter 18, Gebo *et al.*, Chapter 9, this volume). In the character analyses of such poorly documented anatomic systems, missing data may make determination of polarity less certain and inevitably will underrepresent the real extent of parallelism and reversal in the characters under study.

Dental Characters Used in This Study

The dentition is anatomically very complex, providing many characters for phylogenetic studies. In this work, we have selected characters relating to tooth numbers, proportions, occlusal crown morphology, crown margin morphology, and root orientation and number. The criterion for recognition of a "character" is that it show minimal variation within genera or species but substantial variation across higher taxonomic groups and that this variation be to some degree independent of similar characters on other teeth. For example, in a survey of the orientation of the M_1 cristid obliqua across various species, we found that on some M_1s, this crest is directed toward the protoconid (state 0); on others the crest is oriented midway between the protoconid and metaconid (state 1); and on still others the crest is directed towards the metaconid (state 2). A survey of the orientation of this crest on M_2 reveals similar variation. In some taxa, both teeth have the same state (0, 1, or 2). If no other condition were found, we would consider the condition of M_{1-2} together as a single character with three states. However, in some taxa, we find different states on M_1 and M_2, demonstrating that some degree of independence exists for each tooth. When we encounter variation of this kind, we treat the variation on each tooth as a separate character. But the choice does not really matter for the purposes of analysis for the following reason: if we consider this character complex as composed of a single, ordered morphocline with five states (0,0; 0,1; 1,1; 1,2; 2,2), it requires four steps to go from state 0,0 to state 2,2. Alternatively, if we characterize the variation as two characters, each with three states (0, 1, or 2), it still requires a total of four steps to achieve state 2 for both characters.

The problem of weighting of characters has been widely discussed and debated (see Maddison and Maddison, 1992, and references therein). It comes into these analyses in several ways. (1) *A priori* weighting is employed when we choose some characters and disregard others. This relegates the characters not chosen to a weight of zero (Neff, 1986). In her study of postcranial characters in this volume, Ford (Chapter 18) eliminated a number of characters before making her analysis. We do not take this approach here in that we have tried to include all the characters mentioned by other workers as well as many new characters of our own devising. We would argue that all characters, regardless of the amount of homoplasy they display, contain some phylogenetically useful information and should, therefore, be included. (2) In cases in which multistate characters are ordered, the number of states recognized in a morphocline will add weight to a transformation series. In other words, if an "equal weights" option is employed in phylogenetic analyses, this does not necessarily mean that each character has the same influence in descriminating among alternative network topologies. For example, a character with four discrete states will always contribute at least three steps to the network length, whereas a binary character may contribute as few as one step (Swofford, 1993). In the analyses reported below, we have chosen to scale multistate characters in such a way that individual steps in a morphocline count for proportionately less as character states are added (Maddison and Maddison, 1992, pp. 197–199). (3) We have chosen to "atomize" our characters rather than risk the loss of significant information, as in the example discussed above. This has the effect of weighting *character complexes* ("megacharacters" of Wheeler, 1986) more heavily than isolated characters. For example, if three characters used to describe the shape of the lower incisors all covary because they are part of the same character complex, this would be logically equivalent to scoring that complex three times, thereby weighting it more heavily. Separation of various elements of an apparently unified functional complex is favored here because it does not require the assumption that the functional complex evolved full-blown as a unified functional unit, often a dangerous and unwarranted assumption for dental anatomy (Fleagle et al., 1994).

A further point to consider in cladistic analyses is the homoplastic occurrence of traits. The high frequency of homoplasy has led some to challenge the accuracy of phylogenetic claims based on analyses of the dentition (e.g., Novacek, 1992). In this view, the dentition is intrinsically more labile in structure than some other anatomic systems and, as such, rapidly loses its phylogenetic "signal" as current functional or adaptive demands reshape it. Whether this is true or not, and we know of no reliable evidence that it is so, we can only concur with the views of Szalay and colleagues (1987) that the ". . . attributes of primate dentitions are not only our best clues to the feeding preferences of these animals, but they clearly mirror their ancestral morphology in spite of the adaptive plasticity of the dentition" (p. 80). Even if it were shown that teeth evolve more rapidly than some other anatomic parts, this is more than offset by the high-density time sampling of information about the teeth of fossil species.

Analytical Approach

Wagner Networks and the Assessment of Alternative Phylogenies

As noted earlier, the use of outgroups for rooting networks (thus, determining the polarity of character states) establishes an interpretation of cladogenesis. However, by determining the most parsimonious unrooted "Wagner network" for the primates under study without folding the network into a cladogram, we can judge the parsimony of selected alternative hypotheses of primate cladogenesis. If a proposed cladogram is Wagner-equivalent (equally parsimonious) to that of the unrooted tree, then the hypothesis would be consistent with our dental evidence. Alternatively, if the cladogram is not Wagner-equivalent, we can assess the number of added steps required in comparison to our maximum-parsimony Wagner network.

Cartmill (1981) has shown that a Wagner network involving several changes in character states can be folded (transformed) at different points, creating a number of rooted cladograms that are equally parsimonious. For example, Fig. 5a shows four taxonomic units (A, B, C, D) that display six "steps" for six two-state characters (u → U, v → V, w → W, x → X, y → Y, z → Z). The cladogram in Fig. 5b has been folded at point 1, between nodes a′ and b′. The one in Fig. 5c has been folded at point 2, between node b′ and terminal taxon D. These two cladograms are Wagner-equivalent, each taking six steps.

A larger number of less parsimonious cladograms are not Wagner-equivalents. For example, the Wagner network in Fig. 5a cannot be folded in such a way that A and D are sister taxa to the exclusion of B (Fig. 5d). Such a cladogram will be less parsimonious.

Rooting of Wagner Networks to Produce Cladograms

Determining the most parsimonious character distribution, by itself, provides insufficient evidence to determine which Wagner-equivalent cladogram comes closest to representing the "true" phylogeny (i.e., those represented in Fig. 5b or 5c, or some other Wagner-equivalent cladogram). A second piece of information is required that cannot be obtained from the character distributions within the study group, namely, the character states of the ancestor. The equally parsimonious cladograms in Fig. 5b and 5c differ only in the nature of the inferred ancestral morphotype: In cladogram 5b, the morphotype is u,v,w,x,y,z, whereas in 5c, the morphotype is u,V,w,X,y,z. To choose between the two, we need to find the most parsimonious cladogram that is rooted in such a way that the morphotype is identical to the structure of the real ancestor. Here, we examine and compare two approaches used to root our Wagner network and describe the implications of each. (1) We root the Wagner network in such a way as to accept the hypothesis that Haplorhini and Strepsirhini are natural groups and that Adapidae are strepsirhines. The network is then folded to be consistent with this haplorhine/strepsirhine dichotomy, that is, so that *Tarsius* and Anthropoidea are more closely related to each other than either is to Adapidae. (2) An alternative method (more commonly favored by phylogenetic systematists) for rooting a Wagner network, and thus establishing the structure of the morphotype, is to use outgroups (e.g., Wat-

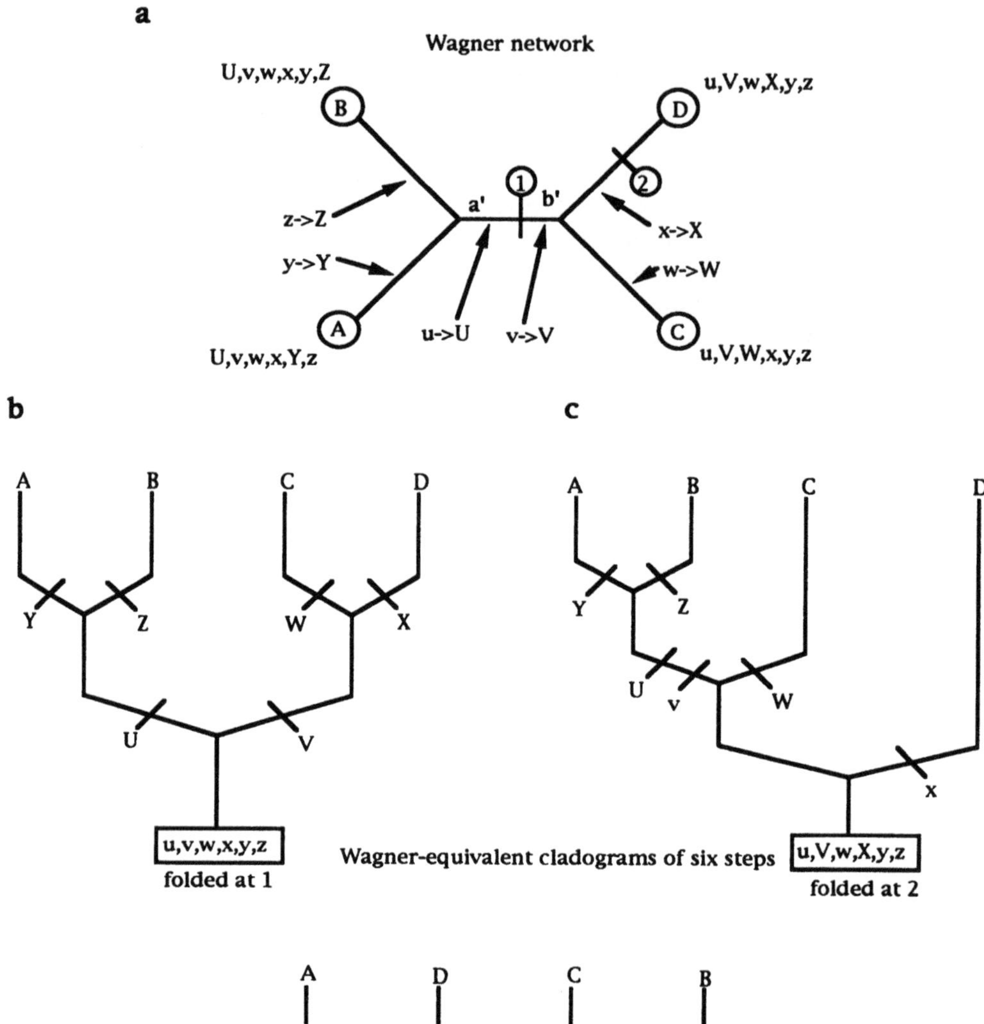

Wagner-equivalent cladograms of six steps

A D C B
U,Y V,X V,W U,Z
d
u,v,w,x,y,z

A non-equivalent arrangement of eight steps

rous and Wheeler, 1981). For this purpose we use *Purgatorius unio* (Paleocene, North America).

Computerized Options Utilized for Parsimony Analysis

By use of the PAUP computer program (Phylogenetic Analysis Using Parsimony) (Swofford, 1993, Version 3.1.1), we have constructed a Wagner network representing the maximum-parsimony hypothesis for the distribution of our characters and their states. Additionally, we explored character-state distributions with MacClade 3.01 (Maddison and Maddison, 1992). Several of the three- and four-state characters that we believe represent morphoclines were identified as "ordered"; all others were identified as "unordered" (see Appendix 1). No *a priori* assumption is made about the polarity of ordered character states. In the case of ordered characters with three states, 0, 1, and 2, for example, the ancestral state at any node could be any of the three states; ordering simply means that to get from state 0 to state 2 (or the reverse), it is necessary to pass through an intermediate state 1. The polarity of various ordered characters is then determined by parsimony and outgroup rooting.

Because the large size of the data set precludes exhaustive searching of all possible trees, "heuristic" algorithms of PAUP were employed to search for optimal (most parsimonious) trees, using randomized, stepwise addition of taxa, repeated ten times, with selection of the "branch swapping by tree bisection–reconnection" option of the program. We also use the "scale" option for ordered characters so that weights are assigned such that the minimum possible length for each ordered character is the same for all characters irrespective of the number of states for that character. PAUP 3.1.1 uses integers to store tree lengths and character weights, so that fractional step weights of 0.50 and 0.33 for three- and four-step ordered characters are unacceptable. Therefore, we used weights of 100, 50, and 33 for 2-state, 3-state, and 4-state characters.

To assess the strength of the most parsimonious Wagner network, we examined how these proposed clusters "decay" as steps are added to the networks. Intuitively, we consider a proposed association to be stronger when it holds together for more steps.

Fig. 5. The concept of maximum-parsimony networks (adapted from Cartmill, 1981). (a) An unrooted or Wagner network with four taxonomic units (A,B,C,D) involving six steps (changes from one character state to another). The two conditions or character states are represented by upper and lower case letters. (b,c) Two Wagner-equivalent cladograms of six steps, folded (i.e., rooted) at different points on the network. Note that the inferred morphotype changes depend on where the network is folded. (d) A non-Wagner-equivalent cladogram of eight steps. This arrangement requires that two character states be independently acquired: states U in taxa A and B, and states V in taxa D and C.

Results

Cladogenesis of the Primates

Most Parsimonious Networks

Without inclusion of outgroups to root the tree, a parsimony analysis using a PAUP heuristic search of 28 taxa yields two maximum-parsimony Wagner networks (Fig. 6). The networks have 39,446 steps with a consistency index of 0.369 (when uninformative characters are excluded). The retention index for the analysis is 0.558, and the homoplasy index is 0.609 (see Maddison and Maddison, 1992, for definitions of these quantities). The two networks differ only in the placement of *Mahgarita* and in the rearrangement of some omomyid taxa. In Network I (Fig. 6), *Mahgarita* is placed with Adapidae; in Network II, this taxon is placed between *Tarsius* and the Omomyidae. Based on features of the cranial anatomy that are not encompassed by this analysis, it is clear that *Mahgarita* is an adapid (see Ross, Chapter 15, this volume; Szalay and Delson, 1979). Therefore, we prefer Network I.

Network I in Fig. 6 was folded to produce two cladograms showing a Halporhine/Strepsirhine dichotomy, that is, so that *Tarsius* links with Anthropoidea (and Omomyidae) to the exclusion of Adapidae. Figure 7 shows the same maximum-parsimony network folded in several other ways. Figure 7A, also consistent with a Halporhine/Strepsirhine dichotomy, shows Omomyidae as a paraphyletic taxon from within which arise *Tarsius* and anthropoids. Figure 7B suggests an alternative view, that the most ancient split among primates occurred between Anthropoidea and all other taxa, living and fossil. Choosing among the alternative foldings of Network I depends on what we assume about the anatomy of the last common ancestor of all primates.

Additional light is shed on our analysis when the networks are subjected to a decay analysis (Fig. 8). Anthropoidea is an exceptionally robust cluster that holds together for the first 81 networks generated. The first four trees retain *Tarsius* as the first outgroup to Anthropoidea.

Outgroup Analysis

When the same 28 primate taxa are rooted using *Purgatorius* as the outgroup, three maximum-parsimony trees of 40,678 steps are found. All trees are rooted among Adapidae with *Cantius* as the sister taxon to all other primates. The majority consensus of these cladograms is illustrated in Fig. 9. Two of the three trees are Wagner-equivalent to Networks I and II illustrated in Fig. 6. The third is identical to Network I except that the positions of two groups are switched: The *Teilhardina–Steinius–Omomys–Ourayia* node is interposed between *Tarsius* and the *Washakius–Nannopithex* node. All trees cluster Anthropoidea together, and all associate *Tarsius* with Anthropoidea. Further, all networks root Anthropoidea within Omomyidae.

In summary of the above, all analyses, whether rooted or unrooted, are consistent with the hypothesis that Anthropoidea is a monophyletic group rooted within a cluster of Omomyidae, the closest relative of which is *Tarsius*. As noted above, taxa for this study were chosen on the basis of their relative primitiveness, degree of completeness, and importance in the literature on anthropoid origins. Relatively complete data were not available for some primitive taxa that are pertinent for understanding early primate cladogenesis or the cladogenesis of Adapidae and Omomyidae (e.g., *Altanius, Donrussellia, Altiatlasius, Loveina, Teilhardina belgica*). Use of derived members of lineages created some complications in our analysis. For example, *Washakius insignis* was included because its dentition is almost completely known (Covert and Williams, 1991). Unfortunately, it represents a derived member of the Washakiini (Honey, 1990). In earlier versions of our analysis, *Washakius* was placed as the outgroup to all other primates. However, subsequent refinement in scoring of our data set indicates that this taxon clusters with other omomyids. A much larger sampling of Eocene taxa is necessary for a fuller understanding of character distributions and cladogensis among omomyids and adapids. Although the intrafamilial relationships of tarsiiformes and adapids are not discussed here, as background for the discussion of anthropoid characters we illustrate the dentitions of representative adapids and omomyids in Figs. 10 and 11.

Anthropoid Origins

The dental evidence for a phylogenetic linking of one or another adapid with anthropoids has been discussed at length by Gingerich (1980) and Rasmussen (1990). Each of these authors calls attention to the dental similarities shared by Anthropoidea and protoadapine adapids, represented in this analysis by *Protoadapis* and *Mahgarita*. As a test of the hypothesis that some or all adapines are the sister group of Anthropoidea, using PAUP 3.1.1, we constrained Anthropoidea to link with (1) *Mahgarita,* (2) *Mahgarita* and *Protoadapis,* or (3) all the adapids used in our study. Linking Anthropoidea with *Mahgarita,* possibility 1, requires 50 added steps above the maximum-parsimony solution of 39,446 steps. Linking of anthropoids with *Protoadapis* and *Mahgarita,* possibility 2, requires 619 steps over the maximum-parsimony solution, whereas linkage of anthropoids with *all* adapids in our study group, possibility 3, requires 317 added steps. To gain an appreciation of how these solutions compare with the maximum-parsimony solution, we found 15 trees shorter than hypothesis 1, but far more than 1000 trees are more parsimonious than the trees proposed in hypotheses 2 or 3. We conclude that hypothesis 1 is plausible only if *Mahgarita* is not an adapid, a view we are not inclined to entertain at the moment (see Ross, Chapter 15, this volume).

Another favored phylogenetic placement of Anthropoidea is as the sister group of (1) *Tarsius,* (2) Omomyidae, or (3) both. Linking anthropoids with

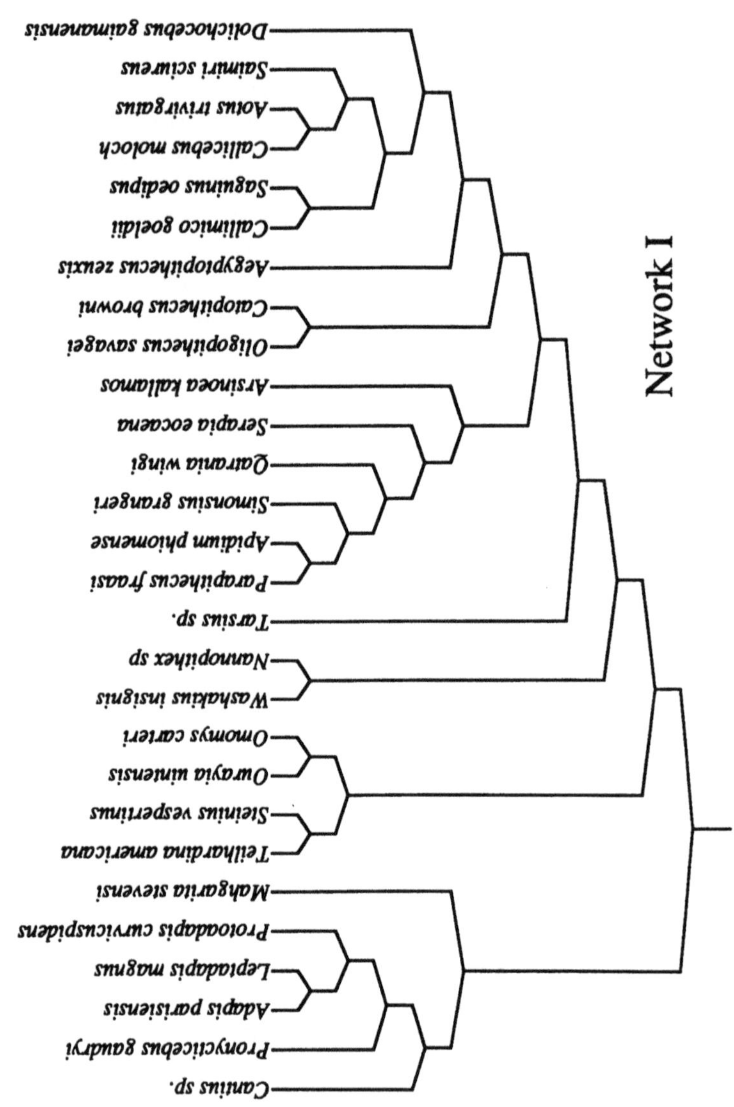
Dolichocebus gaimanensis
Saimiri sciureus
Aotus trivirgatus
Callicebus moloch
Saguinus oedipus
Callimico goeldii
Aegyptopithecus zeuxis
Catopithecus browni
Oligopithecus savagei
Arsinoea kallimos
Serapia eocaena
Qatrania wingi
Simonsius grangeri
Apidium phiomense
Parapithecus fraasi
Tarsius sp.
Nannopithex sp
Washakius insignis
Omomys carteri
Ourayia uintensis
Steinius vespertinus
Teilhardina americana
Mahgarita stevensi
Protoadapis curvicuspidens
Leptadapis magnus
Adapis parisiensis
Pronycticebus gaudryi
Cantius sp.
Network I

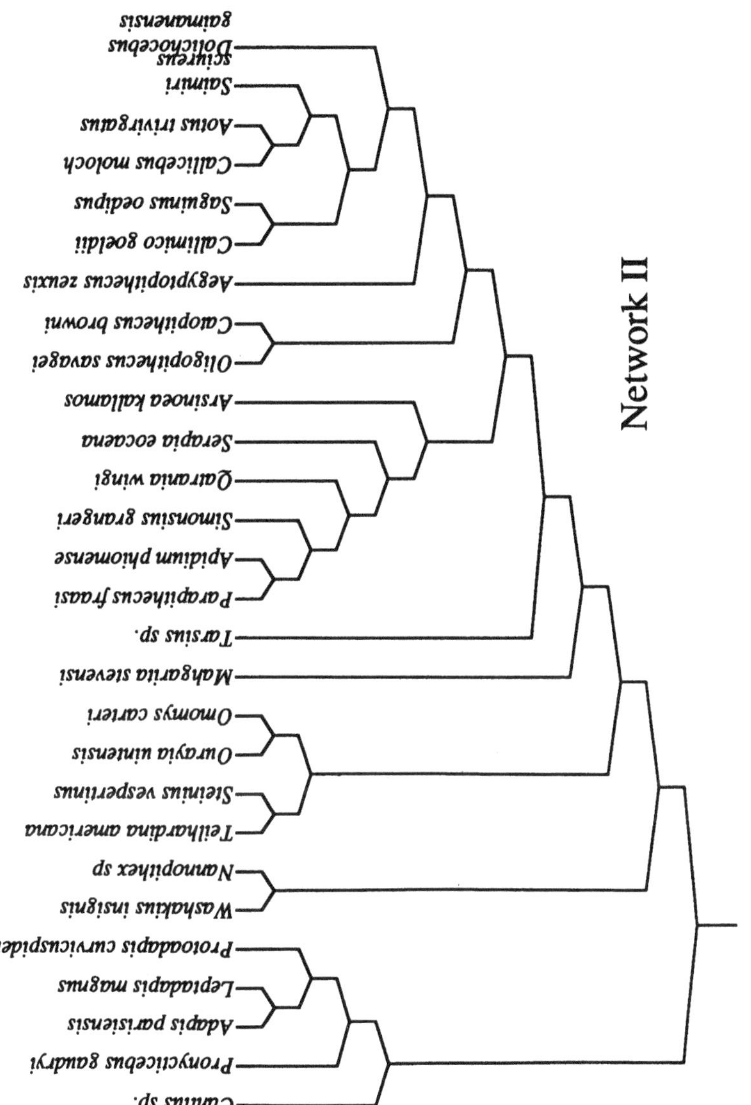

Fig. 6. Two maximum-parsimony networks for primate cladogenesis. Each network of 49,446 steps has been "folded" to produce haplorhine/strepsirhine clades.

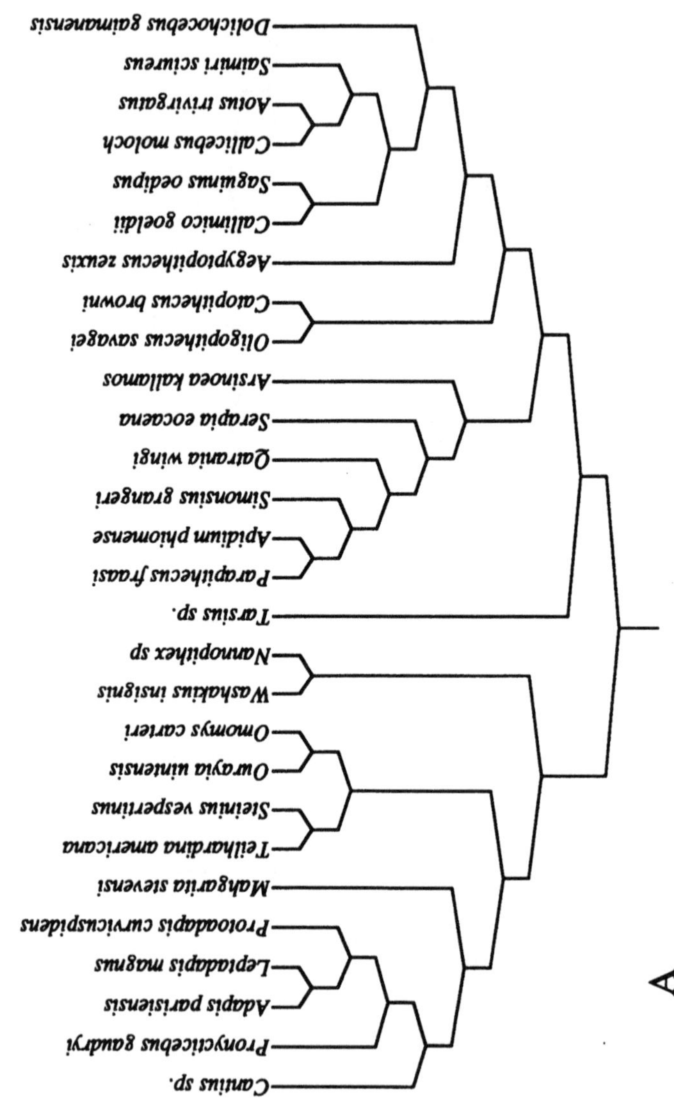
A
Canius sp.
Pronycticebus gaudryi
Adapis parisiensis
Leptadapis magnus
Protoadapis curvicuspidens
Mahgarita stevensi
Teilhardina americana
Steinius vespertinus
Ourayia uintensis
Omomys carteri
Washakius insignis
Nannopithex sp
Tarsius sp.
Parapithecus fraasi
Apidium phiomense
Simonsius grangeri
Qatrania wingi
Serapia eocaena
Arsinoea kallamos
Oligopithecus savagei
Catopithecus browni
Aegyptopithecus zeuxis
Callimico goeldii
Saguinus oedipus
Callicebus moloch
Aotus trivirgatus
Saimiri sciureus
Dolichocebus gaimanensis

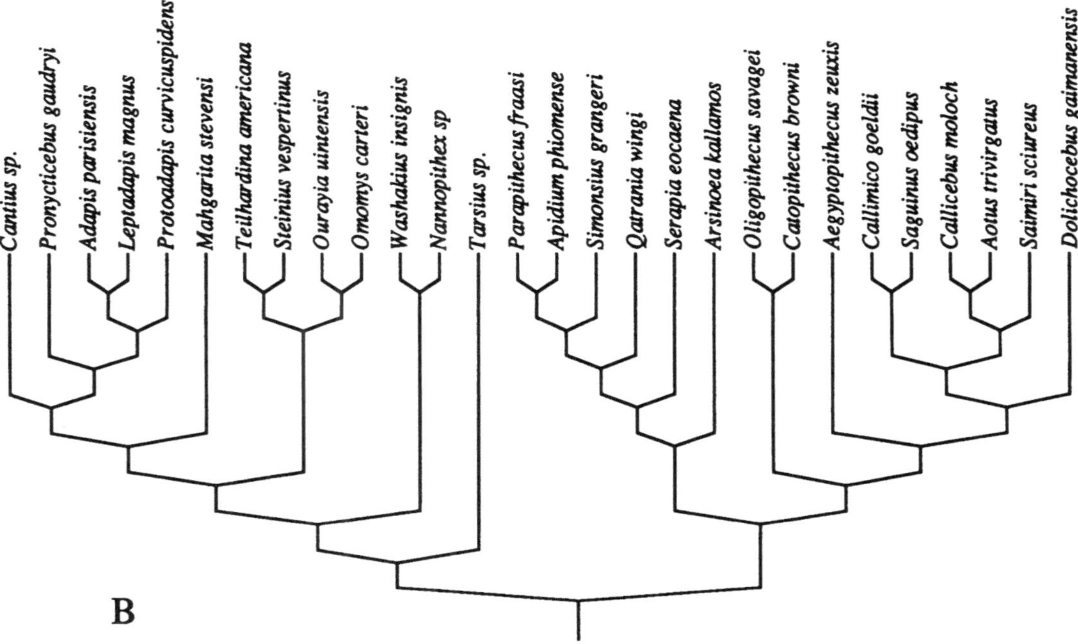

Fig. 7. (A) Maximum-parsimony Network I, from Fig. 6, folded to produce an alternative haplorhine/strepsirhine tree that excludes omomyids. (B) the same network folded to conform to an anthropoid versus "prosimian" tree. Both trees are 49,446 steps.

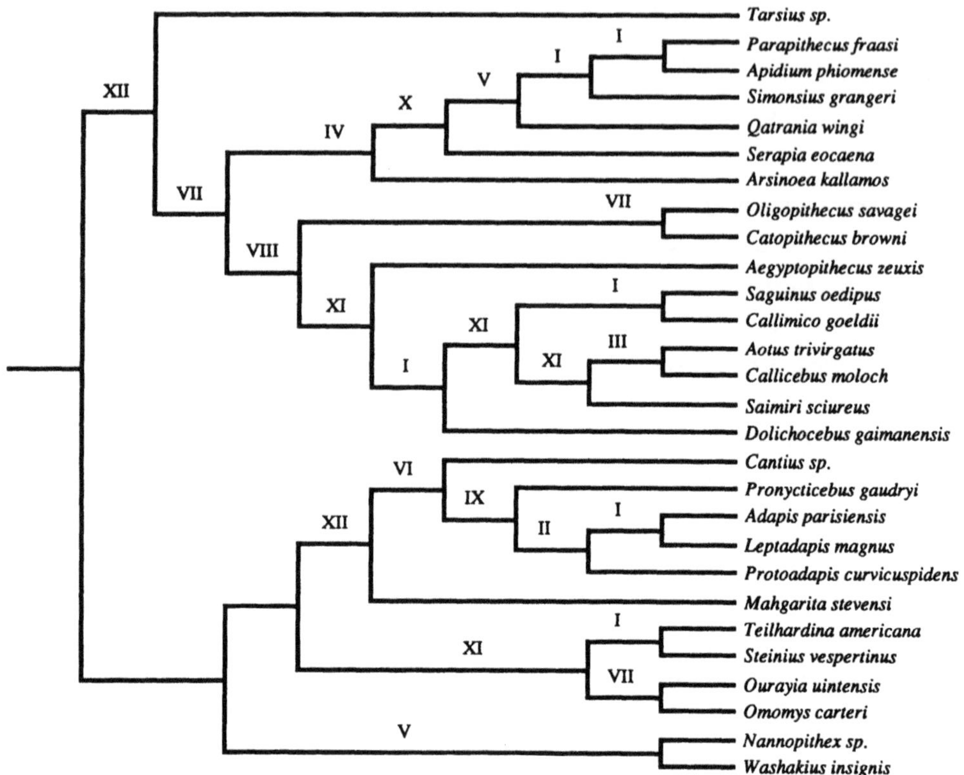

Fig. 8. Robusticity of various linkages in the maximum-parsimony networks of 49,446 steps based on the number of less-optimum trees discovered before the linkage is broken. I, the first 915 trees discovered show this linkage; II, the first 639 trees; III, the first 527 trees; IV, the first 378 trees; V, the first 231 trees; VI, the first 225 trees; VII, the first 81 trees; VIII, the first 48 trees; IX, the first 27 trees; X, the first 15 trees; XI, the first 4 trees; XII, only one of the two maximum-parsimony trees shows this linkage.

Tarsius and omomyids, i.e., Tarsiiformes, agrees with the maximum-parsimony Network I (Fig. 6) and the rooted network (Fig. 9). Others have suggested that anthropoids might be an offshoot of some advanced omomyid such as *Omomys* or its close relative *Chumashius* (Cartmill and Kay, 1978; Hoffstetter, 1980a,b; Kay, 1980; Rosenberger, 1986; Szalay, 1976; Szalay and Delson, 1979) or of a primitive omomyid like *Teilhardina* (Rose *et al.*, Chapter 1, this volume). A linkage with *Omomys* requires 1932 steps above maximum parsimony. Linkage with *Teilhardina* requires 1899 added steps. We found over 1000 trees more parsimonious than either of these proposed trees. From this we conclude that a phyletic linking of anthropoids with Tarsiiformes as a whole, and with *Tarsius* in particular, is likely, whereas linkage of Anthropoidea as the sister group to any omomyid considered here is substantially less probable.

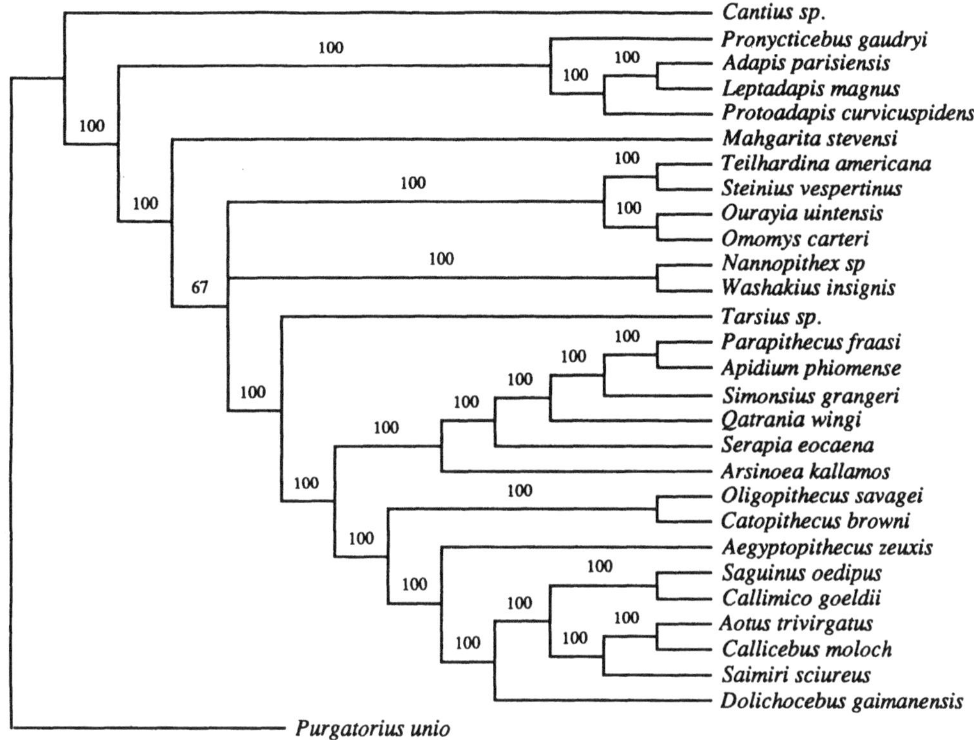

Fig. 9. Majority consensus of three maximum-parsimony trees of 40,678 steps when 28 primate taxa are rooted using *Purgatorius* as the outgroup.

Ancestral Morphotype of Anthropoidea

To assist in what follows, in Fig. 12 we have reproduced our maximum-parsimony tree of Anthropoidea. Key points at which the anthropoid lineage branches, referred to as "nodes," are assigned letters for easy reference. The internal nodes of the tree represent morphotypes. All changes leading up to a node constitute autapomorphies or synapomorphies of that node.

Our maximum-parsimony tree shows a *Tarsius*/Anthropoidea clade, but only a few derived characters support this hypothesis of relationship. On the lower molars, the cristid obliqua is short, reaching only to the base of the trigonid, it is oriented toward the protoconid. The hypoflexids are shallow. After the branching off from *Tarsius,* the remaining taxa (node A, Fig. 12) are anthropoids. Our two maximum-parsimony networks suggest the same

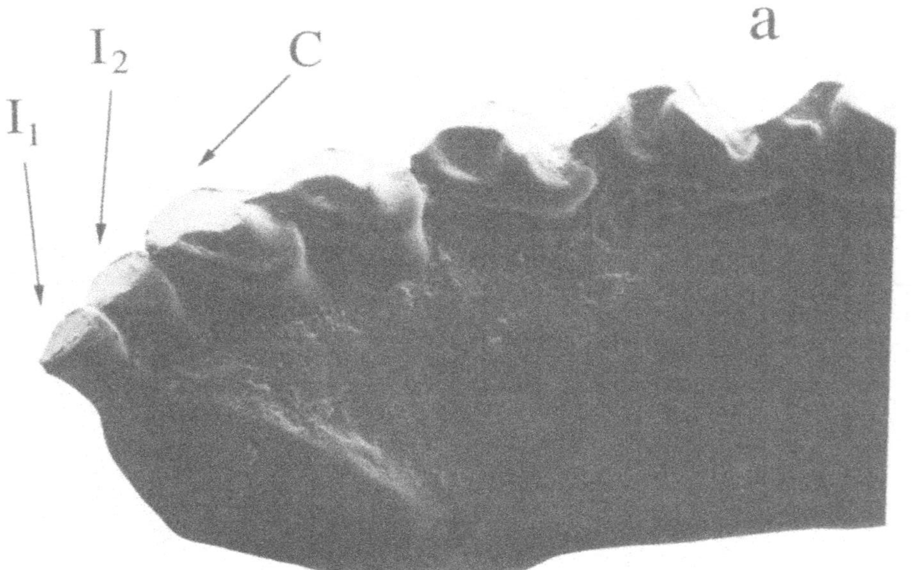

Fig. 10. Dentitions of selected tarsiiform primates. (a) *Washakius insignis,* UW 12580, occlusolingual view of right I_{1-2}, C_1, P_{2-4}; M_1 length of this specimen is 2.4 mm. I_1 and I_2 are labeled. Note that the incisors are small land vertically implanted with I_1 smaller than I_2.
(b) *Washakius insignis,* UW 13426, occlusolingual view of left P_3–M_2; M_1 length 2.5 mm. Note that the M_{1-2} cristid obliquas are long and trenchant and that the M_{1-2} paraconids are large. (c) *Washakius insignis,* UW 10275, occlusal view of right P^2–M^3. M^1 length of this specimen is 2.4 mm. (d) *Teilhardina americana,* UW 6896, occlusolateral view of left P_2–M_3. M_1 length of this specimen is 1.9 mm. (e) *Teilhardina americana,* UW 8961, occlusal view of right P^4–M^3. Note the weakly developed *Nannopithex* fold. M^1 length of this specimen is 1.9 mm. (f,g) *Tarsius* sp. Duke Primate Center skeletal collection. (f) Occlusal view of left I^1–M^3. (g) Occlusolateral view of left I_1–M_3. M_1 length is 2.5 mm; M^1 length is 2.8 mm. Note on lower molars the shallow hypoflexids, and the cristid obliquas, oriented toward the protoconids, reach only to the bases of the trigonids.

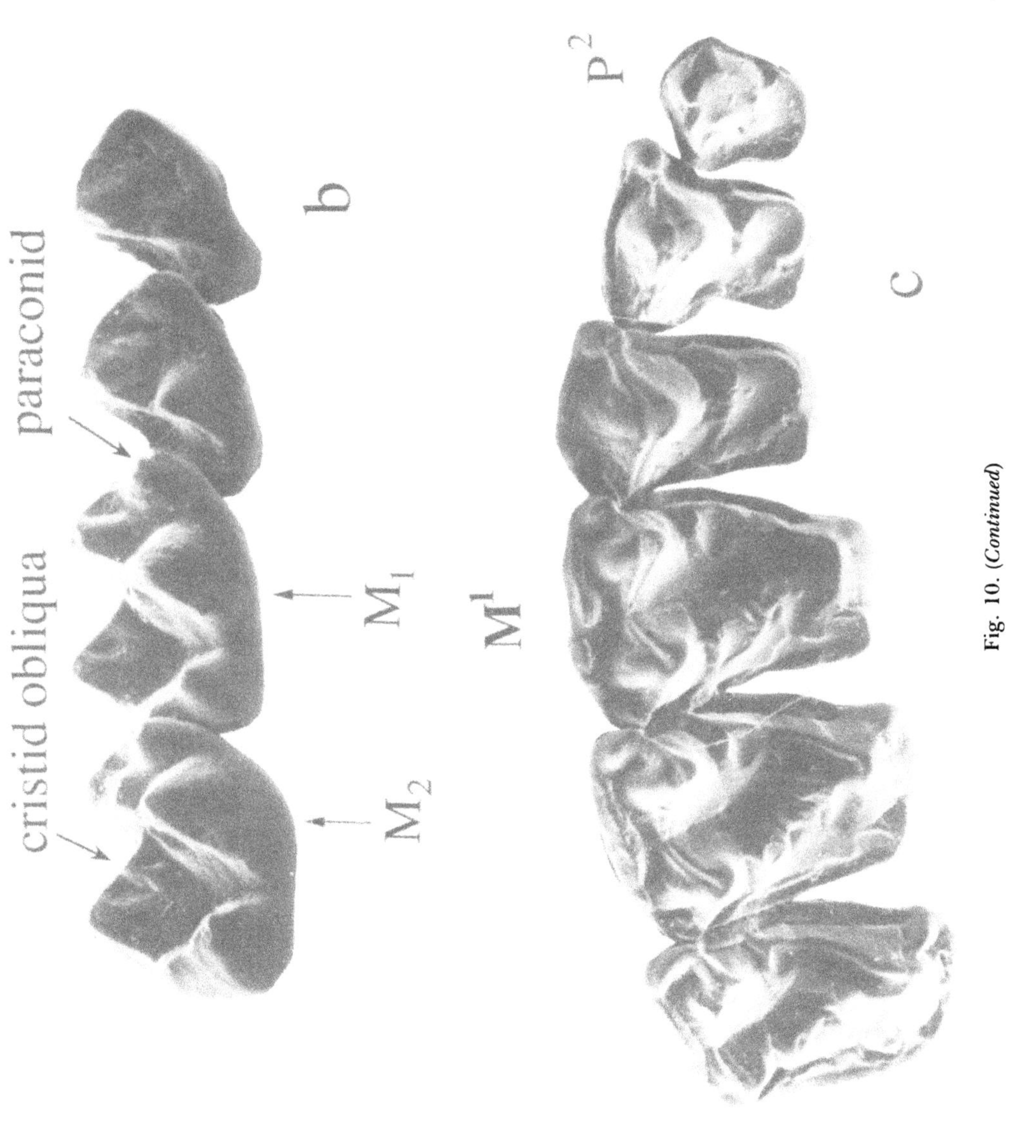

Fig. 10. *(Continued)*

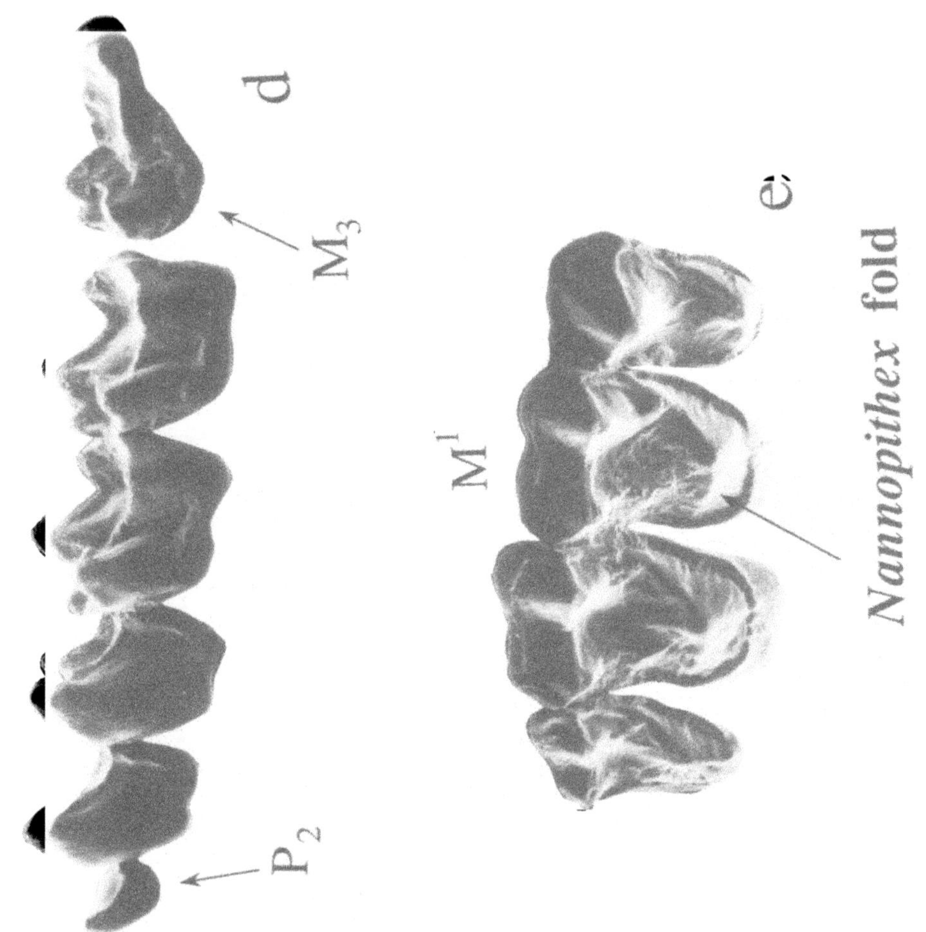

Fig. 10. (*Continued*)

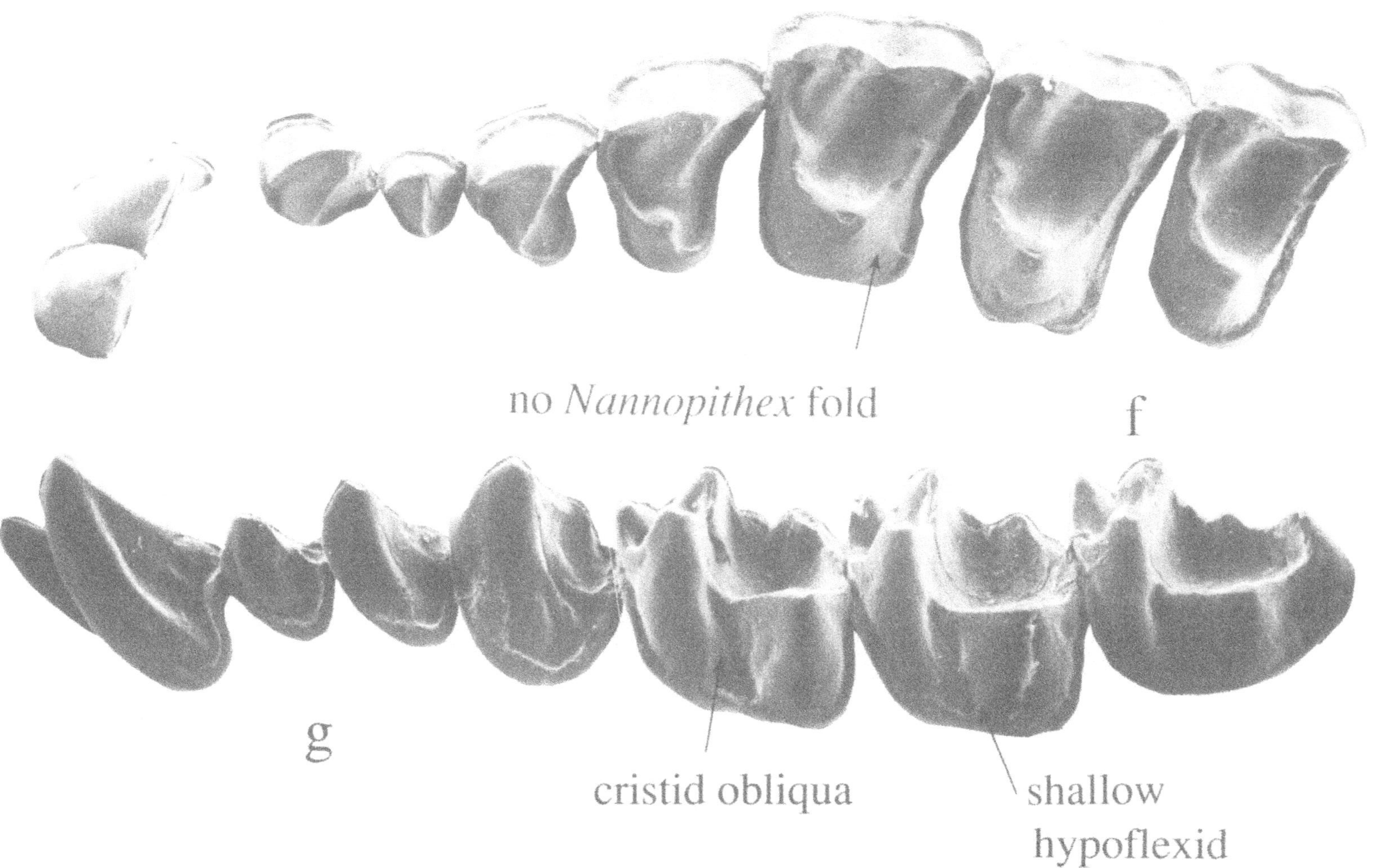

Fig. 10. (*Continued*)

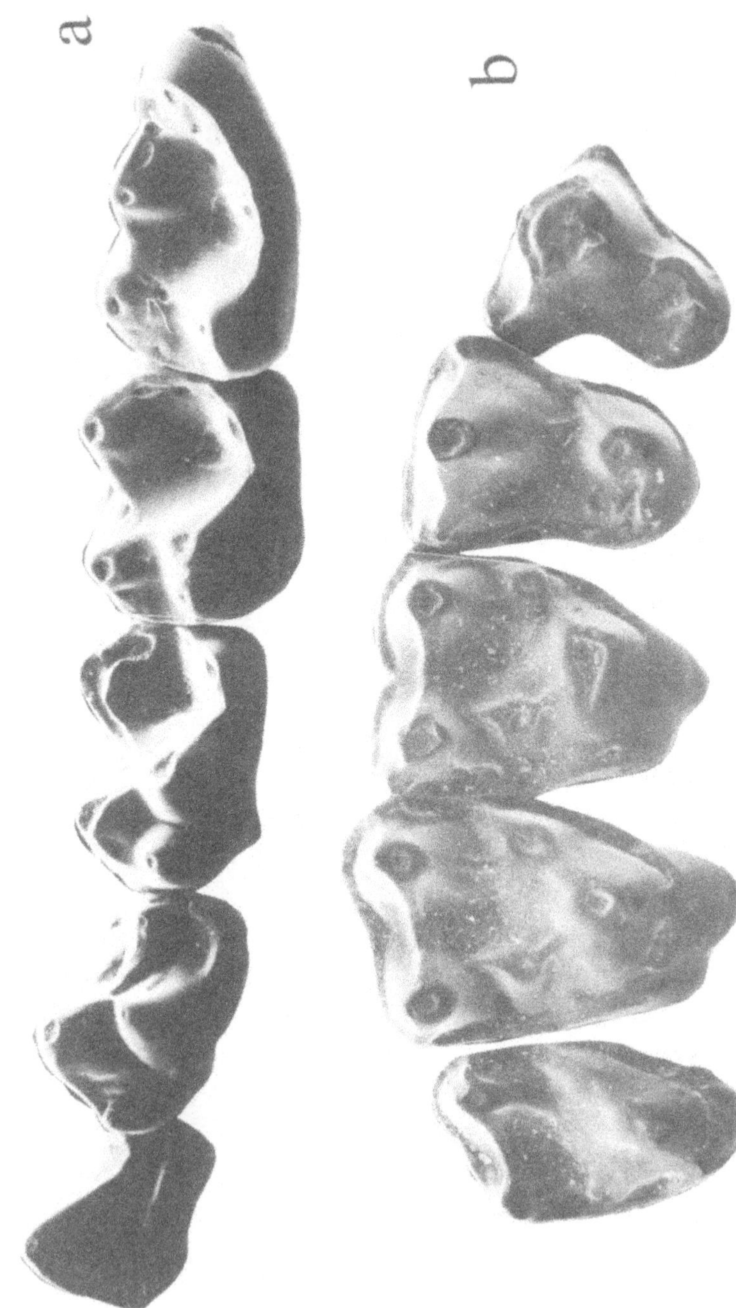

Fig. 11. Dentition of the adapid primate *Cantius eppsi*. (a) BMNH M 13773, occlusolateral view of right P_3–M_3. M_1 length of this specimen is 3.4 mm. (b) BMNH M 15145, occlusal view of right P^3–M^3. M^1 length is 3.4 mm.

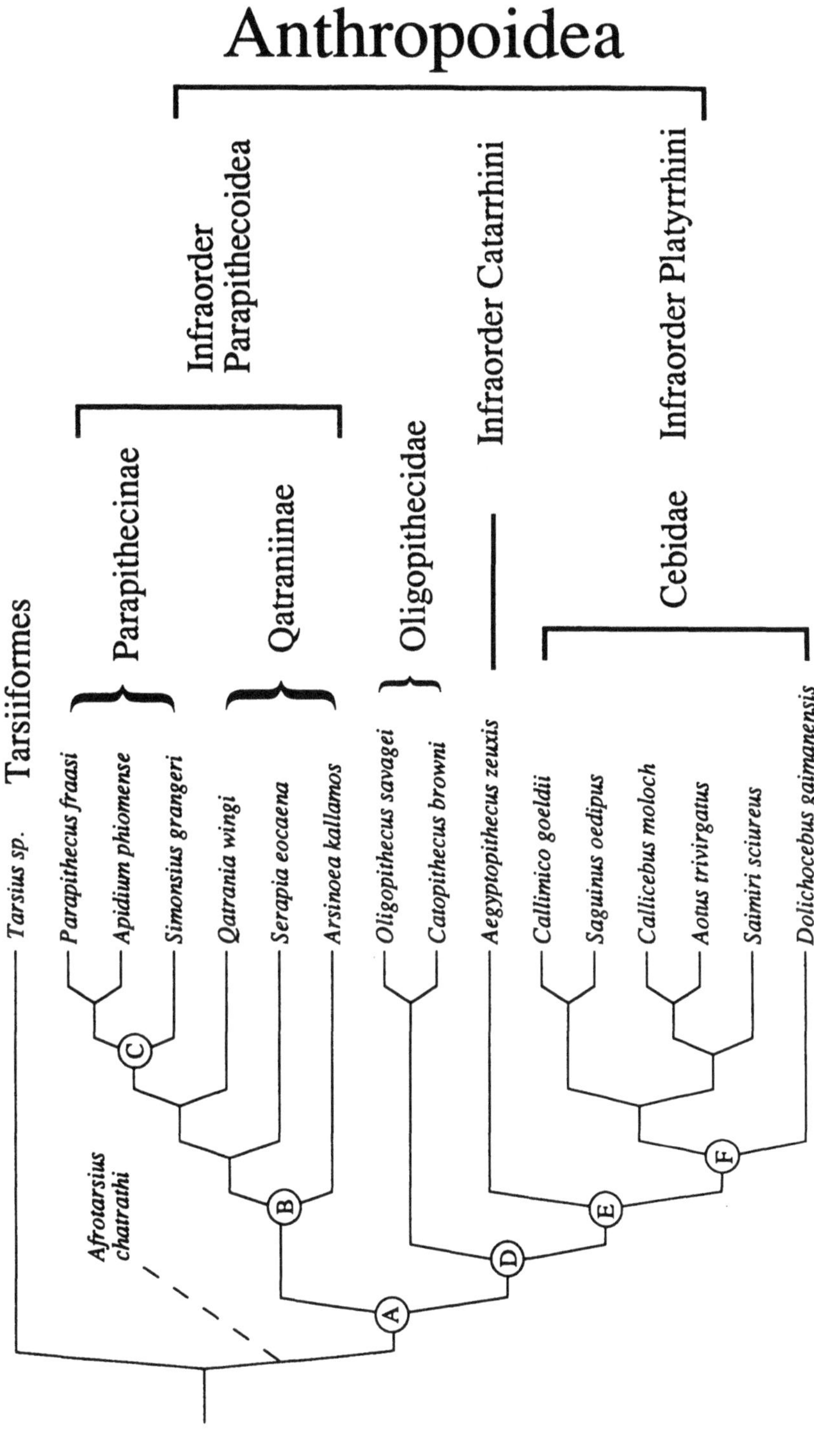

Fig. 12. Maximum-parsimony tree of Anthropoidea extracted from Fig. 6. Key points at which the anthropoid lineage branches or terminates, referred to as "nodes," are assigned letters for easy reference.

morphology for this hypothetical anthropoid ancestor. This morphotype is described below. At several points in this description, we refer to the relative size of a structure, e.g., to the "moderate" size of the lower incisors. Further information about the meaning of these relative terms is supplied in Appendix 1.

Incisors

The proposed ancestral anthropoid, at node A, Fig. 12, had two moderately high-crowned, vertically implanted, spatulate lower incisors arranged in an arcuate battery without intervening spaces (i1, #1; i2, #0; i3, #0; i9, #1; i11, #0; i14, #0). The lower incisors were very small to moderate-sized compared with molar area, with the first incisor smaller than the second (i5, #0 or 1; i6, #1). There was a lingual swelling on the lower second incisor, as well as a strong and complete lingual cingulum (i15, #1; i17, #1). No diastema intervened between the lower lateral incisor and lower canine (i4, #1).

The upper incisor crowns formed a closed arc with contacts occurring between all adjacent teeth (I1, #1; I2, #0). The first upper incisor was spatulate, buccolingually compressed, and much larger than the rounded or slightly compressed second incisor (I4, #2; I6, #1; I7, #0 or 1; I8, #0). The upper central incisor had a simple lingual fovea; its lingual cingulum was moderate and continuous, without a basal lingual cusp (I11, #1; I12, #0). A diastema was present between the upper lateral incisor and upper canine (I3, #0).

Canines

The canines were small to moderate in size, possibly with moderate dimorphism (c1, #1; c2, #1). The lower canine had a rounded oval cross section, and its paracristid was oblique to the occlusal plane (c3, #0; c5, #0). The upper canine was also oval in cross section, and its postparacrista wore against the lower second premolar (C1, #0; C2, #1). A deep groove was present on the mesial aspect of the canine crown (C3, #1). The upper canine lingual cingulum was strong (C4, #1).

Premolars

The three lower premolars of our proposed anthropoid morphotype were slightly basally inflated (p1, #1; p2, #0; p29, #1). The lower second premolar had one root, whereas the third and fourth premolars had two roots (p3, #0; p4, #2). The premolar crowns were slightly to greatly overlapping (crowded) in the mesiodistal axis (p5, #1 or 2). Lower premolars lacked paraconids (p6, #2; p7, #2). The third and fourth lower premolars had weak cristid obliquas (p11, #1). The trigonid of the lower fourth premolar was much longer than the talonid mesiodistally (p12, #0). The second and third premolars lacked metaconids or displayed only a trace of that cusp; the fourth premolar had a small metaconid, positioned close to the protoconid (p14, #0;

p15, #0; p16, #0; p17, #1). The P_4 trigonid was lingually open (p18, #1). The entoconid of P_3 was absent; on P_4 the entoconid was crestlike (p19, #0; p20, #1). On the lower third and fourth premolars the protocristid was distolingually oriented and, on P_4, joined the metaconid (p21, #0; p22, #1; p23, #1). On the talonid of P_3 and P_4 was a small buccally positioned hypoconid (p25, #1; p26, #0). The P_4 lacked a hypocristid (p27).

The upper second premolar had one or two roots; the third and fourth premolars were three-rooted (P1, #1 or 2; P2, #2; P3, #2). The P^2 was smaller than P^3, and P^4 was smaller than M^1 (P4, #0 or 1; P5, #1). Protocones were present on P^{2-4}, with the P^4 protocone low relative to its paracone (P9, #0; P11, #0; P12, #0). The upper premolars lacked hypocones, metacones, and paraconules (P10, #0; P13, #0; P14, #0). P^{2-3} had strong postprotocristae running to the waisted distal crown margins (P17, #0; P18, #1). The lingual cingulum was absent or weakly developed on P^{3-4}; these teeth lacked metastyles (P16, #0; P19, #0).

Molars

There were three two-rooted lower molars, with M_2 larger than M_3 (m1, #0; m2, #1; m3, #1; m4, #1; m5, #0). The trigonid and talonid widths of the second and third lower molars were similar (m6, #1; m7, #1). Paraconids on M_{1-3} were small and positioned mesiolingually between the protoconids and metaconids (m8, #1; m9, #1; m10, #1; m20, #1). The M_3 hypoconulid was situated on a narrow heel (m13, #0; m14, #1). The heights of the molar trigonids and talonids were similar, and cusp relief was moderate to high (m16, #2; m17, #0). The M_1 trigonid was lingually open, with the metaconid positioned distolingual to the protoconid (m18, #0; m19, #1). Wear facet X was absent on the lower molars, and the M_{1-2} lateral protocristids ran towards the metaconids, forming complete distal trigonid walls (m21, #0; m22, #0; m23, #0; m24, #1). First and second lower molars had large entoconids; weak postentoconid sulci were visible, but distal foveae were absent (m25, #3; m26, #1; m41, #0). M_1 and M_2 had moderate-sized hypoconulids positioned near the crown midline (m28, #1; m29, #1; m30, #1). These two molars had strong oblique cristids that were mesiodistally oriented, reaching the trigonid wall at the base of the protoconids and leaving shallow hypoflexids (m31, #1; m32, #0; m33, #0; m34, #0; m35, #0; m46, #1; m47, #1). M_{1-2} hypocristids were weak (m38, #1). The molar talonids were notched lingually but not open (m40, #1). Molar cusps were slightly inflated basally; buccal cingulae were present but incomplete (m44, #1; m45, #1).

The upper molars had three roots (M1, #0; M2, #0). M^{1-2} were quite transverse; that is, they were substantially broader than they were long, with M^1 similar in size to or smaller than M^2 (M3, #1; M4, #2). A *Nannopithex* fold was absent (M5, #0). These teeth had small conules; paraconules were attached to the preprotocristae (M8, #1; M15, #0; M16, #1). Hypocones, possibly large (derived from the cingulum), were present on M^1 and M^2, positioned distally and slightly lingually to the protocones (M6, #0; M10, #0 or 1; M11, #0 or 1; M12, #0). Prehypocristae were absent (M13, #0; M14, #0). The M^{1-2} postprotocristae were strong and ran to the base of the metacone (M18, #0). Mesostyles were absent from the weak buccal cingulae (M17, #0; M21, #1). The upper molars had strong and complete lingual cingulae and lacked pericones (M20, #0; M22, #2). Body size was less than 500 g (M23, #0).

Synapomorphies of Anthropoidea

Changes occurring along the lineage segment between its base and node A in Fig. 12 are the synapomorphies of Anthropoidea. This phylogenetic reconstruction suggests that the following are anthropoid synapomorphies: vertically implanted lower incisor roots and crowns (versus more procumbent) (i11, #1 → #0; i12, #1 → #0); reduction in the size of M_3 compared with M_2 (versus larger M_3) (m5, #2 → #0); reduction in the height of molar trigonids so that they are similar in height to the talonids (versus much higher) (m16, #1 → #2); reduction in the size of the molar paraconids (versus large paraconids) (m20, #2 → #1); appearance of a postentoconid sulcus on M_{1-2} (versus absent) (m26, #2 → #1); development of a moderate-sized M_{1-2} hypoconulid (versus absent) (m27 & 28, #3 → #1); weakening of the M_{1-3} hypocristid (versus strong) (m38 & 39, #2 → #1); development of a P^2 protocone (versus absent) (P12, #1 → #0); and weakening of the upper molar buccal cingulum (versus strong) (M21, #2 → #1).

Anthropoid Cladogenesis

A number of clades are identified in Fig. 12. The first group to cleave from the ancestral primate node is Parapithecidae (node B). Remaining anthropoids divide at their base (node D) into Oligopithecidae on one hand and, on the other, at node E, Propliopithecidae (the sister group of Miocene–Recent catarrhines) plus Platyrrhini.

Parapithecidae

We recognize as parapithecids a collection of taxa including *Apidium, Parapithecus, Simonsius, Qatrania, Arsinoea,* and *Serapia* (Simons, 1992) (Figs. 13–16). All of the lower premolar and molar characters mentioned below are known for all taxa in the above list. Only a single upper molar is known for *Qatrania* (Simons and Kay, 1988), and no upper teeth have been described for *Parapithecus, Arsinoea,* or *Serapia.* As some of these parts of the dentition become better known, it may turn out that what we now consider distinctive characters of parapithecids as a whole are actually characteristic of only a more restricted clade within parapithecids.

A number of parapithecid synapomorphies evolved along the lineage segment between nodes A and B in Fig. 12. A small P_3 metaconid was developed (versus absent) (p16, #0 → 1). The P_{3-4} hypoconids were displaced lingually (versus buccally positioned) (p26, #0 → #1). The M_3 talonid was reduced in breadth compared with the trigonid (versus both having similar widths) (m7, #1 → #0). M_{1-2} had very low cusp relief (versus moderate to high) (m17, #0 → #1). The M_{1-2} cristid obliqua was weakened and rounded (versus strong) (m31, #1 → #0). P^{2-3} had rounded distal crown margins (versus waisted) (P18, #1 → #0). A very large paraconule developed on the upper premolars (versus absent) (P14, #2 → #0). The upper molar paraconules and metaconules were enlarged and "puffy" (M8, #1 → #2; M16, #1 → #2). The molar postprotocristids were rounded and indistinct (M18, #0 → #1).

Cladogenesis within Parapithecidae

Within Parapithecidae we recognize two subfamilies, Parapithecinae (node C, Fig. 12) and Qatraniinae (all taxa between nodes B and C, Fig. 12). Parapithecinae (*Parapithecus, Apidium,* and *Simonsius*) are specialized members of the family sharing a number of derived dental character states. Qatraniinae (*Qatrania, Serapia,* and *Arsinoea*) possess several convincing shared-derived parapithecid dental features. However, overall they are more primitive in their dental structure and may well represent a paraphyletic assemblage from within which parapithecines arose.

Serapia and *Arsinoea,* recently described by Simons (1992) (Fig. 13), are described on the basis of lower teeth. We place them among parapithecids because they possess parapithecid synapomorphies mentioned above such as presence of a small P_3 metaconid (p16, #1), lingual displacement of the P_{3-4} hypoconids (p26, #1), very low cusp relief on M_{1-2} (m17, #1) with weakened and rounded cristid obliquas (m31, #0), and a reduced M_3 talonid breadth (m7, #0).

Serapia and *Arsinoea* lack parapithecine characteristics developed at node C (Fig. 12; see below). They do not possess deeply notched P_{3-4} posterior trigonid walls with discontinuity between the medial and lateral protocristids and deep M_1 hypoflexids. Nor do they have the M_2 trigonids wider than the talonids. Furthermore, wear facet X is absent on the second lower molars, and there is no sulcus between lower molar protoconids and metaconids. Although they are both parapithecids, they share no convincing synapomorphies between them, so *Arsinoea,* first, and then *Serapia* are considered comb-like offshoots of a primitive parapithecid stock.

Arsinoea has a somewhat larger, more lingually placed P_4 metaconid, reminiscent of propliopithecids, oligopithecids, and platyrrhines. It remains a possibility that *Arsinoea* may be closer to the last common ancestor of the latter taxa (at node D, Fig. 12) than to that of parapithecids, further supporting the proposal that the premolar structure of parapithecids (i.e., having a small,

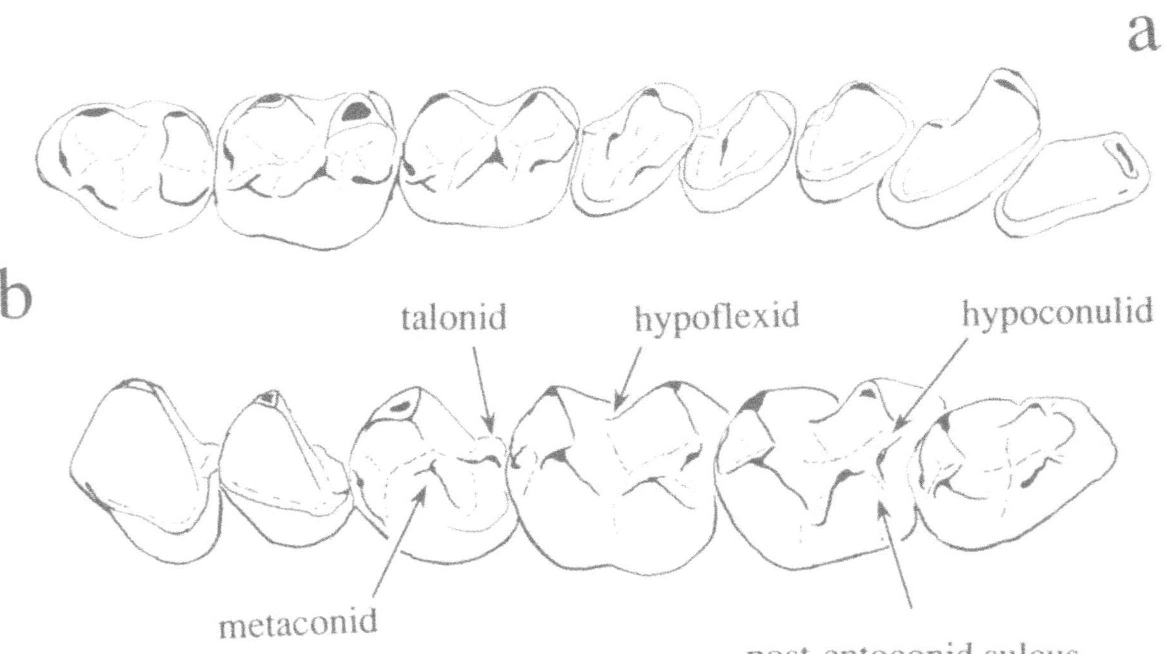

Fig. 13. The mandibular dentition of (a) *Serapia eocaena*, DPC 7340, left I_2–M_3, and (b) *Arsinoea kallimos*, CGM 42310, right P_2–M_3. On M_{1-2}, note the moderately large hypoconulids, postentoconid sulci, and shallow hypoflexids. The premolars have abbreviated talonids. P_4s are small, with distolingually positioned metaconids (*Arsinoea* has a larger one than *Serapia*).

distally positioned P_4 metaconid like Eocene Holarctic primates) is primitive for anthropoids. Another interesting aspect of the anatomy of *Arsinoea* is that, unlike more derived parapithecids such as *Apidium*, the mandibular symphysis was unfused. (Personal observations of the type of specimen indicate a rugose, planar symphyseal plate bordered by a raised lip.) This finding shows that symphyseal fusion, a hallmark of all modern anthropoids, occurred independently in parapithecids on one hand and propliopithecids and platyrrhines on the other. Parenthetically, although some suggestions have cropped up recently that the oligopithecid *Catopithecus* might have an unfused symphysis (Simons, 1989), examination of the available specimens does not convince us of this.

Qatrania wingi (Fig. 14) shares the parapithecid synapomorphies mentioned above. It lacks two important synapomorphies of parapithecines (*Parapithecus, Apidium, Simonsius*): There is no wear facet X on the second lower molars, nor is there a deeply incised sulcus between lower molar protoconids and metaconids. Unfortunately, the anatomy of the upper teeth is poorly known. It would be especially important to know if the upper premolars of *Qatrania* had enlarged paraconules like those of parapithecines.

A single parapithecid tooth is known from a late Eocene or early Oligocene (Rasmussen *et al.*, 1992) locality in Algeria (de Bonis *et al.*, 1988; Godinot, Chapter 10, this volume). The new taxon *Biretia piveteaui* was erected for this lower tooth, believed by de Bonis *et al.* to be an M_1. Comparisons with *Qatrania wingi* were difficult for those authors because of the worn state of the type specimen of *Q. wingi*, CGM 40240 (Simons and Kay, 1983). However, new material of *Q. wingi* is now available for comparison (DPC 3045 and 6125; Simons and Kay, 1988). This Algerian tooth is very similar to the lower second molar of the Fayum species. The M_2 of *Q. wingi* and *Biretia* are of similar size and have trigonids wider than the talonids (m6, #0). As in *Q. wingi* (but unlike

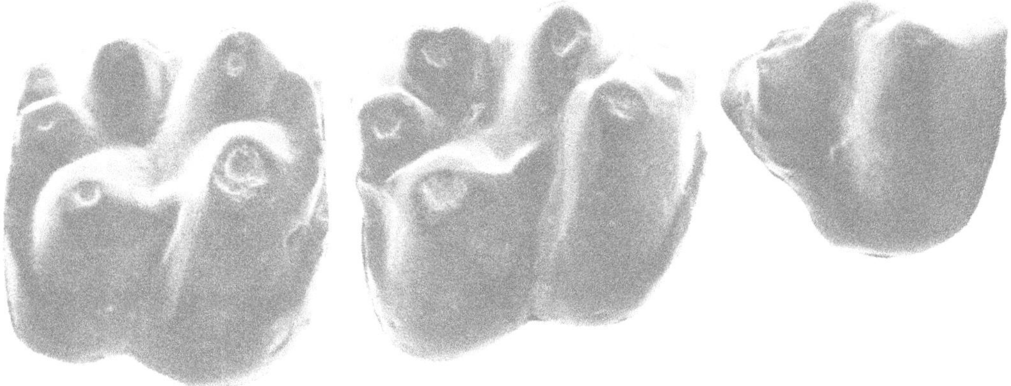

Fig. 14. The dentition of *Qatrania wingi*, DUPC 6125, occlusolateral view of right P_4–M_2. M_1 length of this specimen is 2.8 mm.

in other parapithecid genera), the trigonid is slightly higher than the talonid (m16, #1). The position of the metaconid of *Biretia* is more nearly transverse to the protoconid, as in the M_2, but not the M_1 of *Q. wingii* (m19, #0). The trigonid of *Biretia* is closed lingually as is the M_2 of *Qatrania*. The placement of the cristid obliqua is more buccal, as in a *Qatrania* M_2, not its M_1. De Bonis *et al.* (1983) state that the Algerian taxon "probably" possessed a wear facet X (p. 930). If this is so, that is apparently the only feature in which this specimen differs from *Q. wingii*. Pending recovery of more material, *B. piveteaui* may not be distinguishable from *Q. wingi*.

Parapithecinae (node C, Fig. 12) includes *Apidium, Parapithecus,* and *Simonsius*. *Apidium phiomense* is very closely related to *Parapithecus fraasi* (Fig. 15). The two differ principally in *A. phiomense* having a proportionally larger M_3 and centroconids on its lower molars. *Apidium moustafai,* a geologically older and more primitive species of *Apidium* (Simons, 1962) not analyzed here, resembles *P. fraasi* even more closely. Its M_3 is smaller, and molar centroconids are smaller than those of *A. phiomense* (Gingerich, 1978; personal observations). This increased resemblance between *P. fraasi* and the earlier-occurring species of *Apidium* (*A. moustafi*) further reinforces the hypothesis of relationship between the two.

Two caveats are necessary in consideration of the fragmentary nature of the evidence available for *Parapithecus fraasi*. The type, and only described specimen, is a lower jaw preserving the incisors, canines, and cheek teeth. In the absence of any known upper teeth, the distinctive morphology of the upper cheek teeth of *Apidium,* with its enlarged molar pericones and premolar paracones, cannot be compared with *P. fraasi*. The number of incisors in *P. fraasi* is also in doubt. The only specimen is broken in the symphyseal region with just two incisors preserved (Fig. 15). Much disagreement has been expressed about how many incisors were present before this damage occurred [see Kay and Simons (1983) and Simons and Rasmussen (1991) for reviews]. One view is that *P. fraasi* had just one incisor, possibly a deciduous one, in each jaw half, but others argue that this species had two incisors, the lower central incisors having been lost by postmortem damage. Resolution of this question is not feasible with the present material (Simons and Rasmussen, 1991).

The debate over the number of incisors of *Parapithecus fraasi* has relevance for the classification and phylogeny of *Simonsius* (= *Parapithecus*) *grangeri*. The latter is a very derived species of parapithecid exhibiting a number of autapomorphies, the most important of which is the loss of both lower incisors, a condition unique among primates (Kay and Simons, 1983). The single fact that *Simonsius* lacks all lower incisors as an adult and *Parapithecus* certainly has at least one pair, and perhaps two, should warrant a taxonomic distinctness of the two at the generic level.

Simonsius (Fig. 16) is placed most parsimoniously as a sister taxon to the *Apidium*/*Parapithecus* group within the parapithecine clade (node C, Fig. 12). Among the possible synapomorphies that *Apidium* and *Parapithecus* share to the exclusion of *Simonsius* are the following: the P_{3-4} hypoconids were large

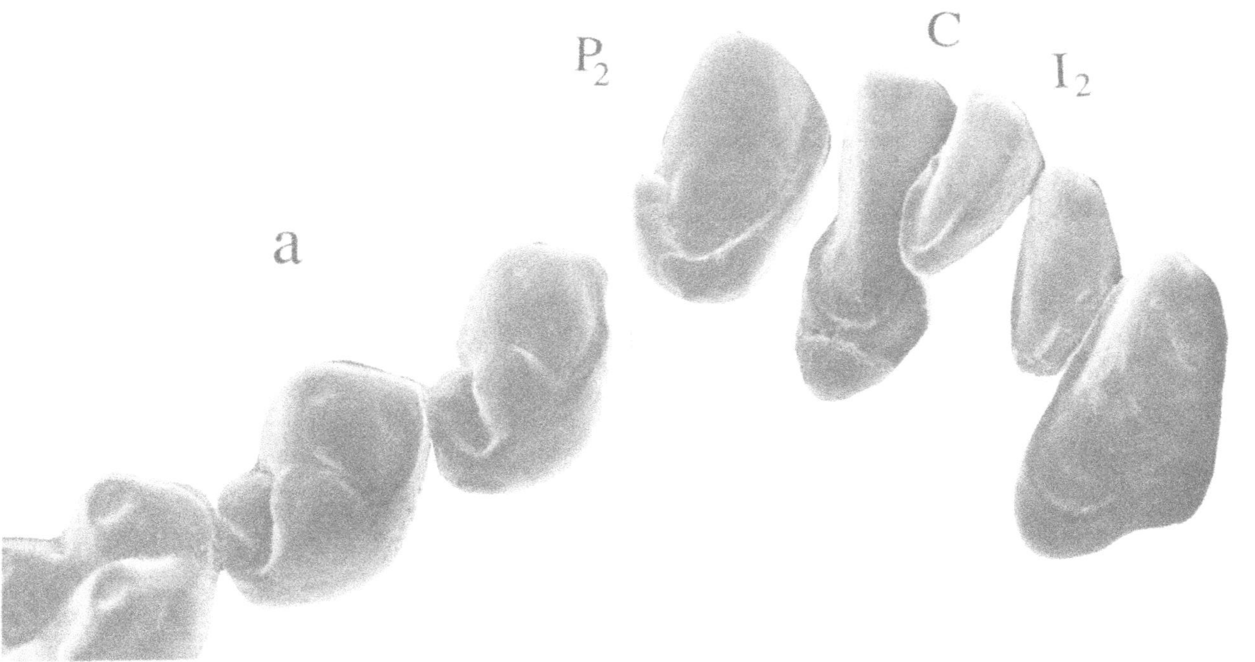

Fig. 15. The dentition of *Parapithecus fraasi* and *Apidium phiomense*. (a) *Parapithecus fraasi,* TYPE, Stuttgart, occlusolingual view of right I_2, C, left I_2, C,P_{2-4}, M_1 (part). M_1 length of this specimen is 3.9 mm. The number of incisors in *Parapithecus fraasi* is in doubt: The specimen is broken in the symphyseal region with just two incisors preserved. Other features of note include premolars with short talonids and small, distolingually positioned metaconids. (b) *Parapithecus fraasi,* TYPE, Stuttgart, occlusolateral view of P_4–M_3. (c) *Apidium phiomense,* DPC 1048, occlusolingual view of left P^{3-4}, M^1. M^1 length of this specimen is 3.8 mm. Features of note include the large P^4 paraconule unattached to preprotocrista and rounded P^{3-4} distal crown margins. (d) *Apidium phiomense,* DPC 1048, occlusal view of left M^2. Note enlarged paraconule and metaconule, indistinct postprotocristid, large hypocone, and pericone. (e) *Apidium phiomense,* DPC 3080, occlusolingual view of right M_{1-2}. Note the large centroconid. M_1 length of this specimen is 3.2 mm.

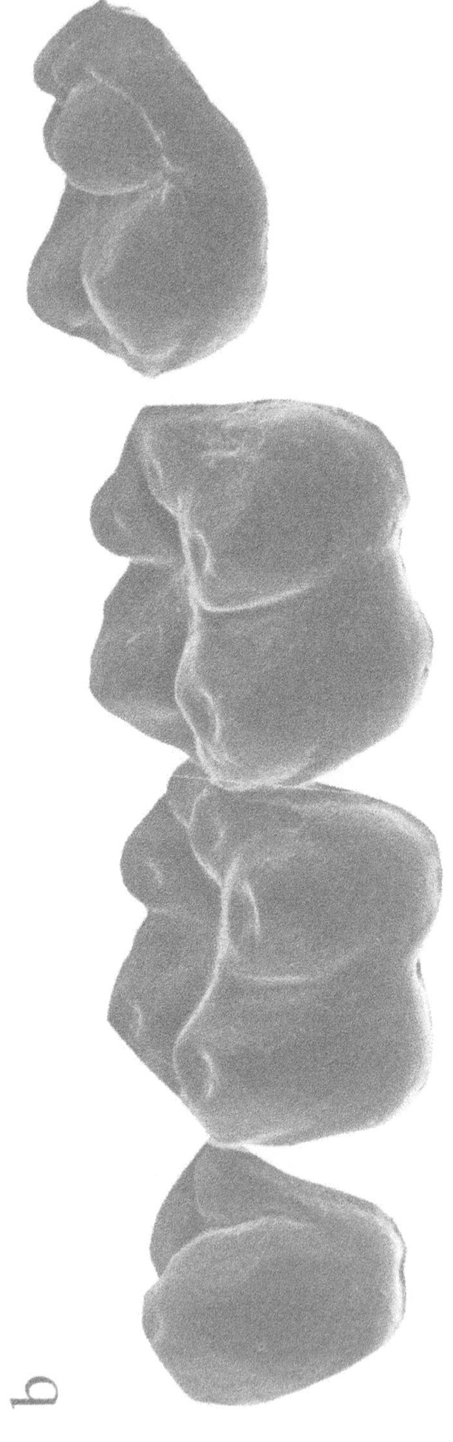

Fig. 15. *(Continued)*

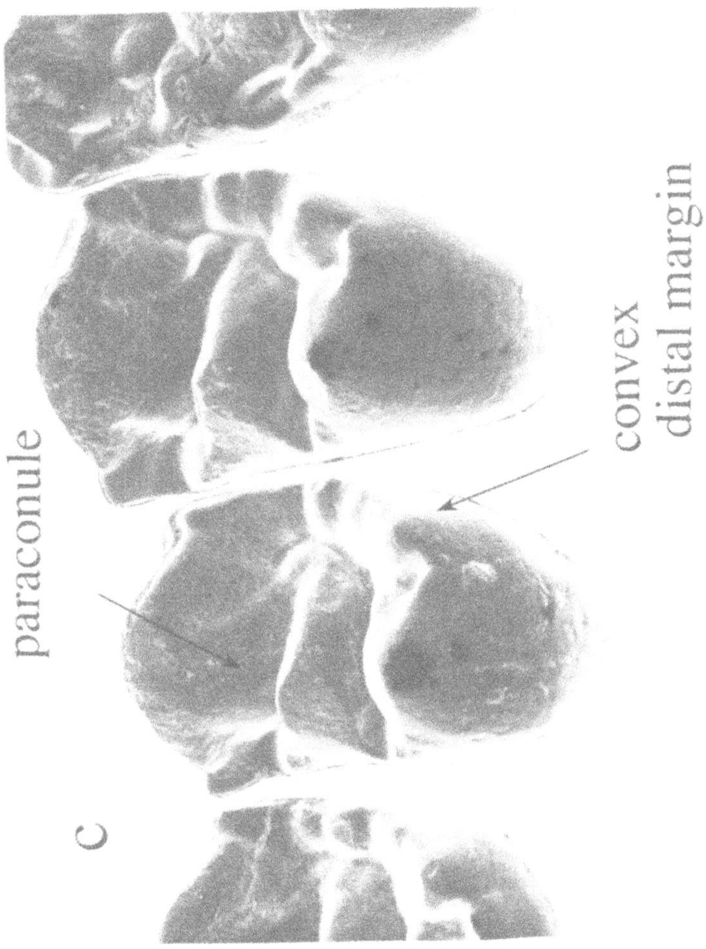

Fig. 15. *(Continued)*

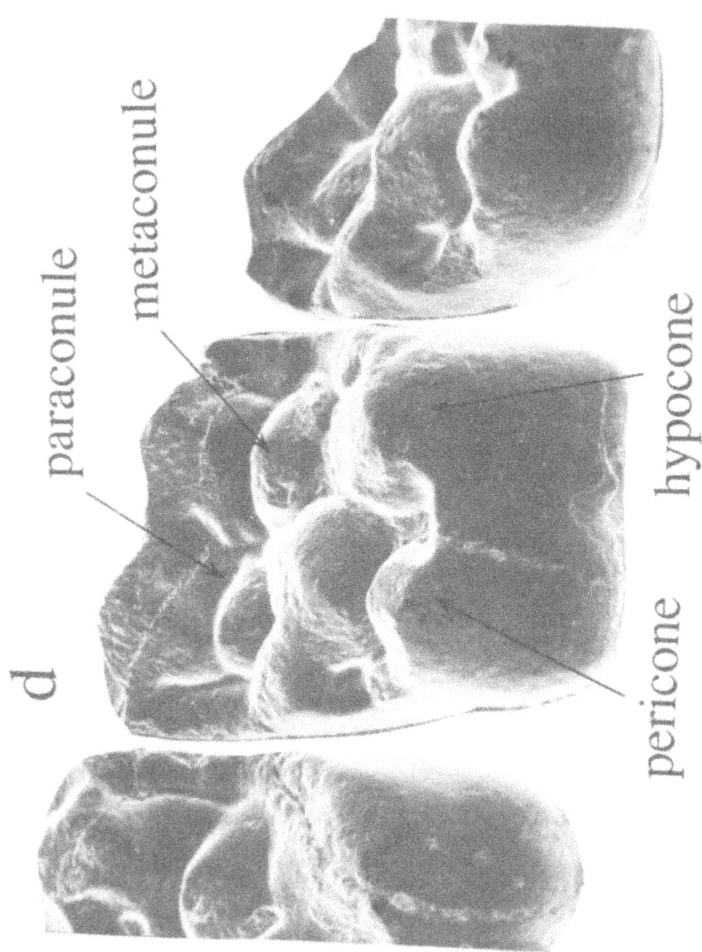

Fig. 15. *(Continued)*

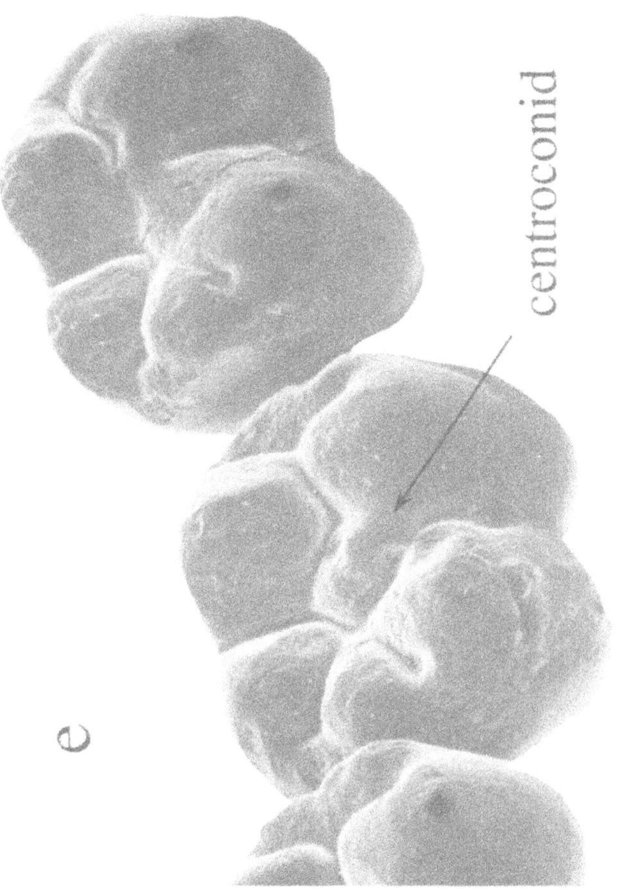

Fig. 15. (*Continued*)

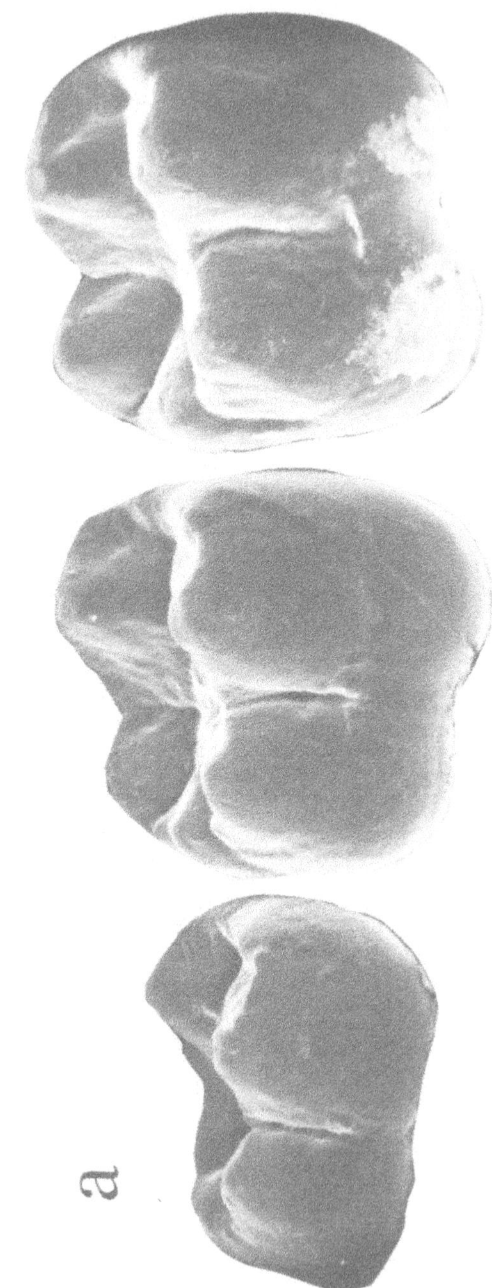

Fig. 16. The dentition of *Simonsius grangeri*. (a) DPC 3110, occlusobuccal view of right M_{1-3}. M_1 length of this specimen is 5.8 mm. (b) DPC 1123, occlusal view of right M^{1-2}. M^1 length of this specimen is 4.44 mm. (c) DPC 2807, occlusolingual view of C_1–M_1. M_1 length of this specimen is 5.8 mm.

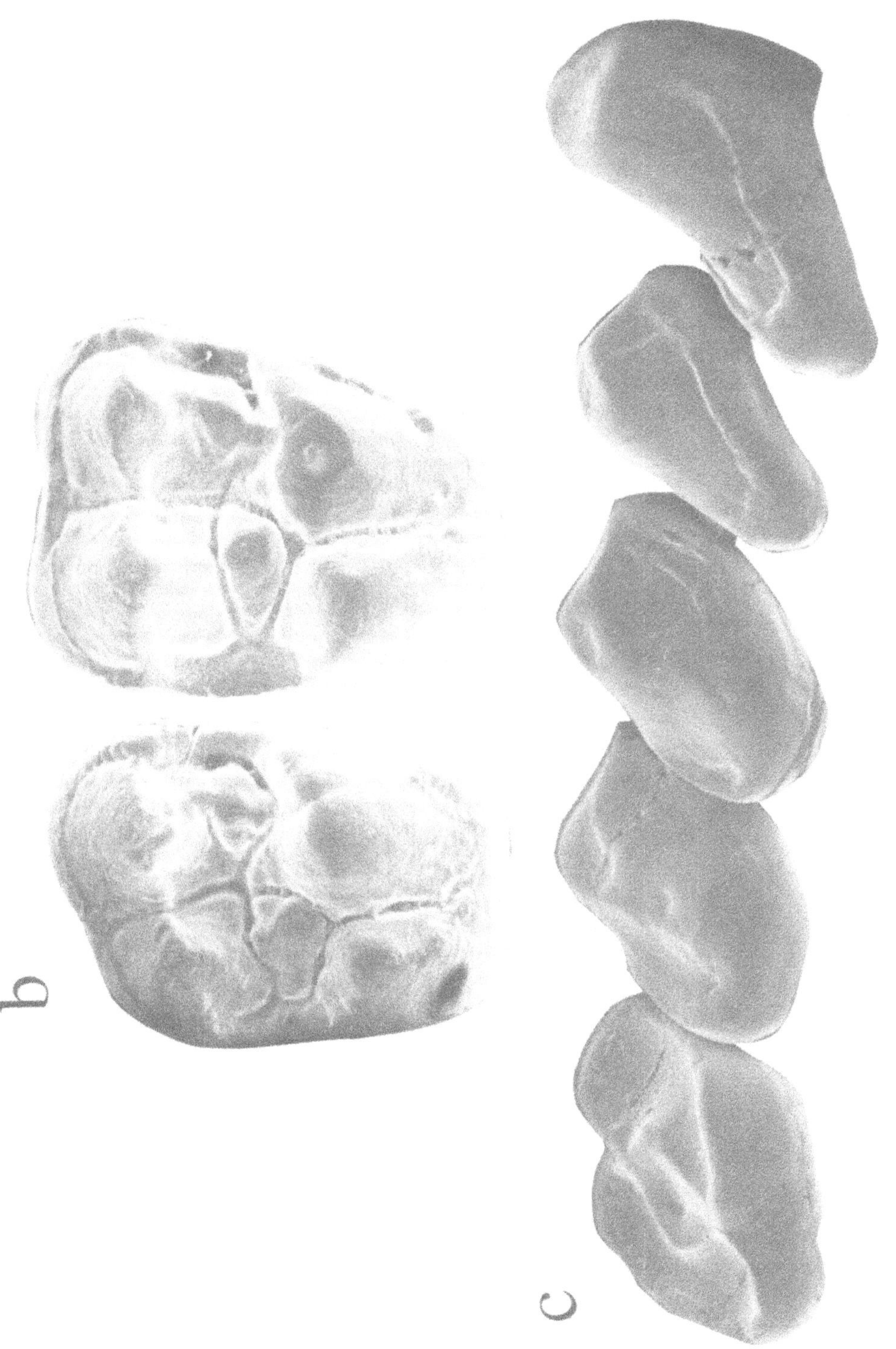

Fig. 16. *(Continued)*

(versus absent in *Simonsius*) (p25, #0 versus #2); the P_2 buccal cingulum was incomplete (versus complete) (p28, #1 versus #2); the M_1 cristid obliqua reached the trigonid at a point distolingual to the protoconid (versus directly distal to the protoconid) (m32, #1 versus #0); the P_4 paraconid was absent (versus reduced in size) (p7, #2 versus #1); and the molar cusps were more inflated (versus less inflated) (m44, #2 versus #1). Our results do not support the hypothesis that *Simonsius* is the sister group of *Parapithecus* or that the two should be considered congeners. Several of the shared similarities of *Simonsius* and *Parapithecus,* on which the proposed congeneric status of the two has been based, are more likely retained primitive features (e.g., M_3 smaller than M_2, absence of a centroconid; Simons, 1974). Of course, if *P. fraasi* could be shown to have had incisor reduction to one tooth foreshadowing the complete loss of lower incisors in *Simonsius,* this would be an important synapomorphy lending support to a hypothesis of sister-group relationship between the two species.

Oligopithecids, Catarrhines, and Platyrrhines

Oligopithecines, catarrhines, and platyrrhines (node D, Fig. 12) constitute the remainder of Anthropoidea after the branching of the Parapithecidae. Dentally these are anthropoids of modern aspect. They share the following synapomorphies: crowding of the premolars was reduced (versus being crowded) (p5, #1 → #0); the P_4 talonid was increased in size so that the trigonid and talonid are similar in mesiodistal length (versus a dominant trigonid) (p12, #0 → #1); the P_4 metaconid was enlarged, medially positioned, and widely separated from the protoconid so that the lateral protocristid was transversely oriented (versus small metaconid, close to the protoconid, and positioned distolingual to it) (p14, #0 → #1; p17, #1 → #2; p23, #1 → #0); similarly, the M_1 metaconid was shifted to a more transverse position relative to the protoconid (versus distolingual to it) (m19, #1 → #0); the molar cusps were more marginally positioned (versus molar sides more sloping, crown constricted) (m44, #1 → #0); the lingual moiety of P^2 was reduced to become a more triangular tooth (versus a more transverse tooth) (P6, #1 → #0); the lingual cingula of the upper premolars were strengthened (versus faint or absent) (P19, #0 → #1); the upper molar metaconules were lost (versus small metaconules) (M16, #1 → #0); and body size was larger than 500 g (versus less than 500 g) (M23, #0 → #1).

Oligopithecidae

Oligopithecus and *Catopithecus* (Fig. 17) are closely related sister taxa separated at node D (Fig. 12). The positioning of oligopithecids before the split between catarrhines and platyrrhines is in agreement with the previously expressed views of Kay (1977) and Szalay and Delson (1979). As intraspecific variation within these species is better documented, it may prove necessary to consider the two congeneric. Oligopithecids depart from the morphotype at

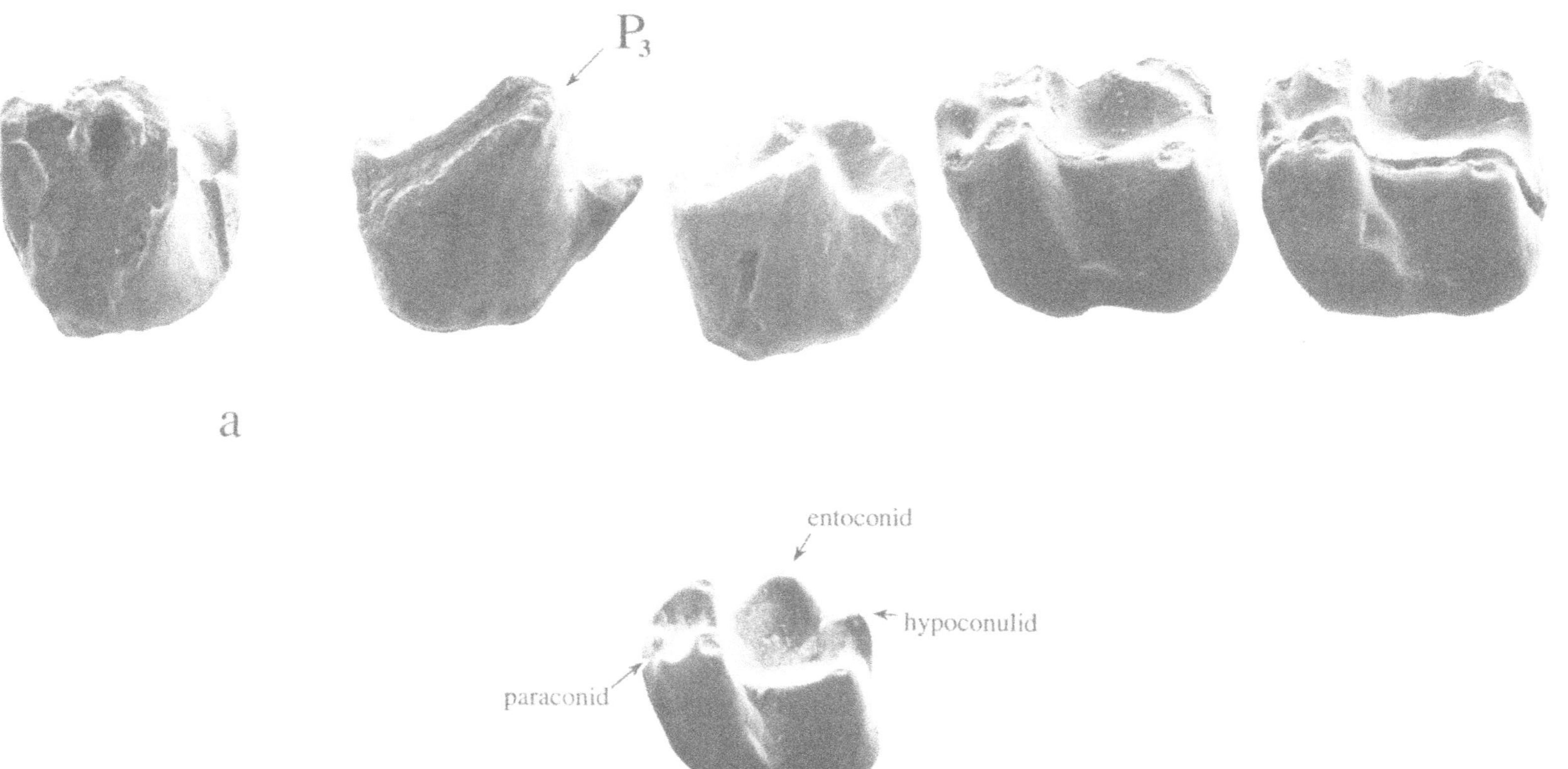

Fig. 17. Dentition of *Oligopithecus savagei.* (a) CGM 29627, occlusolateral view of left C_1–M_2. M_1 length of this specimen is 5.8 mm. Note the enlarged and raised P_3, functionally aligned with upper canine. (b) DPC 6062, occluslingual view of right M_1 (length = 3.2 mm). Note twinned hypoconulid and entoconid.

node D by having lost the $P^2/_2$ so that the upper canine occluded against P_3 (versus retaining three premolars and C–P_2 occlusion) (p2, #0 → #1; C2, #1 → #3). Also, P_3 was enlarged and raised in conjunction with its functional alignment with the upper canine (p10, #1 → #2). On the molars, the hypoconulid and entoconid were twinned (vs. having a centrally positioned hypoconulid) (m30, #1 → #0), and the M^{1-2} paraconule was absent (versus present, small) (M8).

Catarrhini and Platyrrhini

The maximum-parsimony solutions place *Aegyptopithecus* (here taken as representative of Propliopithecidae) as a sister taxon to Platyrrhini. Because *Aegyptopithecus* is the only catarrhine included in this study, the derived characters shared by *Aegyptopithecus* and platyrrhines should be understood as those that unify the Catarrhini and Platyrrhini (node E, Fig. 12). The synapomorphies that unify Catarrhini and Platyrrhini include the closure of the trigonids of P_4–M_1 (p18, #1 → #0; m18, #0 → #1) and loss of the molar paraconids (m20, #1 → #0). P^{2-3} had rounded waisted distal crown margins (versus waisted) (P18, #1 → #0). Hypocones developed on the upper premolars (P13, #0 → #2), and lateral posterior transverse cristae were formed on M^{1-2} (M19, #1 → #0).

The propliopithecid *Aegyptopithecus* (Fig. 18) (and presumably its close relative *Propliopithecus*) is assumed here to be the sister to living Old World monkeys and apes following the views of many workers (e.g., Delson, 1975; Fleagle and Kay, 1983, 1987; Harrison, 1987; Hoffstetter, 1980a,b; Szalay and Delson, 1979). Those authors list a number of convincing dental and postcranial synapomorphies that unite propliopithecids with Miocene–Recent "apes" and Old World monkeys. Distinctive derived features of *Aegyptopithecus* and Miocene catarrhines included premolar reduction so that the upper canine occluded against the P_3 and development of wear facet X on the molars (p2, #1 → #0; C2, #1 → #3; m24, #0 → #1) (the former in parallel with some parapithecids, and the latter with oligopithecids). Recent characterization of other aspects of the dentition of the early cercopithecid *Victoriapithecus* reinforces the interpretation that propliopithecids are the probable sister group of cercopithecids (Benefit, 1992, 1993). For the reasons cited in the above papers, an alternative derivation of cercopithecids from Parapithecinae, as argued by Simons and others (e.g., Gebo and Simons, 1987; Kay, 1977; Simons, 1974), is considered unlikely.

Platyrrhini

Platyrrhini (node F in Fig. 12) is a monophyletic clade unified by synapomorphies. The lower incisor crown height was reduced (i9, #1 → #0). P_{3-4} became single rooted (versus two rooted) (p4, #2 → #0). The P_4 entoconid became a small discrete cusp (versus cristiform) (p20, #1 → #2), and the P_{3-4} hypoconid was lost (versus small) (p25, #1 → #2). The M_3 became single rooted (versus two rooted) (m4, #1 → #0). The M_3 talonid was reduced so that its trigonid was much wider than the talonid (versus similar widths) (m7, #1 → #0), and its heel was lost (versus present, narrow) (m14, #1 → #0). The M_{1-2} postentoconid sulcus was lost (m26, #1 → #2). The M_{1-2} hypoconulid was reduced to a very small cusp and lost on M_3 (versus moderate-sized on all three teeth) (m27, m28, #1 → #2; m29, #1 → #3). The M_{1-2} hypocristid was strengthened (versus weak) (m38, #1 → #2), and the M_{1-2} buccal cingulum was reduced to a trace (versus partial) (m45, #1 → #0). P^{3-4} were reduced to two roots (versus three roots) (P2, #2 → #1; P3, #2 → #1). P^4 was much smaller than M^1 (versus slightly smaller) (P5, #1 → #0). M^2 was reduced to two roots (versus three) (M2, #0 → #1). A weak prehypocrista developed on the upper molars (versus absent) (M13, #0 → #1), and mesostyles were present on their buccal cingula (versus absent) (M17, #0 → #2). Finally, the M^{1-3} lingual cingulum was weak and incomplete (versus strong, complete) (M22, #2 → #1).

Early Miocene *Dolichocebus* departs significantly from the structure of other platyrrhines and greatly resembles the morphotype of platyrrhines (for example, in having a lingually open P_4 trigonid). This taxon may be an early offshoot of the platyrrhine stem (Fleagle and Kay, 1989).

Alternative Proposed Phylogenies

As noted in the introduction, two points of disagreement in anthropoid cladogenesis concern the phyletic position of the Oligopithecidae and Parapithecidae in relation to platyrrhines and various groups of catarrhines (Figs. 2 and 3). To assess these alternatives, we have examined the number of added steps required to produce a tree compared with the maximum-parsimony trees in Fig. 6. In a "heuristic" search using PAUP 3.1.1, less parsimonious trees were identified and saved. Figure 19 is the cumulative distribution of the number of these trees versus tree length for trees equal to or greater than 39,763 steps, at which point we had "discovered" 1014 trees. Tree lengths of an alternative hypothesis can be compared with this distribution to give a visual impression of their plausibility.

Parapithecids. One possibility in particular is that parapithecids are the sister group to catarrhines. This alternative solution can be represented in our study group by setting parapithecids as the sister group to *Aegyptopithecus*. This hypothesis better explains the distribution of some characters of the lower molars. For example, the enlargement and central positioning of the molar hypoconulids and the appearance of a molar distal fovea would not have to be acquired convergently in parapithecids and *Aegyptopithecus* (or lost in oligopithecids and platyrrhines). Evidence against such a hypothesis is that the apparently derived premolar reorganization distinctive of platyrrhines, oligopithecids, and catarrhines would have had to evolve independently twice or been lost convergently twice. Such a solution adds 450 extra steps to our

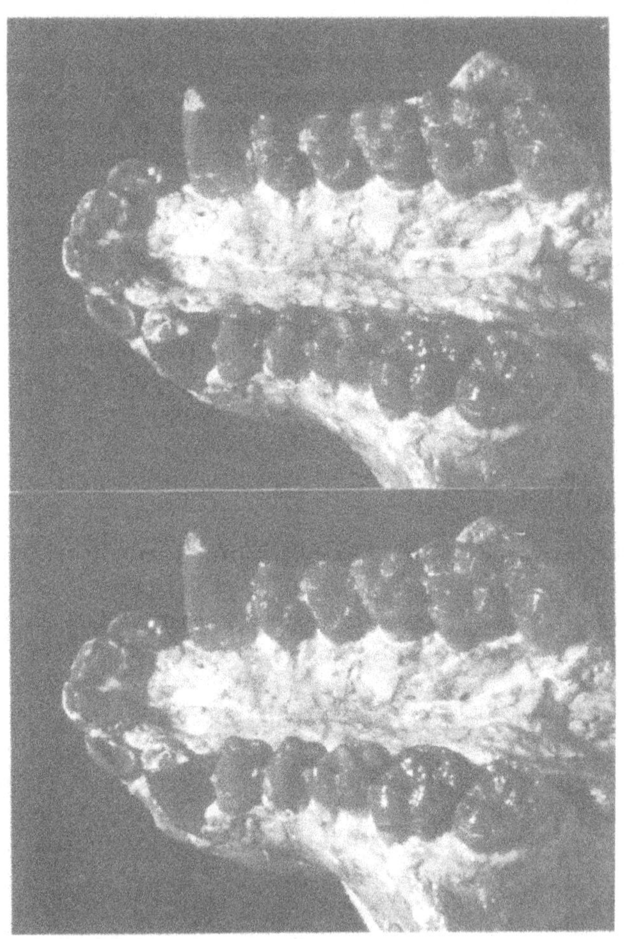

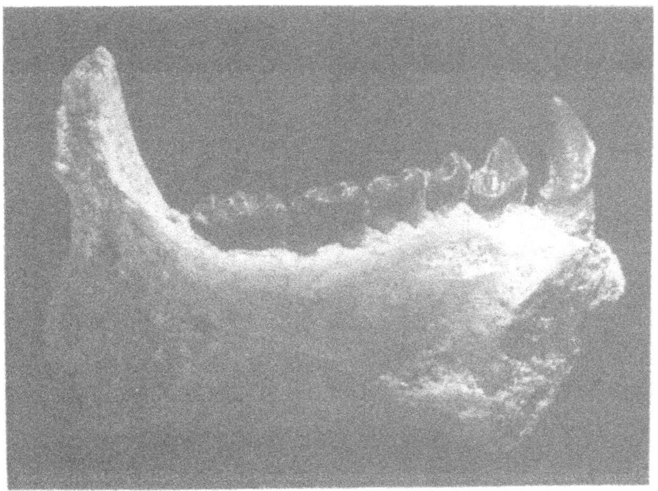

Fig. 18. *Aegyptopithecus zeuxis.* (a) Maxillary dentition from skull CGM 40237 (stereophotograph) (M^1 length = 5.6 mm). (b,c) Mandibular dentition from DPC 1028, left $C–M_3$ (M_1 length = 5.9 mm). b, lingual view; c, buccal view.

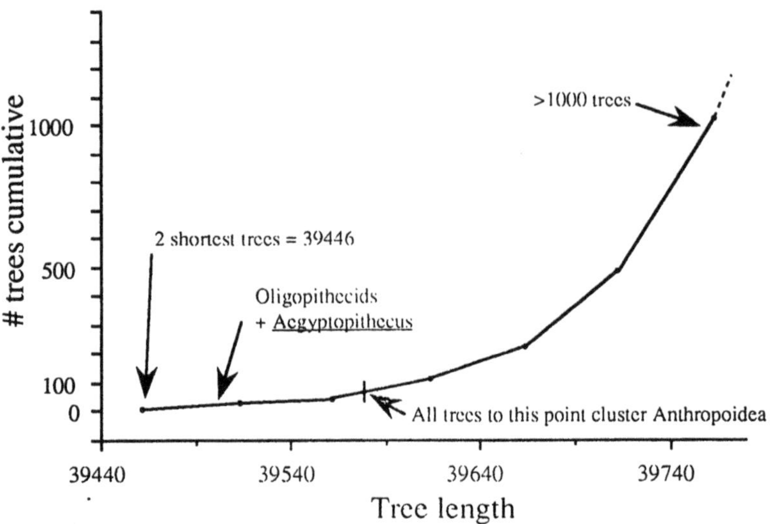

Fig. 19. Bivariate plot of the number of trees found at various less optimal tree lengths. Points are identified along this curve corresponding to the shortest two trees discovered, the length of the tree where oligopithecids and *Aegyptopithecus* are considered sister groups, and the point beyond which not all trees cluster Anthropoidea together.

maximum parsimony cladogram. The number of trees more parsimonious than this solution is far in excess of 10,000 (Fig. 19).

Oligopithecids. Alternative phylogenetic arrangements also have been proposed for the Oligopithecidae. One alternative is that they might be stem anthropoids, i.e., the sister taxon to other living and fossil Anthropoidea, including Parapithecidae (Fleagle and Kay, 1987). This is implausible from the dental evidence. The problem here also stems from the structure of the premolars. In order for oligopithecids to be primitive anthropoids, the apparently primitive parapithecid premolar structure would have had to pass through an advanced condition resembling that in oligopithecids, platyrrhines, and catarrhines before returning to the primitive state. Such a reorganization requires 39,846 steps (versus 39,446 steps). Again, more than 1000 trees were found to be shorter than this (see above).

Gingerich (1977) proposed that *Oligopithecus* may be an adapid. This suggestion is no longer tenable now that we have adequate cranial material for its close relative *Catopithecus* exhibiting postorbital closure similar to other anthropoids (Rasmussen and Simons, 1992; Simons, 1990). Linkage of Oligopithecids with *Mahgarita* requires 40,078 steps, far in excess of parsimony.

Oligopithecids also have been considered primitive relatives of propliopithecids (Simons, 1989, 1990). Premolar loss and development of an enlarged P_3 are shared with *Aegyptopithecus* and *Propliopithecus*. A network linking oligopithecids with *Aegyptopithecus* requires 39,496 steps, just 50 extra steps compared with the most parsimonious network (Fig. 19). Thus, we consider such an alternative as plausible, although not the best fit for our data.

Status of Less-Well-Known Possible Early Anthropoids

Altiatlasius koulchii

We have not examined the fragmentary material of this taxon from the Paleocene of Morocco (Sigé *et al.*, 1990). Its describers place *Altiatlasius* within the Omomyidae, and, believing that omomyids are haplorhines, they suggest it is a sister taxon of Anthropoidea. Gingerich (1990) and Rose *et al.* (Chapter 1, this volume) rightly conclude that until more and better-preserved material is known, its allocation to any primate group is problematic.

Algeripithecus minutus

The new primate *Algeripithecus minutus* recently has been described from the early Eocene of Algeria (Godinot and Mahboubi, 1992). Godinot and Mahboubi suggest that this small primate is an early anthropoid most similar to *Aegyptopithecus*. Just two isolated teeth, an upper second molar and a lower third molar, have been described to date. These new specimens do share several of the autapomorphies of the last common ancestor of anthropoids: The M^2 has a poorly developed buccal cingulum (M21), enlarged hypocones (M11), and lacks a metaconule (M16). The M_3 has a weak hypocristid (M39). The extremely bunodont nature of the upper molars also suggests that this taxon may have shared the additional lower molar autapomorphies of the ancestral anthropoid (if parapithecids are the sister group of other anthropoids) such as low cusp relief (m17) and rounded cristid obliquas and hypocristids (m31 and m38). Godinot (Chapter 10, this volume) has described isolated P_3 and P_4 of this animal as being simple, lacking metaconids. This is very different from what we would reconstruct for a primitive catarrhine, which we suggest had a large lingually positioned P_4 metaconid, but more closely resembles that of parapithecids. More complete material is needed before allocation to Anthropoidea is warranted.

Afrotarsius chatrathi

Afrotarsius chatrathi (Fig. 20) is known from a lower jaw preserving M_{1-3} and lower portions of the crowns of P_{3-4}. Originally described by Simons and Bown (1985) as an African tarsiid, it was subsequently suggested to be a possible sister taxon of Anthropoidea (Fleagle and Kay, 1987; Ginsburg and Mein, 1987).

Enough of the characters (Table I) are available to analyze *Afrotarsius* by

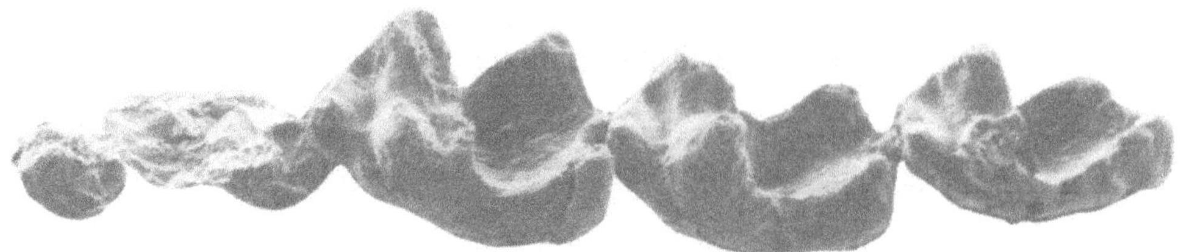

Fig. 20. *Afrotarsius chatrathi*, TYPE, right P_{3-4} (broken), M_{1-3} (M_1 length = 2.6 mm).

mapping it onto our Wagner networks (Fig. 6). The most parsimonious arrangement places *Afrotarsius* as the sister group to all anthropoids. Linking *Afrotarsius* with omomyids or adapids is less parsimonious, as is a unique relationship with *Tarsius*.

The lower molars of *Afrotarsius* share several apparent synapomorphies with anthropoids: the M_1 cristid obliqua was oriented mesiodistally toward the protoconid (actually a synapomorphy of *Tarsius* and Anthropoidea); its metaconid was transverse to the protoconid, and there were small lower molar hypoconulids; and M_2 was larger than M_3 (m5, m19, m27, m34). This taxon differs from the anthropoid morphotype in lacking postentoconid sulci on the molars and in having an M_3 hypocristid stronger than in anthropoids.

Proteopithecus sylviae

Proteopithecus sylviae (Table I) is a Fayum late Eocene/early Oligocene taxon known from a maxilla with P^4–M^2 and another tooth identified as P^3 (Simons, 1990). Simons placed *Proteopithecus* within the Oligopithecidae, possibly because it was a primitive anthropoid but with only two premolars. However, it is more plausible that this upper premolar is a P^2 because the tooth lacks a protocone. Protocones are often absent on P^2 in primates but are rarely absent on P^3. Simons (1992) apparently has reached the same conclusion, for he notes that the P_2 of new *Proteopithecus* material is not enlarged.

Little can be said about the phylogenetic position of this animal on the basis of the scant published remains. Overall, the best fit of the available data places *Proteopithecus* with small-bodied platyrrhines like *Callimico* or *Saguinus,* which have reduced or absent molar hypocones. The possibility that *Proteopithecus* represents an African protoplatyrrhine stock deserves consideration as this animal becomes better known.

Amphipithecus mogaungensis

The Burmese primate *Amphipithecus mogaungensis* was described by Colbert (1937) as an anthropoid. Simons (1963, 1965, 1971) concurred that it was an anthropoid and later even suggested that it might be a hominoid (Simons and Pilbeam, 1965). Szalay (1972; Szalay and Delson, 1979) has suggested

Amphipithecus may be an adapid. Ciochon *et al.* (1985) described a second worn specimen preserving M_{1-2} and attempted to confirm its status as an anthropoid, but Holroyd and Ciochon (Chapter 6, this volume) are noncommittal about its affinities.

Fourteen dental features could be scored (Table I), three of which represent synapomorphies with some or all anthropoids. *Amphipithecus* has an enlarged P_4 metaconid (p17), resembling anthropoids of modern aspect (oligopithecids, platyrrhines, and catarrhines), and first molar trigonid and talonid of similar height as in the above taxa, as well as advanced parapithecids (m16) and weak, rounded hypocristids reminiscent of parapithecids (m38). When mapped onto the networks in Figs. 7 and 8, the most parsimonious tree shows *Amphipithecus* to be the sister group of parapithecids. However, because of the great amount of missing data, we suggest that the affinities of *Amphipithecus* cannot yet be resolved.

Pondaungia cotteri

A second Asian primate from the middle Eocene of Burma often discussed as a possible anthropoid primate is *Pondaungia cotteri*. First described in 1927 by Pilgrim, *Pondaungia* presently is known only from four fragments representing M^{1-2} and M_{2-3} (Ba Maw *et al.*, 1979; Ciochon and Holroyd, Chapter 6, this volume; Pilgrim, 1927) (Table I). It has been variously interpreted as a condylarth (von Koenigswald, 1965), an early anthropoid, possibly a catarrhine (Pilgrim, 1927; Simons, 1972; Szalay and Delson, 1979; Ba Maw *et al.*, 1979), or an adapid (Ciochon and Holroyd, Chapter 6, this volume).

The possibility that *Pondaungia* is related to *Amphipithecus* is impossible to assess directly since only one tooth, the lower second molar (extremely worn in *Amphipithecus*), is known in common between *Amphipithecus* and *Pondaungia*. Possible anthropoid features of this animal are the low-crowned, puffy cheek teeth with trigonids and talonids of similar height, the possible presence of wear facet X, the small size of M_3 relative to M_2, and upper molars with poorly developed buccal cingula and cingulum-derived hypocones. However, the critical aspects of the dentition that are diagnostic of anthropoid status, such as the structure of the incisors and premolars, are unknown at present. Furthermore, the extreme degree of crenulation exhibited by this taxon makes scoring difficult; thus, interpretation of characters has been inconsistent among researchers who have studied this material [some have felt, for example, that the hypocone was attached to a *Nannopithex* fold (Pilgrim, 1927)].

Taking all of this into consideration, it is not surprising that *Pondaungia* fits comfortably with practically any group we have analyzed. When this genus is mapped onto the trees in Fig. 6, it fits best with the adapid *Mahgarita*. It is, however, only slightly less parsimonius to link *Pondaungia* with *Cantius*, as the sister group of Anthropoidea, or with parapithecids. All that can now be said with confidence is that *Pondaungia* appears to be highly autapomorphic, and until the antemolar dentition is known, we believe its phyletic status will remain enigmatic. Certainly, we agree with Ciochon and Holroyd (Chapter 6, this volume) that this taxon does not provide any definitive evidence for an Asian origin for Anthropoidea.

Discussion

As shown in previous studies (e.g., Delson and Rosenberger, 1980; Kay, 1980), no single character or suite of characters represent unique dental apomorphies of Anthropoidea. In fact, homoplasy is all-pervasive in the dental anatomy of primates. This should neither surprise us nor lead us to reject the use of dental traits in the interpretation of the phylogenetic relations of primates. We suspect that as the anatomy of the postcranium and cranium becomes better known in various extinct groups, allowing us to characterize anatomic variation more fully at the generic and specific levels, homoplasy will be found to characterize those anatomic systems to the same or greater degree as the dentition.

We can now propose some answers to the questions in the introduction. Our findings are summarized in the cladograms in Figs. 12 and 21:

1. Evidence of the dental anatomy supports the Haplorhini as a clade that includes Omomyidae, with Anthropoidea and *Tarsius* clustered within it. Thus, Omomyidae is a paraphyletic group from which *Tarsius* and anthropoids arose (Fig. 21). This proposed arrangement resembles most closely that in Fig. 1b,c but differs somewhat in that *Tarsius* and Anthropoidea share a common ancestor after separating from Omomyidae.
2. Anthropoidea is a monophyletic group (Figs. 12 and 21).
3. Our analysis suggests that a natural subdivision occurs between Parapithecidae and another group comprised of Oligopithecidae, Platyrrhini, and Catarrhini.

   We see clearly two stages in the evolution of the anthropoid dentition, corresponding to this dichotomy and documented among the early anthropoids of the Fayum, Egypt. In the first stage, most closely approximated by qatraniine parapithecids (Fig. 12), the majority of the changes have occurred in molar structure. On M_{1-2} the height of molar trigonids was slightly reduced, and the paraconids were reduced but not eliminated. On the M_{1-2} talonids, the hypocristid was weakened, and a postentoconid sulcus appeared along with a moderate-sized M_{1-2} hypoconulid. M_3 was reduced in size compared with M_2. Additionally, the lower incisors became more vertically implanted, and a protocone developed on P^2. The upper molar buccal cingulum was weakened.

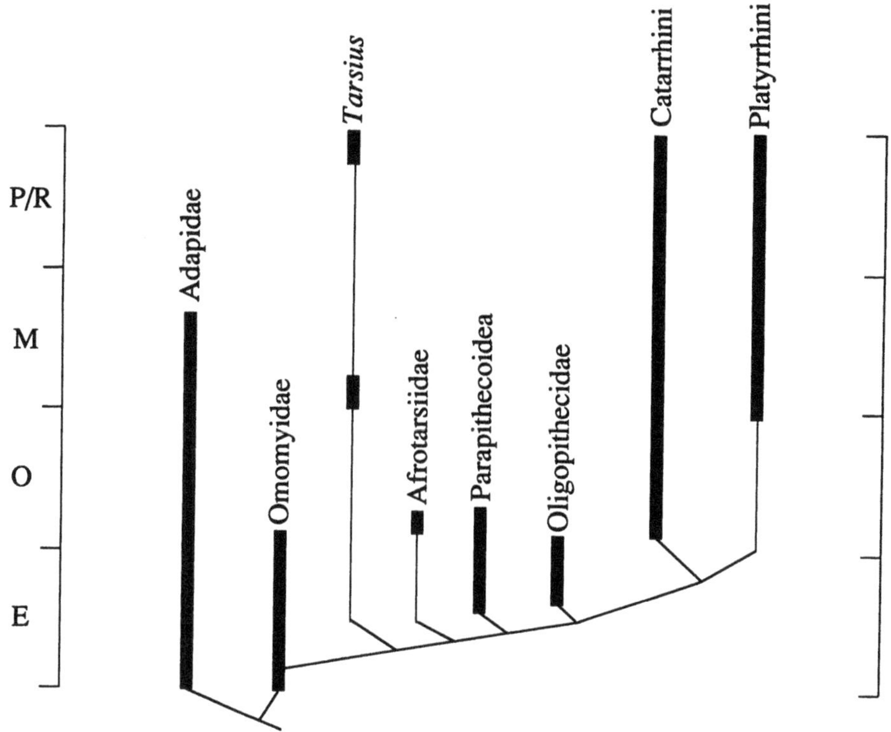

Fig. 21. Known temporal range and suggested phylogenetic relationships of the major groups of primates discussed in the text. Cenozoic epochs at the left are not drawn to scale. E, Eocene; O, Oligocene; M, Miocene; P/R, Pliocene to Recent.

Added to the changes mentioned above, in a second stage in the evolution of the anthropoid dentition, represented by oligopithecids, platyrrhines, and catarrhines (Fig. 12), the majority of changes occurred in the structure of the premolars: the premolars became less crowded; P_4 was structurally reorganized such that its talonid increased in size, becoming similar in mesiodistal length to the trigonid; and the P_4 metaconid was enlarged, repositioned medially, and became widely separated from the protoconid so that the lateral protocristid was transversely oriented. Similarly, the M_1 metaconid was shifted to a more transverse position relative to the protoconid, and the molar cusps were more marginally positioned. The lingual moiety of P^2 was reduced to become a more triangular tooth, and the lingual cingula of the upper premolars were strengthened. Additionally, the upper molar metaconules were lost. Finally, body size increased.

Recently described late Eocene *Arsinoea* resembles parapithecids in most details, but its P_4 (with a strong, more lingually positioned metaconid) is reminiscent of propliopithecids, oligopithecids, and platyrrhines. When it is known from more complete specimens of the upper dentition, *Arsinoea* may require reassignment as a primitive anthropoid of modern aspect.

Our reconstruction showing parapithecids as the sister taxon to other anthropoids finds added support in the postcranial anatomy. Fleagle and Kay (1987) list primitive, otherwise nonanthropoid, postcranial features, especially in the femur, tibia, and cuboid, that support the argument for parapithecids as primitive anthropoids; however, at this point the interpretation of these features is far from clear. The postcranium of Paleogene prosimian primates is poorly known, so that establishing polarities for many anthropoid features is difficult. Moreover, parapithecid postcranial material has been documented only for *Apidium phiomense.* Gebo *et al.* (Chapter 9, this volume) describe new parapithecid postcrania of uncertain allocation. Oligopithecids are poorly represented. Senut and Thomas (1992) note the existence of oligopithecid postcranial remains from Oman, and Gebo *et al.* (Chapter 9, this volume) describe oligopithecid postcranials. Postcranial material of propliopithecids has been described from the Fayum (Fleagle and Simons, 1979, 1982; Gebo, 1987; Gebo and Simons, 1987) and Oman (Senut and Thomas, 1992), but no femoral material has yet been described. As unsatisfying as this may be, we must await further discoveries to resolve this dilemma.

We present information about several fragmentary fossil primates from the Paleogene of Burma and Africa (*Pondaungia, Amphipithecus, Algeripithecus,* and *Proteopithecus*) that have been considered as possible anthropoids. For the moment, we follow Holroyd and Ciochon (Chapter 10, this volume) in assigning the Burmese *Pondaungia* to Adapidae. *Amphipithecus,* also from Burma, may prove to be an anthropoid. *Algeripithecus* also has some possible anthropoid molar synapomorphies and could be a part of the anthropoid radiation, but more information is needed before this can be suggested with confidence. With better material, *Proteopithecus* may prove to belong to an African platyrrhine stock.

As the fossil record of anthropoids becomes better known, it is gradually becoming more and more difficult to define its anatomic synapomorphies. The anatomy of the anthropoidean ear region has several possible synapomorphies, even in its earliest known occurrence in early Oligocene parapithecid *Apidium* and the oligopithecid *Catopithecus* from the late Eocene. However, most of the details of this region are similar to *Tarsius* (Cartmill and Kay, 1978) and seem to be synapomorphies of another larger clade including Anthropoidea and *Tarsius.* Postorbital closure would also be a shared-derived similarity of this larger clade (Cartmill, Chapter 16, this volume). Fusion of the metopic suture between the frontal bones occurs in early anthropoids, although such fusion is common in strepsirhines and *Tarsius* and may not, therefore, be an anthropoid synapomorphy. Fusion of the mandibular symphysis, shared by all living anthropoids, occurs in some adapids such as *Mahgarita* and has been discussed as a possible autapomorphy of the last common ancestor of protoadapines and anthropoids, e.g., by Gingerich and Rasmussen. It is now clear, however, that symphyseal fusion occurred independently in several different anthropoid clades, because *Arsinoea* clearly had an unfused symphysis. The spatulate shape of the incisors and having an I_2 that is larger than the I_1 are characters that have long been considered anthropoid synapomorphies or as synapomorphies of a more inclusive group composed of adapids and anthropoids. However, it is quite possible that these incisor characters are actually synapomorphies for the order Primates (e.g., Covert and Williams, 1991, Chapter 2, this volume).

There remain a number of dental features, especially in the molars (see above), that probably are synapomorphies of anthropoids, although perhaps none of them is a unique synapomorphy. *All* the synapomorphies of Anthropoidea may ultimately be found as convergent synapomorphies in other groups of primates or other mammals.

Many of our colleagues have expressed disappointment that major variation in phylogenetic geometry (that is, phylogenetic rearrangements of the major groups of primates or anthropoids) can be found with the addition of just a small number of steps in our phylogenetic analysis. A possible solution is to impose an explicit *a priori* or *a posteriori* weighting scheme on the characters. This, in turn, might lead to stronger (that is, more consistent) linkages. Anyone is welcome to impose such a scheme on our data, but we hesitate to adopt this "solution." It may be perfectly true that some characters may be of higher "value" within the group studied, but how can we determine objectively which are these characters? Such a solution, although appealing, may boil down to arguments among scholars that "my characters are heavier than yours."

We suspect that the real reason for the instability in the resolution of higher-level phylogeny within primates and anthropoids results not from the characters but from the geometry of cladogenesis itself. The major groups we are studying may have originated in short periods of speciation. The distinctiveness of modern primate groups resulted from a "weeding out" of some intermediates and from morphological reorganization occurring after the initial cladogenesis through longer periods of slowly accumulated change. If so, limited morphological change occurred at the origin of the major groups of primates and anthropoids. It should not then be surprising that most of the synapomorphies of living Anthropoidea will not be found in all of its primitive members as they become better known. Ultimately for us, the solution is to find more fossils closer to the hypothesized times of diversification of Anthropoidea.

Anthropoid Classification

Table III represents our classification of the order Primates. Several aspects of the classification need clarification in light of our phylogenetic conclusions. (1) Because we do not believe that Plesiadapiformes belong within primates or are the sister group to Primates, we do not use the term "Euprimates" for what Simons (1972) called "primates of modern aspect." (2) Although we recognize a division of Primates into Haplorhini and Strepsirhini, this subdivision rests on an insecure foundation. Because we do not know what the primate morphotype looked like, we are uncertain about where to "fold" our maximum-parsimony network. Thus, as illustrated in Fig. 7, our maximum-parsimony networks can be folded in several ways to preserve a primary dichotomy between living haplorhines and strepsirhines. But, depending on how it is folded, Omomyidae can be positioned within either of the two groups (compare Fig. 6, Network I, with Figure 7A). Indeed, rooting the network using the Paleocene plesiadapiform *Purgatorius* as an outgroup (Fig. 10) makes Haplorhini monophyletic but leaves Strepsirhini as a paraphyletic taxon (*Cantius* is the sister-group of all other primates, and Omomyidae arose from within Adapidae). (3) As Szalay and Delson (1979) recognized, there are insufficient numbers of commonly used taxonomic levels to recognize formal dichotomies for (a) Haplorhini versus Strepsirhini, (b) Tarsiiformes versus Anthropoidea, and (c) Platyrrhini versus Catarrhini. Szalay and Delson (1979) resolved this by not recognizing the taxon Anthropoidea. As an alternative solution we classify Haplorhini and Strepsirhini as "semiorders" following the practice of Szalay and Li (1986), Szalay *et al.* (1987), and Sigé *et al.* (1990). The other two dichotomies can then fall in as suborders and infraorders. (4) We have elevated Parapithecoidea to the level of an infraorder coordinate with Platyrrhini and Catarrhini, but we have chosen not to recognize a formal taxonomic level to signify the primary division within Anthropoidea between Parapithecoidea and a group comprising Platyrrhini and Catarrhini. (5) Finally, Oligopithecidae and Afrotarsiidae have been placed as *incertae sedis* within Anthropoidea. We argue here that the former is a sister group to the platyrrhine/catarrhine clade, and the latter should be regarded as a sister group to Anthropoidea (Figs. 12 and 22).

Acknowledgments

We thank Elwyn Simons, Bob Emry, Richard Thorington, Jerry Hooker, Marc Godinot, Bert Covert, and Tom Bown for allowing us to examine specimens in their care. We profited from many hours of discussions with Tom Bown, Bert Covert, Eric Delson, John Fleagle, Pat Holroyd, Tab Rasmussen, Alfie Rosenberger, Elwyn Simons, and many others too numerous to name. This research was supported by grants from NSF and the Smithsonian Institution.

Appendix

Dental characters and character states used in the cladistic analysis. Characters followed by an asterisk are considered "ordered," as explained in the text. Dental terminology follows Kay (1977).

Lower Teeth

Incisors

i1*. Lower incisor number: 0 = three; 1 = two; 2 = one: I_1 present, I_2 absent; 3 = lower incisors absent.
i2. Lower incisor occlusal arrangement: 0 = arcuate battery from lateral perspective; 1 = cusp tips staggered.
i3. Lower incisor crown spacing: 0 = no spaces; 1 = spaces present between crowns.
i4. I_2-C diastema: 0 = present; 1 = absent.
i5*. I_{1-2} size (ratio of I_{1-2} area to M_1 area): 0 = very small (<0.69); 1 = moderate sized (≥0.70, ≤1.07); 2 = large (>1.07).
i6*. I_1:I_2 proportions (ratio of I_1 area to I_2 area): 0 = I_1 much smaller than I_2 (<0.65); 1 = I_1 smaller than I_2 (≥0.65, ≤0.82); 2 = I_1 almost as large as I_2 (>0.83, ≤ = 1.00); 3 $I_1 > I_2$ (>1.0).
i7*. I_1 crown width (spatulate incisors only): 0 = considerably wider (m-d) than root (spatulate): 1 = narrow at apex, wider than root; 2 = "styliform" (crown apex approximately the same width at the cervical margin).
i8. I_2 crown cross-sectional shape (ratio of mesiodistal length to buccolingual breadth): 0 = rounded oval (>0.70); 1 = mesiodistally compressed (<0.64).
i9*. Lower incisor crown height (cementoenamel junction to crown tip on the buccal surface): 0 = low crowned; 1 = moderately high crowned; 2 = high crowned.
i10. I_{1-2} crown buccal outline (spatulate incisors only): 0 = gently curved in lateral perspective; 1 = acutely curved.
i11. Lower incisor roots: 0 = erect or vertical; 1 = procumbent.
i12*. Lower incisor crowns: 0 = erect or vertical; 1 = procumbent; 2 = very procumbent.
i13. Tooth comb: 0 = absent; 1 = with three teeth; 2 = with two teeth.
i14. I_1 crown shape: 0 = spatulate; 1 = lanceolate; 2 = daggerlike.
i15. I_2 heel development (the heel is a lingual swelling at the base of crown): 0 = heel absent; 1 = heel present.
i16. Incisor lingual enamel: 0 = well developed; 1 = poorly developed or absent.
i17*. Lower incisor lingual cingulum: 0 = absent to weak; 1 = strong but incomplete; 2 = strong and complete.

Canines

c1*. Female $C^1/_1$ size (relative to molars): 0 = very small; 1 = small to moderate; 2 = large.

c2*. $C^1/_1$ dimorphism (square root male C_1 area/square root female C_1 area): 0 = low (<1.07); 1 = moderate (≥1.07, <1.17); 2 = high (≥1.17).
c3. C_1 cross-sectional shape: 0 = rounded oval; 1 = mesiodistally compressed.
c4. C_1 lingual crest development: 0 = rounded; 1 = sharp.
c5. Canine paracristid (not scored if species has canine incorporated into a tooth comb): 0 = oblique to occlusal plane; 1 = nearly horizontal to occlusal plane; 2 = forms part of cropping mechanism with I_{1-2}.

Premolars

p1. $P^1/_1$: 0 = present; 1 = absent.
p2. $P^2/_2$: 0 = present; 1 = absent.
p3. P_2 roots: 0 = single; 1 = double.
p4. P_{3-4} roots: 0 = P_3 single, P_4 single; 1 = P_3 single, P_4 double; 2 = P_3 double, P_4 double.
p5*. Premolar crowding (mesiodistal overlapping of crowns): 0 = no crowding; 1 = slightly crowded; 2 = very crowded.
p6*. P_3 paraconid: 0 = large; 1 = small; 2 = absent.
p7*. P_4 paraconid: 0 = large; 1 = small; 2 = absent.
p8. P_3 or P_4 paraconid shape: 0 = conical; 1 = crestlike.
p9. P_4 paraconid position: 0 = mesial to protoconid; 1 = mesiolingual, between protoconid and metaconid; 2 = mesial to metaconid, widely spaced from metaconid; 3 = twinned with metaconid.
p10*. P_3 protoconid height: 0 = lower than P_4 protoconid; 1 = equal to P_4 protoconid; 2 = higher than P_4 protoconid.
p11*. P_{3-4} cristid obliqua: 0 = absent; 1 = weak; 2 = strong.
p12*. P_4 trigonid length: 0 = much longer than talonid in mesiodistal length; 1 = equal to, or slightly longer than, talonid; 2 = shorter than talonid.
p13. P_2 protoconid size and shape: 0 = slender, projects above protoconids of P_{3-4}; 1 = massive, projects above protoconids of P_{3-4}; 2 = not projecting.
p14. P_4 metaconid position: 0 = close to protoconid; 1 = widely separated from protoconid.
p15. P_2 metaconid size: 0 = absent or trace; 1 = small.
p16*. P_3 metaconid size: 0 = absent or trace; 1 = small; 2 = large.
p17*. P_4 metaconid size: 0 = absent or trace; 1 = small; 2 = large.
p18. P_4 trigonid, lingual configuration: 0 = closed; 1 = open.
p19*. P_3 entoconid or lingual talonid crest: 0 = absent; 1 = lingual talonid crest; 2 = entoconid is a small discrete cusp.
p20*. P_4 entoconid and lingual talonid crest: 0 = absent; 1 = lingual

talonid crest present but an entoconid does not stand out above it; 2 = entoconid is a small discrete cusp.

p21. P_4 lateral and medial protocristids: 0 = continuous between metaconid and protoconid; 1 = discontinuous between metaconid and protoconid; 2 = absent.

p22. P_3 lateral protocristid orientation: 0 = transversely oriented; 1 = distolingually oriented; 2 = absent.

p23. P_4 lateral protocristid orientation: 0 = transversely oriented; 1 = distolingually oriented; 2 = absent.

p24. P_{3-4} posterior trigonid wall: 0 = complete (taxa without metaconids are assigned this character state if there is a protocristid); 1 = deeply notched.

p25*. P_{3-4} hypoconid size: 0 = large; 1 = small; 2 = absent.

p26. P_{3-4} hypoconid position: 0 = buccally positioned, distal to protoconid; 1 = lingually positioned, distal to metaconid, or between protoconid and metaconid.

p27*. P_4 hypocristid shearing development: 0 = absent; 1 = weak; 2 = strong.

p28*. P_2 buccal cingulum development: 0 = absent; 1 = incomplete, broken at protoconid and hypoconid; 2 = complete.

p29*. Lower premolar inflation: 0 = not basally inflated; 1 = slightly basally inflated; 2 = very basally inflated.

Molars

m1. $M_{3/3}$: 0 = present; 1 = absent.

m2. M_1 root number: 0 = one; 1 = two.

m3. M_2 root number: 0 = one; 1 = two.

m4. M_3 root number: 0 = one; 1 = two.

m5*. M_2 length : M_3 length (ratio of M_2 mesiodistal to M_3 mesiodistal): 0 = M_2 longer than M_3 (>1.10); 1 = M_2 and M_3 lengths similar (≤1.10, ≥1.00); 2 = M_2 shorter than M_3 (<1.00).

m6*. M_2 trigonid width (ratio of buccolingual breadths of trigonid and talonid): 0 = much wider than talonid (≥1.11); 1 = widths similar (<1.11, >0.95); 2 = much narrower than talonid (≤0.95).

m7. M_3 trigonid width (based on relative buccolingual breadths): 0 = much wider than talonid (≥1.20); 1 = trigonid and talonid widths similar (≤1.20).

m8*. M_1 paraconid position: 0 = buccal (mesial to protoconid); 1 = midline, between protoconid and metaconid; 2 = lingual (mesial to metaconid but widely spaced from it); 3 = lingual (twinned with metaconid).

m9*. M_2 paraconid position: 0 = buccal (mesial to protoconid); 1 = midline, between protoconid and metaconid; 2 = lingual (mesial to metaconid but widely spaced from it); 3 = lingual (twinned with metaconid).

m10*. M_3 paraconid position: 0 = buccal (mesial to protoconid); 1 = midline, between protoconid and metaconid; 2 = lingual (mesial to metaconid but widely spaced from it); 3 = lingual (twinned with metaconid).

m11. M_1 parastylid: 0 = absent; 1 = present.

m12*. Molar metastylids: 0 = absent; 1 = small; 2 = large.

m13. M_3 hypoconulid: 0 = single; 1 = double.

m14*. M_3 heel: 0 = absent; 1 = narrower than talonid; 2 = approximately equal in width to talonid.

m15. Molar enamel surface: 0 = smooth; 1 = crenulated.

m16*. M_1 trigonid height (ratio of trigonid height to talonid height): 0 = higher than talonid (≥ 1.20); 1 = slightly higher than talonid (≥ 1.10, < 1.20); 2 = trigonid and talonid of similar height (< 1.10).

m17. M_{1-2} cusp relief: 0 = moderate to high; 1 = low.

m18. M_1 trigonid lingual configuration: 0 = open; 1 = closed.

m19. M_1 metaconid position: 0 = lingual to protoconid; 1 = distolingual to protoconid.

m20*. M_{1-2} paraconid development: 0 = absent; 1 = small; 2 = large.

m21. M_{1-2} lateral protocristid orientation (see Kay, 1977): 0 = runs toward metaconid; 1 = runs toward hypoflexid; 2 = absent.

m22. M_1 distal trigonid wall: 0 = complete; 1 = deeply notched by protoconid/metaconid sulcus; 2 = medial and lateral protocristid do not meet, but no sulcus is discerned.

m23. M_2 distal trigonid wall: 0 = complete; 1 = deeply notched by protoconid/metaconid sulcus; 2 = medial and lateral protocristid do not meet, but no sulcus is discerned.

m24. M_{1-3} wear facet X: 0 = present; 1 = absent.

m25*. M_{1-2} entoconid: 0 = absent; 1 = crestlike, barely stands out on lingual talonid marginal crest; 2 = a small discrete cusp; 3 = a large cusp.

m26*. M_{1-2} postentoconid sulcus: 0 = prominent; 1 = faintly visible; 2 = absent.

m27*. M_1 hypoconulid size: 0 = large; 1 = moderate; 2 = small; 3 = absent.

m28*. M_2 hypoconulid size: 0 = large; 1 = moderate; 2 = small; 3 = absent.

m29*. M_3 hypoconulid size: 0 = large; 1 = moderate; 2 = small; 3 = absent.

m30*. M_{1-2} hypoconulid position: 0 = twinned to entoconid; 1 = near midline; 2 = slightly buccal to midline.

m31*. M_{1-2} cristid obliqua development: 0 = weak (rounded); 1 = strong (trenchant); 2 = very strong (trenchant).

m32*. M_1 cristid obliqua orientation: 0 = reaches trigonid wall at a point distal to protoconid; 1 = reaches trigonid wall at a point distolingual to protoconid; 2 = reaches trigonid wall at a point distal to metaconid.

m33*. M_2 cristid obliqua orientation: 0 = reaches trigonid wall at a point distal to protoconid; 1 = reaches trigonid wall at a point distolingual to protoconid; 2 = reaches trigonid wall at a point distal to metaconid.

m34*. M_1 cristid obliqua terminus: 0 = runs to base of trigonid; 1 = runs part way up the distal trigonid wall; 2 = connects with protoconid tip or protocristid.

m35*. M_2 cristid obliqua terminus: 0 = runs to base of trigonid; 1 = runs part way up the distal trigonid wall; 2 = connects with protoconid tip or protocristid.

m36*. M_3 cristid obliqua terminus: 0 = runs to base of trigonid; 1 = runs part way up the distal trigonid wall; 2 = connects with protoconid tip or protocristid.

m37*. M_{1-2} centroconid development: 0 = present; 1 = absent but cristid obliqua bends sharply in hypoflexid; 2 = absent.

m38*. M_{1-2} hypocristid development: 0 = absent or seen only as a trace; 1 = weak; 2 = strong.

m39*. M_3 hypocristid development: 0 = absent or seen only as a trace; 1 = weak; 2 = strong.

m40*. M_{1-2} talonid, lingual configuration: 0 = widely open; 1 = notched lingually but not open; 2 = closed.

m41. M_{1-2} distolingual fovea (posterointernal basin): 0 = absent; 1 = present.

m42. M_{1-2} hypocristid configuration: 0 = simple, 1 = with accessory cusp close to hypoconid.

m43. M_{1-2} cristid obliqua: 0 = notched; 1 = straight.

m44*. Molar cusp inflation: 0 = cusps marginally positioned; 1 = slightly inflated; 2 = very inflated.

m45*. M_{1-2} buccal cingulum development: 0 = absent to trace; 1 = partial, broken at protoconid and hypoconid; 2 = complete, low; 3 = complete, gutterlike.

m46*. M_1 hypoflexid: 0 = very shallow; 1 = shallow; 2 = deep.

m47*. M_2 hypoflexid: 0 = very shallow; 1 = shallow; 2 = deep.

Upper Teeth

Incisors

I1*. I^1–I^2 interstitial contact: 0 = absent; teeth widely spaced; 1 = present as narrow contact; 2 = I^2 tightly packed against I^1, I^1 preparacrista abbreviated.

I2. I^1–I^1 interstitial contact: 0 = present; 1 = absent: a wide space occurs in the midline between these teeth.

I3. I^2-C diastema: 0 = present; 1 = absent.

I4*. I^1 area: I^2 area: 0 = areas approximately equal (≤ 1.00); 1 = I^1 slightly larger than I^2 (>1.00, <1.40); 2 = I^1 much larger than I^2 (≥ 1.40).

I5*. I^1 size (I^1 area: M^1 area): 0 = small (≤ 0.50); 1 = moderate (>0.50, <0.56); 2 = large (≥ 0.56).

I6*. I^1 occlusal shape (mesiodistal length/buccolingual breadth): 0 = rounded oval (<1.05); 1 = buccolingually compressed (≥ 1.05, <1.30); 2 = extremely compressed (≥ 1.30).

I7*. I^2 occlusal shape (mesiodistal length/buccolingual breadth): 0 = rounded oval (≤ 1.05); 1 = slightly buccolingually compressed (>1.05, <1.30); 2 = extremely buccolingually compressed (≥ 1.30).

I8. I^1 crown shape: 0 = spatulate; 1 = peglike; 2 = daggerlike.

I9. I^1 lingual fovea: 0 = simple; 1 = dual, with midcrown pillar.

I10. I^1 occlusal edge orientation (spatulate incisors only): 0 = occlusal edge orthogonal to long axis of root; 1 = occlusal edge wears at a steep angle to long axis of root; 2 = crown with pronounced mesial asymmetry (= mesial process) in unworn state.

I11*. I^{1-2} lingual cingulum: 0 = weak, discontinuous; 1 = moderate, continuous; 2 = strong.

I12. I^1 basal lingual cusp: 0 = absent; 1 = present.

I13. I^1–I^2 buccal cingulum: 0 = absent; 1 = present.

Canines

C1. C^1 cross-sectional shape: 0 = oval; 1 = rounded.

C2. Upper canine occlusion: 0 = C^1 wears against P_{1-2}; 1 = C^1 wears against P_2 only; 2 = C^1 wears against P_{2-3}; 3 = C^1 wears against P_3 only.

C3. C^1 mesial groove (females): 0 = shallow or absent; 1 = deep.

C4*. C^1 lingual cingulum: 0 = weak or absent; 1 = strong; 2 = very strong.

Premolars

P1*. P^2 root number: 0 = one; 1 = two; 2 = three.

P2*. P^3 root number: 0 = one; 1 = two; 2 = three.

P3*. P^4 root number: 0 = one; 1 = two; 2 = three.

P4*. P^2 area : P^3 area: 0 = P^2 much smaller (≤ 0.85); 1 = P^2 smaller (>0.85, <0.95); 2 = P^2 equal (≥ 0.95).

P5*. P^4 area : M^1 area: 0 = P^4 much smaller (≤ 0.66); 1 = P^4 smaller (>0.66, ≤ 0.76); 2 = P^4 subequal (>0.76).

P6. P^2 occlusal outline: 0 = triangular; 1 = suboval with the long axis buccolingual; 2 = suboval with the long axis mesiodistal; 3 = round, peglike.
P7. P^4 occlusal outline: 0 = triangular; 1 = suboval, long axis mesiodistal; 2 = squared.
P8. P^{3-4} trigon/talon proportions: 0 = trigon ≥ talon; 1 = trigon < talon.
P9. P^3 protocone: 0 = present; 1 = absent.
P10. P^4 metacone: 0 = absent; 1 = present.
P11. P^4 protocone: 0 = low relative to paracone; 1 = high relative to paracone.
P12. P^2 protocone: 0 = present; 1 = absent.
P13. Premolar hypocones: 0 = absent; 1 = present on P^4 only; 2 = present on P^{3-4}; 3 = present on P^{2-4}.
P14*. P^4 paraconule: 0 = large; 1 = small; 2 = absent.
P15. P^{3-4} parastyles: 0 = present; 1 = absent.
P16. P^{3-4} metastyles: 0 = absent; 1 = present.
P17. P^{3-4} postprotocrista: 0 = strong, runs to distal crown margin; 1 = absent, or weak and short; 2 = strong, short.
P18. P^{2-3} distal crown margin: 0 = smoothly rounded; 1 = waisted between buccal and lingual cusps.
P19. P^{3-4} lingual cingulum: 0 = absent or weak; 1 = strong.

Molars

M1*. M^{1-2} root number: 0 = three, three; 1 = three, two; 2 = two, two.
M2*. M^3 root number: 0 = three; 1 = two; 2 = one.
M3*. M^2 shape (buccolingual breadth/mesiodistal length): 0 = very transverse (≥1.60); 1 = transverse (<1.60, ≥1.30); 2 = squared (<1.30).
M4*. M^1 area : M^2 area: 0 = M^1 much bigger (≥1.40); 1 = M^1 bigger (<1.40, >1.0); 2 = M^1 equal to or smaller (≤1.0).
M5*. M^{1-2} *Nannopithex* fold: 0 = absent; 1 = weak; 2 = strong.
M6*. M^{1-2} pseudohypocone: 0 = absent; 1 = small; 2 = large.
M7. M^{1-2} metaconule: 0 = single; 1 = double.
M8*. M^{1-2} paraconule: 0 = absent; 1 = small; 2 = large.
M9*. M^{1-2} preprotoconule: 0 = absent; 1 = weak; 2 = strong.
M10*. M^1 hypocone: 0 = large; 1 = small; 2 = absent.
M11*. M^2 hypocone: 0 = large; 1 = small; 2 = absent.
M12. M^{1-2} hypocone position: 0 = distal, slightly lingual to protocone; 1 = distal, far lingual to protocone.
M13*. M^{1-2} prehypocrista development: 0 = absent; 1 = weak; 2 = strong, reaches to postprotocrista, encloses the talon lingually.
M14*. M^3 prehypocrista development: 0 = absent; 1 = weak; 2 = strong, reaches to postprotocrista, encloses the talon lingually.

M15. M^1 or M^2 paraconule position: 0 = attached to preprotocrista; 1 = buccal to preprotocrista.
M16*. M^{1-2} metaconule size: 0 = absent; 1 = small; 2 = large.
M17. M^{1-2} mesostyle size: 0 = absent; 1 = present, attached to ectocrista; 2 = present on buccal cingulum.
M18*. M^{1-2} postprotocrista development: 0 = strong, runs to base of metaconule or metacone; 1 = strong but short; does not reach base of metacone; 2 = absent.
M19. M^{1-2} lateral posterior transverse crista development: 0 = sharp; 1 = indistinct.
M20*. P^4–M^1 pericone: 0 = absent; 1 = small; 2 = large.
M21*. P^{3-4}, M^{1-2} buccal cingulum development: 0 = absent; 1 = weak; 2 = strong.
M22*. M^{1-3} lingual cingulum development: 0 = absent; 1 = weak, broken; 2 = strong, complete.
M23. Body size (as inferred from molar size): 0 = small (<500 g); 1 = large (≥500 g).

References

Ba Maw, U., Ciochon, R. L., and Savage, D. E. 1979. Late Eocene of Burma yields earliest anthropoid primate, *Pondaungia cotteri*. *Nature* **282:**65–67.

Beard, K. 1990. Gliding behavior and paleoecology of the alleged primate family Paromomyidae (Mammalia, Dermoptera). *Nature* **345:**340–340.

Benefit, B. R. 1992. The phylogeny and paleodemography of *Victoriapithecus*—new evidence from the deciduous dentition. *Am. J. Phys. Anthrop. Suppl.* **14:**48.

Benefit, B. R. 1993. The permanent dentition and phylogenetic position of *Victoriapithecus* from Maboko Island, Kenya. *J. Hum. Evol.* **25:**83–172.

Butler, P. M. 1980. The tupaiid dentition, in: W. P. Luckett (ed.), *Biology and Evolutionary Relationships of Tree Shrews,* pp. 171–204. New York: Plenum Press.

Cartmill, M. 1981. Hypothesis testing and phylogenetic reconstruction. *Z. Zool. Syst. Evol. Forsch.* **19:**73–96.

Cartmill, M., and Kay, R. 1978. Cranio-dental morphology, tarsier affinities, and primate suborders, in: D. Chivers and J. Joysey (eds.), *Recent Advances in Primatology,* Vol. 3, pp. 205–214. Academic Press, London.

Ciochon, R. L., Savage, D. E., Thaw Tint, and Ba Maw, U. 1985. Anthropoid origins in Asia? New discovery of *Amphipithecus* from the Eocene of Burma. *Science* **229:**756–759.

Colbert, E. H. 1937. A new primate from the upper Eocene Pondaung formation of Burma. *Am. Mus. Novit.* **651:**1–18.

Covert, H., and Williams, B. 1991. The anterior lower dentition of *Washakius insignis* and adapid–anthropoidean affinities. *J. Hum. Evol.* **21:**463–467.

de Bonis, L., Jaeger, J.-J., Coiffait, B., and Coiffait, P.-E. 1988. Découverte du plus ancien primate Catarrhinien connu dans l'Éocène supérieur d'Afrique du Nord. *C. R. Acad. Sci. Paris* **306(II):**929–934.

Delson, E. 1975. Toward the origin of the Old World monkeys. *Coll. Int. C.N.R.S.* **281:**839–850.

Delson, E., and Rosenberger, A. 1980. Phyletic perspectives on platyrrhine origins and anthropoid relationships, in: R. Ciochon and A. B. Chiarelli (eds.), *Evolutionary Biology of the New World Monkeys and Continental Drift,* pp. 445–448. Plenum Press, New York.

Fleagle, J., and Kay, R. 1983. New interpretations of the phyletic position of Oligocene hominoids, in: R. Ciochon and R. Corruccini (eds.), *New Interpretations of Ape and Human Ancestry,* pp. 181–210. Plenum Press, New York.

Fleagle, J., and Kay, R. F. 1987. The phyletic position of the Parapithecidae. *J. Hum. Evol.* **16:**483–532.

Fleagle, J., and Kay, R. F. 1989. The dental morphology of *Dolichocebus gaimanensis,* a fossil monkey from Argentina. *Am. J. Phys. Anthrop.* **78:**221.

Fleagle, J. G., and Simons, E. L. 1979. Anatomy of the bony pelvis in parapithecid primates. *Fol. Primatol.* **31:**176–186.

Fleagle, J. G., and Simons, E. L. 1982. The humerus of *Aegyptopithecus zeuxis:* A primitive anthropoid. *Am. J. Phys. Anthropol.* **59:**175–193.

Fleagle, J. G., Kay, R. F., and Anthony, M. R. L. 1994. Fossil New World monkeys, in: R. F. Kay, R. M. Madden, R. L. Cifelli, and J. J. Flynn (eds.), *Vertebrate Paleontology in the Neotropics: The Miocene Fauna of La Venta, Colombia.* Smithsonian Institution Press, Washington, D.C.

Franzen, J. L. 1987. Ein neuer Primate aus dem Mitteleozän der Grube Messel (Deutschland, S.-Hessen). *Cour. Forsch. Inst. Senskenberg* **91:**151–187.

Gebo, D. 1987. Anthropoid origins—the foot evidence. *J. Hum. Evol.* **15:**421–430.

Gebo, D. L., and Simons, E. L. 1987. Morphology and locomotor adaptations of the foot in early Paleogene anthropoids. *Am. J. Phys. Anthrop.* **74:**83–101.

Gingerich, P. D. 1977. Radiation of Eocene Adapidae in Europe. *Geobios Mem. Spec.* **1:**165–182.

Gingerich, P. D. 1978. The Stuttgart collection of Oligocene primates from the Fayum Province, Egypt. *Paläont. Z. (Stuttgart)* **52:**82–92.

Gingerich, P. 1980. Eocene Adapidae, paleobiology, and the origin of South American Platyrrhini, in: R. Ciochon and B. Chiarelli (eds.), *Evolutionary Biology of the New World Monkeys and Continental Drift,* pp. 123–138. Plenum Press, New York.

Gingerich, P. D. 1990. African dawn for primates. *Nature* **346:**411.

Gingerich, P., and Schoeninger, M. 1977. The fossil record and primate phylogeny. *J. Hum. Evol.* **6:**483–505.

Ginsburg, L., and Mein, P. 1987. *Tarsius thailandica* nov. sp., premier Tarsiidae (Primates, Mammalia) fossile d'Asie. *C. R. Acad Sci. (Paris)* **304 (Ser. II):**1213–1214.

Godinot, M., and Mahboubi, M. 1992. Earliest known simian primate found in Algeria. *Nature* **357:**324–326.

Gregory, W. K. 1922. *The Origin and Evolution of the Human Dentition.* Williams & Wilkins, Baltimore.

Harrison, T. 1987. The phyletic relationships of the early catarrhine primates: A review of the current evidence, *J. Hum. Evol.* **16:**41–80.

Hoffstetter, R. 1977. Phylogenie des primates. *Bull. Mem. Soc. Anthropol. (Paris)* **4(Ser. 13):**327–352.

Hoffstetter, R. 1980a. Los monos platirinos (Primates): Origen, extension, fologenia, taxonomia. *Act. II Cong. Arg. Paleotol. Biostratig.* **II:**291–303.

Hoffstetter, R. 1980b. Origin and deployment of New World monkeys emphasizing the southern continents route, in: R. Ciochon and A. B. Chiarelli (eds.), *Evolutionary Biology of the New World Monkeys and Continental Drift,* pp. 103–122. Plenum Press, New York.

Honey, J. 1990. New washakiin primates (Omomyidae) from the Eocene of Wyoming and Colorado, and comments on the evolution of the Washakiini. *J. Vert. Paleontol.* **10:**206–221.

Kay, R. F. 1977. The evolution of molar occlusion in the Cercopithecidae and early catarrhines. *Am. J. Phys. Anthropol.* **46:**327–352.

Kay, R. F. 1980. Platyrrhine origins: a reappraisal of the dental evidence, in: R. Ciochon and A. B. Chiarelli (eds.), *Evolutionary Biology of the New World Monkeys and Continental Drift,* pp. 159–188. Plenum Press, New York.

Kay, R. F. 1990. The phyletic relationships of extant and fossil Pitheciinae (Platyrrhini, Anthropoidea) *J. Hum. Evol.* **19:**175–208.

Kay, R. F., and Simons, E. L. 1983. Dental formulae and dental eruption patterns in Parapithecidae. *Am. J. Phys. Anthrop.* **63:**353–375.

Kay, R. F., Thorington, R. W., Jr., and Houde, P. 1990. Eocene plesiadapiform shows affinities with flying lemurs not primates. *Nature* **345:**342–344.

Kay, R. F., Thewissen, J. G. M., and Yoder, A. D. 1992. Cranial anatomy of *Ignacius graybullianus* and the affinities of the Plesiadapiformes. *Am. J. Phys. Anthrop.* **89:**477–498.

Koenigswald, G. H. R. von. 1965. Critical observations upon the so-called higher primates from the upper Eocene of Burma. *Proc. Kon. Nederl. Akad. Wet. Ser. B.* **68:**165–167.

Le Gros Clark, W. E. 1934. *Early Fore-Runners of Man.* Ballière, London.

Maddison, W., and Maddison, D. 1992. *MacClade, Analysis of Phylogeny and Character Evolution, Version 3.0.* Sinauer Associates, Sunderland, Massachusetts.

Neff, N. A. 1986. A rational basis for *a priori* character weighting. *Syst. Zool.* **35:**110–123.

Novacek, M. 1992. Mammalian phylogeny: shaking the tree. *Nature* **356:**121–125.

Patterson, B. 1954. The geologic history of nonhominid primates in the Old World. *Hum. Biol.* **26:**191–209.

Pilgrim, G. E. 1927. A *Sivapithecus* palate and other primate fossils from India. *Mem. Geol. Surv. India* **14:**1–24.

Rasmussen, D. T. 1990. The phylogenetic position of *Mahgarita stevensi:* protoanthropoid or lemuroid? *Int. J. Primatol.* **11:**437–467.

Rasmussen, D. T., and Simons, E. 1988. New specimens of *Oligopithecus savagei,* early Oligocene primate from the Fayum, Egypt. *Fol. Primatol.* **51:**182–208.

Rasmussen, D. T., and Simons, E. L. 1992. Paleobiology of the oligopithecines, the earliest known anthropoid primates. *Int. J. Primatol.* **13:**477–508.

Rose, K. D., and Bown, T. M. 1991. Additional fossil evidence on the differentiation of the earliest euprimates. *Proc. Natl. Acad. Sci. U.S.A.* **88:**98–101.

Rosenberger, A. L. 1986. Platyrrhines, catarrhines and the anthropoid transition, in: E. Wood, L. Martin, and P. Andrews (eds.), *Major Topics in Primate and Human Evolution,* pp. 66–88. Cambridge University Press, Cambridge.

Senut, B., and Thomas, H. 1992. First discoveries of anthropoid postcranial remains from Taqah (early Oligocene, Sultinate of Oman), in: *Abstracts XIV Congress, International Primatological Society.* Strasbourg, France.

Sigé, B., Jaeger, J.-J., Sundre, J., and Vianey-Liaud, M. 1990. *Altiatlasius koulchii* n. gen. et sp., primate omomyidé du Paléocène supérieur du Moroc, et les origines des Euprimates. *Paleontographica* **212:**1–24.

Simons, E. L. 1962. Two new primate species from the African Oligocene. *Postilla,* **64:**1–12.

Simons, E. L. 1963. A critical reappraisal of Tertiary primates, in: J. Beuttner-Janusch (ed.), *Evolutionary and Genetic Biology of Primates,* pp. 65–129. Academic Press, New York.

Simons, E. L. 1965. New fossil apes from Egypt and the initial differentiation of Hominoidea. *Nature* **205:**135–149.

Simons, E. L. 1971. Relationships of *Amphipithecus* and *Oligopithecus. Nature* **232:**489–491.

Simons, E. L. 1972. *Primate Evolution.* Macmillan, New York.

Simons, E. L. 1974. *Parapithecus grangeri* (Parapithecidae, Old World Higher Primates): New species from the Oligocene of Egypt and the initial differentiation of Hominoidea. *Postilla* **166:**1–12.

Simons, E. 1989. Description of two genera and species of late Eocene Anthropoidea from Egypt. *Proc. Natl. Acad. Sci. U.S.A.* **86:**9956–9960.

Simons, E. L. 1990. Discovery of the oldest known anthropoidean skull from the Paleogene of Egypt. *Science* **247:**1521–1526.

Simons, E. L. 1992. Diversity in the early Tertiary anthropoidean radiation in Africa. *Proc. Natl. Adad. Sci. U.S.A.* **89:**10743–10747.

Simons, E. L., and Bown, T. M. 1985. *Afrotarsius chatrathi,* first tarsiiform primate (?Tarsiidae) from Africa. *Nature* **313:**475–477.

Simons, E. L., and Kay, R. F. 1983. *Qatrania,* a new basal anthropoid primate from the Fayum, Oligocene of Egypt. *Nature* **304:**624–626.

Simons, E. L., and Kay, R. F. 1988. New material of *Qatrania* from Egypt with comments on the phylogenetic position of Parapithecidae (Primates, Anthropoidea). *Am. J. Primatol.* **15:**337–347.

Simons, E. L., and Pilbeam, D. R. 1965. Preliminary revision of the Dryopithecinae (Pongidae, Anthropoidea). *Fol. Primatol.* **3:**81–152.

Simons, E. L., and Rasmussen, D. T. 1991. The generic classification of Fayum Anthropoidea. *Int. J. Primatol.* **12:**163–178.

Swofford, D. 1993. *PAUP (Phylogenetic Analysis Using Parsimony), version 3.1.1.* Computer program distributed by Illinois Natural History Survey, Champagne, Illinois.

Szalay, F. S. 1972. *Amphipithecus* revisited. *Nature* **236:**179.

Szalay, F. S. 1976. Systematics of the Omomyidae (Tarsiiformes, Primates): taxonomy, phylogeny and adaptations. *Bull. Am. Mus. Nat. Hist.* **156:**157–450.

Szalay, F. S., and Delson, E. 1979. *Evolutionary History of the Primates.* Academic Press, New York.

Szalay, F. S., and Li, C. F. 1986. Middle Paleocene Euprimate from southern China and the distribution of Primates in the Paleogene. *J. Hum. Evol.* **15:**387–398.

Szalay, F. S., Rosengerger, A., and Dagosto, M. 1987. Diagnosis and differentiation of the Order Primates. *Yrbk. Phys. Anthrop.* **30:**75–105.

Watrous, L., and Wheeler, Q. D. 1981. The outgroup comparison method of character analysis. *Syst. Zool.* **30:**1–11.

Wheeler, Q. D. 1986. Character weighting and cladistic analysis. *Sys. Zool.* **35:**102–109.

Function and Fusion of the Mandibular Symphysis in Primates

14

Stiffness or Strength?

MATTHEW J. RAVOSA and WILLIAM L. HYLANDER

Introduction

Over the past 30 years a series of morphological and experimental analyses have attempted to address questions about the functional and evolutionary significance of mandibular symphyseal fusion or complete ossification of the joint between the two dentaries (Scapino, 1965, 1981; Hylander, 1975a, 1977, 1979a,b, 1984, 1985; Beecher, 1977, 1979, 1983; Hirschfeld *et al.*, 1977; Dessem, 1985; Hylander *et al.*, 1987; Greaves, 1988, 1993; Ravosa, 1991; M. J. Ravosa, *unpublished data;* Ravosa and Hylander, 1993; Hylander and Johnson, 1994; Ravosa and Simons, 1994). This work has increased our understanding of the functional morphology of the mammalian masticatory apparatus, in part by highlighting the interaction of jaw mechanics, diet and allometry on

MATTHEW J. RAVOSA and WILLIAM L. HYLANDER • Department of Biological Anthropology and Anatomy, Duke University Medical Center, Durham, North Carolina 27710. *Present address for M. J. R.:* Department of Cell and Molecular Biology, Northwestern University Medical School, Chicago, Illinois 60611-3008, and Department of Zoology, Division of Mammals, Field Museum of Natural History, Chicago, Illinois 60605-2496.

Anthropoid Origins, edited by John G. Fleagle and Richard F. Kay. Plenum Press, New York, 1994.

symphyseal form. Many of these studies have also influenced adaptive explanations for the evolution of anthropoid craniodental morphology and impact directly on hypotheses regarding phylogenetic affinities among certain Eocene and Oligocene primates (Hiiemae and Kay, 1972, 1973; Gingerich, 1977, 1979; Beecher, 1977, 1979; Cachel, 1979a,b; Hylander, 1979a,b; Szalay and Delson, 1979; Rosenberger, 1981, 1986; Rosenberger *et al.*, 1985; Rasmussen, 1986, 1990; Greaves, 1988, 1993; Simons, 1989, 1990, 1992; Ravosa, 1991; M. J. Ravosa, *unpublished data;* Rasmussen and Simons, 1992; Ravosa and Hylander, 1993; Ravosa and Simons, 1994).

Most of the above morphological and experimental studies have emphasized that symphyseal fusion is an adaptation to strengthen the symphyseal joint (Hylander, 1975a, 1977, 1979a,b, 1984, 1985; Beecher, 1977, 1979, 1983; Hirschfeld *et al.*, 1977; Scapino, 1981; Ravosa, 1991; M. J. Ravosa, *unpublished data;* Hylander and Johnson, 1994; Ravosa and Simons, 1994), although there remains some disagreement as to the relative importance of mastication versus incision as determinants of symphyseal fusion. In contrast, Greaves (1988, 1993) has argued recently that symphyseal fusion in anthropoid primates is an adaptation to stiffen, rather than strengthen, the symphysis during the crushing of small, hard objects (e.g., seeds) along the incisors. The purpose of this chapter is to evaluate various explanations for symphyseal form and function in anthropoids. In doing so, we will review available experimental, neontological, and paleontological evidence regarding the significance of jaw mechanics, mechanical properties of the diet, and allometric constraints related to jaw-muscle force production and food-processing behavior, on symphyseal fusion in primates and other mammals.

Discussion

Symphyseal Mobility and Fusion

In 1988, Greaves suggested that symphyseal fusion in anthropoid primates is linked to a pattern of forceful crushing of small, hard objects (e.g., seeds) by large vertically implanted incisors. He notes that "only in those few animals where the incisors have taken on an important crushing function can the costs of fusion be tolerated" (p. 56). Greaves's argument for the functional significance of symphyseal fusion is based largely on the proposition that although an unfused symphysis is able to transmit balancing-side muscle force to an extent equal that of a fused symphysis, an unfused symphysis has the undesirable characteristic of allowing a relatively large amount of independent movement between the two mandibular halves (dentaries) during incision. Therefore, when large masticatory forces are needed to crush small, hard objects along, for example, the right side of the incisor tooth row, bite force generated by muscle force from the left side (the balancing side) is dissipated or wasted along the incisors on the left rather than contributing to the crushing of the small object between the incisors on the right (Fig. 1). This loss of usable force occurs because during crushing of the small, hard objects along the incisors on the right, an unfused and therefore flexible symphysis allows the balancing-side (left) upper and lower incisors to contact one another.

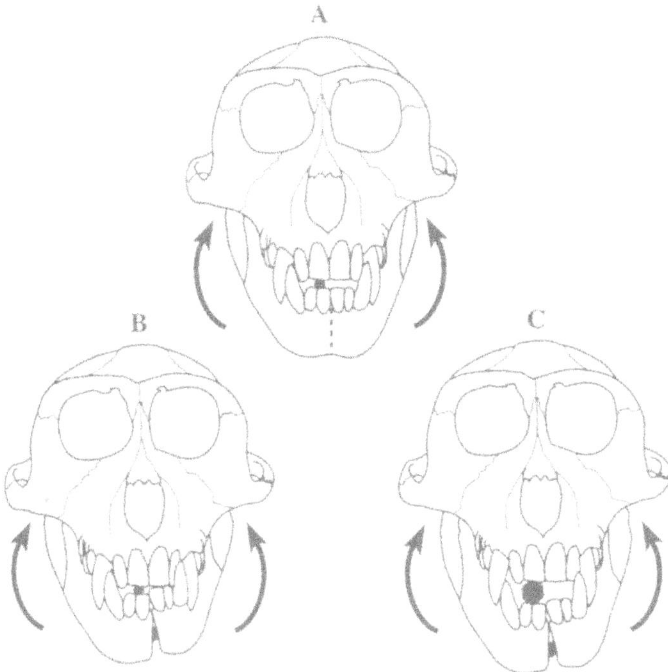

Fig. 1. Frontal views of an imaginary adult primate with a fused symphysis (A) and one with an unfused symphysis (B) during the incisal crushing of a small, hard object such as a seed. During incision the mandibular corpora are twisted about their long axes by a laterally positioned resultant jaw-muscle force and bent in the parasagittal plane. The arrows indicate the direction of twisting; i.e., the lower border of the corpus everts and the postcanine alveolar process inverts. When the animal in A crushes the small, hard object along the incisors on the right, there is no contact between the upper and lower incisors on the left. In this case jaw-muscle force from both the left and right sides contributes to the production of bite force applied to the food object. On the other hand, when the animal in B crushes the small, hard object along the incisors on the right, the incisors on the left come into contact. In this case little or no jaw-muscle force from the left (the balancing side) is applied to the food object. Instead, this force is dissipated uselessly along the contacting balancing-side incisors. When the animal with an unfused symphysis in C crushes a large-diameter food item along the incisors on the right, there is no contact between the upper and lower incisors on the left. Thus, jaw-muscle force from the left side contributes to the production of useful incisal bite force even though the symphysis is unfused (adapted from Greaves, 1988).

Conversely, during incisal biting of much larger objects along the right side, as the diameter of these large items exceeds the amount of independent movement of each dentary at the symphysis, the balancing-side (left) upper and lower incisors do not contact one another. In this case, force associated with the balancing-side jaw muscles is transmitted across the symphyseal joint and therefore contributes to crushing the food item on the right side. In sum, Greaves argues that the primary function of a fused symphysis is to stiffen the symphyseal joint so as to reduce independent movement of each dentary during the vigorous incisal crushing of small, hard objects such as seeds. Reducing this undesirable movement ensures that all of the incisal bite force is applied directly to the small food object, rather than some of it being dissipated uselessly along the directly contacting upper and lower balancing-side incisors.

For the purposes of this section, two major issues require consideration. First, what data exist to support Greaves's model? And second, are there equally or more plausible explanations for the functional significance of a fused mandibular symphysis? Prior to discussing these two issues, we will comment briefly on Greaves's criticisms of previous explanations for symphyseal fusion.

We fully agree with Greaves that having an unfused symphysis allows a greater amount of independent movement between the dentaries (Crompton and Hiiemae, 1970; Kallen and Gans, 1972; Scapino, 1981; Hirschfeld *et al.*, 1977; Hiiemae and Crompton, 1985). However, we differ somewhat with his opinion [and also Rosenberger's (1986)] that there is considerable disagreement among workers regarding the extent to which balancing-side jaw-muscle force can be transmitted across an unfused versus a fused symphysis. That is, no published accounts to our knowledge have suggested that a significant amount of the force produced by the balancing-side jaw muscles is somehow dissipated or lost to the system as a result of the presence of an unfused symphysis (cf. Dessem, 1985, regarding vertical force transmission across the symphyseal joint in three carnivoran genera).

As a point of clarification, we believe that Greaves (1988) has misinterpreted what Beecher (1979) and others have suggested regarding the functional significance of symphyseal fusion. Although Greaves correctly notes that Beecher implied "that a fused symphysis represented the strongest possible connection and would resist mechanical failure," we disagree with his statement that "most would explain symphyseal fusion as a mechanism to *optimize* (italics ours) muscle force transmission from one side of the lower jaw to the other" (p. 32). Other workers (Scapino, 1981; Dessem, 1985; Rosenberger *et al.*, 1985; Rosenberger, 1986) have also apparently misinterpreted what has been said as to the relationship between balancing-side jaw-muscle force transmission and symphyseal fusion (Hylander, 1975a, 1979a,b; Beecher, 1977, 1979). It appears to us that most workers agree that in taxa with unfused symphyses, force is transferred across the symphysis just as efficiently as in those taxa with fused symphyses (cf., Dessem, 1985), as long as the symphyseal tissues do not experience structural failure during these periods of force transfer.

In order to prevent further misunderstanding, we would like to clarify that although Beecher (1977, 1979) and Hylander (1975a, 1979a,b) have suggested that primate taxa with fused symphyses are able to transfer more balancing-side muscle force across the symphyseal joint, this is not because a fused symphyseal joint *optimizes* this force transmission. Rather, it is because a fused symphysis is best able to resist structural failure from the increased symphyseal stress resulting from increased recruitment of balancing-side jaw-muscle force during mastication. Thus, Beecher and Hylander simply suggest that a fused symphysis is less apt to experience structural failure from the repetitious cyclical loads associated with *increased* levels of balancing-side muscle force during molar biting and chewing.

The main aspect of Greaves's model emphasizes the importance of incisal crushing of small, hard objects as the major determinant of symphyseal fusion. Although Greaves did not specify what kind of small, hard objects were crushed, other than to note "(e.g., seeds)," for primates these objects must be mainly or exclusively seeds. Therefore, in testing this hypothesis, it is useful to consider whether there are any primates that have both a fused symphysis and habitually crush small, hard seeds or other small, hard objects along their incisors. Broadly speaking, there are two primate groups that bear on this issue. The most important group for our purposes consists of extant primates, whereas the other group consists of the "subfossil" Malagasy lemurs.

Greaves's model can be tested with available behavioral and morphological data for extant primates. However, one question must be considered most relevant: do living anthropoids, all of which have fused symphyses, habitually crush small, hard seeds or any other small, hard object along their incisors? A review of available data on dietary behavior in anthropoids indicates that the vast majority do not include small seeds or any other small, hard object as a major component of their diet (for summaries see Richard, 1985; Fleagle, 1988). The literature on anthropoid feeding behavior also provides no evidence for the habitual crushing of small, hard objects along the incisors. One such study among five species of free-ranging cercopithecine monkeys indicates that "the molars were always used to open and crush seeds except in 1% of the tested fruits by a *C. aethiops* monkey" (Happel, 1988, p. 319). Additional morphological studies indicate that small-object feeding, such as eating small seeds, is associated with both relatively *small* incisors and *little* incisal preparation of these foods (Jolly, 1970; Hylander, 1975b). Moreover, little evidence exists to suggest that early anthropoids crushed small, hard seeds along the incisors (e.g., Rosenberger, 1986; Fleagle, 1988). Therefore, the morphological and dietary-behavior data for anthropoid primates do not support Greaves's model.

Greaves's model also cannot explain the presence of partial symphyseal fusion in prosimians such as bamboo lemurs (*Hapalemur*), i.e., calcification and/or incipient ossification of the cruciate ligaments, because these primates possess a delicate toothcomb and small incisors, which preclude habitual crushing of small, hard objects along their incisors. Moreover, an analysis of symphyseal morphology and diet in extant prosimians, none of which have fully fused symphyses, demonstrates that those species with partially fused symphyses feed principally on leaves and/or bamboo (Beecher, 1977, 1979; Ravosa, 1991), not on small, hard objects as predicted by Greaves's model. Thus, morphological and dietary-behavior data for both extant primate suborders do not support the incisor-crushing model of symphyseal fusion.

Rather than rely mainly on observations of dietary preference in free-ranging primates, we conducted an additional test of Graves's model to determine what an animal with a fused symphysis and large incisors does when it actually eats a small, hard object. We presented hard, unpopped popcorn kernels* to two adult macaques (*Macaca fascicularis*) and four adult baboons (*Papio anubis*) housed at the Duke University Vivarium. In each case the subject first typically clasped the kernel with its lips and incisors, and then immediately transferred the kernel to the molar region where it was crushed. No incisal crushing of the hard kernels took place. From a biomechanical standpoint these observations are not remarkable since, in chewing small, hard objects, higher bite forces can be generated and dissipated more effectively along the molars than along the incisors. Thus, our behavioral test also does not support the incisor-crushing model of symphyseal fusion.

Greaves's model can also be tested in the five genera of subfossil Malagasy lemurs, all of which developed symphyseal fusion. These lemurs are distributed among three separate subfamilies—the Archaeolemurinae, Megaladapinae, and Palaeopropithecinae. No two of these groups are sister taxa, which indicates that a completely fused symphysis evolved independently at least three times among Malagasy primates.

Only among two species of the Archaeolemurinae (*Archaeolemur edwardsi, A. majori*) is there any evidence to support the hypothesis that habitual incisor use and a fused symphysis are linked functionally. Similar to anthropoids, these primates had both a fused symphysis and large upper and lower incisors. In contrast, although *Hadropithecus stenognathus* (Archaeolemurinae) and all three species of the Palaeopropithecinae (*Palaeopropithecus maximus, P. ingens, Archaeoindris fontoynonti*) had fused symphyses, they had fairly diminutive incisors, indicative of minimal amounts of incisal preparation of food items (cf., Jolly, 1970; Hylander, 1975b). Most importantly, although all three species of the Megaladapinae (*Megaladapis edwardsi, M. grandidieri, M. madagascariensis*) had fused symphyses, *they lacked upper incisors altogether*, demonstrating that symphyseal fusion and incisor crushing cannot possibly be linked in these taxa. Thus, the evidence from subfossil Malagasy lemurs provides no convincing support for the incisor-crushing model of symphyseal fusion.

*We consider unpopped popcorn kernels to be small, hard objects based on Fig. 1 in Greaves's (1988) paper (p. 55). In this figure, a small, hard object is depicted as about one-third the height of the upper central incisor of an adult baboon, which is more or less the same size as a popcorn kernel.

A brief consideration of Greaves's model for members of the Carnivora serves to underscore fundamental and widespread inconsistencies in the purported functional link between incisal crushing of hard, small objects and symphyseal fusion. For instance, there are many carnivorans with fused symphyses, diminutive incisors and no evidence for the presence of routine incisal crushing behaviors, e.g., the ursid species *Ailuropoda melanoleuca* (giant panda), *Helarctos malayanus* (Malayan sun bear), and *Melursus ursinus* (sloth bear) and the mustelid species *Conepatus chinga* (hog-nosed skunk), *Eira barbara* (tayra), *Mellivora capensis* (ratel), *Paraonyx congica* (Zaire clawless otter), *Arctonyx collaris* (hog badger), *Taxidea taxus* (North American badger), *Mydaus javanensis* (sunda stink-badger), and *Suillotaxus marchei* (Philippine badger) (Scapino, 1981; M. J. Ravosa, *unpublished data*). Thus, similar to the case of extant and extinct primates, the comparative evidence from carnivorans also suggests that a fused symphysis and the incisal crushing of small, hard objects are not linked.

Since a specific functional association between incisal crushing of small, hard objects and symphyseal fusion appears unlikely for several reasons (Ravosa and Hylander, 1993), what alternative models exist to explain symphyseal fusion? Perhaps the functional association between symphyseal fusion and increased stiffness of the symphysis is important for molar biting and chewing rather than for incisor biting. For instance, species that routinely chew small, hard objects and/or thin, tough leaves along the postcanine teeth might be subject to problems similar to those envisioned by Greaves for incisal crushing. If so, symphyseal fusion could be an adaptation to stiffen the symphysis so as to reduce the amount of balancing-side tooth contacts during unilateral mastication.

It seems to us, however, that this argument is not overly persuasive for the following reason. Both anisognathic jaws and an unfused mandibular symphysis probably represent the primitive condition for primates and other mammals (cf., Herring and Scapino, 1973; Zingeser, 1976). If so, perhaps there is nothing to be gained by fusing the symphysis to prevent extensive balancing-side tooth contacts during unilateral mastication if the preexisting presence of markedly anisognathic jaws precludes the widespread occurrence of such contacts. Thus, it appears that the development of symphyseal fusion in primates with anisognathic jaws must have occurred for reasons unrelated to preventing undesirable balancing-side tooth contacts during the chewing of small or thin objects along the molars and premolars. However, once symphyseal fusion occurred, this may have facilitated the evolution of the more isognathic jaws typical of anthropoid primates.*

*Isognathy refers to the condition in which the widths of the upper and lower dental arches are about the same. This condition allows the upper and lower postcanine teeth on the left and right sides of the jaws to come into near-functional occlusion simultaneously. On the other hand, anisognathy refers to the condition in which the widths of the upper and lower arcades are markedly different from one another. In these instances, ordinarily the postcanine teeth on only one side can be brought into functional occlusion at any given time. Among primates, an-

Anthropoids do, however, provide evidence for a possible functional link between symphyseal fusion and isognathy simply because all anthropoids are characterized by these two morphological states. In contrast, subfossil Malagasy lemurs provide some evidence for the lack of this link. There are three taxa with fused symphyses for which data are available: *Archaeolemur edwardsi, Megaladapis edwardsi,* and *Palaeopropithecus maximus. Archaeolemur edwardsi* had at the second molars an isognathic upper arch-width/lower arch-width ratio of 1.04, which at the fourth premolars increased to a more anisognathic ratio of 1.21 (slightly higher than the anthropoid mean). *Megaladapis edwardsi* had an arch-width ratio of 1.11 at the second molars, which decreased slightly to a value of 1.09 at the fourth premolars; these ratios are respectively equal to and much less than the anthropoid means. *Palaeopropithecus maximus,* however, retained the primitive condition of anisognathy; in this species the arch-width ratio at the second molars was 1.27, and at the fourth premolars it was 1.30. Thus, one of three subfossil lemurs with symphyseal fusion retained the primitive condition of anisognathic jaws, suggesting that isognathy and symphyseal fusion are not necessarily functionally linked. This differs from the evidence provided by comparisons between extant anthropoids and extant prosimians.

Symphyseal Stress and Fusion

In addition to its stiffening effects, symphyseal fusion also strengthens the symphysis so as to help counter symphyseal stress during incisal biting and unilateral mastication. Following Beecher (1977, 1979), a completely fused mandibular symphysis is stronger than an unfused symphysis of similar dimensions essentially because bone is considerably stronger than the fibrocartilage and ligaments of an unfused symphyseal joint.* The following is a review of the available experimental, behavioral, and morphological evidence regarding the influence of masticatory stress on symphyseal form and fusion.

---

thropoid jaws show a nearly isognathic condition at the second molars with a mean upper arch-width/lower arch-width ratio of 1.11 for 13 genera. Anteriorly at the fourth premolars, anthropoid arcades are significantly more anisognathic (mean arch-width ratio is 1.19) (Mann–Whitney U test, $p < 0.01$). The extant prosimian sample of 12 genera have a mean arch-width ratio of 1.22 at the second molars for 12 genera; this value increases significantly at the fourth premolars to 1.29. In suborder comparisons, extant prosimian jaws are significantly more anisognathic than anthropoid jaws at both points along the dental arcades. Within each suborder these occlusal patterns do not vary allometrically or at the family level (Kruskal–Wallis analysis of variance, $p > 0.05$).

*Sometimes the morphological distinction between a completely fused symphysis and a partially fused symphysis is less significant in a functional sense. For instance, although the symphyseal joint of many large-bodied felids, ursids, and mustelids is only partially fused, from a functional standpoint this type of symphysis may be nearly as strong as a fully ossified one at resisting dorsoventral shear forces. This may also apply to the Late Eocene anthropoids *Catopithecus browni* and *Arsinoea kallimos,* both of which apparently had a partially fused mandibular symphysis (Simons, 1989, 1990, 1992).

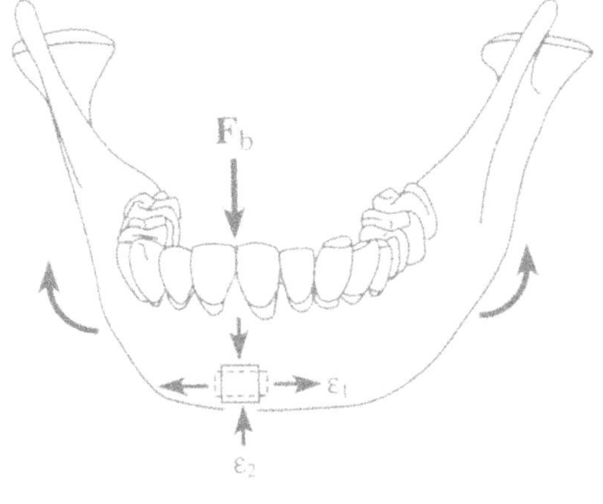

Fig. 2. Frontal view of an adult primate mandible with a fused symphysis. During incisal biting the mandibular corpora are twisted about their long axes by a laterally-positioned resultant of jaw-muscle force. This twisting everts the lower borders of both corpora such that the symphysis is bent vertically. This in turn results in compression along the upper or alveolar surface of the symphysis and tension along the inferior or basal aspect of the symphysis. The bite force ($\mathbf{F}_b$) also causes localized stress concentrations within the alveolar processes of the incisors. Vertical bending stress at the symphysis can be most effectively countered by complete symphyseal fusion simply because bone is much stronger than the ligaments of an unfused joint and, if necessary, by increasing the amount of cortical bone distributed along the symphysis. The arrows labeled ϵ_1 and ϵ_2 indicate the direction of the maximum and minimum principal strains (principal tension and principal compression) (adapted from Hylander, 1984).

Symphyseal Stress and Incision

Hiiemae and Kay (1972, 1973) provide one of the earlier considerations of incisal biting as a determinant of symphyseal fusion in anthropoids. They note that spatulate, vertically implanted incisors and symphyseal fusion are two features common to early anthropoids and suggest that perhaps a fused symphysis in anthropoids functions to strengthen the symphyseal joint against greater incisal loading. A similar argument has been emphasized by Rosenberger *et al.* (1985), who posit that symphyseal fusion in anthropoids counters "powerful shearing forces . . . taken up by the symphysis" (p. 30) during the incisal preparation of "tough-coated fruits and, possibly, fruits with hard edible contents, such as seeds and nuts" (Rosenberger, 1986, p. 77).

It has been noted that during incision (and during mastication), the mandibular corpora of macaques are twisted about their long axes (Hylander, 1979a, 1988). This axial twisting everts the lower borders of both corpora such that the symphysis is bent vertically (Fig. 2). In turn, this type of vertical bending causes compression along the upper or alveolar surface of the symphysis and tension along the inferior or basal aspect of the symphysis,* and this stress can be most effectively countered by complete symphyseal fusion and, if necessary, by increasing the amount of cortical bone along the symphy-

*Twisting or torsion of the mandibular corpus about its long axis results from the overall muscle-force resultant, which lies lateral to the mandible. These twisting moments evert the lower or basal border of the mandibular corpus and invert the upper or alveolar portion (Hylander, 1979a, 1988).

sis, particularly along its basal or inferior aspect. Therefore, symphyseal fusion could represent an adaptive response to counter symphyseal bending stress during powerful incision.

This argument, as well as those by Hiiemae and Kay and Rosenberger, however, is not well supported simply because of the widespread presence of an unfused symphysis in various mammals that are known to engage in powerful incision (Hylander, 1979b). Among primates, *Daubentonia madagascariensis* is one such example.

Symphyseal Stress and Mastication

To date, the majority of studies among primates and other mammals suggest that symphyseal fusion is designed to resist stresses associated with two loading patterns experienced during molar biting and chewing: lateral transverse bending of the mandibular symphysis, or wishboning; and dorsoventral shear of the symphysis. Before continuing, we would like to state that the possible presence of stress regimes associated with dorsoventral shear and wishboning are not mutually exclusive. In fact, both of these loading patterns appear to occur among anthropoids during routine unilateral mastication (Hylander, 1979a, 1984, 1985; Hylander and Johnson, 1994). These two possible causes of symphyseal fusion are considered at greater length below.

Wishboning Stress and Symphyseal Fusion. In vivo bone-strain analyses indicate that during unilateral mastication, the mandibular symphysis of anthropoid primates experiences lateral transverse bending or wishboning (Hylander, 1984, 1985). This wishboning stress results from a laterally directed component of muscle force on the balancing side and a laterally directed component of bite force, and perhaps jaw-muscle force, on the working side (Hylander and Johnson, 1994) (Fig. 3).* Since wishboning results in high stress concentrations and high strain magnitudes at the symphysis, especially along its lingual surface, this stress is best countered by complete symphyseal ossification simply because bone is much stronger than the ligaments of an unfused joint. Thus, perhaps the presence of an unfused symphysis in extant prosimians indicates that wishboning stress along the prosimian symphysis is less than that experienced by anthropoids. Moreover, in taxa with a fused symphysis, increased wishboning is best resisted by adding more bone along the labiolingual axis of the symphysis (Hylander, 1984, 1985) and/or inclining the long axis of the symphysis more horizontally (Ravosa, 1991).

Certain morphological data for primates apparently bear on the issue of wishboning in anthropoids. Comparisons of allometric trajectories of symphysis width indicate that the data for anthropoids are significantly trans-

*The tendency for the balancing-side mandibular corpus to experience wishboning during mastication results in part from the late and pronounced activity of the balancing-side deep masseter muscle during the terminal portion of the power stroke (Hylander *et al.*, 1987; Hylander and Johnson, 1994).

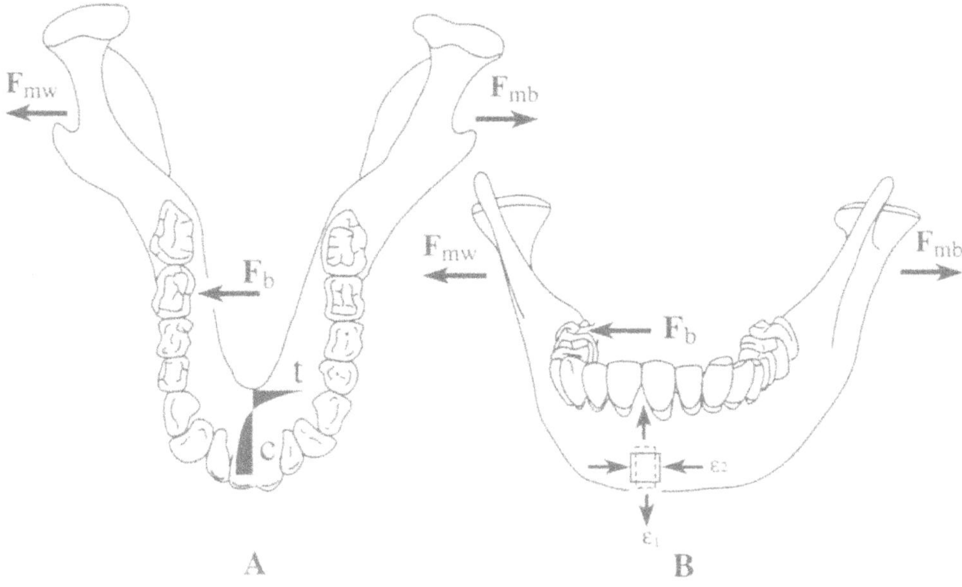

Fig. 3. Superior (A) and frontal (B) views of an adult primate mandible with a fused symphysis. During unilateral mastication the mandibular symphysis of anthropoids experiences lateral transverse bending or wishboning. This wishboning stress results from a laterally directed component of jaw-muscle force on the balancing side ($\mathbf{F}_{mb}$) and a laterally directed component of bite force ($\mathbf{F}_b$) and muscle force on the working side ($\mathbf{F}_{mw}$). Since wishboning results in especially high tensile stresses along its lingual aspect (A), this stress is best countered by complete symphyseal ossification simply because bone is much stronger than the ligaments of an unfused joint. Increased wishboning is best resisted in a fused symphysis by increasing the labiolingual thickness of the symphysis and/or orienting the long axis of the symphysis more horizontally. The relative size of the arrows is not representative of the amount of forces on each side of the jaw (adapted from Hylander, 1984, 1988).

posed above those for recent prosimians with unfused symphyses (Fig. 4) (Ravosa, 1991).* These data indicate that anthropoids have a relatively large symphysis and provide additional support for the hypothesis that compared to prosimians, anthropoid symphyseal morphology is related to resisting increased wishboning stress.

Dorsoventral Shear Stress and Symphyseal Fusion. Bone-strain data (Hylander, 1977, 1979a), as well as histological (Beecher, 1977, 1979) and morphological analyses (Hylander, 1979b; Beecher, 1983; Ravosa, 1991; M. J.

*Ravosa's (1991) study of jaw scaling included a sample of 65 anthropoid taxa (n = 318 adult cases) and 35 recent prosimian taxa with unfused symphyses (n = 220). Here, we have increased the number of species in each suborder to 91 anthropoids (n = 424) and 45 extant prosimians with unfused symphyses (n = 879). Suborder comparisons of least-squares regression lines for symphysis height and width and corpus height and width indicate significant transpositions of the anthropoid scaling trajectories over those for extant prosimians in all four cases (analysis of covariance, $p < 0.01$).

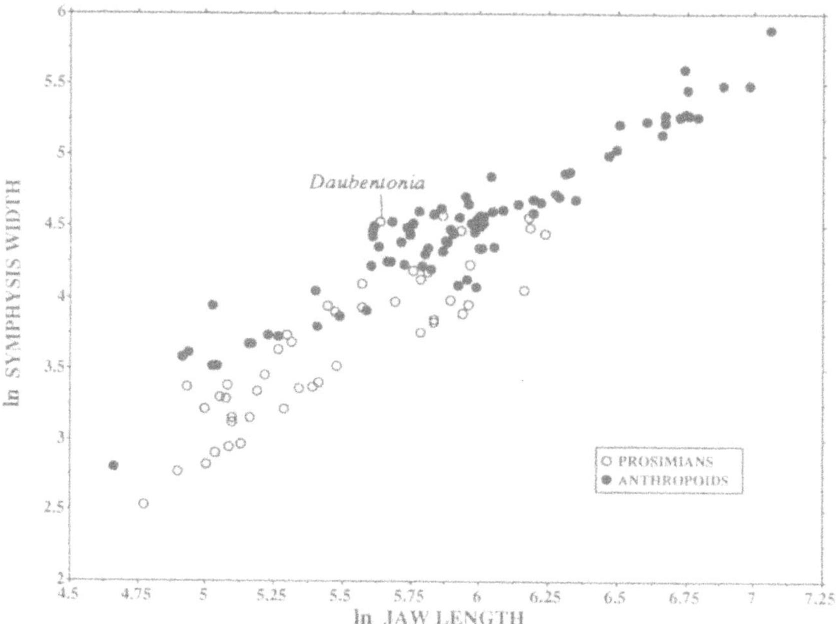

Fig. 4. A plot of ln symphysis width versus ln jaw length in the two living primate suborders. This sample is based on 91 anthropoid and 45 prosimian species; all of the extant prosimians have unfused symphyses. Anthropoid data scatter for symphysis width is transposed above that for extant prosimians, which supports the hypothesis that anthropoids have a relatively large symphysis probably related, in part, to resisting increased wishboning stress. Note that although *Daubentonia*, a prosimian, has an unfused symphysis, it plots relatively high in the anthropoid range for symphysis width; in this case an enlarged symphysis is likely a result of the presence of large, centrally growing incisors. In addition, several indriids and lemurids with partially fused symphyses plot relatively high in the anthropoid range for symphysis width (adapted from Ravosa, 1991).

Ravosa, *unpublished data*), indicate that during unilateral mastication, the balancing-side mandibular corpus of primates experiences parasagittal bending. This in turn is associated with dorsoventral shear stress at the symphysis. The amount of dorsoventral shear stress is directly related to the amount of vertically oriented force transmitted across the symphysis from the balancing to the working side of the mandible (Fig. 5). Thus, an increased recruitment of balancing-side jaw-muscle force causes an increase in dorsoventral shear, and this is best countered by complete symphyseal fusion for reasons noted above (Hylander, 1975a, 1979a,b, 1984; Beecher, 1977, 1979, 1983). Because of the presence of an unfused symphysis in living prosimians, one might predict that dorsoventral symphyseal shear stress and parasagittal bending moments about the balancing-side mandibular corpus are lower than those experienced by anthropoids. Moreover, in taxa with a fused symphysis, increased dorsoventral shear is best resisted by simply increasing the amount of cortical bone distributed along the symphysis.

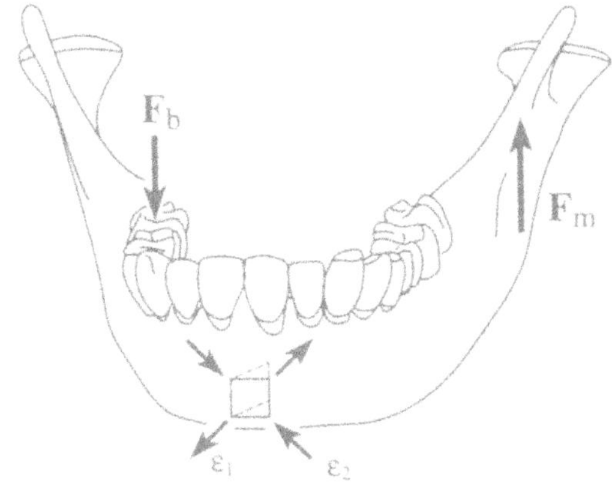

Fig. 5. Frontal view of an adult primate mandible with a fused symphysis ($\mathbf{F}_b$ is the vertical component of molar bite force). During unilateral mastication the balancing-side mandibular corpus of primates experiences parasagittal bending, which in turn causes dorsoventral shear stress at the symphysis. The amount of dorsoventral shear is related to the amount of the vertical component of jaw-muscle force ($\mathbf{F}_m$) transmitted across the symphysis from the balancing to the working side of the mandible. Thus, an increased recruitment of balancing-side jaw-muscle force causes an increase in dorsoventral shear stress, which is best countered by complete symphyseal fusion. Increased dorsoventral shear is best resisted in a fused symphysis by increasing the amount of cortical bone distributed along the symphysis. The arrows labeled ϵ_1 and ϵ_2 indicate the directions of the maximum and minimum principal strains (principal tension and principal compression), respectively (adapted from Hylander, 1984).

Certain morphological data for primates also apparently bear on the issue of dorsoventral shear and parasagittal bending in anthropoids. Comparisons of allometric trajectories between prosimians with unfused symphyses and anthropoids (with fused symphyses) indicate that the anthropoid data scatter for corpus height is also significantly transposed above that for prosimians (Fig. 6), implying that anthropoids have relatively larger corpus *and* symphysis dimensions (Ravosa, 1991). These data provide additional support for the hypothesis that anthropoid corpus and symphysis (Fig. 4) morphologies are possibly related to resisting increased parasagittal bending stress and dorsoventral shear stress, both of which result from a greater recruitment of balancing-side jaw-muscle force.

Of the various loading regimes discussed, wishboning forces initially appear to be the most important determinant of symphyseal fusion in primates simply because the stress concentrations produced during wishboning are considerably higher than those associated with dorsoventral shear forces (Hylander, 1984, 1985; Hylander and Johnson, 1994). However, because living prosimians with partially fused symphyseal joints have calcified or ossified ligaments oriented primarily to resist dorsoventral shear (Beecher, 1977, 1979), and because extant and extinct prosimians with partially fused symphyses have bony rugosities oriented mainly to resist dorsoventral shear (Beecher, 1983; Ravosa and Simons, 1994; M. J. Ravosa, *unpublished data*), perhaps significant levels of wishboning stress occur only among those taxa with complete symphyseal fusion. Thus, symphyseal fusion in early anthropoids may have followed initially from increased dorsoventral shear resulting from the recruitment of relatively greater amounts of vertically oriented balancing-side jaw-muscle force. Once symphyseal fusion was attained, this

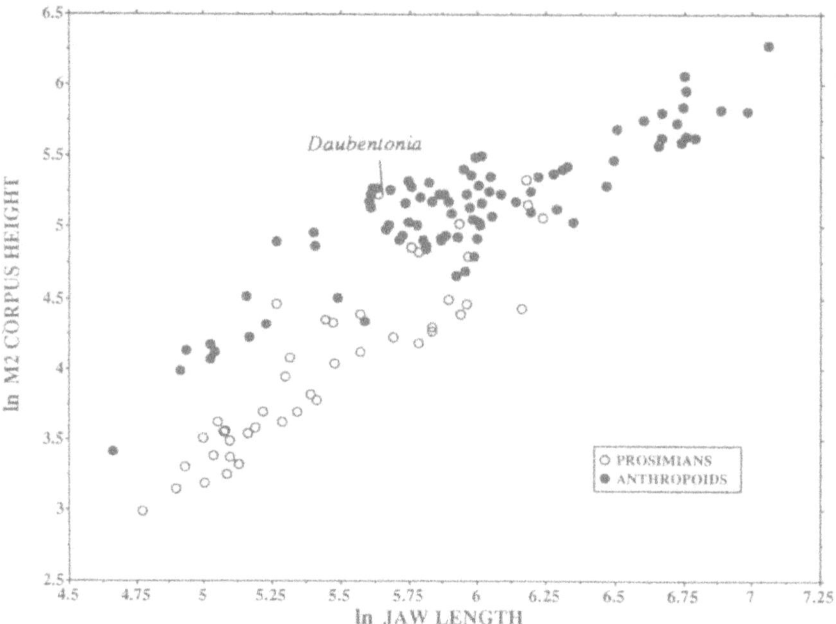

Fig. 6. A plot of ln corpus height versus ln jaw length in the two living primate suborders. This sample is based on 91 anthropoid and 45 prosimian species; all of the extant prosimians have unfused symphyses. Anthropoid data points for both corpus height and symphysis width (see Fig. 4) are transposed above those for extant prosimians, which further supports the hypothesis that anthropoids have more robust symphysis and corpus dimensions than do living prosimians. These more robust corpus and symphysis dimensions may reflect a structural adaptation to counter increased parasagittal bending of the corpus and increased dorsoventral shear of the symphysis with increased recruitment of vertically oriented balancing-side jaw-muscle force. Note again that although *Daubentonia* has an unfused symphysis, it plots relatively high in the anthropoid range for corpus height; this is probably a structural adaptation to counter elevated parasagittal bending stress associated with powerful incisal biting and gnawing. Also, several indriids and lemurids with partially fused symphyses plot relatively high in the anthropoid range for corpus height, indicating that taxa with partial fusion have more robust jaws than taxa with unfused symphyses (adapted from Ravosa, 1991).

may have facilitated the recruitment of increased amounts of horizontally oriented balancing-side jaw-muscle force, which effectively increased grinding forces between the teeth and elevated levels of wishboning stress along the symphysis.

Diet, Allometry, and Symphyseal Fusion. Both diet and allometry may also influence the recruitment of balancing-side jaw-muscle force, and these two factors may therefore influence symphyseal fusion. In primates, comparative histological data (Beecher, 1977, 1979), gross anatomic data (Beecher, 1983; Ravosa, 1991; M. J. Ravosa, *unpublished data*), and feeding-behavior data (Beecher, 1983; Ravosa, 1991) suggest that increased symphyseal stress is associated with the processing of greater amounts of structurally resistant or tougher foods. These foods also entail greater repetitive or cyclical loading of the mandible (Hylander, 1979b, 1988; Bouvier and Hylander, 1981).* In general, mammals with unfused or partially fused symphyses and tougher diets have greater numbers of ligaments at the symphyseal joint oriented to resist dorsoventral shear forces (Beecher, 1977, 1979). Beecher also notes that in prosimians with this morphological pattern, the cruciate ligaments, resisting dorsoventral shear forces, tend to be calcified or ossified; i.e., they exhibit partial symphyseal fusion. It is of interest that primate taxa with tough diets and partially fused symphyses, e.g., Indriinae and Hapalemurinae, have relatively larger symphysis and corpus dimensions, which in fact tend to cluster with similar measures for anthropoids (Ravosa, 1991).†

Comparative work among cercopithecine monkeys suggests that symphyseal stresses are influenced by dietary and allometric factors, both of which covary positively with body size (Hylander, 1985). That is, larger animals tend to eat increasingly tougher or more resistant foods entailing larger loads and greater repetitive loading of the mandible (Scapino, 1981; Beecher, 1983; Hylander, 1985; Ravosa, 1991; M. J. Ravosa, unpublished data). Moreover, larger animals have greater constraints on jaw-muscle force production requiring relatively more balancing-side muscle-force recruitment (Herring and Scapino, 1973; Scapino, 1981; Beecher, 1983; Hylander, 1984, 1985; Ravosa, 1991; M. J. Ravosa, *unpublished data;* Ravosa and Simons, 1994). In fact, Scapino (1981) suggests that the allometric effects of diet and jaw-muscle force production constrain the variability in symphyseal form at larger body sizes, such that the sole morphological option for larger animals is simply to fuse the mandibular symphysis. Similar explanations of symphyseal form and/or fusion have been suggested by Herring and Scapino (1973) for pigs, Beecher (1983) and M. J. Ravosa (*unpublished data*) for adapids and omomyids, Hylander (1985) for Old World monkeys, Ravosa (1991) and Ravosa and Simons (1994) for subfossil lemurs, and Scapino (1981) and M. J. Ravosa (*unpublished data*) for felids, ursids and, mustelids.

In addition, Hylander (1985) suggests that the presence of increased balancing-side jaw-muscle force recruitment among larger primates should result in greater wishboning stress. Beecher (1977, 1979, 1983) likewise notes that dorsoventral shear stress presumably increases with positive allometry since, within a monophyletic group, larger species tend to process tougher food items and have greater numbers of symphyseal ligaments oriented to

*Repetitive or cyclical loading of the mandible (and the increased likelihood of fatigue failure) from increased overall muscle force recruitment may be a more important determinant of symphysis and corpus morphology than high, infrequent, or nonrepetitive loads (Hylander, 1979b, 1988; Bouvier and Hylander, 1981). Although it is not clear whether tougher and more fibrous foodstuffs always require higher bite force magnitudes, such a diet does entail greater repetitive loading of the lower jaw.

†On the basis of molar morphology, *Mahgarita stevensi* also likely had a tougher, more resistant diet (Fleagle, 1988; Rasmussen, 1990), which perhaps explains the presence of symphyseal fusion and robust symphyseal dimensions in this small-bodied North American adapid (Figs. 7 and 8) (M. J. Ravosa, *unpublished data*).

resist shear stresses. Thus, Beecher's comments may also support claims regarding positive allometry of balancing-side muscle-force recruitment as a result of constraints on muscle force production in larger-bodied taxa.

If there are similar constraints on jaw-muscle force production in prosimians as there are in anthropoids, then the relative amount of balancing-side muscle-force recruitment among prosimians may also increase with positive allometry. Analyses of the scaling of jaw dimensions in prosimians indicate that mandibular symphysis height and corpus height (Fig. 6) scale with strong positive allometry relative to jaw length (Ravosa, 1991; M. J. Ravosa and W. L. Hylander, *unpublished data*). The prosimian jaw-scaling patterns may be correlated with positive allometry of balancing-side muscle-force use, which in turn may explain the incidence of symphyseal fusion in all large-bodied subfossil lemurs regardless of dietary preference (Figs. 7 and 8), much as predicted by

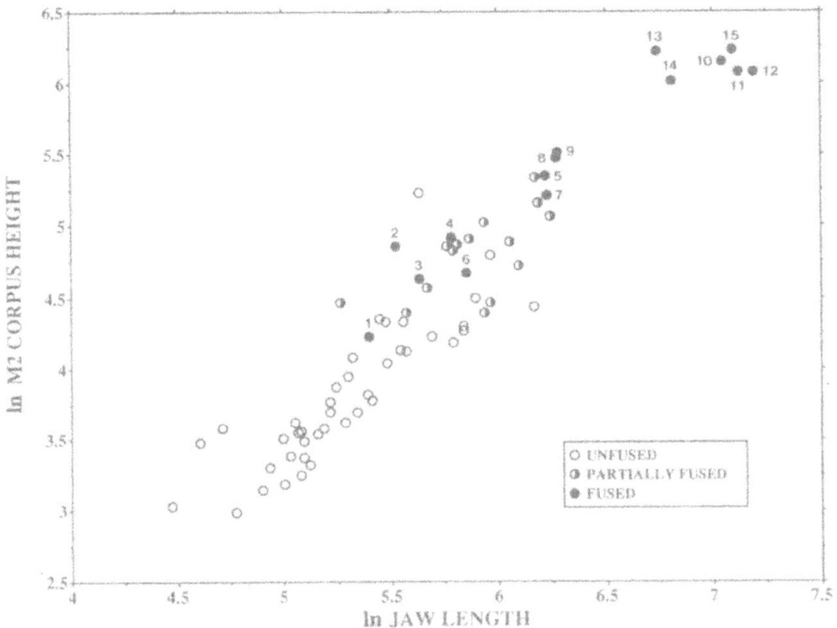

Fig. 7. A plot of ln corpus height versus ln jaw length in 72 extant and extinct prosimians with unfused (open circles), partially fused (half-open circles), and fully fused (numbered solid circles) symphyses. Note that taxa with fully and partially fused symphyses have corpus measures that fall in the middle or at the upper end of the data scatter for prosimians with unfused symphyses. This further supports the hypothesis that primates with partial symphyseal fusion often have more robust jaws than do taxa with unfused symphyses. Note that the pattern of symphyseal fusion is size-related, with larger taxa having greater degrees of fusion. As noted in Fig. 6, although *Daubentonia* has an unfused symphysis, it plots relatively high in the range for corpus height (adapted from M. J. Ravosa, 1991, *unpublished data*). 1, *Mahgarita stevensi;* 2, *Adapis betillei;* 3, *Adapis* sp.; 4, *Adapis parisiensis;* 5, *Leptadapis magnus;* 6, *Notharctus tenebrosus;* 7, *Archaeolemur majori;* 8, *A. edwardsi;* 9, *Hadropithecus stenognathus;* 10, *Megaladapis madagascariensis;* 11, *M. grandidieri;* 12, *M. edwardsi;* 13, *Palaeopropithecus maximus;* 14, *P. ingens;* 15, *Archaeoindris fontoynonti.*

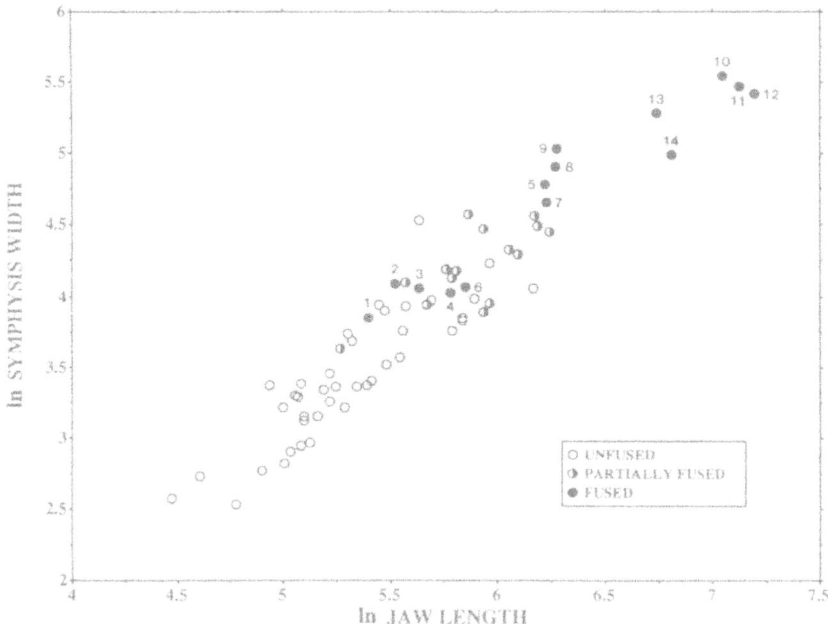

Fig. 8. A plot of ln symphysis width versus ln jaw length in 71 extant and extinct prosimians with unfused (open circles), partially fused (half-open circles), and fully fused (numbered solid circles) symphyses. Note again that taxa with fully and partially fused symphyses have symphysis dimensions that lie in the middle or at the upper end of the data scatter for prosimians with unfused symphyses. This plot also indicates a size-related pattern of symphyseal fusion much as demonstrated previously in Fig. 7 (adapted from M. J. Ravosa, 1991, *unpublished data*). Species numbers same as in Fig. 7.

Beecher, Hylander, and Scapino. Moreover, as first noted by Beecher (1983), a similar argument may apply to the large-bodied adapids such as *Notharctus tenebrosus, Leptadapis magnus,* and *Adapis parisiensis* (Figs. 7 and 8) (M. J. Ravosa, *unpublished data*). Since many of the large-bodied extant prosimians, subfossil lemurs, and adapids have or had tougher or more fibrous diets (Tattersall, 1982; Beecher, 1983; Fleagle, 1988), this may also explain the pattern of positive allometry of corpus and symphysis dimensions among prosimians as well as the development of partial or complete symphyseal fusion among the larger-bodied taxa.

In sum, the analyses above indicate that both dietary and allometric factors may directly affect the extent of mechanical stresses at the symphysis and therefore may also influence the development of mandibular symphyseal fusion. These analyses also support the argument that symphyseal fusion and more robust mandibular dimensions are often part of the same functional pattern (Hylander, 1979b).

There are, however, several issues regarding symphyseal fusion in primates and other mammals that remain unresolved. That is, it appears that not all cases of symphyseal fusion can be fully explained by the allometric and/or dietary arguments. At the subordinal level in mammals, there is considerable body size variability at which adults develop a completely fused symphysis (Scapino, 1981; Beecher, 1983). Moreover, some carnivorans, notably canids, do not exhibit a size-related pattern of mandibular symphyseal fusion; 44,000-g wolves have symphyseal joints very similar in morphology to the simple, unfused joints of 1500-g fennecs (Scapino, 1981; M. J. Ravosa, *unpublished data*). Furthermore, there are some small-bodied carnivorans with rather unremarkable diets and fully fused symphyses, e.g., the Javan mongoose *Herpestes javanicus* (775 g) and the meerkat *Suricata suricatta* (850 g) (Scapino, 1981; M. J. Ravosa, *unpublished data*). Among primates, it is not clear why some small-bodied New World monkeys, also with mechanically unremarkable diets and feeding behaviors, have fused mandibular symphyses, e.g., the golden-lion tamarin *Leontopithecus rosalia* (500 g) and Goeldi's marmoset *Callimico goeldii* (630 g).

Perhaps some of this unexplained variability in symphyseal morphology is related to differences in jaw-muscle recruitment patterns or to differences in the orientation of the jaw muscles, either of which influences the degree of symphyseal stress. Furthermore, there is always the distinct possibility that symphyseal fusion in certain small-bodied carnivorans and many small-bodied New World monkeys may result from phylogenetic effects. For instance, it is possible that a fused symphysis among small-bodied platyrrhines was inherited from a larger-bodied ancestor. Nevertheless, the above counter-examples emphasize the need for additional experimental, morphological, and behavioral investigations into the functional bases of mandibular symphyseal fusion in mammals.

Phylogenetic Analysis and Symphyseal Fusion

One final consideration of our study concerns the importance of symphyseal fusion for discussions of anthropoid origins and primate systematics. To the best of our knowledge, a completely fused or partially fused symphysis has evolved independently at least once in basal anthropoids, at least three times in subfossil lemurs (Archaeolemurinae, Megaladapinae, Palaeopropithecinae), at least five times in Adapinae (*Adapis, Leptadapis, Mahgarita, Caenopithecus, Cercamonius*), at least once in Notharctinae (*Notharctus*), and at least twice in Sivaladapinae (*Sivaladapis, Indraloris*) (Gingerich, 1977, 1980; Szalay and Delson, 1979; Tattersall, 1982; Beecher, 1983; Rosenberger *et al.*, 1985; Rasmussen, 1986, 1990; Ravosa, 1991; M. J. Ravosa, *unpublished data*). Therefore, although the presence of complete or partial symphyseal fusion might be relevant for discussing adaptive scenarios (Hiiemae and Kay, 1972, 1973; Beecher, 1977, 1979, 1983; Hylander, 1979b; Rosenberger *et al.*, 1985; Rosenberger, 1986; Ravosa, 1991; M. J. Ravosa, *unpublished data;* Rasmussen and Simons, 1992), the likelihood of a fused symphysis as a result of functional convergence (i.e., homoplasy) makes symphyseal fusion a poor character for establishing phylogenetic affinities in any mammalian group (Beecher, 1977).

Conclusions

Our analysis of the available experimental, neontological, and paleontological data on symphyseal morphology among primates and other mammals suggests that Greaves's (1988) argument regarding the functional link between the incisal crushing of small, hard objects such as seeds and mandibular symphyseal fusion is not supported (see also Ravosa and Hylander, 1993). Thus, contrary to Greaves's assertion, "the costs of symphyseal fusion" can be tolerated among taxa that do not habitually use their incisors to crush small, hard objects. Although habitual incisal preparation of large food items may have a significant effect on the morphology of the mandibular symphysis and corpus (Hylander, 1979b; Bouvier, 1986a,b; Ravosa, 1991; Daegling, 1992), little evidence exists to support the claim that symphyseal fusion is an adaptation to facilitate the incisal crushing of small, hard objects. This is because there are no data to support the hypothesis that incisal crushing of small, hard objects has ever constituted a frequent behavior among those primate taxa with a fused symphysis. Thus, it is highly unlikely that this model characterizes those factors influencing the evolution of symphyseal fusion in early anthropoids.

Instead, symphyseal fusion in primates appears to be a structural adaptation to resist two types of loading regimes experienced during unilateral mastication: wishboning and dorsoventral shear. Of these two loading regimes, wishboning stress may be the more important determinant of symphyseal form because the stress concentrations associated with wishboning are considerably higher than those associated with dorsoventral shear (Hylander, 1984, 1985; Hylander and Johnson, 1994). However, it is possible that significant wishboning stress occurs only among those taxa that already possess complete symphyseal fusion. Thus, perhaps the initial occurrence of symphyseal fusion in early anthropoids may have been primarily a result of increased dorsoventral shear (Beecher, 1977, 1979, 1983; Ravosa and Simons, 1994).

Repetitive or cyclical loading of the mandible, in part related to the mechanical properties of the diet, may also have an important effect on the degree of symphyseal fusion. On a related note, because there probably are allometric constraints on overall muscle-force production, larger species with their tougher and more resistant diets must recruit relatively more balancing-side muscle force, and this presumably results in increased symphyseal stress that is best countered by varying degrees of symphyseal fusion. However, since the earliest anthropoids were quite small, the allometric argument may not be applicable in this case. Instead, symphyseal fusion in early anthropoids is more likely related, in part, to a diet of unripe or tough-coated fruits or nuts requiring forceful, vigorous mastication (Beecher, 1977, 1979, 1983; see also Cachel, 1979a,b; Rosenberger *et al.*, 1985; Rosenberger, 1986; Rasmussen and Simons, 1992). Most importantly, this in turn is associated with the recruitment of increased amounts of balancing-side jaw-muscle force (Beecher, 1977, 1979, 1983; Hylander, 1977, 1979a,b; Ravosa, 1991; M. J. Ravosa, *unpublished data;* Hylander *et al.*, 1992).

ACKNOWLEDGMENTS

We thank the following curators and staff for providing access to various cranial specimens: E. Simons, P. Chatrath (Duke Primate Center); M. Godinot (Université Montpellier); A. Friday (Cambridge University); J. Wilson, E. Lundelius (Texas Memorial Museum); M. Rutzmoser (Museum of Comparative Zoology); I. Tattersall, E. Delson, G. Musser, W. Fuchs, J. Alexander, S. Anderson, M. Novacek (American Museum of Natural History); B. Patterson, J. Kerbis, W. Stanley, J. Flynn, S. McCarroll (Field Museum of Natural History); R. Thorington, R. Emry, L. Gordon, R. Purdy, C. Anderson, L. Coley (National Museum of Natural History); P. Andrews, P. Jenkins, C. Stringer, J. Hooker, R. Kruszynski, M. Sheridan, M. Sheldrick (British Museum of Natural History); J. Roche, M. Trainer, D. Goujet, F. Petter (Muséum National d'Histoire Naturelle); M. Hoogmoed, C. Smeenk, D. Reider (Rijksmuseum van Natuurlijke Historie); R. Angermann (Museum für Naturkunde); C. Beard, L. Krishtalka (Carnegie Museum of Natural History); B. Latimer, L. Jellema, L. Linden (Cleveland Museum of Natural History); F. Sibley, M. Turner (Peabody Museum of Natural History); T. Daeschler (Academy of Natural Sciences of Philadelphia); M. Coombs (Pratt Museum of Natural History); C. Grigson (Odontological Museum); R. Dubos (University of Connecticut); D. Klingener (University of Massachusetts); and B. Rakotosamimanana (Université de Madagascar). The helpful comments of R. Beecher, M. Cartmill, D. Daegling, B. Demes, J. Fleagle, W. Greaves, R. Kay, and an anonymous reviewer are appreciated greatly. K. Johnson is thanked for technical assistance. This study was supported by NIH grants to WLH (DE-04531) and MJR (DE-05595), NSF grants to WLH (BNS-8711842) and MJR (BNS-8813220), Leakey Foundation, Boise Fund, Field Museum, and American Philosophical Society grants to MJR, and the Department of Biological Anthropology and Anatomy, Duke University Medical Center.

References

Beecher, R. M. 1977. Function and fusion at the mandibular symphysis. *Am. J. Phys. Anthropol.* **47**:325–336.

Beecher, R. M. 1979. Functional significance of the mandibular symphysis. *J. Morphol.* **159:**117–130.

Beecher, R. M. 1983. Evolution of the mandibular symphysis in Notharctinae (Adapidae, Primates). *Int. J. Primatol.* **4:**99–112.

Bouvier, M. 1986a. A biomechanical analysis of mandibular scaling in Old World monkeys. *Am. J. Phys. Anthropol.* **69:**473–482.

Bouvier, M. 1986b. Biomechanical scaling of mandibular dimensions in New World monkeys. *Int. J. Primatol.* **7:**551–567.

Bouvier, M., and Hylander, W. L. 1981. Effect of bone strain on cortical bone structure in macaques (*Macaca mulatta*). *J. Morphol.* **167:**1–12.

Cachel, S. M. 1979a. A functional analysis of the primate masticatory system and the origin of the anthropoid post-orbital septum. *Am. J. Phys. Anthropol.* **50:**1–18.

Cachel, S. M. 1979b. A paleoecological model for the origin of higher primates. *J. Human Evol.* **8:**351–359.

Crompton, A. W., and Hiiemae, K. M. 1970. Molar occlusion and mandibular movements during occlusion in the American opossum, *Didelphis marsupialis* L. *Zool. J. Linn. Soc.* **49:**21–47.

Daegling, D. J. 1992. Mandibular morphology and diet in the genus *Cebus. Int. J. Primatol.* **13:**545–570.

Dessem, D. 1985. The transmission of muscle force across the unfused symphysis in mammalian carnivores. *Fortschr. Zool.* **30:**289–291.

Fleagle, J. G. 1988. *Primate Adaptation and Evolution.* Academic Press, New York.

Gingerich, P. D. 1977. Radiation of Eocene Adapidae in Europe. *Geobios Mem.* **1:**165–182.

Gingerich, P. D. 1979. Phylogeny of middle Eocene Adapidae (Mammalia, Primates) in North America: *Smilodectes* and *Notharctus. J. Paleontol.* **53:**153–163.

Gingerich, P. D. 1980. Dental and cranial adaptations in Eocene Adapidae. *Z. Morphol. Anthropol.* **71:**135–142.

Greaves, W. S. 1988. A functional consequence of an ossified mandibular symphysis. *Am. J. Phys. Anthropol.* **77:**53–56.

Greaves, W. S. 1993. A reply to Drs. Ravosa and Hylander. *Am. J. Phys. Anthropol.* **90:**513–514.

Happel, R. 1988. Seed-eating by West African cercopithecines, with reference to the possible evolution of bilophodont molars. *Am. J. Phys. Anthropol.* **75:**303–327.

Herring, S. W., and Scapino, R. P. 1973. Physiology of feeding in minature pigs. *J. Morphol.* **141:**427–460.

Hiiemae, K. M., and Crompton, A. W. 1985. Mastication, food transport, and swallowing. In: M. Hildebrand, D. M. Bramble, K. F. Liem, and D. B. Wake (eds.), *Functional Vertebrate Morphology,* pp. 262–290. Harvard University Press, Cambridge.

Hiiemae, K. M., and Kay, R. F. 1972. Trends in the evolution of primate mastication. *Nature* **240:**486–487.

Hiiemae, K. M., and Kay, R. F. 1973. Evolutionary trends in the dynamics of primate mastication. In: M. R. Zingeser (ed.), *Symposia of the Fourth International Congress of Primatology, Vol. 3: Craniofacial Biology of Primates,* pp. 28–64. S. Karger, Basel.

Hirschfeld, Z., Michaeli, Y., and Wienreb, M. M. 1977. Symphysis menti of the rabbit: Anatomy, histology, and postnatal development. *J. Dent. Res.* **56:**850–857.

Hylander, W. L. 1975a. The human mandible: Lever or link? *Am. J. Phys. Anthropol.* **43:**227–242.

Hylander, W. L. 1975b. Incisor size and diet in anthropoids with special reference to Cercopithecidae. *Science* **189:**1095–1098.

Hylander, W. L. 1977. *In vivo* bone strain in the mandible of *Galago crassicaudatus. Am. J. Phys. Anthropol.* **46:**309–326.

Hylander, W. L. 1979a. Mandibular function in *Galago crassicaudatus* and *Macaca fascicularis:* An *in vivo* approach to stress analysis of the mandible. *J. Morphol.* **159:**253–296.

Hylander, W. L. 1979b. The functional significance of primate mandibular form. *J. Morphol.* **160:**223–240.

Hylander, W. L. 1984. Stress and strain in the mandibular symphysis of primates: A test of competing hypotheses. *Am. J. Phys. Anthropol.* **64:**1–46.

Hylander, W. L. 1985. Mandibular function and biomechanical stress and scaling. *Am. Zool.* **25:**315–330.

Hylander, W. L. 1988. Implications of *in vivo* experiments for interpreting the functional significance of "robust" australopithecine jaws. In: F. E. Grine (ed.), *Evolutionary History of the "Robust" Australopithecines,* pp. 55–83. Aldine de Gruyter, New York.

Hylander, W. L., and Johnson, K. R. 1994. Jaw muscle function and wishboning of the mandible during mastication in macaques and baboons. *Am. J. Phys. Anthropol.* **94:** 523–547.

Hylander, W. L., Johnson, K. R., and Crompton, A. W. 1987. Loading patterns and jaw movements during mastication in *Macaca fascicularis:* A bone-strain, electromyographic and cineradiographic analysis. *Am. J. Phys. Anthropol.* **72:**287–314.

Hylander, W. L., Johnson, K. R., and Crompton, A. W. 1992. Muscle force recruitment and biomechanical modeling: An analysis of masseter muscle function during mastication in *Macaca fascicularis. Am. J. Phys. Anthropol.* **88:**365–387.

Jolly, C. J. 1970. *Hadropithecus,* a lemuroid small object feeder. *Man* **5:**525–529.

Kallen, F. C., and Gans, C. 1972. Mastication in the little brown bat, *Myotis lucifugus. J. Morphol.* **136:**385–420.

Rasmussen, D. T. 1986. Anthropoid origins: a possible solution to the Adapidae–Omomyidae paradox. *J. Hum. Evol.* **15:**1–12.

Rasmussen, D. T. 1990. The phylogenetic position of *Mahgarita stevensi:* Protoanthropoid or lemuroid? *Int. J. Primatol.* **11:**439–469.

Rasmussen, D. T., and Simons, E. L. 1992. Paleobiology of the oligopithecines, the earliest known anthropoid primates. *Int. J. Primatol.* **13:**477–508.

Ravosa, M. J. 1991. Structural allometry of the mandibular corpus and symphysis in prosimian primates. *J. Hum. Evol.* **20:**3–20.

Ravosa, M. J., and Hylander, W. L. 1993. Functional significance of an ossified mandibular symphysis: A reply. *Am. J. Phys. Anthropol.* **90:**509–512.

Ravosa, M. J., and Simons, E. L. 1994. Mandibular growth and function in *Archaeolemur. Am. J. Phys. Anthropol.* (in press).

Richard, A. F. 1985. *Primates in Nature.* W. H. Freeman, New York.

Rosenberger, A. L. 1981. A mandible of *Branisella boliviana* (Platyrrhini, Primates) from the Oligocene of South America. *Int. J. Primatol.* **2:**1–7.

Rosenberger, A. L. 1986. Platyrrhines, catarrhines and the anthropoid transition. In: B. A. Wood, L. Martin, and P. Andrews (eds.), *Major Topics in Primate and Human Evolution,* pp. 66–88. Cambridge University Press, Cambridge.

Rosenberger, A. L., Strasser, E., and Delson, E. 1985. Anterior dentition of *Notharctus* and the adapid–anthropoid hypothesis. *Fol. Primatol.* **44:**15–39.

Scapino, R. P. 1965. The third joint of the canine jaw. *J. Morphol.* **116:**23–50.

Scapino, R. P. 1981. Morphological investigation into functions of the jaw symphysis in carnivorans. *J. Morphol.* **167:**339–375.

Simons, E. L. 1989. Description of two genera and species of Late Eocene Anthropoidea from Egypt. *Proc. Natl. Acad. Sci. USA* **86:**9956–9960.

Simons, E. L. 1990. Discovery of the oldest known anthropoidean skull from the Paleogene of Egypt. *Science* **247:**1567–1569.

Simons, E. L. 1992. Diversity in the early Tertiary anthropoidean radiation in Africa. *Proc. Natl. Acad. Sci. USA* **89:**10743–10747.

Szalay, F. S., and Delson, E. 1979. *Evolutionary History of the Primates.* Academic Press, New York.

Tattersall, I. 1982. *The Primates of Madagascar.* Columbia University Press, New York.

Zingeser, M. R. 1976. Arch form, tooth size, and occlusomandibular kinesis in the Ceboidea. *Am. J. Phys. Anthropol.* **45:**317–330.

The Craniofacial Evidence for Anthropoid and Tarsier Relationships

15

CALLUM ROSS

Introduction

Monkeys and apes have long been observed to resemble humans in the external appearance of the head, having a globular braincase, a short snout, and forward-facing eyes. In 1864, Mivart grouped them in the suborder, Anthropoidea, distinct from the Lemuroidea, to which he assigned lemurs, lorises, galagos, aye-ayes, and tarsiers (Mivart, 1864, p. 635). Although Mivart later (1873) identified a lengthy list of features distinguishing anthropoids from lemuroids, he maintained that New and Old World anthropoids had evolved in parallel from separate nonprimate ancestors. Mivart's conception of Anthropoidea—a polyphyletic taxon united by numerous distinctive features of the skull—thrived in the intellectual milieu of the "classical primatological synthesis" in which parallelism was seen as a widespread phenomenon (e.g., Le Gros Clark, 1934, 1959; Simpson, 1945, 1961). However, with the adoption by primate systematists of the principles of phylogenetic systematics (Hennig, 1966) and, later, of the parsimony criterion for choosing between competing hypotheses of evolutionary relationships, the assumption of widespread parallelism fell out of vogue. Anthropoids have come to be interpreted as a closed

CALLUM ROSS • Department of Biological Anthropology and Anatomy, Duke University, Durham, North Carolina 27705. *Present address:* Department of Anatomical Sciences, School of Medicine, Health Sciences Center, State University of New York, Stony Brook, New York 11794.
Anthropoid Origins, edited by John G. Fleagle and Richard F. Kay. Plenum Press, New York, 1994.

descent community (*sensu* Ax, 1985), and their distinctive features have been reinterpreted as synapomorphies inherited from an ancestral stem species. However, although anthropoids are currently thought to be a monophylum, there are conflicting opinions as to which group of nonanthropoid primates (prosimians) is their closest relative. Three hypotheses of anthropoid relationships are currently advocated.

Tarsier–Anthropoid Hypothesis

Similarities between extant tarsiers and anthropoids have led many workers to reject Mivart's division of primates into Lemuroidea and Anthropoidea (e.g., Gervais, 1854; Earle, 1897a–c; Elliot Smith, 1919). Similarities in the mode of placentation and the morphology of the fetal membranes led Hubrecht (1896) to propose that tarsiers, their fossil relatives, and anthropoids should be the sole constituents of the order Primates (Hubrecht, 1897a,b). Other soft-tissue features supporting a grouping of tarsiers and anthropoids to the exclusion of lemuroids were presented by Pocock (1918), although he did not advocate the removal of lemuroids from Primates. Instead he divided Primates into Haplorhini, including tarsiers and anthropoids, and Strepsirhini (Geoffroy, 1812) for the lemurs, lorises, and galagos.

However, as has often been noted (e.g., Rasmussen, 1986), even if *Tarsius* is more closely related to anthropoids than any other *extant* primates, this does not preclude the possibility that some *fossil* primates might be more closely related to anthropoids than *Tarsius*. For this reason, the evidence of the fossilizable bony skeleton, especially the skull, is an important source of data for assessing anthropoid relationships.

Cartmill and Kay (1978) observed that tarsiers and anthropoids share several features of the skull that are not found in other primates, and they suggested that *Tarsius* is more closely related to anthropoids than are any fossil prosimians. The features they advanced in support of this argument were a postorbital septum formed at least in part by a flange of the alisphenoid bone, a transversely narrow tympanic cavity, and a cavity within the anterior petrosal separated from the tympanic cavity proper by a transverse septum that contains the internal carotid canal and is perforated by a small foramen. Cartmill and Kay also suggested that fossil omomyids, tarsiers, and anthropoids are more closely related to each other than to adapids and strepsirhines because of their shared possession of an unpneumatized apical interorbital septum and a promontory artery that is larger than the stapedial.

Cartmill *et al.* (1981) later identified a transverse septum and pneumatized anterior accessory cavity in petrosals from the Fayum. They presented definitions of two types of pathway for the internal carotid through the middle ear: the perbullar pathway found in tarsiers and extant and fossil anthropoids and the "transpromontorial" pathway found in other primates. Subsequent studies by these authors revealed that these intrabullar structures develop very similarly in tarsiers and anthropoids, and the tarsier–anthropoid clade was defined on the basis of six features from the otic region as well as the presence of a postorbital septum (MacPhee and Cartmill, 1986) (see Fig. 1A and Appendix A) (see also Schmid, 1981).

Omomyid–Anthropoid Hypothesis

The observation that tarsiers are the closest living relatives of anthropoids has also led some workers to suggest that anthropoids and tarsiers arose independently from different groups of omomyids. This hypothesis is dependent on the demonstration that there are derived features shared by tarsiers, anthropoids, and omomyids to the exclusion of adapids and strepsirhines and that there are features linking tarsiers and anthropoids with different lineages of omomyids. Szalay *et al.* (1987; see also Szalay 1975; Szalay and Delson, 1979) presented five features of the skull suggesting that tarsiers, anthropoids, and omomyids form a monophylum: shortened facial skull, medially positioned posterior carotid foramen, enlarged promontory artery, the passage of the olfactory nerve above the interorbital septum rather than below it, and the formation of the auditory meatus by an ectotympanic that is partly elongated and partly phaneric.

Although explicitly rejecting the notion that tarsiers and anthropoids form a clade exclusive of omomyids (Szalay *et al.*, 1987, p. 94), these authors did not specify which omomyid lineages might have given rise either to tarsiers or to anthropoids. Rosenberger (1985) had previously suggested that the morphology of the choanal region and glenoid fossa indicated the derivation of *Tarsius* from a microchoerine clade, and that *Necrolemur* is its sister taxon (Fig. 1B). Beard *et al.* (1991) have since argued, on the basis of orbit size and features of basicranial anatomy, that *Shoshonius* is a more likely sister taxon for *Tarsius* (Fig. 1C). Like Szalay *et al.* (1987), however, neither Rosenberger (1985) nor Beard *et al.* (1991) hypothesized a specific omomyid group as ancestral to anthropoids.

Adapid–Anthropoid Hypotheses

The hypothesis that an adapid lineage gave rise to anthropoids has its roots in the work of Wortman (1903–04) but has been advocated for many years by Gingerich (1973, 1984; Gingerich and Schoeninger, 1977). Most recently, Simons (1989, 1991) and Rasmussen (1990; Simons and Rasmussen, 1989; Rasmussen and Simons, 1988, 1992) have suggested that protoadapines are the ancestral group for anthropoids. The most detailed discussion of the craniofacial evidence supporting this hypothesis is Rasmussen's (1990) treatment of the phylogenetic position of the protoadapine *Mahgarita.* There he suggests that *Mahgarita* and anthropoids are sister taxa because of dental

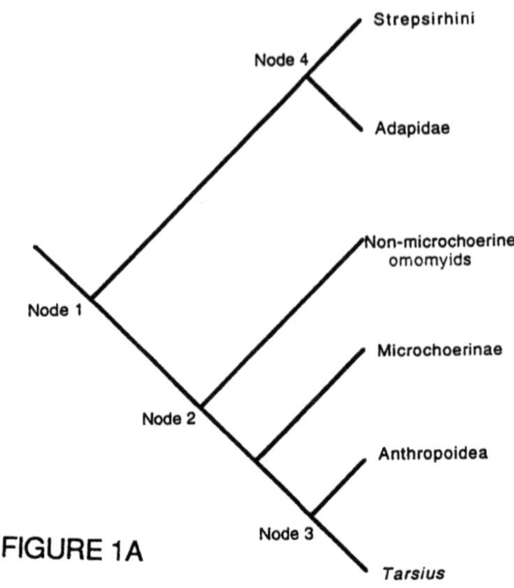

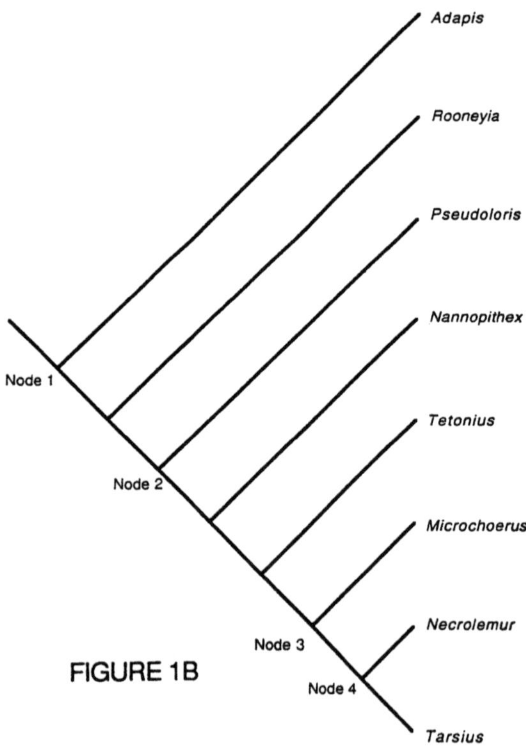

Fig. 1. Cladograms of competing hypotheses of tarsier and anthropoid relationships. (A) Tarsiphile hypothesis of MacPhee and Cartmill (1986). (B) The *Necrolemur–Tarsius* hypothesis of Rosenberger (1985) does not explicitly state how tarsiers are related to anthropoids. (C) The

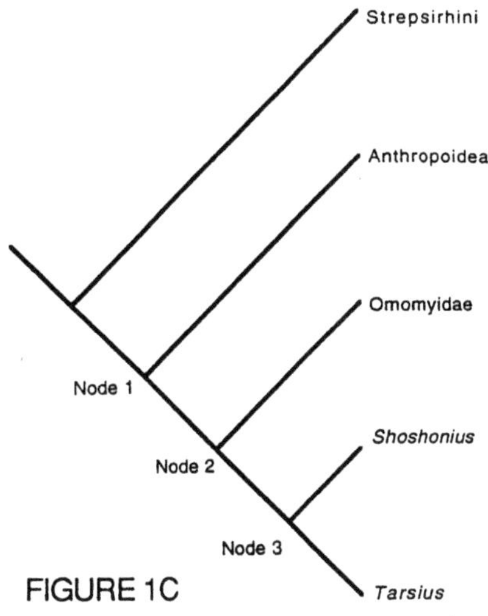

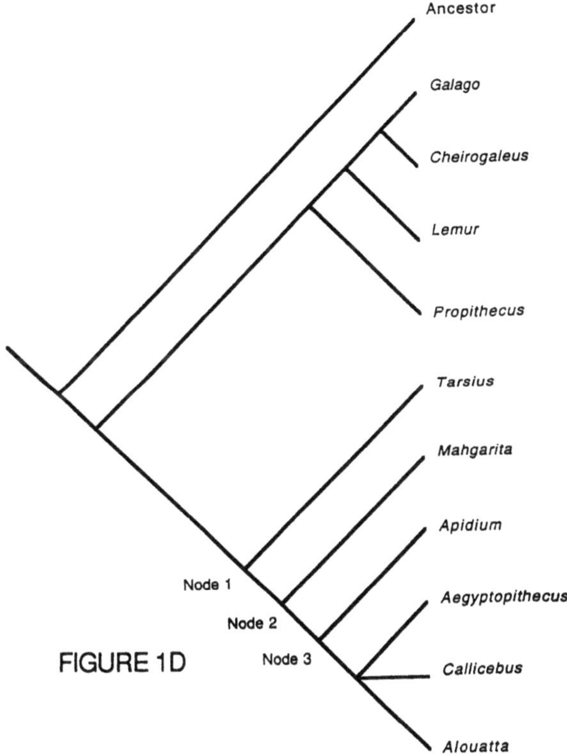

Shoshonius–Tarsius hypothesis of Beard *et al.* (1991). (D) The version of the adapiphile hypothesis shown here is that advanced by Rasmussen (1990). See Appendix A for features claimed by these authors to be present at the labeled nodes.

resemblances (Rasmussen and Simons, 1988) and possession of a fused mandibular symphysis, robust mandibular corpora, a prominent posterior nasal spine, a transverse septum within the middle ear cavity, and a pneumatized mastoid region (Fig. 1D). In a cladistic reincarnation of Wortman's (1903–04) Paleopithecini, Rasmussen argues that *Tarsius* forms a clade with *Mahgarita* and the anthropoids on the basis of their shared possession of a large promontory artery, an extrabullar tympanic bone, and a short and deep facial exposure of the maxilla.

Choosing between the Hypotheses

Debate over the relationships of Anthropoidea persists for two reasons. First, there are conflicting interpretations of the homologies of certain features in the ear region, conflicts arising from semantic confusion over how these features should be described (Cartmill, 1983), erroneous interpretations of the anatomy of certain fossils, and a tendency of some workers to ignore the ontogenetic evidence. Differences of opinion as to whether features in different taxa are "similar enough" to be hypothesized as homologous (and assigned the same character state for phylogenetic analyses) are inevitable and often intractable. It behooves the workers involved to state explicitly the reasons behind their character state assignments so that the assumptions on which phylogenetic hypotheses are contingent are explicit.

The second reason that there is continued debate over anthropoid relationships is that each of the features of the skull traditionally used to define Anthropoidea is also found individually in certain nonanthropoid groups (e.g., a postorbital septum is found in *Tarsius*, a fused metopic suture is seen in *Necrolemur*, and fused mandibular symphyses are exhibited by many adapids), but none of these groups has been definitively shown to be more closely related to anthropoids than any other. Some of these features may be synapomorphies linking anthropoids to nonanthropoid taxa, and some must be convergences or symplesiomorphies, but determining which are which is no easy task. Such character conflicts will only be definitively resolved when (1) one nonanthropoid species is shown to share an overwhelming number of derived features with the earliest members of the anthropoid clade, and (2) the similarities between anthropoids and other nonanthropoid taxa can be convincingly demonstrated to be primitive retentions or convergences.

New fossil discoveries documenting the anatomy of primitive anthropoids can help resolve these questions. For example, the discovery that early anthropoids had unfused mandibular symphyses (Simons, 1991, 1992) suggests that symphyseal fusion in later anthropoids evolved convergently with that of adapids. However, new fossil evidence is not about to resolve all of the conflicts surrounding the evidence for anthropoid relationships, and other means of distinguishing convergences from synapomorphies must be used.

Some workers have used computer-run algorithms to find the most parsimonious arrangement of the data entered (see Rasmussen, 1990; Beard and

MacPhee, Chapter 3, Kay and Williams, Chapter 13, Ford, Chapter 18, this volume). The rationale for use of such algorithms is that chosing the simplest arrangement of the data *given underlying assumptions about how life evolves* is the only way that rational investigation can proceed (Sober, 1988). (The rationale for using parsimony therefore comes from epistemology, not ontology: it is a method for determining the implications of our assumptions about the world.) Once again, therefore, the assumptions underlying character state assignments must be made explicit.

Previous applications of the parsimony criterion to the question of anthropoid relationships have been of limited scope. For example, in analyzing the phylogenetic relationships of the protoadapine *Mahgarita* to anthropoids, Rasmussen (1990) included no omomyid taxa, although omomyids have been hypothesized by other workers to be closely related to anthropoids. Consequently, not only did this analysis not evaluate the possibility that certain omomyids might be more closely related to anthropoids than any adapids, but it also failed to consider the effects that the particular distribution of character states in omomyids might have on the placement of *Mahgarita* relative to strepsirhines and tarsiers. There are similar problems with the use of parsimony by Beard *et al.* (1991) to determine the phylogenetic position of *Shoshonius*. In lumping fossil adapids in with strepsirhines (albeit on the basis of significant postcranial resemblances: Beard *et al.*, 1988), these workers fail to evaluate the possibility that some adapids might be more closely related to anthropoids than are omomyids.

This contribution aims to rectify these problems by evaluating the three principal competing hypotheses of anthropoid relationships in the context of each other. In this chapter, explicit explanations for the character states assigned different taxa are given. This necessitates resolution of semantic confusions that have arisen in regard to some characters and new interpretations of otic structures based on new ontogenetic studies. Because many of the features traditionally used to define Anthropoidea are found in the skull, and because the skull is a rich source of evidence for evaluating relationships, this contribution focuses on the craniofacial evidence for anthropoid relationships. Although this excludes much of the evidence relevant to the question of anthropoid relationships, the postcranial and dental evidence are thoroughly treated elsewhere in this volume (Ford, Chapter 18, Kay and Williams, Chapter 13, this volume).

Materials and Methods

Specimens Examined

Fossil and extant taxa were included in this analysis if they had been specifically hypothesized to be close relatives of anthropoids or tarsiers and if a significant amount of cranial material was available for study. This included

Tarsius syrichta skulls and wet specimens (DUPC*) and fossil skulls of *Mahgarita stevensi* (TMM 41578-9, 41578-20), *Shoshonius cooperi* (CM 31366, 31367, 60493, 60494, 60495), *Microchoerus* sp. (QU 10879), and *Necrolemur antiquus* (BM(NH)-P M3747, M4490; MCZ 8879). Other fossil adapids and omomyids were also included in the analysis in order to sample their major subgroups adequately and thereby to provide a more complete context in which to evaluate the alternate hypotheses of anthropoid relationships. This includes *Notharctus tenebrosus* (AMNH 11466, 23278, 21864), *Adapis* sp. (QU 408, 613), *Rooneyia viejaensis* (TMM 40688-7), and *Tetonius homunculus* (AMNH 4194). Comments on the anatomy of *Leptadapis magnus, Smilodectes gracilis* (USNM-VP 13347, 17994, 17995, 21815), *Pronycticebus gaudryi* (Le Gros Clark, 1934) and *Microchoerus* sp. (QU 10879) are included, although these taxa were not included in the analysis. Data on outgroup taxa were obtained from published descriptions and casts of *Plesiadapis tricuspidens* (Saban, 1963; Russell, 1964; Szalay *et al.*, 1987) and by examination of the skull of *Ignacius graybullianus* (USNM-VP 421608) and skulls of *Tupaia glis* and *T. tana* in the BAACAC and at the USNM-M.

Skulls of lemuroids (*Eulemur fulvus, Avahi laniger, Hapalemur simus, Varecia variegata, Daubentonia madagascariensis, Indri indri, Propithecus verreauxi, P. diadema, Archaeolemur edwardsi,* and *Mesopropithecus pithecoides*), cheirogaleids (*Microcebus murinus, Cheirogaleus medius*), galagos (*Galago senegalensis, Otolemur garnetti*), lorises (*Perodicticus potto, Loris tardigradus*), and anthropoids (*Nasalis larvatus, Presbytis johni, Erythrocebus patas, Macaca fascicularis, Homo sapiens, Brachyteles arachnoides,* and *Pithecia pithecia*) in which auditory bullae had been opened were scored for all characters. Representatives of other haplorhine genera were also scored for extrabullar characters. These skulls are housed at the DUPC, the BAACAC, the USNM-M, and the AMNH.

Five fossil petrosals recovered from the Paleogene deposits of the Fayum were also studied (DUPC 6642, YPM 25972, 25973, 25974, and 23968). Two of these were found in association with anthropoid dentitions: DUPC 6642 in association with a crushed skull and dentition of *Aegyptopithecus,* YPM 23968 in association with a maxillary dentition of *Apidium.* However, the YPM 23968 assemblage also includes a creodont squamosal and petrosal (Cartmill *et al.*, 1981) suggesting that the association of this petrosal with *Apidium* teeth may be by chance. Moreover, in its preserved portions this petrosal closely resembles DUPC 6642 described by Simons and Rasmussen (1989). Thus, the taxonomic status of all four Yale petrosals is in doubt, although they are all

*AMNH, American Museum of Natural History, New York, NY; BAACAC, Duke University Department of Biological Anthropology and Anatomy Comparative Anatomy Collection, Durham, NC; BM(NH)-P, British Museum (Natural History), Palaeontology; CGM, Cairo Geological Museum; CM Carnegie Museum, Pittsburgh, PA; DUPC, Duke University Primate Center, Durham, NC; MCZ, Museum of Comparative Zoology, Boston, MA; QU, Quercy Phosphorites; TMM, Texas Memorial Museum, Balcones Research Center, Austin, TX; USNM, United States National Museum (Smithsonian Institution), USNM-M, Mammalogy, USNM-VP, Vertebrate Paleontology, Washington DC; YPM, Yale Peabody Museum, New Haven, CT.

certainly anthropoid, as the discussion below reveals (and see Cartmill *et al.*, 1981).

Ontogenetic stages of *Callithrix* sp. and *Aotus* sp. were examined and compared with fetal material of *Tarsius syrichta* and *T. bancanus* in the Duke University Comparative Embryology Collection. Additional ontogenetic stages of *Tarsius* were examined and photographed at the Hubrecht Embryological Laboratory, Utrecht, The Netherlands. Computer-assisted three-dimensional reconstructions of the sectioned material of the ear regions of two specimens of *Tarsius* were performed using a computer program, PC3D (Jandel Scientific, San Rafael, CA), in order to document ontogenetic changes in the spatial relationships of the internal carotid artery and the inner ear. Dissections were also performed on two specimens of adult *Tarsius syrichta* from the DUPC in order to examine the morphology of the choanal and pterygoid regions.

Character Analysis

Broadly speaking, a phylogenetic analysis consists of two steps: assignment of character states and searching for the most parsimonious arrangement of these character states. Assignment of character states consists of determining whether structures in different taxa are similar enough to be postulated as homologous *a priori.* Several morphological criteria for determining what "similar enough" means have been proposed, but they are largely modifications on the themes of detailed similarity in shape, size, and position (Wiley, 1981). These are the criteria utilized here.

Ontogenetic evidence has also played an important role in the debate over anthropoid relationships, particularly in MacPhee and Cartmill's (1986) hypothesis that structures of the ear region of tarsiers are homologous with similar structures in anthropoids. Ontogenetic data have often been credited with special weight for resolving questions of homology, but there is enough evidence in the literature to indicate that just because two structures develop in the same way, they need not be homologous (e.g., Alberch, 1985; Northcutt, 1990). Nevertheless, detailed resemblances in the ontogeny of structures in different taxa may be considered as corroborating a hypothesis of homology unless there is good reason to believe otherwise.

Ontogenetic information on the development of the tarsier and anthropoid ear regions is used in this study (1) to determine whether certain otic structures in tarsiers and anthropoids develop in similar ways, (2) to explain why certain structures unique to tarsiers and anthropoids are differently positioned in adult stages of these two taxa, and (3) to hypothesize causal relationships between changes in carotid artery positioning and state changes in other characters. This last point is crucial for ensuring that characters that are causally interrelated are not considered as separate pieces of evidence, a particularly important consideration when parsimony is utilized as a criterion for selecting one pattern of relationships over others (Skelton and McHenry,

1992). In this study, those aspects of morphology that can vary independently of each other between the taxa under consideration were scored as separate characters. For example, the anteroposterior position of the posterior carotid foramen varies independently of its mediolateral or ventrodorsal position among primates.

In order to evaluate the competing hypotheses of anthropoid relationships, the most parsimonious arrangement of the character states described in the text was obtained using the computer program Phylogenetic Analysis Using Parsimony (PAUP, Swofford, 1991). This enables hypotheses regarding the homology of different characters to be evaluated in the context of evidence from other characters. Once again, although the most parsimonious arrangement of the data does not definitively determine which resemblances between anthropoids and other taxa are homologous and which are not, it does enable the implications of the assumptions underlying the character state assignments to be evaluated.

Evidence from the Basicranium: Intrabullar Anatomy

Intrabullar Spaces and Septae

MacPhee and Cartmill (1986) suggest that the cavity lying anteromedial to the tympanic cavity (anterior accessory cavity), the transverse septum separating it from the tympanic cavity, and the small aditus connecting them are derived features shared by tarsiers and anthropoids. Other researchers conclude that these three features are not similar enough in tarsiers and anthropoids to be homologous. Simons and Rasmussen (1989) argue that the "anterior accessory cavity" and the transverse septum separating it from the tympanic cavity cannot be homologous in tarsiers and anthropoids because they are in different positions in the two taxa. They argue that in primitive anthropoids (exemplified by *Aegyptopithecus*) the transverse septum stretches "ventrolaterally from the promontory canal to the lateral bulla wall" and the anterior accessory cavity lies medial to the tympanic cavity, whereas in tarsiers the septum stretches "posteromedially from the carotid canal to the medial bulla wall, and the anteromedial cavity lies largely anterior to the tympanic cavity proper" (Simons and Rasmussen, 1989, pp. 15–16). They also suggest that the apical aditus is not homologous in the two groups. They argue that a constricted opening between the "anteromedial cavity" and the tympanic cavity exists in tarsiers because the anterior and lateral placement of the canal for the internal carotid artery positions it close to the lateral bulla wall, whereas in anthropoids this opening is restricted because of the presence of a laterally directed transverse septum (see also Rosenberger and Szalay, 1980).

The anterior accessory cavity and transverse septum are also claimed to be present in primates other than tarsiers. Simons and Rasmussen argue that an anteromedial expansion of the tympanic cavity is present in all primates (1989), and Rosenberger and Szalay (1980) argue that it is present in omomyids and is probably a synapomorphy of Haplorhini (defined by them to include omomyids, tarsiers, and anthropoids). Simons and Rasmussen (1989) and Rasmussen (1990) suggest that a laterally directed septum arising from the carotid canal in *Mahgarita* is homologous with such a septum in *Aegyptopithecus,* and Beard *et al.* (1991) argue that an "intrabullar septum" containing the carotid canal in *Shoshonius* is homologous with that of both tarsiers and anthropoids.

Ontogeny of the Primate Middle Ear

These arguments are refuted by evidence of otic ontogeny, which reveals that the anterior accessory cavities of tarsiers and anthropoids develop in a very similar fashion and in a way that is unique to them among extant primates so far studied (MacPhee and Cartmill, 1986). Ontogenetic evidence also reveals why the anterior accessory cavity and transverse septum are differently positioned in adults of the two taxa. Moreover, in the context of this developmental evidence, the intrabullar septa of *Mahgarita* and *Shoshonius* can be shown to be morphologically different from those of tarsiers and anthropoids in ways that are probably reflective of different developmental histories.

Anterior Accessory Cavity. In those primates for which developmental evidence is available, the auditory bulla develops as an outgrowth, called the petrosal plate, of a region on the ventral surface of the otic capsule known as the midcapsular arc (MacPhee and Cartmill, 1986) (see Figs. 2B and 3B). In tarsiers and anthropoids for which embryological evidence is available [*Tarsius syrichta* and *T. bancanus, Macaca* sp. and *Leontopithecus rosalia* (MacPhee and Cartmill, 1986); *Tarsius spectrum, Aotus* sp. and *Callithrix* sp. (personal observations)], the anterior portion of the petrosal plate is subjected to "an intense field of pneumatic activity" prior to birth (MacPhee and Cartmill, 1986, p. 238) (Fig. 2C). This pneumatic activity originates in the auditory tube and splits the petrosal plate into medial and lateral lamellae, forming an "anterior accessory cavity" *within* the petrosal plate (Fig. 2D). Expansion of this anterior accessory cavity displaces the medial lamella of the petrosal plate to the medial edge of the cochlear housing. The lateral lamella, containing the internal carotid artery, remains on the cochlear housing's ventral surface, becoming the "transverse septum" of MacPhee and Cartmill (1986) and forming the medial wall of a small tympanic cavity (Fig. 2E). This septum is pierced anteriorly by the apical aditus, representing the place where pneumatic activity originally invaded the petrosal plate and that, in the adult, connects the anterior accessory cavity to the tympanic cavity and auditory tube.

Otic development in those lemuroids for which evidence is available [*Propithecus* sp., *Microcebus murinus, Lemur catta* (MacPhee, 1981)] differs from that of *Tarsius* and anthropoids in two important respects (Fig. 3). First, the inter-

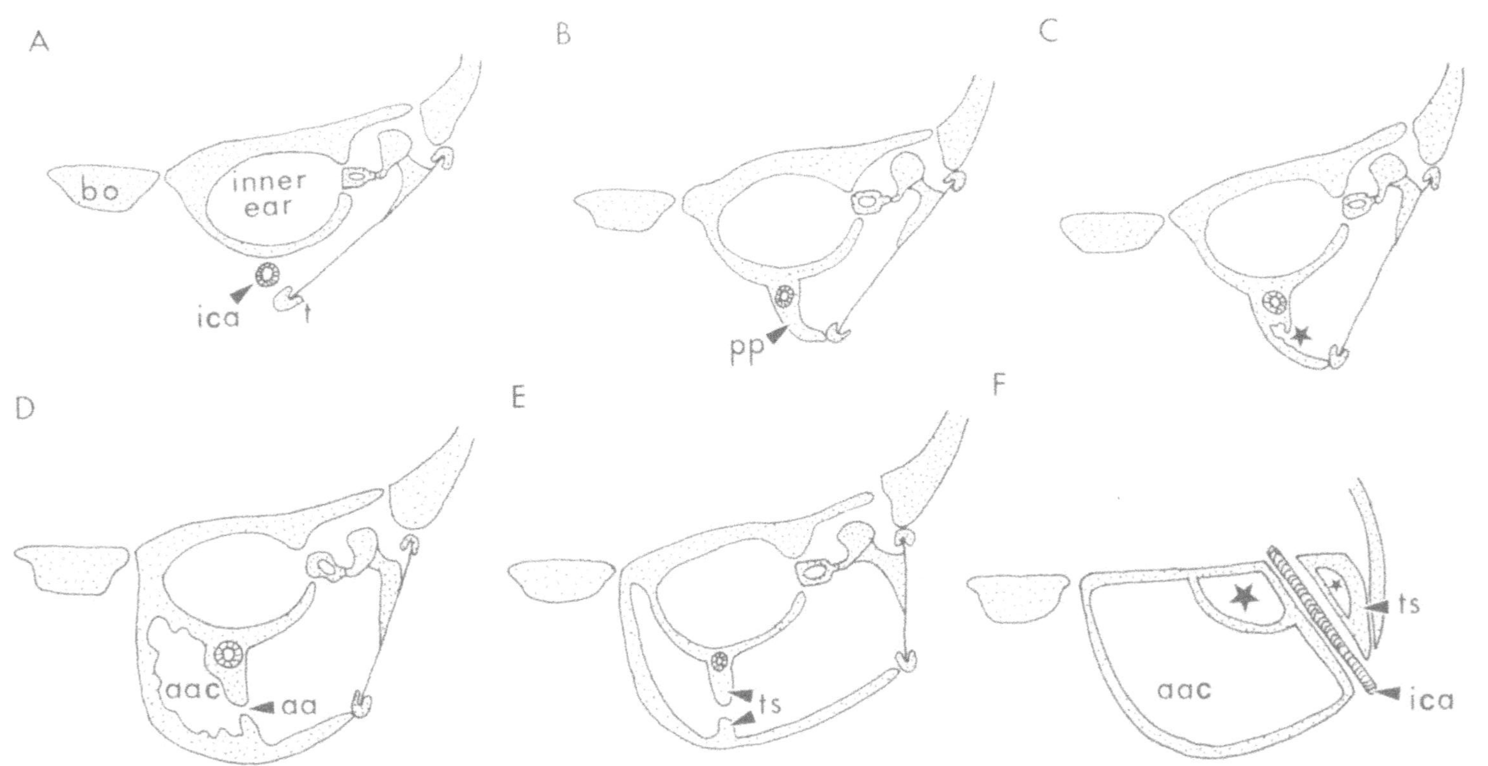

Fig. 2. Diagrammatic cross sections of the ear region illustrating (A through D) pattern of bulla inflation in tarsiers and platyrrhines, (E) condition in adult platyrrhine, and (F) in an adult *Tarsius*. aa, apical aditus; aac, anterior accessory cavity; bo, basioccipital; ica, internal carotid artery; pp, petrosal plate; tc, tympanic cavity; ts, transverse septum; star in (C) marks onset of pneumatization of petrosal plate. The hemisection in F is further forward than A through E—the large star marks the cupula of the cochlea. The internal carotid artery enters the bulla laterally and only grazes the cupula before passing into the braincase. The apical aditus (small star) pierces the transverse septum lying lateral to the internal carotid artery (cf. Fig. 5).

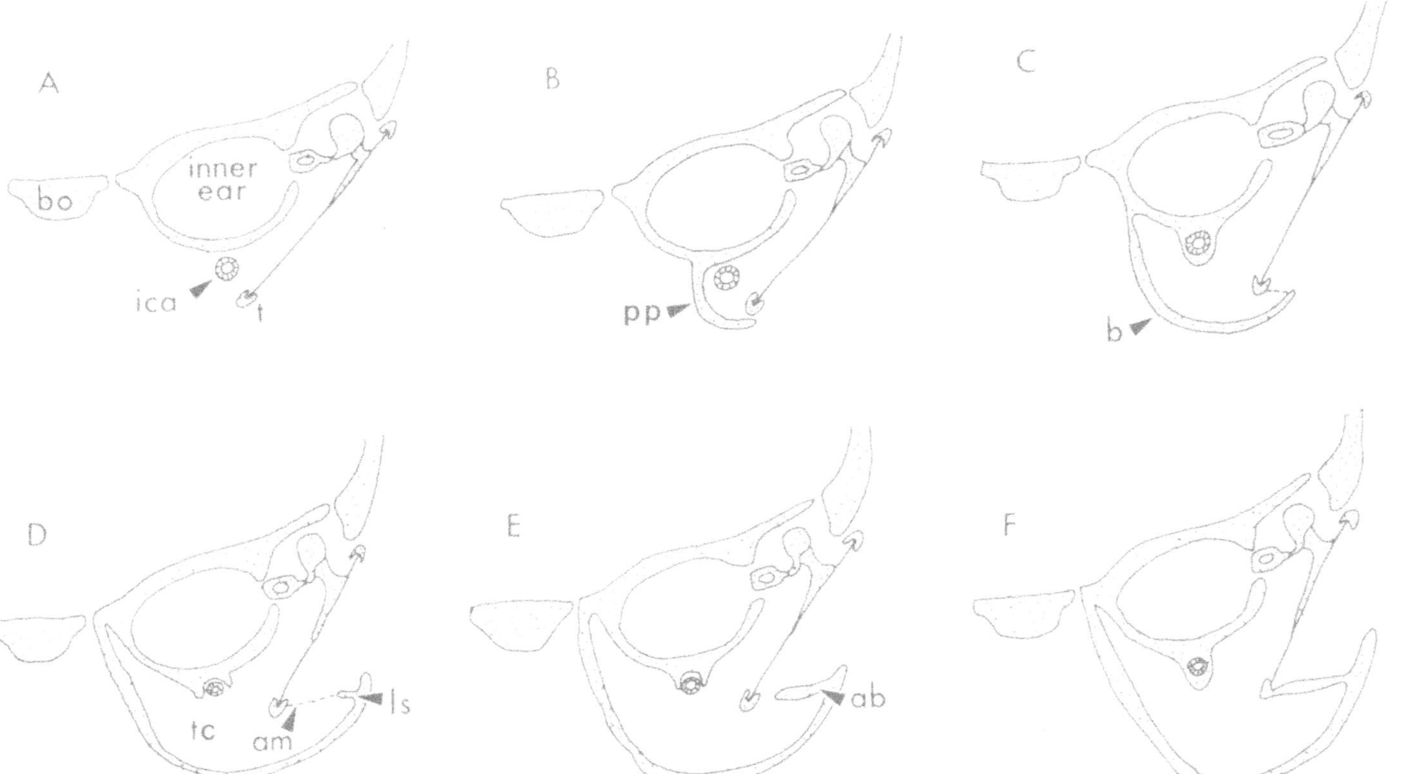

Fig. 3. Diagrammatic cross sections of the ear region illustrating (A through C) pattern of bulla inflation in lemuroid ontogeny; (D) condition in adult lemuroids; (E) condition in adapids; (F) condition in omomyids. *Mahgarita* differs from other adapids in having a complete carotid canal and large internal carotid artery rather than a small artery in an open trough. Abbreviations as in Fig. 2 with the following additions: am, annular membrane; ab, annular bridge; b, bulla; ls, linea semicircularis.

nal carotid artery is not walled up within the petrosal plate as the medial bulla wall develops (Fig. 3B). Second, the petrosal plate is not split into medial and lateral lamellae by pneumatic activity, and no anterior accessory cavity is formed. Rather, bony remodeling shifts the whole petrosal plate medially, there is no bony septum crossing the ventral surface of the cochlear housing in the adult, and the tympanic cavity proper is much larger than that of tarsiers and anthropoids (Fig. 3C). A septum extending rostrally from the promontorium to the anterior bulla wall engulfs the internal carotid artery. This septum develops as the "raised margins of the tegmen tympani and epitympanic wing of the petrosal" that have grown together to obliterate the piriform fenestra in the roof of the tympanic cavity (MacPhee, 1981), not from the lateral lamella of the petrosal plate.

Galagos and lorises [*Galagoides demidoff, Galago senegalensis, Loris tardigradus* (MacPhee, 1981)] resemble lemuroids in that the internal carotid artery is not walled up within the petrosal plate as the medial bulla wall begins to develop. They also bear superficial resemblance to tarsiers and anthropoids in that the petrosal plate is split into medial and lateral lamellae by pneumatic activity. However, in lorises and galagos pneumatization originates in the epitympanic recess and invades the bone dorsal to the tegmen tympani and cochlea before spreading down into the petrosal plate. It does not originate in the anterior portion of the tympanic cavity or auditory tube. This difference suggests that the cavity within the petrosal plate in lorises and galagos is probably not homologous with that found in anthropoids and tarsiers.

There are also differences between tarsiers and anthropoids in the patterns of pneumatization of the anterior accessory cavity. In *Tarsius* the apical aditus opens into the auditory tube *anterior* to the point where the tube widens out to become continuous with the tympanic cavity proper. As a result, the auditory tube of fetal tarsiers possesses two distinct endothelial diverticula, one opening into the tympanic cavity proper and one opening into the anterior accessory cavity. In contrast, in anthropoids "the endothelium-lined sac filling up the anterior accessory cavity is not so obviously a distinct diverticulum of the auditory tube in fetal anthropoids as it is in fetal *Tarsius*. In anthropoids, the sac communicates broadly with the lumina of *both* the auditory tube and the cavum tympani . . ." (MacPhee and Cartmill, 1986, p. 238, emphasis added).

However, for two reasons I think these differences are not sufficient to warrant tarsiers and anthropoids being assigned different character states. First, the tympanic cavity and the bony auditory tube both develop from a single continuous cavity, the tubotympanic recess (or elongated first pharyngeal pouch). The difference between the tympanic cavity and the auditory tube inheres in their relative widths—the tympanic cavity is wider than the auditory tube. In tarsiers the apical aditus opens into a narrow portion of the tubotympanic recess, or auditory tube, whereas in anthropoids it opens into a slightly wider part, the tympanic cavity proper. The narrowness of the anterior portion of the tubotympanic recess in fetal tarsiers is attributable to rapid expansion of the anterior accessory cavity (MacPhee and Cartmill, 1986, Fig. 5A), which restricts the diameter of the tubotympanic recess in the region of the apical aditus.

By adulthood, the morphologies seen in fetal tarsiers and anthropoids are reversed. The apical aditus of *Tarsius* opens directly into the tympanic cavity, whereas that of many anthropoids communicates with the tympanic cavity proper via a small recess at the posterior end of the bony auditory tube (preseptal recess of Packer and Sarmiento, 1985). Packer and Sarmiento (1985) cite this difference as evidence that the transverse septa of tarsiers and anthropoids evolved in parallel. However, variation in the position of the apical aditus relative to the tympanic cavity and auditory tube does not negate the fact that this aditus opens near the junction of the two and is not, I think, sufficient to falsify the hypothesis that the anterior accessory cavity is homologous in the two taxa.

Second, the presence of two endothelial diverticula in fetal tarsiers (one in the anterior accessory cavity and one in the tympanic cavity) and only one in anthropoids is explicable with reference to the ontogenies of the internal carotid canal pathways in the two. In early tarsier embryos, such as Hubrecht 358 (*Tarsius spectrum,* HL 6.7 mm, CRL 13.2 mm) the carotid canal traverses the ventrolateral surface of the cochlear housing, a relationship that is preserved at least until a crown–rump length of 45 mm is achieved (Fig. 4A). However, in tarsier neonates, the internal carotid artery has moved far anteriorly (Fig. 4B), the carotid canal angles posteriorly as it ascends through the tympanic cavity to pass behind, and in close proximity to, the apical aditus, before turning to run anteriorly and medially to the Circle of Willis. This anterior displacement and verticalization of the carotid canal brings it up against the medially directed diverticulum of the auditory tube entering the anterior accessory cavity. Being thus "hung up" on the diverticulum of the auditory tube, this portion of the internal carotid artery cannot move any further forward, and the anterior displacement of the posterior carotid foramen produces the curve in the artery evident in Fig. 4B. Further inflation of the anterior accessory cavity posterior and medial to the carotid emphasizes the distinction between the diverticulae in the tympanic cavity and the anterior accessory cavity, so that by birth the internal carotid artery in tarsiers has the appearance of a wedge driven between two diverticula of the auditory tube. In contrast with *Tarsius,* the carotid artery in anthropoids is not displaced anteriorly during ontogeny to the same extent, and it is not not wedged between the tubotympanic recess and the anterior accessory cavity. Consequently, the anterior accessory cavity opens into the tubotympanic recess at the junction of the auditory tube and the tympanic cavity proper (MacPhee and Cartmill, 1986, Fig. 7B).

Thus, differences in carotid artery position can explain why the auditory tubes of tarsiers have two separate diverticula and those of anthropoids have only one. These differences are correlated with differences in carotid artery position, differences that are captured in Characters 9 and 10, and to score

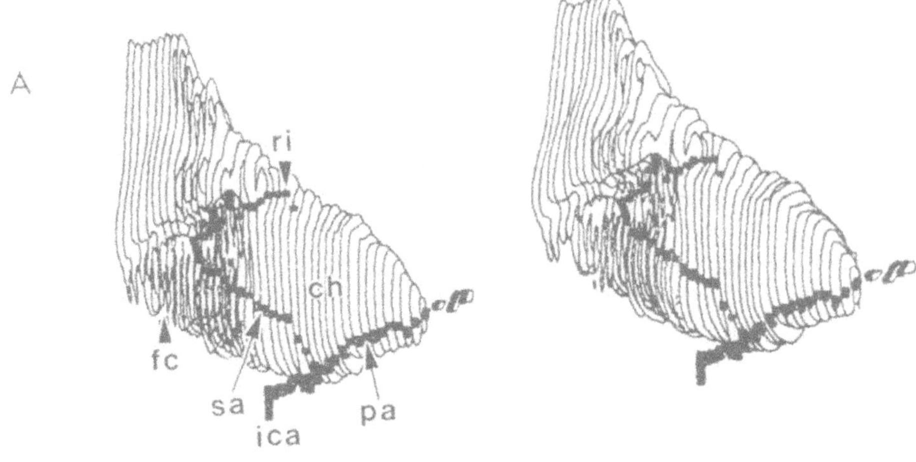

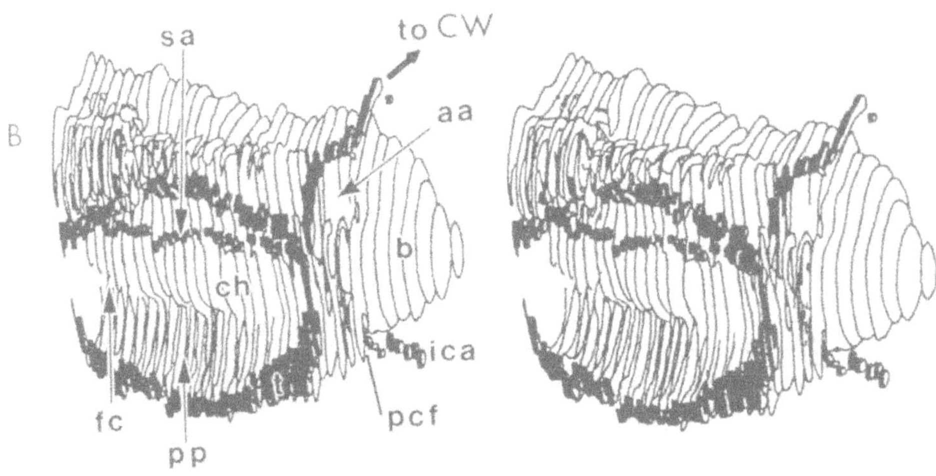

Fig. 4. Stereoscopic three dimensional reconstructions of anterolateral and slightly dorsal views of the ear regions of two ontogenetic stages of *Tarsius*. (A) *Tarsius bancanus* (DUCEC 8402, HL 14.5 mm, CRL 45 mm, section thickness: 20 μm). (B) *Tarsius syrichta* (DUCEC 801, HL 27 mm, CRL 51 mm, section thickness: 16 μm). The younger specimen (A) differs from the older specimen (B) in the following features: there is no petrosal plate (pp) and no anterior accessory cavity causing the bulla (b) to balloon out rostral to the cochlear housing (ch); the internal carotid artery (ica) approaches the cochlear housing (ch) ventral to its midline, well anterior to the fenestra cochleae (fc) (partially obscured in this view); no canal is present around the stapedial artery (sa); and the promontory artery (pa) traverses a large portion of the ventrolateral surface of the cochlear housing before passing dorsally to enter the braincase. In B, the anterior portion of the petrosal plate has been pneumatized through the apical aditus (aa), forming an anterior

these differences here would redundantly score the unique position of *Tarsius* twice.

Other differences between tarsiers and anthropoids in the patterns of pneumatization of the anterior portion of the tympanic cavity inhere in the extent to which this pneumatization invades the petrosal plate. In *Tarsius,* pneumatization splits only the most anterior portion of the petrosal plate, so that in the adult the anterior accessory cavity lies in front of the tympanic cavity proper (Fig. 4B). In anthropoids this pneumatization extends further posteriorly so that the anterior accessory cavity of adult anthropoids lies medial to the anterior portion of the tympanic cavity as well as in front of it.

Notwithstanding these differences between tarsiers and anthropoids in the position of the apical aditus relative to the posterior end of the auditory tube, and in the extent of pneumatization, these animals are unique among extant primates in having a petrosal plate that is pneumatized from the anterior portion of the tympanic cavity.

Transverse Septum. The term "transverse septum" was used by Cartmill and Kay (1978) to describe the sheet of bone separating the tympanic cavity proper from the cavity occupying the anteriormost portion of the petrosal bone in tarsiers and ceboids. As these workers also noted, the internal carotid artery lies within this septum as it traverses the middle ear region. Observations on petrosals from the Fayum led Cartmill *et al.* (1981) to suggest that this pattern was also present in the earliest anthropoids, implying not only that anthropoids are monophyletic but also that they may have close affinities with *Tarsius.* These authors named the cavity occupying the most anterior (or apical) portion of the petrosal the "anterior accessory cavity" and the septum separating it from the tympanic cavity proper the "transverse septum." Subsequent ontogenetic studies revealed that this anterior accessory cavity developed in a very similar fashion in tarsiers and anthropoids, and MacPhee and Cartmill (1986) concluded that these features were synapomorphies of a tarsier–anthropoid clade.

Simons and Rasmussen (1989) argued that the transverse septum of tarsiers is not homologous with that of anthropoids because it lies medial, not lateral, to the internal carotid canal. This assertion is predicated on a restriction of the term "transverse septum" to the sheet of bone extending *either* medially *or* laterally from the carotid canal to the bulla wall and ignores the presence in tarsiers and anthropoids of septa *both* medial *and* lateral to the internal carotid pathway. These septa are not illustrated by Simons and Rasmussen (1989, Fig. 9, p. 14), but the laterally directed septum of *Tarsius* is

---

accessory cavity anterior to the cochlear housing and causing the bulla (b) to balloon anteriorly. The internal carotid artery (ica) has been shunted anteriorly relative to other middle ear structures and is directed caudally as it enters the anteriorly placed posterior carotid foramen (pcf). It then runs straight up through the portion of the transverse septum lying lateral to it and passes up over the top of the anterior accessory cavity to the Circle of Willis (CW). Note that the tympanic bone (t), shown in B but not in A, is extrabullar and has been laterally displaced by the developing bulla (pp).

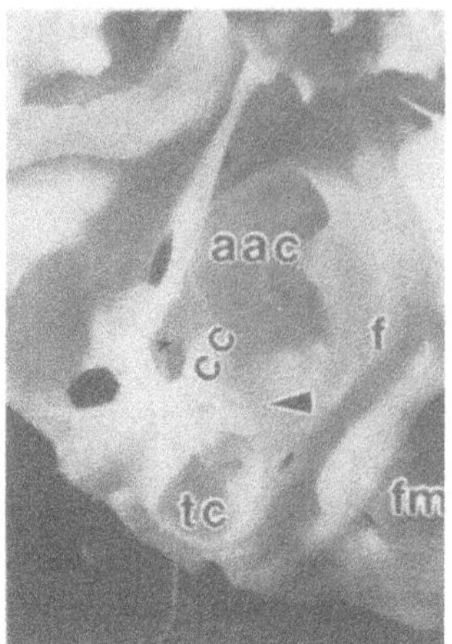

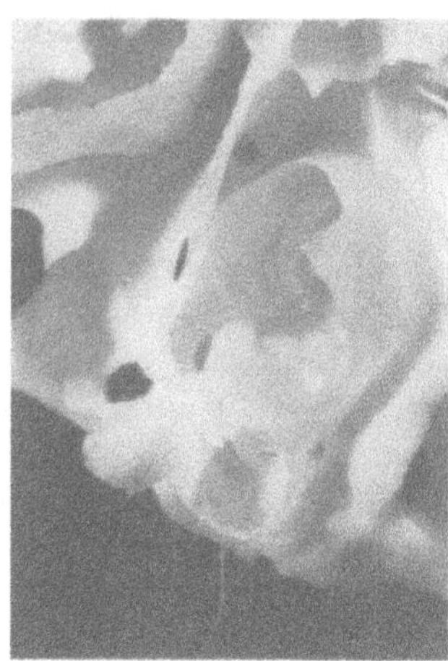

Fig. 5. Ventral view of right ear region of *Tarsius* sp. skull in BACAAC. Rostral is to the top and slightly to the left. The floor of the right bulla has been removed, opening the anterior accessory cavity (aac) and exposing the carotid canal (cc), the lateral portion of the transverse septum (asterisk), the apical aditus (between cc and asterisk), and the medial portion of the transverse septum. The flange of the basioccipital (f) on the medial wall of the bulla is visible. The anterolateral and ventral placement of the posterior carotid foramen can be seen. The foramen magnum (fm) faces directly ventrad. The floor of the tympanic cavity (tc) has also been removed, exposing the fenestra cochleae in the posterior surface of the promontorium.

shown in Fig. 5 as well as by MacPhee and Cartmill (1986, Fig. 3D). The medially directed septum of anthropoids is illustrated in Fig. 6 but is not illustrated by MacPhee and Cartmill (1986, Fig. 3E) because this septum is highly reduced in taxa in which the posterior carotid foramen and the carotid canal are closely apposed to the medial or posterior bulla wall. Nevertheless, it is present in all anthropoids in which there is any separation between the carotid canal and the medial or posterior bulla wall.

Developmental evidence suggests that differences in proportions of these two septae are actually attributable to differences in the position of the internal carotid artery within a single transverse septum. In the early fetal tarsiers studied here, the pathway of the internal carotid artery resembles that of fetal anthropoids in that it pierces the membranous floor of the middle ear cavity at the midline of the ventral surface of the cochlear promontorium, well anterior to the fenestra cochleae, and traverses a large proportion of the promontorium (Fig. 4A). At this stage neither a medially nor a laterally directed septum is present in tarsiers. However, in neonatal tarsiers the posterior carotid foramen is positioned ventrally and laterally away from the ventral surface of the promontorium, and the internal carotid artery is more vertically oriented. As the shift occurs, a septum develops dorsal and medial to the internal carotid canal, and the septum lying lateral to the canal becomes reduced in size so that adult *Tarsius* has a vertically oriented internal carotid artery, a large medially directed septum, and a much reduced lateral transverse septum (Figs. 2F, 4B, and 5).

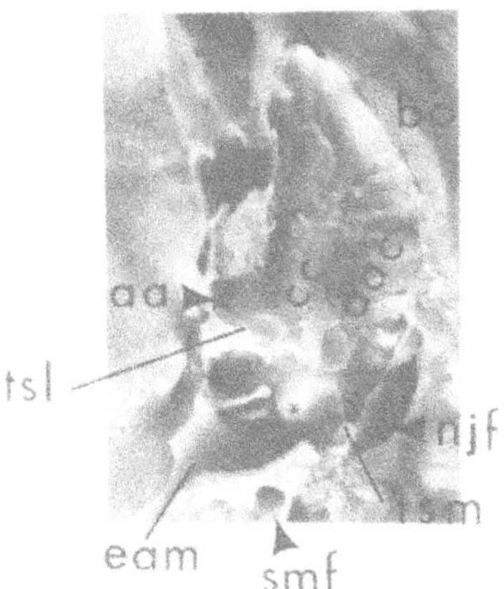

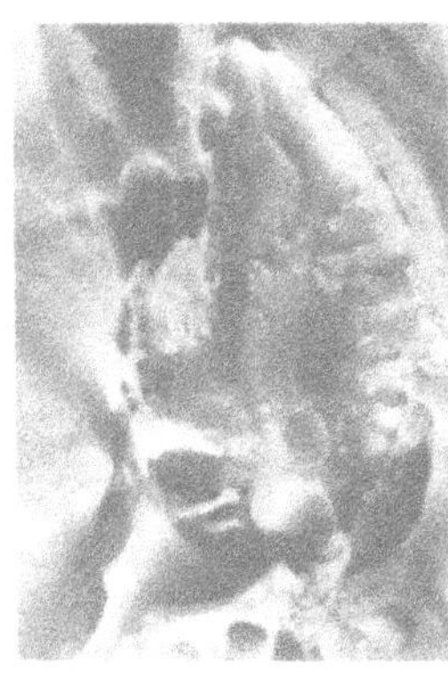

Fig. 6. Ventral view of ear region of *Macaca* sp. (DUPC 075). The midline of the basioccipital (bo) is visible; anterior is obliquely up to the left. Note the tubular external auditory meatus (eam) just anterior to the stylomastoid foramen (smf). Both the medial (tsm) and lateral (tsl) portions of the transverse septum can be seen separating the anterior accessory cavity (aac) from the tympanic cavity (containing the ear ossicles). The tsl is pierced by an apical aditus (aa). The promontorium within the tympanic cavity is labeled with an asterisk. Note that the posterior carotid foramen (broken away) must have lain anterior and medial to the fenestra cochleae, close to the notch in the jugular fossa for the inferior ganglion of IX (njf).

In early anthropoid ontogeny, verticalization of the internal carotid artery does not occur, so an extensive medially directed septum does not develop. In most adult anthropoids the internal carotid artery pierces the bulla's medial wall in close approximation to the jugular foramen and ventral to the midline of the cochlear promontorium. Consequently, there is little room for a septum medial to the internal carotid canal. However, anthropoids with the posterior carotid foramen positioned even slightly ventrally and medially in the bulla wall (as in *Macaca*, Fig. 6; *Aegyptopithecus*, Fig. 8) do have a small septum stretching posteromedially from the carotid canal to the medial wall of the tympanic cavity.

Thus, developmental data suggest that the septa lying medial and lateral to the internal carotid canal in tarsiers and anthropoids are homologous parts of a single septum, and that their differing proportions reflect different positions of the internal carotid artery in adults of the two groups. Developmental evidence corroborates the hypothesis that the anterior accessory cavities, transverse septa, and apical adituses of adult tarsiers and anthropoids are homologous, and it also explains why they are differently positioned.

Discriminating Ontogenetic Pathways in Fossil Taxa

These ontogenetic data can be used to derive criteria for discriminating between anthropoid- and lemuroid-like otic regions in fossils. Anthropoid-like otic regions are distinguished by the presence of a bony transverse sep-

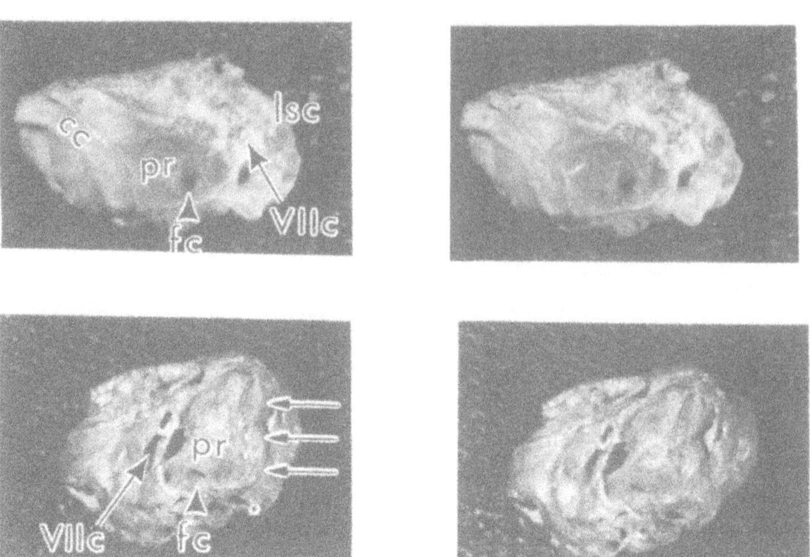

Fig. 7. Ventrolateral and slightly anterior view of Fayum anthropoid petrosals. (A) YPM 25972, isolated left petrosal, anterior to the left, dorsal to the top. The carotid canal (cc) traverses the cochlear housing anterior to the fenestra cochleae (fc). The promontorium (pr) overlies the basal turn of the cochlea which continues anteriorly to the carotid canal. The lateral semicircular canal (lsc) is visible posteriorly. (B) YPM 23968, isolated right petrosal, anterior to right, dorsal to top. Arrows mark transverse septum arising from carotid canal as it crosses the promontorium, well anterior to fenestra cochleae. Asterisk marks remains of bulla defining posterior edge of posterior carotid foramen, also well anterior to the fenestra cochleae. In both (A) and (B) the canal for the facial nerve (VIIc) is broken but visible.

tum (1) arising from the ventrolateral surface of the cochlear housing, (2) containing the internal carotid canal at its base, and (3) forming the lateral wall of an anterior accessory cavity. In two of the petrosals recovered from the Fayum [DPC 2468 (MacPhee and Cartmill, 1986, Fig. 21) and YPM 23968 (see Fig. 7B)], the broken remnants of such a septum can be seen extending ventrolaterally from the portion of the carotid canal lying on the cochlear housing. Unfortunately, however, the relevant portions of the internal carotid canal are broken in DUPC 6642, *Aegyptopithecus zeuxis* (Fig. 8B), and none of the other Fayum petrosals can be definitively assigned to this taxon.

In contrast, no such septum arises from the ventrolateral surface of the promontorium in lemuroids. Lemuroid otic regions sport an anterior septum running from the rostral pole of the promontorium to the anterior bulla wall. This septum is located in the position of the ontogenetically primitive piriform fenestra, where the septum originally formed, not on the ventrolateral surface of the cochlear housing, as would be expected if it had arisen on the midcapsular arc. This septum sports a laterally directed extension that forms the floor of the bony auditory tube and is continuous posterolaterally with the linea semicircularis (Fig. 9A).

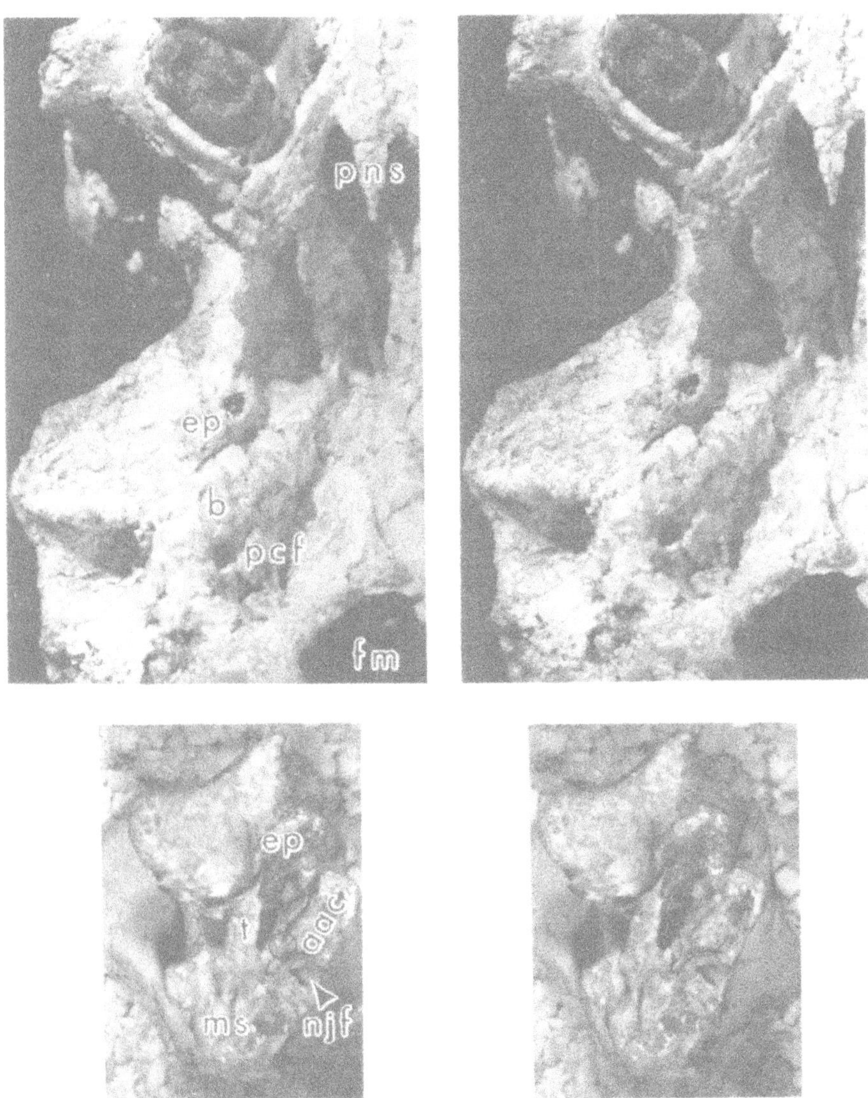

Fig. 8. Ventral views of ear regions of *Aegyptopithecus* fossils from the Fayum. (A) Basicranium of *Aegyptopithecus zeuxis* skull (CGM 40237; photograph courtesy of Dr. R. F. Kay). The posterior carotid foramen (pcf) lies in the midline of the bulla (b), well ventral to the petrosal–basioccipital suture. (B) Petrosal, tympanic, and part of squamosal (DPC 6642; found in association with teeth of *Aegyptopithecus*). Based on position of pcf in A, circle in B marks likely position of pcf. As in *Macaca* (Fig. 6) and other Fayum petrosals, the pcf in DPC 6642 was situated close to notch in jugular fossa for inferior ganglion of IX (njf). The carotid canal is broken, but the groove that remains (marked with arrowhead) delineates the tympanic cavity from the trabeculated ventromedial surface of the cochlear housing that was likely located in the anterior accessory cavity (aac). Compare posterior nasal spine (pns) in A with dissimilar protuberance in Rasmussen (1990, Fig. 2). Note prominent entoglenoid process (ep) at rostral end of strong entoglenoid crest in both specimens and mastoid air cells (ms) in (B).

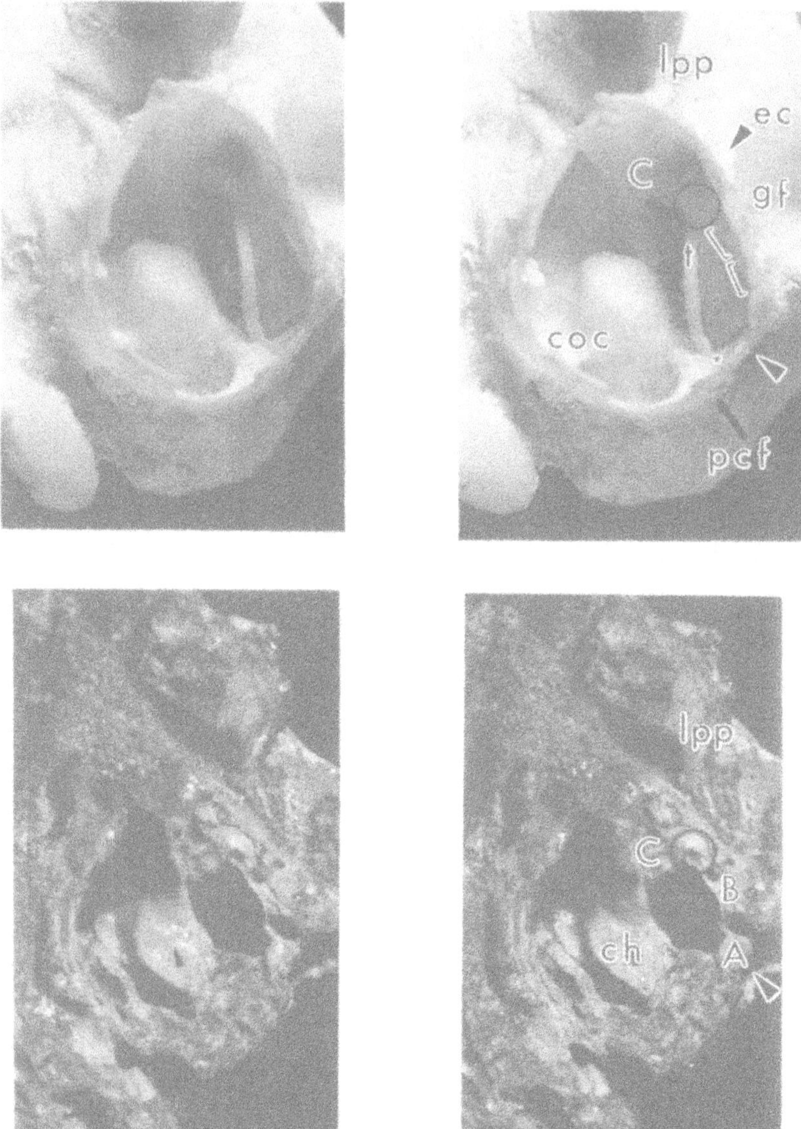

Fig. 9. Ventral views of left ear regions of *Propithecus verreauxi* (DUPC 052) (A) and *Mahgarita stevensi* (TMM 41578-9; type skull) (B). Anterior is to the top. In A the annular membrane can be seen stretching laterally from the tympanic bone (t) to the linea semicircularis (delineated by square brackets). Anterior to the contact between the tympanic bone and the anterior extremity of the linea semicircularis, the anterior canaliculus for the chorda tympani is visible (circled). The asterisk in A marks shelf A, running laterally from the posterior septum and carotid canal to floor the external auditory meatus (unlabeled arrow in A and B). The open trough for the internal carotid artery is marked by two open triangles. The posterior carotid foramen (pcf) is posterolaterally placed; the carotid canal shields the fenestra cochleae from ventral view. The glenoid fossa (gf) is delimited medially by a strong entoglenoid crest (ec). The cochlear canaliculus (coc) is visible medially. (B) The ear region of *Mahgarita stevensi* is very similar. Note that the carotid canal is missing on this side of the specimen. It is, however, preserved in matrix on the right side and in TMM 41578-20. Shelf C (C) floors the bony auditory tube and runs laterally from only that portion of the carotid canal that lies anterior to the cochlea housing (ch).

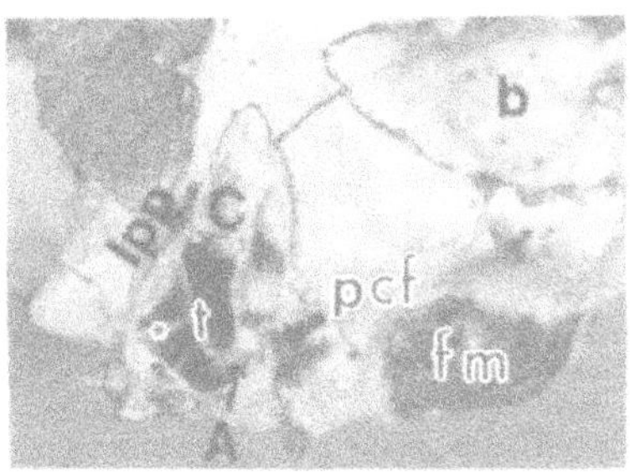

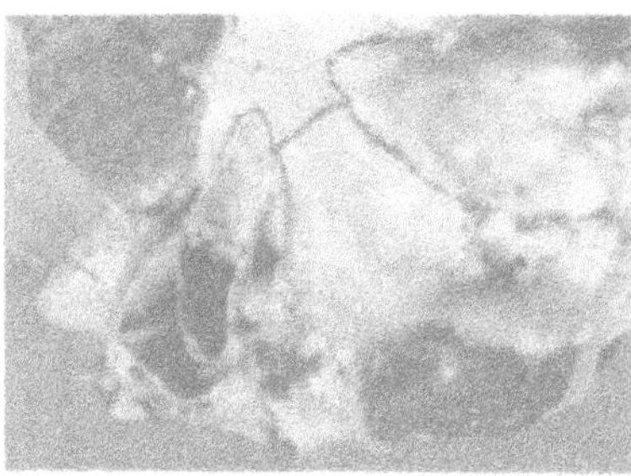

Fig. 10. Ventromedial view of the right ear region of *Rooneyia viejaensis* (TMM 406887). Anterior is obliquely up to the left; the left bulla (b) is visible, and foramen magnum (fm) is visible. The tympanic (t) is fused to the lateral bulla wall via a complete bony annular bridge that sports struts (one of which is marked with an asterisk). Note lateral pterygoid plate (lpp) fused to lateral bulla wall. The posterior carotid foramen (pcf) lies posteromedial, but still anterior to fenestra cochleae. Shelf A (A) extends laterally from the posterior intrabullar septum to the annular bridge.

A lemuroid-like anterior septum is also present in adapids, omomyids, and plesiadapiforms. As in extant lemuroids, this septum sports a laterally directed extension flooring the bony auditory tube—a feature that is visible in *Adapis parisiensis* (Gingerich and Martin, 1981, Fig. 8B; Saban, 1963, Fig. 51; Fig. 13A), *Notharctus tenebrosus* (AMNH 11466, Szalay *et al.*, 1987, Fig. 4), *Smilodectes gracilis* (MPM 2612, in MacPhee and Cartmill, 1986, Fig. 20), *Necrolemur antiquus* (Szalay and Wilson, 1976, Fig. 1), and *Rooneyia viejaensis* (Fig. 10). In none of these fossil taxa does the septum flooring the auditory tube extend back onto the promontorium; in all of them it runs ventrolaterally to the lateral wall of the bulla to become continuous posteriorly with an osseous structure between the lateral bulla wall and the intrabullar tympanic—a partial annular bridge in adapids or a complete annular bridge in omomyids. In *Tetonius* the internal carotid artery runs through a similar septum, which extends ventrolaterally from the internal carotid canal toward the lateral bulla wall but is broken at is lateral extent (Fig. 11B). *Shoshonius* resembles *Tetonius* in lacking the laterally directed portion, also possibly because of breakage, but differs in that the internal carotid artery does not run through the anterior septum but lies medial to it. *Ignacius, Plesiadapis,* and *Phenacolemur* all possess a (presumed) avascular anterior longitudinal septum that lacks the laterally directed portion. In *Ignacius graybullianus* this septum lies lateral to the petrosal–entotympanic suture illustrated by Kay *et al.* (1992), so it is formed from the petrosal bone, as in all extant lemuroids. *Tupaia* and *Ptilocercus* also have such a laterally directed septum, but it is formed from the entotympanic element (MacPhee, 1981).

In light of these comparisons, the "transverse septum" in *Mahgarita* (shelves B and C in Fig. 9B), claimed by Rasmussen (1990) to be anthropoid-like, can be shown to be more lemur- and adapid- than anthropoid-like. Significantly, this septum, lying anterior to the cochlear housing (shelf C in Fig. 9) differs from that of extant anthropoids (and YPM 23968) in not arising from the transpromontorial portion of the internal carotid canal.

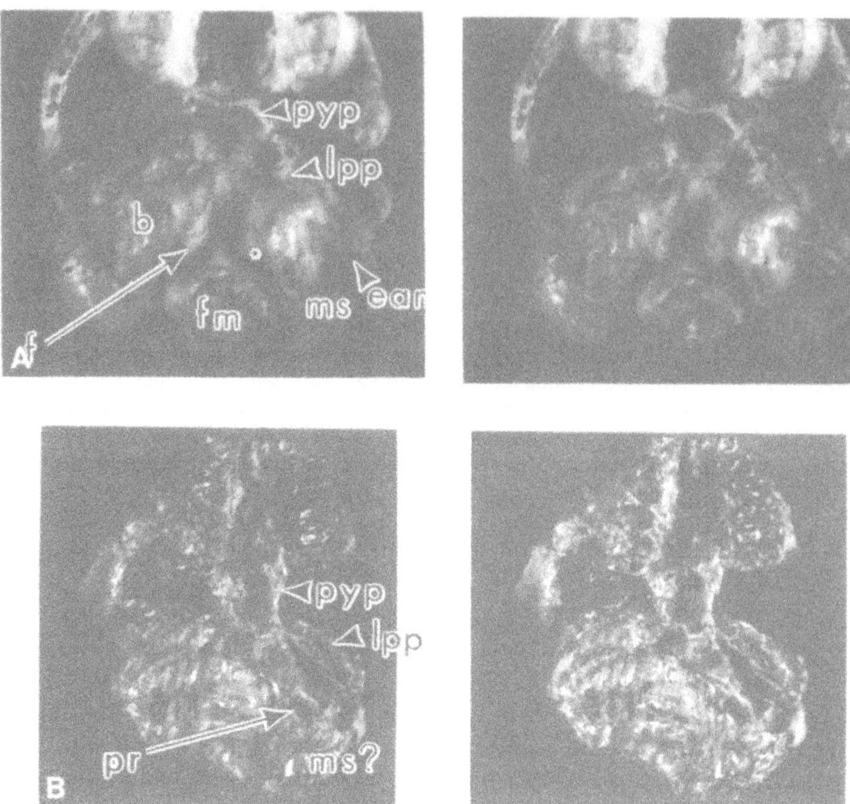

Fig. 11. Ventral views of omomyid basicrania. (A) *Necrolemur* sp. (B) *Tetonius homunculus* (AMNH 4194). Asterisk in A marks position of posterior carotid foramen. In B the promontory canal can be seen on the surface of the anterior portion of the promontorium (pr) and running into the septum lying anterior and medial to it. In both A and B the pyramidal processes (pyp) are medially positioned relative to the toothrow; the lateral pterygoid plate (lpp) is fused to the lateral bulla wall, and mastoid air cells (ms), visible in *Necrolemur,* may also have been present in *Tetonius.* (fm, foramen magnum).

It is possible that this septum is broken in TMM 41578-9 and that it was "perforated by a fairly broad foramen" (Rasmussen, 1990, p. 449), but there are three reasons for thinking this unlikely (Fig. 9B). First, although the posterior edge of this shelf is cracked and distorted, it shows no signs of breakage. Second, it is continuous laterally with the horizontally oriented septum referred to here as shelf B, the medial edge of which reveals no signs of breakage, as would be expected if shelf C extended further posteriorly in life. Third, although the bone that formed the promontory canal is lost bilaterally from the portion of the canal immediately posterior to shelf C in TMM 41578-9, it is clear that the promontory canal on the right side of TMM 41578-20 did not have a septum extending laterally from its surface. Thus, there appears to be no reason to suggest that shelf C ever extended further posteriorly than it does in TMM 41578-9.

Beard *et al.* (1991) suggest that "[t]he internal carotid canal in *Shoshonius* runs dorsally within a bony septum as in *Tarsius* and anthropoids . . ." (p. 2). The septum in *Shoshonius* does not arise from that portion of the internal carotid canal that lies on the cochlear housing (Beard *et al.*, 1991, Fig. 2), and there are no obvious breakages where such a septum could have been. Rather, the artery runs in the ventral edge of a posteriorly directed septum forming the medial wall of the cochlear fossula. Such a septum is found in all adapids, omomyids, extant lemurs, and subfossil Malagasy primates as well as *Plesiadapis, Ignacius,* and *Phenacolemur* (Szalay, 1973; MacPhee, 1981). It carries the internal carotid canal on its ventral edge in lemuroids, adapids, plesiadapiforms, and *Shoshonius,* and it is present in the same position even when the carotid is placed medially, as in *Rooneyia* and *Necrolemur.*

In lemuroids this septum develops via pneumatization from the enlarged hypotympanic sinus of the caudal tympanic process of the petrosal (MacPhee, 1981). Such a septum is seen in anthropoids in which pneumatization of the petrosal plate has extended so far posteriorly as to be medial to the cochlear fossula, but it is not seen in *Tarsius* because pneumatization does not extend this far caudally. It is not clear whether this posterior septum can be homologized in all non-*Tarsius* primates, because in anthropoids the pneumatization of the caudal tympanic process of the petrosal originates from within the anterior accessory cavity, whereas in lemuroids it originates in the hypotympanic extension of the tympanic cavity. In any event, the widespread distribution of this posterior septum suggests that its presence in *Shoshonius* is of little value in assessing its phylogenetic relationships.

Character Analysis

Two characters are needed to capture the different patterns of intrabullar pneumatization found among primates and their outgroup taxa: one to describe the different patterns of pneumatization, and one to describe the differing extent of petrosal plate pneumatization in tarsiers and anthropoids.

Character 1: Transverse septum arising from the cochlear housing. Absent (0); present and forming the lateral wall of an anterior accessory cavity pneumatized from the auditory tube or the auditory tube–tympanic cavity junction (1) (the hypothesis of the homology of these structures in tarsiers and anthropoids is based on the developmental data presented above); present and forming the lateral wall of an anterior accessory cavity pneumatized from the epitympanic recess (2).

Character 2: Extent of pneumatization of anterior accessory cavity. Anterior accessory cavity lies anterior to the tympanic cavity and is not trabeculated (0); anterior accessory cavity extends medial to the tympanic cavity and is trabeculated (1).

Pneumatization of the Mastoid Region

Pneumatization of the mastoid region from the epitympanic recess has been invoked as evidence linking omomyids to anthropoids (Szalay *et al.*, 1987) and linking adapids to anthropoids (Rasmussen, 1990). The presence of such pneumatization in *Necrolemur, Adapis, Mahgarita,* and lorisoids, its probable presence in *Notharctus* and *Tetonius,* and the widespread occurrence of mastoid inflation in eutherians (Klaauw, 1931) suggest that parallel evolution of this feature was common.

Character Analysis

Character 3: Pneumatization of mastoid from epitympanic recess. Pneumatization is present (1); absent (0).

Pathway for the Internal Carotid Artery through the Ear Region

As discussed above, the internal carotid artery in tarsiers and anthropoids runs through the middle ear cavity within a transverse septum separating the anterior accessory cavity from the tympanic cavity proper (Cartmill and Kay, 1978). Because this septum is developmentally derived from the medial bulla wall, the artery is technically perbullar or within the medial bulla wall (Cartmill *et al.*, 1981; MacPhee and Cartmill, 1986).

Wible (1986) documented the distribution of perbullar character states among Eutheria. In his figures describing the perbullar pathway, the artery is shown lying within the medial bullar wall itself, medial to the promontorium; his figure of the extrabullar pathway shows the artery lying medial to the medial bulla wall, outside the tympanic cavity; and his figure of the transpromontorial pathway has the artery lying lateral to the medial bulla wall, within the intrabullar cavity on the ventral surface of the promontorium. However, none of these figures describes the situation seen in anthropoids, in which the internal carotid is both transpromontorial and perbullar, lying on the ventrolateral surface of the cochlear housing and enclosed in a transverse septum derived from the medial bulla wall. This situation arises because the bulla wall develops as an outgrowth of the petrosal precisely at the point where the artery contacts the cochlear housing, a feature unique to tarsiers and anthropoids among eutherians. In fact, although many different eutherians possess perbullar carotid pathways, there is a great deal of variation in the bones making up the perbullar canal.

Thus, there are three independent aspects of the anatomy of the pathway of the internal carotid artery through the middle ear cavity that must be evaluated: whether or not it is perbullar; the bones making up the bulla and perbullar canal; and the position of the artery and its enclosing canal relative to other otic structures. The developmental evidence discussed above demonstrates that the carotid artery lies within a derivative of the medial bulla wall in tarsiers and anthropoids (Character 1). The other two aspects of the carotid pathway through the ear region are discussed here.

Bones Making up the Perbullar Pathway

Simons and Rasmussen have asserted that the "'perbullar' pathway is not a reliable indicator of phyletic affinity [because] it has evolved convergently in other mammalian groups" (Simons and Rasmussen, 1989, p. 13). However, variation within the category "perbullar" provides useful data for evaluating phylogenetic hypotheses. All proboscideans have a bulla formed by an entotympanically derived bone. Because they also have a perbullar pathway, proboscideans have a perbullar pathway formed by the entotympanic bone. All pholidotans have a perbullar canal formed by the petrosal, ectotympanic, and entotympanic bones (Kampen, 1905, cited in Wible, 1986, p. 322). Within each of the five nonprimate orders displaying a perbullar character state for some taxa and an extrabullar or transpromontorial state for others (Xenarthra, Carnivora, Lagomorpha, Rodentia, Chiroptera), the distribution of the perbullar state corroborates "generally accepted hypotheses of phyletic relationships" (Wible, 1986). Within Carnivora, arctoids and cynoids share a perbullar pathway formed by rostral and caudal entotympanics, whereas feloids (where known) share a transpromontorial pathway. Leporid lagomorphs all share a perbullar pathway formed by the tympanic bone, whereas ochotonids have an extrabullar pathway (Wible, 1986). Among Chiroptera, all microchiropterans share a transpromontorial pathway, and all megachiroptera have an anteriorly shifted internal carotid artery that either "runs in a groove on the intratympanic surface of the rostral entotympanic or in a short carotid canal between the rostral entotympanic and the anterior pole of the promontorium" (Wible, 1984, p. 248).

Among xenarthrans the distribution of the perbullar pathway appears not to corroborate a "generally accepted hypothesis of phyletic relationships"—dasypodids and bradypodids share a perbullar pathway that is not found in anteaters (Wible, 1986). However, there is evidence to suggest that the bullae, and therefore the perbullar pathway for the internal carotid artery, may have evolved independently in dasypodids and bradypodids. Although both taxa possess an entotympanic bulla, in *Dasypus* the entotympanic element appears initially in the posterior portion of the tympanic floor and grows anteriorly, whereas in *Bradypus* the element first ossifies anteriorly and then expands posteriorly in the tympanic floor. Wible suggests that this difference may indicate nonhomology of entotympanics in dasypodids and *Bradypus* (Wible, 1984, p. 91).

Thus, although a perbullar pathway *per se* is not a "reliable indicator of phyletic affinities," the distribution of the different types of perbullar pathway *does* corroborate "generally accepted hypotheses of phyletic relationships."

Position of the Internal Carotid Canal within the Middle Ear Cavity

As the internal carotid artery travels through the middle ear cavity, its different parts can vary in relation to other intrabullar structures, and the relations of each of these parts have been utilized in the debate over anthropoid origins. In their original definition of the perbullar pathway, Cartmill *et al.* (1981) assert that (1) the last common ancestor of tarsiers and anthropoids had an anteromedially located posterior carotid foramen, and that the anterolateral position seen in *Tarsius* is an autapomorphy of that lineage; that in tarsiers and anthropoids, including the Fayum petrosals, the internal carotid artery (2) "does not travel near the cochlear window but always well anterior to it"; (3) "travels over only a very small part of the promontorium and in any event does not cross the latter's ventrolateral surface"; (4) is perbullar and therefore not within the tympanic cavity proper; and (5) in *Tarsius* but not in any extant or fossil anthropoids a stapedial artery branches off "beneath the anterior pole of the promontory, and travels backward along the length of the promontory in order to reach the stapes" (Cartmill *et al.*, 1981, p. 17).

In contrast, in noncheirogaleid Malagasy primates the internal carotid artery (1) enters the bulla posterolaterally through the posterior wall, (2) "travels on or near the ventral lip of the cochlear window, (3) traverses "the entire ventral or ventrolateral surface of the promontorium," (4) lies lateral to the medial bulla wall and therefore within the tympanic cavity proper and is enclosed in a canal derived from the promontorium lateral to the petrosal plate, and (5) the stapedial canal, present in all extant non-haplorhines except lorisiforms and cheirogaleids, "travels only a short distance from its origin to the obturator foramen of the stapes" (Cartmill *et al.*, 1981, p. 17).

Simons and Rasmussen take issue with four of these claims, contending that the ear regions of *Alouatta* and *Aegyptopithecus zeuxis* (DPC 6642) exhibit several features claimed by Cartmill *et al.* (1981; MacPhee and Cartmill, 1986) to be characteristic of the transpromontorial pathway. Specifically, they claim that in *Aegyptopithecus* and *Alouatta* (1) the posterior carotid foramen is situated posterolaterally in the bulla, "nearly *ventral* to the cochlear window"; (2) the internal carotid/promontory artery arrives at the promontorium ventral, not anterior, to the fenestra cochleae; (3) it traverses "about half" of the ventrolateral surface of the cochlear promontorium in *Alouatta* but crosses the ventral and anterior surface of the promontoriium of *Aegyptopithecus* (DPC 6642) and the other fossil petrosals from the Fayum; and (4) that "the lateral border of the canal lies *lateral* to the cochlear window and forms a wall of the tympanic cavity proper." On this basis they claim that "[t]he 'perbullar' pathway as described by MacPhee and Cartmill (1986) is a typological construct that does not adequately describe the variation in petrosal morphology that occurs among prosimian and anthropoid primates . . ." (Simons and Rasmussen, 1989, p. 13, quotation marks in original).

These claims regarding variability in the position of the different parts of the intrabullar carotid pathway are examined below, starting at the posterior carotid foramen and working proximally in the direction of blood flow.

Position of the Posterior Carotid Foramen in the Bulla. Tarsius is unique among primates in having the posterior carotid foramen lying anterior to a line joining the midpoints of the ventral rims of the tympanics. The posterior carotid foramen is medially placed in the bulla in lorisoids, *Tupaia, Callithrix, Callicebus, Necrolemur,* and *Rooneyia,* in the midline of the bulla in atelines, *Saguinus, Leontopithecus, Saimiri, Aotus,* pithecines, *Cebus,* most cercopithecines, colobines, hominoids, and *Aegyptopithecus,* and laterally located in plesiadapiforms, cheirogaleids, lemuroids, adapids, *Tarsius, Shoshonius,* and *Alouatta.* The carotid foramen is ventrally located in the bulla in *Tarsius* and in many platyrrhines and catarrhines as well as in *Aegyptopithecus* (see Fig. 8A) but dorsally placed, adjacent to the dorsal edge of the bulla, in other omomyids (including *Shoshonius,* Beard and MacPhee, Chapter 3, this volume), and in adapids. Among the Fayum petrosals the position of the posterior carotid foramen can be determined with confidence only in YPM 25972 and 23968 (see Fig. 7A, B). In both of these specimens the posteromedial edge of the posterior carotid foramen is preserved lying immediately medial to a notch in the jugular fossa (which probably held the inferior ganglion for the glossopharyngeal nerve) and hence posteriorly and only slightly ventrad in the bulla.

Position of the Internal Carotid Pathway Relative to the Fenestra Cochleae. As Simons and Rasmussen point out, *Alouatta* often has the posterior carotid foramen lying ventral to the fenestra cochleae, in apparent violation of MacPhee and Cartmill's definition of the perbullar pathway. However, this is probably an autapomorphic condition among anthropoids, as no platyrrhines other than *Alouatta* have their posterior carotid foramen situated ventral or posterior to the fenestra cochleae. Except for *Alouatta,* all anthropoids examined resemble *Macaca* in having the posterior carotid foramen anteriorly and medially placed relative to the fenestra cochleae (see Fig. 6). However, even though the posterior carotid foramen lies ventral (even posterior in some specimens) to the fenestra cochleae in *Alouatta,* the internal carotid canal curves anteriorly and medially to reach the cochlear housing well anterior to the fenestra cochleae and does not run across its ventral lip. This is also true of all extant anthropoids and *Tarsius.*

In the Fayum petrosals YPM 25972 and 23968, the position of the posteromedial border of the posterior carotid foramen relative to the fenestra cochleae varies with the orientation of the fossils, so its position along the anteroposterior axis cannot be definitively determined in isolated petrosals. However, it is highly unlikely that the posterior carotid foramen was placed ventral to the fenestra cochleae in these specimens (Fig. 7A, B).

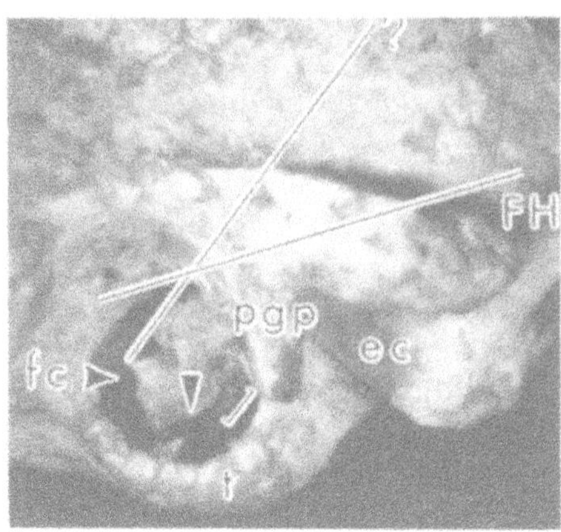

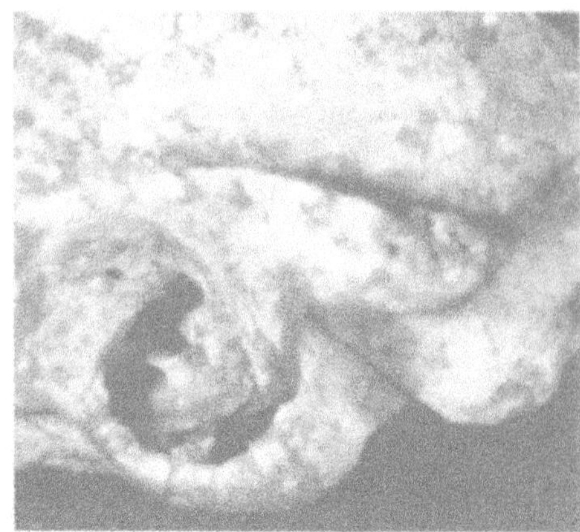

Fig. 12. Lateral view through external auditory meatus of DPC 6642 (*Aegyptopithecus zeuxis*); t, tympanic bone; pgp, postglenoid process; ec, entoglenoid crest. Unlabeled arrow marks most posterior extent of groove representing broken remnant of carotid canal, well anterior to fenestra cochleae (fc). Square bracket marks groove itself, visible in Fig. 11B running anterodorsally. The line marked FH represents probable orientation of Frankfort horizontal as determined from casts and photographs. Line marked "?" represents orientation of Frankfort horizontal if internal carotid arrives at the promontorium ventral to fenestra cochleae. This orientation is unlikely.

Other petrosals (YPM 25973 and 25974) from the Fayum do not preserve the posterior carotid foramen, although they do preserve enough of the carotid canal to determine that it undoubtedly entered the bulla well anterior to the fenestra cochleae and in close approximation to the medial bulla wall and jugular fossa. In all these fossils, the point where the carotid canal first contacts the promontorium lies well anterior to the fenestra cochleae, and it does not travel "on or near the ventral lip of the cochlear window" (Fig. 4A, B). The remnants of the internal carotid canal in DUPC 6642 reveal that this is also true in *Aegyptopithecus* (Fig. 12).

Fossil adapids, where known, have a posterolaterally positioned posterior carotid foramen and a carotid canal that lies near the ventral lip of the fenestra cochleae, obscuring it from ventral view. This is certainly the case in *Adapis parisiensis* (Fig. 13A), *Smilodectes gracilis* (MacPhee and Cartmill, 1986), and *Notharctus* as well as *Mahgarita* (Fig. 13B). As far as can be determined at present, *Shoshonius* probably had its internal carotid canal positioned posterior to the fenestra cochleae, but additional preparation is needed to confirm this. The illustration of *Necrolemur,* Montauban 9, in Szalay (1973, Fig. 12) shows the carotid artery well separated from the fenestra cochleae. *Tetonius* lacks the relevant portion of the internal carotid canal as a result of postmortem breakage (Fig. 12B). Thus, the earliest fossil anthropoids, tarsiers, and extant anthropoids all share a placement of the carotid canal that is not near the ventral lip of the fenestra cochleae. Those omomyids for which the anatomy is known also appear to share this condition.

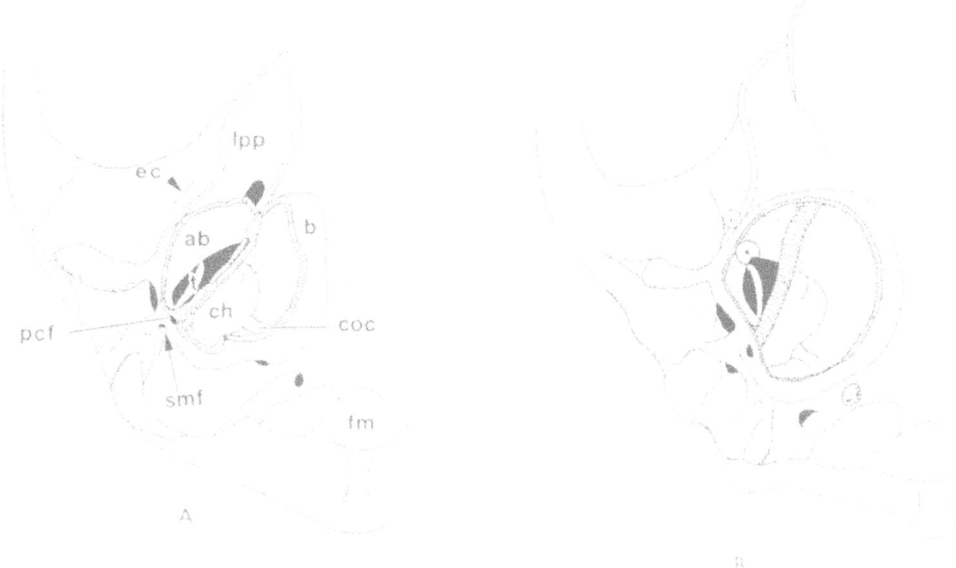

Fig. 13. Reconstruction of ventral view of right ear region of *Mahgarita stevensi* (B) compared with right ear region of *Adapis parisensis* (A) (partly from Stehlin, 1912, and partly from *Adapis* specimen studied at conference). A is labeled, and the same structures are visible in *Mahgarita*. The bulla (b) is broken away to reveal the intrabullar cavity. Note the strong entoglenoid crest (ec), the posterolateral posterior carotid foramen (pcf) adjacent to the stylomastoid foramen (smf), the incomplete annular bridge (ab), the lateral pterygoid plate (lpp) forming a laminar contact with the lateral bulla wall, the intrabullar tympanic (t), and the lemur-like trough for internal carotid artery in *Adapis* in contrast with the large complete canal in *Mahgarita*. fm, foramen magnum; coc, cochlear canaliculus. Circles in B indicate foramina for passage of chorda tympani.

Position of the Internal Carotid/Promontory Artery on the Promontorium. My examination of the original fossils suggests that differences of opinion in the literature (Cartmill *et al.*, 1981; MacPhee and Cartmill, 1986, versus Simons and Rasmussen, 1989, Rasmussen, 1990) regarding the position of the internal carotid artery on the promontorium result largely from differing uses of the term "promontorium."

In human anatomy texts the promontory is referred to as the bony protuberance on the medial wall of the tympanic cavity that overlies the basal turn of the cochlea (Williams *et al.*, 1989, p. 1225). However, Cartmill *et al.* (1981) label all of the cochlear housing lying posterior and lateral to the carotid canal as "promontory," irrespective of the fact that this bone overlies the cupula as well as the basal turn of the cochlea (Cartmill *et al.*, 1981, Figs. 1 and 2). Moreover, Simons and Rasmussen not only use the term "promontorium" to refer to the bony mound lying posterior and lateral to the carotid canal (Simons and Rasmussen, 1989, Fig. 7), but they also use it to refer to that portion of the cochlear housing lying *medial* to the carotid canal (Simons and Rasmussen, 1989, Figs. 6, 8, and 9). This difference of opinion as to what the promontorium is reflects the divergent views of these authors regarding the homologies of intrabullar spaces.

As discussed under Character 1, MacPhee and Cartmill (1986) do not consider the cavity lying medial to the carotid canal to be tympanic cavity proper (they call it the anterior accessory cavity), so the bony cochlear housing protruding into it cannot be called a promontorium as defined in human anatomy texts. On the contrary, Simons and Rasmussen apparently *do* consider the anterior accessory cavity to be part of the tympanic cavity, so it is possible for it to contain a promontorium. Moreover, examination of DUPC 6642, YPM 23968, 25972, 25973, 25974, and two skulls of *Alouatta seniculus* at the DUPC suggests that the bony portion of the cochlear housing lying medial to the carotid canal in all these specimens probably contains the medial portion of the basal turn of the cochlea and hence fits the term promontorium *sensu* human anatomy. If *this* bony protuberance is defined as "promontorium," then the carotid canal does cross its ventrolateral surface in the Fayum petrosals and extant anthropoids, confirming Simons and Rasmussen's assertions that the ear regions of *Alouatta, Aegyptopithecus,* and other early anthropoids resemble the definition of the transpromontorial pathway more than that of the perbullar.

This dispute over whether the cochlear housing lying medial to the internal carotid canal in anthropoids should be called "promontorium" obscures the real question being asked by these workers: how do primates differ in the position of the internal carotid canal on the cochlear housing? Intertaxonomic variability in the different positions of the internal carotid canal was discussed above. When these aspects of carotid canal position are taken into account there is little additional evidence for intertaxonomic differences in the position of the internal carotid pathway on the promontorium. Only tarsiers differ from all other primates in having an anteriorly located transpromontorial portion of the internal carotid artery that only contacts the cochlea at its cupula.

Stapedial Artery Vestigial in the Adult and Runs Posteriorly to Reach the Stapes. The posterior direction of the stapedial artery after its divergence from the promontory artery in adult tarsiers is correlated with the anterior position of the internal carotid artery (Characters 8 and 9) and need not be discussed further.

Character Analysis

Character 4: Presence or absence of perbullar pathway. Perbullar pathway absent (0); perbullar pathway formed exclusively by the petrosal bone (1).

Character 5: Anteroposterior location of posterior carotid foramen in bulla. Posterior to line joining midpoints of tympanic bones (0); anterior to this line (1).

Character 6: Mediolateral position of posterior carotid foramen in bulla (or-

*dered).** Posterior carotid foramen medially placed (0); in the midline of the bulla (1); laterally positioned (2).

Character 7: Ventrodorsal position of the carotid foramen in the bulla. Dorsally positioned, adjacent to basioccipital or mastoid bone (0); ventrally positioned (1).

Character 8: Position of posterior carotid foramen relative to fenestra cochleae (ordered). Posterior to the fenestra cochleae (0); ventral to it (1); anterior to it (2).

Character 9: Position of the internal carotid canal relative to the fenestra cochleae. Internal carotid pathway runs across the ventral lip of the fenestra cochleae, shielding it from ventral view when a canal is present (0); internal carotid canal does not shield the fenestra cochleae from ventral view (1).

Character 10: The position of the portion of the internal carotid/promontory artery (or its accompanying nerves) lying on the promontorium anterior to the fenestra cochleae. The internal carotid artery and its accompanying nerves lies on the ventrolateral surface of the promontorium (0); or anterior to it, contacting only the cupula of the cochlea (1).

Relative Size of the Intrabullar Branches of the Internal Carotid Artery

Szalay (1975) claimed that omomyids, anthropoids, and tarsiers can be distinguished from adapids by the relative size of the promontory and stapedial canals. However, although *Necrolemur* and *Rooneyia* do resemble tarsiers and anthropoids in having relatively larger promontorial canals than stapedial, this is also true of one specimen of *Notharctus* (AMNH 11466) (Gingerich, 1973), some specimens of *Adapis* but not others (Gingerich and Martin, 1981), and both skulls of *Smilodectes gracilis* (MacPhee and Cartmill, 1986). One specimen of *Mahgarita* (TMM 41578-9) lacks a stapedial canal (Rasmussen, 1990), and one has a much reduced canal (TMM 41578-20). Moreover, *Shoshonius* and *Tetonius,* like *Tupaia,* have large stapedial and promontory canals (Beard and MacPhee, Chapter 3, this volume). Some adult *Tarsius* lack a patent stapedial artery (see Diamond, 1992, p. 441, for instances of presence of a stapedial artery stem in adult *Tarsius*), although the stapedial canal is still present, whereas anthropoids have lost all trace of a stapedial canal.

Character Analysis

Character 11: Size of stapedial and promontory canals. Both the stapedial and promontory canals are large (0); the stapedial is slightly smaller than the promontory (1); the stapedial is highly reduced or absent altogether (2); the

*Designates character states ordered for PAUP analysis.

stapedial is larger than the promontory (3); both the promontory and stapedial canals are lacking (4).

Character 12: Morphology of promontory canal. The promontory canal can be absent (?), or if present, can form an open trough (0); or a complete canal (1).

Character 13: Presence or absence of canal for internal carotid artery or nerves. A canal running from the posterior carotid foramen to the promontorium can be present (1) or absent (e.g., lorisids, galagids, and cheirogaleids) (0).

Tympanic Position and Morphology

In tarsiers (MacPhee and Cartmill, 1986, Fig. 4e), anthropoids, and lorisiforms (MacPhee, 1981), the bony tympanic bone is fused to the external surface of the lateral bulla wall and is therefore extrabullar, or "phaneric" (Fig. 2). In tarsiers and extant catarrhines the tympanic has a laterally directed tubular extension not found in platyrrhines. In those platyrrhines with a transversely expanded tympanic bone, the expansion occurs medial to the tympanic membrane, forming a broad, ribbon-like tympanic annulus. *Aegyptopithecus, Apidium* (DU 2468), and *Catopithecus* (Simons, 1989) resemble platyrrhines in having an extrabullar tympanic bone that is not laterally expanded, and *Aegyptopithecus* resembles platyrrhines also in having a medially expanded tympanic annulus (see Fig. 8B).

In lorises and galagos the tympanic is also extrabullar, fused to the lateral surface of the auditory bulla. In lemuroids and cheirogaleids the tympanic bone is intrabullar or "aphaneric" (MacPhee, 1981), lying medial to the lateral bulla wall, within the tympanic cavity. The tympanic is ribbon-shaped and separated from the lateral bulla wall by a gap, the recessus dehiscence, which is bridged by a connective tissue sheet, the annular membrane (see Fig. 9A). A semicircular bony rampart on the internal surface of the lateral bulla wall, the linea semicircularis (Conroy, 1980), is the only bony structure between the lateral bulla wall and the tympanic bone (see Figs. 3 and 9A).

As in lemuroids and cheirogaleids, known adapids [*Smilodectes gracilis* (MacPhee and Cartmill, 1986, Fig. 20), *Adapis parisiensis, Leptadapis magnus* (Stehlin, 1912), and *Pronycticebus gaudryi* (Saban, 1963, Fig. 50)] possess a ribbon-shaped intrabullar tympanic bone and a patent recessus dehiscence. However, the recessus is greatly reduced by the medial extension of a horizontally oriented flange of the petrosal bone, an incomplete bony annular bridge. In Eocene omomyids for which the anatomy is known [*Tetonius* (Beard and MacPhee, Chapter 3, this volume), *Necrolemur, Rooneyia,* and *Shoshonius*] and in *Plesiadapis,* the portion of the tympanic bone that provides attachment for the tympanic membrane is also intrabullar, but a complete bony annular bridge, reinforced by bony struts, extends from the tympanic to the lateral bulla wall, obliterating the recessus dehiscence (see Fig. 10).

MacPhee and Cartmill (1986) suggest that the absence of a bony annular bridge between the tympanic bone and the lateral bulla wall is a synapomorphy of a tarsier–anthropoid clade. This character is better described as a phaneric or extrabullar tympanic bone, differentiating the tarsier–anthropoid condition, where the annular bridge is absent because the tympanic is fused to the lateral surface of the bulla wall, from the lemuroid condition, in which the tympanic lies medial to the bulla wall but a bony annular bridge is absent and the recessus dehiscence is bridged with connective tissue.

Recently, Rasmussen (1990) has suggested that the protoadapine *Mahgarita* may have resembled tarsiers and anthropoids in possessing a phaneric tympanic bone. This hypothesis is tentatively advanced on the basis of the presence on both sides of the type skull of *Mahgarita* (TMM 41578-9) of a horizontally oriented sheet of bone (labeled shelf A in Fig. 9B) lying ventral to the external auditory meatus, which Rasmussen (1990) suggests "lies lateral to the bulla" (p. 449). Rasmussen evaluates four possible interpretations for this shelf of bone, finally concluding: "The possible occurrence of an ectotympanic band, fused to the petrosal bone lateral to the bulla, suggests that the ectotympanic could not have been intrabullar with a free inferior arc. Such a lemuroid pattern is also highly unlikely given the presence of a lateral transverse septum. There simply was not enough space to allow passage of a large, free ring-like structure, as seen in lemuroids . . ." (Rasmussen, 1990, pp. 449–450).

However, the shelf of bone (shelf A) referred to by Rasmussen does not lie lateral to the bulla as he suggests. It is continuous anteriorly and posteriorly with the lateral bulla wall, and its lateral extremity sports a low, broken rampart of bone that is certainly a remnant of the ventral continuation of the lateral bulla wall. In these respects, shelf A of *Mahgarita* closely resembles a similar shelf of bone present in lemuroids, *Rooneyia,* and *Adapis* that runs laterally from the posterior intrabullar septum to the lateral bulla wall and lies ventral to the external acoustic meatus. On the left side of the type skull of *Mahgarita,* shelf A can be seen to be continuous anteriorly with another, narrow, horizontally oriented shelf of bone (shelf B, Fig. 9A) extending medially from the lateral bulla wall. As discussed above, shelf B shows no signs of breakage on its medial edge and is continuous anteriorly with a third bony shelf (shelf C) that provides a bony floor for the auditory tube. In these respects the ear region of *Mahgarita* closely resembles that of *Adapis parisiensis* (Saban, 1963, Fig. 51), *Rooneyia viejaensis,* and extant and subfossil lemuroids (see Fig. 9B). None of these shelves are present in the fossil petrosals from the Fayum, including that of *Aegyptopithecus zeuxis* (DPC 6642).

Rasmussen argues that there was not enough room for a free, intrabullar, ring-like tympanic in *Mahgarita* because of the presence of a laterally directed transverse septum. As discussed above, this septum is unlikely to have been more extensive than it is in TMM 41578-9, so it could not have extended posteriorly far enough to preclude the presence of an intrabullar tympanic bone. Moreover, the arrangement of the promontorium and the annular bridge in *Mahgarita* is very similar to that of *Adapis parisiensis* (Stehlin, 1912; Gingerich and Martin, 1981) and *Leptadapis magnus* (Stehlin, 1912), both of which possessed free, ribbon-like intrabullar tympanics. As in extant lemuroids the tympanic bones were oriented somewhat obliquely, almost parallel to the pathway of the internal carotid artery (Stehlin, 1912, Fig. 161 and 179), fitting easily between the cochlear housing and shelf B. In the extant lemuroids listed above, the anterior limb of the tympanic bone is in contact with the anterior extremity of the linea semicircularis immediately posterior to the anterior canaliculus for chorda tympani, and the posterior limb is in contact with the junction between the linea semicircularis and shelf A (which runs from the posterior septum to the posterior extremity of the linea semicircularis) (Fig. 9A). The tympanic is similarly positioned in *Adapis,* although there is a partial annular bridge in exactly the same position as the linea semicircularis of lemuroids. There was certainly room for a tympanic situated in this way in *Mahgarita,* with the anterior limb in contact with the anterior extremity of shelf B (linea semicircularis) adjacent to the anterior canaliculus for chorda tympani, and the posterior limb in contact with the junction of shelf A and the posterior septum. This reconstruction of *Mahgarita* (shown in Fig. 13B) is supported by the presence in *Mahgarita* of a crest on the "superior border of the internal surface of the meatus . . ." (Rasmussen, 1990, p. 449) in a position identical to that of the fused crura of lemuroids.

However, although the tympanic of *Mahgarita* was almost certainly intrabullar, there is no reason to believe that it was completely fused to shelf B, as is the case in omomyids. In all primates (and *Plesiadapis*) in which the tympanic is fused to the annular bridge, it is discernable as a thickening of the bridge's medial rim (e.g., *Rooneyia, Necrolemur*), a thickening that is not present in *Mahgarita.*

In sum, these observations suggest that *Mahgarita* possessed an intrabullar tympanic bone, as seen in extant lemuroids and as originally suggested by Wilson and Szalay (1976), in conjunction with a very incomplete annular bridge or a very large linea semicircularis. Indeed, on the basis of positional criteria the linea semicircularis can be homologized with the partial annular bridge, differing from it only in size. There is no evidence that the position of the tympanic of *Mahgarita* resembled that of any extant or fossil anthropoid. This tympanic was probably ribbon-like, as in known adapids and lemuroids.

Character Analysis

Three characters are needed to describe the shape and position of the tympanic bone in primates.

Character 14: Position of ventral edge of the tympanic bone. Tympanic bone intrabullar, or aphaneric (0); extrabullar or phaneric (1).

Character 15: The shape of the tympanic bone. Tympanic is a ribbon-like or only slightly expanded annulus (0); tympanic laterally expanded into a collar or tube (1); or, because of fusion with surrounding bones, of unknown shape (?).

Character 16: Morphology of annular bridge. Ossifications between the tympanic and the bulla are not found in those taxa with an extrabullar tympanic, so this character is not analyzable for them (?). Linea semicircularis or partial annular bridge formed on an entotympanic bulla (0); linea semicircularis formed on a petrosal bulla (1); a complete annular bridge (2).

Evidence from the Basicranium: Nonotic Characters

Encroachment of the Auditory Bulla on the Pterygoid Fossa

Rosenberger (1985) notes that in *Tarsius, Necrolemur,* and *Tetonius,* the anteriormost portion of the auditory bulla extends rostrally between the lateral pterygoid plates to encroach on the pterygoid fossa. This is also the case for *Shoshonius* and *Ignacius graybullianus,* but such encroachment is not seen in other primates. This character may be related to the degree of flexion occurring in the basicranium of *Tarsius,* as Rosenberger suggested; however, *Ignacius* exhibits this character and has a flat basicranium. I think the encroachment seen in *Tarsius* is unlikely to have been present in the last common ancestor of *Tarsius, Shoshonius,* and *Necrolemur* because in *Tarsius* the anterior portion of the bulla actually houses an anterior accessory cavity (Fig. 7), whereas such a cavity is not present in these omomyids.

Character Analysis

Character 17: Encroachment of the auditory bulla on the pterygoid fossa. Present and formed by anterior accessory cavity (1); present and formed by the tympanic cavity (2); absent (0).

Pterygoid–Bulla Contact

Extensive contact between the lateral pterygoid plate and the anterolateral surface of the auditory bulla in *Tarsius* and many Eocene prosimians has been noted for many years (e.g., Wortman, 1903–04), but Rosenberger (1985) suggests that the laminar overlap seen in *Rooneyia, Microchoerus, Tarsius,* and *Necrolemur* is significantly different from the abutting contact seen in *Adapis* and *Notharctus.* The difference between "laminar" and "abutting" is that in the latter the bulla exhibits a strong crest articulating with the plate via a clear suture, whereas in the former the plate lies flat against the bulla wall.

Shoshonius also possesses extensive laminar overlap of the lateral pterygoid plate and the anterolateral bulla wall (Beard *et al.,* 1991; Beard and

MacPhee, Chapter 3, this volume),* the extent of which equals that of *Tarsius, Necrolemur, Microchoerus,* and *Rooneyia.* No overlap, laminar or abutting, is evident in *Aegyptopithecus* or any extant platyrrhines, except occasionally in *Aotus. Mahgarita* possesses an extensive laminar contact between the lateral pterygoid plate and the lateral bulla wall (Fig. 9B). Two characters are recognized here: one describing the nature of the contact, and one describing the extent.

Character Analysis

Character 18: Nature of contact between the lateral pterygoid plate and the bulla wall. Contact between the auditory bulla and the lateral pterygoid plate may be absent (0), laminar (1), or abutting (2).

Character 19: Extent of contact between the lateral pterygoid plate and the bulla wall. When present the contact may be slight (0), or very extensive, approaching the external auditory meatus posteriorly (1).

Basioccipital Flange Overlaps Medial Bulla Wall

Shoshonius and *Tarsius* are characterized by an extensive overlap of a flange of the basioccipital bone onto the medial wall of the auditory bulla (Beard and MacPhee, Chapter 3, this volume). Extensive flanges are visible in *Microchoerus* (QU 10879) and two specimens of *Necrolemur antiquus* in the British Museum [BM(NH)P M3747 and M4490] (although M4490 has a less extensive flange on the right side of the bulla than the left, and the flange on the left bulla is less extensive than the flanges in M3747). MCZ 8879 exhibits a flange in the same position as in *Tarsius* and *Shoshonius.* Thus, the size of this flange in *Necrolemur* appears to be variable.

Smilodectes exhibits extensive and very robust flanges of basioccipital overlapping the medial bulla wall, particularly well developed in USNM 17994, 17995, and 17996. Small and robust flanges are also visible in some *Notharctus* skulls (USNM 21864), although they are much less extensive than those in *Smilodectes. Ignacius* possesses a small basioccipital flange on the medial bulla wall, less extensive than those of *Shoshonius* and *Tarsius.*

Dissections of *Tarsius* specimens indicate that although the portion of this flange that lies immediately adjacent to the braincase sports a low protuberance giving origin to tendons of m. rectus capitis anterior, this is not true of the majority of this flange. Therefore, muscle attachment does not account for its large size in *Tarsius,* and Rosenberger (1985) may be correct in hypothesizing that this flange might be needed to stabilize the petrosal of *Tarsius* because it is incompletely attached to the surrounding basicranium. This hy-

*In the caption to Fig. 4 of Beard *et al.* (1991), they state that the "lateral pterygoid wings overlap the anteromedial bullar wall." This is a *lapsus calami* for anterolateral wall.

pothesis does not account for the presence of smaller flanges in *Smilodectes* and *Notharctus*, however.

Character Analysis

Character 20: Flange of basioccipital overlapping medial bulla wall. Extensive (1); minimal or absent (0).

Suprameatal Foramen

Beard *et al.* (1991) illustrate a small foramen situated in the lateral surface of the base of the postglenoid process, which they suggest is homologous with the suprameatal foramen found in *Tarsius* and illustrated by Saban (1963) in *Necrolemur.* The identity of the foramen in *Shoshonius* with the suprameatal foramen in *Tarsius* is dubious. The suprameatal foramen in *Tarsius* is large, two-thirds the size of the external auditory meatus in length, and is situated immediately over the external auditory meatus. The foramen in *Shoshonius* is small and is located anterior to the external auditory meatus in the lateral surface of the postglenoid process. The foramen illustrated by Saban (1963) in the skull of *Necrolemur* approximates the anatomy of the suprameatal foramen in *Tarsius* in its size and position more closely than that of *Shoshonius.* This foramen is not preserved in MCZ 8879, probably because of crushing of the specimen. Kay *et al.* (1992) reported a suprameatal foramen as being present in *Ignacius, Tupaia,* and *Ptilocercus;* however, they did not homologize it with that of *Tarsius.* In *Ignacius* and tree shrews, the "suprameatal foramen" resembles that of *Shoshonius* in its small size and its position in the posterior root of the zygomatic arch.

Character Analysis

Character 21: Suprameatal foramen. Absent (0); present, small and in the posterior root of the zygomatic arch (1); present, large, and above the external auditory meatus (2).

Parotic Fissure

Beard and MacPhee (Chapter 3, this volume) identify a significant resemblance between *Tarsius* and *Shoshonius* in the presence of a groove lodged between the posterolateral corner of the bulla and the mastoid process. In *Tarsius,* this groove contains the stapedius fossa, and a patent parotic fissure opening into it provides passage for the stapedius tendon. Whether the stapedius muscle shares a "primitive stylomastoid foramen" with the facial nerve is not clear from adult skulls, although it does so in fetal tarsiers. Certainly, the groove in the posterolateral corner of the bulla of adult tarsiers also contains the stylomastoid foramen. The pyramidal eminence is visible in the roof of the tympanic cavity, protruding from the ventromedial surface of the bony facial canal.

Beard and MacPhee note the presence in *Shoshonius* of a groove at the posterolateral corner of the bulla closely resembling the groove for the stapedius fossa and stylomastoid foramen in *Tarsius*. They suggest, probably correctly, that there was a patent parotic fissure in *Shoshonius* as well, a notable resemblance between tarsiers and *Shoshonius*.

The distribution of this feature among omomyids and outgroup taxa examined here suggests that a patent parotic fissure is primitive for omomyids and probably for primates as well (as suggested by Beard and MacPhee, Chapter 3, this volume). A groove between the posterolateral corner of the bulla and the mastoid process, very similar to that seen in *Shoshonius* and *Tarsius*, is present in *Tetonius* and *Necrolemur*. In *Necrolemur* (MCZ 8879), the right tympanic cavity also sports a pyramidal eminence on the ventromedial surface of the facial canal in a position identical to that of *Tarsius*. *Tupaia* exhibits a groove between the bulla and the mastoid similar to that of *Tarsius* as well as possessing a patent parotic fissure (MacPhee, 1981). A "dual stylomastoid foramen" in *Ignacius* has previously been hypothesized to give passage to an arterial connection between the posterior auricular and the stapedial arteries (Kay *et al.*, 1992); however, the anatomy of the most inferior of these foramina suggests that it may represent an extrabullar stapedius fossa. On both sides of the skull of *Ignacius* (USNM-VP 421608), a larger foramen lying ventral to the stylomastoid foramen opens into a short anteromedially directed canal. The opening of this canal in the tympanic cavity is not preserved, and the canal is filled with matrix at its anterior end. However, any opening in this canal would be situated immediately medial to the posterior crus of the ectotympanic and posterior to the fenestra vestibuli. A small rugose ridge on the bulla wall immediately outside this foramen defines a small area that may have given origin to the stapedius. Certainly the extrabullar groove closely resembles those of *Tarsius*, *Necrolemur*, and *Tupaia*, into which the stylomastoid foramen and parotic fissure open.

The presence of a dual stylomastoid foramen in *Plesiadapis* (Russell, 1964), well illustrated by Saban (1963), and the probable extrabullar position of the stapedius fossa in *Phenacolemur* (MacPhee, 1981) imply that a patent parotic fissure was present in plesiadapiforms as well. Altogether, these data suggest that a patent parotic fissure and extrabullar stapedius fossa are primitive for primates. No dual stylomastoid foramen or parotic fissure has been reported in adapids, and none is present in extant strepsirhines or fossil or extant anthropoids.

The polarity of this feature among mammals is unclear. A crista parotica posterior, defining the posterior boundary of the stapedius fossa and confining it to the tympanic cavity proper, is present in the early Cretaceous *Vin-*

celestes and has been claimed to be a synapomorphy linking this taxon to marsupials and eutherians (Wible, 1990). However, in adult lipotyphlans, elephant shrews, tree shrews (MacPhee, 1981, p. 13, Fig. 2), chiropterans, carnivores, and artiodactyls (Wible, 1984), a lateral section of the caudal tympanic process of the petrosal lies medial to the stapedius fossa, excluding the stapedius fossa from the tympanic cavity. A lateral moiety of the caudal tympanic process of the petrosal lying medial to the stapedius fossa is also present in some of the eutherian petrosals from the Bug Creek Anthills (Wible, 1990, p. 197, Fig. 7C); the late Createceous *Kennalestes* and *Asioryctes* both possess a "strut" running from the tympanic process to the promontorium, excluding the stapedius fossa from the tympanic cavity (Kielan-Jaworowska, 1981); and the Liassic mammal *Morganucodon* may have had a stapedius fossa lying outside the tympanic cavity, as it exhibits no caudal tympanic processes at all (Kermack *et al.*, 1981). These data suggest that the presence of a parotic fissure and an extratympanic stapedius fossa is probably a primitive feature among eutherians.

Character Analysis

Character 22: Patent parotic fissure. Present (0); absent (1).

Evidence from the Orbital Region

Orbit Size

Beard *et al.* (1991) demonstrated that *Shoshonius* and *Tarsius* share massively hypertrophied orbits compared to those of other primates. This character was scored by examining plots of log orbit diameter against log skull length provided by Martin (1990), Kay and Cartmill (1977), and Beard *et al.* (1991). *Shoshonius* and *Tarsius* fall well above the nocturnal primate regression line and are scored as having extremely large orbits (Beard *et al.*, 1991). *Tetonius, Necrolemur, Microcebus,* and *Galago senegalensis* have large orbits, their values falling on or above the nocturnal primate regression line (Kay and Cartmill, 1977). The remaining taxa examined are scored as having small orbits. Measures of orbit diameter in *Ignacius* (Kay *et al.*, 1992) and *Mahgarita* (unpublished observation) were plotted on Martin's graph and fell well below the diurnal primate line.

Character Analysis

Character 23: Size of orbits (ordered). Extreme enlargements of the orbits (2); large orbits (1); small orbits (0).

Postorbital Septum

Tarsiers and anthropoids are the only vertebrates possessing a postorbital septum, a thin sheet of bone stretching from the postorbital bar to the braincase and separating the orbital contents from the temporal fossa. This septum is formed by expansions of the frontal, zygomatic, and alisphenoid bones. Indeed, *Tarsius* and anthropoids are the only mammals possessing a contact between the zygomatic and alisphenoid bones.

Simons and Rasmussen (1989) suggest that because the zygomatic bone forms a greater proportion of the postorbital septum in anthropoids than in tarsiers, "[t]he key similarity between tarsiers and anthropodieans—direct contact between the alisphenoid and zygomatic bones—has apparently evolved convergently" (p. 9). This argument has been made previously (Simons and Russell, 1960; Simons, 1972) and countered by Cartmill (1980) in the following way:

> [D]ifferences in the relative contribution of various dermal bones to the formation of the braincase's side walls in different anthropoid groups . . . surely do not imply that the last common ancestor of the anthropoids had a partly unossified braincase like that of reptiles, or that different anthropoid groups must be traced back to different premammalian ancestors. It is therefore equally unwarranted to infer from the fact that the frontal, alisphenoid, and zygomatic contributions to the postorbital septum differ in the relative sizes among haplorhines that the last common ancestor among Haplorhini lacked a postorbital septum" (Cartmill, 1980, p. 257).

I concur with this line of reasoning. The observed intertaxonomic differences in the relative proportions of the frontal, alisphenoid, and zygomatic bones that form the postorbital septum may result from parallel evolution of the postorbital septum, but they may also result from changing proportions of the bones after the septum was acquired.

Beard and MacPhee (Chapter 3, this volume) reject the possibility that the postorbital septum of tarsiers is homologous with that of anthropoids *a priori* and do not score the two taxa with the same character state. This is not a position I agree with. I believe it is better to score morphological similarities (i.e., zygomatic–alisphenoid contact) *and* differences (i.e., the differing proportions of the zygomatic in tarsiers and anthropoids) than to reject hypotheses of homology *a priori* on the basis of differences alone. In any event, the discovery of more fossils is the only way that this particular question can be definitively resolved.

Character Analysis

Character 24: Postorbital closure (ordered). None (0); postorbital bar present (1); postorbital septum present (2).

Character 25: Composition of the postorbital septum. Zygomatic forms most of the septum in anthropoids (0), whereas the frontal forms most of the septum in *Tarsius* (1).

Zygomatic–Lacrimal Contact on Inferior Orbital Margin

Cartmill (1978) noted that anthropoids and *Tarsius* share possession of a maxillary contribution to the inferior orbital margin, precluding a zygomatic–lacrimal contact. Some lorises and galagos also lack a zygomatic–lacrimal contact on the inferior orbital rim, but it is usually present in lemuroids. *Shoshonius, Tetonius,* and *Necrolemur* also exhibit a maxillary contribution to the inferior orbital margin preventing any lacrimal–zygomatic contact. On the *Rooneyia* skull (TMM 40688-7) there is a very tiny zygomatic–lacrimal contact on one side of the skull but not on the other. *Mahgarita* resembles tarsiers and anthropoids in lacking a contact between zygomatic and lacrimal bones on the inferior orbital margin (Rasmussen, 1990). *Contra* Wortman (1903–04), zygomatic–lacrimal contact is present in *Notharctus* and *Adapis.*

Character Analysis

Character 26: Zygomatic–lacrimal contact: Present (0), absent (1).

Interorbital Septum

The literature on the question of anthropoid relationships contains much discussion on a structure referred to as the "interorbital septum" (e.g., Cartmill, 1972; Simons and Rasmussen, 1989). Cartmill (1970) noted that in adult tarsiers and small-bodied anthropoids the bony orbits are so closely approximated to each other at a point ventral to the olfactory tract that an extensive unpneumatized interorbital septum is formed, occluding the posterior portion of the nasal fossa. In tarsiers and small-bodied anthropoids there is only a single thin sheet of bone separating the two orbits at this point. The skulls of *Necrolemur, Tetonius, Microchoerus,* and *Pseudoloris* also appear to exhibit an interorbital septum (Cartmill and Kay, 1978), although it is only possible to confirm that this septum is formed by a single lamina of bone in two specimens of *Necrolemur* [MCZ 8879, BM(NH)P M4490] (see Fig. 14).

This interorbital septum has been cited as evidence supporting the notion of a clade consisting of omomyids, tarsiers, and anthropoids (Cartmill and Kay, 1978). However, Simons and Rasmussen (1989) note that such an interorbital septum was not present in *Aegyptopithecus.* Rather, *Aegyptopithecus* exhibits two laminae of bone separated by a narrow space, not a unilaminar interorbital septum as in tarsiers and small-bodied anthropoids. Moreover,

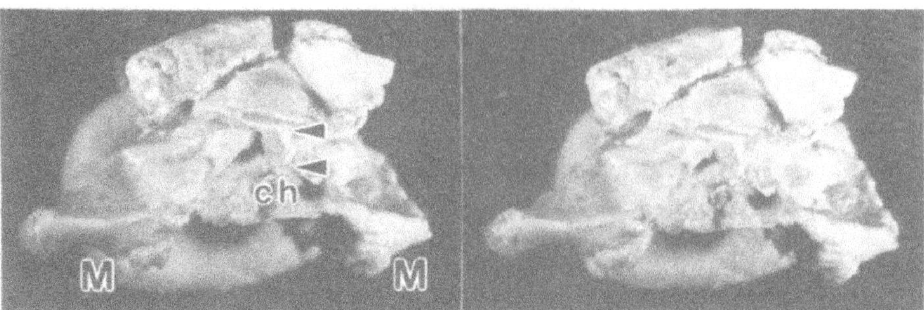

Fig. 14. Posterior view of face of *Necrolemur antiquus* (MCZ 8879). The face is broken off the specimen through the interorbital region, exposing the unilaminar interorbital septum (top and bottom of septum marked by tips of arrows) to posterior view. (Matrix remains in place on either side of the interorbital septum). The frontal bone is collapsed into the anterior cranial fossa above. (The maxillary toothrow on each side is marked with an M.) The peaked shape of the choanal opening (ch) (filled with matrix) is apparent.

Simons and Rasmussen (1989, p. 5) illustrate a specimen of *Galago senegalensis moholi* in which a "single thin lamina of presphenoid bone separates the orbits." They suggest on the basis of the absence of an interorbital septum in *Aegyptopithecus* and the presence of one in *Galago senegalensis* that an interorbital septum cannot be a synapomorphy of a tarsier–anthropoid clade.

The term "interorbital septum" is commonly used in the embryological literature to describe "a vertically arranged, higher than wide, unpaired cartilage plate, which stands as a free connecting stem between the nasal capsule and pars chiasmatica of the orbitotemporal region" (translation of Matthes, 1922, p. 159, quoted in Haines, 1950, p. 596). As Haines points out, the emphasis on lateral compression is important in defining this interorbital septum—if the cartilage is not a thin, "vertically arranged" cartilaginous plate but a thicker rod of cartilage, it is not an interorbital septum (Haines, 1950).

In adult primates the cartilage lying between the orbits and immediately in front of the optic foramina is replaced by bone. The morphology of this bony partition between the orbits can take several forms: a single lamina of bone, as in *Tarsius, Aotus, Saguinus,* and *Callimico;* two laminae of bone extensively fused together in the midline, as in *Pithecia, Callimico, Callithrix argentata,* and some *Leontopithecus;* two laminae of bone that are convex medially and only in contact for a very restricted area, as in many cercopithecines, *Cacajao, Microcebus, Galago senegalensis,* some *Callicebus,* and *Cebus albifrons;* or two laminae of bone completely separated by varying amounts of space, as in *Aegyptopithecus,* atelines, hominoids, many *Cebus,* and some *Leontopithecus.** The first two conditions are often referred to as being "interorbital septa," whereas the second two

*Data for this character were derived from examination of hemisected skulls at the FMNH, many of which are figured in Hershkovitz (1977).

are seldom referred to in this way; Simons and Rasmussen (1989) were the first authors to recognize the limited contact between the medial orbital walls in some cheirogaleids as an interorbital septum.

The varying degrees of separation of the medial orbital walls are probably determined by the relative size and position of the orbits (Haines, 1950; Simons and Rasmussen, 1989). Small animals have relatively larger eyes and orbits and hence narrower interorbital regions, whereas larger animals tend to have relatively smaller, more widely separated orbits and therefore more space between the medial orbital walls. However, as noted by Cave (1967), the interorbital regions of large primates show variability in the nature of the cavity occupying this interorbital space. In large-bodied strepsirhines the interorbital region is occupied by a posterior extension of the nasal fossa into the sphenoid bone, floored by the lamina transversa (Cave, 1967). In large-bodied anthropoids, however, this interorbital space is not occupied by an extension of the nasal fossa but by a sinus of variable origin. In *Alouatta* and *Pongo* the interorbital, intrasphenoidal sinus is an extension of the maxillary sinus (Cave and Haines, 1940; Haines, 1950; Cave, 1967); in African apes and *Homo* the interorbital sinus is formed by an expanded ethmoidal sinus. Observations on one specimen each of *Ateles paniscus, Callicebus torquatus, Lagothrix lagotricha, Cebus albifrons,* and *Cebus apella* suggest that the interorbital, intrasphenoidal sinus is continuous with the frontal sinus in these taxa. In cercopithecines and pitheciines, the intrasphenoidal sinus is fully trabeculated, making it difficult to determine whether it is continuous with other sinuses or not.

This variability in the nature of the cavities occupying the interorbital region in anthropoids suggests that the last common ancestor of all anthropoids had no interorbital space at all and that this space was occupied by an interorbital septum, either bilaminar or unilaminar, as originally suggested by Cave (1967). Although the hypothesis that primitive anthropoids possessed an interorbital septum is not supported by the interorbital morphology of *Aegyptopithecus* (Simons and Rasmussen, 1989), *Aegyptopithecus* is neither the oldest nor the most primitive anthropoid known (Simons, 1991, 1992), so it is unlikely to be morphotypic for that group.

In any event, because the width and morphology of the interorbital region are probably related to relative orbit size, and because relative orbit size has already been included in the data set, to score interorbital morphology as a separate character would introduce redundancy into the data set. Moreover, it would not capture the fact that two profoundly different morphologies are represented by anthropoids, *Tarsius,* omomyids, and cheirogaleids on the one hand and other toothcombed strepsirhines and *Adapis* on the other. In the former group and in *Notharctus,* the orbits are most closely approximated below the olfactory tract, and the smaller members of these groups possess interorbital septa of varying size, whereas in *Adapis* and most noncheirogaleid strepsirhines, the orbits are most closely approximated above the olfactory tract, and none of them possesses an interorbital septum.

Character Analysis

Character 27: Pronounced interorbital constriction. Absent (0); present below olfactory tract (1).

Lacrimal–Palatine Contact on Medial Orbital Wall

Among primates a contact between the palatine and the lacrimal bone on the medial wall of the orbit is common only in *Eulemur* and *Lemur catta,* although it is frequently present in *Hapalemur griseus* and is sometimes present in *Lepilemur, Phaner,* and *Cheirogaleus major* (Cartmill, 1978). This feature is also present in *Tupaia* (Le Gros Clark, 1959). Among fossil primates, *Ignacius, Plesiadapis* (Kay *et al.,* 1992), *Necrolemur, Pronycticebus* (Russell, 1964), and *Rooneyia* have a large frontal–maxillary contact excluding the lacrimal and palatine from contact, a condition they share with most cheirogaleids. This is also the case for most *Adapis,* although the Cambridge specimen of *Adapis parisiensis* exhibits a small os planum of the ethmoid. Anthropoids, *Tarsius,* and lorisoids possess a large os planum of the ethmoid bone that participates in excluding the palatine from a contact with the lacrimal. The medial orbital wall of *Shoshonius* and *Tetonius* is not well enough preserved to determine the arrangement of the bones in the medial orbital walls of these taxa.

Character Analysis

Character 28: Contact between lacrimal and palatine. Present (0); separated by a large frontomaxillary contact (and in some taxa, a small os planum of the ethmoid) (1); presence of a large os planum in the medial orbital wall (2).

Foramen Rotundum

The foramen rotundum is not present as a discrete opening primitively in mammals (Müller, 1934), nor is it present in *Ignacius, Plesiadapis,* or *Tupaia* (Kay *et al.,* 1992). *Necrolemur* (Simons and Russell, 1960), *Lemur catta,* lorisoids, and cheirogaleids also lack a foramen rotundum (Yoder, 1992). A complete foramen rotundum is present in *Indri, Eulemur, Rooneyia,* and *Pronycticebus* (Le Gros Clark, 1934) as well as *Tarsius* and all extant anthropoids. It is not possible to determine whether a distinct foramen rotundum is present in *Tetonius* and *Shoshonius.*

Character Analysis

Character 29: Foramen rotundum. Absent (0); present (1).

Lacrimal Foramen

The lacrimal foramen lies outside the orbital margin in *Plesiadapis, Tupaia, Necrolemur, Rooneyia, Adapis, Mahgarita* (Rasmussen, 1990), *Pronycticebus* (Russell, 1964), extant strepsirhines, and *Tarsius,* on the rim of the orbit in *Notharctus* and some anthropoids, and within the orbit in *Ignacius* and the remaining anthropoids.

Character Analysis

Character 30: Position of lacrimal foramen. Outside orbital margin (0); within the orbit or on the orbital rim (1).

Fusion of the Metopic Suture

In many adult prosimians the two frontal bones are not fused in the midline. In adult anthropoids, however, the metopic suture is fused and obliterated. This has been one of the traditional hallmarks of anthropoids (Simons, 1972), being found in all fossil and extant anthropoids for which frontals are known. However, this feature is also found in *Necrolemur, Tarsius, Tupaia,* and several strepsirhines (*Eulemur, Hapalemur*). An unfused suture is found in most adapids (except some *Adapis*), most strepsirhines (except some lemuroids), *Plesiadapis,* and *Ignacius.*

Character Analysis

Character 31: Metopic suture. Fused in adult (1); unfused in adult (0).

Orbital Margin Convergence

The orbital margins of prosmian primates are more convergent than those of most mammals (Cartmill, 1970), and the orbital margins of many anthropoids are more convergent than those of nonanthropoid primates (Ross, 1993). The orbital margins of all primates are certainly more convergent than those of *Tupaia* and probably *Plesiadapis.* This character was scored by comparing values obtained by measuring casts of fossil taxa with the range of values for orbital convergence in extant primates reported elsewhere (Ross, 1993) (ranging from 34° in *Euoticus elegantulus* to 85° in *Papio hamadryas, Macaca hecki,* and *Rhinopithecus roxellanae*). Fossil taxa with values for orbital convergence falling within the primate range were scored as having primate-like orbital convergence: this included *Ignacius* (37°), *Aegyptopithecus* (71°), *Adapis* (57° and 60°), *Microchoerus* (47°), and *Rooneyia* (63°). The specimens of *Necrolemur* examined by me and the known specimens of *Shoshonius* and *Tet-*

onius are too distorted to be measured, but comparison of preserved portions suggest that these animals had *Microchoerus*-like values for orbital convergence. *Plesiadapis* was not measured because of crushing of the skull.

Character Analysis

Character 32: Orbital convergence. Orbital margins less convergent than primates (0); exhibit primate-like values for convergence (1).

Evidence from the Pterygoid Region

Nasal Spine of Palatine

Rasmussen (1990) suggests that both *Aegyptopithecus* and *Mahgarita* share the possession of a pronounced nasal spine on the posterior palatine torus. Comparisons of the morphology of this region in *Aegyptopithecus* (Fig. 8A) with TMM 41578-20 (Rasmussen, 1990, Fig. 2) reveal that the short, rounded spine in *Mahgarita* bears little resemblance to the long and robust spine of *Aegyptopithecus*. Rather, in its size, the posterior nasal spine of *Mahgarita* more closely resembles that of *Plesiadapis, Tarsius,* lorisids, large galagos (such as *Otolemur crassicaudatus*), and many extant anthropoids. *Ignacius, Tupaia, Rooneyia,* small galagos, cheirogaleids, and most lemurids lack a distinct posterior nasal spine or have a very small one, hardly protruding posteriorly at all. *Adapis* sp. possesses a distinct posterior nasal spine of a similar size and shape to that of *Mahgarita*. The spine of *Leptadapis* most closely resembles that of *Aegyptopithecus* in size. This character is best divided into three ordered character states.

The posterior margin of the palate often exhibits a posterior palatine torus, a ridge of bone lying along the posterior border of the palate (Novacek, 1986; Yoder, 1992). Such a posterior palatine torus is evident in *Ignacius, Plesiadapis, Palaechthon nacimienti, Tupaia, Tarsius, Shoshonius, Necrolemur, Rooneyia, Microchoerus, Adapis, Leptadapis,* lemurids, and cheirogaleids. Extant anthropoids, *Avahi, Daubentonia, Lepilemur,* and lorisines lack such a torus (Yoder, 1992). *Aegyptopithecus* exhibits a derived morphology of the posterior border of the palate (Fig. 8A) in that it protrudes anteriorly on either side of the posterior nasal spine, forming a W-shaped posterior palatal border. Because of this the pyramidal processes are long and oriented anteroposteriorly (forming the lateral branches of the "W"). These pyramidal processes are also quite rugose, as is the case in adapids, including *Mahgarita*. On the basis of the hypothesis that this morphology is homologous with the posterior palatine torus, *Aegyptopithecus* and *Mahgarita* are scored as having a robust torus. This feature may be a primitive eutherian trait, as it is also found in *Asioryctes* and *Kennalestes* (Kielan-Jaworowska, 1981).

Character Analysis

Character 33: Posterior nasal spine (ordered). Reduced or absent (0); small but distinct (1); robust and long (2).

Character 34: Posterior palatine torus. Present (0); absent (1).

Choanal Shape and the Position of the Pyramidal Processes

Rosenberger (1985) noted that in *Tarsius* and *Necrolemur* the choanal openings are peaked, and the pyramidal processes of the palatine bone are situated well medial to the toothrow, in contrast with the condition found in adapids, *Rooneyia,* and all anthropoids in which the choanal openings are more arched and the pyramidal processes are laterally placed. *Shoshonius* resembles *Tarsius* and *Necrolemur* in having sharply peaked choanae and medially placed pyramidal processes. Although the choanal region is not well preserved in *Tetonius* and *Microchoerus,* the pyramidal processes are certainly closely approximated to the midline in these taxa. The presence of this character complex in an anaptomorphine (*Tetonius homunculus*), an omomyine (*Shoshonius cooperi*), and a microchoerine (*Necrolemur*) suggests that it is a retention from the ancestral omomyid stock.

The fact that *Rooneyia* lacks this complex of features and also has relatively small eyes compared with those of other omomyids suggests that orbit size may indirectly account for the unique choanal morphology seen in other omomyids and *Tarsius.* As noted by Spatz (1968), orbital enlargement may displace the palate (and choanae) ventrally and posteriorly relative to the braincase. Posterior displacement of the choanal region in omomyids brings it back into a pterygoid region that is already distinguished by constriction of the pterygoid fossae and medial displacement of the pyramidal processes as a result of orbital enlargement. This combination of factors reduces the area available for attachment of the medial pterygoid muscle while at the same time bringing its inferior head into close proximity with the nasopharynx.

The crowding of this area that results is obvious in *Tarsius.* The superior head of the medial pterygoid muscle takes origin from the medial orbital wall, and the inferior head arises from the most anterior depths of the mesopterygoid fossa. The medial pterygoid plate has the inferior head of the medial pterygoid muscle pressed to its lateral surface and the mucosa of the nasopharynx bound to its medial surface. The medial pterygoid plates thus provide a rigid frame for the portion of the nasopharynx immediately posterior to the choanae, protecting it from contractions of the medial pterygoid muscle.

In most larger primates, with relatively smaller orbits and less palatal kyphosis (e.g., callitrichids), or in smaller primates with relatively large orbits that have a low degree of frontation (i.e., they face dorsally) (e.g., *Loris*), the origin of the medial pterygoid muscle lies well posterior to the choanae, no

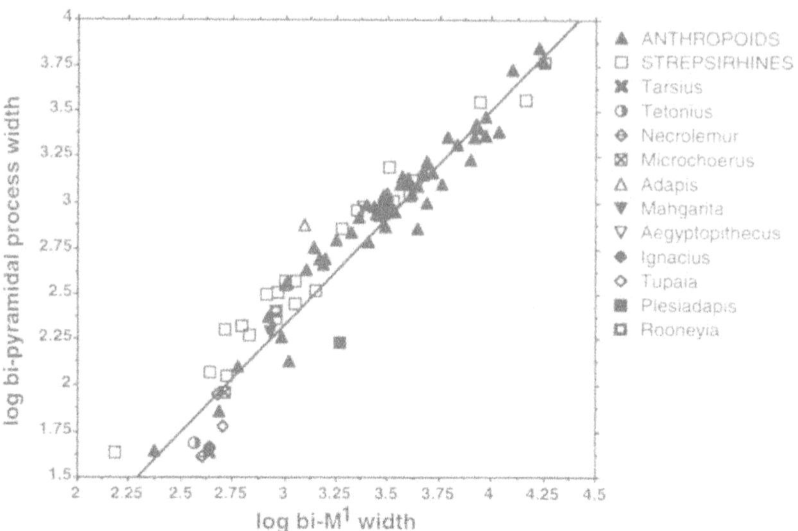

Fig. 15. Log-log plot of bipyramidal process width against bi-M[1] width in primates (data kindly provided Dr. Matt Ravosa). Reduced major axis added. Note that *Plesiadapis, Ignacius,* and *Tupaia* fall well below the reduced major axis with *Tetonius* and *Tarsius.*

protective role for the medial pterygoid plate is required, and this plate is often absent altogether.

Thus, the shape of the choanae in omomyids and *Tarsius* is more than likely correlated with the large orbits in these taxa (Character 23). However, the position of the medially placed pyramidal processes is not related to orbit size because medially placed pyramidal processes are also evident in the outgroup taxa utilized here, which all have relatively small orbits. Figure 15 is a log-log plot of bipyramidal process width against the width of the palate at M1, with the reduced major axis added.* There is some debate as to how to reduce continuously distributed quantitative characters to discrete character states (Chappill, 1989). The method chosen here was to visually compare the values for preorbital rostrum length and skull length against the RMA line for this data set. Those taxa with values falling above the line were scored as having long snouts; those falling below the line as having short snouts. The main problem with this method is that the regression line dividing the two character states is empirically determined, so that addition of taxa to the data set will change the slope of the line, possibly changing the character state assignments.

The values for most primates fall close to the reduced major axis. Notably, however, the values for *Necrolemur, Tetonius, Tarsius, Tupaia, Plesiadapis, Ignacius,* and some anthropoids and strepsirhines fall below the line. (*Shoshonius* was not measured, but it clearly resembles *Tarsius* in the morphology of this region and was scored accordingly.) Medial placement of the pyramidal processes is undoubtedly primitive for eutherians, as this feature is also found in zalambdalestid eutherians (Kielan-Jaworowska, 1984), *Asioryctes,* and *Kennalestes* (Kielan-Jaworowska, 1981).

Character Analysis

Character 35: Pyramidal processes. Medially placed (0); laterally placed (1).

Length of the Medial Pterygoid Plate

Rosenberger (1985) suggests that *Necrolemur* and *Tarsius* share a short medial pterygoid plate. According to Rosenberger, *Microchoerus* has a "moderately short" medial pterygoid plate. Moderately short plates are also found in *Aegyptopithecus,* but many extant anthropoids have medial pterygoid plates consisting merely of rudimentary ridges on the medial surface of the lateral pterygoid plate that are continuous with a small hamular process (e.g., *Saimiri, Lagothrix, Alouatta,* and *Callicebus*). Such highly reduced medial pterygoid plates are also found in *Ignacius* and *Mahgarita.* The medial pterygoid plates in these forms are therefore relatively smaller than those of *Tarsius, Necrolemur,* and *Shoshonius,* which remain well differentiated from the lateral plate, probably because they form a bony frame to protect the nasopharynx. *Notharctus* has well-developed medial pterygoid plates; *Adapis* has highly reduced medial pterygoid plates.

Character Analysis

Character 36: Length of medial pterygoid plate (ordered). Three character states are required to define this character: long medial pterygoid plate extending one-third to one-half of the distance to the anterior surface of the bulla (0); short but well defined from lateral pterygoid plate (i.e., distinct from lateral pterygoid plate for its entire dorsoventral extent, thereby forming a distinct, albeit reduced, pterygoid fossa between the plates) (1); medial pterygoid plate entirely absent or reduced to a low rugosity along medial surface of lateral pterygoid plate (2).

Evidence from the Masticatory Apparatus

Snout Size and Shape

Beard *et al.* (1991; Beard and MacPhee, Chapter 3, this volume) claim that a short snout or preorbital rostrum is a resemblance shared by *Shoshonius* and *Tarsius,* and Rasmussen (1990) suggests that tarsiers, anthropoids, and *Mahgarita* share possession of a short and deep facial exposure of the maxilla. To evaluate these claims, measures of preorbital rostrum length, maxillary depth, and skull length were taken from between one and six skulls of 15 species of extant haplorhines and 19 species of extant strepsirhines, one species of tree shrew, as well as several fossil primates and nonprimates. Preorbital rostrum length was measured as the shortest distance from the orbital margin to the anterior edge of the canine alveolus. (*Necrolemur* and *Rooneyia* were assumed to have three premolars, so the fourth tooth from the molar tooth row was assumed to be the canine in these taxa.) This measure excludes the premaxilla from estimates of snout length; however, this was the only measure that could be taken on most of the fossil specimens examined. Moreover, as Rasmussen (1990) has pointed out, most of the variation in snout length among primates is in the length of the maxilla. Maxillary depth was measured as the distance from the anterior edge of the canine alveolus to the anterior corner of the nasal–maxillary suture. When this suture was not visible, the superolateral corner of the nasal aperture was used. Skull length was measured as the distance from inion to the anterior edge of the canine alveolus.

A log-log plot of the mean values of preorbital rostrum length and skull length is shown in Fig. 16, with the reduced major axis shown. Those taxa with mean values falling below the line were scored as having short snouts; those falling above were scored has having long snouts. When this data set is

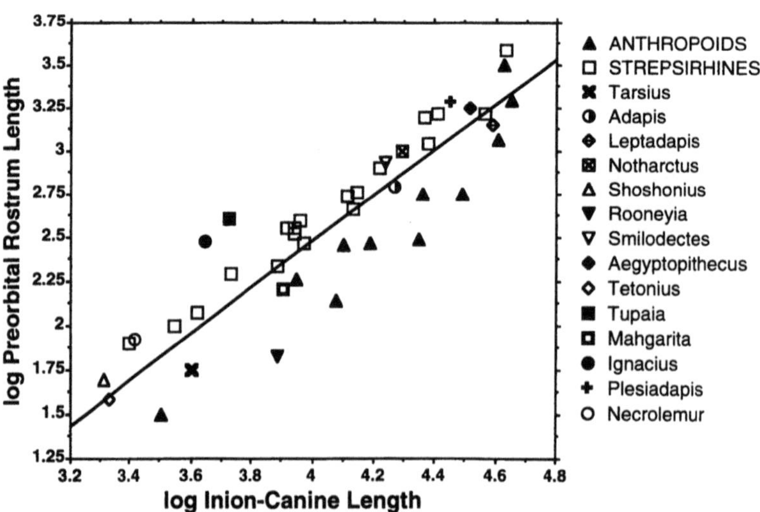

Fig. 16. Log-log plot of preorbital rostrum length against inion–canine length in primates. Reduced major axis added. Note that most anthropoids fall below the line, as do *Tarsius, Tetonius, Mahgarita, Leptadapis,* and *Adapis.*

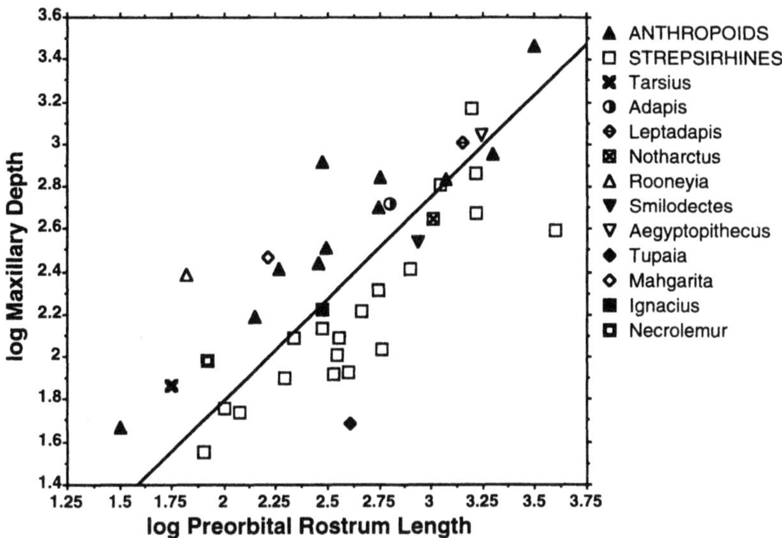

Fig. 17. Log-log plot of maxillary depth against preorbital rostrum length. Note that most anthropoids fall above the line, as do *Necrolemur, Rooneyia, Tarsius, Mahgarita, Adapis,* and *Leptadapis.*

used, *Aegyptopithecus, Necrolemur,* and *Shoshonius* all have long snouts, whereas *Tarsius* and *Tetonius* have short snouts.

It has been suggested that snout length is correlated with orbit size (Cartmill, 1970), a suggestion that is supported by the fact that *Necrolemur* appears to have a slightly longer snout and relatively smaller eyes than *Tarsius* and *Shoshonius.* However, it is clear that snout length and orbit size can vary independently in those taxa, such as *Loris,* that have low degrees of orbital frontation (*sensu* Cartmill, 1972) (i.e., the orbits face dorsally). Moreover, some taxa such as callitrichids and *Saimiri* have short snouts and relatively small orbits.

Character state assignments describing maxillary depth were assessed in a similar way to snout length. Figure 17 is a log-log plot of mean values for maxillary depth and preorbital rostrum length. The deep maxilla hypothesized by Rasmussen (1990) for *Mahgarita* is confirmed, a trait that is also present in *Aegyptopithecus, Rooneyia, Necrolemur, Tarsius, Adapis, Leptadapis, Daubentonia,* and *Propithecus.* Shallow maxillae are seen in most strepsirhines, *Notharctus, Smilodectes, Ignacius, Tupaia,* and *Alouatta.* This trait was not measured in the casts of *Plesiadapis tricuspidens* and *Tetonius* because they are highly deformed.

Character Analysis

Character 37: Snout length. Short snouts (1); longer snouts (0).

Character 38: Maxillary depth. Deep maxilla (0); shallow maxilla (1).

Fusion of the Mandibular Symphysis and Robust Mandibular Corpora

Rasmussen (1990) has suggested that possession of robust mandibular corpora and a fused mandibular symphysis link *Mahgarita* to anthropoids (see also Gingerich, 1980). Symphyseal fusion characterizes the mandibles of all extant anthropoids and several adapid lineages. Robust mandibular corpora are likely to be functionally correlated with fusion at the mandibular symphysis, however. This assertion is based on experimental and comparative studies by Hylander (1977, 1979, 1984, 1985), Beecher (1977, 1979, 1983), and Ravosa (1991) that suggest that vertically deep jaws are adapted for resisting increased sagittal bending stresses incurred on the balancing side during mastication. These stresses are greatest in animals with fused symphyses because fusion allows greater use of balancing-side muscles. Symphyseal fusion also allows the symphysis to resist greater "wishboning" of the mandible during unilateral mastication, and this results in increased transverse bending of the corpus, which is best resisted by increasing mediolateral corpus dimensions. Thus, increased robustness of mandibular corpora in adapids and anthropoids can be hypothesized to be functionally related to symphyseal fusion, and the two features cannot be considered to be separate pieces of evidence for anthropoid relationships.

Neither of these features may be useful for assessing the phylogenetic relationships of anthropoids, however, as the earliest anthropoids appear not to have had fused symphyses (Rasmussen and Simons, 1992; Simons, 1992), suggesting that complete symphyseal fusion almost certainly arose independently in anthropoids and adapids. These characters are fully discussed by Ravosa and Hylander (Chapter 14, this volume).

Character Analysis

Character 39: Complete symphyseal fusion and robust mandibular corpora. Absent (0); present (1).

Temporomandibular Joint

Rosenberger (1985) observed that *Tarsius, Necrolemur,* and *Microchoerus,* have an anteroposteriorly oriented trough-like glenoid fossa with a prominent entoglenoid crest, a condition that is also found in *Shoshonius,* whereas *Rooneyia* has a "shallower glenoid gutter and less differentiation of the crest" (Rosenberger, 1985, p. 188). Rosenberger contrasts this with the biconcave articular fossa found in all other primates sampled by him, a condition he suggests is primitive for euprimates. He also notes that some *Plesiadapis* possess a

> strong, albeit short, entoglenoid crest. However, since the morphology of the glenoid fossa manifests no other important resemblances to tarsiiforms (it is essentially

a flat and expansive surface), it is likely that an entoglenoid crest developed independently in *Plesiadapis* to stabilize the condyle with a large, laterally positioned articular fossa (Rosenberger, 1985, p. 188).

The absence of "other important resemblances to tarsiiforms" is not relevant to the question of whether the entoglenoid crest in *Plesiadapis* is homologous with that of tarsiiforms, and the morphology of the fossa and the morphology of the entoglenoid crest must be considered as separate characters. *Aegyptopithecus, Adapis, Notharctus, Mahgarita,* and extant anthropoids all possess a glenoid fossa that is biconcave and transversely wide compared to the situation in *Necrolemur, Shoshonius,* and *Tarsius.*

The strong entoglenoid process claimed to link *Tarsius* to *Necrolemur* is actually present in numerous taxa, including *Propithecus, Shoshonius,* pitheciines, atelines, and *Aegyptopithecus.* Many taxa lack an entoglenoid process altogether (*Mahgarita*), or it is weak, consisting of only a thin lamina of bone riding up onto the lateral bulla wall.

Character Analysis

Character 40: Temporomandibular joint morphology. Trough-like fossa oriented anteroposteriorly (1); biconcave and transversely wide (0).

Character 41: Entoglenoid process morphology. Entoglenoid process weak (e.g., *Mahgarita*) or absent with the glenoid fossa a flat plane, as in *Ignacius* (0); strong, consisting of a distinct ridge of bone separating the glenoid fossa from the bulla (e.g., *Aegyptopithecus,* see Figs. 8A and 12) (1).

Interincisal Diastema

Beard (1988) has demonstrated that the possession of a relatively broad interincisal diastema, a likely correlate of strepsirhinism, is the primitive condition for primates. His measurements also indicate that two *Notharctus,* one *Smilodectes,* and one *Plesiadapis* specimen have interincisal diastemata that are relatively as narrow as those of tarsiers and anthropoids. Although this does not *necessarily* mean that these taxa possessed haplorhine-like upper lips, it does provide support for Rasmussen's (1986) contention that at least some adapids may have possessed soft-tissue features characteristic of tarsiers and anthropoids. My own estimate of mean relative median gap width (defined in Beard, 1988) for *Rooneyia* gives a value of 0.94, outside the range of extant haplorhines (0.11–0.48).

Character Analysis

Character 42: Interincisor diastema width. Narrow, haplorhine-like interincisal diastema, i.e., falls within the range of extant haplorhines (1); broad diastema, wider than that of extant haplorhines (0).

Table I. Character–Taxon Matrix

		1	2	3	4	5	6	7	8	9	10	11	12	13	14	15	16	17	18	19
1	*Ignacius*	0	?	?	0	0	2	0	0	0	0	4	?	1	1	1	?	2	0	?
2	*Plesiadapis*	0	?	?	0	0	2	0	0	0	0	4	?	1	0	?	2	0	?	?
3	*Tupaia*	0	?	0	0	0	0	0	0	0	0	0	1	1	0	0	0	0	0	?
4	*Tetonius*	0	?	?	0	0	?	0	?	?	0	0	?	?	0	?	?	2	1	1
5	*Necrolemur*	0	?	1	0	0	0	0	0	1	0	1	1	1	0	?	2	2	1	1
6	*Shoshonius*	0	?	0	0	0	2	0	0	?	0	0	1	1	0	?	2	2	1	1
7	*Rooneyia*	0	?	0	0	0	0	0	0	1	0	1	1	1	0	?	2	0	1	1
8	*Adapis*	0	?	1	0	0	2	0	0	0	0	1,3	0	1	0	0	1	0	2	0
9	*Notharctus*	0	?	?	0	0	2	0	0	0	0	1,3	0	1	0	0	1	0	2	0
10	*Mahgarita*	0	?	1	0	0	2	0	0	0	0	1,2	1	1	0	0	1	0	1	1
11	*Aegyptopithecus*	1	0	1	?	0	1	1	2	1	0	2	1	1	1	0	?	0	0	?
12	*Tarsius*	1	1	0	1	1	2	1	2	1	1	2	1	1	1	1	?	1	1	1
13	Platyrrhines	1	0	1	1	0	0,1,2	0,1	1,2	1	0	2	1	1	1	0	?	0	0,1	?
14	Catarrhines	1	0	1	1	0	1	1	2	1	0	2	1	1	1	1	?	0	0	?
15	*Galago senegalensis*	2	?	1	0	0	0	0	0	0	?	4	1	0	1	?	?	0	1	0
16	Lemur	0	?	0	0	0	2	0	0	0	0	3	0	1	0	0	1	0	1	0
17	Microcebus	0	?	0	0	0	2	0	0	0	0	4	?	0	0	0	1	0	1	0

PAUP Analysis and Results

A branch-and-bound analysis was performed on the character–taxon matrix shown in Table I using PAUP (Swofford, 1991).* Three taxa were designated as outgroups because use of several outgroup taxa is more likely to result in accurate estimates of character polarities than use of only one. *Ignacius, Plesiadapis,* and *Tupaia* were chosen as outgroups because they represent groups commonly believed to be closely related to primates and because the two fossil taxa provide some measure of stratigraphic polarity.

Characters 6, 8, 23, 24, 33, and 36 were designated as ordered. The characters were all scaled equally (1000) regardless of the number of character states to prevent multistate characters from biasing the results. The initial PAUP run, without any topological constraints, produced two equally parsimonious trees. The strict consensus of these trees is shown in Fig. 18, with a consistency index (CI) of 0.565 and a rescaled CI of 0.363. This tree shows anthropoids to be monophyletic with *Tarsius* as their sister taxon and with a tarsier–anthropoid clade forming a monophyletic group with omomyids. *Mahgarita* is the sister taxon to this omomyid–anthropoid–tarsier clade.

However, this tree suggests that the Malagasy primates do not form a monophyletic clade. This result is unlikely to be correct. Yoder's (1992) phylogenetic analysis of morphological and molecular evidence for cheirogaleid relationships found Malagasy primates to form a monophyletic clade with a confidence interval, as established by a bootstrap analysis, of 95%. It therefore

*Version 3.1 licensed to R. F. Kay.

20	21	22	23	24	25	26	27	28	29	30	31	32	33	34	35	36	37	38	39	40	41	42
0	1	0	0	0	?	1	0	1	0	1	0	1	0	0	0	2	0	1	0	0	0	0
0	?	0	0	0	?	?	0	1	0	0	0	?	1	0	0	?	0	?	0	0	1	1
0	1	0	0	1	?	0	0	0	0	0	1	0	0	0	0	0	0	1	0	0	0	0
?	?	0	1	1	?	1	1	?	?	?	?	1	?	?	0	?	1	?	0	?	?	?
1	2	0	1	1	?	1	1	1	0	0	1	1	1	0	0	1	0	0	0	1	1	0
1	1	0	2	1	?	1	1	?	?	?	?	1	1	0	0	1	0	?	0	1	1	?
0	0	?	0	1	?	0,1	1	1	1	0	0	1	0	0	1	?	1	0	0	0	0	0
0	0	1	0	1	?	0	0	1	?	0	0,1	1	1	0	1	2	1	0	1	0	0	?
0	0	1	0	1	?	0	1	?	?	1	0	1	?	?	1	0	0	1	1	0	0	1
0	0	1	0	1	?	1	0	?	?	0	?	1	1	0	1	2	1	0	1	0	0	?
0	0	1	0	2	0	1	1	?	?	1	1	1	2	0	1	1	0	0	1	0	1	1
1	2	0	2	2	1	1	1	2	1	0	1	1	1	0	0	1	1	0	0	1	1	1
0	0	1	0,1	2	0	1	1	1,2	1	1	1	1	0,1	1	0,1	1,2	0,1	0,1	1	0	0,1	1
0	0	1	0	2	0	1	1	2	1	1	1	1	0,1	1	1	0,1	0,1	0	1	0	1	1
0	0	1	1	1	?	0	1	2	0	0	0	1	0	1	1	1	0	1	0	0	0	0
0	0	1	0	1	?	0	0	0	1	0	1	1	0	0	1	1	0	1	0	0	0	0
0	0	1	1	1	?	0	1	2	0	0	0	1	0	0	1	1	0	1	0	0	0	0

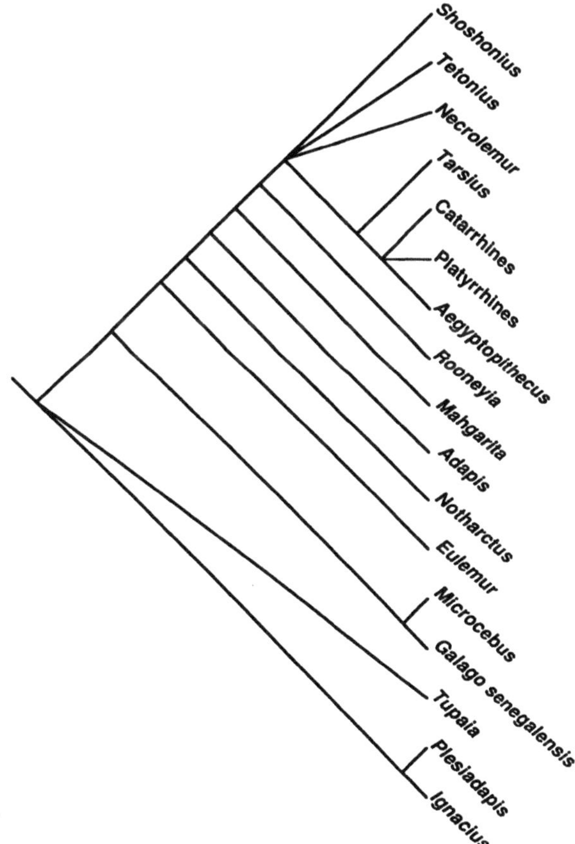

Fig. 18. Most parsimonious unconstrained tree.

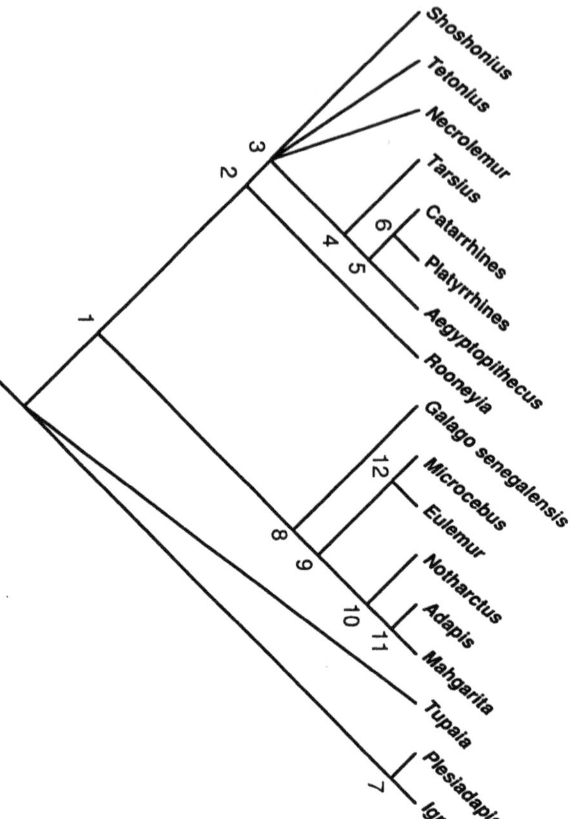

Fig. 19. Most parsimonious tree with the constraint of Malagasy-primate monophyly imposed. See Appendix B for details of character states at nodes.

seems unlikely that Malagasy primates are not monphyletic. Consequently, the data set for this study was run through PAUP again with the constraint of Malagasy primate monophyly imposed. The branch-and-bound analysis produced four equally parsimonious trees, and Fig. 19 is the strict consensus of those trees. The consistency index for the four trees (and for the consensus tree) is 0.551 for all characters but, when rescaled to exclude uninformative characters, is only 0.343.

To assess the confidence limits on the clades within this tree a general heuristic bootstrap analysis (Felsenstein, 1985) was performed using PAUP. A bootstrap analysis randomly samples with replacement characters from the original data set, creating a series of bootstrap character–taxon matrices of the same size as the original matrix (in this case 42 characters). A heuristic analysis was then performed on each of these matrices to find the most parsimonious tree. In this case 100 bootstrap character–taxon matrices were generated, and a heuristic search was performed on each of these. The "majority-rule" consensus tree (Margush and McMorris, 1981) of the resulting 100 trees was then obtained, and the number of times each of the clades appeared in these trees is presented as a percentage in Appendix B. The "majority-rule" consensus tree is made up of those clades that appear in the majority of these 100 trees.

The decay indices for each of the clades within the most parsimonious tree were also found, and these are also given in Appendix B. A decay index is the number of unit steps that parsimony must be relaxed to cause a particular clade to loose its support. This index is obtained by calculating the semistrict consensus tree for all trees that are one step longer than the most parsimonious, two steps longer, three steps longer, etc. (Mishler *et al.*, undated).

Discussion

The parsimony analysis performed here corroborates the hypothesis that tarsiers and anthropoids shared a common ancestor more recently than either of them did with any known primate. However, the bootstrap analysis puts very low confidence limits on this tree. Nodes 4, 5, and 6 were found in only 59%, 90%, and 60% of trees in the bootstrap analysis, respectively (Appendix B). Felsenstein (1985) suggests that we should accept only those clades with a confidence limit of 95% or higher. If these confidence limits were applied to this tree (Fig. 19), none of these clades would be acceptable. The decay indices also suggest that there is little support for many of the clades in Fig. 19. The clades descended from nodes 1–3, 6–9, and 11 all collapse when parsimony is relaxed one step. Nodes 4 and 5 are more stable, with the tarsier–anthropoid clade having a decay index of three steps and the anthropoid clade having a decay index of five steps. Therefore, although the craniofacial evidence provides some corroboration for the notion of a tarsier–anthropoid clade, the decay indices and the bootstrap analysis suggest that the craniofacial evidence for anthropoid relationships is either too incomplete to be of use or that there is too much homoplasy in the anatomy of the head for it to be a reliable source of data. Because the skulls of so few Eocene prosimians and anthropoids are known, the former is probably the case. With this in mind, it is important when evaluating the craniofacial evidence for anthropoid relationships to discriminate between hypotheses of homology that fail morphological tests and those that fail phylogenetic tests of homology (Wiley, 1981).

Adapid–Anthropoid Hypothesis

The adapid–anthropoid hypothesis is not supported by the evidence from the skull. Of the four features claimed by Rasmussen (1990) to support a *Tarsius–Mahgarita*–Anthropoidea clade (Fig. 1D, node 1), one fails morphological tests of homology, and three are most parsimoniously interpreted as convergent.

(1) Rasmussen's (1990) hypothesis that *Mahgarita* might have possessed an extrabullar tympanic bone is not substantiated by my examination of the fossil evidence. Gingerich and Martin's (1981) claim that the "free-fused annulus" of early anthropoids resembles the "free annulus" of adapids is also not correct. The tympanic in *Aegyptopithecus* is not an annulus and is not free (Fig. 11), the tympanic of *Catopithecus* is also fused to the lateral bulla wall (Simons, 1991), and that of *Mahgarita* was almost certainly intrabullar as in all adapids for which this anatomy is known.

(2) A large promontory and reduced or absent stapedial canal are features shared by *Tarsius, Mahgarita,* and anthropoids. However, although the anatomy of these canals does a good job of grouping extant taxa into "logical" groups, this is certainly not true of Eocene primates. Even if the specimens of *Adapis* and *Mahgarita* that show variability in relative size of these canals are actually separate species or even genera, there still has to have been copious convergence and reversal among adapids. It is unparsimonious to assume that the last common ancestor of adapids and anthropoids had a stapedial canal that was reduced relative to that of the promontory.

(3) Rasmussen's (1990) claim that *Mahgarita* shares a short, deep facial exposure of the maxilla with tarsiers and some anthropoids is confirmed by the quantitative data presented here. However, these features are not most parsimoniously considered to be synapomorphies of a *Tarsius–Mahgarita–*anthropoid clade. Indeed, both these features provide support for a *Mahgarita–Adapis* clade.

(4) The lack of a zygomatic–lacrimal contact on the inferior orbital margin in *Mahgarita* is an interesting resemblance to *Tarsius,* anthropoids, and omomyids. The lack of this contact in some lorises and galagos, its presence on one side of the *Rooneyia* skull, and its presence in *Ignacius* suggest that this could be a primitive primate feature. It is equally parsimonious to assume that it evolved convergently among all of these taxa.

Two of the five features hypothesized by Rasmussen (1990) to support a *Mahgarita*–anthropoid clade (Fig. 1D, node 2) fail morphological tests of homology. Consequently, these features were not scored as potentially homologous for the parsimony analysis reported above. (1) The "lateral transverse septum" claimed by Rasmussen (1990) to be present in *Mahgarita* and anthropoids is actually more lemur- than anthropoid-like and therefore fails morphological tests of its status as a *Mahgarita*–anthropoid synapomorphy. (2) In my opinion, the posterior nasal spine of *Mahgarita* claimed to resemble that of *Aegyptopithecus* bears little morphological resemblance to it and cannot be used as evidence linking this protoadapine to anthropoids.

The remaining three features fail phylogenetic tests of homology (i.e., in the context of the other cranial data reported here, it is unparsimonious to suggest that these features were present in the last common ancestor of adapids and anthropoids). (1) Mastoid pneumatization is present at least in lorisoids, *Necrolemur,* and *Adapis* and probably in *Tetonius* and *Notharctus* and must therefore have been lost and/or evolved several times within primates.

(2 and 3) Similarly, the fused mandibular symphysis and robust mandibular corpora (which can be treated as a single character), traditionally cited as evidence linking adult anthropoids to some adapids, are known to have evolved in parallel numerous times within Adapidae and among other mammals. More importantly, they are also probably absent in *Catopithecus* (Rasmussen and Simons, 1992). It is most parsimonious, therefore, to assume that these features evolved in parallel in adapids and anthropoids.

Thus, the evidence from the skull does not support the adapid–anthropoid hypothesis, and the best evidence for a link between adapids and anthropoids remains the similarities between the molars of *Oligopithecus* and protoadapines outlined by Rasmussen and Simons (1988). The evidence from the postcranium does not argue definitively for or against the adapid–anthropoid hypothesis: some workers note similarities between the postcrania of various adapids and strepsirhines (Beard *et al.,* 1988; Beard and Godinot, 1988) that suggest the monophyly of a strepsirhine–adapid clade to the exclusion of other primates, whereas other workers suggest that postcranial features argue for a link between *Europolemur* and anthropoids (Franzen, 1987).

One way to resolve the conflict between the dental and postcranial data while explaining the close neontological similarities between tarsiers and anthropoids to the exclusion of strepsirhines, is to assume that adapids are the primitive euprimates and that strepsirhines and anthropoids both arose from adapid-like animals (Rasmussen, 1986). However, if the adapid–strepsirhine postcranial synapomorphies are primitive, then the alternate condition found in omomyids, tarsiers, and anthropoids must be reversals to a primitive condition and interpreted as synapomorphies of a clade consisting of these taxa, dispelling the notion of an adapid–anthropoid clade.

Notably, none of the adapid–strepsirhine postcranial synapomorphies has yet been confirmed in any protoadapine, unless the Messel adapid can be referred to this group. If protoadapids possessed a postcranium that lacked the distinctive strepsirhine synapomorphies, and if adapids are in fact a paraphyletic taxon, then Rasmussen's (1986) phylogeny may yet be correct. The craniofacial evidence, however, suggests that this is not the case.

Omomyid–Anthropoid Hypothesis

Both the omomyid–anthropoid and the tarsier–anthropoid hypotheses claim that tarsiers, omomyids, and anthropoids form a monophyletic group to the exclusion of other primates. Unlike the tarsier–anthropoid hypothesis, however, the omomyid–anthropoid hypothesis proposes that anthropoids and tarsiers evolved from separate omomyid stem groups. The claim for a tarsier–anthropoid–omomyid clade is discussed before the conflicting hypotheses of tarsier relationships proposed by different advocates of the omomyid–anthropoid hypothesis are evaluated.

Is There a Tarsius–Anthropoidea–Omomyidae Clade (Haplorhini Sensu Szalay et al., 1987)?

The most explicit statement of the evidence for the monophyly of a omomyid–tarsier–anthropoid clade is presented by Szalay *et al.* (1987), but three of the cladograms in Fig. 1 (Fig. 1A, node 2; Fig. 1B, between nodes 1 and 2 if anthropoids are nested within the omomyids; Fig. 1C, node 1) hypothesize the existence of such a clade. Szalay *et al.* present 13 features alleged to define that clade, six of which are postcranial characters likely to be primitive for euprimates (Dagosto, 1990), and two of which are features of the soft tissues of the eye and brain. The remaining five features were redefined and included in the analysis presented here.

Although the most parsimonious tree obtained here verifies the monophly of the Haplorhini *sensu* Szalay *et al.*, the bootstrap analysis did not find this clade. In the most parsimonious tree only three characters (excluding equivocal characters) provide support for this clade—the absence of ventral shielding of the fenestra cochleae by the internal carotid canal, a stapedial canal that is slightly smaller than the promontory canal, and the presence of an extensive contact between the lateral pterygoid plate and the lateral bulla wall—and all of these characters require some assumptions if they are to be interpreted as synapomorphies of a tarsier–omomyid–anthropoid clade.

The evolutionary polarity of the first of these three characters is uncertain. A shielded fenestra cochleae is seen in Eocene insectivores *Palaeoryctes* and *Pararyctes* (Gunnell, 1989) as well as in tree shrews, lipotyphlans, and microchiropterans (MacPhee and Cartmill, 1986, p. 261), whereas a posterolateral posterior carotid foramen position and shielded fenestra cochleae are seen in *Ignacius, Phenacolemur, Plesiadapis tricuspidens,* and *P. cookei* (even though the carotid arterial system in these taxa is much reduced: MacPhee *et al.,* 1983; Kay *et al.,* 1990). If the slightly reduced stapedial canal is a synapomorphy of a tarsier–omomyid–anthropoid clade, then the stapedial canal must have been reenlarged in *Tetonius* and *Shoshonius.* If the extensive contact between bulla and lateral pterygoid plate is a synapomorphy of this clade, it must have been lost subsequently in anthropoids.

None of the characters proposed by Szalay *et al.* provide definitive support for this clade; all are more parsimoniously interpreted in other ways.

(1) A shortened facial skull cannot be definitively claimed to characterize the last common ancestor of Haplorhini (*sensu* Szalay *et al.*). *Aegyptopithecus, Shoshonius,* and *Necrolemur* have long snouts as measured and scored in this study, and some extant anthropoids have long snouts. Better fossil evidence of early anthropoids is required if this feature is to be demonstrated to be morphotypic for this group.

(2) The presence of a pronounced interorbital constriction below the olfactory tract cannot be used to support the notion of Haplorhini advanced by Szalay, *et al.* because it may well have been present in the last common ancestor of all primates, subsequently to be lost in adapid, lemuroid, and some

anthropoid lineages. Certainly, this feature is present in all the omomyids for which skulls are known. Moreover, given that small-bodied anthropoids (*Saimiri,* callitrichids, *Callicebus, Aotus*) possess a pronounced interorbital constriction below the olfactory tract, and that the earliest known anthropoids were fairly small, it seems likely that this feature was present in the anthropoid ancestor. Although it is not certain that this feature is a haplorhine synapomorphy, I share the view of Szalay *et al.* (1987, p. 95) that this is "a feature of exceptional significance"; it may have established the necessary preconditions for the evolution of the anthropoid and tarsier postorbital septum (see below).

(3) The polarity of posterior carotid foramen positioning in primates and their relatives is uncertain. Some think the posterolateral position characterized the primate ancestor (Szalay *et al.,* 1987); others think the posteromedial position is primitive for primates (Schmid, 1981; MacPhee and Cartmill, 1986). In the former case the posteromedial position is seen as derived and can be assumed to define Haplorhini (*sensu* Szalay *et al.*), but only if the anterolateral position seen in *Tarsius,* the posterolateral position seen in *Shoshonius,* and the midline position seen in many anthropoids are seen as secondarily derived states. Determination of the polarity of this feature among primates must await the discovery of more fossils.

(4) Enlargement of the promontory canal at the expense of the stapedial is also not a good feature for defining a tarsier–anthropoid–omomyid clade, as it is widespread in adapids, and equally sized canals are present in *Shoshonius* and *Tetonius* (Beard and MacPhee, Chapter 3, this volume).

(5) Framing of the external auditory meatus by the ectotympanic may link omomyids to tarsiers and anthropoids, but it is impossible to tell because in every omomyid skull I have seen the sutures between the bulla and the tubular structures framing the external auditory meatus are obliterated.

It has often been argued that omomyids are too derived dentally (e.g., Gingerich and Martin, 1981; Simons, 1989) and postcranially (Gingerich, 1980) to have given rise to anthropoids. This argument is based on the observation that omomyids have reduced canines, an I_1 that is larger than I_2, a fused tibiofibula, and anterior calcaneal lengthening. The washakiin omomyid *Washakius insignis* has an unreduced canine and a lower lateral incisor that is larger (in root size) than the mesial incisor (Covert and Williams, 1991); *Steinius vespertinus* retains an unreduced canine, although it has an enlarged I_1 (Rose and Bown, 1991); and *Teilhardina belgica* and possibly *T. americana* have a canine that is still large and an I_1 that is only slightly larger than I_2 (Bown and Rose, 1987, p. 28). Dagosto (1988) observes that tibio-fibular fusion is not in fact found in all omomyids (*Hemiacodon* lacks it) and that the anterior calcaneal lengthening is not too extreme. In any event, anterior calcaneal length is related to body size (Gebo and Dagosto, 1988), and a transition from a small omomyid to a large anthropoid would be expected to result in a decrease in calcaneal length. An omomyid origin for anthropoids is therefore not ruled out by the dental and postcranial evidence.

Necrolemur–Tarsius Hypothesis

Rosenberger's (1985) hypothesis that *Necrolemur* and *Tarsius* form a monophyletic group nested within an omomyid clade that excludes anthropoids is not supported by the parsimony analysis presented here. His claim that *Necrolemur* and *Tarsius* are linked to the exclusion of other omomyids by the shared possession of four characters (Fig. 1B, node 4) is weakened by the observation that all of these features are present in *Shoshonius* and, where preserved, in *Tetonius* as well, suggesting that they were present in the last common ancestor of all omomyids. If this is the case, these features cannot be synapomorphies of a *Tarsius–Necrolemur* clade but of a clade consisting of tarsiers, anthropoids, and all non-*Rooneyia* omomyids.

I suggested that the sharply peaked choanae seen in *Tarsius, Shoshonius,* and *Necrolemur* are correlated with the large orbits of these taxa. Orbit size in *Tarsius* actually resembles that of *Shoshonius* more than that of *Necrolemur* [if the estimate of orbit size on *Shoshonius* by Beard *et al.* (1991) is correct], although the orbits of all these animals, and probably those of *Tetonius,* are large enough to have produced peaked choanae. Such orbital enlargement with concomitant constriction of the choanal region was probably present in the last common ancestor of all omomyids except *Rooneyia,* suggesting that reduction in orbit size played a significant part in the origin of anthropoids (Ross, 1993).

The shortening of the medial pterygoid plate in omomyids is a common phenomenon among primates, including *Mahgarita* among the adapids, many extant strepsirhines, and platyrrhines. There is therefore no reason to suggest that it links *Tarsius* and *Necrolemur* or *Tarsius* and *Shoshonius* to the exclusion of other primates.

The suggestion that extreme encroachment by the bulla onto the pterygoid fossa is a synapomorphy of a *Tarsius–Necrolemur* clade is probably not true because the anterior cavity in the bulla of these two taxa fails morphological tests of homology. *Necrolemur* does not exhibit the laterally directed transverse septum of *Tarsius* that indicates the presence of an anterior accessory cavity forming the anterior portion of the intrabullar cavity. Consequently, this anterior intrabullar space cannot be hypothesized to be homologous in *Necrolemur* and *Tarsius,* so its encroachment on the pterygoid fossa cannot be homologous either. Moreover, even if the encroachment were hypothesized to be homologous in these two taxa, the presence of this feature in *Tetonius* and *Shoshonius* would suggest that this too is a primitive omomyid trait.

The presence of an anteroposteriorly guttered glenoid fossa in *Tarsius, Shoshonius,* and *Necrolemur* is noteworthy; however, it is unparsimonous to assume that this is a synapomorphy of a clade consisting of two or three of these animals to the exclusion of anthropoids.

The presence of a strong entoglenoid crest between the glenoid fossa and the bulla is characteristic not only of *Tarsius* and *Necrolemur* but also of *Shosh-*

onius and numerous anthropoids, including *Aegyptopithecus.* According to the parsimony analysis presented here, this is best interpreted as a synapomorphy of a clade consisting of anthropoids, *Tarsius, Shoshonius,* and *Necrolemur.*

Thus, the characters presented in favor of the *Tarsius–Necrolemur* hypothesis either fail morphological tests of homology or are likely to be primitive for omomyids, as indicated by their presence in *Shoshonius* and, when preserved, *Tetonius.*

Tarsius–Shoshonius Hypothesis

The suggestion by Beard *et al.* (1991) that *Tarsius* and *Shoshonius* form a monophyletic group that is the sister taxon to other omomyids is not corroborated by the evidence examined here. Two of these features, (reduced preorbital rostrum and orbit size) have already been discussed. The remainder are examined here.

Like Rosenberger (1985), Beard *et al.* suggest that omomyids and *Tarsius* are linked by the shared possession of an extensive overlap of the lateral bulla wall by the lateral pterygoid plate (Fig. 1C, node 2). Some kind of contact is probably primitive for primates, but an extensive contact does appear to characterize all omomyids (although Rosenberger's claim that the "laminar" overlap seen in omomyids is different from that of adapids is falsified by the case of *Mahgarita*). However, the results of this analysis suggest that it is most parsimonious to assume that anthropoids lost extensive laminar overlap of the lateraly pterygoid plate and the lateral bulla wall in their descent from an ancestral lineage that had such a contact.

The hypothesis of homology between the suprameatal foramen of *Shoshonius* and that of *Tarsius* (Fig. 1C, node 2) is not supported by the details of its size and position, and it is most parsimonious to assume that this feature evolved in parallel in *Necrolemur* and *Tarsius.*

The position of the posterior carotid foramen in the bulla in *Shoshonius* resembles that of *Tarsius* only in being laterally and ventrally placed. It is unparsimonous to assume that the lateral position in these taxa is a synapomorphy of a *Tarsius–Shoshonius* clade. The ventral position of the posterior carotid foramen links these two taxa to the anthropoid clade to the exclusion of all other primates, in which it is dorsally placed.

The basioccipital flange on the medial bulla wall in *Tarsius* and *Shoshonius* is a feature they share with *Necrolemur,* and it may be primitive for omomyids (the relevant anatomy being missing in *Tetonius*). It is morphologically very similar in the three taxa, if somewhat more anteriorly placed in *Tarsius,* but is most parsimonously seen as a primitive omomyid feature lost in anthropoids.

The claim that the internal carotid artery of *Shoshonius* resembles that of *Tarsius* in running through an intrabullar septum is incorrect. This feature fails morphological tests of homology.

Thus, the features claimed to be synapomorphies of a *Tarsius–Shoshonius* clade are either too dissimilar in the two taxa to be hypothesized as homologous or are most parsimoniously interpreted as primitive for omomyids or for primates.

Tarsier–Anthropoid Hypothesis

Tarsiers and anthropoids share three features of the skull that are unique to them in the history of life and therefore are probably homologous. These features were all cited by MacPhee and Cartmill (1986) in support of their tarsier–anthropoid hypothesis (Fig. 1C, node 3): (1) a postorbital septum that includes a contact between the zygomatic and alisphenoid bones; (2) an anterior accessory cavity formed within the petrosal plate by pneumatization from the tympanic cavity and separated from the tympanic cavity by a transverse septum arising from the cochlear housing; and (3) a perbullar pathway for the internal carotid artery formed entirely by the petrosal bone. The second of these is developmentally correlated with possession of a small tympanic cavity, so these cannot be cited as separate pieces evidence for a tarsier–anthropoid clade. The third represents a refinement of the definition of the perbullar pathway advanced by MacPhee and Cartmill (1986) in that it specifies that the perbullar pathway is formed entirely by the petrosal bone.

Tarsiers and anthropoids also share several features that, if present in their last common ancestor, must have evolved convergently in other groups or have been lost among some anthropoids. Three of these were treated by MacPhee and Cartmill (1986). (1) They originally suggested that tarsiers and anthropoids shared a last common ancestor with an anteromedially located posterior carotid foramen and that the anterolateral position of *Tarsius* is autapomorphic. This may be the case, but only the discovery of more fossils will document the evolutionary polarity of carotid foramen position in primates. Certainly tarsiers and most anthropoids share the positioning of the posterior carotid foramen anterior to the fenestra cochleae. (2) Tarsiers and anthropoids also share an extrabullar tympanic bone with concomitant loss of the annular bridge, a feature that is most parsimoniously interpreted as convergent with *Ignacius* and lorisoids. (3) Tarsiers and anthropoids are distinguished by a marked reduction of the stapedial artery and canal relative to the promontory. This feature is most parsimoniously considered convergent in *Mahgarita*.

The remaining feature that is parsimoniously interpreted as a tarsier–anthropoid synapomorphy was not mentioned by MacPhee and Cartmill (1986). Its presence or absence is not yet documented in *Shoshonius* and *Tetonius*, and it may well turn out to be a synapomorphy linking all omomyids to a tarsier–anthropoid clade. (4) A narrow interincisal diastema, probably correlated with the haplorhine condition of the upper lips, is a feature that tarsiers and anthropoids share with *Notharctus*, *Smilodectes*, and *Plesiadapis*.

Is the postcranial evidence compatible with the tarsier–anthropoid hypothesis? The postcranial evidence for anthropoid relationships has been thoroughly documented by several workers (Dagosto, 1988; Szalay and Dagosto, 1988; Gebo, 1986, 1989), and its implications explored in an excellent review by Dagosto (1990). From their earliest appearance in the fossil record the postcrania of all prosimians show similarities that are not found often in other mammals, indicating that they are likely to be synapomorphies of a clade including all prosimians. Anthropoids tend to exhibit the primitive conditions of these traits, suggesting that anthropoids either retained these features in a direct lineage from the euprimate ancestor or reversed to this "primitive" condition after passing through a prosimian-like stage of postcranial evolution. Dagosto (1990) favors the latter model on the grounds that it does not require the convergent evolution of neontological or cranial similarities between anthropoids and certain prosimian taxa. The tarsier–anthropoid hypothesis advocated here is not incompatible with this model, as it not only interprets the extremely rare tarsier–strepsirhine–omomyid–adapid postcranial similarities as synapomorphies, but it flags the equally rare tarsier–anthropoid cranial similarities as synapomorphies as well. Moreover, the anthropoid reversal to primitive mammalian postcranial morphology can be explained as an adaptive shift to increased dependence on above-branch quadrupedalism (Gebo, 1986, 1989).

The Evolution of the Postorbital Septum

I have suggested elsewhere, following Cartmill (1980), that the postorbital septum of tarsiers and anthropoids may have evolved to insulate the orbital contents from contractions of the temporalis muscle occurring during the power stroke of mastication (Ross, 1991, 1993). This insulation is necessary, I think, not solely because of the presence of a fovea in tarsiers and anthropoids (Cartmill, 1980) but because of anterior displacement of the origin of the temporalis muscle concomitant with increased frontation of orbits that are already highly convergent.

Dissections of tarsiers and numerous anthropoids indicate that the anteriormost fibers of the temporalis muscle curve around the postorbital septum between their origin on the frontal bone and their insertion on the medial surface of the ascending ramus of the mandible, a curvature suggesting that were the postorbital septum not present, the anteriormost fibers of temporalis would transect the orbital boundary and displace the orbital contents during mastication (Ross, 1993, 1994).

This prompts two further questions. Tarsiers do not have orbits that are as frontated as those of anthropoids of equal orbital convergence, and some lemuroids have more frontated orbits (Cartmill, 1970); so (1) can this explanation for the origin of the postorbital septum apply to tarsiers, and (2) why do anthropoids have a combination of highly frontated and convergent bony orbits? In fact, given their degree of orbital convergence, tarsiers have orbits that are more frontated than all strepsirhines except the lemurids *Eulemur fulvus* and *Lemur catta* and the indriids *Propithecus* and *Indri*. These large-bodied animals are relatively small-eyed, suggesting that the relatively small eye is not in close physical proximity to the temporalis. Certainly, dissections indicate that the temporalis of *Eulemur fulvus* and that of *Propithecus* do not impinge on the orbital contents; they run directly from origin to insertion without curving around the eye. *Tarsius,* although it has a low degree of orbital frontation, has enormous eyes that crowd back on the temporal fossa, and the fact that the temporalis does curve around the septum in tarsiers suggests that its decreased orbital frontation associated with orbital enlargement (Cartmill, 1972) does not serve to alleviate the danger of the temporalis impinging on the orbital contents.

The answer to the second question comes from a study of basicranial flexion in primates carried out by Matt Ravosa and me (Ross and Ravosa, 1993). We found that the angle of basicranial flexion covaries with orbital axis orientation (as measured from lateral radiographs) in haplorhines but not in strepsirhines. This suggests that the extreme orbital approximation in tarsiers and anthropoids (but not strepsirhines) has integrated the bony orbit and the anterior limb of the basicranium into a single structural unit such that increased basicranial flexion in tarsiers and anthropoids must result in increased orbital frontation. Two possible causes of increased basicranial flexion, and therefore increased orbital frontation in haplorhines, are increased brain size relative to basicranial length and increased frontal lobe dimensions. Ross and Ravosa (1993) found that as neurocranial volume relative to basicranial length increases, the basicranium becomes more flexed.

In light of this, it is significant that although both adapid and omomyid endocasts reveal an increase in the size of the occipital and temporal lobes of the brain relative to other Eocene mammals, only omomyids show any signs of basicranial flexion (Rosenberger, 1985) or anteriorly positioned foramina magna (Simons and Russell, 1960; Szalay, 1976). *Tetonius, Necrolemur,* and *Rooneyia* all exhibit a partial overlap of the occipital lobe over the cerebellum (Radinsky, 1970, 1979), a feature that is not reported in the two adapids for which endocasts are known: *Adapis* and *Smilodectes.* If omomyids have large brains relative to basicranial length, as well as having some degree of basicranial flexion, these may be features that they share with tarsiers and anthropoids but not with adapids or most extant strepsirhines. Determining whether or not this was actually the case must await data on neurocranial volume, basicranial length, and the degree of basicranial flexion in the relevant fossils.

If this argument regarding the context and functional significance of the origin of the postorbital septum is correct, then adapids do not qualify as anthropoid (or tarsier) ancestors on two counts: they do not exhibit the high degree of orbital approximation seen in omomyids, nor do they show any signs of the brain enlargement necessary to account for the basicranial flexion and orbital frontation that forced the evolution of a postorbital septum. Thus, the most compelling argument against the adapid–anthropoid hypothesis for anthropoid origins is that there is no reason for any known adapid lineage to have evolved a postorbital septum, whereas there are several reasons to think that a tarsier-like omomyid lineage might have needed one.

Which omomyid lineage might this have been? There is little evidence supporting the hypothesis of a closer relationship of a tarsier–anthropoid clade to *Shoshonius* than to any other omomyids; the sole feature is a posterior carotid foramen that is ventrally placed in the auditory bulla, a feature found in *Shoshonius, Tarsius,* and numerous anthropoids. (The carotid canal in *Shoshonius* is probably anteriorly located relative to the fenestra cochleae, but further preparation of the specimens is required to confirm this.)

Indeed, a *Shoshonius–Tarsius*–Anthropoidea clade is, I think, unlikely, considering the derived dental specializations of the Washakiini, including *Shoshonius* (Honey, 1990). The origin of the tarsier–anthropoid clade from an as yet unknown omomyid lineage is perhaps a more accurate reflection of the current state of our knowledge, as shown in Fig. 19.

That at least some kind of non-*Rooneyia* omomyid gave rise to anthropoids and *Tarsius* is indicated by the short snouts and strong entoglenoid processes of microchoerines and *Shoshonius* that are widely distributed among extant and fossil anthropoids. The presence in *Necrolemur* of a fused metopic suture is also suggestive, although this area is badly preserved in *Tetonius* and *Shoshonius.*

Tarsius certainly shares numerous features with microchoerines and with *Shoshonius* that must have subsequently been lost in anthropoids if they have their origins among such forms: extremely large orbits with correlated medial displacement of pyramidal processes and peaked choanae; encroachment of the auditory bulla on the pterygoid fossa; a trough-like glenoid fossa; a flange of the basioccipital on the medial bulla wall; and possibly a suprameatal foramen. However, as suggested above, the skulls of these omomyids may be the best examples of the context in which a postorbital septum might be expected to arise.

Definition of Anthropoidea

These results suggest that the definition of Anthropoidea on the basis of craniofacial features cannot invoke two features that are traditionally used: a postorbital septum and fused metopic suture. Rather, they are defined by the possession of a trabeculated anterior accessory cavity extending medially to the tympanic cavity; the combination of a large promontory and absent stapedial artery in the adult; and a postorbital septum formed predominantly by the zygomatic bone. Anthropoids are also differentiated from some other primates by features that have likely evolved convergently elsewhere within the order: a pneumatized mastoid region, possibly evolved convergently in *Necrolemur, Tetonius, Mahgarita,* and lorisoids; loss of the contact between the lateral pterygoid plate and the bulla wall, seen in some lemuroids; a lacrimal foramen lying inside or on the rim of the orbit, the latter condition also being found in *Notharctus,* the former in *Ignacius;* and a fused mandibular symphysis with correlated robust mandibular corpora, evolved convergently among adapids.

Thus, perhaps the most informative definition of the Anthropoidea would be one that captures the evolutionary history that constrained them to be what they are today—tarsiers with small eyes. Conversely, tarsiers may be small anthropoids with extremely large eyes. Certainly, many of the craniofacial features traditionally used to define anthropoids are also present in tarsiers; hence, the anthropoid or "human-like" appearance of their heads.

Acknowledgments

I thank the Drs. Fleagle, Kay, and Simons for inviting me to participate in the Anthropoid Origins Conference on which this book is based, and the L.S.B Leakey Foundation and the Wenner-Gren Foundation for supporting the conference. Drs. Matt Cartmill, Richard F. Kay, Tab Rasmussen, Matt Ravosa, Hans Thewissen, Kathleen Smith, and one anonymous reviewer provided detailed comments on earlier drafts of this manuscript. Drs. Elwyn Simons, Jack Wilson, Rich Kay, Farish Jenkins, and L. K. Krishtalka very generously granted permission for me to study and photograph fossils in their care. The following people were most helpful and hospitable during my studies, and I thank them for their many kindnesses: Charles Schaff at the MCZ, Cambridge, MA; Melissa Winans and Bob Rainey at the Balcones Research Center, Austin, TX; Dr. Liza Shapiro in Austin, TX; Dr. Chris Beard at the CM; Dr. Betsy Dumont in Pittsburgh, PA; Drs. Ross MacPhee and Wolfgang Fuchs at the AMNH; Dr. Richard Thorington at the USNM; Prithijit Chatrath at the DUPC; and Dr. J. G. E. Thewissen in the Department of Biological Anthropology and Anatomy at Duke. I particularly thank Dr. Kathleen Smith for her assistance with PC3D and for numerous discussions regarding ontogeny and phylogeny. This work was supported by a Dissertation Travel Award from the Duke University Graduate School, a Collection Study Grant from the AMNH, a Grant-in-Aid of Research from Sigma Xi, and a Dissertation Improvement Grant from NSF (BNS-9100523).

Appendix A: Features of the Skull Cited in Support of Hypotheses of Anthropoid and Tarsier Relationships Shown in Fig. 1

Tarsiphile Hypothesis

These are from MacPhee and Cartmill (1986) (Fig. 1A).

Node 1. Annular bridge present and incomplete
Anterior accessory cavity absent
Tympanic cavity large

Posterior carotid foramen posteromedial
Carotid pathway transpromontorial
Stapedial and promontorial canals large and subequal
Node 2. Stapedial canal smaller than promontory
Node 3. Annular bridge absent
Anterior accessory cavity present
Tympanic cavity small
Posterior carotid foramen anteromedial
Carotid pathway perbullar
Stapedial artery vestigial, promontory artery unreduced
Postorbital septum
Node 4. Posterior carotid foramen posterolateral (subsequently lost with reduction of carotid system in lorisiforms and cheirogaleids)

Omomyophile Hypotheses

Features suggested by Szalay *et al.* (1987) as supporting a *Tarsius*–Anthropoidea–Omomyidae clade (Haplorhini *sensu* Szalay *et al.*) and compatible with Fig. 1A–C, although the authors do not support the tarsiphile hypothesis (Fig. 1B, C).

Shortened facial skull
Olfactory process above the interorbital septum
Carotid artery enters skull medially
Promontory artery hypertrophied and at least its bony canal is absolutely larger than the stapedial one
Auditory meatus formed by ectotympanic which is elongated and partly outside of the auditory bulla or "phaneric"

Necrolemur–Tarsius Hypothesis

Rosenberger (1985) does not present a specific hypothesis regarding anthropoid relationships to these taxa except to exclude them from the *Tarsius–Necrolemur* clade. (Fig. 1B).

Node 1. Moderately flexed basicranium
Node 2. Choanae probably peaked
Medially placed pyramidal processes
Node 3. Choanae probably peaked
Moderately short medial pterygoid plate
Bulla moderately encroaching on pterygoid fossa
Basicranium moderately flexed
Laminar contact between lateral pterygoid plate and lateral bulla wall
Glenoid fossa moderately guttered with moderate entoglenoid crest
Node 4 Peaked choanae
Short medial pterygoid plate
Extreme encroachment by bulla on pterygoid fossa (convergent with *Tetonius*)
Deeply guttered glenoid fossa with marked entoglenoid crest

Tarsius autapomorphy: extreme basicranial flexion.

Tarsius–Shoshonius Hypothesis

These are from Beard *et al.* (1991) (Fig. 1C).

Node 1. The authors cite Cartmill and Kay (1978), Cartmill *et al.* (1981), MacPhee and Cartmill (1986), Rosenberger and Szalay (1980), Schmid (1981), and Gebo (1986) in support of a *Tarsius*–anthropoid–omomyid clade.
Node 2. Lateral pterygoid wings overlap anteromedial (*sic*, anterolateral) bullar wall
Suprameatal foramen present
Node 3. Posterior carotid foramen posterolateral in position
Reduced snout
Basioccipital flange on medial bullar wall
Enlarged orbits relative to skull length
Internal carotid artery in intrabullar septum (convergent with anthropoids)
Branches 4 and 5 (convergences shared by tarsiers and anthropoids).
Loss of ectotympanic annular bridge
Partial postorbital closure

Mahgarita–Anthropoidea Hypothesis

These are from Rasmussen (1990) (Fig. 1D).

Node 1. Promontory canal enlarged (stapedial absent or reduced)
Tympanic extrabullar
Maxillary facial exposure short and deep (convergent in *Propithecus*)
Lack of zygomatic–lacrimal contact on inferior orbital margin (convergent in *Galago*)
Node 2. Lateral transverse septum of middle ear present
Mastoid region pneumatized (convergence with Lorisidae)
Prominent posterior nasal spine
Fused mandibular symphysis
Robust mandibular corpora (convergent in *Propithecus*)
Node 3. Intrabullar latticework present
Postorbital plates of zygomatic closing eye socket
Zygomatic–alisphenoid contact (convergent in *Tarsius*)
Lacrimal canal opens within orbit

Appendix B: Confidence Limits, Decay Indices, and Character States at Nodes in Most-Parsimonious Constrained Tree

The tree described is shown in Fig. 19. Data are presented in the following sequence: Node, confidence limits for node from bootstrap analysis; decay index for node (number of unit steps parsimony must be relaxed to cause this clade to lose support); (? equivocal) derived character states supporting nodes (Character: state, description).

1; clade not found by heuristic bootstrap analysis; 1 step; 18: 1, laminar contact between lateral bulla wall and lateral pterygoid plate; ?21: 0 or 1, if derived relative to outgroup then 0, absence of suprameatal foramen; 27: 1, pronounced orbital constriction present below olfactory tract (lost in *Adapis–Mahgarita* clade); ?35: 0 or 1, if derived relative to outgroup then 1, pyramidal process laterally placed; 36: 1, medial pterygoid plate short but well defined from lateral pterygoid plate.

2; clade not found by heuristic bootstrap analysis; 1 step; 9: 1, internal carotid canal does not shield fenestra cochleae from ventral view; 11: 1, stapedial canal slightly smaller than promontory (stapedial reenlarged in *Shoshonius* and *Tetonius* and further reduced in *Tarsius*–Anthropoidea clade); 19: 1, extensive contact between lateral pterygoid plate and bulla wall (convergent in *Mahgarita;* lost in anthropoids); ?27: 0 or 1, if derived relative to outgroup then 1, loss of zygomatic-lacrimal contact on inferior orbital margin (equivocal because *Rooneyia* exhibits contact on only one side); 29: 1, foramen rotundum present (unknown for *Shoshonius, Tetonius,* and *Aegyptopithecus*) (convergent in *Eulemur*).

3; clade not found by heuristic bootstrap analysis; 1 step; ?20: 0 or 1, if derived relative to outgroup then 1, extensive flange of basioccipital overlaps medial bulla wall (lost in anthropoids, unknown for *Tetonius*); ?23: 0, 1 or 2, if derived relative to outgroup then 1 or 2, large or extremely large orbits; 26: 1, zygomaticolacrimal contact on inferior orbital margin absent (convergent in *Mahgarita* and *Ignacius*); ?31: 0 or 1, if derived relative to outgroup then 1, metopic suture fused in adult (convergent in *Eulemur* and *Tupaia;* equivocal because unknown for *Shoshonius* and *Tetonius*); 34: 0 or 1, if derived relative to outgroup then 1, posterior nasal spine small but distinct (unknown in *Tetonius,* further enlarged in *Aegyptopithecus;* convergent in *Adapis–Mahgarita* clade and *Plesiadapis*); ?40: 0 or 1, if derived relative to outgroup then 1, trough-like glenoid fossa (unknown in *Tetonius,* lost in anthropoids); ?41: 0 or 1, if derived relative to outgroup then 1, strong entoglenoid process (unknown in *Tetonius,* polymorphic in platyrrhines).

4; 59%; 3 steps; 1: 1, transverse septum arising from the cochlear housing and forming lateral wall of an anterior accessory cavity pneumatized from the tympanic cavity; 4: 1, perbullar pathway formed exclusively by the petrosal bone; 7: 1, posterior carotid foramen ventrally placed; 8: 2, posterior carotid foramen located anterior to fenestra cochleae (ventral to fenestra cochleae in some *Alouatta*); 11: 2, stapedial artery highly reduced or absent altogether (convergent in one specimen of *Mahgarita*); 14: 1, ventral edge of tympanic bone extrabullar or phaneric (convergent in *Galago senegalensis* and *Ignacius graybullianus*); 24: 2, postorbital septum present; 28: 2, presence of large os planum in medial orbital wall (unknown in *Aegyptopithecus*); 42: 1, narrow, haplorhine-like interincisor diastema (convergent in *Notharctus* and *Plesiadapis,* unknown for *Mahgarita, Shoshonius,* and *Tetonius*).

5; 90%; 5+ steps; 2: 1, anterior accessory cavity extends medial to tympanic cavity and is trabeculated; 3: 1, pneumatization of mastoid from epitympanic recess present (convergent in *Necrolemur, ?Tetonius, Adapis–Mahgarita* clade and *Galago senegalensis*); ?21:0, reduction of orbit size to small (reversal); 6: 1, posterior carotid foramen in the midline of the bulla; 18: 0, no contact between auditory bulla and lateral pterygoid plate (reversal); 20: 0, no flange of basioccipital riding up on to the lateral bulla wall (reversal); 21: 0, suprameatal foramen absent (reversal); 22: 1, parotic fissure not patent (convergent in strepsirhines and adapids); ?23: 0 or 1, if derived relative to omomyid condition then 0, small orbits; 25: 0, postorbital septum formed predominantly by zygomatic bone; 30: 1, lacrimal foramen positioned inside or on orbital margin (convergent in *Notharctus* and *Ignacius*); ?35: 0 or 1, if derived relative to omomyid condition then 1, laterally placed pyramidal processes (reversal); 39: 1, complete symphyseal fusion and robust mandibular corpora in adults (convergent in adapids); ?40: 0 or 1, if derived relative to omomyid condition then 0, glenoid fossa biconcave and transversely wide (reversal).

6; 60%; 1 step; 36: 1, posterior palatine torus absent.

7; clade not found by heuristic bootstrap search; 1 step; 6: 2, carotid foramen laterally placed; (convergent in *Tarsius, Shoshonius,* strepsirhines, and adapids); 11: 4, lack of stapedial and promontory canals (retained or convergent in *Microcebus* and *Galago senegalensis*); 16: 2, presence of a complete anular bridge (retained in omomyids); 24: 0, no postorbital closure; 29: 1, lacrimal and palatine separated by a large frontomaxillary contact (convergent in *Adapis,* retained in *Rooneyia* and *Necrolemur*).

8; clade not found by heuristic bootstrap search; 1 step; 19: 0, slight contact between lateral pterygoid plate and lateral bulla wall (very extensive in *Mahgarita*); ?21: 0, suprameatal foramen absent (primitive condition for primates equivocal); 22: 1, closure of parotic fissure (convergent in anthropoids); 35: 1, pyramidal processes laterally placed (primitive condition for primates equivocal).

9; clade not found by heuristic bootstrap search; 1 step; 6: 2, posterior carotid foramen laterally positioned (convergent in *Plesiadapis–Ignacius* clade, *Tarsius,* and *Shoshonius*); 12: 0, promontory canal forms an open trough on the promontory (reversion to complete canal in *Mahgarita*; 16: 1, linea semicircularis on petrosal bulla.

10; clade not found by heuristic bootstrap analysis; 1 step; 3: 1, stapedial canal slightly smaller than promontory (polymorphic in all three adapids); 39: 1, fused mandibular symphyses and robust mandibular corpora in adult (convergent in anthropoids).

11; 51%; 1 step; 3: 1, pneumatization of mastoid from epitympanic recess present; 27: 0, lack of pronounced interorbital constriction; 33: 1, small but distinct posterior nasal spine present (convergent in *Plesiadapis,* extant haplorhines, *Necrolemur* and *Shoshonius*); 36: 2, highly reduced or absent medial pterygoid plate (convergent in *Ignacius*); 37: 1, short snout relative to skull length (convergent in *Tarsius, Rooneyia,* and some extant anthropoids); 38: 0, deep maxilla (convergent in omomyids, *Tarsius, Aegyptopithecus,* and some extant anthropoids).

12; 100% (clade enforced via topological constraints; 3: 0, no pneumatization of mastoid from epitympanic recess (probably a primitive retention).

References

Alberch, P. 1985. Problems with the interpretation of developmental sequences. *Syst. Zool.* **34:** 46–58.

Ax, P. 1985. Stem species and the stem lineage concept. *Cladistics* **1:**279–287.

Beard, K. C. 1988. The phylogenetic significance of strepsirhinism in Paleogene primates. *Int. J. Primatol.* **9:**83–96.

Beard, K. C., and Godinot, M. 1988. Carpal anatomy of *Smilodectes gracilis* (Adapiformes, Notharctinae) and its significance for Lemuriform phylogeny. *J. Hum. Evol.* **17:**71–92.

Beard, K. C., Dagosto, M., Gebo, D. L., and Godinot, M. 1988. Interrelationships among primate higher taxa. *Nature* **331:**712–714.

Beard, K. C., Krishtalka, L., and Stucky, R. K. 1991. First skulls of the Early Eocene primate *Shoshonius cooperi* and the anthropoid–tarsier dichotomy. *Nature* **349:**64–67.

Beecher, R. M., 1977. Function and fusion at the mandibular symphysis. *Am. J. Phys. Anthrop.* **47**:325–336.
Beecher, R. M., 1979. Functional significance of the mandibular symphysis. *J. Morphol.* **159**:117–130.
Beecher, R. M., 1983. Evolution of the mandibular symphysis in Notharctinae (Adapidae, Primates). *Int. J. Primatol.* **4**:99–112.
Bown, T. M., and Rose, K. D. 1987. Patterns of dental evolution in early Eocene anaptomorphine primates (Omomyidae) from the Bighorn Basin, Wyoming. *Paleont. Soc. Mem.* **23.** *J. Paleont.* **61**(5):suppl.
Cartmill, M. 1970. *The Orbits of Arboreal Mammals.* Unpublished Ph.D. dissertation, University of Chicago.
Cartmill, M. 1972. Arboreal adaptations and the origin of the Order Primates. In: R. H. Tuttle (ed.), *The Functional and Evolutionary Biology of Primates,* pp. 97–122. Aldine, Chicago.
Cartmill, M. 1978. The orbital mosaic in prosimians and the use of variable traits in systematics. *Folia Primatol.* **30**:89–114.
Cartmill, M. 1980. Morphology, function and evolution of the anthropoid postorbital septum. In: R. L. Ciochon and A. B. Chiarelli (eds.), *Evolutionary Biology of the New World Monkeys and Continental Drift,* pp. 243–274. Plenum Press, New York.
Cartmill, M. 1983. Assessing tarsier affinities: Is anatomical description phylogenetically neutral? *Geobios Mem. Spec.* **6**:153–161.
Cartmill, M., and Kay, R. F. 1978. Craniodental morphology, tarsier affinities, and primate suborders. In: D. J. Chivers and K. A. Joysey (eds.), *Recent Advances in Primatology, Vol. 3,* pp. 205–214. Academic Press, London.
Cartmill, M., MacPhee, R. D. E., and Simons, E. L. 1981. Anatomy of the temporal bone in early anthropoids with remarks on the problem of anthropoid origins. *Am. J. Phys. Anthrop.* **56:** 3–22.
Cave, A. J. E. 1967. Observations on the platyrrhine nasal fossa. *Am. J. Phys. Anthrop.* **26**:277–288.
Cave, A. J. E., and Haines, R. W. 1940. The paranasal sinuses of the anthropoid apes. *J. Anat.* **74**:493–523.
Chappill, J. A. 1989. Quantitative characters in phylogenetic analysis *Cladistics* **5**:217–234.
Conroy, G. C. 1980. Ontogeny, auditory structures and primate evolution. *Am. J. Phys. Anthrop.* **52**:443–451.
Covert, H. H., and Williams, B. A. 1991. The anterior dentition of *Washakius insignis* and adapid–anthropoidean affinities. *J. Hum. Evol.* **21**:463–467.
Dagosto, M. 1988. Implications of postcranial evidence for the origin of Euprimates. *J. Hum. Evol.* **17**:35–56.
Dagosto, M. 1990. Models for the origin of the anthropoid postcranium. *J. Hum. Evol.* **19**:121–139.
Diamond, M. K. 1992. Homology and evolution of the orbitotemporal venous sinuses of humans. *Am. J. Phys. Anthrop.* **88**:211–244.
Earle, C. 1897a. On the affinities of *Tarsius:* A contribution to the phylogeny of the Primates. *Am. Naturalist* **31**:569–687.
Earle, C. 1897b. Relations of *Tarsius* to the lemurs and apes. *Science* **5**:258–260.
Earle, C. 1897c. Further considerations on the systematic position of *Tarsius. Science* **5**:657–658.
Elliot Smith, G. 1919. On the zoological position and affinities of *Tarsius. Proc. Zool. Soc. Lond.* **1919**:465–475.
Felsentein, J. 1985. Confidence limits on phylogenies: An approach using the boostrap. *Evolution* **39**(4):783–791.
Franzen, J. L. 1987. Ein neuer Primate aus dem Mitteleozän der Grube Messel (Deutschland, S-Hessen). *Cour. Forsch. Inst. Senckenberg* **91**:151–187.
Gebo, D. L. 1986. Anthropoid origins—the foot evidence. *J. Hum. Evol.* **15:** 421–430.
Gebo, D. L. 1989. Locomotor and phylogenetic considerations in anthropoid evolution. *J. Hum. Evol.* **18**:191–223.
Gebo, D. L., and Dagosto, M. 1988. Foot anatomy, climbing and the origin of the Indriidae. *J. Hum. Evol.* **17**:135–154.

Geoffroy Saint-Hilaire, E. 1812. Tableau des Quadrumanes, Ou des animaux composant le premier Order de la Classe des Mammifères. *Ann. Mus. Hist. Nat.* **19:**85–122, 156–170.

Gervais, P. 1854. *Histoire Naturelle des Mammifères.* Paris.

Gingerich, P. D. 1973. Anatomy of the temporal bone in the Oligocene anthropoid *Apidium* and the origin of the Anthropoidea. *Fol. Primatol.* **19:**329–337.

Gingerich, P. D. 1976. Cranial anatomy and evolution of the early Tertiary Plesiadapidae. *Univ. Michigan Pap. Paleontol.* **15:**1–141.

Gingerich, P. D. 1980. Eocene Adapidae, paleobiogeography, and the origin of South American Platyrrhini. In: R. L. Ciochon and A. B. Chiarelli (eds.), *Evolutionary Biology of the New World Monkeys and Continental Drift,* pp. 123–138. Plenum Press, New York.

Gingerich, P. D. 1984. Primate evolution: Evidence from the fossil record, comparative morphology and molecular biology. *Yearb. Phys. Anthropol.* **27:**57–72.

Gingerich, P. D., and Martin, R. D. 1981. Cranial morphology and adaptations in Eocene Adapidae. II. The Cambridge skull of *Adapis parisiensis. Am. J. Phys. Anthrop.* **56:**235–258.

Gingerich, P. D., and Schoeninger, M. 1977. The fossil record and primate phylogeny. *J. Hum. Evol.* **6:**483–505.

Gunnell, G. F. 1989. *Evolutionary History of Microsyopoidea (Mammalia, ?Primates) and the Relationship between Plesiadapiformes and Primates. University of Michigan Papers on Paleontology No. 27.* The University of Michigan, Ann Arbor.

Haines, R. W. 1950. The interorbital septum in mammals. *J. Linn. Soc. Lond. Zool.* **41:**585–607.

Hennig, W. 1966. *Phylogenetic Systematics.* University of Illinois Press, Urbana.

Hershkovitz, P. 1977. *Living New World Monkeys (Platyrrhini). Volume 1.* Chicago: University of Chicago Press.

Honey, J. G. 1990. New washakiin primates (Omomyidae) from the Eocene of Wyoming and Colorado, and comments on the evolution of the Washakiini. *J. Vert. Paleont.* **10:**206–221.

Hubrecht, A. A. W. 1896. Die Keimblase von *Tarsius. Festschr. Gegenbaur.* Leipzig.

Hubrecht, A. A. W. 1897a. *The Descent of the Primates.* New York: Charles Scribner's Sons.

Hubrecht, A. A. W. 1897b. Relations of *Tarsius* to the lemurs and apes. *Science* 5:550–551.

Hylander, W. L. 1977. *In vivo* bone strain in the mandible of *Galago crassicaudatus. Am. J. Phys. Anthrop.* **46:**309–326.

Hylander, W. L. 1979. An experimental analysis of temporomandibular joint reaction force in macaques. *Am. J. Phys. Anthrop.* **51:**433–456.

Hylander, W. L. 1984. Stress and strain in the mandibular symphysis of primates: A test of competing hypotheses. *Am. J. Phys. Anthrop.* **64:**1–46.

Hylander, W. L. 1985. Mandibular function and biomechanical stress and scaling. *Am. Zool.* **25:**315–330.

Kampen, C. J. van. 1905. Die Tympanalgegend des Säugetierschädels. *Gegenbaurs Morphol. Jahrb.* **34:**321–722.

Kay, R. F., Thewissen, J. G. M., and Yoder, A. D. 1992. The cranial anatomy of *Ignacius graybullianus* and the affinities of the Plesiadapiformes. *Am. J. Phys. Anthrop.* **89:**477–498.

Kay, R. F. and Cartmill, M. 1977. Cranial morphology and adaptations of Palaechthon nacimienti and other Paromomyidae (Plesiadapoidea, ? Primates), with a description of a new genus and species. *J. Hum. Evol.* **6:**19–53.

Kay, R. F., Thorington, R. W. Jr. and Houde, P. 1990. Eocene plesiadapiform shows affinities with flying lemurs not primates. *Nature* **345:**342–344.

Kermack, K. A., Musset, F. and Rigney, H. W. 1981. The skull of *Morganucodon. Zool. J. Linn. Soc.* **71:**1–158.

Kielan-Jaworowska, Z. 1981. Evolution of the therian mammals in the Late Cretaceous of Asia. Part IV. Skull structure in *Kennalestes* and *Asioryctes. Palaeont. Pol.* **42:**25–78.

Kielan-Jaworowska, Z. 1984. Evolution of the therian mammals in the Late Cretaceous of Asia. Part V. Skull structure in Zalambdalestidae. *Palaeont. Pol.* **46:**107–117.

Klauuw, C. J. van der. 1931. On the auditory bulla in some fossil mammals with a general introduction to this region of the skull. *Bull. Am. Mus. Nat. Hist.* **62:**1–352.

Le Gros Clark, W. E. 1934. On the skull structure of *Pronycticebus gaudryi. Proc. Zool. Soc. Lond.* **1934:**19–27.

Le Gros Clark, W. E. 1959. *The Antecedents of Man.* Harper, New York.
MacPhee, R. D. 1981. Auditory regions of primates and eutherian insectivores: Morphology, ontogeny and character analysis. *Contrib. Primatol.* **18:**1–282.
MacPhee, R. D. E., and Cartmill, M. 1986. Basicranial structures and primate systematics. In: D. R. Swindler and J. Erwin (eds.), *Comparative Primate Biology, Vol. 1: Systematics, Evolution and Anatomy,* pp. 219–275. Alan R. Liss, New York.
MacPhee, R. D. E., Cartmill, M. and Gingerich, P. D. 1983. New Palaeogene primate basicrania and the definition of the order Primates. *Nature* **301:**509–511.
Margush, T., and McMorris, F. R. 1981. Consensus *n*-trees. *Bull. Math. Biol.* **43:**239–244.
Martin, R. D. 1990. *Primate Origins and Evolution. A Phylogenetic Reconstruction.* London: Chapman and Hall.
Mishler, B. D., Donoghue, M. J., and Albert, V. A. Undated. The decay index as a measure of robustness within a cladogram.
Mivart, St. G. J. 1864. Notes on the crania and dentition of the Lemuridae. *Proc. Zool. Soc. Lond.* **1864:**611–648.
Mivart, St. G. J. 1873. On *Lepilemur* and *Cheirogaleus,* and on the zoological rank of the Lemuroidea. *Proc. Zool. Soc. Lond.* **1873:**484–510.
Müller, J. 1934. The orbitotemporal region of the skull of the Mammalia. *Arch. Neerl. Zool.* **1:**118–259.
Northcutt, R. G. 1990. Ontogeny and phylogeny: A re-evaluation of some conceptual relationships and some applications. *Brain Behav. Evol.* **36:**116–140.
Novacek, M. J. 1986. The skull of leptictid insectivorans and the higher-level classification of eutherian mammals. *Bull. Am. Mus. Nat. Hist.* **183:**1–112.
Pocock, R. I. 1918. On the external characters of lemurs and of *Tarsius. Proc. Zool. Soc. Lond.* **1918:**19–53.
Radinsky, L. B. 1970. The fossil record of prosimian brain evolution. In: C. Noback and W. Montagna (eds.), *The Primate Brain. Advances in Primatology, Vol. 1,* pp. 209–224. Appleton-Century-Crofts, New York.
Radinsky, L. B. 1979. *The Fossil Evidence of Primate Brain Evolution (James Arthur Lecture).* American Museum of Natural History, New York.
Rasmussen, D. T. 1986. Anthropoid origins: A possible solution to the Adapidae–Omomyidae paradox. *J. Hum. Evol.* **15:**1–12.
Rasmussen, D. T. 1990. The phylogenetic position of *Mahgarita stevensi:* Protoanthropoid or lemuroid? *Int. J. Primatol.* **11:**439–469.
Rasmussen, D. T., and Simons, E. L. 1988. New specimens of *Oligopithecus savagei,* early Oligocene primate from Egypt. *Fol. Primatol.* **51:**182–208.
Rasmussen, D. T., and Simons, E. L. 1992. Paleobiology of the Oligopithecines, the earliest known anthropoid primates. *Int. J. Primatol.* **13:**1–32.
Ravosa, M. J. 1991. Structural allometry of the prosimian mandibular corpus and symphysis. *J. Hum. Evol.* **20:**3–20.
Rose, K. D. and Bown, T. M. 1991. Additional fossil evidence on the differentiation of the earliest euprimates. *Proc. Natl. Acad. Sci. USA* **88:**98–101.
Rosenberger, A. L. 1985. In favor of the *Necrolemur*–tarsier hypothesis. *Fol. Primatol.* **45:**179–194.
Rosenberger, A. L., and Szalay, F. S. 1980. On the tarsiiform origins of Anthropoidea. In: R. L. Ciochon and A. B. Chiarelli (eds.), *Evolutionary Biology of the New World Monkeys and Continental Drift,* pp. 139–157. Plenum Press. New York.
Ross, C. F. 1991. Muscle attachment and the function of the haplorhine postorbital septum. *Am. J. Phys. Anthrop.* [*Suppl.*] **12:**154.
Ross, C. F. 1993. *The Functions of the Postorbital Septum and Anthropoid Origins.* Unpublished Ph.D. thesis, Duke University.
Ross, C. F. 1994. The muscular and osseous anatomy of the primate anterior temporal fossa and the functions of the postorbital septum. *Am. J. Phys. Anthrop.* (in review).
Ross, C. F., and Ravosa, M. 1993. Basicranial flexion, relative brain size and facial kyphosis in anthropoid primates. *Am. J. Phys. Anthrop.* **91:**305–324.

Russell, D. E. 1964. Les mammifères paléocène d'Europe. *Mem. Mus. Natl. Hist. Nat. Paris [C]* **13:**1–321.

Saban, R. 1963. Contribution a l'étude de l'os temporal des primates. Déscription chez l'homme et les prosimiens. Anatomie comparée et phylogenie. *Mem. Mus. Natl. Hist. Nat. Paris [A]* **29:**1–378.

Schmid, P. 1981. Comparison of Eocene nonadapids and *Tarsius.* In: A. B. Chiarelli and R. S. Corrucini (eds.), *Primate Evolutionary Biology. Selected Papers (Part A) of the VIIIth Congress of the International Primatological Society,* pp. 6–13. Springer-Verlag, Berlin.

Simons, E. L. 1972. *Primate Evolution. An Introduction to Man's Place in Nature.* Macmillan, New York.

Simons, E. L. 1989. Description of two genera and species of Late Eocene Anthropoidea from Egypt. *Proc. Natl. Acad. Sci. USA* **86:**9956–9960.

Simons, E. L. 1991. Additional evidence on the differentiation of the earliest euprimates. *Proc. Natl. Acad. Sci. USA* **88:**98–101.

Simons, E. L. 1992. Diversity in the early Tertiary anthropoidean radiation in Africa. *Proc. Natl. Acad. Sci. USA* **89:**10743–10747.

Simons, E. L., and Rasmussen, D. T. 1989. Cranial anatomy of *Aegyptopithecus* and *Tarsius* and the question of the tarsier–Anthropoidean clade. *Am. J. Phys. Anthrop.* **79:**1–23.

Simons, E. L., and Russell, D. 1960. Notes on the cranial anatomy of *Necrolemur. Breviora* **127:** 1–14.

Simpson, G. G. 1945. The principles of classification and a classification of mammals. Bull. Am. Mus. Nat. Hist. **85:**1–350.

Simpson, G. G. 1961. *Principles of Animal Taxonomy.* Columbia University Press, New York.

Skelton, R. R., and McHenry, H. 1992. Evolutionary relationships among early hominids. *J. Hum. Evol.* **23:**309–349.

Sober, E. 1989. *Reconstructing the Past: Parsimony, Evolution, and Inference.* MIT Press, Cambridge, Massachusetts.

Spatz, W. B. 1968. Die Bedeutung der Augen fur die sagittale Gestaltung des Schädels von *Tarsius* (Prosimiae, Tarsiiformes). *Fol. Primatol.* **9:**22–40.

Stehlin, H. G. 1912. Die Säugetiere des schweizerischen Eoceäns. Critischer Catalogen der Materialen. VII(I): *Adapis. Abh. Schweiz. Palaont. Ges.* **38:**1165–1298.

Swofford, D. L. 1991. PAUP, Phylogenetic Analysis Using Parsimony, Version 3.1. Computer program distributed by the Illinois Natural History Survey, Champaign, Illinois.

Szalay, F. S. 1973. New Paleocene primates and a diagnosis of the new suborder Paomomyiformes. *Fol. Primatol.* **19:**73–87.

Szalay, F. S. 1975. Phylogeny of primate higher taxa: The basicranial evidence. In: W. P. Luckett and F. S. Szalay (eds.), *Phylogeny of the Primates: A Multidisciplinary Approach,* pp. 91–125. Plenum Press, New York.

Szalay, F. S. 1976. Systematics of the Omomyidae (Tarsiiformes, Primates): Taxonomy, phylogeny and adaptations. *Bull. Am. Mus. Nat. Hist.* **156:**157–450.

Szalay, F. S., and Dagosto, M. 1988. Evolution of hallucial grasping in the primates. *J. Hum. Evol.* **17:**1–33.

Szalay, F. S., and Delson, E. 1979. *The Evolutionary History of the Primates.* Academic Press, New York.

Szalay, F. S., and Wilson, J. A. 1976. Basicranial morphology of the early Tertiary tarsiiform *Rooneyia* from Texas. *Fol. Primatol.* **25:**288–293.

Szalay, F. S., Rosenberger, A. L., and Dagosto, M. 1987. Diagnosis and differentiation of the Order Primates. *Yearb. Phys. Anthropol.***30:**75–105.

Wible, J. R. 1984. *The Ontogeny and Phylogeny of the Mammalian Cranial Arterial Pattern.* Ph.D. dissertation, Duke University.

Wible, J. R. 1986. Transformations in the extracranial course of the internal carotid artery in mammalian phylogeny. *J. Vert. Paleontol.* **6:**313–325.

Wible, J. 1990. Petrosals of late Cretaceous marsupials from North America and a cladistic analysis of the petrosal in therian mammals. *J. Vert. Paleontol.* **10:**183–205.

Wiley, E. O. 1981. *Phylogenetics: The Theory and Practice of Phylogenetic Systematics.* John Wiley & Sons, New York.
Williams, P. L., Warwick, R., Dyson, M., and Bannister, L. H. 1989. *Gray's Anatomy,* 37th ed. Churchill Livingstone, London.
Wilson, J. A., and Szalay, F. S. 1976. New adapid primate of european affinities from Texas. *Fol. Primatol.* **25:**294–312.
Wortman, J. L. 1903–04. Studies of Eocene Mammalia in the Marsh Collection, Peabody Museum. Part II. Primates. *Am. J. Sci.* **15:**163–176, 399–414, 419–436; **16:**345–368; **17:**23–33, 133–140, 203–214.
Yoder, A. D. 1992. *The Phylogenetic Affinities of the Cheirogaleidae: A Molecular and Morphological Analysis.* Unpublished Ph.D. thesis, Duke University.

Anatomy, Antinomies, and the Problem of Anthropoid Origins

16

MATT CARTMILL

Introduction

At the moment, nearly all students of primate evolution agree that modern anthropoids comprise a holophyletic clade with respect to other living primates, i.e., that there was once a single species from which all anthropoids, and no other extant animals, are descended. The 1992 conference at Duke University upon which this volume is based demonstrated that there is not much current agreement on other points concerning anthropoid origins, no matter whether cladistic questions (which primates are the phyletic sister group of the Anthropoidea?) or gradistic questions (which characteristic traits of modern anthropoids did the ancestral anthropoid species possess?) are at issue.

In this chapter, I wish to look at some of the cladistic questions concerning anthropoid relationships and to suggest some reasons why they have proved so intractable. To a large extent, our persistent inability to answer these questions to everybody's satisfaction reflects the equivocal nature of the relevant evidence, which has always confronted us with a seeming conflict between neontological and paleontological data. I think, however, that our failure also reflects some underlying defects in the traditions of comparative morphological argument and discourse.

MATT CARTMILL • Department of Biological Anthropology and Anatomy, Duke University, Durham, North Carolina 27710.
Anthropoid Origins, edited by John G. Fleagle and Richard F. Kay. Plenum Press, New York, 1994.

Neontology versus Paleontology

The neontological evidence available today points fairly clearly to a tarsier–anthropoid clade that excludes other living primates. Although that cladistic hypothesis is not contradicted by the primate fossil record, it also receives no support from it. This incongruity and the arguments that it has engendered have persisted now for almost a century, and new fossils and new sources of data have tended to reinforce it rather than resolve it.

The first sustained comparative anatomic argument for a tarsier–anthropoid clade was put forward by A. A. W. Hubrecht (1897), who observed that tarsiers and anthropoids resembled one another embryologically (and differed from more primitive mammals) in the early formation of the chorionic cavity and the correlated exclusion of the yolk sac from the process of placentation. In toothcomb prosimians (Luckett, 1975, 1976), the allantois and yolk sac are large, and a transitory choriovitelline placenta is formed between the epithelium of the uterine wall and the combined chorion and yolk-sac splanchnopleure (Fig. 1A). As development proceeds, the yolk sac dwindles, and the allantois spreads around the embryo to form a diffuse, chorioallantoic (but still epitheliochorial) secondary placenta. In tarsiers and anthropoids, on the other hand, the reduced yolk sac is separated from the blastocyst very early on by precocious expansion of the chorionic cavity (Fig. 1B,C). No choriovitelline placenta is formed. The allantois is vestigial, but its blood vessels vascularize a hemochorial placenta. Hemochorial placentation, which is a feature shared by many nonprimate mammals, is not in itself compelling evidence for a tarsier–anthropoid clade; however, the suppression of the yolk sac's role in placentation, the early vascularization of the placenta by the allantoic arteries, and the reduction of the allantoic sac to a vestige are plausibly interpreted as haphlorhine synapomorphies (Martin, 1990, p. 457).

Such features of placentation were among the anthropoid traits of *Tarsius* that Pocock (1918) pointed to in erecting the taxon Haplorhini to encompass tarsiers and anthropoids. The second haplorhine trait that Pocock designated was the structure of the nose. In strepsirhine ("twisty-nosed") primates, the nostrils point laterally, and are separated by the rhinarium, a moist, naked patch of skin around the nostrils bearing a median cleft that extends back onto the palate. This extension of the perinasal skin onto the palate creates a gap between the median upper incisors. In haplorhine ("single-nosed") primates—i.e., tarsiers and anthropoids—there is no rhinarium and no gap.

Pocock's third haplorhine characteristic was the postorbital septum. In lemurs and lorises, as in mammals generally, the periorbita has a broad lateral and posterior contact with the temporal musculature (Fig. 2). In *Tarsius* and anthropoids, the zygomatic, frontal, and alisphenoid send bony processes into this region to form a composite wall between the orbit and the temporal muscles (Fig. 3). In a 1980 article, I argued that this walling-off of the orbit is correlated with the presence of a fovea centralis in the extant haplorhines,

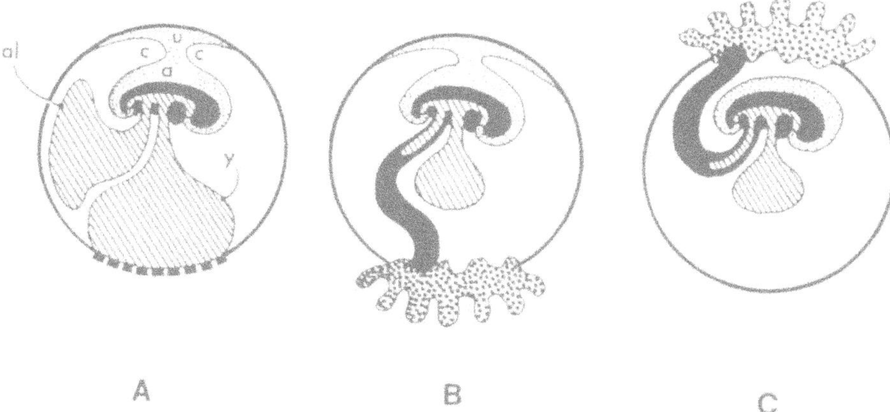

Fig. 1. Comparative placentation in primates (diagrammatic). In typical strepsirhines (A), as in most other amniotes, the allantois (al) and yolk sac (y) are large, and chorioamniotic folds (c) fuse over the dorsal surface of the embryo to enclose the amniotic cavity (a) and pinch it off from the uterine lumen (u). The initial placenta (heavy dashed line) is a choriovitelline organ, vascularized by the vessels of the yolk sac. In *Tarsius* (B) and anthropoids (C), the yolk sac and allantois are reduced, a choriovitelline placenta never forms, and the definitive chorioallantoic placenta (heavy stipple) is a hemochorial organ vascularized by the allantoic arteries via the body stalk (black). In anthropoids, the amniotic cavity is never confluent with the uterine lumen; it appears precociously as a cavitation in the ectoderm (as a necessary corollary of implantation and placenta formation at the embryonic pole of the blastocyst). White areas in each diagram represent the chorionic cavity (extraembryonic celom); light stipple, the amniotic cavity and uterine lumen; diagonal hachure, the gut and the associated endoderm-lined extraembryonic membranes.

and I endorsed Martin's (1973) conclusion that the fovea and the haplorhines' shared lack of a retinal tapetum lucidum point to their common descent from a more anthropoid-like diurnal ancestor.

Two additional cranial features shared by tarsiers and anthropoids have been proposed as synapomorphies of Pocock's Haplorhini. One is the extreme approximation of the orbits seen in tarsiers and small anthropoids, in which the interorbital region is compressed into a thin but extensive sheet of bone (the interorbital septum) and the olfactory connections between the nasal fossa and the braincase are displaced forward and reduced to the dimensions of a slender canal (Cave, 1967; Cartmill, 1972). The other derived cranial feature linking anthropoids to *Tarsius* is the pattern of pneumatization of the middle ear. In the developing tarsier fetus, a pneumatic diverticulum grows medially from the posterior end of the auditory tube and expands medially into the front end of the petrosal bulla, inflating the petrosal plate and ballooning up into a secondary cavity enclosed on all sides by the petrosal. This "anterior accessory cavity" lies anteromedial to the tympanic cavity proper, which is correspondingly reduced and separated from the anterior cavity by a bony septum. The internal carotid artery follows a "perbullar" course (Cart-

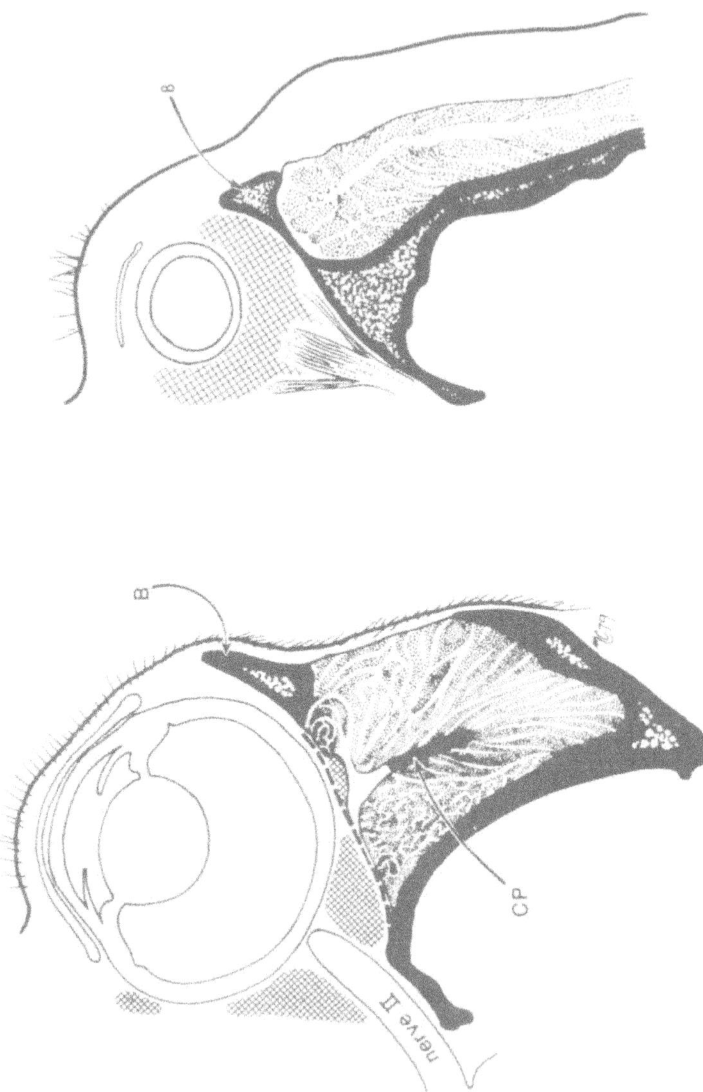

Fig. 2. Diagrammatic horizontal section through the orbits of *Lemur catta* and *Homo sapiens*. In *Lemur* (left), there is a wide expanse of free periorbita (heavy dashes) across the orbital surface of the temporalis muscle; in *Homo* (right), the muscle is separated from the periorbita by a bony septum. B, postorbital bar; CP, coronoid process. Cross-hatching represents fat. After Cartmill (1980) and McGrath and Mills (1984).

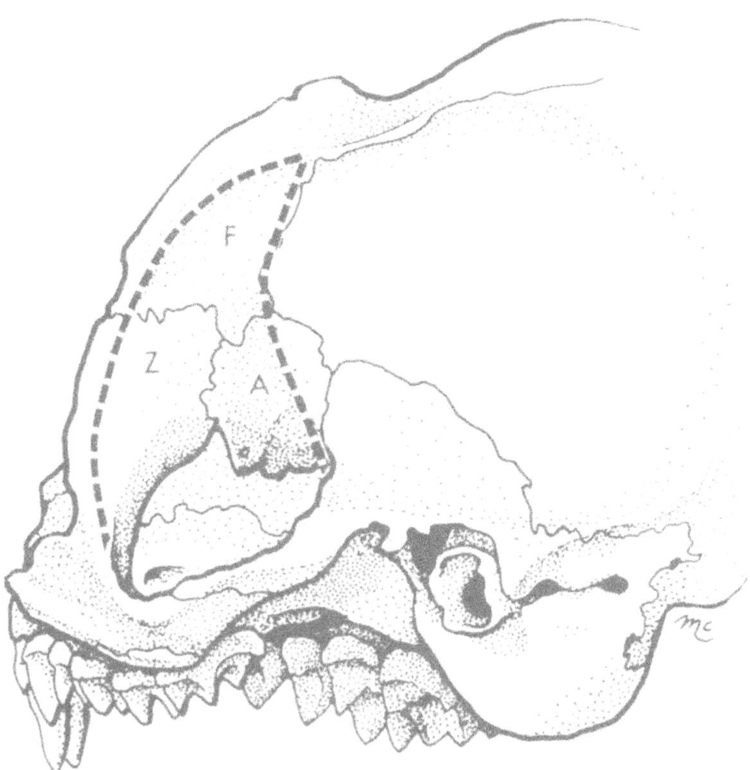

Fig. 3. Posterolateral view of left orbitotemporal region of the skull of an adult *Tarsius syrichta* with persistent sutures (Field Museum No. 56741), showing contributions of frontal (F), zygomatic (Z), and alisphenoid (A) to the postorbital septum (dashed outline). After Cartmill (1980).

mill *et al.*, 1981) through this septum in adult tarsiers, running vertically upward between the tympanic and anterior accessory cavities to enter the braincase. Similar ontogenetic events occur during fetal life in the anthropoids that have been studied. The principal difference in this regard between *Tarsius* and anthropoids is that the anthropoid anterior accessory cavity develops into a mass of interconnected air cells separated by bony trabeculae, rather than a single large cavity like that of *Tarsius* (MacPhee and Cartmill, 1986).

As far as I know, these morphogenetic processes and the otic morphologies they produce are uniquely shared by tarsiers and anthropoids. Other cranial traits that have been proposed as synapomorphies of a tarsier–anthropoid clade (presence of a separate foramen rotundum, reduction of the stapedial artery, anterior and/or medial repositioning of the posterior carotid foramen, loss of the subtympanic extension of the tympanic cavity) have almost certainly evolved in parallel in other primate lineages, and their phylogenetic significance is accordingly ambiguous.

The biochemical data bearing on tarsier affinities also present some ambiguities and contradictions, but the bulk of the evidence supports the tarsier–anthropoid clade. Yoder's (1992) phylogenetic analysis of nucleotide sequence data from the mitochondrial cytochrome B gene shows *Tarsius* clustering with strepsirhines. However, this conclusion is suspect, since the same analysis also assigns *Homo* to the strepsirhine clade rather than pairing it with *Saimiri* (Yoder, 1992, p. 126). The currently available data on nucleotide sequences from the nuclear part of the genome, and on peptide sequences in proteins that it codes for, provide generally consistent if not overwhelming support for the proposition that *Tarsius* is the sister group of anthropoids among living primates (Beard and Goodman, 1976; Baba *et al.,* 1982; Czelusniak *et al.,* 1988; de Jong and Goodman, 1988; Koop *et al.,* 1989a,b).

A striking metabolic trait linking tarsiers to anthropoids is the need for dietary vitamin C. Almost all mammals that have been tested, including stepsirhine primates, retain the primitive tetrapod ability to synthesize ascorbic acid in their livers. However, anthropoids and tarsiers (like guinea pigs and certain bats) lack a crucial enzyme that catalyzes the final step in the synthesis of ascorbic acid from glucose. These animals accordingly require vitamin C in their diets to prevent scurvy (Elliot *et al.,* 1966; Pollock and Mullin, 1987).

All these facts support Pocock's concept of Haplorhini as a clade including anthropoids and tarsiers but excluding other living primates. The collective force of these various lines of evidence has been sufficient to convince most investigators over the course of the last 20 years or so that anthropoids and tarsiers are more closely related to each other than either is to any of the extant lemurs and lorises. There have of course been prominent dissenters, and different investigators have propounded widely differing lists of synapomorphies linking tarsiers to anthropoids and other primate groups. Pettigrew *et al.* (1989) infer from the neuroanatomy of the visual system that *Tarsius* is the sister group of all other living primates—a hypothesis formerly urged on other grounds, but subsequently abandoned, by Gingerich (1978) and Schwartz (1978). Schwartz and Tattersall (1987) now contend that tarsiers are deeply embedded in the strepsirhine clade as the sister group of Lorisiformes. In support of this contention, they point to various dental resemblances between tarsiers and lorisiforms and to such galago-like features of *Tarsius* as orbital enlargement, elongation of the calcaneus and navicular, and supposed homologies of chromosomal G-banding (Poorman *et al.,* 1985). Whereas Schwartz and Tattersall admit the anterior accessory cavity and the perbullar course of the internal carotid as possible haplorhine synapomorphies but dismiss tarsier–anthropoid resemblances in placentation as probable convergences (on the grounds that *Tarsius* is nonanthropoidlike in its modes of implantation and amnion formation), Rasmussen (1986; Simons and Rasmussen, 1989) rejects the supposedly anthropoid-like features of the tarsier skeleton as false homologies while accepting narial anatomy, placentation, and biochemistry as evidence for a tarsier–anthropoid clade excluding toothcomb prosimians (but including most Eocene primate genera). Rosenberger (1985), on the other hand, recognizes a narrow interorbital region and small nasal cavity as probable synapomorphies of Pocock's Haplorhini (including omomyids) but regards the haplorhine condition itself as a primitive feature within primates and therefore rejects it as evidence for a sister-group relationship between tarsiers and anthropoids (Rosenberger and Strasser, 1985).

Despite these and other sources of persistent disagreement, it seems clear on balance that neontological data reveal several plausible molecular, physiological, and anatomic apomorphies that link anthropoids to tarsiers. Candidate synapomorphies linking either tarsiers or anthropoids to any toothcomb prosimians are far fewer and more dubious. There are real morphological resemblances between tarsiers and certain toothcomb prosimians (especially galagos), but they are unlikely to represent synapomorphies. The supposed resemblances between the chromosomal banding patterns of galagos and tarsiers have proved to be insubstantial (Healy, 1992). What we know about living primates thus supports the cladistic reality of Pocock's Haplorhini.

What we know about fossil primates does not. The difficulties of applying the haplorhine–strepsirhine distinction to early fossil primates have been remarked by many paleoprimatologists, from G. G. Simpson in 1945 (p. 182) down to the present. Of the cranial features that can be reasonably interpreted as tarsier–anthropoid synapomorphies, the only one that seems to be present in any Paleogene prosimians is the apical interorbital septum, a smaller version of which is seen in certain microchoerines and perhaps in *Tetonius* (Cartmill, 1975; Cartmill and Kay, 1978; Szalay *et al.*, 1987). No early prosimians exhibit a postorbital septum or an anterior accessory cavity of either the tarsier or anthropoid type. Dagosto and her co-workers (Dagosto, 1985; Szalay *et al.*, 1987) have enumerated several postcranial features shared by tarsiers, primitive anthropoids, and at least some of the so-called Eocene "tarsioids," but the only one of these that we can be reasonably sure is not primitive for euprimates is the rigid inferior tibiofibular joint, a trait that has evolved in parallel in many mammalian lineages and furnishes correspondingly weak evidence for the reality of the haplorhine clade.

Some adapids are thought to share apomorphies, including various derived features of the anterior dentition (vertically implanted incisors, fusion of the mandibular symphysis, $I_2 > I_1$, premolar–canine honing) with anthropoids but not with tarsiers (Gingerich, 1977, 1984; Rasmussen, 1986; Simons, 1989). However, it is becoming increasingly difficult to interpret all of these anterior dental features as adapid–anthropoid synapomorphies, since some omomyids also had anthropoid-like lower incisors (Covert and Williams, 1991), and certain early anthropoids probably retained unfused mandibular symphyses (Simons, 1990, 1992). Furthermore, adapids also appear to share several apomorphies of the postcranial skeleton with toothcomb prosimians but not with anthropoids (Dagosto, 1988).

An impressive list of candidate synapomorphies (fused tibiofibula, tarsal elongation, gutter-like glenoid fossa, tubular ectotympanic, peaked choanae, extensions of the basioccipital and pterygoid plates onto the bulla) link modern tarsiers to certain omomyids but not to anthropoids (Gingerich, 1981; Rosenberger, 1985; Beard and MacPhee, Chapter 3, this volume). These traits provide evidence supporting the view that the phyletic sister group of Anthropoidea is a tarsiiform grouping including *Tarsius* and the Omomyidae *sensu lato.* However, the distribution of these tarsier-like traits among early "tarsioids" is spotty and inconsistent (*Necrolemur,* for example, has a tubular ectotympanic but no bullar flange of the basioccipital, whereas the reverse is true of *Shoshonius*), and so only a small percentage of them can be consistently interpreted as synapomorphies of a broadly defined tarsiiform group within Haplorhini (Beard and MacPhee, Chapter 3, this volume).

On Constructing and Deconstructing Morphology

All of this can be summarized by saying that the early fossil primates are not telling us the same phylogenetic story that the living primates tell us. How, in practice, do evolutionary biologists resolve dilemmas engendered by this sort of conflicting evidence? When we are dealing with only three taxa—call them A, B, and C—the currently popular procedures are simple and easily understood. We begin by making a list of all the things that A and B have in common but C does not, and go on to make similar lists for the A–C and B–C pairs. We then purge each list of any items that we have reason to think were present in the last common ancestor of all three taxa. Finally, we count the remaining (nonprimitive) traits in each of the three lists. The longest list wins the contest, and the two taxa it represents are united in a clade that excludes the third.

There are some problems with this procedure, the most fundamental one being that it becomes unworkable as the number of taxa increases and the number of possible clades skyrockets. But apart from this purely practical difficulty, this procedure presents some interesting theoretical difficulties. They can be summed up by saying that the determination of what will count as a morphological similarity is not adequately constrained by the facts of biology.

The number of resemblances we can find between any two real objects—particularly objects as complex as organisms—is practically limitless. A sufficiently ingenious or perverse investigator can therefore go on for a surprisingly long time seeking out such resemblances and adding them to a favored list to make sure of securing a desired outcome. For instance, if we wished to argue for an elephant–human clade that excludes chimpanzees, we might advance such candidate elephant–human synapomorphies as a projecting nose, a bony chin, hairlessness, and an inability to gallop. To take a less fanciful example, if we wish to conclude that Megachiroptera and Primates form a clade that excludes Microchiroptera, we can point with Pettigrew *et al.* (1989, p. 492) to the fact that both megabats and primates flex their necks while resting (though of course megabats do so while hanging upside down).

A desired conclusion can also be bolstered by subtraction as well as addition—that is, by subtracting items from the lists that one dislikes before the listed items are tallied. One way to do this is to invert the morphocline polarity of inconveniently distributed characters, thereby changing unwanted synapomorphies into irrelevant symplesiomorphies to be purged from the list. Such a polarity reversal can often be accomplished by selection of an appropriate outgroup (see Beard and MacPhee, Chapter 3, this volume, on the tubular acoustic meatus). It can also be accomplished by reorganizing a multistate morphocline into a dichotomized variable. For example, the postorbital septum of *Tarsius* can be dichotomized out of existence by lumping the tarsier and lemur conditions together into a "nonanthropoid" or "primitive" character state that contrasts with a "fully anthropoid" condition.

A more common strategy is to begin by dismissing unwanted resemblances as convergences, which can then be evicted from the lists on grounds of *prima facie* nonhomology before undertaking any phylogenetic analysis. This is commonly done by seeking out and emphasizing either morphological or functional differences between the structures involved. A purely morphological example of this sort of *a priori* homology judgment is furnished by the argument of Simons and Rasmussen (1989) that the postorbital septum of *Tarsius* cannot be homologous with that of anthropoids because the zygomatic component of the septum is relatively smaller in tarsiers than in anthropoids.

Interactions between morphology and function are also sometimes invoked in making *a priori* judgments of nonhomology. This has been done in two opposite ways by primatologists seeking to dismiss resemblances between tarsiers and anthropoids. The first way is that adopted by Packer and Sarmiento (1984), who argue that tarsier–anthropoid similarities in middle-ear morphology are not homologous because they are functionally similar but morphologically different. The second way is to argue the converse: that similar structures in tarsiers and anthropoids can be dismissed as nonhomologous if they are *morphologically* similar but *functionally* different. For example, Schwartz and Tattersall (1987, p. 28) contend that the apical interorbital septa of tarsiers and anthropoids are nonhomologous despite their morphological similarities because they have different functional correlates (being side effects, in the view of these authors, of optic hypertrophy in tarsiers and of olfactory reduction in anthropoids). Similar arguments involving the postorbital septum are elaborated by Simons and Rasmussen (1989), who point to the unique periorbital processes of the maxilla in *Tarsius* as an indication that the periorbital ossifications in tarsiers and anthropoids serve different functions and must therefore have evolved independently.

But perhaps the most common, subtle, and rhetorically effective way of eliminating unwanted morphological resemblances from consideration is to define them out of existence by reconceptualizing the anatomy of the region in question. This covert objective underlies a great deal of comparative anatomic description and helps to explain why new concepts and terms are perpetually being advanced in redescribing anatomic regions that were originally described in considerable detail by 19th-century investigators.

Many clear instances of this sort of deconstruction of unwanted similarities can be found in the 1989 article by Simons and Rasmussen, which defines away most of the special cranial resemblances between *Tarsius* and anthropoids that other investigators have posited. The postorbital septum, for example, is reconceptualized by Simons and Rasmussen into postorbital processes of the zygomatic and of the frontal, thus reducing the similarities between tarsiers and anthropoids that reside in the concept of the septum as an unanalyzed whole. The anterior accessory cavity of the middle ear is redefined, more or less according to Szalay's (1975) usage, as synonymous with van Kampen's (1905) "hypotympanic sinus," whereby it ceases to be a peculiarity of the extant Haplorhini. The "perbullar" pathway of the internal carotid artery as conceptualized by MacPhee and Cartmill (Cartmill *et al.*, 1981; MacPhee and Cartmill, 1986) is rejected by adopting Wible's (1984) much broader conception of the same term, whereupon it ceases to be a special resemblance between tarsiers and anthropoids. The apical interorbital septum is similarly rejected as a tarsier–anthropoid similarity on the grounds that *Galago senegalensis* shows such a septum in the conceptually different sense in which the term was originally used by Haines (1950), namely, a small patch of unpneumatized orbitosphenoid about the same size as the optic foramen and directly in front of it.

I do not intend by these observations to protest Simons and Rasmussen's reconceptualizations of various anatomic regions to support their own phylogenetic scheme. My own (Cartmill, 1972) redefinition of Haines's concept of the apical interorbital septum, for example, represented exactly the same sort of maneuver two decades previously as that carried out by Simons and Rasmussen. But though I will present some reasons for rejecting certain aspects of the particular picture painted by Simons and Rasmussen (see also Ross, Chapter 15; Beard and MacPhee, Chapter 3, this volume), I cannot discern any overarching principles that provide more general grounds for preferring one such conceptual scheme to another. What principles of anatomic conceptualization, for example, might be invoked in adopting Luckett's (1975, Table 1) conception of amniogenesis in *Tarsius* as a two-stage process, of which the first stage is homologous to the cavitation event in anthropoids, rather than Schwartz's (1978) conception of it as a single global event that is unlike the anthropoid process as a whole? What general rules warrant accepting my conception of the bony septum between the tympanic and anterior accessory cavities of anthropoids and tarsiers as a unitary structure (within which the carotid canal may assume various positions), as opposed to Simons and Rasmussen's analysis of that septum into nonhomologous precarotid and postcarotid components? In the absence of any general principles for settling such conceptual issues, I am unable to place any ultimate faith in the objectivity of anatomic description or of any technology of phylogenetic reconstruction that begins by dissecting complex morphologies into verbal descriptions.

The resemblances we perceive between objects are a function of the concepts we bring to them. The definition of morphological characters and their states, and therefore all cladistic analysis of morphology, depends on the anatomic concepts we adopt. Those concepts are not forced upon us by the facts of biology. Deciding what concepts and boundaries to apply in clustering a number of dissimilar morphologies under a smaller number of descriptive headings is not purely a matter of objective description; it also involves what can only be described as an act of poetic imagination. Unfortunately, just as different poets describe things in different ways, so do different anatomists, and the differences are not always phylogenetically neutral.

In a 1982 paper, I demonstrated this by using two different systems of anatomic terminology to describe the primate ear region. Both systems had approximately the same information content. Each sufficed to distinguish the various basic morphologies of the primate ear region, and neither system contained any redundancies: that is, all possible combinations of character states in each system described possible morphologies. I then applied these systems of descriptors to the ear regions of five primate genera and showed that the two systems yielded different maximally parsimonious unrooted trees. In one system, the sister group of *Tarsius* was *Necrolemur;* in the other, it was *Cebus.* I concluded from this that anatomic terms themselves may incorporate phylogenetic bias and that "parsimony assessments based on morphological [characters] cannot reliably yield neutral judgments of conflicting phylogenetic hypotheses."

Schoch (1986), noting that this conclusion "undercuts much of the basis of phylogenetic analysis," attempted to refute it by reanalyzing my examples. He concluded that I had counted wrong and that neither of my systems of descriptors yields a single most parsimonious cladogram. But he counted things up differently because he adopted different assumptions in enumerating character-state transitions. I had arrayed my multistate characters in linear morphoclines, in which some states were further apart than others; Schoch assumed that all character states are equidistant. I had equalized character weighting (i.e., morphocline lengths) by weighting the transitions between adjacent states more heavily for two-state than for multistate characters; Schoch adopted unitary weights for all transitions. Since there are no reasons for preferring his assumptions to mine, Schoch's solution does not bolster our confidence in the objectivity of this sort of exercise.

I continue to be persuaded that complex morphologies can often be described in different ways that may convey equivalent anatomic information but have different phylogenetic implications. This fact may help to explain how different investigators applying identical cladistic methods to essentially similar bodies of data can nevertheless arrive at very divergent phylogenetic conclusions, as the chapters in this volume bear witness.

Some Possible Constraints on the Morphological Imagination

Is there any way out of this dilemma? I have no general solution to offer, but I would like to offer one recommendation: that we stop making *a priori* judgments of homology. Homology judgments should emerge as fruits of a phylogenetic analysis at the end, not dictate the data on which that analysis operates from the start (Cartmill, 1992, 1994). The postorbital septum provides a case in point.

There is a broad consensus among primatologists today that the postorbital septa of tarsiers and anthropoids cannot be homologous. Such a judgment would be warranted if we could infer from the entire body of relevant evidence, *including the evidence of the postorbital septum itself,* that the last common ancestor of tarsiers and anthropoids lacked the septum. But many students of primate phylogeny have ruled the postorbital septum out of court from the very beginning as a candidate tarsier–anthropoid synapomorphy, insisting that the morphological or functional differences (or both) between the septa of tarsiers and anthropoids make it impossible for them to be homologous. The arguments that have been advanced for this conclusion deserve to be examined critically.

The postorbital septum is a bony wall interposed between the temporalis muscle and the periorbita, eliminating the primitive (for mammals) direct contact between them. As noted above, three bones—the frontal, alisphenoid, and zygomatic—contribute to this wall in both tarsiers and anthropoids. In many anthropoids, the zygomatic contribution is by far the largest; in tarsiers, the septal flanges of the three bones are roughly equal in size (Fig. 3). The postorbital septum is never entirely complete; the orbit still communicates with the infratemporal fossa via the inferior orbital fissure, through which the infraorbital branches of the maxillary nerve and blood vessels pass. This fissure is large in *Tarsius,* moderate in *Homo* and *Aotus,* and small to tiny in other anthropoids (Cartmill, 1980). No other primates have a bony wall between the periorbita and temporalis. Although a robust postorbital bar like that of *Hadropithecus* may give a false impression of partial postorbital closure (Tattersall, 1973), a broad communication still persists in such cases between the orbit and the upper part of the temporal fossa.

The unique structural similarities between the postorbital septa of tarsiers and anthropoids seem on the face of it to argue strongly for their homology. Yet many investigators have denied not only that homology but the similarities themselves. Simons and Russell (1960) argued that

> the manner of postorbital closure in *Tarsius* (insofar as the malar and frontal are concerned) is distinct from that seen in catarrhines and platyrrhines. Closure in this area in *Tarsius* is chiefly effected by an outward and downward growth of a flange of the frontal (with relatively little malar expansion), while in higher Primates the greater part of the dorsolateral area of enclosure is contributed by the development of a dorsal plate of the malar. These differences strongly imply that the partial postorbital closure of *Tarsius* only parallels that of the Anthropoidea and is not a character of their common inheritance.

These assertions have been repeated frequently by various investigators during the ensuing 30 years and were recently reiterated with particular insistence by Simons and Rasmussen (1989). Nevertheless, the assertions are incorrect; and even if they were correct, they would be irrelevant.

As Fig. 3 demonstrates, the zygomatic (malar) contribution to the bony wall between the temporal fossa and the orbit in at least some tarsiers is fully as large as the frontal contribution. (Figure 3, which is reprinted from my 1980 paper on the postorbital septum, incidentally obviates Simons and Rasmussen's complaint that my perceptions of the postorbital septum were distorted because I "approached the orbit from the front" rather than viewing it "from the temporal fossa at an oblique posterolateral angle.") The relative extent of the frontal, zygomatic, and alisphenoid contributions to the postorbital septa of tarsiers is impossible to assess in almost all adult skulls because of their complete sutural closure. In fact, the specimen illustrated in Fig. 3 is the only skull of an adult *Tarsius* I know of that retains discernible sutures in this area. This fact casts doubt on the accuracy and significance of Simons and Rasmussen's measurements of the width of the zygomatic postorbital processes in *Tarsius*.

In assessing the relative size of the postorbital septum's components in various primates, it is important to bear in mind that the posterior surface of the septum proper is delimited by the temporal line. Periorbital processes of the frontal and zygomatic that extend above or anterior to this line do not lie between the periorbita and temporalis and therefore do not contribute to the postorbital septum. The striking frontal flanges that raise the superior edges of the orbits up above the braincase in *Tarsius bancanus* may contribute to the impression of expansive periorbital growth of the frontal, but they contribute nothing to postorbital closure. Neither do the expanded postorbital processes of the frontal and zygomatic in lorises, which lie mainly in front of the temporal line and are therefore not morphologically comparable to the flanges of these bones that extend into the postorbital septa of *Tarsius* (*contra* Rosenberger and Szalay, 1980, p. 151) and anthropoids.

It cannot be denied that the zygomatic part of the postorbital septum in tarsiers is relatively smaller than its counterpart in most anthropoids. But it is hard to see why these differences in proportion are supposed to represent evidence against the homology of the septa in two groups. One might argue with equal justice that the last common ancestor of *Tarsius* and anthropoids must have lacked a postorbital *bar* because the zygomaticofrontal suture lies further up on the bar in platyrrhines than it does in *Tarsius*. The zygomatic component in the human postorbital septum is also relatively smaller than that of platyrrhines (Fig. 4), but that fact does not furnish grounds for suspecting that hominids attained postorbital closure independently of other Anthropoidea.

Functional arguments have also been put forward to support *a priori* rejection of the postorbital septum as a potential tarsier–anthropoid synapomorphy. Simons and Rasmussen argue that the hypertrophied eyeballs of tarsiers are peculiarly vulnerable to injury because they protrude far beyond

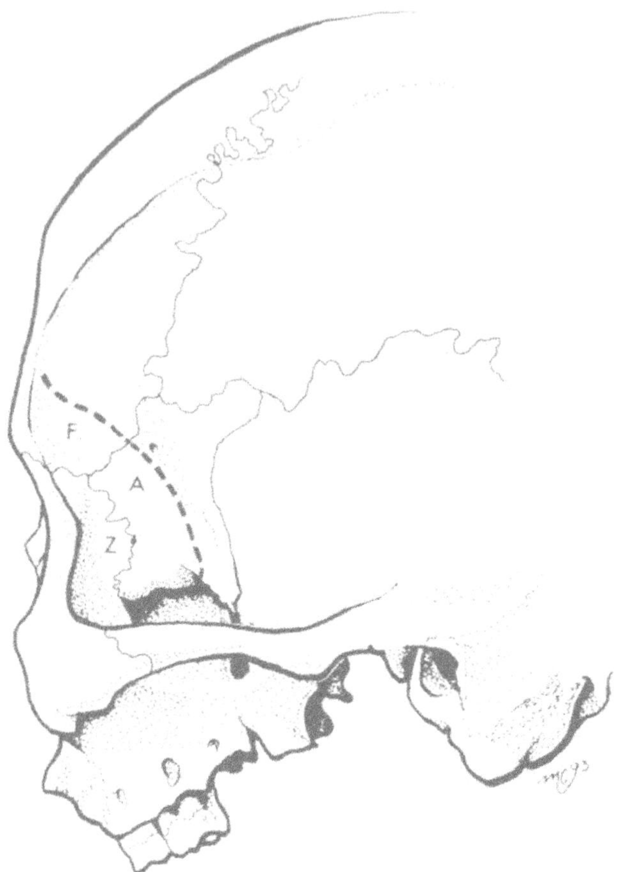

Fig. 4. Posterolateral view of left orbitotemporal region of an adult human skull with persistent sutures (author's collection), showing contributions of frontal (F), zygomatic (Z), and alisphenoid (A) to the postorbital septum. The heavy dashed line indicates the approximate position of the septum's attachment to the braincase.

the orbital margins. They also contend that the leaping locomotor habits of tarsiers imperil the enlarged eyes by subjecting them to "sudden jarring stops and starts in dense foliage." The postorbital septum of *Tarsius,* they conjecture, somehow serves to support the eyeball and protect it against these dangers. Since anthropoids do not have enormous protruding eyeballs, Simons and Rasmussen conclude that the postorbital septa of tarsiers and anthropoids must have evolved separately, for different functional reasons.

The functional account proposed here is implausible: how could the spread of small bony flanges across the periorbital surface of the temporalis muscle function to protect a protruding eyeball from the effects of "sudden jarring stops and starts"? But even if we accept this account, we are not obliged to accept the homology judgment that goes with it, because functional differences between two structures ordinarily have no bearing on their homology. Functional considerations can play important roles of other sorts in phylogenetic reconstruction, particularly in establishing the polarity of character states. When two states of the same character serve identical functions, the one that does so less efficiently is likely to be primitive. But function can tell us nothing about the homology or nonhomology of shared derived states proposed as synapomorphies. Even if the postorbital septa of tarsiers and anthropoids serve entirely different functional purposes—a point that surely remains to be demonstrated—the septum might have evolved originally to serve one of those purposes and have readapted later on to serve the other.

We may argue on various grounds that a postorbital septum like that of *Tarsius* is unlikely to have evolved into one like that of *Callithrix,* or vice versa. (Probably each is apomorphous in various respects relative to the other.) But there is nothing about the morphology of either septum that would make such a transformation impossible. There are therefore no morphological grounds for rejecting the hypothesis of their homology *a priori.* If and when we are compelled to conclude that the two are not homologous, it will only be because a convincing analysis of haplorhine phylogeny has given us convincing reasons for thinking that the last common ancestor of tarsiers and anthropoids lacked a postorbital septum. And to make our analyses and reasoning convincing to others, we need to weight all the evidence and not throw genuine similarities out of the scales because we feel that the phylogenies they imply are unworthy of serious consideration. Like it or not, the postorbital septum is a genuine and apomorphic (nonprimitive) similarity of tarsiers to anthropoids. There are others. Like it or not, there are other apomorphic similarities that link tarsiers to various omomyids, anthropoids to various adapids, adapids to toothcomb prosimians, and so on. Most of these resemblances must represent convergences, but we cannot determine which ones they are in advance of our phylogenetic analyses (Cartmill, 1994). If we admit all this and stop excluding one another's pet facts and concepts from our analyses because we find them uncongenial, we may discover generally agreeable solutions to such intractable phylogenetic problems as that of anthropoid origins. If we do not discover a solution, we may at least determine that we do not know enough yet to answer the question. Either outcome would represent a step in the right direction.

Acknowledgments

Much of the foregoing reflects the thinking and research of Kim Coleman Healy, Callum Ross, and Anne Yoder, whose work I have had the privilege of helping to supervise during the past several years at Duke, and from whom I have learned at least as much as they have learned from me. I am grateful to Drs. J. G. Fleagle, R. F. Kay, and E. L. Simons for inviting me to participate in the 1992 conference on anthropoid origins on which this volume is based, and to Dr. Kaye Brown for helpful comments on earlier drafts of this manuscript.

References

Baba, M. L., Weiss, M. L., Goodman, M., and Czelusniak, J. 1982. The case of tarsier hemoglobin. *Syst. Zool.* **31**:156–165.

Beard, J. M., and Goodman, M. 1976. The hemoglobins of *Tarsius bancanus.* In: M. Goodman and R. E. Tashian (eds), *Molecular Anthropology,* pp. 239–255. Plenum Press, New York.

Cartmill, M. 1972. Arboreal adaptations and the origin of the order Primates. In: R. H. Tuttle (ed.), *The Functional and Evolutionary Biology of Primates,* pp. 97–122. Aldine-Atherton, Chicago.

Cartmill, M. 1975. Strepsirhine basicranial structures and the affinities of the Cheirogaleidae. In: W. P. Luckett and F. S. Szalay (eds.), *Phylogeny of the Primates: A Multidisciplinary Approach,* pp. 313–354. Plenum Press, New York.

Cartmill, M. 1980. Morphology, function, and evolution of the anthropoid postorbital septum. In: R. L. Ciochon and A. B. Chiarelli (eds.), *Evolutionary Biology of the New World Monkeys and Continental Drift,* pp. 243–274. Plenum Press, New York.

Cartmill, M. 1982. Assessing tarsier affinities: Is anatomical description phylogenetically neutral? *Geobios Mem. Spec.* **6**:279–287.

Cartmill, M. 1992. Homology as a morphological concept. *Am. J. Phys. Anthrop. [Suppl.]* **14**:57–58.

Cartmill, M. 1994. A critique of homology as a morphological concept. *Am. J. Phys. Anthropol.* **94**:115–123.

Cartmill, M., and Kay, R. F. 1978. Cranio-dental morphology, tarsier affinities, and primate suborders. In: D. J. Chivers and K. A. Joysey (eds.), *Recent Advances in Primatology, Vol. 3,* pp. 205–214. Academic Press, London.

Cartmill, M., MacPhee, R. D. E., and Simons, E. L. 1981. Anatomy of the temporal bone in early anthropoids, with remarks on the problem of anthropoid origins. *Am. J. Phys. Anthrop.* **56**:3–21.

Cave, A. J. E. 1967. Observations on the platyrrhine nasal fossa. *Am. J. Phys. Anthrop.* **26**:277–288.

Covert, H. H., and Williams, B. A. 1991. The anterior lower dentition of *Washakius insignis* and adapid–anthropoidean affinities. *J. Hum. Evol.* **21**:463–467.

Czelusniak, J., Koop, B., Tagle, D., Shoshani, H., Goodman, M., Braunitzer, G., Kleinschmidt, T., de Jong, W. W., and Matsuda, G. 1988. Perspectives from amino acid and nucleotide sequences on cladistic relationships among higher taxa of Eutheria. In: H. H. Genoways (ed.), *Current Mammalogy,* pp. 545–572. Plenum Press, New York.

Dagosto, M. 1985. The distal tibia of primates with special reference to the Omomyidae. *Int. J. Primatol.* **6**:45–75.

Dagosto, M. 1988. Implications of postcranial evidence for the origin of euprimates. *J. Hum. Evol.* **17**:35–36.

de Jong, W. W., and Goodman, M. 1988. Anthropoid affinities of *Tarsius* supported by lens alpha-A crystallin sequences. *J. Hum. Evol.* **17**:575–582.

Elliot, O., Yess, N. J., and Hegsted, D. M. 1966. Biosynthesis of ascorbic acid in the tree shrew and slow loris. *Nature* **212**:739–740.

Gingerich, P. D. 1977. Radiation of Eocene Adapidae in Europe. *Geobios, Mem. spec.* **1**:165–182.

Gingerich, P. D. 1978. Phylogeny reconstruction and the phylogenetic position of *Tarsius.* In: D. J. Chivers and K. A. Joysey (eds), *Recent Advances in Primatology, Vol. 3,* pp. 249–256. Academic Press, London.

Gingerich, P. D. 1981. Early Cenozoic Omomyidae and the evolutionary history of the tarsiiform primates. *J. Hum. Evol.* **10**:345–374.

Gingerich, P. D. 1984. Primate evolution: Evidence from the fossil record, comparative morphology, and evolutionary biology. *Yearb. Phys. Anthropol.* **27**:57–72.
Haines, R. W. 1950. The interorbital septum in mammals. *J. Linn. Soc. Lond. (Zool.)* **41**:585–607.
Healy, K. C. 1992. *Chromosome Banding as a Source of Phylogenetic Information: A Critique Based on the Case of* Tarsius. Ph.D. dissertation, Duke University.
Hubrecht, A. A. W. 1897. *The Descent of the Primates.* Charles Scribner's Sons, New York.
Kampen, P. N. van. 1905. Die Tympanalgegend des Säugetierschädels. *Geg. Morph. Jahrb.* **34**:321–722.
Koop, B. F., Tagle, D. A., Goodman, M., and Slighton, J. L. 1989a. A molecular view of primate phylogeny and important systematic and evolutionary questions. *Mol. Biol. Evol.* **6**:580–612.
Koop, B. F., Siemieniak, D., Slighton, J. L., Goodman, M., Dunbar, J., Wright, P. C., and Simons, E. L. 1989b. *Tarsius* β- and δ-globin genes: Conversions, evolution, and systematic implications. *J. Biol. Chem.* **264**:68–79.
Luckett, W. P. 1975. Ontogeny of the fetal membranes and placenta: Their bearing on primate phylogeny. In: W. P. Luckett and F. S. Szalay (eds.), *Phylogeny of the Primates: A Multidisciplinary Approach,* pp. 157–182. Plenum Press, New York.
Luckett, W. P. 1976. Cladistic relationships among primate higher categories: Evidence of the fetal membranes and placenta. *Fol. Primatolol.* **25**:245–276.
MacPhee, R. D. E., and Cartmill, M. 1986. Basicranial structures and primate systematics. In: D. R. Swindler and J. Erwin (eds.), *Comparative Primate Biology, Vol. 1,* pp. 219–276. Alan R. Liss, New York.
Martin, R. D. 1973. Comparative anatomy and primate systematics. *Symp. Zool. Soc. Lond.* **33**:301–337.
Martin, R. D. 1990. *Primate Origins and Evolution: A Phylogenetic Reconstruction.* Princeton University Press, Princeton.
McGrath, P., and Mills, P. 1984. *Atlas of Sectional Anatomy: Head, Neck, and Trunk.* Karger, Basel.
Packer, D., and Sarmiento, E. E. 1984. External and middle ear characteristics of primates with reference to tarsier–anthropoid affinities. *Am. Mus. Novit.* **2787**:1–23.
Pettigrew, J. D., Jamieson, B. G. M., Robson, S. K., Hall, L. S., McNally, K. I., and Cooper, H. M. 1989. Phylogenetic relations between microbats, megabats and primates (Mammalia: Chiroptera and Primates). *Phil. Trans. R. Soc. Lond. [B]* **325**:489–559.
Pocock, R. I. 1918. On the external characters of the Lemurs and of *Tarsius. Proc. Zool. Soc. Lond.* **1918**:19–53.
Pollock, J. I., and Mullin, R. J. 1987. Vitamin C biosynthesis in prosimians: Evidence for the anthropoid affinity of *Tarsius. Am. J. Phys. Anthrop.* **73**:65–70.
Poorman, P. A., Cartmill, M., MacPhee, R. D. E., and Moses, M. 1985. The banded karyotype of *Tarsius bancanus* and its implications for primate phylogeny. *Am. J. Phys. Anthrop.* **66**:215.
Rasmussen, D. T. 1986. Anthropoid origins: A possible solution to the Adapidae–Omomyidae paradox. *J. Hum. Evol.* **15**:1–12.
Rosenberger, A. L. 1985. In favor of the necrolemur–tarsier hypothesis. *Fol. Primatol.* **45**:179–194.
Rosenberger, A. L., and Strasser, E. 1985. Toothcomb origins: Support for the grooming hypothesis. *Primates* **26**:73–84.
Rosenberger, A. L., and Szalay, F. S. 1980. On the tarsiiform origins of Anthropoidea. In: R. L. Ciochon and A. B. Chiarelli (eds.), *Evolutionary Biology of the New World Monkeys and Continental Drift,* pp. 139–157. Plenum Press, New York.
Schoch, R. M. 1986. *Phylogeny Reconstruction in Paleontology.* Van Nostrand Reinhold, New York.
Schwartz, J. H. 1978. If *Tarsius* is not a prosimian, is it a haplorhine? In: D. J. Chivers and K. A. Joysey (eds), *Recent Advances in Primatology, Vol. 3,* pp. 195–204. Academic Press, London.
Schwartz, J. H., and Tattersall, I. 1987. Tarsiers, adapids and the integrity of Strepsirhini. *J. Hum. Evol.* **16**:23–40.
Simons, E. L. 1989. Description of two genera and species of late Eocene Anthropoidea from Egypt. *Proc. Natl. Acad. Sci. USA* **86**:9956–9960.
Simons, E. L. 1990. Discovery of the oldest known anthropoidean skull from the Paleogene of Egypt. *Science* **247**:1567–1569.

Simons, E. L. 1992. Diversity in the early Tertiary anthropoidean radiation in Africa. *Proc. Natl. Acad. Sci. USA* **89:**10743–10747.

Simons, E. L., and Rasmussen, D. T. 1989. Cranial morphology of *Aegyptopithecus* and *Tarsius* and the question of the tarsier–anthropoid clade. *Am. J. Phys. Anthrop.* **79:**1–24.

Simons, E. L., and Russell, D. E. 1960. Notes on the cranial anatomy of *Necrolemur. Mus. Comp. Zool. Brev.* **127:**1–14.

Simpson, G. G. 1945. The principles of classification and a classification of mammals. Bull. Am. Mus. Nat. Hist. **85:**1–350.

Szalay, F. S. 1975. Phylogeny of primate higher taxa: The basicranial evidence. In: W. P. Luckett and F. S. Szalay (eds.), *Phylogeny of the Primates: A Multidisciplinary Approach,* pp. 91–125. Plenum Press, New York.

Szalay, F. S., Rosenberger, A. L., and Dagosto, M. 1987. Diagnosis and differentiation of the order Primates. *Yearb. Phys. Anthrop.* **30:**75–105.

Tattersall, I., 1973, Cranial anatomy of the Archaeolemurinae (Lemuroidea, Primates). *Anthropol. Papers Am. Mus. Nat. Hist.* **52:**1–110.

Wible, J. R. 1984. *The Ontogeny and Phylogeny of the Mammalian Cranial Arterial Pattern.* Ph.D. dissertation, Duke University.

Yoder, A. D. 1992. *The Phylogenetic affinities of the Cheirogaleidae: A Molecular and Morphological Analysis.* Ph.D. dissertation, Duke University.

Postcranial Anatomy and the Origin of the Anthropoidea

17

MARIAN DAGOSTO and DANIEL L. GEBO

Introduction

One of the major unresolved problems in the phylogeny of primate higher taxa is the origin and relationships of Anthropoidea. Several hypotheses are currently in contention, two of which entail the idea of descent from the Eocene euprimate families Adapidae and Omomyidae. The idea that omomyids or tarsiers are the stem group for anthropoids is based primarily on characters of the narial region and placentation shared by *Tarsius* (a presumed descendant of omomyids) and anthropoids (Hill, 1919; Luckett, 1975; Pocock, 1918); features of middle ear construction and arterial circulation (Cartmill and Kay, 1978; Cartmill et al., 1981; Szalay, 1975); and the relationship of the interorbital septum to the nasal fossa (Cartmill, 1972; Cave, 1967) shared by tarsiers, omomyids, and anthropoids. This hypothesis will be abbreviated as OA. The evidence for a phyletic relationship between adapids and anthropoids (AA) rests primarily on features of the dentition shared by

---

MARIAN DAGOSTO • Department of Cell and Molecular Biology, Northwestern University Medical School, Chicago, Illinois 60611, and Department of Mammalogy, American Museum of Natural History, New York, New York 10024. DANIEL L. GEBO • Department of Anthropology, Northern Illinois University, DeKalb, Illinois 60115.
Anthropoid Origins, edited by John G. Fleagle and Richard F. Kay. Plenum Press, New York, 1994.

adapids and anthropoids (Gingerich, 1980; Rasmussen and Simons, 1988; Simons and Rasmussen, 1989) and shared features of the basicranial region of *Mahgarita* and anthropoids (Rasmussen, 1990).

Features of the postcranial skeleton have played a relatively minor role in this debate but have been used to construct three different types of systematic arguments. The first and strongest argument establishes shared-derived characters that unite omomyids, adapids, or tarsiers with anthropoids. The second type of argument asserts that one or more of the proposed fossil or living ancestors of anthropoids is actually too derived to have given rise to anthropoids. The third argument is related to the second; it establishes that the derived features of the proposed ancestor are shared with primates other than anthropoids, and thus the group is more closely related to these other taxa than to anthropoids. In this chapter we will discuss and analyze the characters used to support or reject OA and AA, using these three types of arguments. We will also discuss the results of a computer-assisted phylogenetic analysis of these characters.

Although our purpose in this chapter is only to examine the contribution of postcranial evidence to this problem, it is our belief that any good phylogenetic hypothesis must consider evidence from multiple systems. We also feel strongly that any phylogenetic analysis is only as good as the characters it is based on. Determinations of homology, polarity, and weighting, as subjective as they may often seem to be, are crucial to phylogeny reconstruction (Neff, 1986; Szalay and Bock, 1991). Different decisions can and do result in different outcomes. It is therefore our intent here to make our reasoning for decisions on polarity, homology, and weighting as explicit as possible.

For the determination of homology we follow the guidelines laid down by Remane (1956) and Simpson (1961). Our determinations of polarity are primarily based on outgroup analysis; plesiadapiforms, dermopterans, and tree shrews are considered to be the closest relatives of euprimates. As will become obvious in the discussion below, determination of polarity for some features is problematic; this is especially the case in those instances where adapiforms and lemuriforms have one condition and tarsiiforms and anthropoids another, neither of which is represented in any outgroups. Although many previous analyses have automatically assumed that the strepsirhine condition is primitive, we believe that this broad assumption is unwarranted. We treat these cases as polarity unknown.

Schemes for weighting characters have been proposed by Hecht and Edwards (1977) and Neff (1986), among others. Our weights incorporate the ideas of both of these works and involve a combination of both *a priori* and *a posteriori* weighting (Neff, 1986). Criteria for *a priori* weighting include the confidence we have in our hypothesis of homology and the nature of the character state (i.e., losses and reductions are of lower weight than acquisitions). *A posteriori* criteria include an assessment of the degree of homoplasy of the character after phylogenetic analysis; i.e., if there are no known instances of homoplasy, the feature is given high weight; if there are many known instances of homoplasy, it is given low weight.

Character Analysis

Characters Supporting Adapid–Anthropoid or Omomyid–Anthropoid Monophyly

Shared-derived characters uniting anthropoids with either group of Eocene primates could provide strong corroborating evidence for one of the competing hypotheses. Unfortunately, proponents of the major schools of thought have been unable to identify many such features. The few that have been proposed are problematic in that homology and polarity are difficult to determine, or the features are considered to be of low weight.

The proponents of AA have not cited any shared-derived postcranial traits to support their arguments, but our analysis suggests that, depending on one's interpretation of the primitive condition in euprimates, reduction in the length of the tarsals (features C1 and N1, see Tables I and II and Fig. 1A) may be a derived feature shared by strepsirhines and anthropoids. However, this feature is of very low weight, since it has evolved in parallel numerous times (see discussion below). The angle of the neck and head of the femur (F2) is generally high in prosimians (more perpendicular neck) and low in anthropoids (more angled neck); the anthropoid condition is probably derived within euprimates, although plesiadapiformes also have a low neck angle (Fig. 1A; M. Dagosto and P. Schmid, *unpublished observations*). Some adapids (*Adapis, Smilodectes*) also have a relatively low neck angle, like anthropoids.

The OA hypothesis is also supported by a few characters shared by tarsiiforms and anthropoids (Fig. 1B). In most strepsirhines the distal edge of the humeral trochlea is perpendicular to the long axis of the humerus, whereas in anthropoids and omomyids the distal edge of the trochlea is at an angle to the long axis (H1) (Szalay and Dagosto, 1980). Exceptions to this general scheme include some individuals of *Notharctus* and *Adapis* and lorisines, which have an angled distal edge. Among haplorhines callitrichines have a perpendicular edge. Although we believe that a perpendicular edge is primitive for Strepsirhini, and an angled edge is primitive for Haplorhini, we are unsure of the primitive condition for euprimates. In any case, this feature is clearly subject to some homoplasy.

Anthropoids are distinguished from prosimians in having a long femoral neck (F1) and a more posteriorly located lesser trochanter (F3); these are likely to be derived conditions within primates, although they have evolved many other times in other mammalian orders (M. Dagosto and P. Schmid, *unpublished observations*). In contrast to other prosimians (including other omomyids), *Necrolemur* exhibits both of these features, and *Microchoerus* exhibits a posteriorly located lesser trochanter (the femoral neck is not preserved). In addition, the femur of *Necrolemur* (but not *Microchoerus*) is remarkably reminiscent of the parapithecid material from the Fayum in the presence of an intertrochanteric crest connecting the greater and lesser trochanters, posteriorly yielding a walled-off surface (F10).

Table I. Description of Characters and Character States Used in the Phylogenetic Analysis[a]

Femur

F1. length of femoral neck (o,1) index N2/BSTD in M. Dagosto and P. Schmid, unpublished observations)
- 0, ≤75 (short)
- 1, 75–120
- 2, >120 (long)

F2. angle of femoral neck (o,2)
- 0, <60
- 1, 60–70
- 2, >70

F3. angle of lesser trochanter (r,2)
- 0, medial (0–30°)
- 1, posterior (>30°)

F4. size of third trochanter (o,1)
- 0, large
- 1, small
- 2, crest or absent

F5. knee index (o,1)
- 0, <90 (shallow knee)
- 1, 90–100
- 2, >100 (deep knee)

F6. femoral head shape (o,2)
- 0, spherical
- 1, semicylindrical
- 2, cylindrical

F7. anterior extension of greater trochanter (r,2)
- 0, no extension
- 1, extension

F8. anterior bend of proximal femur (r,2)
- 0, none
- 1, bent

F9. relative length of trochanteric fossa (o,2)
- 0, long >125
- 1, moderate 110–125
- 2, very short <100

F10. presence of "intertrochanteric crest" (r,3)
- 0, no crest
- 1, crest

Tibia

T1. fusion (o,2)
- 0, not fused, facet present
- 1, close apposition, facet present

Talus

A1. flexor fibularis groove (r,2)
- 0, medial
- 1, lateral

A2. slope of fibular facet (r,2)
- 0, flat
- 1, sloped

A3. length of ast–tib articulation (r,2)
- 0, short
- 1, long

A4. size of posterior trochlear shelf (o,1)
- 0, none
- 1, small
- 2, large

A5. talar neck length (o,2)
- 0, <100 (short)
- 1, >100 (long)

Calcaneus

C1. rel length of anterior part (o,1)
- 0, <40 (short)
- 1, 40–45
- 2, >45 (long)

C2. position of peroneal tubercle (o,3)
- 0, distal to joint
- 1, at joint
- 2, proximal to joint

C3. calcaneal bowing (r,3)
- 0, none
- 1, present

Navicular

N1. length relative to width (o,2)
- 0, <90 (short)
- 1, 100–150
- 2, >150 (long)

N2. navicular cuboid facet (r,3)
- 0, lateral
- 1, plantar

Entocuneiform/MT1 articulation

1. shape of joint (u,3)
- 0, mammal condition
- 1, prosimian condition
- 2, anthropoid condition

Foot

O1. axis (u,2)
- 0, mesaxonic
- 1, paraxonic
- 2, ectaxonic

(*continued*)

Table I. (*Continued*)

2, close apposition, no facet
3, fused
T3. shape of distal surface (r,3)
0, square/parallel
1, trangular
T4. rotation of medial malleolus (o,3)
0, none
1, slight medial
2, strong medial
T5. shape of medial malleolus (u,3)
0, flat
1, anteriorly convex, posteriorly flat
2, all convex
T6. shape of distal tibial shaft (r,2)
0, no compression, round shaft
1, compression ant–post
T7. position of tibialis posterior groove (r,2)
0, medial side of malleolus
1, posterior side of malleolus

O2. toilet claw (r,2)
0, absent
1, present
O3. prehallux (r,2)
0, present
1, absent

Humerus
H1. shape of trochlea (u,2)
0, conical
1, cylinder, medial edge at angle to shaft
2, cylinder, perpendicular distal edge
H2. dorsoepitrochlear pit (u,1)
0, present
1, small, shallow
2, absent

Wrist
W1. facet for triquetrum on pisiform (r,3)
0, large
1, reduced

[a]Abbreviations in parentheses: o, ordered; r, reversible; u, unordered; 1, low weight; 2, medium weight; 3, high weight. The character state matrix is given in Table II.

The characters cited as potential shared-derived features by each hypothesis are few, and most are problematic in that they are subject to homoplasy and/or difficult to polarize. The strongest features for OA (the similarity of proximal femoral anatomy of *Necrolemur* and anthropoids) are found in the very group of omomyids most specialized in other postcranial and dental features and thus least likely to have given rise to (but not necessarily least likely to have been the sister group of) anthropoids.

Characters in Which the Fossil Groups Are Claimed to Be Too Derived to Have Given Rise to Anthropoids

In the absence of strong derived characters demonstrating clear links between groups, both sides have relied on a different type of argument: demonstrating that the unfavored Eocene fossil taxon has derived features of the postcranium not shared by anthropoids, making it unlikely that the fossil group could have been ancestral to the anthropoids. Thus, AA argues that omomyids have derived postcranial features such as an elongate tarsus and fused tibia–fibula, that are not characteristic of anthropoids (Gingerich,

Table II. Character State Matrix

Taxa	Characters[a]																																		
	F1	F2	F3	F4	F5	F6	F7	F8	F9	F10	T1	T3	T4	T5	T6	T7	A1	A2	A3	A4	A5	C1	C2	C3	N1	N2	O1	O2	O3	E1	H1	H2	W1
Primitive	1	0/1	0	0	0	0	0	0	1	0	0	0	0	0	0	0	0/1	0	0	0	0	0	0	0	0	0	?	0	0	0	0	0/2	0
Anaptomorphine	?	?	?	?	?	?	?	?	?	0	2	0	1	1	0	1	0	0	1	1	1	2	1	0	2	0	?	?	1	1	1	1	?
Omomyine	0	2	0	0	2	1	1	1	1	0	2	0	1	1	0	1	0	0	1	1	1	2	1	0	2	0	?	?	1	1	1	1	?
Microchoerine	2	1	1	0/2	2	0	1	1	0/1	1	3	0	1	1	1	0	0	0	1	2	0	2	1	1	?	?	?	?	1	1	1	0	?
Tarsius	0	2	0	0	2	2	1	1	2	0	3	0	1	1	1	0	0	0	1	0	0	2	1	1	2	0	2	1	1	1	1	2	0
Notharctine	1	1	0	0	1	0	0	0	0	0	0	1	2	2	0	1	1	1	1	2	1	1	1	0	1	1	0	0	1	1	2	2	0
Protoadapine	1	1	0	0	?	0	0	0	?	0	0	?	?	?	?	?	1	1	1	2	1	1	1	0	1	?	1	0	1	1	2	2	?
Adapinan	1	0	0	0	0	0	0	0	1	0	0	1	2	2	0	1	1	1	1	2	1	0	2	0	?	?	?	?	1	1	0	0	1
Lemuriform	1	1	0	0	1	0	0	0	0	0	0	1	2	2	0	1	1	1	1	2	1	1	2	0	1	1	2	1	1	1	2	2	1
Platyrrhine	2	0	1	2	0	0	0	0	0	0	1	0	1	1	0	0	0	0	0	0	1	1	1	0	0	0	1	0	0	2	1	0	0
Apidium	2	0	1	2	1	0	0	0	0	1	1	0	1	1	0	0	0	0	0	0	1	0	1	0	0	0	?	?	?	2	1	0	?

[a]Characters are defined in Table I.

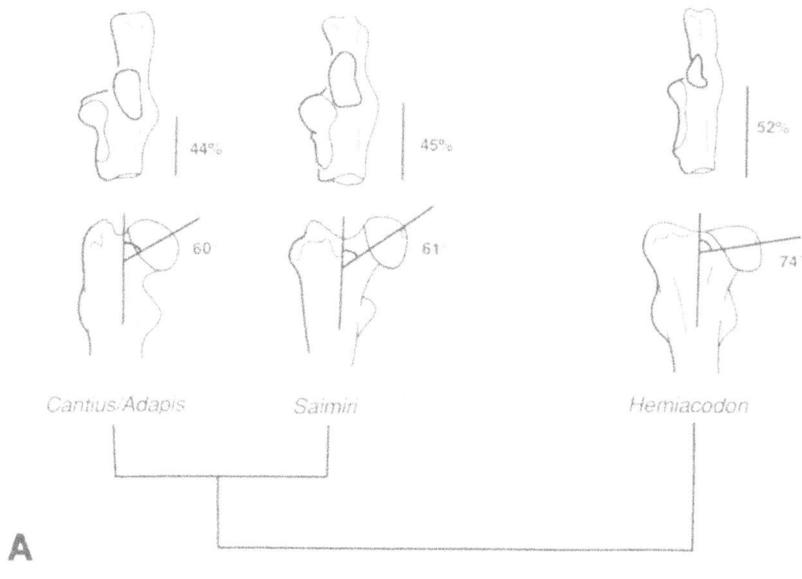

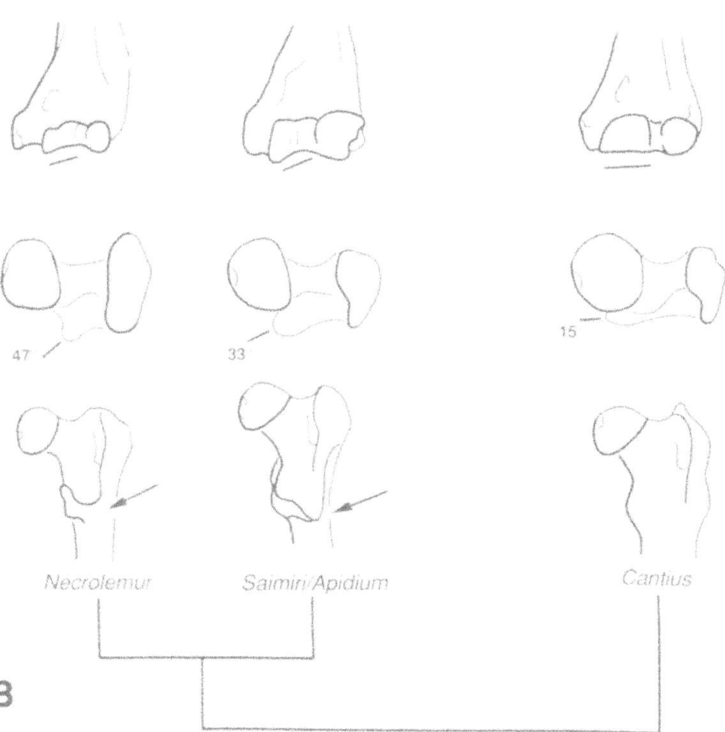

Fig. 1. Potential shared-derived features linking adapids (A) or omomyids (B) to anthropoids. (A) Top, Length of the distal portion of the calcaneus relative to total length. Bottom, angle of the femoral neck and head relative to the long axis of the shaft. (B) Top, angle of trochlea relative to the long axis of the humerus. Center, position of the lesser trochanter of the femur. Bottom, formation of the intertrochanteric crest in *Necrolemur* and *Apidium* (center). Also note that *Necrolemur* and Apidium have relatively longer, more slender necks than *Cantius*.

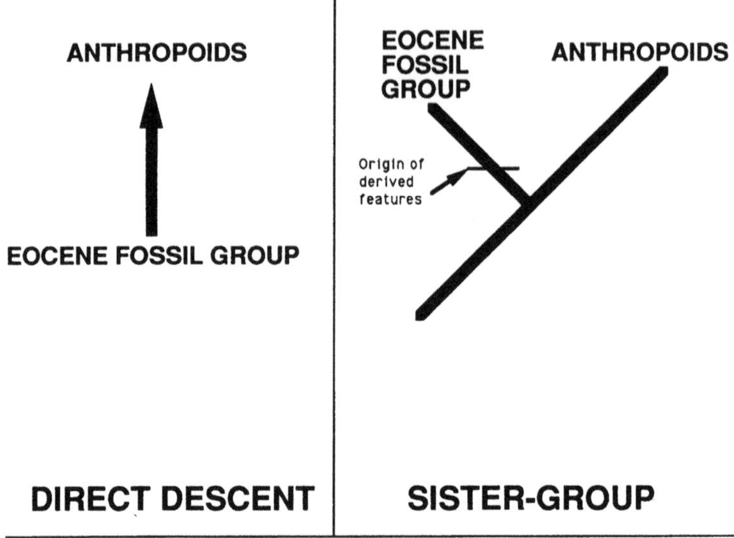

Fig. 2. Discovering derived features in a potentially ancestral taxon that are not found in Anthropoidea does disprove a hypothesis of direct descent (left) but does not disprove the more general hypothesis of a sister-group relationship between the taxa (right), since the derived features could have originated after the split between anthropoids and the fossil group.

1980). The OA proponents argue that adapids have derived features in the structure of the talus, navicular, and carpus that unite them with lemuriforms; anthropoids do not share these features (Beard *et al.,* 1988).

The "too derived" type of argument can be decisive, but it only eliminates from consideration one type of hypothesis: that of *direct descent* from the group (Szalay, 1977). The possibility of a *sister-group* relationship between the derived fossil taxon and anthropoids is not disproved, since the derived characters could have developed within the fossil group after the split with anthropoids (Fig. 2). Therefore, it is also necessary to determine whether AA, OA, or their variants argue for direct descent or simply a sister-group relationship between anthropoids and the fossil taxon. The AA proponents have usually argued for some form of direct descent, citing either adapids or protoadapids in general as the source for anthropoids or identifying specific potential ancestors such as *Cercamonius* (e.g., Gingerich, 1975). The OA proponents are less explicit about identifying ancestors but also usually imply that omomyids are the source group of anthropoids (e.g., Rosenberger and Szalay, 1980). Thus, both hypotheses, as currently stated, are vulnerable to the "too derived" argument.

The ability of such arguments to decide in favor of OA versus AA depends on the number of taxa within the fossil group that can be eliminated from consideration as direct ancestors (without proposing a reversal in the character). If the derived feature can be shown to be present in the morphotype of the fossil group, all taxa are eliminated, and the hypothesis of descent from the group is disproved. If the derived feature is present in only some subtaxa of the fossil group and is not thought to be primitive for the whole group, then only those subtaxa are eliminated from consideration: anthropoids could be descended from the other, more primitive members of the group.

Even though derived features not found in anthropoids are present in some or all members of the fossil group, one can always save the favored hypothesis with resort to a *post-hoc* hypothesis of reversal. That is, even though derived features are present in the proposed ancestor, they could have disappeared in the ancestral anthropoid. Thus, it is also necessary to evaluate the probability of reversal in the character.

In this section we will evaluate the characters used by proponents of AA and OA to show that the unfavored fossil group is too derived to have given rise to anthropoids. As discussed above, two criteria are used to evaluate the argument. (1) Is the character present in the morphotype of the fossil group? If not, in which subtaxa is it present (which groups can be eliminated from consideration as potential ancestors)? (2) What is the probability of reversal?

Features Cited by AA

Long Tarsals (C1, N1). All omomyids currently known, including *Teilhardina,* the earliest recognized omomyid, possess calcanea in which the length of the portion distal to the posterior talocalcaneal facet is at least 50% of total length (Gebo, 1988; Szalay, 1976). Thus, relative distal calcaneal length of at least 50% was probably primitive for omomyids. This exceeds the values found in anthropoids and adapids. However, the differences among early adapids such as *Cantius mckennai* (44%), small platyrrhines such as *Saimiri* and *Aotus* (45%), and omomyids (52%) are small (see Fig. 1) and, in fact, do not exceed the range of variation that can be demonstrated within many living species (Dagosto, 1986).

Long calcanea may be a tarsiiform synapomorphy; however, it is also entirely possible that the basal euprimate was characterized by this feature. Early Eocene adapids such as *Cantius mckennai,* which are smaller than middle Eocene notharctines, also have relatively longer distal calcanea (44% for *C. mckennai* versus 37–39% for *Notharctus* and *Smilodectes:* Gebo *et al.,* 1991). If long (i.e., anterior length > 50%) calcanea are primitive for euprimates, reduction in adapids and anthropoids could be considered a synapomorphy, but clearly one of low weight, as discussed below.

Omomyids and tarsiers are also characterized by relatively long naviculars compared to those of adapids, lemurs, or anthropoids (Fig. 3A; Dagosto, 1990). In contrast to anterior calcaneal length, the differences between omomyids and adapids are more marked. Naviculars as long as those in omomyids are probably not primitive for euprimates.

No matter what phylogenetic scheme one adopts, it is clear that tarsals

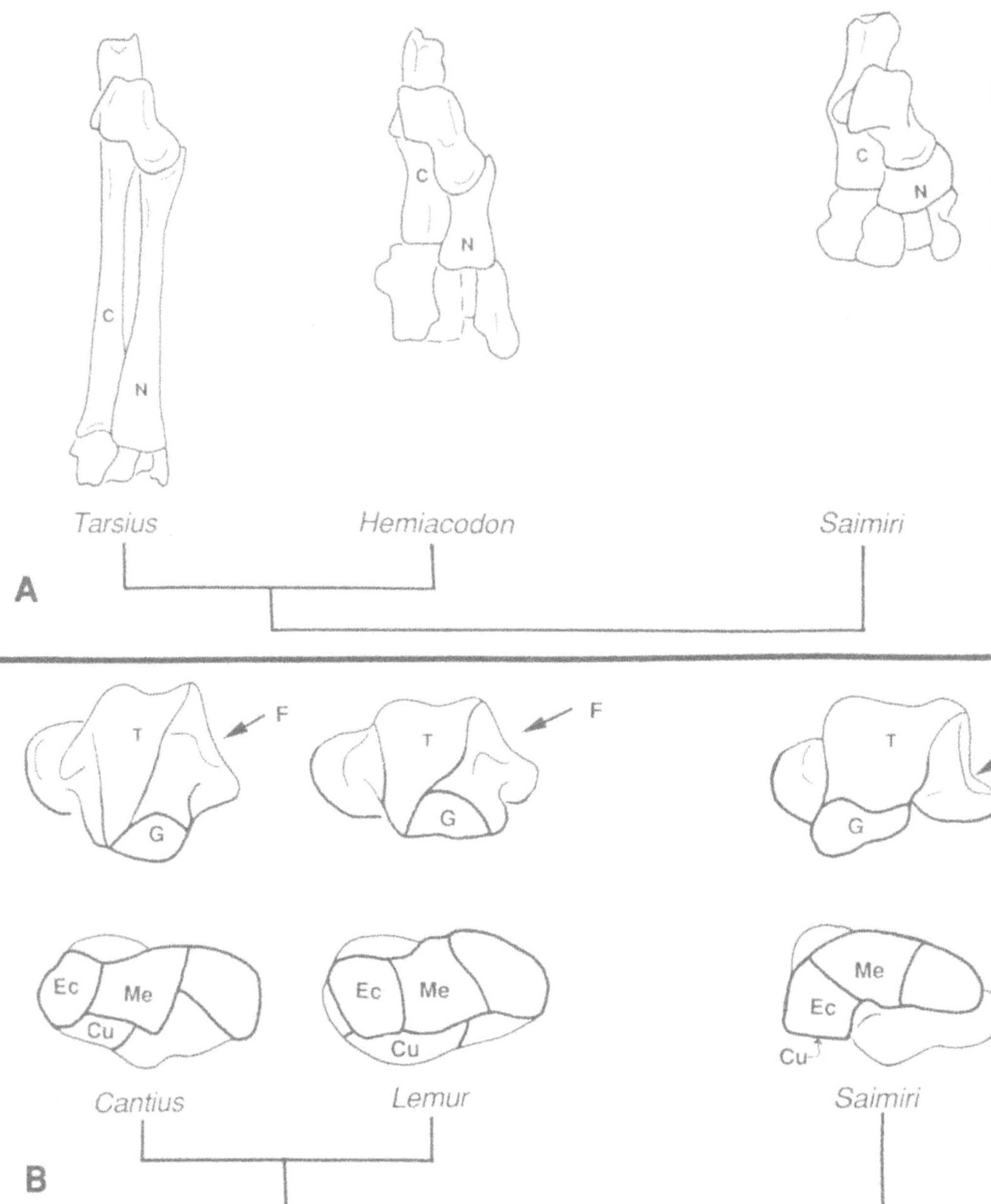

Fig. 3. Some features for which researchers have claimed that either omomyids (A) or adapids (B) are too derived to be ancestral to anthropoids. (A) Relative length of the distal portion of the calcaneus (C) and length of the navicular (N). Specimens are drawn to same talar width. (B) Top, Posterior view of the right talus illustrating the slope of the fibular facet (F) and the relative position of the tibial facet (T) and groove for flexor fibularis (G). Bottom, Distal view of the navicular illustrating the relative position of the articular facets for the cuboid (Cu), the ecto-cuneiform (Ec) and the mesocuneiform (Me) bones.

have increased and decreased in length a number of times in primate evolution. Probable examples of reduction include protoadapine (~40%) to adapinan (32%), primitive lorisid (~50%) to lorisines (39–45%), early notharctine (44%) to late notharctine (38%), and lemurid (47%) to indriid (45%). Probable examples of increase include ? to *Tarsius* (70–75%), primitive anaptomorphine (~50%) to microchoerine (>60%), primitive cheirogaleid (~50%) to *Microcebus* and *Phaner* (60%), and primitive lorisid (~50%) to galagos (65–75%). Thus, we consider this character to be of low weight and therefore of limited value in phylogenetic analyses.

Tibio-Fibular Fusion (T1). This is clearly a derived trait, but it is found in only some members of the Omomyidae. It is characteristic of *Necrolemur* and *Tarsius,* but the known omomyines (*Hemiacodon, Shoshonius*) and anaptomorphines (*Nannopithex, Absarokius*) have unfused bones (K. C. Beard, *personal communication;* Covert and Hamrick, 1993; Dagosto, 1985; Schlosser, 1907; Schmid, 1992). It is very unlikely that fusion was present in the morphotype of Omomyidae. However, as discussed below, close apposition of the tibia and fibula and even partial fusion is characteristic of omomyids, many small platyrrhines, and early anthropoids (*Apidium*) and may very well be a shared feature of this group.

Tibio-fibular fusion has occurred independently in many mammalian lineages (Barnett and Napier, 1953; Carleton, 1941); thus, it is of low weight. However, we know of no cases where it has been established that a fully fused tibia-fibula of an ancestor has unfused in a descendant. We consider reversal in this feature to be highly unlikely.

Characters Cited by OA

Lateral Talus Slope (A2). A gradual lateral slope of the talofibular joint surface on the talus is present in living lemuriforms and all known adapids (Fig. 3B), representing both subfamilies (Beard *et al.*, 1988; Dagosto, 1988; Gebo, 1988). This feature is already present in the very early Eocene taxon *Cantius ralstoni.* Therefore, it is most likely that this derived feature was part of the adapid morphotype. No other mammals except marsupials exhibit this condition, and in notharctines, the flaring is not as great as in later strepsirhines. Therefore, we believe that it is not likely to have been primitive for euprimates, although this is a possibility. If so, the alternate condition seen in haplorhines supports OA monophyly.

There are some instances of partial reversal. The closest approach to the nonstrepsirhine condition (nearly vertical, nonflaring tibiofibular joint surface on talus, with a small process distally) among strepsirhines is among galagos and indriids, but even here the condition is not identical to that of haplorhines. Reversal is therefore possible but never occurs to the point that it perfectly mimics the alternate state.

Position of Talotibial Facet and Groove for Flexor Fibularis (A1). In all known adapids and lemuriforms, the groove is lateral to the posterior part of the facet; therefore, we consider this trait to have been part of the adapid morphotype. In omomyids, tarsiers, and anthropoids (OTA) the groove is plantad and central to the facet. Although the haplorhine condition is typical of most mammals we examined, the morphotype primate and euprimate conditions are difficult to determine. In view of the distribution in other mammals, we interpreted the OTA condition to be primitive for euprimates (Beard *et al.*, 1988). However, paromomyids and dermopterans also have a laterally positioned groove, indicating that this condition may in fact be primitive for primatomorphs (Beard, 1991). If the adapiform–lemuriform condition proves to be primitive for euprimates, the alternate state of OTA provides evidence for OTA monophyly.

Navicular–Cuboid Articulation (N2). In adapids and lemuriforms, the naviculocuboid articulation is large, and on the navicular, the facet for the cuboid is contiguous with and plantad to the navicular–ectocuneiform and navicular–mesocuneiform facets. In haplorhines, the cuboid facet is only contiguous with the ectocuneiform facet. The strepsirhine condition is unique among mammals; there is very little doubt that this is a derived character uniquely shared by these taxa. Because among adapids the navicular is known only in notharctines, there is less strong evidence that this is a feature of the strepsirhine or adapid morphotype. However, the earliest known strepsirhine navicular, that of *Cantius ralstoni,* does exhibit this feature.

There are a few instances in which some reversal of this feature has occurred. In *Daubentonia* and *Archaeolemur,* the cuboid facet is more laterally placed than in other tooth-combed lemurs, so that its medial edge lies just at the junction of the mesocuneiform and ectocuneiform facets. The same is true in galagines and lorisines, but in neither of these cases do the taxa revert completely to the omomyid–anthropoid condition of the trait. Therefore, the probability that anthropoids reversed from a strepsirhine-like condition is very low.

Relative Size of Triquetral Facet on Pisiform (W2). In *Adapis* and living lemuriforms, the size of the triquetral facet on the pisiform bone is reduced (Beard *et al.,* 1988; Beard and Godinot, 1988). Among mammals, this occurs only in these taxa. However, since notharctines do not exhibit this trait, it is not likely to have been part of the adapid morphotype. Thus, only *Adapis* can be excluded from anthropoid ancestry. No instances of reversal in this trait have been reported.

Conclusion

The features cited by AA to say that omomyids are too derived to have given rise to anthropoids are either not present in the omomyid morphotype or are subject to substantial homoplasy (low weight). Length of the navicular is the strongest character supporting this hypothesis. The features cited by OA to say that adapids are too derived to have given rise to anthropoids are more likely to have been present in the adapid morphotype (and therefore all known adapids can be eliminated as ancestors) and, although subject to some homoplasy, are not very likely to be reversed. For most of these characters, a hypothesis of reversal is only necessary if one adopts AA.

Derived Characters Linking Eocene Euprimates with Taxa Other Than Anthropoids

Another way the "too derived" argument has been used in this debate is as follows. If the derived characters of the fossil group are shared with primates other than anthropoids, they can provide strong evidence that this other primate group is more closely related to the fossil group than are anthropoids. The derived features of adapids discussed by Beard *et al.* (1988) are shared with lemuriforms, supporting strepsirhine monophyly. Similarly, AA has sometimes argued that omomyids are more closely related to tarsiers than they are to anthropoids. Like the general form of the "too derived" argument, this argument disproves *direct descent* from the fossil group but cannot disprove a *sister-group* relationship between anthropoids and the (Eocene + living) group (Fig. 4). However, since both OA and AA have generally phrased their arguments such that anthropoids descend from the fossil group, they are vulnerable to this type of argument.

In this context, we can cite other features that support strepsirhine monophyly, although in these cases it is more difficult to reconstruct the

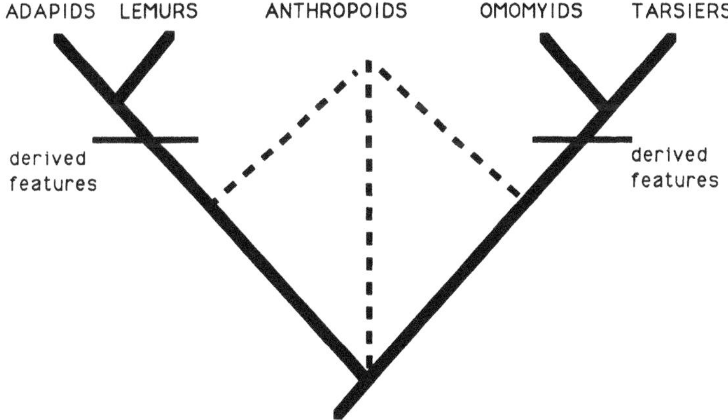

Fig. 4. Demonstrating that a putative ancestral fossil group shares derived characters with a group of primates other than anthropoids indicates that the fossil taxon is more closely related to this other taxon than to anthropoids. Many features can be cited to indicate that omomyids and tarsiers form a monophyletic group and that lemuriforms and adapids form a monophyletic group. However, demonstrating such ties is not informative about the phylogenetic position of anthropoids: anthropoids may still be the sister of either of these groups or a separate lineage, as indicated by the dotted lines.

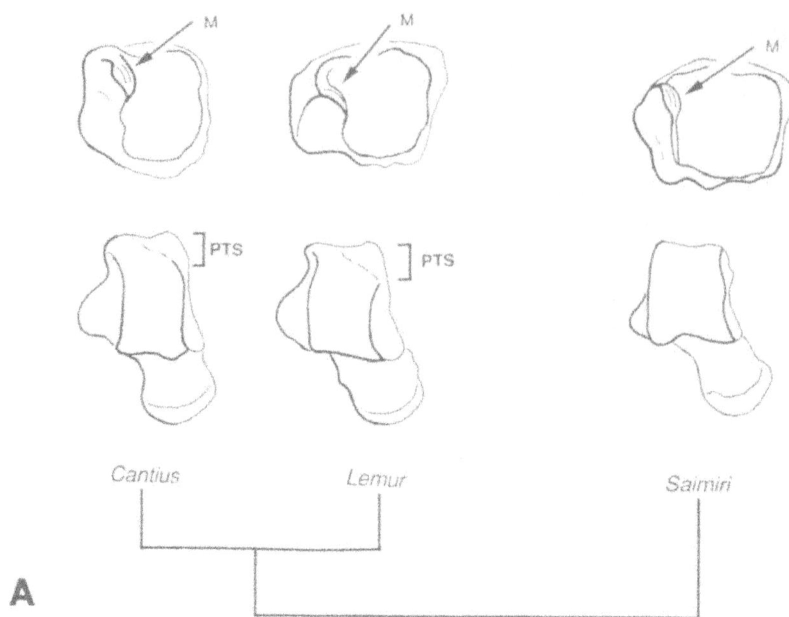
M
M
M
PTS
PTS
Cantius
Lemur
Saimiri
A

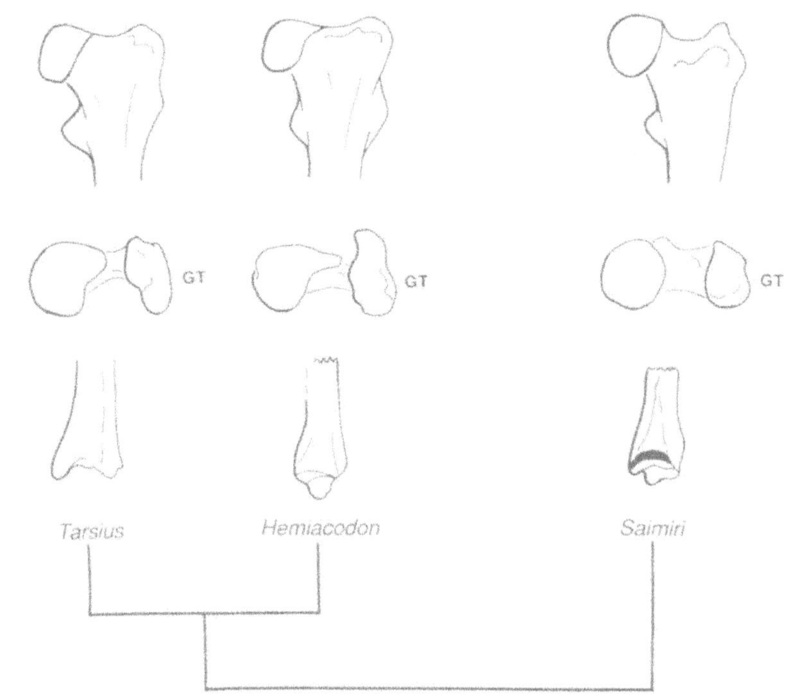
GT
GT
GT
Tarsius
Hemiacodon
Saimiri
B

primitive euprimate condition (Fig. 5A). In all of these cases, Lemuriformes and Adapiformes (LA) share one morphology, and omomyids, *Tarsius*, and anthropoids (OTA) share another.

Distal Tibia (T1–4)

In LA, the distal tibia is characterized by a set of traits that allow a considerable degree of rotation between the tibia and fibula during flexion and extension of the upper ankle joint (Dagosto, 1985). The inferior surface of the tibia has a triangular outline, the medial malleolus is strongly rotated medially, and its surface is completely convex. The distal tibio-fibular joint facet is quite large (high) and is normally anteriorly restricted. In contrast, in OTA, the inferior tibial facet is square in outline, the medial malleolus is only moderately rotated, and its posterior surface is flat, although the anterior surface is convex. In omomyids an articular facet between the distal tibia and fibula is apparently lacking, but all anthropoids (including specimens of *Apidium* where this area is visible) have a small, low, usually anteroposteriorly elongated articular facet (Ford, 1980, 1988). Whether LA conditions represent strepsirhine synapomorphies, or if OTA similarities represent OTA synapomorphies, is difficult to determine unless the ancestral euprimate condition can be reconstructed. At the present time, it appears that the large degree of malleolar rotation, the wholly convex malleolar articular surface, and the triangular outline of the inferior tibial surface are high-weight, derived features, since they occur in no other mammals (Covert, 1988). The large, anteriorly restricted distal tibio-fibular facet is also probably derived (Hershkovitz, 1988), but both dermopterans and paromomyids have smaller, similarly placed facets.

Posterior Trochlear Shelf (A4)

Adapids, lemuriforms,and omomyids exhibit a posterior trochlear shelf on the talus; anthropoids and tarsiers (and other mammals including tree shrews, plesiadapids and dermopterans) have none. In omomyids (except *Necrolemur*) this shelf is very small, but in LA the shelf is quite large. There are at least two ways to interpret this trait. If primitive euprimates had a large shelf, then reduction links OTA, and loss links TA. If a small shelf is primitive,

Fig. 5. Additional features supporting strepsirhine (A) and tarsiiform (B) monophyly. (A) Top, Distal view of tibia illustrating the rotation of the articular surface of the medial malleolus (M). Bottom, Dorsal view of the talus indicating the posterior trochlear shelf (PTS) in *Cantius* and *Lemur* and its lack in anthropoids. (B) Top, anterior view of proximal femur illustrating the cylindrical femoral head and the short, perpendicular neck of omomyids and *Tarsius* compared to anthropoids. Center, Superior view of the femur illustrating the extreme anterior extension of the greater trochanter (GT) in tarsiiforms. Bottom, *Tarsius* and *Necrolemur* have fused tibiofibulas (shown in anterior view); other omomyids (like *Hemiacodon;* lateral view of distal tibia) lack the articular facet between the two bones that is present in anthropoids (colored black in *Saimiri;* lateral view).

then a large shelf links LA (Covert, 1988). We accept the second alternative on the basis that a small shelf is a likely intermediate condition between no shelf and a large shelf and thus is more likely to be a primitive condition.

Dorsoepitrochlear Fossa (H2)

This fossa on the medial side of the posterior surface of the distal humerus is pronounced in Fayum anthropoids (*Apidium, Aegyptopithecus,* and the L-41 taxa; Gebo *et al.*, Chapter 14, this volume) and most platyrrhines and Miocene hominoids; thus, it has been cited as a potential anthropoid synapomorphy (Ford, 1988; Harrison, 1987). However, the fossa is also present in *Microchoerus* and *Necrolemur,* moderately developed in *Hemiacodon* and anaptomorphines (contra Dagosto, 1990), but absent in *Tarsius* (Szalay and Dagosto, 1980). Plesiadapiformes generally have a well-developed fossa, but tupaiids and dermopterans do not. It is usually very shallow in lemuriforms and adapids other than adapinans. An absent or shallow fossa may thus be a potential strepsirhine synapomorphy, although this feature is clearly subject to substantial homoplasy.

Several features can also be cited to support tarsier–omomyid monophyly (Fig. 5B). In addition to long tarsals (especially a long navicular), the absence of a distal tibio-fibular facet or fusion of the tibia and fibula (T1) characterize all omomyids and *Tarsius.* In contrast to adapids and most anthropoids, femora of omomyiform primates and *Tarsius* have a very anteriorly extensive greater trochanter, an anteriorly bowed proximal femur, a very short trochanteric fossa, a very short femoral neck, and a very deep knee (F5, 7-9) (M. Dagosto and P. Schmid, *unpublished observations*). Omomyines (but not microchoerines) have a semicylindrical femoral head (F6). All of these features have, however, evolved convergently in galagos.

Conclusion

There seems to be significant evidence that Strepsirhini (adapids + lemuriforms) is a monophyletic group. OTA also share features; however, at the present time we believe the OTA similarities represent the primitive euprimate condition. Therefore, these similarities cannot support OTA monophyly. However, if we are wrong, and the LA condition is primitive, then these features provide significant support for OA. Among OTA, tarsiers and omomyids share several probably derived features, supporting tarsier–omomyid monophyly.

Phylogenetic Reconstruction

Using the features and assumptions (homology, weighting, polarity) outlined above, we performed a computer-assisted phylogenetic analysis using

PAUP (version 3.0, Swofford, 1991). Our data matrix is given in Table II. The terminal taxa used for analysis are families or subfamilies of fossil primates, since the entire suite of features discussed here is known in so few fossil species. In most cases, the data matrix for the fossil groups is derived from elements that may come from different species within the group. Our experience with both fossil and living primates shows that these features are very stable at the family or subfamily level; thus, we feel secure characterizing subfamilies by the morphology of one or a few members. In the rare cases where features are known to be polymorphic within the group, we have indicated this. Anaptomorphine character states are based on *Absarokius, Arapahovius, Tetonius,* and other unallocated material (Covert and Hamrick, 1993; Savage and Waters, 1978; Szalay, 1976; Szalay and Dagosto, 1980). Omomyine anatomy is based on *Hemiacodon, Omomys,* and *Shoshonius* (K. C. Beard, *personal communication;* Dagosto, 1985; Rosenberger and Dagosto, 1992; Simpson, 1940; Szalay and Dagosto, 1980). Microchoerines are based on *Necrolemur* and *Microchoerus* (Dagosto, 1985; Godinot and Dagosto, 1983; Schlosser, 1907; Schmid, 1979; Szalay and Dagosto, 1980). Notharctine anatomy is abstracted from *Cantius, Notharctus,* and *Smilodectes* (Beard and Godinot, 1988; Covert, 1985; Gebo *et al.*, 1991; Gregory, 1920); adapines from *Adapis* and *Leptadapis* (Beard and Godinot, 1988; Dagosto, 1983); and protoadapines from the Messel adapids, *Europolemur koenigswaldi,* and *Pronycticebus neglectus* (Franzen, 1987, 1988; von Koenigswald, 1979; Thalmann *et al.*, 1989).

The character states for lemuriforms and platyrrhines are based on an assessment of the most likely primitive condition for these groups (Dagosto, 1986; Ford, 1986, 1988; Gebo, 1986). Features were polarized based on the character states present in the outgroup, which is based on plesiadapiforms, dermopterans, and tupaiids.

We used the branch and bound procedure, which is guaranteed to find all most-parsimonious trees. Bootstrap analysis (Felsenstein, 1985) was employed to test the significance of the groupings observed. Statistics (length, CI, tree geometry) for the trees produced under various assumptions are given in Table III.

In view of the previous discussion, the results of this analysis are not surprising. Since anthropoids share few clearly established shared-derived characters with any prosimian taxa and exhibit several features that are also seen in the outgroups but not commonly in prosimians (F5, A3, A4, N1, O3), the most parsimonious hypotheses are those that establish prosimian monophyly (Fig. 6A; Table III). The bootstrap found this grouping in 81% of 500 replications; in contrast, trees supporting haplorhine or simiolemuriform monophyly were only found in 13% and 4% of these replications, respectively. Although the debate on anthropoid origins has usually been reduced to a choice between tarsiiform and adapid ancestors, an alternate hypothesis that must be seriously considered is that neither of these groups is ancestral to or even the exclusive sister group of anthropoids but that prosimians (adapids plus lemuriforms and omomyids plus *Tarsius*) are monophyletic (holophyletic) and that anthropoids are the sister group of this clade. This hypothesis is the

Table III. Summary of PAUP Runs[a]

Constraints	O	W	All characters	Tree geometry	Excluding prosimian monophyly	Tree geometry
None	O	W	155, 0.794	Prosimian monophyly	133, 0.82	Unresolved trichotomy
None	U	U	71, 0.803	Prosimian monophyly	57, 0.825	Haplorhine monophyly
Adapid–anthropoid	O	W	160, 0.769	[(A + L)An]	133, 0.820	[(A + L)An]
Adapid–anthropoid	U	U	73, 0.781	[(A + L)An]	58, 0.810	[(A + L)An]
Omomyid–anthropoid	O	W	160, 0.792	[(O + T)An]	133, 0.820	[(O + T)An]
Omomyid–anthropoid	U	U	72, 0.792	[(O + T)An]	57, 0.825	[(O + T)An]
Tarsius–anthropoid	O	W	178, 0.685			
Anaptomorphine–anthropoid	O	W	168, 0.726[b]			
Omomyine–anthropoid	O	W	172, 0.709			
Microchoerine–anthropoid	O	W	169, 0.722			
Notharctine–anthropoid	O	W	179, 0.692			
Protoadapine–anthropoid	O	W	165, 0.739[b]			
Adapinan–anthropoid	O	W	175, 0.697[b]			
Lemuriform–anthropoid	O	W	183, 0.667			

[a]Constraints, the constraint tree in effect for the run; O, ordered; W, weighted; U, unordered, unweighted; All characters, all characters were included in the analysis; the numbers are the number of steps in the resulting tree and the consistency index of the tree; Excluding prosimian monophyly, characters that support prosimian monophyly were removed from the analysis; Tree geometry, geometry of the resulting consensus tree; A, adapids; L, lemuriforms; An, anthropoids; O, omomyids; T, Tarsius.
[b]Tree length may be artifically low because of missing data.

one most strongly supported by postcranial evidence (Ford, 1988). However, it does contradict the cranial, dental, and soft anatomic evidence supporting OA and AA. If the postcranial characters supporting prosimian monophyly are removed from the analysis or recoded as new character states exhibited only by anthropoids, the result is an unresolved trichotomy of anthropoids, strepsirhines, and tarsiiforms (Fig. 6B; Table III).

The hypothesis that lemuriforms are more closely related to adapids than either group is to anthropoids is supported by the available postcranial evidence (nine mainly high-weight consistent features; Table IV), although this group was reproduced in only 74% of the bootstrap replications. During the conference, the possibility was raised that adapids and lemurs share features because of similarity in body size. Although we do not doubt that change in size is an important contributing factor in the development of these features (Dagosto, 1988), we believe that the probability is low that adapids and lemurs acquired this set of features independently because of parallel acquisition of large body size. We base our conclusion on several lines of evidence: (1) anthropoids, which also become large, do not exhibit any of these features; (2) the characteristic strepsirhine or haplorhine expression of the form of these features does not vary greatly despite a large size range within each group; (3)

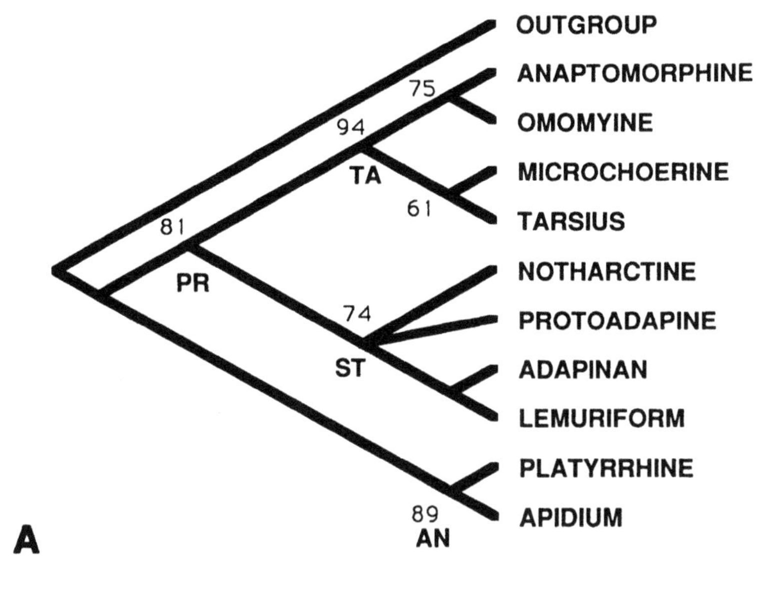

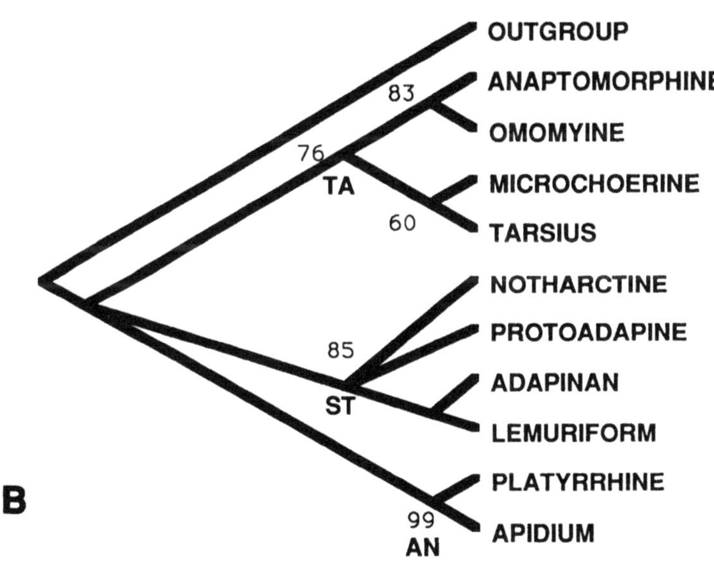

Fig. 6. The most parsimonious trees resulting from the analysis of the complete data set (A) and when the characters supporting prosimian monophyly are removed from the analysis (B). The number at each node is the percentage of times the monophyletic group was repeated in the bootstrap analysis. Only groups found in more than 50% of replications are shown. Synapomorphies for the labeled nodes are given in Table IV; other tree statistics are given in Table III.

Table IV. Lists of Apomorphies at Selected Nodes of the Cladograms Shown in Figs. 6 and 7

	Apomorphies				Apomorphies		
Node	Character	Change	Consistency	Node	Character	Change	Consistency
The strict consensus of the three most parsimonious trees (Fig. 6A)							
PR	F2	0 → 1	0.6	ST	T1	1 → 0	0.75
	T7	0 → 1	0.5		T3	0 → 1	1.0
	A3	0 → 1	1.0		T4	1 → 2	1.0
	A4	0 → 1	0.5		T5	1 → 2	1.0
	N1	0 → 1	1.0		A1	0 → 1	1.0
	O1	0 → 1	0.67		A2	0 → 1	1.0
	O3	0 → 1	1.0		A4	1 → 0	0.5
	H2	0 → 1	0.6		N2	0 → 1	1.0
	Mean		0.73		H1	1 → 2	0.67
					Mean		0.88
TA	F1	1 → 0	0.5				
	F2	1 → 2	0.6	AN	F1	1 → 2	0.5
	F5	1 → 2	0.5		F3	0 → 1	0.5
	F6	0 → 1	0.67		F4	0 → 2	1.0
	F7	0 → 1	1.0		E1	1 → 2	1.0
	F8	0 → 1	1.0		Mean		0.75
	F9	0 → 1	0.6				
	T1	1 → 2	0.75				
	C1	1 → 2	0.5				
	N1	1 → 2	1.0				
	O2	0 → 1	0.5				
	Mean		0.69				
The strict consensus of the six most parsimonious trees after the feature supporting prosimian monophyly were excluded from the analysis (Fig. 6B)							
TA	F1	1 → 0	0.5	ST	T1	1 → 0	0.75
	F2	1 → 2	0.6		T3	0 → 1	1.0
	F6	0 → 1	0.67		T4	1 → 2	1.0
	F7	0 → 1	1.0		T5	1 → 2	1.0
	F8	0 → 1	1.0		T7	0 → 1	0.5
	F9	0 → 1	0.6		A1	0 → 1	1.0
	T1	1 → 2	0.75		A2	0 → 1	1.0
	O1	1 → 2	0.67		N2	0 → 1	1.0
	O2	0 → 1	0.5		H1	1 → 2	0.67
	Mean		0.69		H2	0 → 2	0.6
					Mean		0.85
AN	F1	1 → 2	0.5				
	F2	1 → 0	0.5				
	F3	0 → 1	0.5				
	F4	0 → 2	1.0				
	E1	1 → 2	1.0				
	Mean		0.70				

(*continued*)

Table IV. (*Continued*)

Node	Apomorphies			Node	Apomorphies		
	Character	Change	Consistency		Character	Change	Consistency
The strict consensus of three trees if tarsiiforms and anthropoids are constrained to be a monophyletic group (Fig. 7A)							
TA	F1	1 → 0	0.5	ST	T3	0 → 1	1.0
	F2	1 → 2	0.6		T4	1 → 2	1.0
	F5	1 → 2	0.5		T5	1 → 2	1.0
	F6	0 → 1	0.67		T7	0 → 1	0.5
	F7	0 → 1	1.0		A1	0 → 1	1.0
	F8	0 → 1	1.0		A2	0 → 1	1.0
	F9	0 → 1	0.6		A4	1 → 2	0.4
	C1	1 → 2	0.5		N2	0 → 1	1.0
	N1	1 → 2	0.67		H1	1 → 2	0.67
	O1	1 → 2	0.67		H2	0 → 2	0.6
	O2	0 → 1	0.5		Mean		0.82
	Mean		0.65				
AN	F1	1 → 2	0.5	HA	T1	0 → 1	1.0
	F2	1 → 0	0.6				
	F3	0 → 1	0.5				
	F4	0 → 2	1.0				
	A3	1 → 0	0.5				
	A4	1 → 0	0.4				
	N1	1 → 0	0.667				
	O3	1 → 0	0.5				
	E1	1 → 2	1.0				
	Mean		0.63				
The strict consensus of three trees if strepsirhines and anthropoids are constrained to be a monophyletic group (Fig. 7B)							
TA	F1	1 → 0	0.5	ST	T1	1 → 0	0.75
	F2	1 → 2	0.6		T3	0 → 1	1.0
	F5	1 → 2	0.5		T4	1 → 2	1.0
	F6	0 → 1	0.67		T5	1 → 2	1.0
	F7	0 → 1	1.0		T7	0 → 1	0.5
	F8	0 → 1	1.0		A1	0 → 1	1.0
	T1	1 → 2	0.75		A2	0 → 1	1.0
	C1	1 → 2	0.5		A4	1 → 2	0.4
	N1	1 → 2	0.67		N2	0 → 1	1.0
	O1	1 → 2	0.67		H1	1 → 2	0.67
	O2	0 → 1	0.5		H2	0 → 2	0.6
	Mean		0.67		Mean		0.81
AN	F1	1 → 2	0.5	SL	F9	1 → 0	0.75
	F2	1 → 0	0.6				
	F3	0 → 1	0.5				
	F4	0 → 2	1.0				
	A3	1 → 0	0.5				
	A4	1 → 0	0.4				
	N1	1 → 0	0.67				
	O3	1 → 0	0.5				
	E1	1 → 2	1.0				
	Mean		0.63				

the detailed, extensive similarity between adapids and lemurs is strong evidence that these character states are homologous. For most of these features, these animals are not just similar, they are the same.

Tarsiiform monophyly is supported by 11 characters and was found in 94% of replications; this is the only grouping that approaches statistical significance (Felsenstein, 1985). However, most of these features are of lower weight and less consistent than those supporting strepsirhine monophyly, since they have evolved convergently elsewhere in primates, either in galagos or anthropoids.

Forcing anthropoids to be the sister group of strepsirhines or tarsiiforms results in longer trees, but these two trees are not different in length (Fig. 7; Table III). In each case, only one feature was recognized as a synapomorphy of the forced clades (Table IV). Trees linking anthropoids exclusively with any of the terminal taxa, living or extinct, are even more costly (Table III).

The results of this analysis are in general agreement with those of Ford (Chapter 13, this volume), who analyzed a largely different set of features in an only slightly overlapping set of taxa. In particular, both analyses support prosimian monophyly, strepsirhine monophyly, tarsiiform monophyly, and anthropoid monophyly in that *Apidium* always groups with other anthropoids rather than with any prosimians. (The new taxa from L-41 are too incomplete to be included in this analysis, but they too show no strong similarities to any prosimian group: Gebo *et al.*, Chapter 14, this volume.) In neither analysis are omomyid–anthropoid, tarsier–anthropoid, or adapid–anthropoid groups the most parsimonious arrangements of taxa.

Thus, *if only postcranial evidence is considered,* it is difficult to envision either adapids or omomyids as *ancestral* to anthropoids, since both groups have a number of derived characteristics that are not exhibited by anthropoids. Anthropoids may be the sister group of tarsiiformes, the sister group of strepsirhines, or an independent lineage of euprimates. The last hypothesis is the most strongly supported, and the first two possibilities are equally likely.

We reiterate here our point that phylogenies based on only one character system are incomplete; in effect, characters not included are given zero weight. The conclusion of prosimian monophyly yielded by this analysis of postcranial characters is clearly at odds with information from cranial, facial, otic, visual, reproductive, and molecular systems that favor tarsier–anthropoid and potentially omomyid–anthropoid ties (e.g., Cartmill and Kay, 1978; Ross, Chapter 15, this volume; Rosenberger and Szalay, 1980) and dental features that favor adapid–anthropoid ties (Gingerich, 1980; Rasmussen and Simons, 1988; Simons and Rasmussen, 1989). In our opinion, many of the features supporting tarsiiform–anthropoid ties are of higher weight than the postcranial features supporting prosimian monophyly for reasons given elsewhere (Dagosto, 1990). Although the postcranial data do not provide strong support for this hypothesis, based on our assessment of this other evidence, and the fact that alternate polarities for many of our

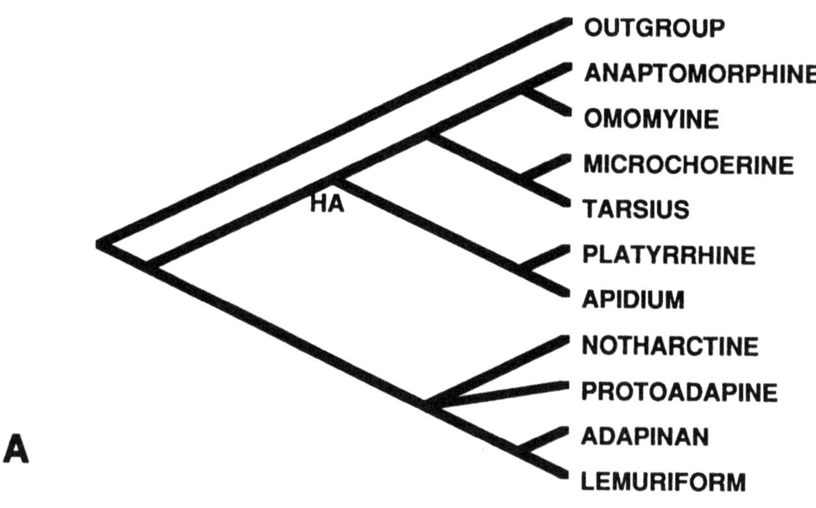

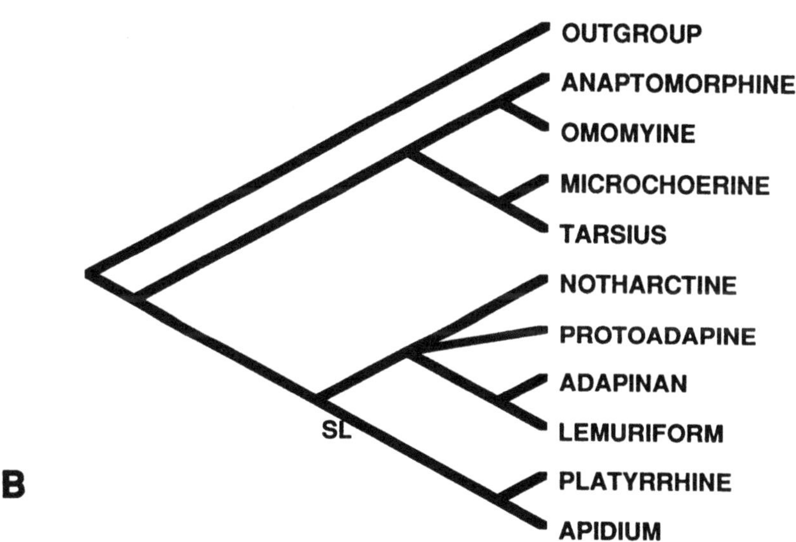

Fig. 7. Most parsimonious trees resulting when anthropoids are constrained to group with tarsiiforms (A) and with strepsirhines (B). In both cases, the tree length is 160 steps (Table III). Both hypotheses are supported by only one synapomorphy, haplorhine monophyly by the relatively close apposition of the distal tibio-fibular joint [T1(1)] and simiolemuriform monophyly by a relatively long intertrochanteric crest on the femur [F9(0)] (Table IV). Nodes AN, TA, and ST are as in Fig. 6.

features would support tarsiiform–anthropoid ties, we believe that haplorhine monophyly is still the best hypothesis of primate relationships.

These disjunct character distributions force a choice between at least two interpretations of primate evolution:

1. Anthropoids originated from some primitive euprimate much earlier than is supposed by the current incarnations of the OA or AA models. This is necessary if anthropoids are the sister group of prosimians as a whole or if they branched off at the base of the strepsirhine or tarsiiform radiations. This hypothesis has previously been considered unlikely because of the lack of fossil evidence for such an early derivation, a position that can no longer be maintained (Beard *et al.,* 1994; Godinot and Mahboubi, 1992). This scenario, however, involves considerable parallel evolution between tarsiiforms (especially *Tarsius*) and anthropoids in the systems mentioned above (except in the case where haplorhine monophyly is accepted and these features are considered basal for all haplorhines).
2. Anthropoids are derived from some known prosimian lineage (the split could have been early or late), but the transformation in the postcranium was quite profound. Deriving anthropoids from either fossil group or from the tarsier lineage necessitates a considerable number of evolutionary reversals in the postcranium (in addition to all the features that must reverse if any prosimian is ancestral: Ford, 1988, Dagosto, 1990).

The choice between these scenarios ultimately involves accepting the homology of one set of features over another; disagreements between workers stem partly from the use of different criteria to judge homology (Dagosto, 1990). Given the probable antiquity of the prosimian–anthropoid split and/or the extensive nature of the transformation, anthropoid origins may well be another example of a case where the fossil record will be particularly useful and probably necessary to the determination of polarity and homology and thus a more robust phylogenetic hypothesis (Donoghue *et al.,* 1989).

Acknowledgments

We thank J. Fleagle and R. Kay for the invitation to participate in the symposium, and the Leakey and Wenner-Gren Foundations for support of the meeting. We also extend our thanks to the staffs of the many museums at which we have collected data relevant to this project. The National Science Foundation and the Wenner-Gren Foundation have supported much of this research. We thank the reviewers for their careful reading of the manuscript and their useful comments.

References

Barnett, C. H., and Napier, J. R. 1953. The rotatory ability of the fibula in eutherian mammals. *J. Anat.* **87**:11–21.

Beard, K. C. 1991. Vertical postures and climbing in the morphotype of primatomorpha: Implications for locomotor evolution in primate history. In: Y. Coppens and B. Senut (eds), *Origine(s) de la Bipedie chez les Hominides (Cahiers de Paléoanthropologie)*, pp. 79–87. CNRS, Paris.

Beard, K. C., Dagosto, M., Gebo, D. L., and Godinot, M. 1988. Interrelationships among primate higher taxa. *Nature* **331**:712–714.

Beard, K. C., and Godinot, M. 1988. Carpal anatomy of *Smilodectes gracilis* (Adapiformes, Notharctinae) and its significance for lemuriform phylogeny. *J. Hum. Evol.* **17**:71–92.

Beard, K. C., Qi, T., Dawson, M. R., Wang, B., and Li, C. 1994. A diverse new primate fauna from middle Eocene fissure-fillings in southeastern China. *Nature* **368**:604–609.

Carleton, A. 1941. A comparative study of the inferior tibio-fibular joint. *J. Anat.* **76**:45–55.

Cartmill, M. 1972. Arboreal adaptations and the origin of the Order Primates. In: R. Tuttle (ed.), *The Functional and Evolutionary Biology of Primates*, pp. 97–122. Aldine-Atherton, Chicago.

Cartmill, M., and Kay, R. F. 1978. Cranio-dental morphology, tarsier affinities and primate suborders. In: D. J. Chivers and K. A. Joysey (eds), *Recent Advances in Primatology*, pp. 205–213. Academic Press, New York.

Cartmill, M., MacPhee, R. D. E., and Simons, E. L. 1981. Anatomy of the temporal bone in early anthropoids. *Am. J. Phys. Anthropol.* **56**:3–21.

Cave, A. J. E. 1967. Observations on the platyrrhine nasal fossa. *Am. J. Phys. Anthropol.* **26**:277–288.

Covert, H. H. 1985. Adaptations and evolutionary relationships of the Eocene primate family Northarctidae. Ph.D. dissertation, Duke University.

Covert, H. H. 1988. Ankle and foot morphology of *Cantius mckennai:* Adaptations and phylogenetic implications. *J. Hum. Evol.* **17**:57–70.

Covert, H. H., and Hamrick, M. W. 1993. Description of new skeletal remains of the Early Eocene Anaptomorphine Primate *Absarokius* (Omomyidae) and a discussion of its adaptive profile. *J. Hum. Evol.* **25**:351–362.

Dagosto, M. 1983. Postcranium of *Adapis parisiensis* and *Leptadapis magnus* (Adapiformes, Primates). *Fol. Primatol.* **41**:49–101.

Dagosto, M. 1985. The distal tibia of primates with special reference to the Omomyidae. *Int. J. Primatol.* **6**:45–75.

Dagosto, M. 1986. The joints of the tarsus in the Strepsirhine primates. Ph.D. dissertation, City University of New York.

Dagosto, M. 1988. Implications of postcranial evidence for the origin of Euprimates. *J. Hum. Evol.* **17**:35–56.

Dagosto, M. 1990. Models for the origin of the anthropoid postcranium. *J. Hum. Evol.* **19**:121–140.

Donoghue, M. J., Doyle, J. A., Gauthier, J., and Kluge, A. G. 1989. The importance of fossils in phylogeny reconstruction. *Annu. Rev. Ecol. Syst.* **20**:43–60.

Felsenstein, J. 1985. Confidence limits on phylogenies: an approach using the bootstrap. *Evolution* **39**:783–791.

Ford, S. M. 1980. A systematic revision of the Platyrrhini based on features of the postcranium. Ph.D. dissertation, University of Pittsburgh.

Ford, S. M. 1986. Systematics of the New World Monkeys. In: D. R. Swindler and J. Erwin (eds.), *Comparative Primate Biology*, pp. 73–135. Alan R. Liss, New York.

Ford, S. M. 1988. Postcranial adaptations of the earliest platyrrhine. *J. Hum. Evol.* **17**:155–192.

Franzen, J. L. 1987. Ein neuer Primate aus dem Mitteleozän der Grube Messel (Deutschland, S.-Hessen). *Cour. Forsch. Inst. Senckenberg* **91**:151–187.

Franzen, J. L. 1988. Ein weiterer Primatenfund aus der Grube Messel bei Darmstadt. *Cour. Forsch. Inst. Senckenberg* **107**:275–289.

Gebo, D. L. 1986. The anatomy of the prosimian foot and its application to the primate fossil record. Ph.D. dissertation, Duke University.

Gebo, D. L. 1988. Foot morphology and locomotor adaptation in Eocene Primates. *Fol. Primatol.* **50:**3–41.

Gebo, D. L., Dagosto, M., and Rose, K. D. 1991. Foot morphology and evolution in early Eocene *Cantius. Am. J. Phys. Anthropol.* **86:**51–73.

Gingerich, P. D. 1975. A new genus of Adapidae (Mammalia, Primates) from the late Eocene of Southern France, and its significance for the origin of higher primates. *Contrib. Mus. Pal. Univ. Michigan* **24:**163–170.

Gingerich, P. D. 1980. Eocene Adapidae, paleobiogeography, and the origin of South American Platyrrhini. In: R. L. Ciochon and A. B. Chiarelli (eds.), *Evolutionary Biology of the New World Monkeys and Continental Drift,* pp. 123–138. Plenum Press, New York.

Godinot, M., and Dagosto, M. 1983. The astragalus of *Necrolemur* (Primates, Microchoerinae). *J. Paleontol.* **57:**1321–1324.

Godinot, M., and Mahboubi, M. 1992. Earliest known simian primate found in Algeria. *Nature* **357:**324–326.

Gregory, W. K. 1920. On the structure and relations of *Notharctus:* An American Eocene primate. *Mem. Am. Mus. Nat. Hist.* **3:**51–243.

Harrison, T. 1987. The phylogenetic relationships of the early catarrhine primates: A review of the current evidence. *J. Hum. Evol.* **16:**41–80.

Hecht, M. K., and Edwards, J. L. 1977. The methodology of phylogenetic inference above the species level. In: M. K. Hecht, P. Goody, and B. Hecht (eds.), *Major Patterns of Vertebrate Evolution,* pp. 3–51. Plenum Press, New York.

Hershkovitz, P. 1988. The subfossil monkey femur and subfossil monkey tibia of the Antilles: A review. *Int. J. Primatol.* **9:**365–384.

Hill, J. P. 1919. The affinities of *Tarsius* from the embryological aspect. *Proc. Zool. Soc. Lond.* **1919:**476–491.

Koenigswald, W. von. 1979. Ein Lemurenreste aus dem eozänen Ölschiefer der Grube Messel bei Darmstadt. *Palaont. Z.* **53:**63–76.

Luckett, W. P. 1975. Ontogeny of the fetal membranes and placenta: Their bearing on primate phylogeny. In: W. P. Luckett and F. S. Szalay (eds.), *Phylogeny of the Primates,* pp. 157–182. Plenum Press, New York.

Neff, N. A. 1986. A rational basis for *a priori* character weighting. *Syst. Zool.* **35:**110–123.

Pocock, R. I. 1918. On the external characters of the lemurs and of *Tarsius. Proc. Zool Soc. Lond.* **1918:**19–53.

Rasmussen, D. T. 1990. The phylogenetic position of *Mahgarita stevensi:* Protoanthropoid or lemuroid. *Int. J. Primatol.* **11:**439–470.

Rasmussen, D. T., and Simons, E. L. 1988. New specimens of *Oligopithecus savagei,* early Oligocene primate from the Fayum, Egypt. *Fol. Primatol.* **51:**182–208.

Remane, A. 1956. *Die Grundlagen des natürlichen Systems, der vergleichenden Anatomie und der Phylogenetik.* Geest und Portig, Leipzig.

Rosenberger, A. L., and Dagosto, M. 1992. New craniodental and postcranial evidence of fossil tarsiiforms. In: S. Matano, R. H. Tuttle, H. Ishida, and M. Goodman (eds.), *Topics in Primatology, Vol. 3,* pp. 37–51. University of Kyoto Press, Kyoto.

Rosenberger, A. L., and Szalay, F. S. 1980. On the tarsiiform origins of the anthropoidea. In: R. L. Ciochan and A. B. Chiarelli (eds.), *Evolutionary Biology of New World Monkeys and Continental Drift,* pp. 139–157. Plenum Press, New York.

Savage, D. E., and Waters, B. T. 1978. A new omomyid primate from the Wasatch formation of southern Wyoming. *Fol. Primatol.* **30:**1–29.

Schlosser, M. 1907. Bietrag zur Osteologie und systematischen Stellung der Gattung *Necrolemur,* sowie zur Stammesgeschicte der Primaten überhaupt. *Neues Jb. Miner. Geol. Palaont. Mh.* **1907:**199–226.

Schmid, P. 1979. Evidence of microchoerine evolution from Dielsdorf (Zurich region, Switzerland)—a preliminary report. *Fol. Primatol.* **31:**301–311.

Schmid, P. 1992. Leaping adaptation in Eocene primates. In: *Abstracts of the XIVth Congress of the International Primatological Society,* p. 71. International Primatological Society, Strasbourg, France.

Simons, E. L., and Rasmussen, D. T. 1989. Cranial morphology of *Aegyptopithecus* and *Tarsius* and the question of the Tarsier–Anthropoidean clade. *Am. J. Phys. Anthropol.* **79:**1–23.

Simpson, G. G. 1940. Studies on the earliest primates. *Bull. Am. Mus. Nat. Hist.* **77:**185–212.

Simpson, G. G. 1961. *Principles of Animal Taxonomy.* Columbia University Press, New York.

Swofford, D. L. 1991. PAUP: Phyogenetic Analysis Using Parsimony, Version 3.0., Champaign, Illinois.

Szalay, F. S. 1975. Phylogeny of primate higher taxa: The basicranial evidence. In: W. P. Luckett and F. S. Szalay (eds.), *Phylogeny of the Primates,* pp. 91–125. Plenum Press, New York.

Szalay, F. S. 1976. Systematics of the Omomyidae (Tarsiiformes, Primates): Taxonomy, phylogeny, and adaptations. *Bull. Am. Mus. Nat. Hist.* **156:**157–450.

Szalay, F. S. 1977. Ancestors, descendants, sister groups and testing of phylogenetic hypotheses. *Syst. Zool.* **26:**12–18.

Szalay, F. S., and Bock, W. J. 1991. Evolutionary theory and systematics: relationships between process and patterns. *Z. Zool. Syst. Evol. Forsch.* **29:**1–39.

Szalay, F. S., and Dagosto, M. 1980. Locomotor adaptations as reflected on the humerus of Paleogene primates. *Fol. Primatol.* **34:**1–45.

Thalmann, U., Haubold, H., and Martin, R. D. 1989. *Pronycticebus neglectus*–an almost complete adapid primate specimen from the Geiseltal (GDR). *Palaeovertebrata* **19:**115–130.

Primitive Platyrrhines? Perspectives on Anthropoid Origins from Platyrrhine, Parapithecid, and Preanthropoid Postcrania

18

SUSAN M. FORD

Introduction

Despite much effort over many years, we still have only a poorly resolved and hotly contended understanding of the relationships among major primate groups, both fossil and living (see recent review, Martin, 1993) (Fig. 1). Anthropoids have long been assumed to have originated from a known North American Eocene group, either among the Omomyidae (e.g., MacPhee and Cartmill, 1986; Rosenberger and Dagosto, 1992; Rosenberger and Szalay, 1980; Szalay, 1975; Wortman, 1904) or the Adapidae (Gidley, 1923; Gingerich, 1980, 1984; Rasmussen, 1986, 1990; Rasmussen and Simons, 1992). New discoveries of early anthropoids, possible anthropoids, or preanthropoids from middle Eocene to early Oligocene beds in North Africa, the Arabian peninsula, and Asia (de Bonis *et al.*, 1988; Ciochon and Holroyd, Chapter 6, this volume; Ciochon *et al.*, 1985; Godinot and Mahboubi, 1992; Pickford, 1986; Culotta, 1992; Sigé *et al.*, 1990; Simons, 1990, 1992; Thomas *et al.*,

SUSAN M. FORD • Department of Anthropology, Southern Illinois University, Carbondale, Illinois 62901.
Anthropoid Origins, edited by John G. Fleagle and Richard F. Kay. Plenum Press, New York, 1994.

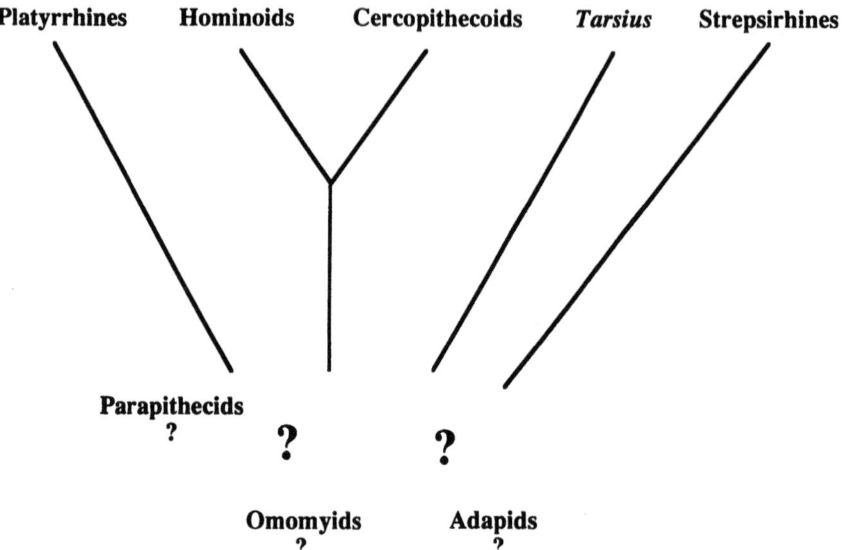

Fig. 1. Current views on anthropoid origins.

1988, 1989; and perhaps the enigmatic new find from Jebel Chambi, Tunisia: Court, 1993), while exciting, have acutally raised more questions than they have answered, reopening in particular the possibility that the anthropoid lineage predates any of the known North American Eocene primate groups (Hershkovitz, 1974; Hoffstetter, 1974a,b, 1980).

Most chapters in this volume and, indeed, most studies attempting to attain a clearer understanding of anthropoid origins and primate relationships concentrate on dental features, in large part because of the scarcity of early anthropoid and preanthropoid cranial and postcranial remains. However, skeletal elements are available for members of many of the groups that are critical for a study of the monophyly and origins of Anthropoidea. Systematic studies concentrating on many functional regions run less risk of being biased by unrecognized convergences in any single complex. In addition, there is much important information on phylogenetic relationships in the postcranial skeleton (Dagosto, 1986, 1988, 1990; Fleagle and Kay, 1987; Ford, 1980a,b, 1986; Strasser and Delson, 1987; Szalay *et al.*, 1987).

In the recent literature, the "ancestral primate morphotype" or immediately "preanthropoid morphotype" for skeletal features has been variously described as more platyrrhine-like (Ford, 1986, 1988, 1990; Godinot, 1992), more consistent with a cheirogaleid-like behavioral pattern (Gebo, 1986, 1988), more "prosimian" and, in specific features, adapid-like (Dagosto, 1986, 1990) or omomyid-like (Dagosto, 1986, 1990), or more parapithecid-like (Fleagle and Kay, 1987). Perhaps not surprisingly, in most cases, the ancestral "model" is generally the group that is the focus of that researcher's primary interest and work; in addition, the major thrust of these recent works has been to determine the behavioral repertoire of some hypothetical earliest primate, with the anthropoids treated as an afterthought.

Here, an appraisal of the evidence from several regions of the postcranium is made specifically to address key issues in anthropoid origins, using varying conceptions of the ancestral morphology. The range of variability found both within and between major primate groups is considered in order to address some of the outstanding issues surrounding Anthropoidea, including: (1) Is it a monophyletic group? (2) What was the likely ancestral morphotype (and associated behavioral repertoire) of anthropoids? (3) What were the most likely phylogenetic and geographic "sources" of the first anthropoids, including whether they were closely related to *Tarsius* or to extant strepsirhines, and if either omomyids or adapids were the "source" group for the first anthropoids? And (4) what were the emerging lineages in the basal radiation of anthropoids (or at least those for which postcrania are known)?

Methods and Materials

General Approach

There are a number of potential methods for addressing questions of anthropoid origins (or any systematic issue). In a very simplistic sense, these can be divided into two approaches. The first involves detailed biological, functional, and morphological analyses of a handful of traits, often related in a large complex (e.g., ear region, molar dentition, orbit, upper ankle joint); this is in fact the approach of several chapters in this volume. I agree that there is no substitute for a careful analysis and understanding of traits, whether for a study of that complex alone or as part of a larger data set for, e.g., a numerical cladistic analysis (see Mickevich and Weller, 1990; Neff, 1986; Pogue and Mickevich, 1990). However, functional complexes can and frequently have developed convergently, even in closely related animals, and looking at one complex in isolation makes it impossible to choose between equally plausible hypotheses of homologous or convergent attainment of these features. The ongoing debates on the homologies of primate orbital closure and basicranial structures are graphic illustrations of this problem. Although one can envision "logical transformations" between morphs of a complex (as in Szalay *et al.*, 1987), often more than one transformational sequence can be conceived as plausible, each supporting contradictory results. In addition, evolution has not always proceeded in entirely logical (by human standards) pathways.

A second approach is an analysis of many traits, preferably crossing numerous functional regions and complexes. As one adds characters and complexes as well as taxa, this type of analysis quickly outstrips the ability of any human to manage without the aid of some computer algorithm. Increasingly, particularly if one accepts an emphasis on shared and derived features as the prime indicators of relationship, this has led to the development of numerical cladistic approaches (see, e.g., Farris, 1970, 1983, 1989; Farris *et al.*, 1970; Felsenstein, 1983, 1985; Swofford, 1991a; Wiley *et al.*, 1991). A problem with numerical cladistic approaches that has left them open to criticism is the danger that they can become simply a compilation of character state trees with little consideration for or relevance to the biology of those traits.

There is a possible rapprochement: a combination of these two types of analysis. As Mickevich and Weller (1990) and Neff (1986) have pointed out, a cladistic study (including a numerical cladistic study) is only as good as the characters and character analysis that underlie it. This implies a judicious and deliberate selection of characters used in a numerical analysis, combined with a careful biologically based evaluation of their interaction in functional complexes. In addition, it requires the use of independent and sound bases both for assigning polarity and transformation sequences (morphoclines) and for assuming homology or homoplasy. It is argued here that this combined approach will lead to the most informative, potentially most accurate, and most biologically sound hypotheses on evolutionary relationships. However, it is also more time consuming, depending on the number of characters analyzed, and considerably more difficult to summarize briefly.

Character Choice and Coding

This study is based on a cladistic analysis of 52 postcranial features drawn from the shoulder, elbow, hip, knee, and upper and transverse ankle joints and incorporates data on four bones: the humerus, femur, tibia, and calcaneus. This step is the ultimate "weighting" of any data set: the heavy weight given characters that are included, the zero weight given all those that are excluded (see, e.g., Swofford and Olsen, 1990). The most variable characters and those that were clearly primarily the result of size differences (allometry) were excluded. The characters used here are discussed more fully elsewhere (Ford, 1980a, 1986, 1990); this data set represents a judiciously pruned subset of those I have used earlier, lacking many features that appear to contribute primarily "noise" and that prior study has shown to be prone to extremely high levels of homoplasy or variability. However, as discussed in my earlier works, features that may represent "character complexes" in some taxa are nonetheless coded separately as discrete characters here. Although this runs the risk of overweighting the "character complex" in those taxa in which the features covary and cofunction, prior analyses have indicated that most components of functional complexes do not covary across all primates but only in those animals in which the components serve a particular unified function to fulfill a particular anatomic/biological role (e.g., aspects of the proximal femur in small-bodied leapers). In other primates doing other things, the complex may essentially "come apart," and the different components vary independently. This independence, and indeed the covariance as a functional complex, will become fully appreciated evolutionarily only if the traits are originally analyzed as independent (see also Kay and Williams, Chapter 13, this volume). All data were collected and coded by myself, with the single exception of some data on *Adapis* taken from Dagosto (1983) and Godinot (1991). It is impossible given the confines of space here to elaborate on the deliberations entered into over the choice and coding of each character; some of this is presented elsewhere (see especially Ford, 1980a). However, the analytical processes involved are described. These, and the various permutations of analyses described, are important in both empowering and limiting the resulting conclusions; thus, they are discussed here in some detail.

The characters included 33 qualitative features and 19 quantitative features, all of which were coded into discrete character states (see Appendix A). Coding, of course, is essential for any cladistic analysis, but it was a difficult first step. Nearly all of the qualitative features were traits with more than two morphs (i.e., more than simply presence/absence). Deciding on cutoff points between, e.g., "medium" and "large" protuberances was considerably more difficult than coding a feature such as the connection or lack thereof between two facets. In some cases, although several qualitative morphs were originally scored, these were collapsed into fewer, broader categories because of considerable variability within genera and species.

Even more difficult was determining discrete categories for quantitative features. However, comparisons of analyses done on qualitative features alone to those done on a combined data set make it clear, as it has been to researchers for a very long time, that critical information is contained in features that are best and most accurately assessed by measurement. To exclude these data from cladistic analyses simply because it is difficult to divide continuous variability into discrete categories, as suggested by Farris (1990; see also discussion in Chappill, 1989), severely limits the information content of the analysis, and the resulting cladogram and can yield erroneous results. Both angles and indices of linear measurements were used; indices were chosen over regressions or residuals primarily because of the inclusion of fossil taxa. With incomplete and often broken fossils, it is risky to attempt to regress variables against some estimate of body weight, and any single estimator of body size may not be present in all fossils. In addition, discovery of each new fossil or extant specimen would require reanalyzing the entire set and perhaps recoding the character. Indices allow more flexible comparisons across taxa, across data sets, and across time in different studies, yet they provide some measure of control for size (but see Sokal and Rohlf, 1981). In most cases, the denominator of the index was some variable that was shown to have a close relationship to body weight across the extant taxa, at least within platyrrhines (for which a larger sample was available; Ford *et al.*, 1991). A modified and less rigorous version of Mickevich's (1982) gap coding was used to identify discrete character states for the quantitative features. In all cases, a "coarse-grained" rather than a "fine-grained" approach was used; i.e., some taxa showed wide ranges of variability for many traits, and there were often large areas of overlap. A deliberate choice was made to recognize fewer, broader categories, thus erring on the side of underestimating true diversity in favor of not assuming too much distinctiveness for taxa with small sample sizes or represented by single specimens.

Taxa Studied

Thirty-six taxa were included in the study (see Appendix B); all taxa were examined on the generic level, with subgeneric variation pooled as polymorphism. These included all 16 extant platyrrhine genera plus two extinct forms (*Cebupithecia* and *Homunculus*); *Apidium*, the best-known parapithecid postcranially; three adapid genera—*Adapis parisiensis* and *Leptadapis magnus* from Europe and *Notharctus* (*N. robustior, N. tenebrosus, N. venticolis*) from North America, which is very similar to the postcranium of the earlier *Cantius* (Covert, 1986, 1988; Gebo *et al.*, 1991); and one omomyid, *Hemiacodon gracilis*. Although other omomyids were not included, despite some diversity within this group, there is considerable similarity in many aspects of their postcranium (Dagosto, 1986, 1988; Gebo, 1986, 1988). Also included were six extant strepsirhines representing some of the considerable diversity in this group (*Otolemur crassicaudatus, Cheirogaleus medius, Eulemur macaco, Daubentonia madagascariensis,* and *Avahi laniger*); *Tarsius* (*T. spectrum* and *T. syrichta*); three cercopithecoids (*Presbytis melalophos, Nasalis larvatus,* and *Rhinopithecus roxellanae*); two hominoids (*Pan troglodytes* and *Hylobates lar*); the Miocene catarrhine *Pliopithecus vindabonensis;* and the Oligocene propliopithecid *Aegyptopithecus zeuxis*. Originally, several other taxa, particularly fossil taxa for which there were very little data, were included. However, the analytical program used, PAUP 3.0s, was found to be highly sensitive to taxa with very large amounts of missing data, so these were deleted from the analysis.

Polymorphisms, Polarity, and Weights

A decision was made to recognize and code polymorphic taxa, whether because of species differences or other intrageneric variability, as truly polymorphic. Frequent alternatives include dropping polymorphic (variable) characters from analysis or coding polymorphic taxa as monomorphic for the most common morph or monomorphic for the putatively most primitive morph present in the taxon (see Dickenson, 1993; Mabee and Humphries, 1993). Individuals working with electrophoretic and DNA data sets have been more willing than morphologists to grapple with polymorphic data sets because they cannot easily argue them away; however, polymorhisms are a reality and actually are far more common than monomorphisms for most morphological features in most taxa. Recognition of polymorphisms creates a considerably messier but more biologically accurate data set and allows one to incorporate variability directly into an analysis. PAUP 3.0s permits the incorporation of polymorphic traits and their analysis, eased particularly by a user-defined stepmatrix detailing allowable transformations between states in more complex transformation series (Swofford, 1991a; Mabee and Humphries, 1993; MacClade 3.0 will now also handle polymorphic taxa as something other than just "uncertainty," although this program was not yet available when these analyses were run; Maddison and Maddison, 1993). Coding polymorphic taxa as such and frequently using stepmatrices appeared the most effective way, at present, to confront the issue of variability, which is frequently overlooked or avoided in systematic studies, particularly cladistic analyses.

Key to any cladistic analysis is the determination of polarity of traits. Although I favor the outgroup method to polarize traits (Watrous and Wheeler, 1981; Maddison *et al.,* 1984), choice of an appropriate outgroup has proven a particularly thorny issue for any study of early primate radiations. As anthropoid origins are pushed back in time, the anthropoid lineage (or lineages) must be increasingly accepted as a product of one of the earliest primate divergence events. Unfortunately, despite a burgeoning fossil record of early anthropoids and preanthropoids, particularly from middle Eocene to early Oligocene beds in North Africa and the Arabian peninsula (de Bonis *et al.,* 1988; Godinot and Mahboubi, 1992; Pickford, 1986; Sigé *et al.,* 1990; Simons, 1990, 1992; Thomas *et al.,* 1988, 1989; and perhaps Court, 1993) and possibly Asia (Culotta, 1992), very little is yet known of the postcranium of the earliest anthropoids prior to the divergence into clear catarrhine and platyrrhine clades. The primary exception is a now large sample of parapithecid postcrania, most believed to have belonged to *Apidium phiomense* (Fleagle and Kay, 1987; Gebo *et al.,* Chapter 9, this volume). Although some have rooted a tree by the use of the earliest members of a group (e.g., parapithecids as the root for anthropoids; Fleagle and Kay, 1987), I consider that the use of a single taxon, even if closely related and quite old, such as *Plesiadapis* for primates, is not a sound biological technique for rooting a phylogenetic tree. It is not only possible but highly likely that many known mammals from the Paleocene of North America and Europe are themselves derived and distinctive forms (see, e.g., Beard, 1990; Kay *et al.,* 1990, 1992); this is all the more likely for any extant form. Likewise, despite its antiquity, *Apidium* cannot automatically be assumed to represent the ancestral anthropoid morphotype, particularly should the ancestry of the Anthropoidea stretch back into the Eocene and out of northeast Africa.

In this study, a primitive ancestral morphotype for a preprimate mammal was constructed by the combined consideration of the morphology of generalized mammals (e.g., some insectivores, tupaiids), bats, dermopterans, and Paleocene and Mesozoic mammals, including plesiadapiforms. The most widespread morphs present in these mammals, particularly if they were present in a number of early mammals, were considered primitive. As a last resort, traits with the most widespread distribution within primates were coded as primitive when reference to the outgroup was not conclusive (see detailed discussion in Ford, 1980a,b, 1990; see also Maddison *et al.*, 1984). This preprimate eutherian morphotype was defined as ANCSTATES in PAUP (see Appendix A).

Almost all included traits are multistate. Transformation series were determined largely on the basis of logical functional and anatomical transitions (not unlike the approach described in Szalay *et al.*, 1987). In some cases, the transformation was a simple graded series in size, e.g., large ↔ medium ↔ small. In others, the transformations appeared more complex, and change was determined to proceed in more than one direction from the ancestral state. In any characters that included a state of absence of a feature, a matrix was defined that allowed absence to occur by loss from any other character state, or any state to be gained from absence. One effect of this method of coding (with a user-defined "stepmatrix," Swofford, 1991a) is that loss of a feature is weighed less heavily than attainment of some other specific morph of that trait; this was considered a desirable effect (see, e.g., Hecht and Edwards, 1977). The use of a stepmatrix rather than "unordered" polarity retains information on the polarity/transformation sequence of the other states for that character. Other than the lowered weight for characer absence (= presumed loss) and the weight implied by inclusion of traits, no other weighting was explicitly applied to individual characters in this study.

Some runs were performed with scaled weighting in effect (see Swofford, 1991a), which sets the weight of each character to 1 despite the number of character states. One reason for scaled weighting is that, otherwise, multistate traits may be more heavily weighted than two- or three-state traits. Scaled weighting is most effective and helpful if states are arbitrary rather than legitimate morphs (D. L. Swofford, *personal communication*). In this study, both qualitative and quantitative traits are likely to represent legitimate morphs because of the coarse-grained approach to coding quantitative data. Therefore, the information content of a character change is no less simply because the character was altered at some previous time or will change again at some later point, particularly if each change or new state is unique. Moreover, rapidly evolving (= multistate) traits are not necessarily subject to higher degrees of homoplasy; if they are, then this should be reflected by the inconsistency of their pattern of change with those of other traits. In fact, if one's ingroup is inclusive enough (e.g., all mammals, all animals, all life), it is highly likely that nearly all characters would be multistate. Thus, the appearance of two-state characters (present/absent) is likely an artifact of the size of the group under study. The scaled-weight runs in general did not provide any significant new insights into the data set or the issues of anthropoid relationships and origins over that provided by equal weighting (the default of PAUP 3.0s); thus, the scaled-weight run will be reported for only one set of options (see below).

PAUP Options and Constraints

The resulting data matrix was analyzed using PAUP 3.0s (Swofford, 1991a). Exhaustive searches for all most-parsimonious trees could not be performed because of the large number of taxa. Therefore, the heuristic branch-swapping technique (options TBR/MULPARS, rooted) was utilized, which approximates and approaches a most parsimonious solution. There is no guarantee that this approach will find all most-parsimonious cladograms (see Swofford, 1991a; Swofford and Olsen, 1990, for discussion); however, in most cases where comparisons have been done, the heuristic approach has performed extremely well (D. L. Swofford, *personal communiation*). As Maddison (1991) and Swofford (1991a) have emphasized, however, whole "islands" of trees might be missed by this method, one of which could include the most parsimonious solution. Therefore, it is important to do multiple searches in order to reduce this possibility of error. Random addition of taxa was used for several replications; this varied with run, depending on limitations on storage. In several cases, runs were repeated several days or weeks apart and on different machines; very similar although not always identical results were obtained, but the 100% and Adams congruence trees were identical, raising confidence that the heuristic technique was providing an acceptable search.

One of the most powerful analytical aspects of a program such as PAUP is that it allows one to explore the effects of different assumptions about evolutionary trajectories of character change and about relationships. Because many trees of equal or nearly equal length are generated, it is impossible to argue on the basis of the PAUP analysis alone that any one tree is the best and truest resolution of relationships. However, it does allow us to exclude some highly unlikely relationships and encourages us to explore more fully the characters that would support one arrangement over another.

With this goal in mind, the "outgroup" for polarity determination was varied to ascertain the effect of assumptions about the primitive primate morphotype; five different options were tried. Most runs were done using the defined preprimate, eutherian morphotype ANCSTATES. Additionally, separate runs were done excluding this "ancestor" and defining each of the following as the "outgroup": *Notharctus* (an adapid); *Hemiacodon* (the best-known omomyid); *Apidium* (the best-known parapithecid); and *Callicebus* (as a generalized platyrrhine; see Ford, 1986, 1988, 1990). These runs therefore examined the assumptions that each of these four taxa, in turn, retains a morphology most like that of the ancestral primate (see Table I). For the run using ANCSTATES as the outgroup and no constraints, both equal and scaled weightings were used.

The ability of PAUP to constrain certain taxa to hold together and form monophyletic groups was used to explore various hypotheses about primate interrelationships in more detail and to track character change given various historically hypothesized clades. These runs were all initiated using ANCSTATES to root the tree (see Table I). In all cases, no assumptions about

Table I. Lengths and Indices of Adams-Consensus Trees[a]

	Length[b]	Consistency index	Homoplasy index	Retention index
Hypothetical ancestor, equal weights	1028+ (1004+)	0.755	0.878	0.544
Hypothetical ancestor, scaled weights	1033+ (1011+)	0.756	0.878	0.546
Hypothetical ancestor, "Major" constraint[c]	1038+ (1016+)	0.745	0.880	0.519
Hypothetical ancestor, Platyrrhine constraint	1034+ (1013+)	0.750	0.879	0.531
Hypothetical ancestor, "Catarrhines + Platyrrhines" constraint[d]	1037+ (1012+)	0.753	0.879	0.537
Hypothetical ancestor, Platyrrhines + *Apidium* constraint[e]	1030+ (1007+)	0.748	0.880	0.526
Hypothetical ancestor, Haplorhine constraint[f]	1045+ (1013+)	0.744	0.880	0.515
Hemiacodon as outgroup	1043+	0.740	0.883	0.480
Notharctus as outgroup	1031+	0.747	0.882	0.496
Apidium as outgroup	1031+	0.747	0.882	0.496
Callicebus as outgroup	1024+	0.752	0.881	0.511
Hypothetical ancestor, "Anthropoids + omomyids" constraint[g]	1035+ (1006+)	0.756	0.878	0.546
Hypothetical ancestor, "Anthropoids + adapids" constraint[g]	1044+ (1022+)	0.738	0.881	0.499

[a]Lengths and indices are reported for the Adams-consensus trees. Only the Adams- and strict-consensus trees were considered in evaluating hypotheses of relationships (see text). Estimates for these parameters do not include changes in characters with user-defined stepmatrices of transformation sequences (11 of 52 characters); thus, length is indicated with a "+" to indicate that there is additional, unmeasured length. These values should not be used in any strict sense as indicators of the accuracy or "value" of one tree over another (see text).

[b]Length for those trees using a defined hypothetical ancestor (the first seven and the last two) includes changes from the "hypothetical ancestor" (defined in ANCSTATES) to the earliest primate. Lengths with a specific primate defined as "outgroup" do not include any changes from a preprimate eutherian mammal; therefore, by default, these four trees should be shorter. The lengths given in parentheses for those using ANCSTATES to root the tree are the lengths minus the steps required to go from ANCSTATES to the earliest primate.

[c]The "Major" constraint forced the program to hold all catarrhines together (including *Pliopithecus* and *Aegyptopithecus* but not including *Apidium*) and to hold all extant strepsirhines together.

[d]The "Catarrhines + Platyrrhines" constraint forced the program to hold all catarrhines (including *Pliopithecus* and *Aegyptopithecus*) and platyrrhines together as a large group, not including *Apidium*.

[e]The "Platyrrhines + *Apidium*" constraint forced the program to hold all platyrrhines plus *Apidium* together.

[f]The "Haplorhine" constraint forced the program to hold all anthropoids, *Tarsius*, and *Hemiacodon* together. However, it made no assumptions about relationships within this group or about the affinities of any other primates.

[g]The "Anthropoids + omomyids" constraint forced the program to hold all anthropoids and *Hemiacodon* together but excluded *Tarsius*. The "Anthropoids + adapids" constraint forced the program to hold all anthropoids and *Adapis*, *Leptadapis*, and *Notharctus* together, making no assumptions about individual relationships.

relationships within the constrained group or between any other primates were made. These included the following constraints: (1) The "major" constraint forced PAUP to accept the integrity of both the Catarrhini (including *Pliopithecus* and *Aegyptopithecus*) and the extant Strepsirhini, since numerous features not included in this study give almost overwhelming support to the monophyly of both groups (e.g., features of the ear, orbit, nose, and dentition for catarrhines, and the toothcomb and grooming claw for strepsirhines; see review in Fleagle, 1988). This constraint was the only one of all introduced that was felt to be well supported by independent evidence. (2) The "platyrrhine" or "Plat" constraint forced PAUP to recognize the monophyly and integrity of the Platyrrhini, a group that is also nearly uniformly accepted yet almost impossible to define (see Delson and Rosenberger, 1980; Fleagle and Kay, 1987; Ford, 1986, 1990; Kay and Williams, Chapter 13, this volume). (3) The "catarrhine + platyrrhine" or "Cat/Plat" constraint forced all platyrrhines and catarrhines, including *Pliopthecus* and *Aegyptopithecus* but excluding *Apidium* (parapithecids), to form a monophyletic group. (4) The "platyrrhines + *Apidium*" or "Plat + parapith" constraint held the platyrrhines and parapithecids (*Apidium*) together as a monophyletic clade, in keeping with suggestions by Hoffstetter (1974b, 1980) and Ford (1988). (5) The "Haplorhine" constraint forced the program to hold all anthropoids, *Tarsius*, and *Hemiacodon* (representing omomyids) together. (6) The "anthropoids + omomyids" or "Anth+omomyid" constraint forced the program to hold all anthropoids and *Hemiacodon* together but excluded *Tarsius*. And finally, (7) the "anthropoids + adapids" or "Anth+adapid" constraint forced the program to hold all anthropoids and *Adapis, Leptadapis,* and *Notharctus* together. The results of these many runs will be discussed where appropriate.

Tree Conflict and Consensus

All runs produced a number of nearly equally parsimonious trees, often well over 100. The number of trees, however, is a direct effect of the large number of taxa included in the study. In fact, nearly all trees were relatively minor variants of one another. The lengths of these trees were very similar despite differences in outgroup, applied constraints, etc. (see Table I). The minor differences in tree length are not sufficient, in my view, given the large overall length, to argue for the supremacy of one hypothesis of relationships over another on the basis of parsimony alone. In order to maximize confidence in the relationships being examined, attention was focused on those clades that appeared in all the most- and the nearly most parsimonious trees produced by a run, shown in the strict-consensus trees, or in relationships (nested taxa) consistent with all trees produced by a run, shown in the Adams-consensus trees. Use of consensus trees must be approached with caution (see, e.g., Barrett *et al.*, 1991, 1993; Dickenson, 1993; Nelson, 1993; and especially Swofford, 1991b). These are not actual phylogenies; the polychotomies most likely represent unresolved relationships and not true polychotomies (but see also Gould, 1989; Corruccini, 1990). Indices computed on these consensus trees therefore should not be interpreted as representing the consistency or retention index of any of the trees that were used to compute the consensus; these constituent trees will generally have much higher consistency indices. However, here I am interested in larger evolutionary events that are broadly supported. Examining the more conservative but likely more robust consensus trees is more appropriate, and if used cautiously, indices computed on these trees can provide some basis for comparison.

Choices made at earlier branching points for recognition of clades ("Haplorhini" or "Prosimii," for example) and character changes have dramatic effects on the hypothesized patterns of subsequent evolution. Accordingly, although everyone may agree that two taxa share a certain trait, under some assumptions that trait may be seen as a primitive retention and thus uninformative for systematic studies. Likewise, in taxa that exhibit variability (such as almost all taxa included here, from generic to ordinal level), assumptions about character polarity at preceding nodes will directly determine which character state is "pulled out" in an analysis to represent the ancestral morph for that taxon. These complications are inherent in any systematic study; the advantage of a study of this nature is that they are more apparent and less hidden; therefore, they are more easily confronted. The implications of hypotheses of evolution of both taxa and characters (not necessarily the same sets of hypotheses) for later stages will be considered in detail throughout the analysis.

Results

Measures of Parsimony

Table I presents various measures of length, consistency, homoplasy, and retention of the Adams-consensus trees for various runs. It is apparent that the various permutations examined yield consensus trees of fairly similar measure. Lengths vary from 1024 steps (using *Callicebus* as the outgroup) to the longest by 21 additional steps, at 1045 steps (using *Hemiacodon* as the outgroup, approached closely by the tree produced using ANCSTATES and the "Haplorhine" constraint). It should be noted that the lengths of those trees using "ANCSTATES" as a root are inflated by the addition of this hypothetical ancestor and the steps required to proceed from this ancestor to the common primate ancestral node. In general, this involved between 21 and 32 additional steps. When these additional steps are subtracted, all trees using "ANCSTATES" to determine character polarity were considerably shorter than all trees using a known primate to root (by 20+ steps), with the shortest being that with no phylogenetic constraints (length 1004+), closely approached by the "Anth + omomyid" and the "Plat + parapith" trees.

Consistency indices varied from 0.740 to 0.756; these indices did not include the stepmatrix characters. As Swofford (1991b) discussed, consistency indices are heavily influenced by the number of taxa and characters examined (more taxa, less consistency), which is constant across the runs presented here. The least "consistent" was the consensus tree produced with *Hemiacodon* as the outgroup; the most consistent were those with "ANCSTATES" and no additional internal constraints and also with the "Anth + omomyid" constraint. All trees had very high homoplasy indices in a narrow range (0.878–0.883); however, polymorphism (which is widespread in this data set) by its very nature implies greater homoplasy and inflates the homoplasy index. Although these trees contain considerable homoplasy, they may be the most consistent trees allowable for these characters. Finally, the retention indices range from 0.480 to 0.546. Again, the highest are those runs with ANCSTATES and no internal constraints or with the "Anth + omomyid" constraint, followed by either the "Plat" constraint or the "Cat/Plat" constraint (holding the catarrhines and platyrrhines together as a monophyletic clade to the exclusion of *Apidium* and nonanthropoids). The lowest values are those using any one of the known primates as the outgroup, with *Hemiacodon* as outgroup the lowest and *Callicebus* as outgroup the highest of the four. Although individual trees in any particular trial would have much "better" values, the relationship between those in various trials is consistent with the measures reported for the Adams-consensus trees.

These parameters (length and consistency, homoplasy, and retention indices) strongly suggest that use of the hypothetical preprimate ancestor (ANCSTATES) to root the tree provides more parsimonious arrangements than does the use of a "typical" omomyid or adapid, a parapithecid, or a generalized platyrrhine (although the last comes closest). To the degree that parsimony can be used as an evaluator of the biological accuracy of a phylogenetic hypothesis, this has important implications, which will be pursued below. Because of this fact, further analyses examining the effects of forcing or constraining certain clades to be monophyletic were all performed using ANCSTATES to root the tree. Beyond this statement, these parameters are similar enough that they do not readily support any meaningful choice between the various possible hypotheses of relationships, either between trees in an individual run or between different choices of weighting or determining polarity, in those analyses using ANCSTATES but applying varying possible phylogenetic constraints. The effect of these options on differing patterns of character change and differing hypotheses of relationship in light of other evidence must be examined in order to choose between them (see below).

Clade Recognition

Table II indicates which major primate groups were recognized as discrete clades in each trial. As is readily apparent, a number of classically recognized primate groups were clearly recognized as clades based on postcranial

Table II. Groups Recognized in Various PAUP Runs[a]

Groups recognized as clades	Hypoth. ancestor, equal wts.	Hypoth. ancestor, scaled wts.	Hypoth. ancestor, "Major" constraint	Hypoth. ancestor, Platyrrhine constraint	Hypoth. anc. "Cats. + Plats." constraint	Hypoth. anc., Plats. + *Apidium* constraint
Anthropoidea	√	√	√	√	√	√
Haplorhini						
Prosimii	√	√	√	√	√	√
Strepsirhini			[by default]			
Adapidae						
Tarsioidea (*Tarsius* + omomyids)	√		√	√		√
Platyrrhini				[by default]		
Catarrhini			[by default]	Plus *Apidium*	√	√
Hominoidea	√	√	√	√	√	√
Cercopithecoidea	√	√	√	√	√	√
Apidium with ?[b]	*Aegyptopithecus* (+ OWMs)	Sis. to *Aegyptopithecus* + OWMs	*Saimiri* (*Aotus* + *Callicebus*)	*Aegyptopithecus* + catarrhines	Sister to all anthropoids	*Saimiri* (*Aotus* + *Callicebus*)
Aegyptopithecus with ?[b]	*Apidium* (+ OWMs)	OWMs (+ *Apidium*)	*Pliopithecus* (catarrhines)	*Apidium* + catarrhines	*Saimiri* (*Aotus* + *Callicebus*)	*Pliopithecus* (hominoids)

Groups recognized as clades	Hypoth. ancestor, Haplorhine constraint	Hyp. Anc., "Anthrops + omomyids" constraint	Hyp. anc., "Anthrops + adapids" constraint	*Hemiacodon* as outgroup	*Notharctus* as outgroup	*Apidium* as outgroup	*Callicebus* as outgroup
Anthropoidea	(√) (-*Aegypto.*)	√	(+ *Leptadapis*)	(√) (-*Aegypto.*)	√		
Haplorhini	[by default]						
Prosimii			√			√	(plus *Aegyptopithecus*)
Strepsirhini							
Adapidae							
Tarsioidea (*Tarsius* + omomyids)			√				
Platyrrhini							
Catarrhini			(√) (-*Aegypto.*)	(√) (-*Aegypto.*)			√
Hominoidea	√	√	√	√			√
Cercopithecoidea	√	√	√	√	√	√	√
Apidium with ?[b]	Sister to all anthropoids (trichotomy)	*Aegyptopithecus* (+ OWMs)	*Aegyptopithecus* + *Chiropotes*	Sis. to pitheciines/ catarrhines/atelines	*Aegyptopithecus* (nested in NWMs)	Sister to all primates (trichotomy)	Sis. to pitheciines/ catarrhines/atelines
Aegyptopithecus with ?[b]	"Tarsioids" as sis. to anthros. (trichotomy)	*Apidium* (+ OWMs)	*Apidium* + *Chiropotes*	Sister. to almost all primates	*Apidium* (nested in NWMs)	Sister to all primates (trichotomy)	Sister to *Tarsius* + *Cheirogaleus*

[a]PAUP runs (top row) were varied to examine the effects of different assumptions about (1) character weighting (equal versus scaled; those not indicated were equally weighted); (2) constraining the program to recognize certain clades (see text and Table I for membership of constrained groups); and (3) primitive character states (a hypothetical early placental form versus selected taxa as outgroup) (see text).

[b]Taxa in parentheses are the next closest sister group to the clade of the Fayum genus and the first taxon listed.

data in all or nearly all runs; others never were (unless forced by constraints on the program). Two important effects of PAUP may directly affect the placement of cerain taxa and must be held in mind in interpreting these results. The first is that taxa with seriously incomplete data sets may not be placed as accurately as those with complete information; in this study, this is particularly true for *Aegyptopithecus*. Other taxa with even less complete data sets were originally included but were dropped from the study for this reason. The second is that extremely divergent taxa, those with very long "branch lengths," may also be more difficult to place. Homoplasies (derived traits developed convergently, parallelisms, and reversals) become less easy to recognize as such on very long unbranched lines (Swofford and Olsen, 1990). A group like the Old World monkeys is a good example; they are quite distinctive postcranially and here are represented by only a few extant colobines and with no early members that might "bridge the gap" and break up the long line. These potential confounding effects of the program will be considered where appropriate.

The "Anthropoidea" formed a monophyletic group in all analyses except those using a particular anthropoid (*Callicebus* or *Apidium*) to serve as the outgroup to all other primates (see, e.g., Fig. 2a, the Adams-consensus tree using ANCSTATES, equal weights, and no constraints). The Anthropoidea always included *Apidium* and *Aegyptopithecus* except in the run constraining the haplorhines to hold together, where *Aegyptopithecus* clustered with the tarsioids.

However, there was never a distinct "Haplorhini" in any analysis, except the one run that constrained (forced) them to cluster together (see Table II). Therefore, with this particular group of postcranial features, no even "near"-parsimonious arrangement of these taxa supports the monophyly of the "Haplorhini," regardless of what other taxonomic constraints are placed on the program or what assumptions are made about polarity of traits (i.e., choice of outgroup or root for tree). And as discussed above, the tree produced under the "haplorhine" constraint had among the lowest consistency and retention indices of all permutations examined (see Table I).

In contrast, all runs recognized a distinctive "Prosimii," including strepsirhines, *Tarsius*, the adapids, plus the omomyids, except those runs forcing an omomyid (*Hemiacodon*) or an adapid (*Notharctus*) to serve as outgroup or that were constrained to break apart these taxa (the "Haplorhine" and "Anth + omomyid" constraints). This clade was remarkably consistent across literally hundreds of trees produced in various types of analyses (see Table II). Lack of recognition of a distinct "Haplorhini" does not necessitate the finding of a monophyletic "Prosimii"; thus, the consistent occurrence of this clade despite varying options and constraints was surprising and intriguing. This may reflect striking and widespread convergences in features, or it may indicate that the assumptions on ancestral form (using ANCSTATES) are faulty and that in fact a primate ancestor more like *Hemiacodon* or *Notharctus* is appropriate (see Dagosto, 1990; Dagosto and Gebo, Chapter 17, this volume). However, as discussed above, there is little in this analysis that would support the latter finding. Repeated recognition of a monophyletic "Prosimii" calls into serious question the reality of a haplorhine clade (despite the shared soft tissue resemblances between *Tarsius* and the anthropoids; see Szalay *et al.*, 1987); this will be discussed further below.

The extant strepsirhines never formed a distinct group separate from all others except under the "Major" constraint, which of course forced them to form a clade by default. It would appear that there are no postcranial traits among those examined here that would join the toothcomb and grooming claw as synapomorphies defining the strepsirhine clade.

Noteworthy also is the total absence of a distinct "adapid" clade, which would have to include the North American *Notharctus* along with the European *Leptadapis* and *Adapis* of the taxa examined here. *Notharctus* almost always fell close to some or most of the extant strepsirhines, particularly *Avahi*. This is consistent with other work indicating that the skeleton of *Notharctus* bears significant similarities to that of extant indriids (Gregory, 1920; Alexander, 1992). The two adapini were less consistent, not always closet to each other, and each generally fell as a more distant sister group to the other "prosimians" (see Figs. 2 and 3a,b) or to all primates (Fig. 3c,d). This reflects critical and pervasive systematic and/or functional differences. Kay and Williams (1992; Chapter 1, this volume) found evidence from the dentition to suggest that the adapiformes may not be a monophyletic group. And a number of others, particularly Franzen (1987; Chapter 4, this volume; see also Martin, 1993), have suggested that the Adapini (*Adapis* and *Leptadapis*) may be very different from other adapines, which are much more similar to *Notharctus* in their postcranium (Dagosto, 1983; Franzen, 1987; Gebo, 1988; Thalmann *et al.*, 1989). The position of the two adapini here, as outgroups in most cases, suggests an early divergence and may support the view of Godinot and others (1991, 1992; Godinot and Jouffroy, 1984; Godinot and Beard, 1991; Jouffroy *et al.*, 1991) that these two, particularly *Adapis*, are independently derived from an early primate ancestor, retaining more generalized primate traits than other Eocene primates and also converging some on quadrupedal anthropoid forms (in contrast to the view of Dagosto, 1983, 1990, that these are very derived).

Tarsius and the omomyid *Hemiacodon* were always very close to one another except when *Hemiacodon* was placed as the outgroup to all primates as a default. Although they were usually each other's closest sister taxon, *Avahi* sometimes fell within the group, always as a sister to *Hemiacodon* (e.g., Fig. 2b). Even more frequently, *Cheirogaleus* fell as a sister to the *Tarsius*/*Hemiacodon* clade (as in Figs. 2a,c and 3a,b). The relationships of *Cheirogaleus* and *Avahi* are clearly confusing and suggest interesting patterns of convergence and adaptation despite their differences in size and anatomy (see, e.g., Dagosto, 1986, 1988; Gebo, 1986, 1988). It is particularly noteworthy that no similar "confusions" of relationship resulted from the morphology of *Otolemur*, which along with the other bushbabies is often considered generally similar to *Cheirogaleus*

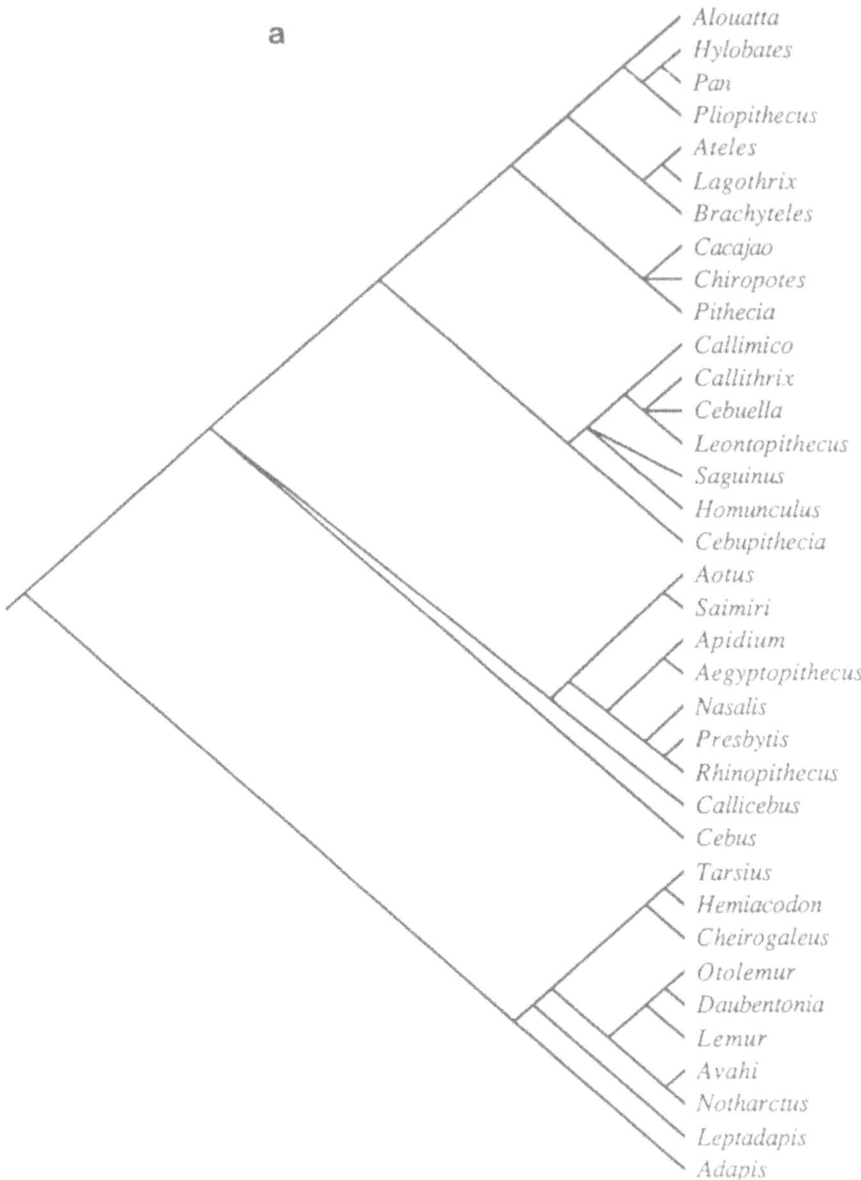

Fig. 2. Adams-consensus trees resulting from PAUP analysis with a hypothetical early eutherian as the outgroup (defined as ANCSTATES), equal character weighting, and the following constraints (see text, Tables I and II): (a) no constraints; (b) "Major" constraint, forcing catarrhines (including *Pliopithecus* and *Aegyptopithecus*, but excluding *Apidium*) and extant strepsirhines to form monophyletic clades; (c) "platyrrhine" constraint, holding fossil and extant platyrrhines together; and (d) "haplorhine" constraint, holding all anthropoids and *Tarsius* plus *Hemiacodon* together.

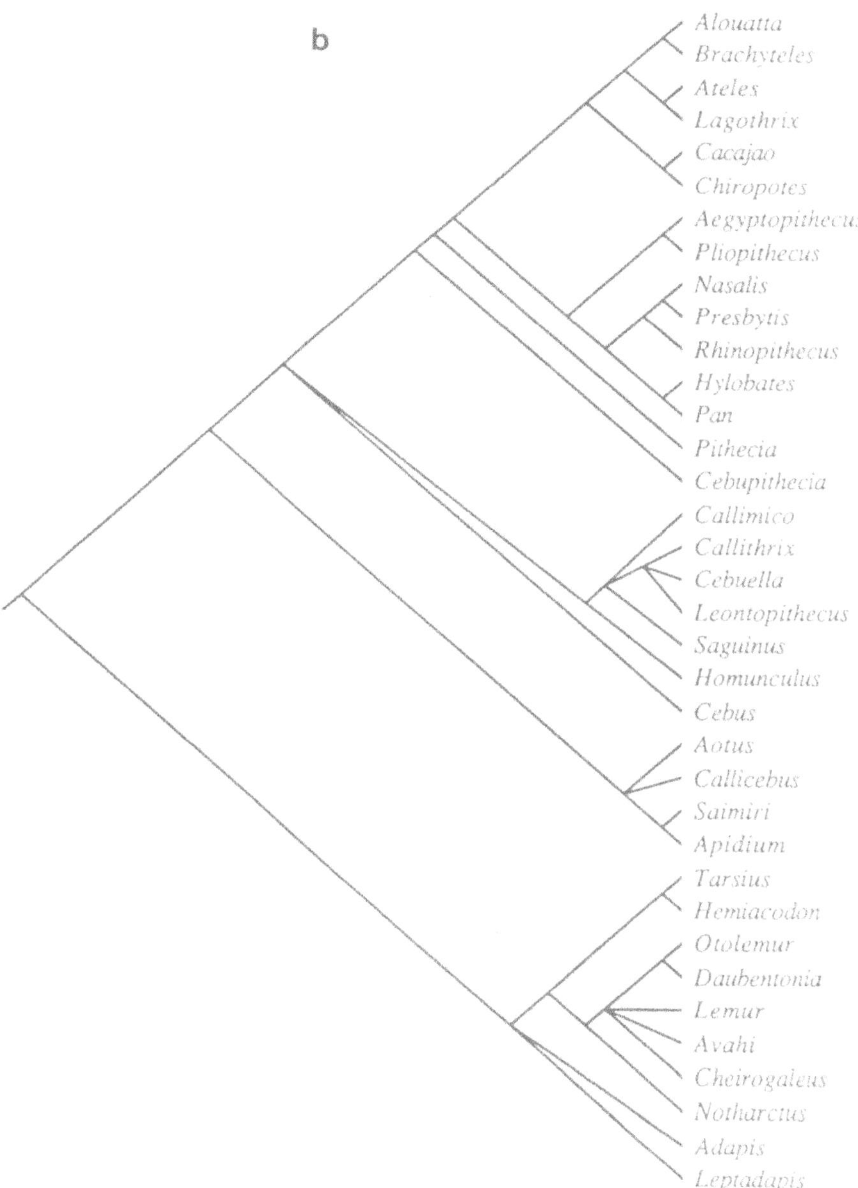

Fig. 2. (*Continued*)

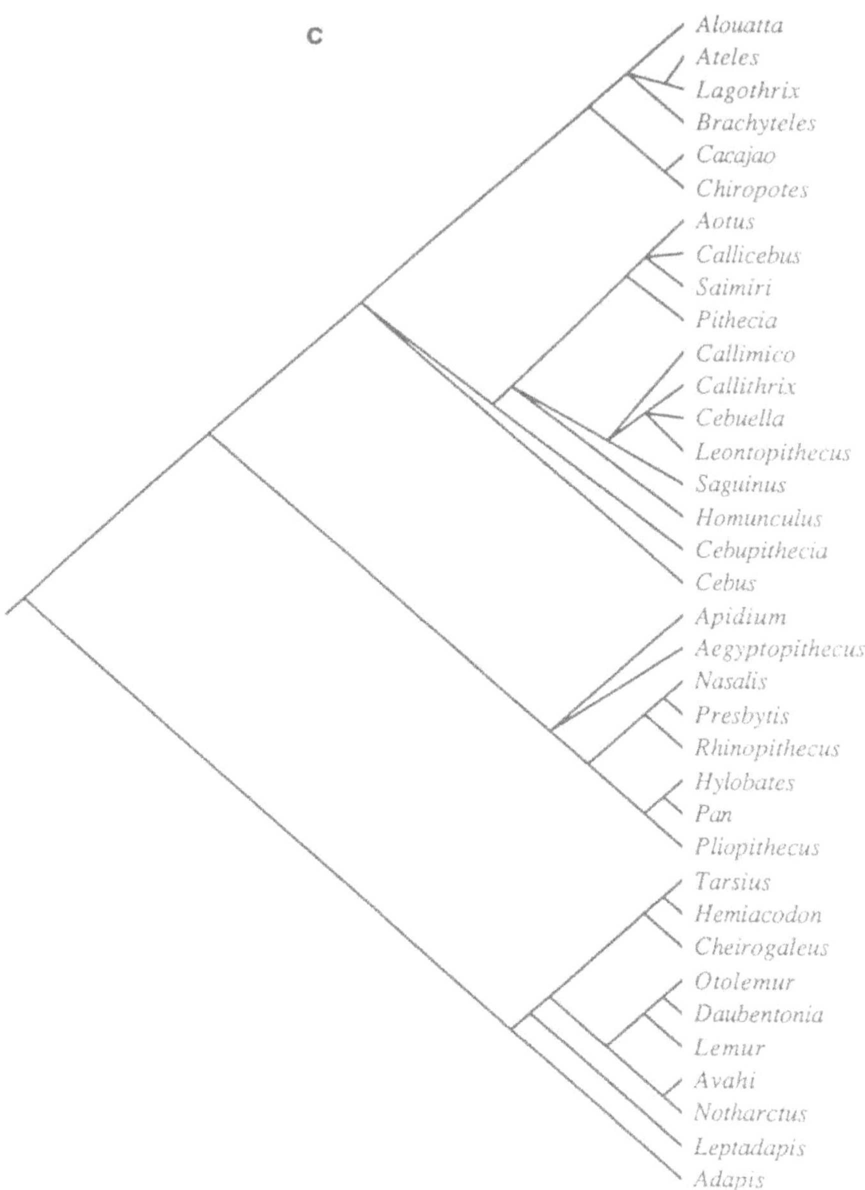

Fig. 2. (*Continued*)

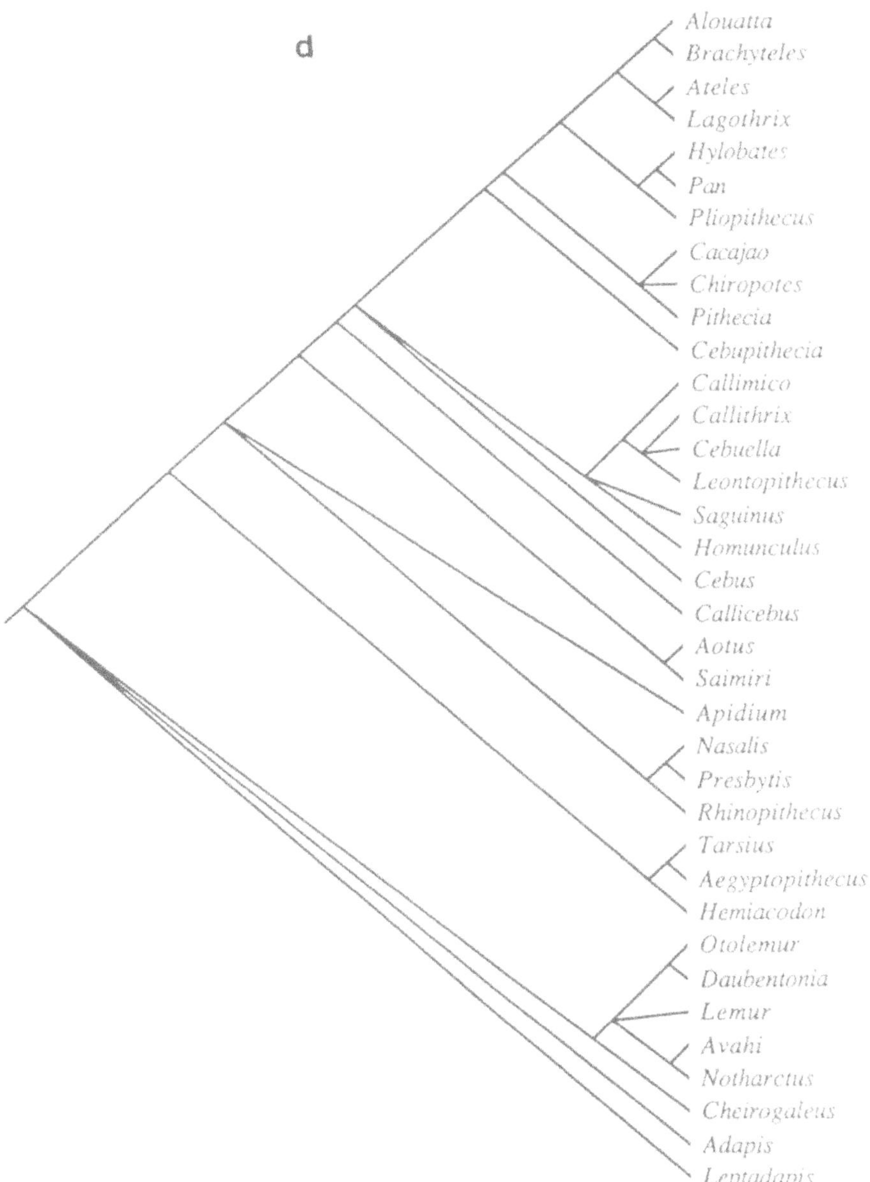

Fig. 2. (*Continued*)

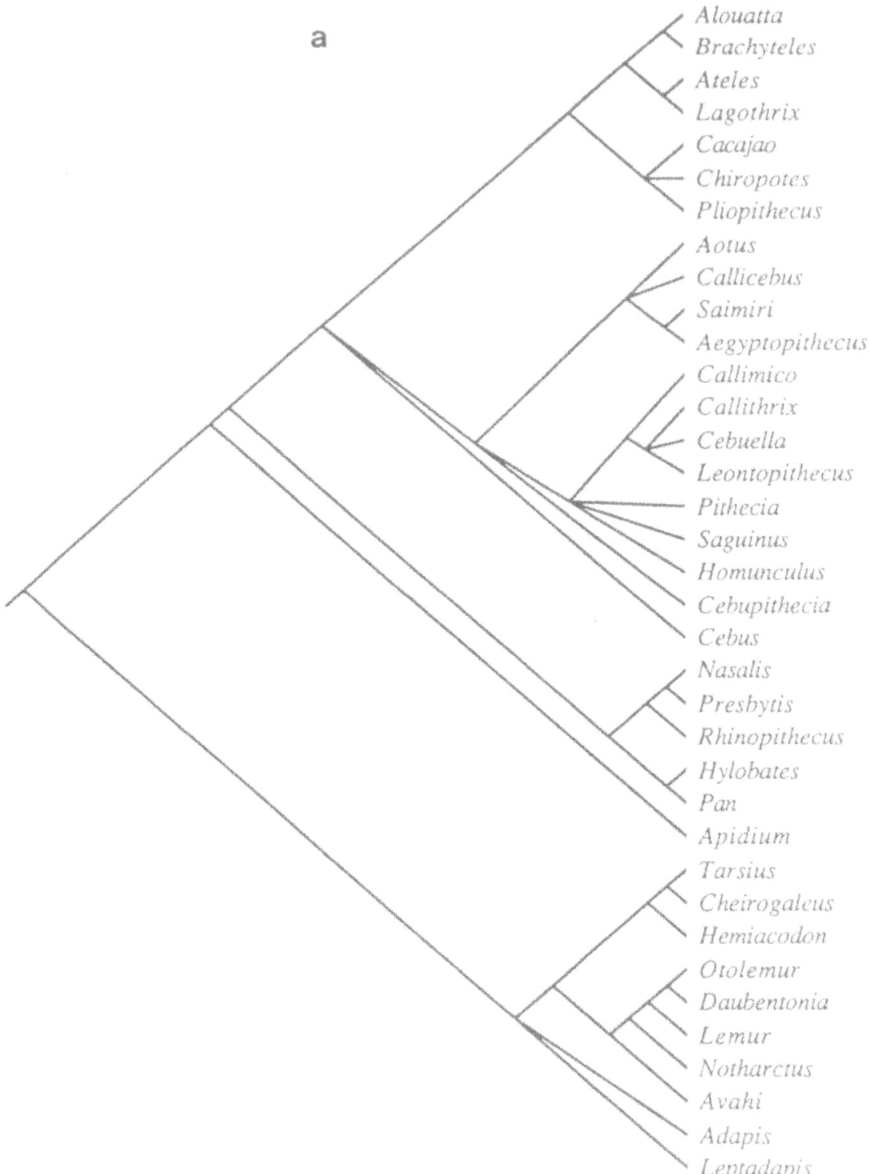

Fig. 3. Adams-consensus trees resulting from PAUP analysis with a hypothetical early eutherian as the outgroup (defined as ANCSTATES), equal character weighting, and the following constraints (see text, Tables I and II): (a) "cat/plat" constraint, forcing catarrhines (including *Pliopithecus* and *Aegyptopithecus*, but excluding *Apidium*) and fossil and extant platyrrhines together but with no assumptions about relationships within this larger group; (b) "plat. + *Apidium*" constraint, forcing *Apidium* (parapithecids) and fossil and extant platyrrhines together; (c) "anth + omomyid" constraint, forcing all "anthropoids" plus *Hemiacodon* to form a monophyletic clade, but with no assumptions about relationships within this group.

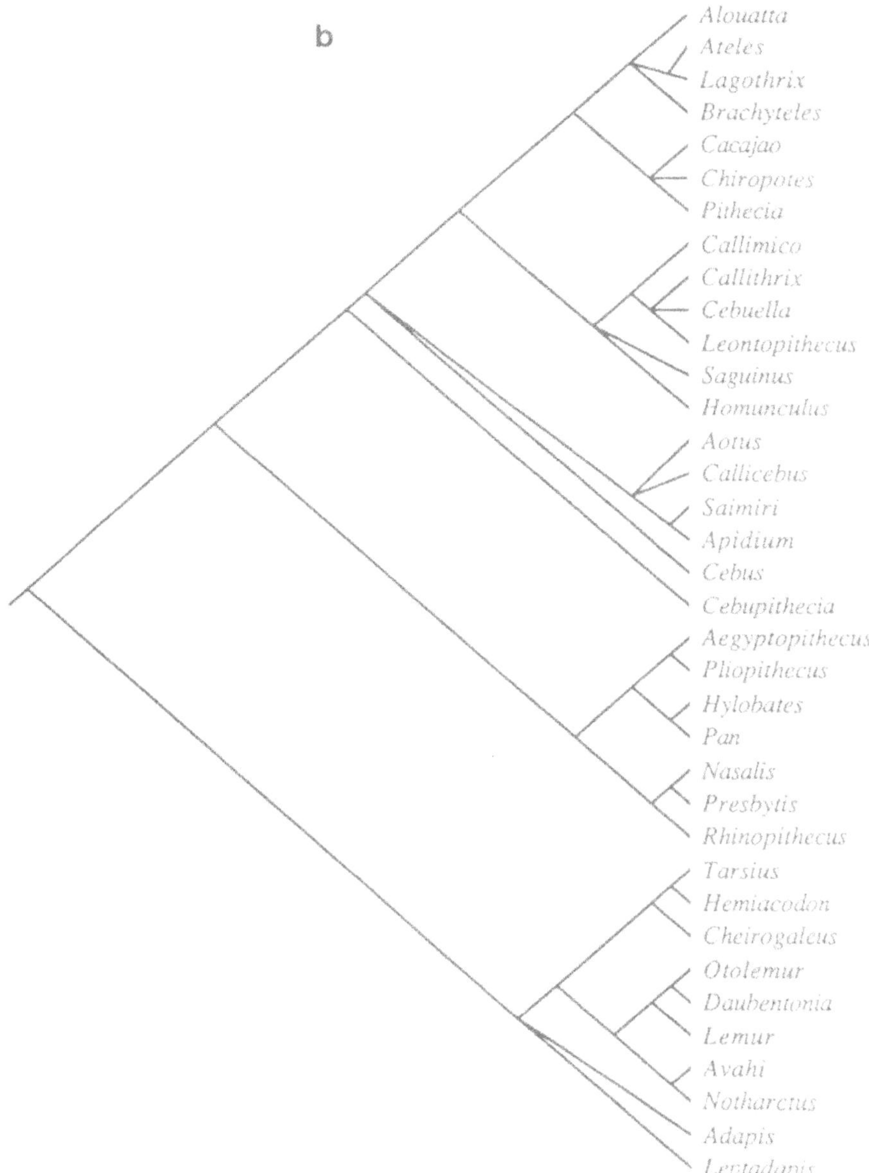

Fig. 3. (*Continued*)

Fig. 3. (*Continued*)

and also as having morphological parallels in the hindlimb to *Tarsius*. The consistently close ties of *Tarsius* and omomyids, usually exclusive of other primates, support the hypothesis of a phylogenetic tie between these taxa.

Within the anthropoids, the cercopithecoids included here (all colobines) always formed a well-defined clade (see Table II). The hominoids, *Hylobates* and *Pan,* also almost always grouped together. It is intriguing that frequently the Miocene *Pliopithecus* was closest to these extant hominoids (as in Figs. 2a,c,d and 3c), although occasionally it fell as sister to all other catarrhines (sometimes with *Aegyptopithecus*). A distinct "Catarrhini" was identified (1) when the program was forced to hold any subset of the anthropoids together (the "Plat," "Cat/Plat," and "Plat + parapith" constraints); (2) when either *Hemiacodon* or *Callicebus* was used as the outgroup to root the analysis; and (3) by default (the "Major" constraint) in recognition of their many other shared derivations (see Fig. 2b). Otherwise, in the most parsimonious arrangement overall (ANCSTATES, no constraints) and the remaining trials, the cercopithecoids and hominoids joined a distinct anthropoid branch separately (e.g., Figs. 2a,d and 3c), with the hominoids as a close sister group to the South American atelines. Even when a monophyletic "catarrhine" clade is identified, the entire clade generally falls closest to the atelines among primates. Clearly, this grouping reflects convergent adaptations, not shared ancestry. Nonetheless, the findings are striking, pointing out the strong similarities in adaptations of hominoids and atelines despite the fact that the neotropical monkeys are much smaller than the hominoids and that at least two, *Lagothrix* and especially *Alouatta,* are not considered to be highly suspensory in their locomotor habits. Indeed, *Lagothrix* and *Ateles* were each other's closest sister taxon in almost every single run, indicating that these groupings are based on morphologies indicative of something other than simply classic forelimb suspension of the type displayed frequently by *Ateles, Brachyteles,* and, especially, *Hylobates.*

The platyrrhines, like the strepsirhines, never formed a distinctive clade unless forced to do so by constraint (see Table II); this was true even when both the extant strepsirhines and the catarrhines were forced to form clades (see Fig. 2b). Rather, there was tremendous fluidity in the placement of the various platyrrhine groups within a nearly universally recognized anthropoid clade. Although the generally recognized ateline, callitrichid, and pitheciine clades usually grouped together, even this was not always the case. *Pithecia* frequently did not cluster with the other pitheciines (e.g., Figs. 2b,c and 3a,c), and the callitrichid clade often included *Homunculus* (Figs. 2a,b,d and 3b) or excluded one callitrichid (usually *Saguinus;* see, e.g., Fig. 2c). A brief glance at some of the strict consensus trees illustrates that these generally indicated a nearly totally unresolved polychotomy of platyrrhines plus separate cercopithecoid and hominoid clades (see Fig. 4).

Critical to an understanding of basal anthropoid relationships are the affinities of the parapithecids, represented here by *Apidium,* and the propliopithecids, represented here by *Aegyptopithecus.* As indicated in Table II, the affinities of these two varied depending on the different assumptions

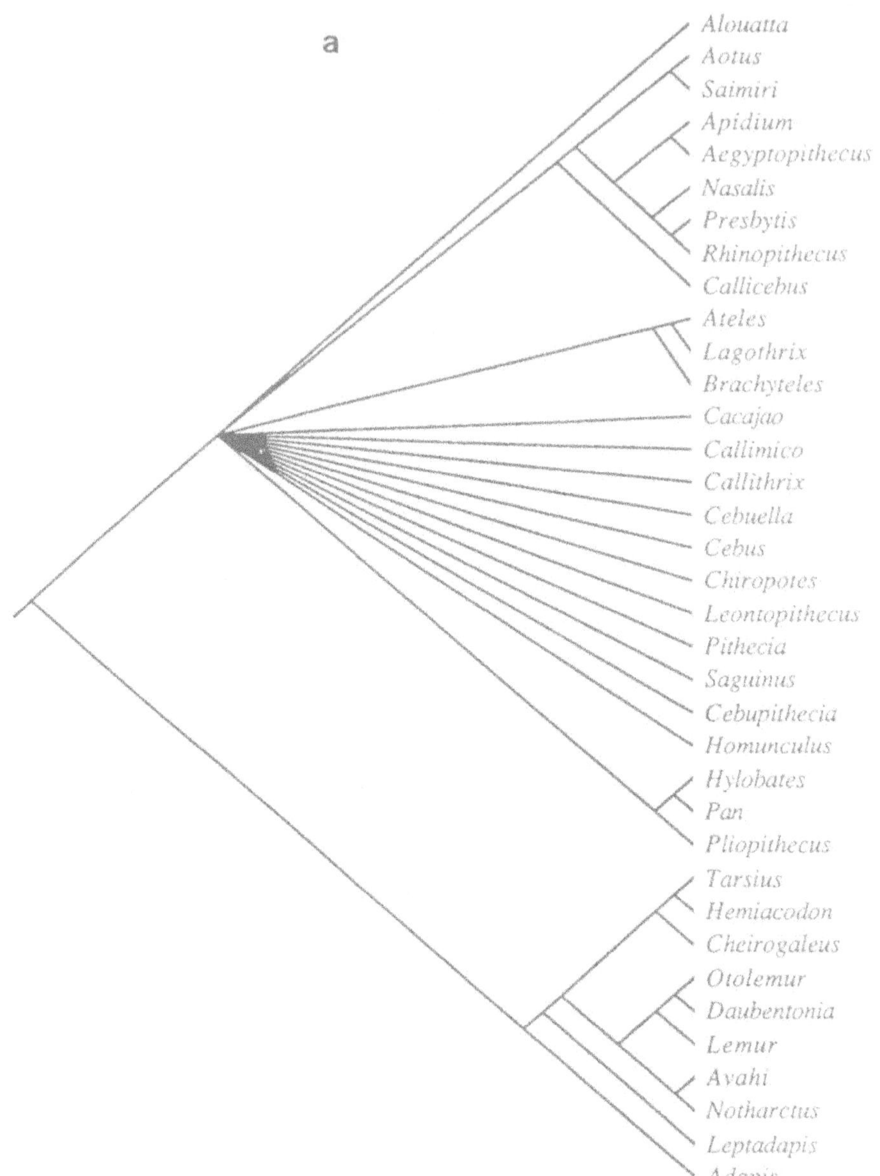

Fig. 4. Strict (100%) consensus cladograms resulting from PAUP analysis with a hypothetical early eutherian as the outgroup (defined as ANCSTATES) and equal weighting of characters: (a) no constraints; (b) "Major" constraint, forcing catarrhines (including *Pliopithecus* and *Aegyptopithecus*, but excluding *Apidium*) and extant strepsirhines to form monophyletic clades. (see text, Tables I and II).

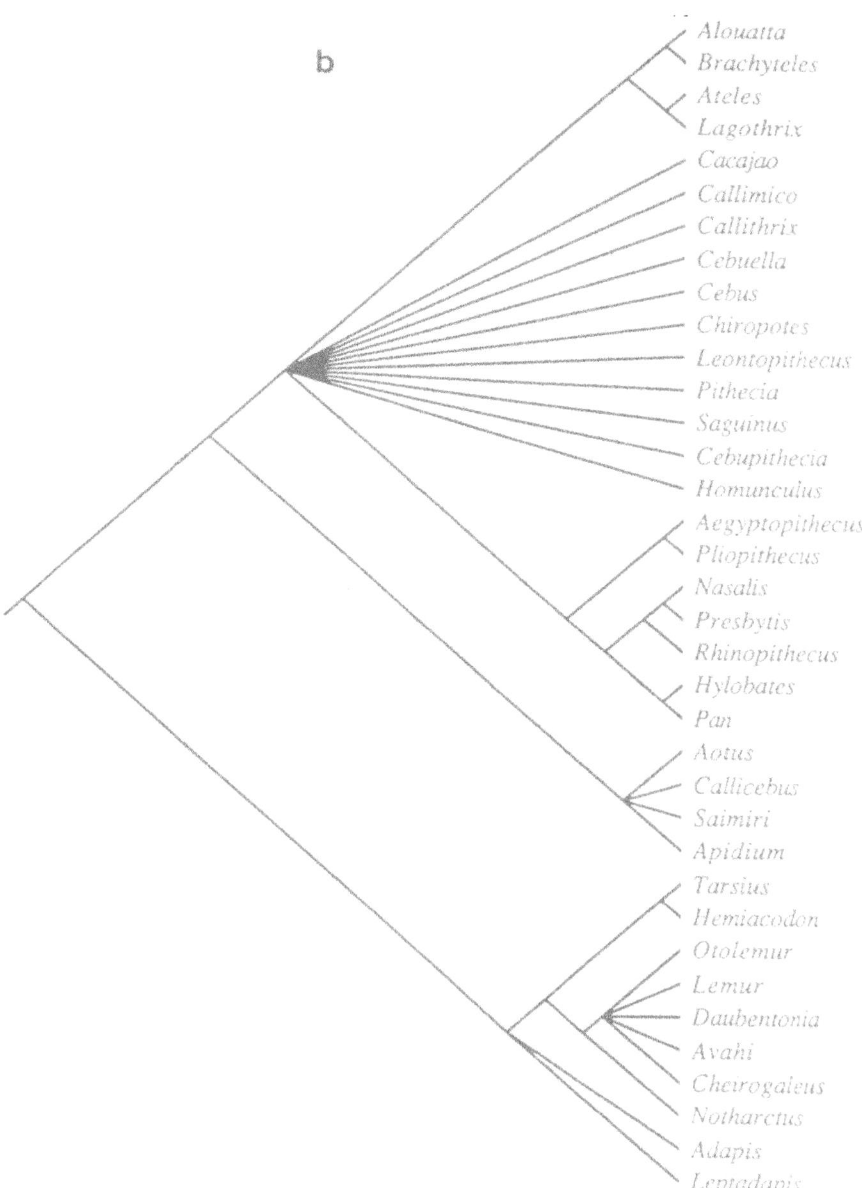

Fig. 4. (*Continued*)

inherent in different analyses; no consistent pattern of relationships was indicated throughout all analyses. In some analyses, they were closest to each other and, as a pair, to either cercopithecoids alone or some larger group of anthropoids. In most, *Aegyptopithecus* fell in a variety of positions; this uncertainty may result from the very incomplete data set for *Aegyptopithecus*. When not with *Aegyptopithecus, Apidium* most consistently clustered with *Saimiri* (and *Aotus* and *Callicebus*) or as a sister to most or all other anthropoids (with the exception of the run in which it was held as outgroup to all primates).

Character Evolution with Alternate Phylogenies

The key to using numerical cladistics as a solid analytical tool is a careful character analysis, both prior to analysis in constructing the original data set and subsequent to any numerical analysis, in order to understand better the implications of the various affinities indicated in different analyses. Certainly, it is only by returning to the original data, the morphology and adaptations of the individual primates under study, and carefully weighing various contradictory (or crossing) potential homologies indicated on the many different, nearly equally parsimonious trees produced that we can hope to gain a clearer and sounder understanding of primate relationships, particularly as they illuminate anthropoid origins.

One real concern is the changes in character transformations as different "phylogenetic choices" are made in the reconstruction of relationships (tree formation). The basal topology of any tree and determinations of early splits have resonating effects on the topology of the rest of the cladogram (see also Kay and Williams, Chapter 1, this volume, for examples of this effect using dental traits). A prime example is the effect on trait changes and polarities at *later* stages by the differing options of recognizing a basal "haplorhine"/"strepsirhine" vs. "anthropoid"/"prosimian" split (see Tables III, V, and VII). As one can see from the few examples on these tables, the effect of these two different basal splits is to produce numerous differences in features characterizing the ancestral primate, "prosimian," and "anthropoid" morphotypes (see also below). There are then rippling effects on all subsequent groupings. Thus, although nearly all runs yielded a "prosimian" group, this group was based on different sets of characters in different trees (see Table V). And in most cases, each of the "prosimian" synapomorphies was subsequently lost or reversed in one or more descendent lineages. The same is true for "Anthropoidea," "Catarrhini," or any other larger clade. Some of these varying interpretations of character changes under different assumptions are discussed here.

Earliest Primate

In every run using ANCSTATES, PAUP identified changes from the preprimate ancestor to a shared ancestor for all primates included in the

Table III. Ancestral Primate Traits[a]

Trait	Hapl/ other[b]	Anth/prosim Any	Major	Cat/Plat	Plat	Plat + Apid.	Anth + omomyid/ other	Hemia. outgroup	Anth + adapids/ other
1	**2**	**2**	**2**	**3**	**2**	**3**	**5?**	5?	**2**
2	**3**	**3**	**3**	**3**	**3**	**3**	**4**	3	**3**
3	1	1	1	1	1	1	1	4	**3**
4	3	3	3	**2**	3	3	3	2	**2**
5	2	2	2	2	2	2	2	2	2
6	0	0	0	0	0	0	0	1	**1**
7	0	0	0	0	0	0	0	0	0
8	**1**	2	2	2	2	2	**1**	1	2
9	1	1	1	1	1	1	1	1	1
10	1	1	1	1	1	1	1	1	1
11	0	0	0	0	0	0	0	0	**1**
12	**1**	3	3	3	3	3	3	1	3
13	**4**	**4**	**4**	**4**	**4**	**4**	**4**	4	**4**
14	**0**	**0**	2	**0**	**0**	**0**	2	2	2
15	**1**	**1**	**1**	**1**	**1**	**1**	**2**	2	**1**
16	**2**	**2**	**2**	**2**	**2**	**2**	**2**	2	**1**
17	0	0	**1**	**1**	0	0	0	1	0
18	0	0	0	0	0	0	0	0	0
19	1	1	1	1	1	1	1	2	**2**
20	**1**	**1**	**1**	**1**	**1**	**1**	**1**	1	**1**
21	**1**	**1**	**1**	0	0	**1**	**1**	2	**1**
22	0	0	0	0	0	0	0	0	0
23	0	0	0	0	0	0	0	0	0
24	1	1	1	1	1	1	1	1	1
25	1	**2**	**2**	**2**	1	1	**2**	2	1
26	0	0	0	0	0	0	0	0	0
27	2	2	2	2	2	2	2	2	2
28	**3**	**2**	**2**	**2**	**2**	**2**	**3**	3	**2**
29	**1**	**1**	**1**	**1**	**1**	**1**	**0**	0	**0**
30	**1**	**1**	**1**	**1**	**1**	**1**	**1**	1	**1**
31	**0**	**1**	**1**	**1**	**1**	**1**	**0**	0	**0**
32	0	0	0	0	0	0	0	0	0
33	**2**	**2**	1	**2**	**2**	**2**	**2**	2	1
34	1	1	1	1	1	1	1	1	1
35	1	1	1	1	1	1	1	0	1
36	1	1	1	1	1	1	1	0	1
37	1	1	1	1	1	1	1	1	1
38	1	1	1	1	1	1	1	1	1
39	**2**	1	1	**2**	1	1	**2**	2	**2**
40	**1**	**2**	**2**	**1**	**1**	**2**	**1**	4	**3**
41	**4**	3	3	3	3	3	3	4	3
42	1	1	1	1	1	1	1	0	1
43	0	**1**	0	**1**	**1**	**1**	0	0	**1**
44	2	2	2	2	2	2	2	2	2
45	**1**	2	2	2	2	**1**	**0**	0	**0**
46	0	0	0	0	0	0	0	0	0

(continued)

Table III. (*Continued*)

Trait	Hapl/ other[b]	Anth/prosim Any	Major	Cat/Plat	Plat	Plat + Apid.	Anth + omomyid/ other	Hemia. outgroup	Anth + adapids/ other
47	1	**0**	**0**	**0**	**0**	**0**	1	0	**0**
48	2	2	2	2	2	2	2	2	2
51	1	1	1	1	1	1	1	2	**2**
52	**2**	**2**	**2**	**2**	**2**	**2**	**1**	1	**2**
53	**1**	**1**	**1**	**1**	**1**	**1**	**1**	1	**1**
54	**2**	**1**	**1**	**1**	**1**	**1**	**1**	1	**1**

[a]Trait numbers in **bold** exhibit different states in different runs. (Trait names in Appendix A.) Character states in **bold** are derived in the ancestral euprimate morphotype; all others are primitive retentions.
[b]Groups are those constrained (haplorhine versus all other primates split; anthropoid versus prosimian split; anthropoid + omomyid versus all others). Any clades other than "anthropoids" and "prosimians" rarely or never occurred unless constrained by PAUP (see text and Tables 1 and 2).

study (an ancestral primate morphotype); the changes were not always precisely the same, however (see Table III). Even when a particular known primate was used as the outgroup, an ancestral primate morphotype differing somewhat from that outgroup was indicated (e.g., see results for *Hemiacodon* in Table III) and was often strikingly similar to the morphotype from other runs. All runs indicated that primates share the following traits (as derived in at least some runs; for primitive retentions see Table III and Appendix A): *calcaneal*—peroneal tubercle reduced and more centered or posteriorly placed (as opposed to anterior) (Characters 1:2 or 3, 2:3, 52:2 or 1), posterior articular facet for the astragalus less elongated (Char. 53:1), and calcaneal heel or posterior portion longer (Char. 54:1 or 2); *tibial*—distal facet for the fibula proximodistally deeper (Char. 13:4), posterior trochlear border becomes less flat, more distinct and rounded (Char. 15:1 or 2), anterior trochlear border becomes sharp (Char. 16:2), and medial malleolus is taller relative to shaft width (Char. 20:1); *femoral*—patellar groove becomes at least moderately deep (Char. 28:2 or 3), lateral lip of the patellar groove becomes less sharp (Char. 29:1 or 0), the facetal surface extends out onto the lateral surface of the lip (Char. 30:1), and the femoral neck becomes wider (Char. 31:1 or 0); and *humeral*—trochlear–capitular ridge becomes somewhat distinct (degree is variable) (Char. 40:1–2–3).

Other characters may have also changed at the origin of the primates to characterize this clade. Most, but not all, runs indicate the following traits as also characterizing the ancestral primate: *femoral*—extension of the femoral head onto the posterior neck increases to moderate amount (Char. 21:1), smooth facet-like area on lesser trochanter narrowed to an oval shape (Char. 25:2), third trochanter shifted distally to about one-sixth to one-fifth of the way down shaft (Char. 33:2); and *humeral*—bicipital groove deeper and narrower (moderately so) (Char. 43:1), and trochlea more "waisted" with a smaller minimum diameter relative to the maximum diameter (Char. 47:0).

In general, this postcranial reconstruction is very similar to the skeleton of modern *Callicebus* or *Cebus.* This can perhaps be best visualized by looking not at a cladogram but at a phenogram in which the branch lengths reflect degree of divergence (e.g., Fig. 5a, no constraints; Fig. 5b, "Major" constraint). In these figures, it is apparent that these platyrrhines have diverged less from the ancestral eutherian and primate morphotypes than have other primates, with the exception of *Adapis* and *Leptadapis.* This is consistent with the arguments of Ford (1986, 1988, 1990), Godinot (1992), and Jouffroy *et al.* (1991) that the ancestral primate was fairly quadrupedal and far less specialized in its activities than, for example, the more leaping notharctines, omomyids, and strepsirhines.

The problem of subsequent reversal, loss, or major alteration of hypothesized synapomorphies is often overlooked in cladistic (or other) analyses and can be marked. For example, if we accept a proximodistally much deeper distal fibular facet on the tibia in the ancestral primate (Char. 13:4), then this is partly reversed in some individuals of many platyrrhine genera (to more intermediate depth); it also becomes less deep in *Cheirogaleus, Lemur,* and particularly in *Hemiacodon* and *Tarsius,* and the facet is much shallower or lost entirely in all catarrhines (including *Pliopithecus*). The posterior and anterior trochlear borders on the tibia (Chars. 15 and 16) reverse to a flatter morphology in many ateline and pitheciine individuals (although there is considerable variability within genera), *Pliopithecus, Hylobates, Daubentonia,* and *Notharctus.* The patellar groove (Char. 28) becomes much less deep in pitheciines, *Pliopithecus,* and *Pan.* Although there are other instances of reversals, they follow these patterns of being relatively few and scattered throughout the order, often in polymorphic taxa; these do not seriously contradict the picture of the ancestral primate derived from these analyses.

A slightly alternative view is presented under certain assumptions. Perhaps the most distinct differences occur under the "Major" constraint (forcing monophyly of both the strepsirhines and the catarrhines) or if *Hemiacodon* is used as the outgroup (see Table III). Under both of these options, a slight "squatting" facet on the tibia (i.e., extension of the trochlear facet onto the anterior shaft) would be hypothesized as present in the ancestral primate morphotype (Char. 17:1). However, this turns out to be a highly variable trait both within and across taxa, so its presence in a hypothetical primate ancestor has little meaning. However, under the "Major" constraint, the ancestral primate would lack derivations found in most other runs of the following features: *tibial*—the distal fibular/tibial facet remains present and anteriorly placed on the shaft (14:~~0~~, remains 2); *femoral*—the third trochanter remains more proximally positioned (33:~~2~~, remains 1); and *humeral*—the bicipital groove remains shallow and wide (43:~~1~~, remains 0). These changes in the ancestral primate derived morphotype do not contradict the model of a predominantly quadrupedal locomotor repertoire that emerges from consider-

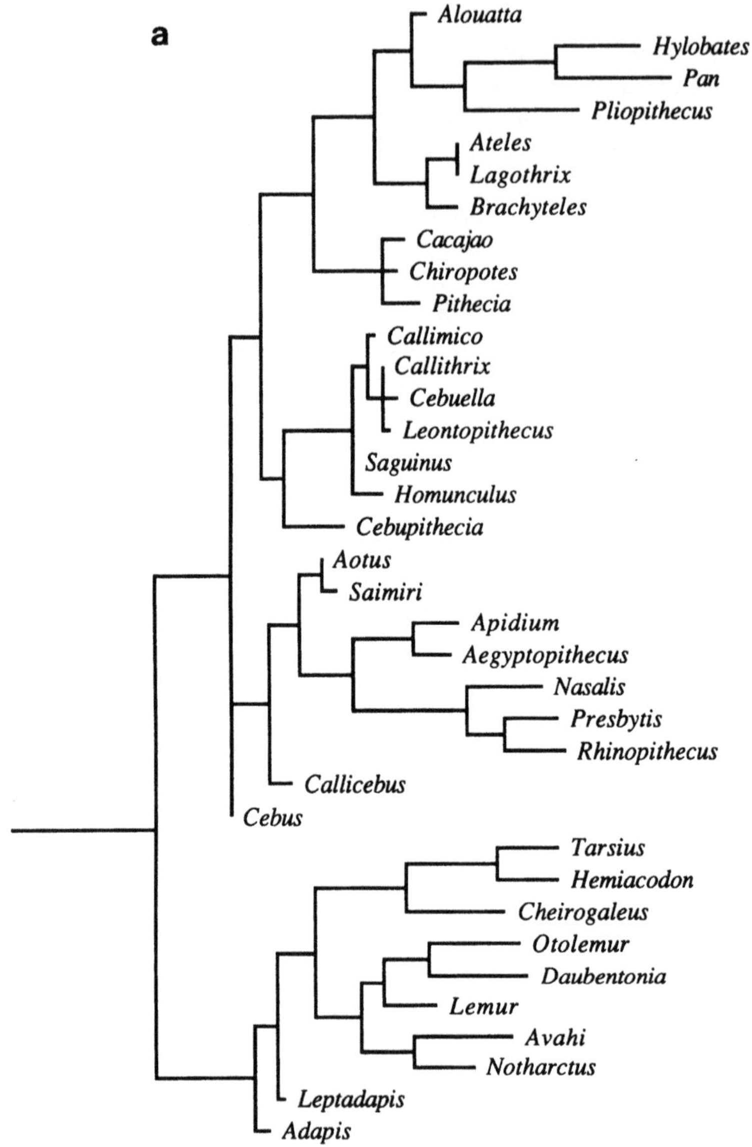

Fig. 5. Phenograms resulting from PAUP analysis with a hypothetical early eutherian as the outgroup (defined as ANCSTATES) and equal weighting of characters: (a) no constraints; (b) "Major" constraint, forcing catarrhines (including *Pliopithecus* and *Aegyptopithecus*, but excluding *Apidium*) and extant strepsirhines to form monophyletic clades (see text, Tables I and II).

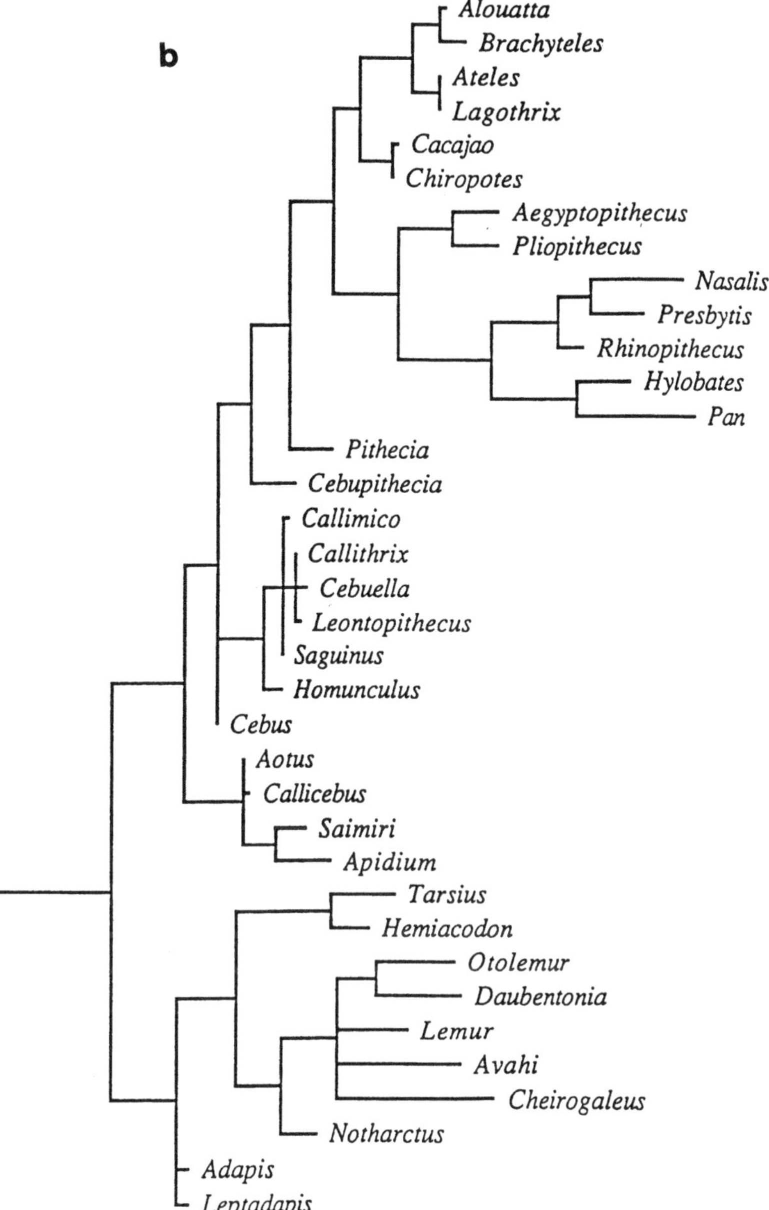

Fig. 5. (*Continued*)

ation of the other traits listed in the primate morphotype. In fact, one change partly resolves an otherwise nonsensical pattern of evolution in other runs—changes in the morphology of the distal tibial facet for the fibula (see below).

Other traits are suggested as primate synapomorphies both with *Hemiacodon* as the outgroup or under the "Haplorhine" constraint: *calcaneal*—posterior articular facet more closely parallels the long axis of the bone (lower angle) (Char. 8:1); *tibial*—distal tibio-fibular facet fused (Char. 12:1) (but see below); *femoral*—patellar groove becomes fairly deep (Char. 28:3), neck becomes very wide (Char. 31:0); and *humeral*—supinator crest becomes very wide (Char. 41:4). These five traits are more consistent with greater leaping activities and would tend to lend some support to the primate ancestor envisaged by Dagosto and others (e.g., Dagosto, 1986, 1988, 1990; Gebo, 1986; Szalay and Dagosto, 1980). However, these would be in addition to those traits listed above, which are not indicative of strong leaping tendencies. Also, when one examines them character by character, the strength of either their placement as eutherian synapomorphies or as indications of leaping become suspect. The more parallel posterior articular facet of the calcaneus (Char. 8:1) must be reversed to a more angled facet in the "ancestral anthropoid," *Daubentonia,* variably in *Notharctus,* and most strikingly in *Tarsius* (where it is highly angled). In addition to a highly angled facet in the saltatorial *Tarsius,* there is polymorphism in the atelines, with many of these largey suspensory primates exhibiting the more parallel facet of many strepsirhines.

The fairly deep patellar groove (Char. 28:3) would be reversed to a more moderate (or even shallower) groove (Char. 28:2) not only in the ancestral anthropoid (being moderately or less deep in nearly all anthropoids) but also in *Notharctus* and *Avahi* (generally considered a strong leaper). The European Adapini frequently fall as outliers to all other primates; thus, their morphology for this and other features would appear pivotal in any attempt at using parsimony (or most other methods) to determine patterns of character evolution. Dagosto (1983) has described the groove as shallow in one specimen and moderately deep in another. Although a deep patellar groove is generally considered a good indicator of leaping propensities (see review, Ford, 1988), in fact there is marked polymorphism between the nominal categories scored here as "shallow," "moderate," and "fairly deep" (states 1–3) in many taxa such as *Adapis* and most platyrrhines with widely varying behavioral repertoires. Although I believe that these different morphs can be distinguished, it appears unlikely that the differences between these intermediate states have any strong impact on behavioral abilities or propensities (unlike the extremes of a nearly flat or extremely deep patellar groove); thus, widespread polymorphism can be maintained. The very wide supinator crest on the humerus (Char. 41:4) would also have to be lost/reversed to a more moderate crest in all anthropoids as well as several strepsirhines (e.g., *Avahi, Daubentonia*), and its morphology is unknown in *Leptadapis.* Likewise, according to the PAUP analysis, under the haplorhine constraint the wide femoral neck (Char. 31:0) could as parsimoniously be construed as derived only in those lineages exhibiting it, including most strepsirhines, *Tarsius* and *Hemiacodon, Nasalis,* and variably in the callitrichids and *Aotus, Callicebus,* and *Pithecia.* This metric trait was particularly difficult to code discretely, given widely overlapping ranges of variation within taxa. Although intuitively it would seem that neck width should have strong biomechanical implications for an animal's support, it appears that far more data on various primates are needed before a realistic sense of meaningful differences can be obtained.

Almost solely under an assumption of *Hemiacodon* as outgroup do the following become added to the primate morphotype (although the primate ancestor is still considerably less specialized than *Hemiacodon*): *calcaneal*—anterior articular (sustentacular) facet for the astragalus does not reach to anterior end of bone (Char. 3:4), anterior and medial sustentacular facets are sharply angled to each other (Char. 4:2), a navicular facet is present, anterior to the anterior sustentacular facet for the astragalus, thus separating this latter facet from the cuboidal facet (Char. 6:1); the anterior calcaneus is somewhat elongated (Char. 51:2), and the peroneal tubercle is much shorter (Char. 52:1); *tibial*—posterior trochlear border becomes sharp (Char. 15:2), tibial trochlea is deeper (anteroposteriorly) than broad (mediolaterally) (Char. 19:2); *femoral*—there is an increase to a broad posterior extension of the femoral head, merging into the neck (Char. 21:2), the lateral margin of the patellar groove becomes quite rounded (Char. 29:0), distal epiphysis is particularly deep (anteroposteriorly) and narrow (mediolaterally) (Char. 35:0), the greater trochanter projects much higher than the head (Char. 36:0); and *humeral*—the trochleocapitular ridge becomes very distinct to sharp-edged (Char. 40:3–4), ventral trochlea is more cylindrical (Char. 42:0), and there is loss of the epicondylar fossa (Char. 45:0). These additions to the primate morphotype would particularly suggest specializations for more leaping activities (see Dagosto, 1986; Ford, 1988, 1990). However, they are so specialized and would require reversal or alteration in so many major lineages (e.g., reacquiring an epicondylar fossa) that even supporters of a more leaping (or "grasp-leaping"—see Szalay and Dagosto, 1980) early primate allow that most of these particular morphologies are likely derivations in those primate lineages in which they occur.

The major exception is the controversy on the polarity of the elongated anterior calcaneus (see, e.g., Anemone, 1990; Dagosto, 1988, 1990; Gebo, 1988; Szalay and Dagosto, 1980). This trait (Char. 51) is generally gauged by an index of length of the anterior calcaneus (with or without all or part of the posterior articular facet) relative to total calcaneal length. Although many faults can be found with this measure, for the purposes of this study I have continued to consider anterior calcaneal length in the manner consistent with the majority of other studies (including my own earlier work). However, as a quantitative feature, the problem of discrete coding or categorization ("short," "elongated," etc., whether numerically coded into states or simply described in prose) makes comparisons between studies difficult. Gebo has generally concluded that a cheirogaleid-like morph is primitive; this would be coded as state 3 in my schema. However, an anterior calcaneus this elongated actually occurs only in cheirogaleids, galagines, omomyids, and *Tarsius,* a highly select group that share, among other things, primarily small body size. Larger strepsirhines along with nearly all anthropoids (including the small callitrichids) exhibit only a moderately elongated anterior calcaneus (state 2) compared with the short one found in early eutherians. However, most runs (see Table III) indicated that the ancestral primate retained this shorter anterior calcaneus (state 1), primarily driven by the shorter anterior calcaneus, as measured by this index, found in all adapids included in this study (which did not include any of the very small European adapids). In view of the very limited distribution of a very short anterior calcaneus in primates (along with the adapids, it also occurs in *Pan* and other very large hominoids, and in a few ateline specimens), it seems much more likely that the earliest primate had a moderately elongated calcaneus (state 2), much like that seen in most anthropoids, *Apidium,* and many strepsirhines. Both the elongated anterior calcaneus of the small-bodied cheirogaleid–galagine–omomyid–*Tarsius* "group" (likely independent derivations in at least two of these taxa) and the short anterior calcaneus of adapids are probable derivations from this ancestral primate morphology.

Thus, even under assumptions of parsimony, individual characters must be examined carefully for implied patterns of evolution, and homoplasy is widespread in some of the best. Additionally, on occasion if characters that coincide or covary in some taxa are coded independently (as here), the priority of parsimony in PAUP and most other numerical cladistic programs may "drive" a nonsensical state to primitive status. For example, because aspects of the alterations in the shape of the distal tibial facet for the fibula are here coded separately (Chars. 12, 13, 14), several biologically unlikely or even impossible combinations can pop up as primitive. Under assumptions of a monophyletic Haplorhini or a *Hemiacodon* outgroup, character 12 indicates that this facet is fused in the ancestor (Char. 12:1) (highly unlikely by anyone's reconstruction of changes in this character; see, e.g., Dagosto, 1985; Fleagle and Simons, 1983; Ford, 1988), whereas characters 13 and 14 continue to show an unfused joint in the ancestral primate morphotype. Also, under most (but not all) assumptions, including the most parsimonious solutions resulting from no assumptions, character 14 indicates that the facet is lost (Char. 14:0), and characters 12 and 13 describe various morphologies of a very visible facet. Of course, the facet is "regained" in most lineages at later points. This is a limitation of the coding process and the program and statistical assumptions made on the basis of parsimony, not a limitation or oddity of nature. However, the implied changes in this joint are among the best examples of why any numerical analysis, cladistic or otherwise, requires judicious and careful study of the character changes implied by any "run."

The morphology of the distal tibia is critical because of its possible role in defining key primate clades. The nature of the distal tibial/fibular articulation in the ancestral primate is problematic, as is the shape of the tibial trochlear facet for articulation with the astragalus. A synovial distal tibio-fibular joint was undoubtedly characteristic of early primates (Dagosto, 1985; Fleagle and Simons, 1983; Ford, 1990). However, the size and position of the facet is less certain (Chars. 12–14). Dagosto and Gebo (Chapter 17, this volume; Dagosto, 1985) and Hershkovitz (1988) claim the proximodistally tall facet characteristic of strepsirhines is derived in that group. However, the small anteroposterior width and anterior placement of this facet in stepsirhines is shared with dermopterans, tree shrews, and paromomyids (Dagosto and Gebo, Chapter 17, this volume). Hershkovitz (1988) describes and figures a "split" facet in *Cercopithecus* (present on both the anterior and posterior corners of the lateral tibial surface); however, the catarrhines I have examined exhibit either no facet (*Hylobates, Presbytis, Rhinopithecus, Cercopithecus mitis, C. aethiops,* and a *Macaca mulatta*) or a very tiny anteriorly placed facet (*Pan, Nasalis, M. fuscata,* and *M. tibetana*). Thus although there is apparently some variability in catarrhines, none of them exhibit the anteroposteriorly long central or posteriorly placed facet seen in platyrrhines and *Apidium* (see also Fleagle and Kay, 1987; Ford, 1990). Unfortunately, this region is unknown for propliopithecids, and the joint is fused in *Tarsius* and some omomyids (Dagosto, 1985). In other omomyids, it is unfused, but apparently the two bones were very closely appressed with a syndesmotic joint; Dagosto (1985) found no evidence of a subchondral contact area on the tibiae she examined, and none is readily apparent in the newly described *Absarokius* tibia (Covert and Hamrick, 1993). These distributions and the morphology in other early mammals has led me to the determination here and elsewhere that the aspects of the "strepsirhine" morph of an anteroposteriorly short anteriorly placed fibular facet are the retained ancestral primate morphotype (see especially Ford, 1980a). Obviously, this is critical in defining either one or another major primate clade. The great proximodistal extension of the facet in strepsirhines, however, is almost certainly derived in that group; this was not included in this analysis.

The most conservative and consistent picture of the derivations in the ancestral primate morphotype would therefore include the features listed in Table IV. Where marked inconsistencies had to be resolved (features not universally recognized as primate in all runs), I opted for those that made the most biological sense and that generally were present under the "Major" constraint except as discussed above. The results of examining character changes under different assumptions largely supports the examination of measures of parsimony in suggesting that choosing an ancestral morphotype modeled after a specific primate (especially *Hemiacodon*) as the outgroup rather than ANCSTATES leads to considerably less viable reconstructions of evolutionary change.

"Haplorhine" Monophyly

As discussed above, a monophyletic "Haplorhini" consisting of the anthropoids, *Tarsius,* and the omomyids (i.e., *Hemiacodon*) never occurred spontaneously in any run. However, in order to examine the effect on character

Table IV. Possible Shared Derived Primate Traits[a]

1. Anterior calcaneus somewhat elongated (Char. 51:2)
2. Calcaneal peroneal tubercle slightly reduced, to moderate size (Chars. 1:2–3 and 52:2)
3. Calcaneal peroneal tubercle more centered or posteriorly placed (as opposed to anterior) (Char. 2:3)
4. Calcaneal posterior articular facet for the astragalus less elongated (Char. 53:1)
5. Longer calcaneal heel (Char. 54:1 or 2)
6. Tibial distal facet for the fibula deeper (proximodistally) (Char. 13:4)
7. Tibal posterior trochlear border becomes less flat, more distinct and rounded (Char. 15:1 or 2)
8. Tibial anterior border becomes sharp (Char. 16:2)
9. Tibial medial malleolus taller relative to shaft width (Char. 20:1)
10. Femoral patellar groove becomes at least moderately deep (Char. 28:2–3)
11. Lateral border of femoral patellar groove becomes somewhat less sharp (Char. 29:1–0)
12. Lateral facetal surface of femoral patellar groove extends out onto the lateral surface of the lip (Char. 30:1)
13. Femoral neck becomes somewhat wider (Char. 31:1–0)
14. Humeral trochlear–capitular ridge becomes somewhat distinct (degree is variable) (Char. 40:1–2–3)

Probable additional features

15. Femoral head develops a moderate extension onto the posterior neck (Char. 21:1)
16. Femoral lesser trochanter narrowed to an oval shape (Char. 25:2)
17. Humeral trochlea more "waisted" with a smaller minimum diameter relative to the maximum diameter (Char. 47:0)

[a]Not all traits are of equal value, and all exhibit some homoplasy.

transformations, one run was performed under a "haplorhine constraint," forcing these taxa to hold together. Very few traits could be found to support this clade, and none of them strongly or with any confidence.

Several traits were identified by the program that, in reality, only united the anthropoids. *Tarsius* exhibited a totally different and reversed or unique form of these characters, which was either shared or unknown in *Hemiacodon* (characters 23, 25, 39, 45, and 46; see Appendices A and B). These are features on the proximal femur (lesser trochanter and the crest running distally from the greater trochanter) and several aspects of the distal humerus in which *Tarsius* and the omomyids share morphologies that are very different from those seen in most anthropoids.

Adapis and *Leptadapis,* as sisters to all primates in this particular analysis, drove two characters to have primitive states lacking in almost all other primates. Thus, the state "shared" by "haplorhines" is also shared by most non-"haplorhines." These are a more pulley-like, less cylindrical humeral trochlea (Char. 47:0) and a slightly to greatly longer anterior calcaneus in comparison to the quite short anterior calcaneus of many early eutherian mammals and these two adapines (Char. 51:2). Two potential synapomorphies are actually so variable within and between taxa as to be of highly suspect value at this or any stage—the interconnection between the anterior and medial (sustentacular) articular facets on the calcaneus (Char. 4:1) and the presence of a "squatting" facet on the anterior face of the distal tibia (Char. 17:1).

However, Dagosto and Gebo (Chapter 8, this volume) suggest that the closely appressed tibio-fibular distal joint with highly limited mobility found in omomyids, *Apidium*, and some *Saimiri*, and one could add in some *Saguinus* (see Fleagle and Kay, 1987; Davis *et al.*, 1993), is possibly primitive for "haplorhines" (Chars. 12–14). I would argue that the omomyid joint is highly derived in its unique combination of a lack of any clear evidence of a synovial joint and close appression of the two bones. This coupled with the highly limited distribution of a closely appressed distal tibio-fibular joint, seen otherwise in only a few anthropoids that appear to share numerous hindlimb features but not forelimb and certainly not dental features with one another, all suggests that the superficial similarities in this joint represent strong convergence in association with specializations for increased leaping, least marked in the platyrrhines (see Ford, 1990).

After discounting the above features, there remain absolutely no shared-derived traits linking the prospective "Haplorhini" together and lending any support whatsoever to the naturalness of this clade.

"*Prosimian*" *Monophyly*

As discussed above, under all options except those that force a split of the "prosimians," a monophyletic "Prosimii" was recognized, including extant strepsirhines, adapids, omomyids, and *Tarsius*. This is consistent with the results of the analysis of Dagosto and Gebo (Chapter 17, this volume), using a somewhat different postcranial data set. In all of these cases, the clade was characterized by the following synapomorphies (see Table V): *calcaneal*—posterior articular facet for the astragalus more closely paralleling the long axis of the bone (lower angle) (Char. 8:1), a much shorter peroneal tubercle (Char. 52:1), calcaneal heel much longer (Char. 54:2); *femoral*—a moderate extension of the femoral head onto the posterior surface of the neck (Char. 21:1), lateral margin of the patellar groove becoming rounded (Char. 29:0), neck becoming very wide (Char. 31:0); and *humeral*—epicondylar foramen displaced closer to middle of shaft, so it lies above the trochlea dorsally (Char. 39:2), supinator crest becoming very wide (Char. 41:4), and epicondylar fossa absent (Char. 45:0). Some of these traits would be considered primitive primate traits under the assumption of a monophyletic "Haplorhini" or a *Hemiacodon*-like early primate (see above).

Under most assumption sets, the following features would also be prosimian synapomorphies: *tibial*—posterior border of the tibial trochlea sharp (Char. 15:2); tibial trochlea much deeper (anteroposteriorly) than broad (mediolaterally), thus forming an anteroposteriorly directed rectangle (Char. 19:2); *femoral*—patellar groove fairly deep (Char. 28:3), and greater trochanter extending much higher than the femoral head (Char. 36:0).

Table V. Ancestral "Prosimian" Traits[a]

	Anth/Prosim				
Trait	Any	Major	Cat/Plat	Plat	Plat + Apid.
1	2 (1)	2 (1)	**2 (1)**	2 (1)	**1**
2	3 (4)	3	3	3 (4)	**4**
3	1 (4)	1 (3)	1 (3)	1 (4)	1 (4)
4	3 (2)	3 (1)	2	3 (2)	3 (2)
5	2	2	2	2	2
6	0	0 (1)	0 (1)	0	0
7	0	0	0	0	0
8	**1**	**1**	**1**	**1**	**1**
9	1	1	1	1	1
10	1	1	1	1	1
11	0	0	0	0	0
12	3	3	3	3	3
13	4	4	4	4	4
14	0	2	0	0	0
15	**2**	1	**2**	**2**	**2**
16	2	2	2	2	2
17	0	1	1	0	0
18	0	0	0	0	0
19	**2**	1	**2**	**2**	**2**
20	1	1	1	1	1
21	1	1	**1**	**1**	1
22	0	0	0	0	0
23	0	0	0	0	0
24	1	1	1	1	1
25	2	2	2	1	1
26	0	0	0	0	0
27	2	2	2	2	2
28	**3**	**3**	2	**3**	**3**
29	**0**	**0**	**0**	**0**	**0**
30	1	1	1	1	1
31	**0**	**0**	**0**	**0**	**0**
32	0	0	0	0	0
33	2	1	2	2	2
34	1	1	1	1	1
35	1	1	**0**	1	1
36	**0**	1	**0**	**0**	**0**
37	1	1	1	1	1
38	1	1	1	1	1
39	**2**	**2**	2	**2**	**2**
40	2 (4)	2 (4)	1 (4)	1 (4)	2 (4)
41	**4**	**4**	**4**	**4**	**4**
42	1	1	1 (0)	1	1
43	1	0	1	1	1
44	2	2	2	2	2
45	**0**	**0**	**0**	**0**	**0**
46	0	0	0	0	0

(continued)

Table V. (*Continued*)

Trait	Anth/Prosim Any	Major	Cat/Plat	Plat	Plat + Apid.
47	0	0	0	0	0
48	2	2	2	2	2
51	1 (2)	1 (2)	1 (2)	1 (2)	1 (2)
52	**1**	**1**	**1**	**1**	**1**
53	1	1	1	1	1
54	**2**	**2**	**2**	**2**	**2**

[a]Trait numbers in **bold** exhibit different states in different runs. Character states in **bold** are derived in the ancestral "prosimian" morphotype; all others are primitive retentions. States in parentheses are shared "prosimian" traits after *Adapis* and *Leptadapis* branch off. Other constrained groupings in Tables III and VII are not included here, because, by default, they precluded recognition of the clade "Prosimii."

Without *Adapis* and *Leptadapis,* which generally fall as sister taxa to the other "prosimians" or even to all primates (see Table II), five other synapomorphies mark the remaining "prosimians": *calcaneal*—peroneal tubercle reduced to a very tiny bump (Char. 1:1), anterior articular (sustentacular) facet for the astragalus not reaching the anterior (cuboid) end of bone (Char. 3:3–4), anterior and medial sustentacular facets sharply angled to each other or indicating astragalocalcaneal contact across the full sustentaculum only in some articulated positions (see Ford, 1988) (Char. 4:1), anterior calcaneus somewhat elongated (Char. 51:2); and *humeral*—the trochleocapitular ridge becoming very sharp-edged (Char. 40:4).

Again, however, two problems must be confronted: (1) the algorithm producing "nonsense" or unlikely evolutionary character sequences, which are only caught if one goes back and examines posited changes character by character, and (2) homoplasy (reversals and convergences). In this case, some runs indicated loss of the peroneal tubercle on the calcaneus as synapomorphic for "prosimians" (Char. 1:0), necessitating its redevelopment in several descendent lineages. Although numerous reacquisitions of this protuberance in various lineages are possible, without strong evidence it is sounder to consider its extreme reduction primitive for the group (as a "prosimian" synapomorphy), with individual lineages independently losing any visible tubercle.

The problem of homoplasy is more critical and less easily resolved. As can be seen in Table VI, many of these features are subject to large amounts of homoplasy. For example, some are simply highly variable across primates, including reduced or angled connection between sustentacular facets on the calcaneus (Char. 4:1); variation among no, slight, and moderate extensions of the femoral head onto the posterior neck (Char. 21:0–2); width of the femoral neck (Char. 31); patellar groove depth (Char. 28); and shape of margins of the femoral patellar groove (Char. 29) (see Ford, 1988, 1990, for discus-

Table VI. Tarsier Affinities

Possible shared-derived haplorhine traits

None

Possible shared-derived prosimian traits

1. Calcaneal peroneal tubercle reduced, short (Char. 52:1)
   But reversed in *Cheirogaleus*; paralleled variably in atelines; tubercle is lost or extremely low in *Tarsius* and omomyids; therefore, may be a strepsirhine trait
2. Calcaneal posterior articular facet less angled, more closely parallels long axis (Char. 8:1)
   But reversed in *Tarsius, Daubentonia*; independently derived in *Aegyptopithecus,* variably in some atelines and *Cebuella*
3. Tibial trochlear facet anteroposteriorly deep, mediolaterally compressed (Char. 19:2)
   But reversed in *Tarsius*
4. Femoral greater trochanter projects further proximally than does the head, by a considerable degree (Char. 36:0)
   But reversed in *Tarsius* and lorisines, paralleled in some cercopithecoids
5. Humeral entepicondylar foramen shifted more toward midshaft, so located above dorsal portion of trochlea (Char. 39:2)
   But paralled in *Apidium, Pliopithecus,* and variably in *Cebus* and some pitheciines
6. Humeral medial epicondyle flat dorsally, no concavity (Char. 45:0)
   But slight concavity in *Cheirogaleus*; flat in *Pliopithecus* and variably in callitrichids and some atelines

Of weaker weight (subject to more homoplasy)

7. Calcaneal heel longer (Char. 54:2)
   But reversed in several strepsirhines; independently derived in *Aegyptopithecus* and variably in *Apidium, Saguinus,* and *Alouatta*
8. Posterior border of tibial trochlear facet is sharp (Char. 15:2)
   Paralleled in many anthropoids
9. Femoral head with moderate extension onto posterior neck (Char. 21:1)
   But variably present in many anthropoids; states 0–2 highly subject to polymorphism
10. Femoral patellar groove lateral margin rounded (Char. 29:0)
    But highly variable in many "prosimians" and anthropoids, not a good indicator of anything
11. Femoral neck wide (Char. 31:0)
    But reversed in *Daubentonia* and some *Lemur,* paralleled variably in many anthropoids, including *Apidium*
12. Humeral supinator crest very wide (Char. 41:4)
    But reversed to ancestral moderately wide in many strepsirhines; most anthropoids are much narrower (states 2 or less); *Apidium* and *Aegyptopithecus* are not really known for this feature, and it is variably wide in some key platyrrhines (*Cebus, Callicebus,* and *Aotus*)
13. Femoral patellar groove fairly deep (Char. 28:3)
    Paralleled in a huge number of anthropoids; much too variable to be good indicator

If Adapis and Leptadapis excluded

14. Calcaneal peroneal tubercle reduced to tiny bump (Char. 1:1)
15. Calcaneal anterior articular facet for astragalus does not reach anterior (cuboid) end of bone (Char. 3:3–4)
    But independently developed in many anthropoids; much homoplasy
16. Calcaneal anterior and medial sustentacular facets sharply angled to one another or do not connect in all positions (Char. 4:1)
    But *Tarsius* and galagos highly specialized anterior calcaneus; convergent in most cercopithecines (further derived), Fayum primates, and variably in many platyrrhines
17. Humeral trochleocapitular ridge becomes sharp-edged (Char. 40:4)
    But independently as polymorphic and highly variable in many platyrrhines

sions of these features). Others would be reversed (not present) in *Tarsius* (although not *Hemiacodon*), seriously weakening their positions as "prosimian" synapomorphies, including the more parallel posterior articular facet on the calcaneus (Char. 8:1); tibial trochlear facet anteroposteriorly deep (Char. 19:2); and femoral greater trochanter projecting further than the head (Char. 36:0).

An anteroposteriorly deeper, mediolaterally compressed distal tibial facet for the astragalus (Char. 19:2) is here seen as a "prosimian" trait but is reversed in *Tarsius*. Dagosto (1985) also examined this feature but limited her coding to those with an (AP)/(ML) index less than 100 (mediolaterally broader) versus those with an index greater than 100 (those anteroposteriorly expanded). Given this coding, she then noted that primates are derived in falling in the latter category, with anteroposteriorly expanded, mediolaterally short facets. Although I concur with her finding that most early eutherians and many other mammals fall in the category of an index around or slightly below 100, I found that most anthropoids (which she did not include in her earlier study) showed considerable variation, with indices hovering around 100 and varying widely (within genera) between 86 and 122, even falling below 86 in some atelines. The nonprimates both Dagosto and I have examined fall into this broader category, and it is this state (Char. 19:1) that I define as ancestral for primates (as retained from a hypothetical ANCSTATES). Only a few specimens of *Saguinus, Cebus,* and *Apidium* approach the high indices (indicating a deep facet anteroposteriorly) that both she and I find characterizing all strepsirhines and adapids, including *Adapis* (see Dagosto, 1983). *Tarsius* does not fall in this category but has states 0–1 instead, and Dagosto's indices for omomyids indicate some exhibiting my state 1 (ancestral) but several exhibiting the derived state 2. Either this is independently derived (and yet highly unusual among mammals) or it represents a shared omomyid ("tarsioid"?)–adapid–strepsirhine trait that is reversed in some omomyids and the highly derived *Tarsius*.

The ancestral "prosimian" that emerges from this analysis is one that would have been marked by buttressed and strengthened joints, especially the elbow, hip, and knee (see Table VI). Changes in the angle of the posterior facet of the calcaneus and the shape of the tibial trochlea suggest less flexibility and more stability in flexion/extension at the tibiotalar and subtalar joints. The possible addition of a deep patellar groove and high greater trochanter suggest increased ability for powerful leaping. When *Adapis* and *Leptadapis* are excluded from the remaining "prosimians," additional synapomorphies appear. Some, such as the elongation of the anterior calcaneus (Char. 51:2), suggest a longer foot and greater power for more leaping behaviors.

Anthropoid Monophyly

As discussed above, the Anthropoidea was recognized as a distinct clade in all analyses except those forcibly separating the group, by pulling one

Table VII. Ancestral Anthropoid Traits[a]

		Anth/prosim							
Trait	Hapl/ other[b]	Any	Major	Cat/Plat	Plat	Plat + Apid.	Anth + omomyid/ other	Hemia. outgroup	Anth + adapids/ other
1	2	2	2	5	2	3	**2**	**2**	2
2	3	3	3	3	3	3	3	**3**	3
3	4	1	1	1	1	1	4	4	3
4	1	3	3	2	3	3	2	2	2
5	2	2	2	2	2	2	2	2	**3**
6	0	0	0	0	0	0	1	0	1
7	0	0	0	0	0	0	0	0	0
8	**2**	2	2	2	2	2	**2**	**2**	2
9	1	1	1	1	1	1	1	1	**2**
10	1	1	1	1	1	1	1	1	1
11	0	0	0	0	0	0	0	0	1
12	**4**	**4**	**4**	**4**	**4**	**0**	4	**4**	**4**
13	4	4	4	4	4	4	4	4	4
14	0	**4**	**3**	**4**	**4**	0	3	**3**	**4**
15	1	1	1	1	1	1	2	2	1
16	2	2	2	2	2	2	2	2	1
17	1	0	1	1	0	0	1	1	0
18	0	0	0	0	0	0	0	0	0
19	1	1	1	1	1	1	1	**1**	**1**
20	1	1	1	1	1	1	1	1	1
21	**0**	1	1	0	0	1	**2**	2	1
22	**1**	**1**	**2**	**1**	**1**	0	**1**	2	0
23	**2**	**1**	**1**	**2**	**1**	**1**	**1**	**1**	**1**
24	**2**	**2**	**2**	**2**	**2**	**2**	1	**2**	**2**
25	2	2	2	2	1	1	2	2	1
26	**1**	0	0	**1**	0	0	0	0	0
27	**1**	**1**	2	2	**1**	**1**	**1**	**1**	**1**
28	3	2	2	2	2	2	**2**	**2**	2
29	1	1	1	1	1	1	0	1	**1**
30	1	1	1	1	1	1	1	1	1
31	**1**	1	1	1	1	1	0	0	**1**
32	0	0	**1**	0	**1**	**1**	1	**1**	**1**
33	2	2	1	2	2	2	2	2	1
34	1	1	1	1	1	1	1	1	**2**
35	**2**	**2**	**2**	1	**2**	**2**	2	**2**	2
36	1	1	1	1	1	1	1	1	1
37	1	1	1	1	1	1	1	1	1
38	1	1	1	1	1	1	1	1	1
39	1	1	1	2	1	1	1	**1**	2
40	2	2	2	1	1	2	3	**3**	3
41	**2**	**2**	**2**	**2**	**2**	**2**	2	**2**	**2**
42	1	1	1	1	1	1	0	0	1
43	0	1	0	1	1	1	0	1	1
44	**3**	2	2	2	2	2	3	2	2

(*continued*)

Table VII. (Continued)

Trait	Hapl/ other[b]	Anth/prosim Any	Major	Cat/Plat	Plat	Plat + Apid.	Anth + omomyid/ other	Hemia. outgroup	Anth + adapids/ other
45	2	2	2	2	2	1	1	1	0
46	1	**1**	**1**	**1**	**1**	0	1	**1**	**1**
47	0	0	0	0	0	0	0	0	0
48	2	2	2	2	2	2	2	2	2
51	2	**2**	**2**	**2**	**2**	**2**	**2**	2	2
52	2	2	2	2	2	2	2	2	2
53	1	1	1	1	1	1	1	1	1
54	**1**	1	1	1	1	1	1	1	1

[a]Trait numbers in **bold** exhibit different states in different runs. (Traits 49 and 50 were excluded). Character states in **bold** are derived in the ancestral anthropoid morphotype; all others are primitive retentions.
[b]Groups are those constrained (haplorhine versus all other primates split; anthropoid versus prosimian split; anthropoid + omomyid versus all others). Any clades other than "anthropoids" and "prosimians" rarely or never occurred unless constrained by PAUP (see text and Tables I and II).

member out as the outgroup to all primates (i.e., *Callicebus* or *Apidium* as outgroup). The monophyly of this group seems certain, particularly as it is well supported by dental and cranial and soft tissue features as well (see others in this volume).

However, even a brief glance at Table VII is sufficient to show that as one examines smaller and smaller clades, assumptions made early in the analysis as to the character state present at any earlier nodes have dramatic effects on the presumed states at later nodes. Thus, there is considerable variation between runs as to which character states are present in the earliest anthropoid and also as to whether those states are primitive retentions or derivations. No single feature is present as an anthropoid synapomorphy, as opposed to a symplesiomorphy, under all assumptions. Several, however, are always indicated as present in the ancestral anthropoid, and usually as a derivation. These include: *calcaneal*—anterior portion moderately elongated (Char. 51:2); *tibial*—distal facet for fibula moderately elongated (anteroposteriorly) (Char. 12:4), distal facet for fibula more central or even posteriorly located (varying by constraints) (Char. 14:3–4); *femoral*—at least a low mound of bone on the posterior surface of the neck (Char. 22:1), intertrochanteric crest absent or slight extension of crest distal to greater trochanter [Char. 23:1(→2)], lesser trochanter displaced posteromedially on shaft (or only slightly more mediad) (Char. 24:2), third trochanter reduced to only a slight rise or rugose area, no distinct trochanter present (Char. 27:1), longer neck (Char. 32:1), distal epiphysis broader mediolaterally and less deep anteroposteriorly (Char. 35:2); and *humeral*—supinator crest moderately wide (Char. 41:2), medial epicondyle slightly angled dorsally to trochlea (Char. 46:1).

Some features are problematic. The extension of the femoral head onto the neck posteriorly (Char. 21) demonstrates a particularly interesting pattern of evolution. As with depth of the patellar groove (Char. 28, discussed above), although morphs are discrete and easily coded, some appear to coexist readily in polymorphic condition in any given genus. A "cylindrical" femoral head is quite distinct and seen throughout the bushbabies (including the fairly quadrupedal *Otolemur* included here), *Tarsius, Hemiacodon,* to a slightly lesser degree in two *Cheirogaleus* specimens examined (see also Anemone, 1990; Dagosto, 1986; Ford, 1980b; Gebo, 1986, 1988; Napier and Walker, 1967), and rarely otherwise. Absolutely no extension is found in most catarrhines and *Apidium*. However, most ateline genera, for example, are polymorphic for states 0 and 1 (no to moderate extension), and callitrichids, *Aotus, Callicebus,* and *Cebus* are broadly polymorphic for states 0, 1, and 2 (no to moderate extension, which merges gradually with neck). *Notharctus* is polymorphic for states 1, 2, and 3. It is difficult to know how to proceed to interpret a character such as this. Intuitively and biomechanically, the differences appear to have great functional implications (see, e.g., discussion in Ford, 1988, 1990). The lack of variability in some taxa, such as the galagos and *Tarsius* on the one hand and the catarrhines on the other, if real, should be accepted as an important difference. This morphological/genetic constraint to a very narrowly defined range of variability suggests that some very different adaptations are occurring than in those taxa that maintain significant amounts of polymorphism. However, not all primate genera are included here, and samples of many are extremely small (some including only one or two specimens); the lack of variability may be simply a function of small sample sizes. Or perhaps more significantly, the presence of polymorphism in a genus may be the result of pooling varying species that are themselves highly constrained. At this time, there are insufficient data on the majority of primates to ascertain how much interspecific variability exists within genera for both postcranial morphology and locomotor behavior. These data are beginning to be compiled (e.g., Davis *et al.*, 1993; Fleagle, 1977; Ford and Hobbs, 1994; Garber and Preutz, 1993; Nisbett, 1993; Rodman, 1979; Ward and Sussman, 1979), but there is still too little information to confront this issue in a broad study such as this, comparing many primate genera. Thus, although features such as femoral head extension and patellar groove depth are likely highly informative, we currently lack critical data to fully appreciate and interpret the information they relay.

A few of the major differences between runs are particulary worth investigating. Again, these seem to focus on marked differences when *Hemiacodon* is defined as the outgroup or between the "Major" constraint (the run constraining catarrhines and strepsirhines) and the other runs recognizing a monophyletic Anthropoidea.

With *Hemiacodon* as outgroup or sister group to anthropoids, several distinctly different things would occur, some of which also occurred when *Hemiacodon* (omomyids), but not *Tarsius,* was constrained to group with anthropoids. Features that otherwise are seen simply as primitive retentions would be interpreted as derived anthropoid synapomorphies (including Chars. 1:2, 2:3, 8:2, 19:1, 28:2, 39:1). This is simply the result of different polarities being determined for these traits when a primate ancestor is defined by this method, as discussed in the "Earliest Primate" section above. Although this affects the number of derivations or alterations occurring with the origin of anthropoids, it does not change which morphologies are seen to characterize the earliest anthropoid.

However, under the *Hemiacodon*-outgroup assumption, an entirely different state would characterize the ancestral anthropoid for some characters (Chars. 3:4, 15:2, 21:2, 22:2, 31:0, 40:3, 42:0). These differences are driven both by assumptions of states at earlier nodes and by the clades that are defined in the analysis. For example, for the position of the anterior articular surface on the calcaneal sustentaculum (AAS, Char. 3), state 4 of AAS not extending to the anterior end of the bone is defined for the anthropoid ancestor, rather than state 1 (broad extension to the anterior end), for two reasons. (1) There is sufficient polymorphism in many anthropoids so that the most parsimonious "decision" is to retain the state present in the closest previous node. And (2) with *Hemiacodon* held as an outgroup (which actually exhibits state 3), *Tarsius* and *Aegyptopithecus* also become sister taxa to all primates (in an unresolved polychotomy), and most of the strepsirhines and the adapids become a sister group to the anthropoids. Since the majority of these nonanthropoids exhibit state 4, this is then seen as the "primitive condition" for primates that is simply retained in the ancestral anthropoid. In contrast, if ANCSTATES is used to define a eutherian ancestral condition, then state 1 is retained in the primitive primate and the early side branch of the anthropoids in most runs. The "anthropoids + omomyid" constraint is instructional here. Using ANCSTATES, it also resulted in the ancestral primate retaining state 1 (see Table III) because *Leptadapis* forms a sister to all primates and also exhibits state 1. However, at the next node state 4 would evolve, similar to the "scenario" with *Hemiacodon* as outgroup, because of the widespread presence of state 4 in the nonanthropoids that branch off somewhat sequentially (rather than forming a discrete clade). This is then naturally retained in the earliest anthropoid.

This analysis of anterior articular facet position (Char. 3) illustrates the kinds of conclusions one is driven to by accepting both parsimony and various assumptions about character state polarity. This is not meant to disparage parsimony as a working hypothesis but to illustrate how essential it is to explore carefully the validity and consequences of that hypothesis on a character-by-character basis after completing an analysis. Most of the other characters with altered state under the "*Hemiacodon*-outgroup" assumption fall prey to a similar sequence of reasoning, including sharpness of the posterior trochlear border of the tibia (Char. 15), femoral neck diameter (Char. 31), and shape of the ventral trochlea of the humerus (Char. 42). These four traits exhibit sufficient variability within large clades (e.g., within Catarrhini, Platyrrhini, Strepsirhini, and Adapidae) to make any determinations of the ancestral condition for each of these groups or for primates suspect and of limited value in determining relationships between these groups (see also discussions above). The other characters that differ here (Chars. 21, 22, and 40), as discussed in sections above, are also ones that are difficult to determine likely patterns of change in because of variability, despite their seeming to be important functionally.

There are striking differences between the results accepting the "Major" constraint from many of the other analyses for the characterization of the ancestral anthropoid morphotype (see especially Table VII; Chars. 1, 4, 17, 21, 26, 27, 33, 39, 40, and 43). These are important because the few clades within primates that would be widely accepted as well established include the extant strepsirhines and, particularly, the catarrhines (see e.g., Fleagle, 1988; Szalay *et al.*, 1987). These features differ across analyses because of the implications of assuming different patterns of radiation within the Anthropoidea. Depending on which subgroups are hypothesized to have branched off first or separately, very different patterns of evolution appear for a number of key traits. These are explored in the following section.

Thus, the traits that can be reasonably hypothesized as characterizing an ancestral anthropoid, as derivations, are listed in Table VIII. These are a series of traits that are not very well understood in terms of biomechanical implications for locomotor behavior and that occur in primates of widely varying habits. However, among living primates, they occur most frequently in combination in medium-sized (1–5 kg) arboreal quadrupedal monkeys that focus on horizontal or mildly angled supports in the middle or upper canopy (e.g., *Aotus, Cebus, Callicebus,* smaller colobines).

The Anthropoid Radiation

Although there are a growing number of early anthropoid or possible anthropoid taxa, e.g., *Beretia, Algeripithecus, Amphipithecus,* and the oligopithecines, most of these are not currently securely known by any postcranial remains (Gebo et al., Chapter 9, this volume; Godinot and Mahboubi, 1992; Rasmussen and Simons, 1992), and thus they cannot be included in the course of this particular study. Four clear clades (or possible clades) of anthropoids remain: the parapithecids (primarily known from the large sample of *Apidium* postcrania, many of which were examined), the propliopithecids (primarily known from the sample of *Aegyptopithecus* postcrania, some of which were examined for this study), the platyrrhines (of probable but unproven monophyly), and the catarrhines (certainly monophyletic, consisting of two major subgroups—the hominoids and the cercopithecoids—both also well established as distinct clades). There is also the problematic *Pliopithecus* from the Miocene of Europe. Although it is generally considered highly probable that both the propliopithecids and *Pliopithecus* are catarrhines (see, e.g., Andrews, 1985; Gebo *et al.*, Chapter 9, this volume; Groves, 1991; Harrison, 1987;

Table VIII. Possible Shared-Derived Anthropoid Traits

1. Tibial fibular facet elongated to moderately long (Char. 12:4)
   But reversed in catarrhines
2. Tibial fibular facet shifts to more posterior position (Char. 14:4)
   But reversed in catarrhines
3. Femoral neck with raised area of bone on posterior surface (not a fully developed ridge or "crista paratrochanterica," which develops in some) (Char. 22:1)
   But reversed in hominoids and many cercopithecoids
4. Femoral lesser trochanter less medial, more posteromedial in placement on shaft (Char. 24:2)
5. Femoral third trochanter reduced to a slight rise or rugosity (Char. 27:1)
   But reversed in *Apidium* and variably in a few platyrrhines
6. Femoral distal epiphysis more mediolaterally expanded, anteromedially compressed (Char. 35:2)
   But reversed in *Apidium* and independently derived in lorisids
7. Humeral supinator crest less wide (Char. 41:2)
   But unknown in many fossil primates, esp. *Apidium* and *Aegyptopithecus*
8. Humeral medial epicondyle slightly displaced dorsally, rather than straight medial projection (Char. 46:1)
   But reversed in hominoids, variably in atelines and callitrichids

Of weaker weight (subject to more homoplasy)

9. Some extension of ridge below trochanteric fossa of femur develops (not yet true intertrochanteric crest) (Char. 23:1)
   But variably lost in many platyrrhines
10. Femoral neck longer (Char. 32:1)
    But tremendous variability within anthropoids and nonanthropoids; useless trait at this level or as currently coded

Simons and Rasmussen, 1991), these assumptions were not made *a priori* in this study. Therefore, based on the presumed clade of a monophyletic Anthropoidea but not a monophyletic Haplorhini (although not necessarily a monophyletic Prosimii either), and associated character changes discussed above and given in Tables IV, VI, and VIII, one can examine the likelihood of differing branching patterns within Anthropoidea.

Numerous features of other portions of the animals clearly unite hominoids and cercopithecoids into a single clade, the Catarrhini (see, e.g., Fleagle, 1988; Groves, 1991; Harrison, 1987; Martin, 1990), but they rarely formed a clade based on postcranial features alone (see Table II). However, this is because the two groups have diverged considerably in the postcranium, each developing very different habits and associated traits. While the cercopithecoids have become quadrupeds *par excellence* (Strasser and Delson, 1987) who often descend to the ground for significant portions of time, hominoids have become large-bodied suspensors and climbers (Fleagle, 1988), sharing a significant suite of derived characters with the atelines, albeit independently derived. However, whenever the catarrhines did form a distinct clade, a number of postcranial synapomorphies were recognized; nearly all were subject to high levels of homoplasy within primates. These include:

calcaneus—the anterior articular facet broadly reaching the anterior or distal end of the bone (Char. 3:1), variably present in many platyrrhines as well as discussed in the previous section; a much wider posterior articular facet (Char. 10:2), also independently developed in a few platyrrhines, *Tarsius*, and *Cheirogaleus; humerus*—loss of the epicondylar foramen (Char. 39:0), which is also lost in several different platyrrhine lineages. Other characters are even weaker and subject to widespread homoplasy across primates and reversals among catarrhines.

However, there is one good catarrhine synapomorphy, and possibly three or four. The posterior articular facet of the calcaneus becomes more squat in shape and less anteroposteriorly elongated (Char. 53:2). This is a unique attribute shared by all of the extant catarrhines among all primates included in this study. The change in shape is not simply a result of widening of the facet, as facet width measured separately against a measure of body size exhibits a parallel increase in several other primates (see above). A concomitant shortening of the facet must also be occurring, although again it is not detectable as a unique occurrence (Char. 9). This results in a less elongate, less helical, and presumably less mobile joint but one that is more stable and may be associated with increased body size in extant catarrhines. However, a similar change in facet shape is not seen in the moderately large atelines, *Aegyptopithecus,* or *Pliopithecus.* For three other traits, catarrhines exhibit the morphology presumed to be primitive for primates, but platyrrhines and *Apidium* share an altered morphology. The catarrhine morphologies include a short and anteriorly placed or unique dual distal tibial facet for the fibula (Chars. 12:2–3, 14:2) and the absence (or loss?) of any bony elaboration (mound or ridge, "crista paratrochanterica") on the posterior femoral neck (Char. 22:0) except slightly in *Presbytis.* For this last, however, Hershkovitz (1988) reports the presence of a crista paratrochanterica in some catarrhine specimens he has examined. If the states for these traits, which are shared uniquely by *Apidium* and the platyrrhines, are considered primitive for anthropoids (see Table VIII and below), then the catarrhines share the unifying features of reversal in some or all of these several femoral and tibial traits. However, if the shared platyrrhine–*Apidium* morphology is considered a synapomorphy of those two, then the catarrhines share very few other good derived postcranial features beyond a more stable, less mobile subtalar joint. Rather, it would appear that as the catarrhine lineage diverged, it was marked by a conservative, fairly unaltered postcranium coupled with well-documented cranial and dental derivations from earlier anthropoids. It was only on the major divergence(s) within Catarrhini, particularly of the Cercopithecoidea and the Hominoidea, that postcranial specializations occurred, separately within each group.

Pliopithecus is frequently grouped with the hominoids as a sister to the atelines; however, when the catarrhines held together as a clade (see Table II), *Pliopithecus* invariably did *not* group with the hominoids but rather with the atelines or with *Aegyptopithecus* (see, e.g., Figs. 2 and 3) as sister to the extant catarrhines. Thus, when *Pliopithecus* was grouped with the hominoids it was largely driven by features that both also shared with the atelines and that, for all three, are most likely convergences superimposed on other highly derived morphologies. These include traits such as a large tuber calcanei (Char. 7:2); an anteromedially rounded distal tibial shaft (Char. 11:1); a slight anterior bowing of the femoral shaft (Char. 26:1), which is not seen in *Hylobates* but is present in *Pan* and only variably in atelines; loss of the epicondylar foramen (Char. 39:0); a moderate to narrower bicipital groove (Char. 43:1–2); and a medial epicondyle that projects in line with the trochlea in at least some individuals (Char. 46:0). In fact, *Pliopithecus* lacks many of the possible catarrhine synapomorphies listed above, including the anterior position of the calcaneal anterior articular facet (Char. 3), loss of the entepicondylar foramen (Char. 39:0), which it retains, and the critical and perhaps best catarrhine synapomorphy, the squarer posterior articular facet on the calcaneus (Char. 53). However, it does share with the other catarrhines a wider posterior articular facet on the calcaneus (Char. 10:2) and the possibly primitive anthropoid trait of a mound or ridge of bone on the posterior femoral neck (Char. 22:0). In addition, it has no visible facet for the fibula on the distal tibia (Chars. 12–13–14:0), a condition otherwise seen only in various catarrhines, although independently lost more than once (in *Hylobates* and various cercopithecoids). These last traits are suggestive of a catarrhine link, but a distant one, lacking key synapomorphies of later catarrhine lineages. Thus, in keeping with Andrews (1985), Groves (1991), and others, it would seem that the postcranial evidence supports the place of *Pliopithecus* with the Anthropoidea and Catarrhini, but as a primitive sister taxon to later Cercopithecoidea and Hominoidea, lacking the full development of the derived subtalar joint and the bony ectotympanic tube and having generally primitive teeth but displaying the autapomorphic development of shearing crests and a derived facial morphology (see, e.g., Benefit and McCrossin, 1991).

Aegyptopithecus is not as well known postcranially as many other extinct primates included here. Most often, and especially under the "Major" constraint, *Aegyptopithecus* fell as a sister taxon to the Catarrhini within a monophyletic Anthropoidea, often closest to *Apidium* or *Pliopithecus* (see Table II); under some assumptions, it fell as a sister to all anthropoids or even all other primates (e.g., under the "*Callicebus*-outgroup" assumption). There are few consistent postcranial traits that would support a catarrhine affinity; most are subject to marked homoplasy—either frequent independent development in other lineages or reversal in a major catarrhine group (hominoids or cercopithecoids)—making their value as indicators of relationship highly suspect. Catarrhines and *Aegyptopithecus* do share a wider posterior articular facet on the calcaneus (Char. 10:2); however, this is variably present in *Cebus, Aotus,* many callitrichids, and some "prosimians." Many cercopithecoids have separate anterior and posterior facets on the calcaneal sustentaculum (e.g., Gebo and Simons, 1987), which *Aegyptopithecus* lacks. But hominoids and some Old World monkeys have connected facets, and separate facets are variably present in many platyrrhines and some strepsirhines (e.g., *Lemur, Otolemur*) and adapids (*Notharctus*), making this feature uninformative about *Aegyptopithecus*'s affinities (see Ford, 1980a, 1988, 1990). The lack of catarrhine synapomorphies supports the views of Fleagle and Simons (1982; Fleagle, 1988; Simons, 1987) that *Aegyptopithecus,* although a catarrhine dentally, is basically very primitive and conservative postcranially. Thus, although *Pliopithecus* and *Aegyptopithecus* can broadly be considered catarrhines, they fall as more primitive sister taxa to the later monkeys and apes. *Pliopithecus* may share a more recent common ancestor with extant forms, but it also appears to be more autapomorphic. *Aegyptopithecus* is much more conservative in its anatomy, although there are some autapomorphies that converge with morphologies seen in various other primates.

Monophyly of the platyrrhines, as discussed above, is not at all well supported. The features Ford (1980a, 1986) had suggested as platyrrhine synapomorphies in the postcranium, primarily on the distal tibia, were later found to be present in the postcranium of the parapithecid *Apidium* as well (Fleagle and Kay, 1987; Ford, 1990). These will be discussed in more detail below; however, if it is determined that these features are independently derived in the platyrrhine and parapithecid lineages, then they would remain as indicators of platyrrhine monophyly, but no longer unique to this group. The only remaining possible platyrrhine synapomorphy from this study would be the development of a more cylindrical, less cone-shaped or hourglass-shaped trochlea on the humerus (Char. 42:0). This must be seen, at best, as an extremely weak and unconvincing synapomorphy; in fact, it characterizes only seven of the 18 platyrrhine genera included here (particularly the callitrichids) and is only variably present in another eight. In addition, it would develop convergently in *Tarsius, Cheirogaleus,* and *Avahi,* opening the possibilities that it is either the primitive primate condition or highly labile. In any case, there appear to be few, if any, convincing and strong platyrrhine postcranial synapomorphies. This is mirrored in both the strict-consensus trees (see Fig. 4), which show a largely unresolved multichotomy for almost all platyrrhine genera, and the phylograms (Fig. 5), which show almost no divergence from a primitive anthropoid (or primitive primate) for such platyrrhines as *Cebus, Aotus,* and *Callicebus.* Nonetheless, there are good biogeographic reasons to continue to consider the Platyrrhini as a monophyletic clade, and Kay and Williams (Chapter 13, this volume) argue that there are at least some features that would support this view.

Therefore, if both *Aegyptopithecus* and *Pliopithecus* are accepted as catarrhines, and the Platyrrhini as monophyletic, the anthropoid radiation becomes a tripartite mystery of "where go (or went) the parapithecids?" Figure 6 presents the four primary possibilities that are even remotely consistent with the various forms of evidence (including geography and stratigraphy). These are: (a) parapithecids were an early side branch, with platyrrhines and catarrhines being each others' closest sister taxon; (b) platyrrhines and parapithecids shared a more recent common ancestor than either did with catar-

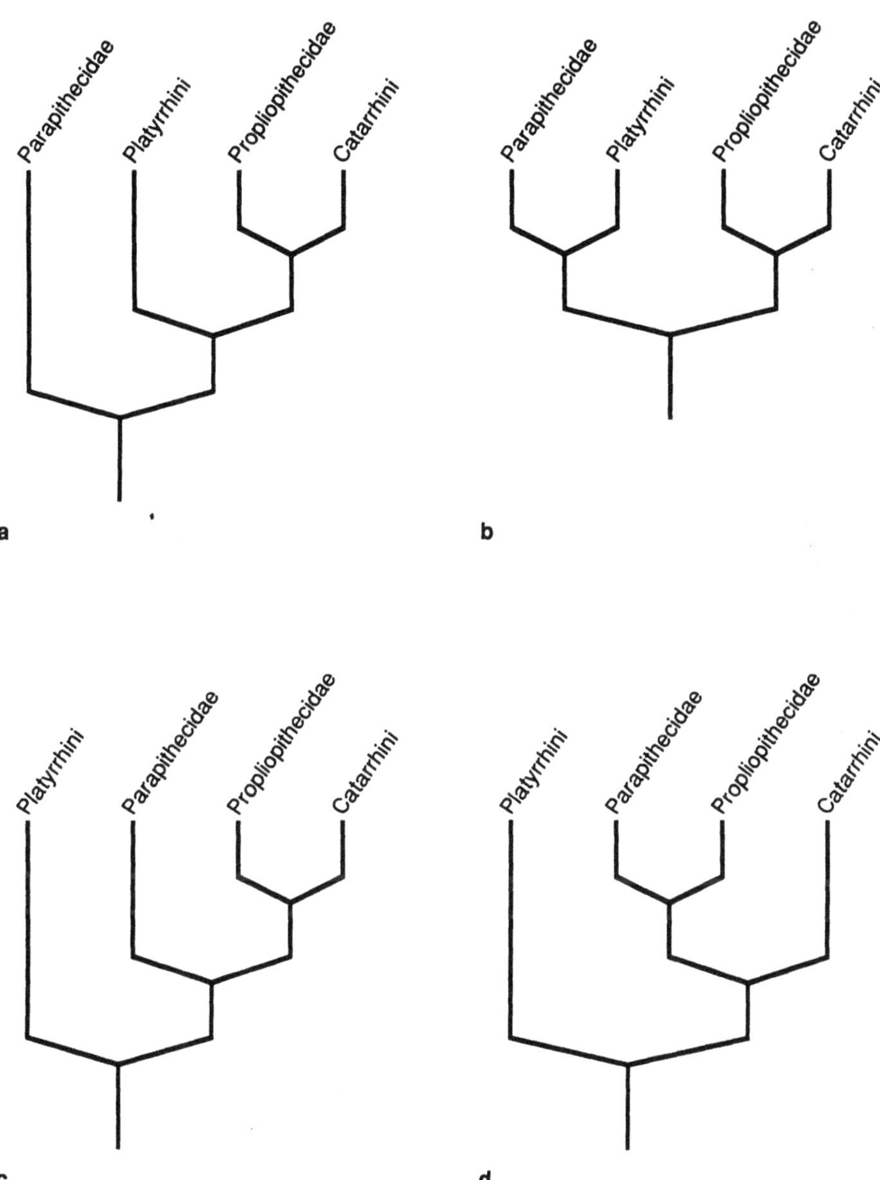

Fig. 6. Possible anthropoid radiations: (a) parapithecids as an early side-branch; (b) platyrrhines and parapithecids share a more recent common ancestor (would be Eocene); (c) catarrhines (including propliopithecids) and parapithecids share a more recent common ancestor (would be Eocene); (d) propliopithecids and parapithecids share a more recent common ancestor and then are closest to catarrhines, with platyrrhines the earliest sidebranch (would be Eocene).

Table IX. Potential Parapithecid Ties

Possible shared-derived Parapithecid–Propliopithecid traits

1. Calcaneal peroneal tubercle becomes very long and low (Char. 1:5)
   But may be primitive eutherian/primate or else independently attained in many platyrrhines, some catarrhines, and tarsioids
2. Calcaneal heel becomes quite long (Char. 54:2)
   But may be primitive primate trait (see Table IV) or independently acquired in many strepsirhines, *Tarsius*, omomyids, and adapids
3. Humeral trochleocapitular ridge much less distinct (Char. 40:1)
   But independently in *Adapis* and variably in many platyrrhines, especially atelines and pitheciines
4. Humeral ventral trochlea closer in width to ventral condyle, at least polymorphically in genus (Char. 48:1 or 1–2)
   But independently in hominoids, some platyrrhines, and some strepsirhines

Possible shared-derived parapithecid–catarrhine traits[a]

1. Calcaneal anterior and medial articular (sustentacular) facets have a break so a continuous contact connection exists only in certain positions (Char. 4:1)
   But reversed in extant hominoids; independent in some strepsirhines and many platyrrhines
2. Femoral head restricted, little to no extension posteriorly onto neck (Char. 21:0)
   But may be retained primitive primate–eutherian trait under some assumptions
3. Femoral intertrochanteric crest present (Char. 23:2)
   But independently also present variably in *Cebus*, *Chiropotes*, and many atelines
4. Femoral shaft slightly bowed anteriorly (Char. 26:1)
   Very weak, variably present in many platyrrhines also

Possible shared-derived parapithecid–platyrrhine traits[b]

1. Tibial distal fibular facet very long anteroposteriorly (Char. 12:4)
   But reversed (to polymorphic for states 3–5) in *Callicebus*, *Saimiri*, *Cebus*, *Saguinus*, and *Lagothrix*
2. Tibial distal fibular facet central to posteriorly displaced (Char. 14:3)
   But reversed to polymorphic for states 2/3 in *Callicebus*; possibly independently in some cercopithecoids

Of weaker weight (subject to more homoplasy)

3. Femoral neck with raised area of bone (possible precedent condition for ridge or "crista paratrochanterica"?) (Char. 22:1)
   But variably reversed to polymorphic for states 0/1 or 0/1/2 in atelines, pitheciines, *Aotus* and *Cebus*; independently occurs in *Presbytis* and *Cheirogaleus*; critically the morphology is unknown in *Aegyptopithecus*
4. Femoral lesser trochanter narrowed to oval shape (Char. 25:2)
   But possibly primitive primate retention (see Table IV) and states 0–2 highly variable (polymorphic) within genera and larger clades; critically, *Apidium* highly autapomorphic and *Aegyptopithecus* and adapids unknown
5. Humeral epitrochlear fossa present (Char. 45:2)
   But would be lost in most callitrichids, only variably present in most NWMs, and independently acquired in *Aegyptopithecus*
6. Humeral medial epicondyle displaced at least slightly dorsally (Char. 46:1)
   But would be variably reversed in many platyrrhines, esp. atelines and callitrichids; independently acquired in cercopithecoids, *Otolemur*, and *Daubentonia*

(*continued*)

Table IX. (*Continued*)

Possible shared-derived platyrrhine–catarrhine traits[c]

1. Femoral third trochanter reduced to a slight rise or rugosity (Char. 27:1)
   But would reverse to variably present in some *Aotus, Callicebus, Saimiri,* and *Callimico*
2. Femoral distal epiphysis mediolaterally broad, anteroposteriorly compressed (Char. 35:2)
   Independently derived in lorisids

Of weaker weight

3. Humeral epicondylar foramen located over trochlea ventrally (but not dorsally; displaced medially) (Char. 39:1)
   Probably primitive eutherian condition (see Tables III and IV); also polymorphism in some platyrrhines; foramen lost from unknown preceding position in catarrhines
4. Humeral trochleocapitular ridge becomes moderately distinct (Char. 40:2)
   Probably primitive eutherian condition (see Tables III and IV); also polymorphism in many platyrrhines

[a]A parapithecid (*Apidium*)–catarrhine clade was only recognized in the run enforcing platyrrhine monophyly (the "plat" constraint).
[b]A parapithecid (*Apidium*)–platyrrhine clade was only recognized in the run enforcing this monophyly (the "plat + apid." constraint).
[c]A platyrrhine–catarrhine clade was only recognized in the run enforcing this monophyly (the "cat/plat" constraint).

rhines; (c) catarrhines (including propliopithecids) and parapithecids shared a more recent common ancestor, and platyrrhines branched off prior to the parapithecid divergence; or (d) propliopithecids and parapithecids share a more recent common ancestor and then are closest to catarrhines, with platyrrhines again the earliest side branch. Although this last is the least likely alternative, as it would necessitate that the catarrhine-like features of the cranium and dentition of propliopithecids be independently evolved, it is included because the postcranial evidence analyzed here frequently suggested this alternative. With the discovery of parapithecids in some of the lowest beds in the Fayum, of earliest Oligocene or even late Eocene age (Simons, 1992; Gebo *et al.*, Chapter 9, this volume), schemata b–d would require the platyrrhines to have diverged from other anthropoids as a distinct clade sometime during the Eocene, long before their first appearance in the fossil record (e.g., MacFadden, 1990).

Table IX presents the putative synapomorphies of these four major potential branching patterns during the basal anthropoid radiation. As can be seen, there is very little of a strong nature to support a parapithecid/catarrhine clade (Fig. 6c) or a parapithecid–propliopithecid clade (Fig. 6d). Most features that might support these groups are likely to be either retained ancestral primate traits or highly subject to homoplasy. Although dental features lend some support to a parapithecid–catarrhine clade, Kay and Williams (Chapter 13, this volume) note that this arrangement would also require numerous reversals in dental traits in the catarrhine lineage from the ancestral anthropoid or else independent derivations in nonparapithecid catarrhines and platyrrhines and possibly in oligopithecines.

The two primary alternatives, a platyrrhine–parapithecid clade (Fig. 6b) or a catarrhine–platyrrhine clade (Fig. 6a), are less easily dismissed. Some discussion of the implications of these alternatives is provided above under "Anthropoid Monophyly" (see Table VIII). A number of traits linking platyrrhines and parapithecids are subject to high levels of homoplasy (Table IX); however, the distal tibial morphology with an anteroposteriorly elongated and central-to-posteriorly displaced fibular facet is not (Chars. 12:4 and 14:3). As discussed above, this morphology is indeed expressed only in *Apidium* and all platyrrhines and is clearly derived within primates. The issue becomes a balancing of the likelihood of this fairly distinctive feature (1) developing convergently in these two lineages, (2) reversing to the more primitive condition in catarrhines, or (3) developing only once as a synapomorphy of parapithecids and platyrrhines. Two pieces of evidence would help resolve this dilemma; neither is available at present. A clear understanding of the function of this altered morphology and thus the likely impetus for its possible independent derivation would help; however, as discussed in Ford (1990), the functional significance is unclear at present. The second critical piece of evidence would be the morphology of this region in other parapithecids and oligopithecines; recent discoveries (Gebo *et al.*, Chapter 9, this volume; Senut and Thomas, 1992) suggest that more information may be available in the near future. Should platyrrhines and parapithecids be sister taxa, then the cranial suture pattern they also share may be an additional synapomorphy (Fleagle and Kay, 1987).

These potential synapomorphies must be weighed against the number of traits that would unite platyrrhines and catarrhines and in which parapithecids (at least *Apidium*) retain the primitive primate morphology. Although the list again is not long, it does include the altered shape of the distal femur so that it is broader mediolaterally with much less deep femoral condyles (Char. 35:2) and reduction of the third trochanter to a slight rise or rugosity (Char. 27:1). Despite differences in coding and terminology between my own descriptions of the third trochanteric area and those of Hershkovitz (1987, 1988), it seems apparent that we are seeing the same general morphologies (especially as we have examined many of the same specimens). A rise but not a distinct, shaped tubercle is widely present in platyrrhines; a distinct, "knobby" tubercle is not, but it is variably present in *Aotus*, *Callicebus*, *Saimiri*, and *Callimico*. Whether this polymorphic occurrence of a third trochanter in some critically unspecialized platyrrhines indicates that they have reversed the reduction from their ancestor or that the ancestor of these and other platyrrhines had not lost it is controversial. However, in the absence of compelling evidence otherwise, and with the knowledge that variability in third trochanter form is widespread in mammals, including many cases of independent loss of this feature, I opt for the interpretation that retention of the genetic code for presence of a "knobby" trochanter in the population, expressed polymorphically, is the ancestral condition retained in these platyrrhines. Loss or reduction of the trochanter then occurred independently in catarrhines and likely in two or more lineages of platyrrhines.

The lesser trochanter has also figured importantly in discussions of parapithecid relationships (Fleagle and Kay, 1987; Ford, 1986, 1988, 1990). Qualitatively, platyrrhines and catarrhines appear to share a smaller lesser trochanter, one that does not extend as far as the head. However, when this feature is analyzed quantitatively, as here, there is sufficient overlap for some taxa that it is impossible to clearly identify more than three morphs. At this low level of discrimination, most primates fall in a nebulous intermediate zone. More data and perhaps a better method of quantifying lesser trochanter size/projection than that used here (see Appendix A; Ford, 1988) may allow better discrimination so that the *appearance* of a shorter lesser trochanter can be tested and perhaps supported quantitatively. One factor is that qualitative assessments of lesser trochanter projection are dependent on head projection/neck length as the point of comparison. The quantitative measure used here, diameter at the lesser trochanter minus shaft diameter immediately below the trochanter, is independent of neck length. Prior observations of the "short" lesser trochanter of platyrrhines and catarrhines may in fact be wrong and the result of confounding lesser trochanter projection with the independent variation in neck length.

Kay and Williams (Chapter 13, this volume) argue that dental features do not fully allow a clear choice; they favor a parapithecids-as-primitive scenario with a platyrrhine–catarrhine clade as advocated earlier by Fleagle and Kay (1987). However, as is clear from Kay and co-workers' careful studies of parapithecid dental characteristics, this group is highly autapomorphic. The same is true of a number of postcranial features such as the shape of the lesser trochanter (Char. 25:5), some intriguing similarities they share with microchoerines (Dagosto and Gebo, Chapter 17; Gebo *et al.*, Chapter 9, this volume), and the nature of the distal tibio-fibular joint (see above). Thus, in a taxon that is known to be highly autapomorphic in a number of features and to have developed some striking convergences with other taxa somewhat specialized for leaping, it is impossible at this time to argue that either the distal femoral morphology or the distal tibial morphology is more likely to have been subject to homoplasy in the evolution of *Apidium,* the only well-known parapithecid postcranium. Resolution of these issues simply must await knowledge of the postcranial anatomy of other, earlier, and/or less specialized parapithecids such as *Qatrania.* For now, an unresolved trichotomy is the best we can arrive at, one that is in need of study and not necessarily indicative of a legitimate bush.

Anthropoid Origins and the Basal Primate Divergence

One key issue surrounding anthropoid origins remains to be expressly addressed: their placement and affinities during the basal primate divergence. As discussed above, prior work has largely focussed on an early primate radiation into the two well-known Holarctic Eocene groups, the Adapidae and the Omomyidae (including the Microchoerinae, the Omomyinae, and the Anaptomorphinae), and later origins of the Anthropoidea, Strepsirhini, and *Tarsius* lineage from one or another of these two groups. However, clearly, the evidence presented here from the postcranium does not support a derivation of the Anthropoidea from either of these two groups.

The "anthropoids + adapids" constraint was among the least parsimonious (Table I); in addition, the extant Strepsirhini fell not as a sister to this group, as might be expected if they are also closely related to or derived from adapids, but rather as a sister to *Tarsius* and the omomyids (Table II). An adapid–anthropoid clade would share only two potential synapomorphies. One would be the shortening of the peroneal tubercle on the calcaneus (Char. 1:2), a highly variable development in primates with polymorphism for states 2–5 widespread in anthropoids and polymorphism for states 1–2 characteristic of adapids and many strepsirhines. The other would be the attainment of a mediolaterally broader, anteroposteriorly shallow distal femoral epiphysis (Char. 35:2). As discussed above, the polarity for this trait in anthropoids is highly suspect, as *Apidium* lacks this derivation in form. It appears as a potential adapid–anthropoid synapomorphy because *Notharctus* specimens approach the anthropoid condition, falling at the high end of the anthropoid range (the end closest to the primitively narrow and deep distal femur). This feature is unknown in the European adapini. The general morphology of the *Notharctus* distal femur is so distinctive in other ways from that of anthropoids that it is suspect whether this one feature is homologous rather than convergent. Particularly as this is a continuously varying trait, an index that may reflect several independent changes, and one that is coded into rather broad categories in which *Notharctus* falls at the extreme end of the range, it cannot be considered a sound basis for an argument for an adapid origin for anthropoids.

A critical aspect of the hypothesis, however, is whether or not adapids are too derived to serve as anthropoid ancestors. It is somewhat complicated by the fact that we are in fact dealing with two different adapid "morphs": the "typical" adapid morph, if you will, evidenced in *Notharctus* (Covert, 1986, 1988; Gebo *et al.*, 1991) and the very different adapini morph found in *Adapis* and *Leptadapis* (allowing for some differences between genera in each category). Under most assumptions, the ancestral primate morphology would be quite different from that seen in *Notharctus,* but not extremely dissimilar from that seen in *Adapis.* However, under the "anthropoid + adapid" constraint, the ancestral primate would hypothetically include several very different morphologies (see Table III). The adapids would not be highly derived from this ancestral primate, particularly the Adapini. If, as under most other options and largely accepted here (see discussion above), these features are *not* considered characteristic of the common primate ancestor, then attainment of these features must be seen as specialized derivations of the adapids, taking them away from the anthropoid lineage and making them too specialized for possible ancestry, whether convergently with strepsirhines or homologously (perhaps as the antecedent clade). This interpretation is favored here not only because of the reasons given above for favoring a different hypothesis about the ancestral primate but also because accepting a more adapid-like ancestral primate would sever any specific tie between strepsirhines and adapids, which appears contrary to current theories on strepsirhine origins and affinities based on both postcranial and craniodental evidence (see others in this volume).

An omomyid origin is the other widely touted potential anthropoid source. Although this option is quite parsimonious (Table I), it again raises problems. First, there are no shared-derived characters of *Hemiacodon* (the best known omomyid and the one included here) and anthropoids in the data set analyzed here beyond something as poor as increased femoral neck length (Char. 32:1), which at the low level of discrimination of morphs adopted here shows widespread homoplasy and polymorphism throughout primates.

Additionally, again, it would drive some reconsideration of the ancestral primate condition that, as discussed above, is seen as less likely to be correct although not impossible. But this solution does not place known omomyids in a position to be ancestral; they themselves (as represented by *Hemiacodon*) would be derived in several significant ways even accepting a redefined ancestral primate. Autapomorphies would include: *calcaneal*–a reduced, almost nonexistent peroneal tubercle (Char. 1:5); a much smaller posterior articular facet, both shorter (Char. 9:0) and narrower (Char. 10:0); and *femoral*–reduction to a tiny, facetal-like area on the lesser trochanter (Char. 25:3); development of a very deep patellar groove (Char. 28:4); and the greater trochanter projecting much higher than the head (Char. 36:0). In addition, *Tarsius* would have to develop independently most of these traits, which it shares with omomyids. Of course, this does leave open the potential that some unknown, more primitive omomyid that totally lacked these derived traits (largely shared by *Hemiacodon* and other known omomyids; see, e.g., Szalay, 1976; Dagosto, 1986, 1988; Dagosto and Gebo, Chapter 17, this volume) could serve as the ancestor for both anthropoids and *Tarsius*. But would this be an omomyid or simply an earlier, less specialized primate?

This leads us directly into a third scenario, the one favored here. Indeed, any primate that could reasonably be reconstructed as an anthropoid progenitor, given our current knowledge of fossil primate morphology, would be less specialized than either known omomyids or known adapids. These northern, Eocene groups are already too specialized in their own unique directions. Of these Holarctic taxa, *Adapis* and *Leptadapis* appear to differ least from a projected early primate. However, the cause of this similarity is at present shrouded in mystery. Are the morphologies primitive retentions? Are these two genera unrelated to other "adapids," a late Eocene migrant from some other place (Franzen, Chapter 4, this volume; Godinot, 1991)? Or are they late adapids highly derived from the ancestral adapid form in ways that converge on the putative primate ancestor? Their size is intriguing here: like anthropoids, they appear to be larger than earlier primates (Ford and Davis, 1992; Gebo, 1987). Are the morphologies and their presumed associated quadrupedal behaviors a result of this size increase, developed convergently in the both adapini and the anthropoids? Or are they primitive for primates and unrelated to body size? At any rate, *Adapis* and *Leptadapis,* in the late Eocene, are too late to be ancestral to anthropoids themselves.

Discussion and Conclusions

This analysis and comparison of a seemingly innumerable group of trees resulting from the exploration of different assumptions about the trait polarity and of constraining monophyly of various possible clades does point to some clear conclusions about the pattern of primate evolution, all based on this data base.

(1) There are substantial biological as well as analytical reasons to prefer a hypothetical ancestor defined by the careful analysis of the form and distribution of character states in a variety of "outgroups" (both fossil and living) as well as within the group under study, rather than the determination of polarity on the basis of some putatively most generalized or primitive member of the clade under study or even a single outgroup taxon, whether that member is fossil or extant. It is *not* sufficient simply to assume that the morphology present in the earliest known members of a group is the most primitive. This point has been made before by many others, but it is well worth repeating, as the assumption of polarity based on early or "generalized" group members is still frequently done.

(2) Strepsirhines share derived attributes in addition to toothcombs and grooming claws, as other recent studies have also begun to document (Dagosto, 1988; Gebo, 1988). None found here is unique to the group, and many are subsequently lost in some members of the clade (as is also the case for the toothcomb; witness *Daubentonia*). *Cheirogaleus* in particular has a tendency not to group with other strepsirhines because of some marked parallels with *Tarsius* and the omomyids; this is in agreement with the work of Gebo (1986, 1988), who has also found that omomyids are more like cheirogaleids than like galagines. The "typical" adapids, such as *Notharctus,* share these postcranial traits and are clearly already too derived postcranially to be ancestral to anthropoids.

The concern here is anthropoid origins and radiation, not details of evolution in other lineages. In addition, the sample size here is much too small to allow any conclusive analysis of details of relationships among non-anthropoid primates. Nonetheless, it is interesting to note the marked distinction drawn repeatedly between *Notharctus,* which consistently falls within the "strepsirhine" clade, and *Adapis* and *Leptadapis,* which usually fall as much more distant outliers. This supports the suggestions of Franzen (Chapter 4, this volume) and Godinot (Chapter 10, this volume) that these late Eocene adapini result from a very different evolutionary history than the other adapids (see also Kay and Williams, Chapter 13, this volume). This cannot be explained simply as the result of their adaptation for increased quadrupedalism (versus more leaping), as they are very different morphologically from some of the more quadrupedal strepsirhines such as *Daubentonia* and *Otolemur.* Their specific type of quadrupedalism is uncertain, as they are also distinct from quadrupedal anthropoids.

(3) The omomyid included in this study, *Hemiacodon,* was almost always closest to *Tarsius.* This relationship is supported by a number of traits, particularly of the calcaneus and femur, of fairly high consistency. This is extremely strong support for the omomyids (many others of which share a number of the features seen in *Hemiacodon;* see Gebo, 1986, 1988; Dagosto and Gebo, Chapter 17, this volume), being derived relatives of *Tarsius* rather than united with other Holarctic Eocene primates in some undifferentiated primordial primate soup out of which later groups evolved. However, this study also showed that *Tarsius* and, separately, the omomyids have each developed numerous, different autapomorphies (see Fig. 5), again removing either from a position as "ancestral to" or close to a "hypothetical ancestor" of other primates. This would include any omomyid similar to *Hemiacodon* postcranially or even further derived as being directly ancestral to the anthropoids. Although a much less differentiated "omomyid" could have filled this role, it is uncertain that anything that generalized would or should be classified as an omomyid.

(4) There is absolutely no support for a haplorhine clade in the features examined here. On the other hand, the consistency with which *Tarsius* and a representative omomyid grouped with other nonanthropoid primates, the strepsirhines and adapids (particularly *Notharctus*), as a derived clade cannot be easily waved aside. None of the several postcranial traits uniting the "prosimians" would appear to be as convincing a synapomorphy as the various cranial and soft tissue traits that have been put forward to support a monophyletic Haplorhini (see Dagosto, 1990; Szalay *et al.*, 1987); but even these supposedly strong haplorhine traits have come under question recently (Simons, 1990, and others in this volume). The possibility that not only is Haplorhini not a valid taxon but that the largely discounted Prosimii incorporating *Tarsius,* the strepsirhines, and these Eocene Holarctic groups may in fact be a valid monophyletic clade deserves serious consideration.

(5) The Anthropoidea do form a monophyletic clade, unified by a number of traits, although only a few are of high consistency. Many features are, in fact, present in some combination of two of the three major anthropoid groups included here: the platyrrhines, catarrhines, and parapithecids. This makes the determination of relationships within the Anthropoidea and polarity of traits more difficult to determine, and the behavioral reconstruction of the ancestral morphotypic anthropoid becomes dependent on interpretations of these intraanthropoid relationships (see below).

(6) The Catarrhini, although well defined both cranially and dentally, are only loosely supported by characters of the postcranium, most with high

homoplasy indices. Cercopithecoids and hominoids each form highly distinctive, derived clades. The Miocene *Pliopithecus* and the Oligocene propliopithecids most likely are early offshoots of the catarrhine clade, with *Pliopithecus* exhibiting more autapomorphies than *Aegyptopithecus*. The hominoids, in particular, and *Pliopithecus* demonstrate numerous convergences with atelines.

(7) Critical to our understanding of the early anthropoid morphotype, the platyrrhines do not form a distinctive clade in any analysis unless forced to do so. A platyrrhine clade is supported by only a small number of traits, generally with very high homoplasy (or low consistency) values. In fact, few clades within the Platyrrhini were strongly supported, and some platyrrhines frequently fall near the "base" of the anthropoid branch with extremely few derivations, including *Cebus, Aotus, Saimiri,* and, less often, *Callicebus* (see Fig. 5). This strongly suggests that any generalized platyrrhine morphology that could be identified would be nearly identical to the basal anthropoid form and possibly only slightly modified from the basal primate morphology. Similarities of *Aegyptopithecus* skeletal elements to those of many platyrrhines, particularly quadrupedal Neotropical monkeys, serve as strong support for this conclusion. From this, numerous platyrrhine groups have diversified in various ways, some of which converge on other primate groups.

(8) The postcranium provides strong support for cranial and dental features clearly demonstrating the anthropoid nature of parapithecids. Postcranial traits also suggest that parapithecids and catarrhines (including proliopithecids) do not share a closer relationship than either does to platyrrhines. However, neither of the alternative possibilities of a parapithecid–platyrrhine clade or platyrrhine–catarrhine clade (with parapithecids a third, independent lineage with some convergences to both) can be seen as more strongly supported by these data. Any interpretation is complicated, of course, by the difficulties in defining just what is a platyrrhine.

Apidium is clearly highly derived in its own right in a number of traits. Thus, known parapithecids are certainly not representative of "ancestral" anthropoids; rather, they represent a uniquely derived early anthropoid clade that did not survive much beyond the early Oligocene, unless they lingered in yet unsampled areas of Africa. Therefore, the few clearly derived traits (as opposed to primitive retentions, which are also present) that *Apidium* shares with nonanthropoids such as some microchoerines, must be interpreted as convergences.

(9) On the basis of this analysis of these postcranial traits, no cogent argument can be made for a special relationship, much less an ancestor–descendent link, between either adapids or omomyids and anthropoids. This reinforces the growing evidence from Asia and North Africa that anthropoids are a separate and early branch within the primate radiation (see Fig. 7). The lack of a clear relationship of anthropoids to either of the major western Holarctic groups that emerged in the Eocene, coupled with the now growing evidence for the presence of anthropoids in the Eocene of Africa and possibly of eastern Asia, strongly suggests that the Anthropoidea represent an inde-

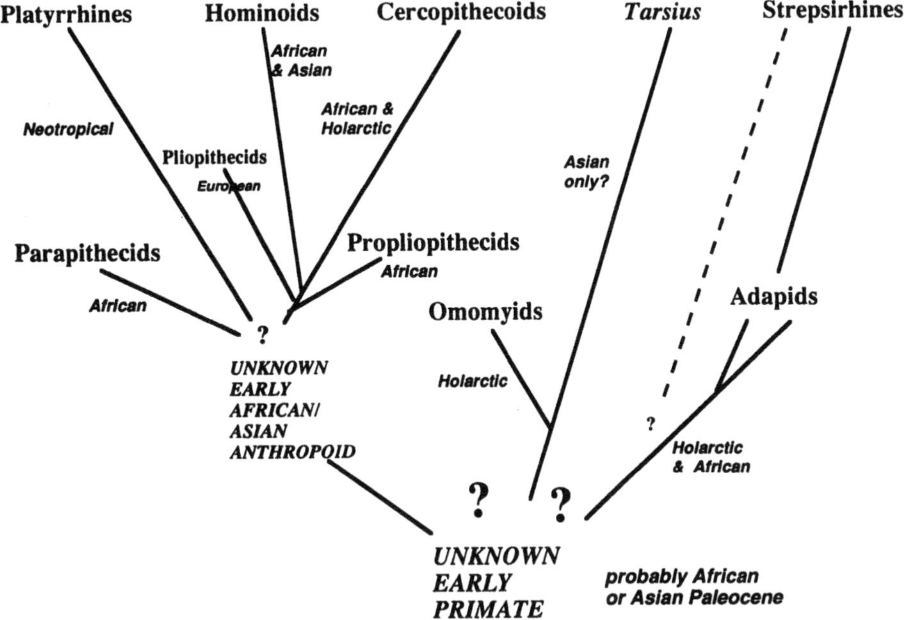

Fig. 7. Hypothesized pattern of primate radiation.

pendent primate lineage of great antiquity within the order. Anthropoids may be a result of the earliest divergence of the primates, arising either in Asia or Africa but certainly not in Europe or North America or from any primate group native to those geographic regions. Their relationship to other, still poorly known early Gondwanan primates, such as *Altiatlasius* and *Afrotarsius,* remains to be clarified.

This leads to the following reconstruction of primate evolutionary events, which must begin with an as yet undiscovered early primate, either Asian or African, in the Paleocene. Recently reported early Eocene placentals from Australia (Godthelp *et al.,* 1992) give further evidence of a still only poorly sampled placental radiation in the southern continents in the early Tertiary. At least three branches evolved from this ancestral primate group.

Two descendent clades migrated into Holarctica, diversifying and becoming very successful in the Eocene—the adapids (which may not be a monophyletic group) and the omomyids. Both of these groups likely originated in the late Paleocene in southern or eastern areas (Africa or Asia). Relatives of the adapids (possibly descendents) either stayed in Africa or returned there by the Miocene, giving rise to the many strepsirhine groups. A late Eocene European clade, including *Adapis* and *Leptadapis,* is of very uncertain origin and affinity to the rest of the group. A relative (but not a descendent) of the omomyids either stayed in or migrated to Asia, leading to the tarsier lineage. Questions about the supposed tarsier affinities of the Fayum *Afrotarsius* (see e.g., Godinot, Chapter 10, Kay and Williams, Chapter 13, this volume) and the discoveries of omomyids and fossil tarsiers in Asia (Beard and Wang, 1991; Culotta, 1992) lend support to the idea that the omomyid–tarsier group originated there, with only the omomyids leaving Asia for other northern continents. It is possible that the two basal groups (the adapid–strepsirhine group and the omomyid–tarsier group) shared a common ancestor more recently than either did with the third primate lineage, in which case the nomen "Prosimii" would be valid.

The third group, the anthropoids, would have differentiated at least in the middle Eocene, and possibly in the latest Paleocene, as a separate branch. Although a remote tie to the omomyid–tarsier group remains possible, given shared soft tissue traits (as argued by Dagosto and Gebo, Chapter 17, this volume), such a link would have to reach back at least into the late Paleocene if not earlier. During the Eocene, anthropoids appear to have been evolving in both Africa and Asia, although evidence is sparse, and the nature of these forms only dimly understood.

Some time in the late Eocene or earliest Oligocene, the extant anthropoid lineages (the Catarrhini and Platyrrhini) and the parapithecids diversified, although the early separation of these three groups was marked by conservation of the ancestral locomotor behaviors and skeletal features. However, quickly after the divergence of the major groups, additional radiations occurred, giving rise to groups that were highly distinctive both craniodentally and postcranially: the cercopithecoids, the hominoids, and the parapithecids. The platyrrhines were least distinctive on emergence, although subsequent to their migration to and isolation in South America, many lineages developed unique anatomies and behaviors.

The earliest anthropoid was most likely primarily quadrupedal, favoring horizontal and mildly angled branch supports in the middle and upper canopy. Although it has been assumed that a key element of the evolution of this adaptation and associated morphologies was increased body size (see Gebo, 1987; Ford, 1988; Ford and Davis, 1992), the moderately small to diminutive size of many of the putative Eocene anthropoids from North Africa (e.g., Godinot and Mahboubi, 1992; Simons, 1990, 1992) seriously challenges that assumption. As yet, it is uncertain that the inflated molar cusps of *Algeripithecus* are true indicators of its anthropoid affinity rather than the convergent adaptations of, say, a small gummivore. Likewise, certainty of the anthropoid affinities of some of the newest finds from Quarry L-41 of the Fayum deposits must await more complete specimens. Others, however, are indisputably anthropoid and just as indisputably smaller than any other anthropoids except the highly specialized callitrichids from South America. Unfortunately, no directly associated postcranial remains of these early, small, and controversial taxa have yet been discovered, although new finds are very promising (Gebo *et al.*, Chapter 9, this volume).

Two final questions remain, one methodological and one practical. The first: Why bother with all the trouble of doing a numerical cladistic analysis? Life, and reconstructing its past patterns, is certainly much simpler if one concentrates on just a handful of favored traits, and the results are much cleaner (if not more accurate). Ultimately, PAUP (and similar programs) is only an analytical tool. Like a pair of calipers, it does not miraculously provide correct solutions to problems, but it does allow us to explore new dimensions of the problem and to approach issues from new directions. Calipers are only as useful as their accuracy, the skill of the user, and the use to which the results are put. Likewise, numerical cladistic programs are only as good as the accuracy and validity of the assumptions they make, the skill of the user (including skill at inputting "good" data based on sound character analyses), and the use made of the multitude of potential output. PAUP both allows and forces one to investigate his/her characters and phylogenetic assumptions far more rigorously than might otherwise occur. In the final analysis, however, one opts for the hypotheses most consistent with the evidence at hand, including characters not incorporated in the computer analysis, time and biogeography, and biological concerns (a not-so-subtle form of weighting). As long as one makes these final arbiters clear, this is not a misuse of these programs but a rational and balanced use of a tremendous tool.

And finally, where do we go from here in order to resolve those glaring question marks (Fig. 7) and to find the great unknowns? The answer to gaining a better understanding may lie in some combination of more and more complete fossils, more critical analysis of characters and character states, refined programs of analysis to allow better inclusion of critical and informative quantitative traits and polymorphic taxa, and the development of an acceptable technique for weighting characters somewhere during the analytical stages. As with prior work on the postcrania by Ford (e.g., 1986) and work on the dentition by Kay and Williams (1992; Chapter 13, this volume), careful examinations of large suites of characters across many primate taxa result in the recognition of large-scale homoplasy within the primate order. Many clades are based on small numbers of low-confidence characters. In addition, even with small sample sizes, many genera demonstrate marked polymorphism for traits, at least within narrow ranges of potential character states. Thus, there is often partial overlap between taxa rather than either total congruence or total disjunction. This reflects the reality of nature; populations across both space and time are varying and polymorphic for a large percentage of features. Eliminating all such features from study may leave researchers with an uninformatively small number of traits to study; including them is messy and difficult but more closely reflects the real world, a world of variability that new fossil finds make only more apparent and more wonderful.

Acknowledgments

I wish to thank Drs. John Fleagle, Richard Kay, and Elwyn Simons for inviting me to participate in the symposium and workshop that led to this volume, which critically affected my own thinking on the topic. Discussions over methodology with Brenda Benefit, Robert Corruccini, Charlene Dickinson, Richard Kay, and especially David Swofford vastly improved this study; in addition, I am particularly grateful to John Fleagle for both the opportunity to study the *Apidium* postcrania and for many long discussions on characters, analysis, and phylogeny. Any errors in method or flaws in logic remain mine alone. David Hobbs ran several programs for me, and he, Lesa Davis, Marian Dagosto, Richard Kay, Elizabeth Strasser, and an anonymous reviewer provided critical readings of the manuscript, for which I thank them.

And finally, I must thank the many individuals who have generously provided me access over the years to materials that were incorporated into this study, including Dr. Richard Thorington, Smithsonian Institution; Drs. Anderson and Delson, American Museum of Natural History; Dr. Mary Dawson, Carnegie Museum of Natural History; Drs. Timm and Hershkovitz, Field Museum of Natural History; Drs. Biegert and Schmid, Anthropologische Institüt, Zurich; Dr. Pru Napier, British Museum (Natural History); Florida State Museum; Museum of Vertebrate Paleontology, UC Berkeley; Philadelphia Academy of Science; São Paulo and Rio de Janeiro museums; and Drs. Herbert Covert, John Fleagle, Alfred Rosenberger, Elwyn Simons, and Frederick Szalay for permission to study specimens in their care.

Appendix A: Characters and Character States Used in PAUP Analysis

Character states are listed as coded, in order from state 0 to state *n*. If a character is labeled with "(stepmatrix)," then a special matrix of transformations between states was defined; otherwise, it is assumed that states transform along the series: $0 \leftrightarrow 1 \leftrightarrow 2$. The state numbers were assigned based on ease of definition of the transformation series; it is not assumed that state 0 was the primitive condition for any group. As often as not, some state other than 0 is actually assumed to represent the primitive condition for early eutherians (as defined through the option ANCSTATES in PAUP 3.0s; see Swofford, 1991a).

For quantitative features, measurements are illustrated and described in Ford (1988; see also 1980a,b, 1986). Indices describe either the shape of a facet or the length of some feature relative to a measure that has a close relationship to body size. Divisions between states for quantitative features were determined by plotting ranges, means, and one standard deviation and seeking natural "gaps" in the data, where there was little overlap between taxa. Since there was a great deal of overlap between and wide variation within primate genera, states were defined conservatively, with the fewest states that seemed to fit the major gaps in the data. A more finely scaled analysis with larger samples analyzed on the species level may well indicate more states can be accurately and clearly identified within primates.

The state defined as ancestral eutherian (in the ANCSTATES option of PAUP, see text) is given in parantheses after the abbreviated name of the character.

Calcaneal Characters

1. Size of the peroneal tubercle (CPT size)(4): 0, absent; 1, tiny; 2, short; 3, moderate; 4, lrg_defined; 5, low_long; 6, doubled (stepmatrix: all states can transform to 0)
2. Location of peroneal tubercle (CPTloc)(1): 0, absent; 1, ant_end; 2, ant_half; 3, centered; 4, post_half (stepmatrix: all states can transform to 0)
3. Position of anterior articular facet for the astragalus (on anterior sustentaculum) (CAASpos)(1): 0, absent; 1, ant(erior end)_broad; 2, ant_point; 3, navic(ular facet); 4, not_ant (not to anterior end) (stepmatrix: all states can transform to 0)
4. Connection between anterior and medial articular (sustentacular) facets (CAMAS)(3): 0, separate; 1, some_posit. (see Ford, 1980a, 1986); 2, angled (sharply); 3, confluent
5. Degree of concavity of cuboid articular facet (CCASconc)(2): 0, flat; 1, cuboid_pivot; 2, slight; 3, moderate; 4, deep; 5, convex_ltpl (lateral/plantar portion) (stepmatrix)
6. Navicular articular facet (CNAS)(0): 0, absent; 1, antAAS (anterior to anterior articular facet for the astragalus); 2, medAAS; 3, latAAS; 4, double_convex (stepmatrix: all states can transform to 0)
7. Tuber calcanei (CTuber)(0): 0, absent; 1, small_mod; 3, large
8. Angle of posterior articular surface to long axis of bone (CANGLPAS)(2): 0, $<3°$; 1, $3° \leq x < 8°$; 2, $8° \leq x < 24°$; 3, $\geq 24°$
9. Length of posterior articular surface relative to maximum length of cuboid articular surface (Cpasl)(1): 0, <90; 1, $90 \leq x < 110$; 2, $110 \leq x < 128$; 3, ≥ 128
10. Width of posterior articular facet relative to maximum length of cuboid articular surface (Cpasw)(1): 0, <48; 1, $48 \leq x < 70$; 2, ≥ 70
51. Length of anterior calcaneus relative to maximum calcaneal length (CANT)(1): 0, ≤ 28; 1, $28 < x \leq 33$; 2, $33 < x \leq 48$; 3, $48 < x \leq 60$; 4, >60
52. Length of peroneal tubercle relative to maximum length of cuboid articular surface (CTPL)(3): 0, absent; 1, <68; 2, 68–168; 3, >168
53. Width to length of posterior articular facet for the astragalus (CPASshp)(0): 0, <42; 1, $42 \leq x \leq 76$; 2, >76
54. Length of posterior calcaneus relative to maximum length of cuboid articular surface (CPOST)(0): 0, ≤ 61; 1, $61 < x < 107$; 2, ≥ 107

Tibial Characters

11. Shape of distal shaft (Tshaft)(0): 0, triangular; 1, antmed_rnded(rounded); 2, lat_rnded; 3, medlat_comprsd(compressed)
12. Anteroposterior length of distal facet for the fibula (TFFL)(3): 0, absent; 1, fused; 2, tiny; 3, short; 4, mod.(erate); 5, long (stepmatrix: all states can transform to 0)
13. Proximodistal depth of distal facet for the fibula (TFFD)(3): 0, absent; 1, fused; 2, tiny; 3, shallow; 4, med; 5, deep (stepmatrix: all states can transform to 0)
14. Position of distal facet for the fibula (TFFpos)(2): 0, absent; 1, fused;

2, ant(erior); 3, central; 4, post(erior); 5, ant_curves_out (stepmatrix: all states can transform to 0)

15. Posterior border of trochlear facet for astragalus (Tpstbord)(0): 0, flat; 1, rounded; 2 sharp
16. Anterior border of trochlear facet for astragalus (Tantbord)(0): 0, flat; 1, rounded; 2, sharp; 3, high and sharp
17. "Squatting" facet on anterior face of distal tibia (Tsqt)(0): 0, absent; 1, slight; 2, moderate; 3, large
18. Horizontal groove on posterior malleolus (Thorizgr)(0): 0, absent; 1, slight; 2, marked
19. Depth (AP) to breadth (ML) of trochlear facet for astragalus (TFAP_FML)(1): 0, <86; 1, $86 \leq x \leq 122$; 2, $122 < x < 150$; 3, ≥ 150
20. Medial malleolar height relative to anterior–posterior diameter of distal shaft (TMMH_SAPD) (0): 0, ≤ 68; 1, $68 < x \leq 101$; 2, >101

Femoral Characters

21. Extension of head onto posterior neck (Fhdpext)(0): 0, sl_none (slight to none); 1, mod(erate); 2, merges_neck; 3, general_ext(ension, or "cylindrical")
22. Presence of ridge on posterior neck ("crista paratrochanterica," Hershkovitz, 1988) (Fridge)(0): 0, flat; 1, low_mound; 2, ridge
23. Intertrochanteric crest (Ftrochcrst)(0): 1, absent; 2, distal_ext (below trochanteric fossa); 3, distinct_crst(crest); 4, turns_lat (stepmatrix: all states can transform to 0)
24. Position of lesser trochanter (FLTpos)(1): 0, medial; 1, sl_med(slightly medial); 2, posteromed(ial); 3, sl_post(slightly posterior); 4, posterior
25. Shape of smooth (subchondral) bone on lesser trochanter (FLTshape)(1): 0, round; 1, fat_oval; 2, oval; 3, tiny_oval; 4, long and thin; 5, double (stepmatrix: states 3, 4, and 5 can all come from state 2 with only one step)
26. Anterior bowing of shaft (Fbow)(0): 0, straight; 1, slight_bow; 2, bow
27. Third trochanter (note: coding of states differs significantly from those of Hershkovitz, 1987, 1988; thus, the apparent differences in coding of individual taxa. In fact, the same basic morphologies are described for the same taxa) (F3tr)(2): 0, absent; 1, sl_rugose/rise; 2, distinct
28. Depth of patellar groove (Fpatgrdep)(1): 0, flat_veryshal; 1, shallow; 2, moderate; 3, fairly_deep; 4, deep
29. Shape of lateral margin of patellar groove (Fpatlatshp)(2): 0, rounded; 1, mod_sharp; 2, sharp
30. Extension of subchondral bone beyond lateral margin of patellar groove ("curled over" onto shaft; Fpatlatext)(0): 0, absent; 1, curls_over
31. Neck diameter relative to medial–lateral proximal shaft diameter, below lesser trochanter [PSTD] (FND_PSTD)(2): 0, >100; 1, 75–100; 2, <75
32. Neck length relative to PSTD (FNL_PSTD)(0): 0, <150; 1, ≥ 150
33. Position of third trochanter, measured as length from greater trochanter to top of third trochanter relative to lateral length of femur (F3tgtLpos)(1): 0, absent; 1, <12; 2, 12–20; 3, >20 (stepmatrix: all states can transform to 0)

34. Superior patellar width relative to PSTD (FSPW_PSTD)(1): 0, <84; 1, 84–115; 2, $115< x \leq 150$; 3, >150
35. Distal epiphysis depth (AP) relative to biepicondylar breadth (FDED_BEB)(1): 0, >107; 1, 107–99; 2, $99> x >71$; 3, ≤71
36. Proximal head projection relative to greater trochanter projection (FHP_GTP)(1): 0, <80; 1, 80–125; 2, $125< x \leq 180$; 3, >180
37. Lesser trochanter projection relative to PSTD (FLTPROJ)(1): 0, <46; 1, 46–88; 2, >88
38. Head shape (FHDSHP)(1): 0, <96; 1, 96–109; 2, >109

Humeral Characters

39. Position of epicondylar foramen (Hepifor)(1): 0, absent; 1, medial; 2, above_venttroch (ventral trochlea); 3, above_dorstroch (stepmatrix: all states can transform to 0)
40. Trochleocapitular ridge (Htrcprid)(0): 0, absent; 1, sl_distinct; 2, mod_distinct; 3, very_distinct; 4, sharp_edged
41. Supinator crest (Hsupcrst)(3): 0, rounded_edge; 1, sharp_edge; 2, moderate wide; 3, very_wide
42. Shape of ventral trochlea (Hventtr)(1): 0, cylinder; 1, cone; 2, spool
43. Bicipital groove (Hbicipgr)(0): 0, shallow_wide; 1, moderate; 2, deep_narrow
44. Shape of deltopectoral crest (Hdelpeccrst)(2): 0, smooth_shaft; 1, rounded; 2, crest_modflare; 3, flat_sup(flattens superiorly); 4, intermed_crest
45. Epicondylar fossa (Hepifos)(2): 0, absent; 1, concave; 2, pit_fossa
46. Dorsal displacement of medial epicondyle (Hmedepi)(0): 0, parallel; 1, slight_dorsangl; 2, large_dorsangl
47. Minimum trochlear diameter (anteroposterior) relative to maximum trochlear diameter (HMinTD_MaxTD) (1): 0, >70; 1, ≤70
48. Ventral condylar width relative to ventral torchlear width (HVCW_VTW)(2): 0, <100; 1, $100 \leq x < 140$; 2, $140 \leq x < 200$; 3, ≥200

49–50. Excluded from analysis; see Appendix B.

51–54. Listed with calcaneal characters.

Appendix B: Data Matrix for PAUP (Version 3.0s)

```
#NEXUS

[primate pc data]
[data for anthropoid origins paper]

begin data;
   dimensions ntax=46 nchar=54;
   format missing=? interleave symbols="0~9" equate="a={0,1}" equate="b={1,2}"
   equate="c={0~3}" equate="d={2,3}" equate="e={3,4}" equate="f={3~5}"
   equate="g={4,5}" equate="h={2,4}" equate="j={0~2}" equate="k={1~3}"
   equate="l={1~4}" equate="m={1,3}" equate="n={5,6}" equate="o={2~5}"
   equate="p={2~4}" equate="q={1,4}" equate="r={0~4}" equate="s={0,2}";

   charlabels [see APPENDIX A]
      CPTsize CPTloc CAASpos CAMAS CCASconc CNAS CTuber CANGLPAS Cpasl
         Cpasw
      Tshaft TFFL TFFD TFFpos Tpstbord Tantbord Tsqt Thorizgr TFAP_FML TMMH_SAPD
      Fhdpext Fridge Ftrochcrst FLTpos FLTshape Fbow F3tr Fpatgrdep Fpatlatshp
         Fpatlatext
      FND_PSTD FNL_PSTD F3tgtLpos FSPW_PSTD FDED_BEB FHP_GTP FLTPROJ FHDSHP
         Hepifor Htrcprid
      Hsupcrst Hventtr Hbicipgr Hdelpeccrst Hepifos Hmedepi HMinTD_MaxTD
         HVCW_VTW Hmeas Hmeas
      CANT CTPL CPASshp CPOST;
   statelabels
      [SEE APPENDIX A]
         ;
   matrix
```

[note: chars. 49&50 excluded]

```
[ 0 =======+      ......  ....1   ......  ....2   ......  ....3   ......  ....4   ......  ....5]
Alouatta          5e13o   jbbd1   mgggj   j00a1   jaj2j   aldja   1a2bd   k1101   baa12   a02??
Aotus             ndlmd   cj2bb   0ggfj   kcjbb   jjabh   ajpj1   aabb2   bb1a1   kaaej   aak??
Ateles            oelc5   jbkk1   1gg50   cc0aa   ajjkj   aapja   1a3bd   ka101   ba21j   a0k??
Brachyteles       14135   02221   15550   ba01b   010ba   aa2a1   112bd   d110b   10b10   002??
Cacajao           be32d   1j2b1   jgfgs   aa01b   aja2j   aaaaa   1a1b3   2a121   bab1j   2a2??

[char. #'s        12345   6789.   12345   6789.   12345   6789.   12345   6789.   12345   6789.]
[ 6 =======+      ......  ....1   ......  ....2   ......  ....3   ......  ....4   ......  ....5]
Callicebus        ode2d   aj2b1   sfeda   kc01a   jba2h   0bba1   aabb2   b110b   11a4b   1a2??
Callimico         pdlmd   j02b1   ag5es   2sb1a   j2a2e   jbdj1   01212   111bp   b003a   102??
Callithrix        2dlmd   002bb   agge2   bj210   j2jb4   aadj1   aa212   ab10h   200da   a0d??
Cebuella          oeqdd   00bb1   agf32   2ss10   22a1e   aa2j1   0a212   b2104   d002a   a02??
Cebus             oercc   jjdbb   affej   cca11   cjjbj   aacj1   aabjd   b11j1   1aabj   b02??

[char. #'s        12345   6789.   12345   6789.   12345   6789.   12345   6789.   12345   6789.]
[11 =======+      ......  ....1   ......  ....2   ......  ....3   ......  ....4   ......  ....5]
Chiropotes        oepck   ja2b1   agee0   2a011   jjbdj   aaaaa   11bbd   b11b1   kab12   212??
Lagothrix         d4mj5   j2bb1   bfg5s   c00a1   aj2dl   aamba   ba3b3   b110d   b0bb2   b02??
Leontopithecus    o31d2   002bb   sg5e2   2aj10   j21d4   a1b11   aad12   1b10b   ba0ka   b0d??
Pithecia          23433   112b1   0geeb   21011   ba112   0a001   a112d   1111e   10???   ?0b??
Saguinus          ddbmk   002bb   0ffp2   kcjba   j2abp   aakj1   aa2a2   bb1j1   1aapj   a0b??

[char. #'s        12345   6789.   12345   6789.   12345   6789.   12345   6789.   12345   6789.]
[16 =======+      ......  ....1   ......  ....2   ......  ....3   ......  ....4   ......  ....5]
Saimiri           6e410   412k1   0feeb   21a1a   b212h   ab3j1   1a212   11112   21a1b   bad??
Cebupithecia      33134   20221   04441   20011   12120   ?1211   ????2   1?113   22122   002??
Homunculus        ?????   ?????   ?????   ?????   22122   01321   01112   11113   30??2   102??
Tarsius           54402   10332   0111?   31001   30303   02301   00110   12224   40020   002??

[char. #'s        12345   6789.   12345   6789.   12345   6789.   12345   6789.   12345   6789.]
[20 =======+      ......  ....1   ......  ....2   ......  ....3   ......  ....4   ......  ....5]
Hemiacodon        53322   10100   ?????   ?????   30313   ?2401   01???   0????   ?????   ?????
Apidium           523a2   20211   05441   2d0ba   0b225   12d11   aa?11   11b31   ?1??2   1ab??
   [18 could be a, 1 specimen]
Aegyptopithecus   53404   0?132   ?????   ?????   ?????   ?????   ?????   ??11    ?1??2   101??
```

```
[char. #'s          12345  6789.   12345  6789.    12345  6789.    12345  6789.    12345  6789.]
[23 =======+        ......  ....1  ......  .....2  ......  ....3   ......  ....4   ......  ....5]
Nasalis             63102  01202   02222  22010    00241  10301    00012  01102    2123?  203??
Presbytis           13102  00212   00002  21110    01241  10321    10012  10202    1111?  202??
Rhinopithecus       13113  30202   00001  21111    00231  10211    10012  11102    1103?  202??

[char. #'s          12345  6789.   12345  6789.    12345  6789.    12345  6789.    12345  6789.]
[26 =======+        ......  ....1  ......  .....2  ......  ....3   ......  ....4   ......  ....5]
Hylobates           13120  00302   10000  12011    00220  01220    11233  31103    12211  011??
Pan                 33121  02312   ?3322  20011    002ha  2ajba    10323  b1a04    22211  010??
Pliopithecus        63314  12232   10000  10011    ?0121  11110    11122  21123    21110  002??
Otolemur            144a1  10111   23522  20031    30002  02301    00211  31134    31120  101??
Lemur               14402  001a1   33322  22020    10012  02301    aa201  01124    42140  001??

[char. #'s          12345  6789.   12345  6789.    12345  6789.    12345  6789.    12345  6789.]
[31 =======+        ......  ....1  ......  .....2  ......  ....3   ......  ....4   ......  ....5]
Daubentonia         14411  00221   33420  00031    10000  02301    10202  11224    32020  101??
Avahi               00322  10110   33522  20022    10010  12201    01210  01124    30020  003??
Cheirogaleus        14423  00122   0130j  j0021    32011  02311    0021a  02034    40b21  01a??

[char. #'s          12345  6789.   12345  6789.    12345  6789.    12345  6789.    12345  6789.]
[34 =======+        ......  ....1  ......  .....2  ......  ....3   ......  ....4   ......  ....5]
Adapis              b3???  ???1?   ?????  ?????    ?001?  ?????    ?????  ???21    41??0  012??
Leptadapis          24131  001k1   ?????  ?????    ?????  ?????    ?????  ?????    ?????  ?????
Notharctus          b230b  10bk1   33521  1b0da    m0a1?  02201    0011b  1?b24    41120  002??
Smilodectes         ?????  ?????   ?????  ?????    00014  02201    00200  122??    ?????  ?????
   [Adapis 19:2, 28:b, all from Dagosto 1983]

[characters 51 on]
[char. #'s          12345  6789.   12345  6789.]
[ 0 =======+        ......  ....6  ......  ....7]
Alouatta            b21b
```

```
Aotus               2d11
Ateles              2b1a
Brachyteles         b1a1
Cacajao             221a

[char. #'s          12345  6789.    12345  6789.]
[ 6 =======+        ......  ....6   ......  ....7]
Callicebus          2211
Callimico           b211
Callithrix          2b11
Cebuella            2211
Cebus               2d11

[char. #'s          12345  6789.    12345  6789.]
[11 =======+        ......  ....6   ......  ....7]
Chiropotes          2211
Lagothrix           2b11
Leontopithecus      2b1a
Pithecia            2b11
Saguinus            2d1j

[char. #'s          12345  6789.    12345  6789.]
[16 =======+        ......  ....6   ......  ....7]
Saimiri             2d11
Cebupithecia        2211
Homunculus          ????
Tarsius             4?12

[char. #'s          12345  6789.    12345  6789.]
[20 =======+        ......  ....6   ......  ....7]
Hemiacodon          3?1?
Apidium             221b
Aegyptopithecus     ??12
```

```
[char. #'s          12345   6789.     12345   6789.]
[23 =======+        ......   ....6    ......   ....7]
Nasalis             2221
Presbytis           2121
Rhinopithecus       2121

[char. #'s          12345   6789.     12345   6789.]
[26 =======+        ......   ....6    ......   ....7]
Hylobates           2121
Pan                 0221
Pliopithecus        2310
Otolemur            4112
Lemur               2112

[char. #'s          12345   6789.     12345   6789.]
[31 =======+        ......   ....6    ......   ....7]
Daubentonia         2111
Avahi               2101
Cheirogaleus        3311

[char. #'s          12345   6789.     12345   6789.]
[34 =======+        ......   ....6    ......   ....7]
Adapis              1??2
Leptadapis          1112
Notharctus          1112
Smilodectes         ????
   ;
end;
   typeset MyFavorites = ord: all, a:23, b:25, c:1, e:2, f:5, g:6, h:3, j:12 13
                    14, k:33;
ANCSTATES ANC = 0:all, 1:2 9 10 19 24 25 28 33 34 35 36 37 38 39 42 47 51,
                      2:5 8 14 27 29 31 44 45 48, 3:4 12 13 41 51, 4:1;
                              [20-0?, 34-0?, 48-?];
end;
```

References

Alexander, J. P. 1992. Alas, poor *Notharctus*. *Nat. Hist.* **8/92:**54–58.

Andrews, P. 1985. Family group systematics and evolution among catarrhine primates. In: E. Delson (ed.), *Ancestors: the Hard Evidence,* pp. 14–22. Alan R. Liss, New York.

Anemone, R. L. 1990. The VCL hypothesis revisited: Patterns of femoral morphology among quadrupedal and saltatorial prosimian primates. *Am. J. Phys. Anthropol.* **83:**373–393.

Barrett, M., Donoghue, M. J., and Sober, E. 1991. Against consensus. *Syst. Zool.* **40:**486–493.

Barrett, M., Donoghue, M. J., and Sober, E. 1993. Crusade? A reply to Nelson. *Syst. Biol.* **42:**216–217.

Beard, K. C. 1990. Gliding behavior and paleoecology of the alleged primate family Paromomyidae (Mammalia, Dermoptera). *Nature* **345:**340–341.

Beard, K. C., and Wang, B. 1991. Phylogenetic and biogeographic significance of the tarsiiform primate *Asiomomys changbaicus* from the Eocene of Jilin Province, PRC. *Am. J. Phys. Anthropol.* **85:**159–166.

Benefit, B. R., and McCrossin, M. L. 1991. Ancestral facial morphology of Old World higher primates. *Proc. Natl. Acad. Sci. USA* **88:**5267–5271.

Bonis, L. de, Jaeger, J.-J., Coiffait, B., and Coiffait, P.-E. 1988. Découverte du plus ancien primate Catarrhinien connu dans l'Éocène supérieur d'Afrique du Nord. *C.R. Acad. Sci. Paris, [II]* **306:**929–934.

Chappill, J. A. 1989. Quantitative characters in phylogenetic analysis. *Cladistics* **5:**217–234.

Ciochon, R. L., Savage, D. E., Thaw, T., and Maw, B. 1985. Anthropoid origins in Asia? New discovery of *Amphipithecus* from the Eocene of Burma. *Science* **229:**756–759.

Corruccini, R. S. 1990. Review of *A Wonderful Life. Hum. Evol.* **5:**579–587.

Court, N. 1993. An enigmatic new mammal from the Eocene of North Africa. *J. Vert. Paleontol.* **13:**267–269.

Covert, H. H. 1986. Biology of early Cenozoic primates. In: D. R. Swindler and J. Erwin (eds.), *Comparative Primate Biology, Vol. 1: Systematics, Evolution, and Anatomy,* pp. 335–359. Alan R. Liss, New York.

Covert, H. H. 1988. Ankle and foot morphology of *Cantius mckennai:* Adaptations and phylogenetic implications. *J. Hum. Evol.* **17:**57–70.

Covert, H. H. and Hamrick, M. W. 1993. Description of new skeletal remains of the early Eocene anaptomorphine primate *Absarokius* (Omomyidae) and a discussion about its adaptive profile. *J. Hum. Evol.* **25:**351–362.

Culotta, E. 1992. A new take on anthropoid origins. *Science* **256:**1516–1517.

Dagosto, M. 1983. Postcranium of *Adapis parisiensis* and *Leptadapis magnus* (Adapiformes, Primates). *Fol. Primatol.* **41:**49–101.

Dagosto, M. 1985. The distal tibia of primates with special reference to the Omomyidae. *Int. J. Primatol.* **6:**45–75.

Dagosto, M. 1986. The joints of the tarsus in the strepsirhine primates: Functional, adaptive, and evolutionary implications. Ph.D. dissertation, City University of New York.

Dagosto, M. 1988. Implications of postcranial evidence for the origin of primates. *J. Hum. Evol.* **17:**35–56.

Dagosto, M. 1990. Models for the origin of the anthropoid postcranium. *J. Hum. Evol.* **19:**121–140.

Davis, L. C., Ford, S. M., and Garber, P. A. 1993. Functional anatomy and positional behavior in three *Saguinus* species. *Am. J. Phys. Anthropol. [Suppl.]* **16:**78–79.

Delson, E., and Rosenberger, A. L. 1980. Phyletic perspectives on platyrrhine origins and anthropoid relationships. In: R. L. Ciochon and A. B. Chiarelli (eds.), *Evolutionary Biology of the New World Monkeys and Continental Drift,* pp. 445–458. Plenum Press. New York.

Dickenson, C. 1993. The phylogeny of New World primates (Platyrrhini, Primates) assessed by high-resolution two-dimensional protein electrophoresis. Ph.D. dissertation, University of Chicago.

Farris, J. S. 1970. Methods of computing Wagner trees. *Syst. Zool.* **19:**83–92.

Farris, J. S. 1983. The logical basis of phylogenetic analysis. In: N. I. Platnick and V. A. Funk (eds.), *Advances in Cladistics, 2,* pp. 7–36. Columbia University Press, New York.

Farris, J. S. 1989. *Hennig86.* Port Jefferson Station, New York.

Farris, J. S. 1990. Phenetics in camouflage. *Cladistics* **6:**91–100.

Farris, J. S., Kluge, A. G., and Eckhart, M. J. 1970. A numerical approach to phylogenetic systematics. *Syst. Zool.* **19:**172–189.

Felsenstein, J. 1983. Parsimony in systematics: Biological and statistical issues. *Annu. Rev. Ecol. Syst.* **14:**313–333.

Felsenstein, J. 1985. Confidence limits on phylogenies: An approach using bootstrap. *Evolution* **39:**783–391.

Fleagle, J. G. 1977. Locomotor behavior and skeletal anatomy of sympatric leaf-monkeys (*Presbytis obscura* and *Presbytis melalophos*). *Am. J. Phys. Anthropol.* **20:**440–453.

Fleagle, J. G. 1988. *Primate Adaptation and Evolution.* Academic Press, San Diego.

Fleagle, J. G., and Kay, R. F. 1987. The phyletic position of the Parapithecidae. *J. Hum. Evol.* **16:**483–532.

Fleagle, J. G., and Simons, E. L. 1982. The humerus of *Aegyptopithecus zeuxis,* a primitive anthropoid. *Am. J. Phys. Anthropol.* **59:**175–193.

Fleagle, J. G., and Simons, E. L. 1983. The tibio-fibular articulation in *Apidium phiomense,* an Oligocene anthropoid. *Nature* **301:**238–239.

Ford, S. M. 1980a. A systematic revision of the Platyrrhini based on features of the postcranium. Ph.D. dissertation, University of Pittsburgh.

Ford, S. M. 1980b. Phylogenetic relationships of the Platyrrhini: The evidence of the femur. In: R. L. Ciochon and A. B. Chiarelli (eds.), *Evolutionary Biology of the New World Monkeys and Continental Drift,* pp. 317–330. Plenum Press, New York.

Ford, S. M. 1986. Systematics of New World monkeys. In: D. R. Swindler and J. Erwin (eds.), *Comparative Primate Biology Vol. 1: Systematics, Evolution, and Anatomy,* pp. 73–135. Alan R. Liss, New York.

Ford, S. M. 1988. Postcranial adaptations of the earliest platyrrhine. *J. Hum. Evol.* **17:**155–192.

Ford, S. M. 1990. Locomotor adaptations of fossil platyrrhines. *J. Hum. Evol.* **19:**141–174.

Ford, S. M., and Davis, L. C. 1992.Systematics and body size: implications for feeding adaptations in New World monkeys. *Am. J. Phys. Anthropol.* **88:**415–468.

Ford, S. M., and Hobbs, D. G. 1994. Species differentiation in the postcranial skeleton of *Cebus. Am. J. Phys. Anthropol. [Suppl.]* **18:**88.

Ford, S. M., Davis, L. C., and Kay. R. F. 1991. New platyrrhine astragalus from the Miocene of Colombia: *Am. J. Phys. Anthropol. [Suppl.]* **12:**73–74.

Franzen, J. L. 1987. Ein neuer Primate aus dem Mitteleozän der Grube Messel (Deutschland, S.-Hessen). *Cour. Forsch. Inst. Senckenberg* **91:**151–187.

Garber, P. A., and Preutz, J. D. 1993. Effect of forest structure on positional behavior in moustached tamarin monkeys. *Am. J. Phys. Anthropol. [Suppl.]* **16:**91.

Gebo, D. L. 1986. The anatomy of the prosimian foot and its application to the primate fossil record. Ph.D. dissertation, Duke University.

Gebo, D. L. 1987. Anthropoid origins—the foot evidence. *J. Hum. Evol.* **15:**421–430.

Gebo, D. L. 1988. Foot morphology and locomotor adaptation in Eocene primates. *Fol. Primatol.* **50:**3–41.

Gebo, D. L., and Simons, E. L. 1987. Morphology and locomotor adaptations of the foot in early Oligocene anthropoids. *Am. J. Phys. Anthropol.* **74:**83–101.

Gebo, D. L., Dagosto, M., and Rose, K. D. 1991. Foot morphology and evolution in early Eocene *Cantius. Am. J. Phys. Anthropol.* **86:**51–73.

Gidley, J. W. 1923. Paleocene primates of the Fort Union, with discussion of relationships of Eocene primates. *Proc. U.S. Nat. Mus.* **63:**1–38.

Gingerich, P. D. 1980. Eocene Adapidae, paleobiogeography and the origin of the South American Platyrrhini. In: R. L. Ciochon and A. B. Chiarelli (eds.), *Evolutionary Biology of the New World Monkeys and Continental Drift,* pp. 123–138. Plenum Press, New York.

Gingerich, P. D. 1984. Primate evolution: Evidence from the fossil record, comparative morphology, and molecular biology. *Yearb. Phys. Anthropol.* **27**:57–72.

Godinot, M. 1991. Toward the locomotion of two contemporaneous *Adapis* species. *Z. Morph. Anthrop.* **78**:387–405.

Godinot, M. 1992. Early primate hands in evolutionary perspective. *J. Hum. Evol.* **22**:267–283.

Godinot, M., and Beard, K. C. 1991. Fossil primate hands: A review and an evolutionary inquiry emphasizing early forms. *Hum. Evol.* **6**:307–351.

Godinot, M., and Jouffroy, F. K. 1984. La main d'*Adapis*. In: E. Buffetaut, J.-M. Mazin, and E. Salmon (eds.), *Actes du Symposium Paléontologique D. Cuvier*, pp. 221–241. Le Serpentaire, Montbéliard.

Godinot, M., and Mahboubi, M. 1992. Earliest known simian primate found in Algeria. *Nature* **357**:324–326.

Godthelp, H., Archer, M., Cifelli, R., Hand, S. J., and Gilkeson, C. F. 1992. Earliest known Australian Tertiary mammal fauna. *Nature* **356**:514–516.

Gould, S. J. 1989. *A Wonderful Life: The Burgess Shale and the Nature of History.* W. W. Norton, New York.

Gregory, W. K. 1920. On the structure and relations of *Notharctus*, an American Eocene primate. *Mem. Am. Mus. Nat. Hist. N.S.* **3**:51–243.

Groves, C. P. 1991. *A Theory of Human and Primate Evolution.* Clarendon Press, Oxford.

Harrison, T. 1987. The phylogenetic relationships of the early catarrhine primates: A review of the current evidence. *J. Hum. Evol.* **16**:41–80.

Hecht, M. K., and Edwards, J. L. 1977. The methodology of phylogenetic inference above the species level. In: M. K. Hecht, P. C. Goody, and B. M. Hecht (eds.), *Major Patterns of Vertebrate Evolution*, pp. 3–51. Plenum Press, New York.

Hershkovitz, P. 1974. A new genus of late Oligocene monkey (Cebidae: Platyrrhini) with notes on postorbital closure and platyrrhine evolution. *Fol. Primatol.* **21**:1–35.

Hershkovitz, P. 1987. The taxonomy of South American sakis, genus *Pithecia* (Cebidae, Platyrrhini): A preliminary report and critical review with the description of a new species and subspecies. *Am. J. Primatol.* **12**:387–468.

Hershkovitz, P. 1988. The subfossil monkey femur and subfossil monkey tibia of the Antilles: A review. *Int. J. Primatol.* **9**:365–384.

Hoffstetter, R. 1974a. Phylogeny and geographic deployment of the Primates. *J. Hum. Evol.* **3**:327–350.

Hoffstetter, R. 1974b. *Apidium* et l'origine des Simiiformes (= Anthropoidea). *C.R. Acad. Sci. Paris. [D]* **278**:1715–1717.

Hoffstetter, R. 1980. Origin and deployment of New World monkeys emphasizing the southern continents route. In: R. L. Ciochon and A. B. Chiarelli (eds.), *Evolutionary Biology of the New World Monkeys and Continental Drift*, pp. 103–138. Plenum Press, New York.

Jouffroy, F. K., Godinot, M., and Nakano, Y. 1991. Biometrical characteristics of primate hands. *Hum. Evol.* **6**:269–306.

Kay, R. F., and Williams, B. B. 1992. Dental evidence for anthropoid origins. *Am. J. Phys. Anthrop. [Suppl.]* **14**:98.

Kay, R. F., Thorington, R. W., Jr., and Houde, P. 1990. Eocene plesiadapiform shows affinities with flying lemurs not primates. *Nature* **345**:342–344.

Kay, R. F., Thewissen, J. G. M., and Yoder, A. D. 1992. Cranial anatomy to *Ignacius graybullianus* and the affinities of the Plesiadapiformes. *Am. J. Phys. Anthropol.* **89**:477–498.

Mabee, P. M., and Humphries, J. 1993. Coding polymorphic data: Examples from allozymes and ontogeny. *Syst. Biol.* **42**(2):166–181.

MacFadden, B. J. 1990. Chronology of Cenozoic primate localities in South America. *J. Hum. Evol.* **19**:7–21.

MacPhee, R. D. E., and Cartmill, M. 1986. Basicranial structures and primate systematics. In: D. R. Swindler and J. Erwin (eds.), *Comparative Primate Biology, Vol. 1: Systematics, Evolution, and Anatomy*, pp. 219–275. Alan R. Liss, New York.

Maddison, D. R. 1991. The discovery and importance of multiple islands of most-parsimonious trees. *Syst. Zool.* **40:**315–328.

Maddison, W. P., and Maddison, D. R. 1993. *MacClade: Analysis of Phylogeny and Character Evolution, Version 3.* Sinauer Associates, Sunderland, MA.

Maddison, W. P., Donoghue, M., and Maddison, D. R. 1984. Outgroup analysis and parsimony. *Syst. Zool.* **33:**83–103.

Martin, R. D. 1990. *Primate Origins and Evolution. A Phylogenetic Reconstruction.* Princeton University Press, Princeton.

Martin, R. D. 1993. Primate origins: Plugging the gaps. *Nature* **363:**223–234.

Mickevich, M. F. 1982. Transformation series analysis. *Syst. Zool.* **31:**461–478.

Mickevich, M. F., and Weller, S. J. 1990. Evolutionary character analysis: Tracing character change on a cladogram. *Cladistics.* **6:**137–170.

Napier, J. R., and Walker, A. C. 1967. Vertical clinging and leaping—a newly recognized category of primate locomotion. *Fol. Primatol.* **6:**204–219.

Neff, N. A. 1986. A rational basis for *a priori* character weighting. *Syst. Zool.* **35:**110–123.

Nelson, G. 1993. Why crusade against consensus? A reply to Barrett, Donoghue, and Sober. *Syst. Biol.* **42:**215–216.

Nisbett, R. A. 1993. The functional ecology of howling monkey positional behavior in the lowland Pacific dry forests of Costa Rica. *Am. J. Phys. Anthropol. [Suppl.]* **16:**151.

Pickford, M. 1986. Première découverte d'une faune mammalienne terrestre paléogène d'Afrique sub-saharienne. *C.R. Acad. Sci. Paris [II]* **302**(19)**:**1205–1210.

Pogue, M. G., and Mickevich, M. F. 1990. Character definitions and character state delineation: The *bête noire* of phylogenetic inference. *Cladistics* **6:**319–361.

Rasmussen, D. T. 1986. Anthropoid origins: A possible solution to the Adapidae–Omomyidae paradox. *J. Hum. Evol.* **15:**1–12.

Rasmussen, D. T. 1990. The phylogenetic position of *Mahgarita stevensi:* Protoanthropoid or lemuroid? *Int. J. Primatol.* **11:**439–469.

Rasmussen, D. T., and Simons, E. L. 1992. Paleobiology of the oligopithecines, the earliest known anthropoid primates. *Int. J. Primatol.* **13:**477–508.

Rodman, P. S. 1979. Skeletal differentiation of *Macaca fascicularis* and *Macaca nemestrina* in relation to arboreal and terrestrial quadrupedalism. *Am. J. Phys. Anthropol.* **51:**51–62.

Rosenberger, A. L., and Dagosto, M. 1992. New craniodental and postcranial evidence of fossil tarsiiforms. In: S. Matano, R. H. Tuttle, H. Ishida, and M. Goodman (eds.), *Topics in Primatology, Vol. 3: Evolutionary Biology, Reproductive Endocrinology, and Virology,* pp. 37–51. University of Tokyo Press, Tokyo.

Rosenberger, A. L., and Szalay, F. S. 1980. The tarsiiform origins of Anthropoidea. In: R. L. Ciochon and A. B. Chiarelli (eds.), *Evolutionary Biology of the New World Monkeys and Continental Drift,* pp. 139–157. Plenum Press, New York.

Senut, B., and Thomas, H. 1992. First discoveries of anthropoid postcranial remains from Taqah (early Oligocene, Sultinate of Oman). In: *Abstracts XIV Congress of the International Primatology Society,* Strasbourg, France.

Sigé, B., Jaeger, J. J., Sudre, J., and Vianey-Liaud, M. 1990. *Altiatlasius koulchii* n. gen. et sp., primate Omomyide du Paléocène Supérieur du Maroc, et les origines des primates. *Palaeontograhica [A]* **214:**1–56.

Simons, E. L. 1987. New faces of *Aegyptopithecus* from the Oligocene of Egypt. *J. Hum. Evol.* **16:**273–289.

Simons, E. L. 1990. Discovery of the oldest known anthropoidean skull from the Paleogene of Egypt. *Science* **247:**1567–1569.

Simons, E. L. 1992. Diversity in the early Tertiary anthropoidean radiation in Africa. *Proc. Natl. Acad. Sci. USA* **89:**10743–10747.

Simons, E. L., and Rasmussen, D. T. 1991. The generic classification of Fayum Anthropoidea. *Int. J. Primatol.* **12:**163–178.

Sokal, R., and Rohlf, F. 1981. *Biometry. Second Edition.* W. H. Freeman, New York.

Strasser, E., and Delson, E. 1987. Cladistic analysis of cercopithecid relationships. *J. Hum. Evol.* **16:**81–99.

Swofford, D. L. 1991a. *PAUP: Phylogenetic Analysis Using Parsimony, Version 3.0s.* Computer Program distributed by the Illinois Natural History Survey, Champaign, Illinois.

Swofford, D. L. 1991b. When are phylogeny estimates from molecular and morphological data incongruent? In: M. M. Miyamoto and J. Cracraft (eds.), *Phylogenetic Analysis of DNA Sequences,* pp. 295–333. Oxford University Press, New York.

Swofford, D. L., and Olsen, G. J. 1990. Phylogeny reconstruction. In: D. M. Hillis and C. M. Moritz (eds.), *Molecular Systematics,* pp. 411–501. Sinauer Associates, Sunderland, MA.

Szalay, F. S. 1975. Phylogeny of primate higher taxa: The basicranial evidence. In: W. P. Luckett and F. S. Szalay (eds.), *Phylogeny of the Primates,* pp. 91–125. Plenum Press, New York.

Szalay, F. S. 1976. Systematics of the Omomyidae (Tarsiiformes, Primates): Taxonomy, phylogeny, and adaptations. *Bull. Amer. Mus. Nat. Hist.,* **156**(3)**:**157–450.

Szalay, F. S., and Dagosto, M. 1980. Locomotor adaptations as reflected on the humerus of Paleogene primates. *Fol. Primatol.* **34:**1–45.

Szalay, F. S., Rosenberger, A. L., and Dagosto, M. 1987. Diagnosis and differentiation of the order Primates. *Yearb. Phys. Anthropol.* **30:**75–105.

Thalmann, U., Haubold, H., and Martin, R. D. 1989. *Pronycticebus neglectus*—an almost complete adapid primate specimen from the Geiseltal (GDR). *Palaeovertebrata* **19:**115–130.

Thomas, H., Roger, J., Sen, S., and Al-Sulaimani, Z. 1988. Découverte des plus anciens ⟨⟨Anthropoïdes⟩⟩ du continent arabo-africain et d'un Primate tarsiiforme dans l'Oligocène du Sultanat d'Oman. *C.R. Acad. Sci. Paris [II]* **306:**823–829.

Thomas, H., Roger, J., Sen, S., Bourdillon-de-Grissac, C., and Al-Sulaimani, Z. 1989. Découverte de Vertébrés fossiles dans l'Oligocène inférieur de Dhofar (Sultanat d'Oman). *Geobios* **22:**101–120.

Ward, S. C., and Sussman, R. W. 1979. Correlates between locomotor anatomy and behavior in two sympatric species of *Lemur. Am. J. Phys. Anthropol.* **50:**575–590.

Watrous, L., and Wheeler, Q. D. 1981. The outgroup comparison method of character analysis. *Syst. Zool.* **35:**102–109.

Wiley, E. O., Siegel-Causey, D., Brooks, D. R., and Funk, V. A. 1991. *The Compleat Cladist: A Primer of Phylogenetic Procedures.* University of Kansas Museum of Natural History, Special Publication 19, pp. 1–158.

Wortman, J. L. 1904. Studies of Eocene Mammalia in the Marsh collection, Peabody Museum. Part II. Primates, suborder Anthropoidea. *Am. J. Sci. [ser. 4]* **17:**204–250.

Anthropoid Origins

Past, Present, and Future

19

JOHN G. FLEAGLE and
RICHARD F. KAY

Introduction

The papers collected in this volume summarize the state of our knowledge about the origins of Anthropoidea—monkeys, apes, and humans. Considering how important this group is for understanding human evolution, and the enormous literature that has grown up around the questions of where anthropoids may have originated and from what group, there is remarkably little consensus on any aspect of this topic. As will be seen in the following review, there are proponents for an African, Asian, or, ultimately, North American or European origin for anthropoids. And different scientists believe the closest relatives of anthropoids to be the extinct Eocene groups Omomyidae or Adapidae or living *Tarsius* from South Asia. In this final chapter we try to pull together the major arguments put forth in earlier chapters, to summarize what we see as the major points of agreement and disagreement over anthropoid origins, to identify major gaps or confusing lines of evidence in current knowledge, and finally to lay out what we see as the most fruitful directions for future research in this area. As we hope to make clear, just because the problem of anthropoid origins remains unresolved, it is not from a lack of

JOHN G. FLEAGLE • Department of Anatomical Sciences, School of Medicine, Health Sciences Center, State University of New York, Stony Brook, New York 11794. RICHARD F. KAY • Department of Biological Anthropology and Anatomy, Duke University, Durham, North Carolina 27710.

Anthropoid Origins, edited by John G. Fleagle and Richard F. Kay. Plenum Press, New York, 1994.

data relevant to the issues being addressed. Quite the contrary, we know a lot more about this topic than we did a decade ago, or even 5 years ago. Indeed, as is often the case in paleontology, the wealth of new information has complicated the problem by adding even more characters to the script and confounded what in retrospect was a very simplified list of alternatives. In addressing anthropoid origins, we will begin with the present.

Anthropoids and Haplorhines Today

What are anthropoids? One issue about which there is general agreement among most current authorities is that anthropoids—New World monkeys, Old World monkeys, and hominoids—are a distinctive natural group that stands apart form other taxa of primates on the basis of numerous derived features. Biomolecular studies (e.g., Miyamoto and Goodman, 1990) invariably find anthropoids to be a distinctive clade among primates, and there are numerous soft anatomic features of the placenta that separate anthropoids from other primates (e.g., Luckett, 1975). In addition, there are many osteological and craniodental characteristics of anthropoids that distinguish this group and potentially can be used to identify fossil anthropoids and track the evolution of the group in the fossil record.

There is also little debate over the basic outline of anthropoid phylogeny. Living anthropoids clearly can be divided into two main groups—New World monkeys (platyrrhines) and Old World higher primates (catarrhines); the latter are further divided into Old World monkeys (cercopithecoids) and apes and humans (hominoids). Aside from occasional suggestions that some features shared by platyrrhines and catarrhines may have been acquired independently and a persistent difficulty in identifying any derived featue to unite all platyrrhines, there is near universal acceptance of this scheme.

Moreover, there is little doubt among students of extant mammals, at least, that the living sister taxon of anthropoids is the genus *Tarsius* (Cartmill 1980, Chapter 16, this volume; Martin 1990; Ross, Chapter 15, this volume). The consensus from molecular phylogeny is that *Tarsius* is the sister taxon of anthropoids (Miyamoto and Goodman, 1990), and to these may be added metabolic similarities and numerous soft anatomic features, including loss of a cleft naked rhinarium and details of placentation, and many features of the anatomy of the eye, including loss of a tapetum and presence of a retinal fovea, that unite tarsiers and anthropoids. In the cranium, tarsiers share with anthropoids a simplified internal carotid artery (part of the blood supply to the brain) lacking the stapedial arterial branch, an anterior accessory air-filled cavity of the middle ear, fusion of the metopic suture between the frontal bones, a reduced nasal turbinate system and absence of the sphenoethmoid recess, an apical interorbital septum, and substantial degree of postorbital closure (Cartmill 1980, Chapter 16; Ross, Chapter 15, this volume). On the basis of these similarities, tarsiers and anthropoids are commonly joined in the infraorder Haplorhini (Pocock, 1918; Martin, 1990). In the face of all these shared-derived features uniting tarsiers and anthropoids, it is perhaps surprising that there is any doubt at all about the origin of anthropoids. Indeed, why have more workers have not simply placed tarsiers among anthropoids and focused on the issue of haplorhine origins?

The Problem

The confusion surrounding anthropoid origins comes from two sources. First, there is the specialized anatomy of tarsiers themeselves. Although tarsiers indeed share a large number of presumably derived features with anthropoids, in many other anatomic features, the extant genus *Tarsius* is a very odd primate by any measure. It has a reduced number of lower incisors, canine-like upper central incisors, very simple premolars, and molars with extraordinarily well-developed cusps, especially the paraconids. The cranium has a number of striking features, some probably associated with the gigantic orbits, including overlap of pterygoid plates with the petrosal bulla, a ventrolateral entrance of the internal carotid artery into the middle ear, and a gutter-like glenoid fossa, that are unknown in any other extant primate. In the postcranial skeleton, tarsiers have extraordinarily large hands, a fused tibia and fibula, and extremely long tarsal bones in the ankle that gave rise to the generic name. In addition, tarsier chromosomes are unusual among primates, and there is continued debate over the phylogenetic signifiance of the tarsier karyotype (Dutrillaux and Rumpler, 1988). Obviously, these striking anatomic specializations preclude a primate morphologically similar to *Tarsius* from any place in the direct ancestry of later, more generalized anthropoids and may well indicate a very early divergence of this genus from all other living primates. Still, this should not obscure the fact of the numerous and trenchant shared-derived similarities with anthropoids. They simply show that *Tarsius* has accumulated many of its own derived features in the long period since it separated from the anthropoid stem.

The evidence that provides a real challenge to the status of tarsiers as the sister taxon of anthropoids is the information provided by the fossil record that expands our knowledge of both anthropoids and fossil "prosimians." The fossil evidence that bears directly on the issue of anthropoid origins concerns three separate issues: the evolutionary record of tarsier-like primates, the fossil record and nature of other "fossil prosimians," and the anatomic structure of early anthropoids. It is not unfair to say that as presently known and generally interpreted, all of this evidence tends to weaken rather than strengthen the case for *Tarsius* as the sister taxon of anthropoids (Cartmill, Chapter 16, this volume). At the very least, it has become increasingly more difficult to draw the boundary between anthropoids and nonanthropoid primates. Thus, before we address current views of anthropoid origins, we must first summarize recent advances and current interpretations concerning primate origins and the early evolution of prosimians (nonanthropoids) as well as new discoveries of early anthropoids.

Primate Origins

Until most recently, any discussion of anthropoid origins involved consideration of three groups of early Tertiary mammals known primarily from North America and Europe: the predominantly Paleocene Plesiadapiformes and two groups of mainly Eocene prosimians—the "lemur-like" adapids and the "tarsier-like" omomyids. In the last decade, however, it has become almost uniformly accepted that only the latter two groups can realistically be considered primates in any phylogenetic sense. Adapids and omomyids share many derived features with living primates and may well be in the direct ancestry of most later taxa (e.g., Wible and Covert, 1987). In contrast, plesiadapiforms are, at best, the most primate-like of known Paleocene mammals. There are very few, if any, shared features to link them only with primates to the exclusion of other mammalian orders such as Dermoptera (flying lemurs) or Scandentia (tree shrews), and all known plesiadapiforms are either too specialized to be ancestral to any later primates or, in the case of *Purgatorius,* too generalized or poorly known to share any particular features with primates to the exclusion of many other orders. Moreover, not only do many plesiadapiforms seem to show more convincing shared features with other orders, such as Dermoptera (e.g., Kay *et al.,* 1992), there are other mammalian orders, including tree shrews and megabats, that have been purported to share equally convincing shared features with primates (Wible and Covert, 1987). At this time, all that seems certain is that primates, plesiadapiforms, tree shrews, and bats seem to share a common ancestry relative to other orders, but the branching sequence among these groups, usually grouped in the superorder Archonta, is unresolved (see MacPhee, 1993; Simmons *et al.,* 1991). The nomenclatorial implication of this is that the term Primates equates to "primates of modern aspect," *sensu* Simons (1972). There is no longer any need to use the term "Euprimates" for this group since Euprimates = Primates.

Fossil Prosimians

With the elimination of plesiadapiforms from consideration as the unique primate ancestor, the early Eocene adapids and omomyids are both the earliest prosimians and the earliest primates (Rose *et al.,* Chapter 1, this volume). These groups appear rather suddenly in North America and Europe at the beginning of the Eocene epoch and have no clearly documented antecedents on other continents (but see Gingerich *et al.*, 1991). They differ from other mammals and resemble later primates in virtually all aspects of their cranial and skeletal anatomy, including diagnostic features of the auditory region (e.g., MacPhee and Cartmill, 1986; Dagosto, 1988, 1993; Dagosto and Gebo, Chapter 17, this volume). Adaptively they share with living prosimians dental features suggestive of more frugivorous–faunivorous or frugivorous–folivorous habits (Williams and Covert, 1994; Strait, 1991), reduced olfactory lobes of the brain, reduced tactile sensitivity of the snout, and enlarged, convergent eyes with a postorbital bar, all associated with an important shift in the mode by which they must have perceived their environment. They resemble later primates in their hands and feet: the cheridea appear to have had enlarged tactile pads and nails rather than claws (e.g., Godinot, 1992), and they had an opposible big toe, suggesting a specialized form of arboreality (Cartmill, 1992). Ecologically, early prosimians seem to replace some of the disappearing plesiadapiforms (Maas *et al.*, 1988), but their geographic and phylogenetic origins are obscure. Not surprisingly, current hypotheses look to more poorly known continents as the ultimate source areas, either India, which was beginning to dock with Asia after a long period of isolation (Krause and Maas, 1990), Africa (Sigé *et al.*, 1990), or Asia (Gingerich *et al.*, 1991). Although both adapids and omomyids share many general features with later primates, and there is increasing evidence of remarkable dental similarity among the earliest members of each (Rose and Bown, 1991; Rose *et al.*, Chapter 1, this volume), the two groups seem to be clearly distinct throughout their known records.

Adapids

As a group, the adapids are larger, probably mostly diurnal, less faunivorous, and more folivrous than the omomyids. In dental anatomy, primitive adapids may have closely resembled the primate common ancestor in some respects; many have a complete dental formula with four premolars; they usually have small incisors and larger, sometimes sexually dimorphic, canines. Some omomyids may also possess these primitive dental features (Rose *et al.*, Chapter 7, this volume). Many advanced adapids have a fused mandibular symphysis, a feature aliging them with living anthropoids. Adapids have traditionally been divided into two geographically distinct groups, the North American notharctines and the European adapines [not considering the late Miocene sivaladapines (Gingerich and Sahni, 1984)]. However, most current authorities recognize three distinct groups (at either family or subfamily levels) among the Eocene adapids: the mostly North America notharctines and two groups among the European genera—the earlier, more diverse cercamoniines (Rose *et al.*, Chaper 1, this volume) or protoadapines of Franzen (1987), and the late, abundant, but less speciose adapines. It is also evident that the geographic distinctions are not exact—the earliest notharctine is also found in Europe, and there is a late Eocene cercamoniine from North America (Wilson and Szalay, 1976) and a recently described species from Asia (Beard *et al.,* 1994).

The North American notharctines are a very abundant and relatively well-known group of fossil primates including a large number of time-successive species (Gingerich and Simons, 1977). They flourished in earliest Eocene but are unknown after the middle Eocene. They are thought to preserve the ancestral morphology for the adapid radiation.

Like notharctines, cercamoniines appear in the earliest Eocene, but unlike the notharctines, they are considerably more generically diverse and lasted to the end of the Eocene (e.g., Gingerich, 1977; Franzen, 1987). Most taxa are known only from teeth—a few from skulls and limbs. Like notharctines, with whom they are frequently grouped, they seem to retain many features that are primitive for both adapids and primates as a group. In addition to the European genera, there is one cercomoniine (*Mahgarita*) from the late Eocene of Texas (Wilson and Szalay, 1976), several poorly known species from the Eocene of Pakistan (Russell and Gingerich, 1987), and one new, unnamed species from China (Beard *et al.,* 1994). Several new primates have been described from North Africa that have cercamoniine affinities. These include *Djebelemur* from Tunisia (Hartenberger and Marandat, 1992), new fossils from Oman (Gheerbrandt *et al.,* 1994), and a new species from the late Eocene of the Fayum.

The better-known adapines appear later in Europe and are in most aspects of cranial and skeletal anatomy (except dental formula) more specialized than the cercamoniines or notharctines (Dagosto, 1983; Franzen, 1987; Godinot, 1984). They are also known from the middle Eocene of China (Beard *et al.,* 1994).

In addition to the groups described above and the late Miocene sivaladapines from Asia, there are various African and Asian genera that are frequently linked with adapids but not clearly with any of the accepted subfamilies (e.g., Sudre, 1975).

Although adapids as a group show striking phentic similarities to extant prosimians in both craniodental and postcranial anatomy (see Gregory, 1920; Tattersall and Schwartz, 1985; Gingerich and Martin, 1981; Rose and Walker, 1985), attempts to find unique shared-derived features linking modern strepsirhines with either adapids as a whole or specific lineages have yielded remarkably little detailed evidence. For example, the toothcomb characteristic of all lemurs and lorises was not present in any known Eocene adapid, and many features they share in common, such as enlarged stapedial branch of the internal carotid artery or ring-like tympanic bone "within" the middle ear cavity, are commonly regarded as the primitive condition for primates or difficult to interpret precisely in most fossils. Indeed, the best evidence linking adapids uniquely with strepsirhines (and presumably farther from anthropoid ancestry than *Tarsius*) are lateral flare on the talus and several other ankle features (found in all adapids) and details of the wrist of *Adapis* but not notharctines (Beard *et al.*, 1988; Beard and Godinot, 1988). However, these features have been disputed.

Some authors note dental features that seem to link cercamoniine adapids as a group with anthropoids—fused mandibular symphysis, small, erect incisors, a projecting canine, the arrangement of molar talonid cusps, etc. (e.g., Gingerich, 1980). This view finds support among several of the contributors to this volume, especially Gingerich, Franzen, Rasmussen, and Simons. There have also been attempts to link specific adapid taxa with anthropoids. Thus, Rasmussen (1990) argued that the cercamoniine *Mahgarita* shows many similarities uniquely related to oligoithecine anthropoids and has further suggested that some European cercamoniines known only from dental remains and currently recognized as cercamoniines may actually turn out to be anthropoids when more of their anatomy becomes known. Gingerich has argued that *Cercamonius* itself was closely related to anthropoids.

Omomyids

Like adapids, omomyids are known primarily from North America and Europe, where they appear in the earliest Eocene. Compared with adapids, omomyids are smaller animals, had larger brains, and contain many more faunivorous–frugivorous and probably nocturnal species (Beard *et al.*, 1991; Covert and Williams, Chapter 2, this volume; Gingerich, 1980; Strait, 1991; Williams and Covert, 1994). Many omomyids are derived in having large procumbent incisors; reduction of other anterior teeth also is common. Others have small vertically implanted incisors and an unreduced premolar number (Kay and Williams, Chapter 13; Rose *et al.*, Chapter 1, this volume; Williams, 1994). Cranially and skeletally, they seem to share some features with other Eocene prosimians and some with tarsiers and anthropoids (Dagosto, 1985; Szalay and Dagosto, 1980; Beard *et al.*, 1988). As in adapids, there are different but not totally distinct radiations of omomyids in North America and Europe, with a few poorly known taxa from other continents.

North American omomyids appear to have been much more diverse than their adapid contemporaries both adaptively as determined by body size and reconstructed dietary habits (Strait, 1991; Williams, 1994; Williams and Covert, 1994) and also in number of genera with more sympatric and synchronic genera in faunas containing only one or two adapids. Traditionally, they have been divided into two subfamilies, the more primitive, smaller, and more faunivorous anaptomorphines and the more advanced, more frugivorous, and often larger omomyines, which also survive much later. Several genera of North American omomyids (*Tetonius, Shoshonius*) have been shown to have striking cranial similarities to extant *Tarsius* (Beard *et al.*, 1991; Beard and MacPhee, Chapter 3, this volume). However, recent studies have demonstated that both the phylogenetic and morphological view of these two subfamilies are overly simplistic (Rose and Bown, 1991, Bown and Rose, 1987, 1991; Williams and Covert, 1994; Williams, 1994). Some cranial and dental similarities link some omomyids to *Tarsius* (e.g., Beard *et al.*, 1991; Gingerich, 1981; Kay and Williams, Chapter 13; Ross, Chapter 15, this volume). Several recent studies have also noted that this is a very diverse group and that several taxa show dental proportions and cranial features more comparable to those of adapines and/or anthropoids (Rosenberger 1986; Covert and Williams 1991; Beard and MacPhee, Chapter 3, this volume).

European omomyids, the microchoerines, are much less diverse both taxonomically and adaptively. There are only four genera, probably derived from an anaptomorphine ancestry. Two taxa, *Necrolemur* and *Pseudoloris*, show cranial, dental, and postcranial similarities to tarsiers (Simons and Russell, 1960; Rosenberger, 1985; Beard and MacPhee, Chapter 3, this volume).

There are a number of poorly known taxa from Africa that have been identified as omomyids, including some that precede the appearance of omomyids in Europe and North America (Sigé *et al.*, 1990; Gingerich, 1990; Simons *et al.*, 1986). However, the affinities of these have been debated. Recently, Beard and colleagues have reported the occurrence of midle Eocene Chinese omomyids with affinities to late Eocene North America taxan *Stockia* and *Macrotarsius*. (Beard *et al.*, 1994). This further supports the view that there were connections between these two continents later in the Eocene. The affinities of earlier possible omomyids from Asia are debated (Rose *et al.*, Chapter 1, this volume; Szalay and Li, 1986).

Just as adapids traditionally have been allied with strepsirhines because of general phenetic similarties, omomyids have traditionally been linked with the genus *Tarsius* on the basis of their small size, large orbits, and elongate tarsals (Gingerich, 1981). Similarly, just as the adapid–strepsirhine link has been subject to more careful phylogenetic scrutiny and has come to be supported cladistically (e.g., Beard *et al.*, 1988, so has the omomyid–tarsier link. This association between the tarsiers and omomyids has been scrutinized and supported by the cladistic analyses of Kay and Williams (Chapter 13, this volume) on the dental evidence, Ross (Chapter 15, this volume) and Beard and MacPhee (Chapter 3, this volume) on the cranial evidence, and Dagosto and Gebo (Chapter 17, this volume) from the postcranial anatomy.

On the other hand, there are only a few features that seem to link omomyids as a group with both tarsiers and anthropoids—shapes of the ankle and elbow (e.g., Beard *et al.*, 1988), cranial features such as the apical interorbital septum (Ross, Chapter 15; Beard and MacPhee, Chapter 3, this volume), and a few morphologically labile dental features mentioned by Kay and Williams (Chapter 13, this volume). There have also been attempts to link specific omomyid taxa with tarsiers or anthropoids. Thus, Simons and Russell (1960) and Rosenberger (1985; also Rosenberger and Dagosto 1992) argued that the microchoerine *Necrolemur* was uniquely related to *Tarsius* independent of other taxa. Most recently, Beard *et al.* (1991; also Beard and MacPhee, Chapter 3, this volume) have argued that *Shoshonius*, *Tetonius*, and *Necrolemur* are all uniquely related to *Tarsius* on cranial and basicranial anatomy. Most significant for understanding anthropoid origins is that in many of these studies different omomyid taxa show different similarities with tarsiers, and the morphological features they share with tarsiers are not features that tarsiers share with anthropoids. For example, *Shoshonius* lacks postorbital closure, a feature present in both *Tarsius* and anthropoids (Beard *et al.*, 1991). Therefore, if *Shoshonius* is the sister taxon to *Tarsius* but not anthropoids, as Beard *et al.* claim, then postorbital closure cannot be a shared-derived character of *Tarsius* and anthropoids.

Some authors have also identified similarities between specific omomyid taxa and anthropoids. Following Simons (1961), Rosenberger (Rosenberger and Szalay, 1980; Rosenberger and Dagosto, 1992) has noted that *Ourayia* has relatively broad, spatulate upper incisors. Covert and Williams (1991) noted that *Washakius* had similar-sized lower incisors rather than an enlarged central incisor as in most omomyids, and Beard and MacPhee (Chapter 3, this volume) found basicranial similarities between *Rooneyia* and anthropoids. Most of these authors used these similariies to argue that omomyids were a relatively diverse group of primates from which anthropoids could have arisen.

Early Anthropoids

The fossil record of early anthropoids has increased dramatically in the last decade (Fig. 1), with new finds from the Fayum, Egypt, and many new sites in Oman and North Africa (e.g., Rasmussen and Simons, 1992; Simons, 1992; Thomas *et al.*, 1989; Godinot, Chapter 10, this volume) and Asia (Beard *et al.*, 1994). Early anthropoids were first found in the Fayum of Egypt in the early part of this century, and because of the large number of specimens and completeness of much of the material, the Fayum primate fauna still provides

Fig. 1. Fossil localities yielding early anthropoids.

the baseline for evaluating other early anthropoid finds. The exact ages of the primate-bearing desposits from the Fayum remain a source of debate, but all probably date from no older than late Eocene to no younger than early Oligocene, or between 35 and 31 MA (Rasmussen *et al.*, 1992; Kappelman, 1992; Gingerich, 1993). Fayum anthropoids are currently placed in three major groups—propliopithecines, oligopithecines, and parapithecoids (Simons *et al.*, Chapter 8, this volume). As they become better known, some taxa, *Proteopithecus* and *Plesiopithecus* (Simons 1992), may be assignable to other, less-well-documented groups.

Propliopithecines

When it was first described over 80 years ago, *Propliopithecus* was identified as an early ape, probably ancestral to the European *Pliopithecus*, generally recognized as an ancestral gibbon. Likewise, the closely related *Aegyptopithecus* was originally regarded as a fossil ape, i.e., the sister group of the living great apes and humans. However, with increased information about the cranial and postcranial anatomy of these taxa, it now understood that the propliopithecines are better considered generalized ancestral catarrhines. With the notable exception of their dental formula, 2.1.2.3/2.1.2.3, propliopithecines lack virtually all of the diagnostic features that unite extant Old World monkeys and apes and more closely resemble platyrrhines in most aspects of their cranial and postcranial anatomy (Fleagle and Kay, 1983; Fleagle, 1988). Indeed, studies of propliopithecines demonstrate that what were previously regarded as platyrrhine cranial and skeletal features are also primitive for catarrhines and thus living anthropoids as a group, and that the acquisition of catarrhine features was a slow and mosaic process (e.g., Fleagle, 1986). As basal catarrhines that are undoubted anthropoids in all respecs, propliopithecines have played a limited role in questions of anthropoid origins. Their significance lies primarily in two respects. First, the fact that they appear to be primitive catarrhines with many platyrrhine-like anatomic features supports the view that many aspects of platyrrhine cranial and skeletal anatomy are also basal anthropoid features; thus, they suport the monophyly of platyrrhines and catarrhines. Second, Rasmussen and Simons have recently argued that the poorly known *Propliopithecus* (or *Moeripithecus*) *markgrafi* is in some respects an intermediate form that links the clearly anthropoid, and catarrhine, propliopithecines with a more primitive, confusing group of early anthropoids, the oligopithecines.

Oligopithecines

The genus *Oligopithecus* was first described in 1962 by Simons, who identified it as a catarrhine on the basis of its having two premolars resembling those of *Propliopithecus.* However, the group remained poorly known until the recovery of abundant new fossils including cranial and skeletal remains of a new genus, *Catopithecus,* in the late 1980s from a new quarry (L-41) near the base of the Jebel Qatrani Formation (Simons 1989, 1990; Rasmussen and Simons, 1992). The new material further highlights the morphological paradox that had been recognized from the fragmentary remains (e.g., Gingerich, 1980). It showed that these taxa were definitely anthropoid in having postorbital closure, a platyrrhine-like ear structure, and a catarrhine dental formula, but at the same time the molar dentition showed similarities to European adapids, particularly cercamoniines. These morphological similarities have thus led Rasmussen and Simons (1992; Rasmussen, Chapter 4, this volume; following earlier arguments by Gingerich, 1980) to argue that anthropoids are derived from a cercamoniine adapid. Although Kay and Williams (Chapter 13, this volume) argue that they are distinctive enough from later catarrhines to be placed in a separate family, the Oligopithecidae, Simons and colleagues (Chapter 8, this volume) place the primitive oligopithecines with propliopithecids. This raises the obvious question of the proper phyletic position of the other group of Fayum anthropoids, the parapithecoids.

Parapithecoids

Parapithecids (*Apidium* and *Parapithecus*) were also first described in the early part of this century, but the true phylogenetic position of this group and even the correct dental formula of some genera remain to be firmly established (e.g., Kay and Simons, 1983; Simons, 1986; Fleagle and Kay, 1987). Although parapithecids clearly show some striking autapomophies, overall, they share many of the "platyrrhine" features of the cranium and postcranial skeleton found in propliopithecines and are more primitive than propliopithecines in other features, including premolar number (three rather than two), mandible shape, and several postcranial features. Moreover, in some features, including premolar shape and several aspects of femoral structure, they appear to be even more primitive than platyrrhines and resemble prosimians. There is little doubt that parapithecids are more primitive than either propliopithecids or any group of living Old World anthropoids, but authorities differ in their views as to whether parapithecids precede or follow the platyrrhine–catarrhine divergence (Fleagle and Kay, 1987; Godinot, Chapter 10; Kay and Williams, Chapter 13; Ford, Chapter 18; this volume). As with the propliopithecines, the parapithecids demonstate that the earliest Old World higher primates were clearly platyrrhine-like in much of their morphology and that in many features, platyrrhines almost certainly preserve the primitive anthropoid condition.

Although *Catopithecus* from Quarry L-41 in the Fayum served to confirm the confusing mix of adapid/anthropoid features of the oligopithecines, the most surprising and less anticipated discoveries from the lower level of Fayum were several new genera of early anthropoids that seem to be loosely allied with the parapithecids (Simons, 1992; Gebo *et al.*, Chapter 9, this volume) and have been placed by Simons *et al.* (Chapter 8, this volume) in the superfamily Parapithecoidea. Although only the type specimens and a few unassigned limb bones have been described so far, these new fossils appear to be critical for understanding the early evolution of anthropoids in several respects. First of all, they demonstate an extraordinary adaptive radiation of anthropoids from the late Eocene that preceded the previously known radiation from the upper, presumably early Oligocene, part of the Fayum. Whereas this Oligocene radiation was one of platyrrhine-like anthropoids (e.g., Fleagle and Kay, 1985; Fleagle, 1988), the earlier radiation consisted of animals that adaptively resemble modern prosimians in their small size and faunivorous/frugivorous dentitions. Second, these new taxa provide further insights into primitive anthropoid features and possible relationships among the Fayum groups. One recently described taxon, *Arsinoea* (Simons, 1992), further reinforces the overall impression of primitiveness for parapithecid group. It definitely has an unfused mandibular symphysis, suggesting that several different anthropoid groups may have acquired this feature independently. Another taxon, *Serapia* (Simons, 1992), offers the first hint of a possible transitional form between the primitive parapithecid premolars and the more derived condition found in oligopithecines, platyrrhines, and catarrhines. *Serapia* has premolars that are intermediate in shape between the two groups but otherwise resembles such primitive parapithecids as *Qatrania* and *Arsinoea* (Kay and Williams, Chapter 13, this volume). Moreover, if primitive anthropoids resembled primitive parapithecoids such as *Arsinoea* and *Serapia*, then their teeth were quite unlike those of cercamoniines, so the resemblance between oligopithecines and cercamoniines may be homoplasies. The alternative view, that parapithecoids and oligopithecines were independently evolved from different prosimian ancestors, has not yet been advanced.

New Asian ?Anthropoids

As originally presented at the Anthropoid Origins conference (Culotta, 1992) and recently published in *Nature*, Beard and colleagues (1994) have reported a new group of anthropoids from the middle Eocene of Shanghuang, China. Identified as anthropoids on the basis of their dental formula, enlarged canines, and several premolar and molar features, the Chinese fossilis are much more primitive than any other taxa allocated to Anthropoidea and are very different from any of the Fayum taxa (except possibly *Afrotarsius*). Their anthropoid or even primate affinities are likely to be debated until additional, especially cranial, material becomes available (see Culotta, 1992). If the new Chinese fossils are indeed basal anthropoids as proposed, it is particularly notable that they show none of the dental similarities to adapids found in the oligopithecines. Indeed, among potential anthropoid sister taxa, they show the greatest similarities to tarsiers, a group that is also represented in the Shanghuang fauna (Beard *et al.*, 1994).

Anthropoid Origins

The increased fossil record of early anthropoids has given tremendous insight into some aspects of early anthropoid evolution and remarkably little into others. The new finds in North Africa have revealed an extraordinary adaptive array of Eocene anthropoids and have also filled in many holes in the differentiation of catarrhines from a primitive ancestor. However, we still know relatively little about either the origin of platyrrhines or the origin of anthropoids as a group. Thus, the fossil record shows that platyrrhine-like morphological features were widespread in Africa 35 million years ago and the sequence in which catarrhines acquired their distinctive features (e.g., Fleagle, 1986). It also appears that most of the dental and gnathic features (premolar cusps, fused mandibular symphysis) that separate living anthropids and "prosimians" do not distinguish the earliest anthropoids. Even though we now know quite a lot about early anthropoid anatomy, none of this helps us much when it comes to determining the source of anthropoids or its sister-group relationship with a known Eocene group. This uncertainty stems from several reasons.

(1) All of the known early anthropoids appear to possess a suite of derived orbital and basicranial features. This leaves a considerable anatomic gap between them and any of the known Eocene prosimians. On the basis of what we know about the evolution of suites of adaptations in other primate groups such as hominids, we might hypothesize that the characteristic anthropoid traits must have evolved in a mosaic way, not all at once. Thus, in human evolution, bipedalism evolved well before brain enlargement. Likewise, in anthropoid evolution, the development of postorbital closure may not have been synchronous with changes in the middle ear. But we have no fossils that document this morphological gap.

(2) We are uncertain about which morphological features of the common ancestor of early anthropoids are simply primitive retentions from the earliest primates. The arguments about whether small erect lower incisors are features linking anthropoids to adapids, to some omomyids, or are simply primitive retentions is a case in point.

(3) We don't know which of the few shared features that are found to link anthropoids with either omomyids or adapids are shared-derived features and which are homoplasies. Particularly perplexing are the similarities oligopithecines show to propliopithecines in number and morphology of premolars in contrast to their apparently more primitive, somewhat adapid-like molars. In contrast, parapithecids seem to be more primitive than oligopithecids in premolar number and shape (as well as various postcranial features), but some

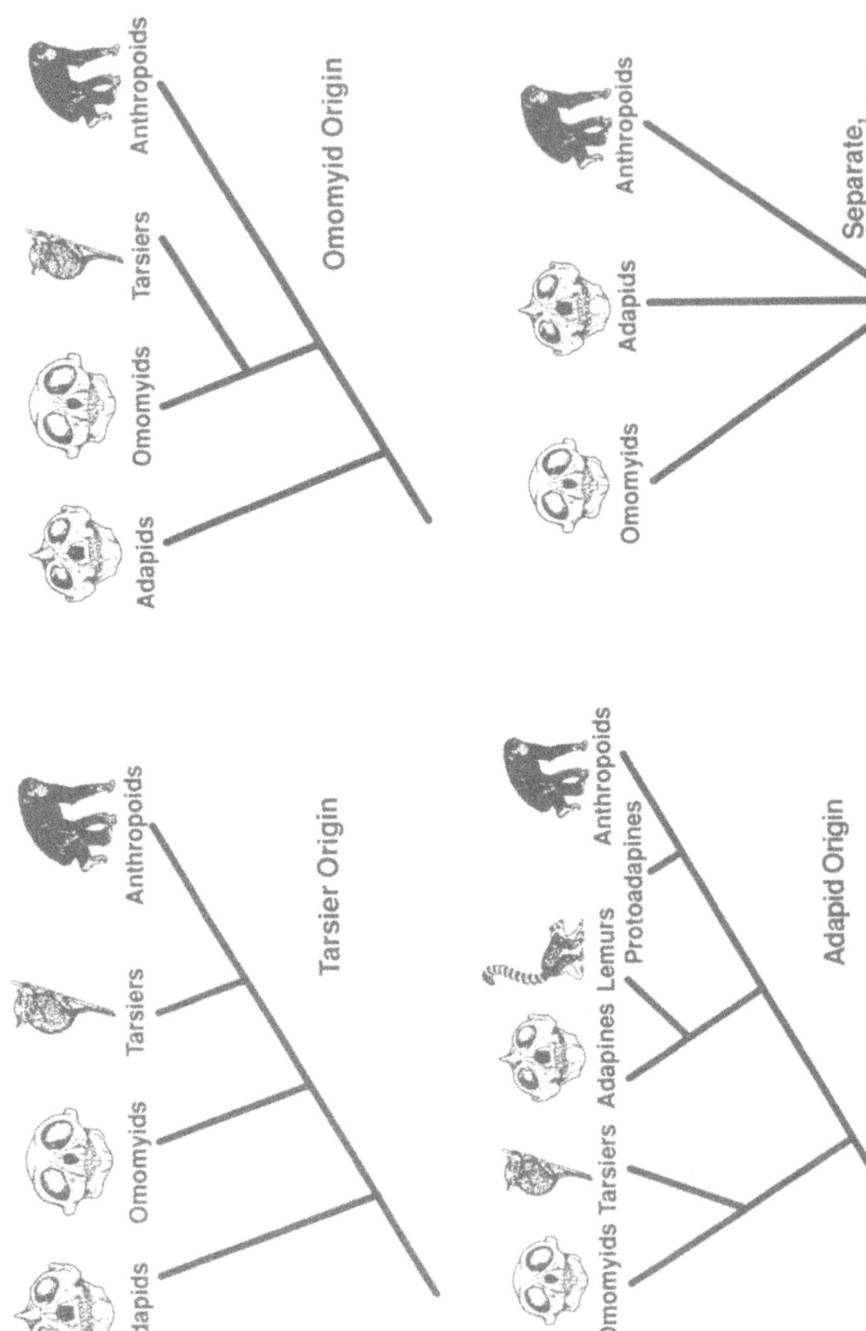

Fig. 2. Four theories of the phyletic relationships among anthropoids and other primate groups.

have more catarrhine-like molars. Did the apparent "catarrhine" molar features of parapithecids evolve in parallel with catarrhines? Both parapithecids and oligopithecines have postorbital closure, and its seems unlikely that they evolved this anthropoid feature independently. However, it is not clear what their common ancestor, and hence the common ancestor of all known later anthropoids, looked like. Moreover, the features shared by all of the early anthropoids, postorbital closure and the anterior accessory chamber of the middle ear, are apparently not present in any of the fossil prosimian groups usually put forth as ancestral anthropoids.

(4) Thus far, the postcranial evidence is of little help. As in cranial morphology, there is little overlap in the postcranial anatomy of anthropoids and fossil prosimians. Phylogenetic analyses of postcranial anatomy (Dagosto and Gebo, Chapter 17; Ford, Chapter 18; this volume) yield a prosimian clade. In view of this confusing evidence it is not surprising that there are four competing hypotheses for anthropoid origins, each with different strengths and weaknesses (Fig. 2).

Tarsier Origin

As noted above, tarsiers share many similarities in reproductive anatomy, eye structure, and cranial anatomy as well as biochemical similarities with anthropoids not found in other living primates. Moreover, the features of postorbital closure and development of an anterior accessory chamber of the middle ear that unite tarsiers and anthropoids are unique features among primates or even among mammals rather than similarities that appear to have evolved in numerous groups (Ross, Chapter 15, this volume). Although the cranial similarities linking tarsiers with anthropoids have been questioned (Simons and Rasmussen, 1989; Beard and MacPhee, Chapter 3, this volume), there seems little reason to do so on morphological grounds (Cartmill, Chapter 16, this volume). Indeed, perhaps the most striking result from all the new remains of fossil primates that have been recovered in recent decades has been the demonstration that postorbital closure and middle ear morphology are the main features distinguishing anthropoids from prosimians. Only tarsiers among all known fossil and extant prosimians share either of these traits with anthropoids.

Nevertheless, the argument that one or more Eocene omomyids without postorbital closure shares unique cranial features with tarsiers alone suggests to many that these tarsier–anthropoid similarities may indeed be homoplasies (see above; Beard *et al.*, 1991; Beard and MacPhee, Chapter 3, this volume). Likewise, to some, the mainly dental and gnathic similarities between adapids and early anthropoids support an other-than-tarsier origin for anthropoids (Rasmussen, Chapter 12, this volume). On dental evidence, the new Chinese fossils could be interpreted as supporting possible tarsier-anthropoid similarities; however, little is known of their cranial anatomy at present.

Omomyid Origin

Many authorities have argued that the sister taxon of anthropoids is not the genus *Tarsius* but a more generalized omomyid. In their view both tarsiers and anthropoids are decended from some common tarsiiform or omomyid ancestry (e.g., Rosenberger and Szalay, 1980; Rosenberger, 1986; Rosenberger and Dagosto, 1992). In this view the haplorhine features uniting *Tarsius* and anthropoids were presumably present in the omomyid ancestor of anthropoids, but not the unique features that are characteristic of *Tarsius*. This view has the advantage of being in accordance with notions of a haplorhine–strepsirhine dicotomy among primates, with molecular studies linking *Tarsius* with anthropoids, and with studies that argue for the origin of *Tarsius* from particular omomyid taxa.

As many have pointed out, even though it would be tidy to be able to apply a haplorhine–strepsirhine division to fossil prosimians, the division of fossil prosimians into haplorhines and strepsirhines is not as clear as among extant primates. It may well be that both omomyids and adapids as well as tarsiers would cluster with anthropoids, and the living strepsirhines are the odd group (Rasmussen, 1986). Moreover, the biggest weakness of this scenario is that the cranial features that most clearly link tarsiers with anthropoids (postorbital closure and anterior accessory chamber of the middle ear) are not present in known omomyids, including the taxa that share various other cranial features with *Tarsius*. Thus, there are very few derived features (possibly an apical interorbital septum and reduced nasal region and a few postcranial characters) that link generalized omomyids with anthropoids. Moreover, if tarsiers are indeed derived from any of the taxa most frequently proposed (*Tetonius, Shoshonius,* and *Necrolemur*), then the unique tarsier–anthropoid cranial featues have evolved independently in tarsiers and anthropoids, and the morphological evidence linking these two groups as haplorines becomes substantially weaker and very difficult to identify among fossils.

A major stumbling block to any of these arguments is the extremely poor fossil record of omomyids. There are only a handful of skulls that document either the orbital anatomy or the ear region, and practically nothing is known about the postcranial anatomy of most taxa (Covert and Williams, Chapter 2; Dagosto and Gebo, Chapter 17, this volume).

Adapid (Cercamoniine or Protoadapine) Origin

Largely on the basis of similarities in the anterior dentition (small incisors, large canines, and frequently fused mandibular symphysis), many workers have long argued for an adapid–anthropoid relationship (e.g., Gingerich, 1980). Supporters of an adapid–anthropoid relationship have noted that omomyids tend to be too specialized dentally (but see Covert and Williams, 1991) and that many of the haplorhine features do not necessarily exclude adapids, as they are not clearly strepsirrhine (Rasmussen, 1987). The postcanine similarities between the oligopithecines and cercamoniines have strengthened the adapid–anthropoid link and have led Rasmussen (Chapter 12, this volume; see also Rasmussen and Simons 1988, 1992) to predict that some of the taxa currently recognized as cercamoniines on the basis of dental remains are actually anthropoids.

One major weakness of the adapid–anthropoid hypothesis is that there are still a number of possibly derived features that adapids share with strepsirhines (Beard *et al.,* 1988; Covert and Williams, Chapter 2, this volume). In addition, many of the features that adapids share with anthropoids are almost certainly primitive primate features, whereas the anthropoid features shared by tarsiers (postorbital closure) and perhaps omomyids (apical interorbital septum, reduced nasal region) are almost certainly derived features. Finally, as with omomyids, there is no indication among any of the known fossils that any adapid ever even approximated the diagnostic anthropoid features of the orbit and ear.

Ancient or Other Origin

The seemingly endless frustration and difficulty of deriving anthropoids from any known group of Eocene prosimians from Europe or North America has led many to argue that the divergence of anthropoids from any known group of prosimians was an ancient event, and the anthropoid ancestor was an unknown Paleogene group inhabiting some relatively unknown continent, more probably Africa or perhaps Asia (see Conroy 1981; Sigé *et al.,* 1990, Beard and MacPhee, Chapter 3, this volume). This position is consistent with the fact that the earliest appearance of anthropoids is in Africa and that many of the Eocene/Oligocene mammals that first appear with anthropoids, such as hyraces and proboscideans, also seem to have no clear holarctic ancestors.

Nevertheless, arguing for a nonadapid, nonomomyid, nontarsier origin for anthropoids is just an expression of ignorance regarding the origin of the group. It is certainly reasonable to expect that the ancestors of this group are not among currently known primate taxa, given how little we know about the fossil record from many parts of the world and the increasing evidence for numerous Eocene anthropoids in Africa (Godinot, Chapter 10, this volume), perhaps Asia (Beard *et al.,* 1994), and even South America, but it is not very satisfying. Furthermore, a relatively ancient split among adapids, omomyids, tarsiers, and anthropoids does not mean that one of these three are not more closely related to anthropoids than the others, only that we might expect identification of their correct phylogenetic relationships more difficult to reconstruct, as each would have had even more time to develop parallelisms and autapomorphies. However, only when these new Eocene taxa are better known will we have a clearer idea of what they really tell us about the antiquity and origins of anthropoids.

Solving Anthropoid Origins

Anthropoids certainly had nonanthropoid ancestors. The origin of this group, just like the evolution of any other group of organisms, is a biological phenomenon that took place. It is also almost certain that most of the hypotheses regarding anthropoid origins from tarsiers, omomyids, adapids, or some unknown group are not correct. Anthropoids most probably had a single origin from a single "prosimian" ancestor. How will we get past the current impasse and resolve the apparently contradictory and mutually exclusive scenarios implied by the current evidence?

It seems most unlikely that phylogenetic analyses of currently known taxa will resolve this problem. In recent years (and in this book) there have been a number of very thoughtful and thorough studies of anthropoid origins examining many anatomic regions for broad samples of extant and fossil anthropoids. These studies are invaluable for clarifying precisely which alternative morphological changes are indicated by different phylogenies and providing a quantitative estimate of how much "morphological change" is required by different phylogenies. However, they have not resolved the problem of anthropoid origins for several obvious reasons. First, the results of any analysis are obviously very dependent on the coding of characters, *a priori* determination of character homology, and the nature of the samples being studied (e.g., Cartmill 1982, Chapter 16, this volume). As a result, whether by "intent," different personal histories, or whatever, different studies often reach different conclusions from analyses of the same material. Developmental studies and extensive comparative samples can help clarify some of the homology problems, but only partially.

A second problem that plagues all phylogenetic analyses is determination of what is primitive. The situation has grown much worse for the primates. A common approach for making such determinations of what is primitive is to examine the morphology of a close relative of the group to be analyzed. For many years it was possible to use a perfectly good outgroup, the Paleocene plesiadapiformes as a basis for determining which characteristics of primates are primitive. They were ideal because it was accepted that they not only were very closely related to primates but also may be near in time to the hypothesized last common ancestor (so that little time would have been available for anatomic changes to have accumulated in the outgroup). Now, however, we have reason to believe that plesiadapiformes may not be the sister group of primates. Other available outgroups within the Archonta are either too specialized (bats) or too far separate in time to be a reliable guide to the morphology of the common ancestor (tree shrews, with practically no fossil record, may have departed substantially from the morphology of the last common ancestor in the last 65–70 milion years). Without appropriate outgroups, it is difficult to determine precisely what was the morphology of the ancestral primate. Without knowing that, it is practially impossible to reconstruct the cladogenesis of early primates, which may be all important for understanding anthropoid origins.

A third problem is the obvious fact that there are no clear answers. To paraphrase Woody Allen, time may fly like an arrow, but evolution flies like a banana. Virtually all phylogenetic analyses find 30–50% homoplasy in the most parsimonious solutions. With so many features evolving and reevolving or reverting many times, it is difficult to have much confidence in the most parsimonious solution when it is clear that parsimony is not obviously a common result of phyletic evolution by natural selection.

As trite as it sounds, resolution can only come from the recovery of new fossils that bridge the current morphological and temporal gaps between currently known anthropoids and prosimians and clarify which of the apparent prosimian–anthropoid similarities reflect real phylogenetic similarities and which are parallelisms and convergences. As Szalay has argued, we need to know the actual morphological transformations involved in anthropoid origins to clarify many of the morphological problems that underlie the current impasse. There are many specific inconsistencies that need to be addressed. For example, it is particularly striking that no fossil prosimians show postorbital closure, yet all early anthropoids show a walled-off orbit. Where did the anthropoid condition come from? Or the tarsier condition, for that matter? Oligopithecines show a dental formula and premolar morphology similar to those of propliopithecines but molar similarities to adapids, suggesting that catarrhines are derived from an adapid ancestry. But, what does this imply about parapithecids, which have three premolars more primitive than those of any other anthropoids but good anthropoid or even catarrhine molar features? Although there are clear dental similarities between oligopithecines and adapids, adapid skeletal anatomy is very clearly not similar to that of anthropoids but very similar to strepsirhines.

Will new fossils resolve these issues or just generate new unanticipated problems? Undoubtedly some of both. Certainly, the new fossil anthropoids from Egypt (Simons, 1992) are vastly different from anything anyone had any reason to anticipate in early anthropoid evolution. And possible new anthropoids from China, as they become better documented, should add surprises (Beard *et al.*, 1994). It seems that as we get closer to the base of the anthropoid radiation, the tree gets very bushy. It is hard to sort out the relationships when you don't even know who the players are. Early parapithecoids such as *Serapia* and *Arsinoea* are helping to sort out the relationships among platyrrhines, parapithecids, and propliopithecines, but other north African Eocene primates provide hints of an anthropoid radiation that precedes the one documented in the Fayum. In addition, the skeletal anatomy of many early anthropoids and fossil prosimians remains largely undocumented (but see Fleagle and Kay, 1987; Gebo *et al.*, Chapter 9; Dagosto and Gebo, Chapter 17, this volume). This is another sort of evidence that will not eliminate current contradictions in the dental and cranial evidence but may throw support to one hypothesis over others.

Moreover, the phyletic issue is just one of the many outstanding problems concerning anthropoid origins. The current evidence seems to favor Africa as the geographic site of anthropoid origins on the basis of the many primitive or perhaps even "transitional" taxa. However, the new Chinese "anthropoids" are more primitive, and the site of origin and the site of the first major adaptive radiation of a group need not be the same. The aspect of anthropoid origins that was least addressed by the authors is that of the adaptive correlates of the initial appearance and radiation of the group (Cartmill, 1980; but see Simons *et al.*, Chapter 8; Ravosa and Hylander, Chapter 14, Ross, Chapter 15, this volume). Most significant in this regard is the realization that the earliest anthropoids were both more prosimain-like and more diverse in many aspects of their biology (size, molar structure) than we had thought before (Simons *et al.*, Chapter 8, this volume). At the moment, Anthropoidea appears to be very bushy at its base; only when we can more clearly sort through these early branches will have a clearer view of the roots of the group.

ACNOWLEDGMENTS

This paper reflects the contributions of all the authors in the volume as well as those who participated in the Duke Conference but did not write a paper for this volume. In particular, we thank B. A. Williams, D. T. Rasmussen, K. C. Beard, K. E. Reed, T. M. Bown, W. Hartwig, and T. C. Rae for extensive discussion of the topics discussed therein. K. E. Reed and J. Kelly provided editorial assistance, and Luci Betti provided artistic support. This work was funded in part by grants for the NSF, the LSB Leakey Foundation, and the Wenner-Gren Foundation.

References

Beard, K. C., and Godinot, M. 1988. Carpal anatomy of *Smilodectes gracilis* (Adapiformes, Notharctinae) and its significance for lemuriform phylogeny. *J. Hum. Evol.* **17**:71–92.

Beard, K. C., Dagosto, M., Gebo, D. L., and Godinot, M. 1988. Interrelationships among primate higher taxa. *Nature* **331**:712–714.

Beard, K. C., Krishtalka, L., and Stucky, R. K. 1991. First skulls of the early Eocene primate *Shoshonius cooperi* and the anthropoid–tarsier dichotomy. *Nature* **349**:64–67.

Beard, C., Qi, T., Dawson, M. R., Wang, B, and Li, C. 1994. A diverse new primate fauna from middle Eocene fissure-fillings in southeastern China. *Nature* **368**:604–609.

Bown, T. M., and Rose, K. D. 1987. Patterns of dental evolution in early Eocene Anaptomorphine primates (Omomyidae) from the Bighorn Basin, Wyoming. *J. Paleontol.* **61**:5.II, suppl.

Bown, T. M., and Rose, K. D. 1991. Evolutionary relationships of a new genus and three new species of Omomyid primates (Willwood Formation, Lower Eocene, Bighorn Basin, Wyoming). *J. Hum. Evol.* **20**:465–480.

Cartmill, M. 1980. Morphology, function, and evolution of the anthropoid postorbital septum.

In: R. L. Ciochon and A. B. Chiarelli (eds.), *Evolutionary Biology of the New World Monkeys and Continental Drift,* pp. 243–274. Plenum Press, New York.
Cartmill, M. 1982.Assessing tarsier affinities: Is anatomical description phylogenetically neutral? *Geobios Mem. Spec.* **6:**279–287.
Ciochon, R. L., and Corruccini, R. S. 1983. *New Interpretaions of Ape and Human Ancestry.* Plenum Press, New York.
Conroy, G. C. 1981. Review of evolutionary biology of the New World monkeys and continental drift. *Fol. Primatol.* **36:**155–156.
Covert, H. H., and Williams, B. 1991. The anterior lower dentition of *Washakius insignis* and adapid–anthropoidean affinities. *J. Hum. Evol.* **21:**463–467.
Culotta, E. 1992. A new take on anthropoid origins. *Science* **256:**1516–1517.
Dagosto, M. 1983. Postcranium of *Adapis parisiensis* and *Leptadapis magnus* (Adapiformes, Primates): Adaptational and phylogenetic significance. *Fol. Primatol.* **41:**49–101.
Dagosto, M. 1985. The distal tibia of Primates with special reference to the Omomyidae. *Int. J. Primatol.* **6:**45–75.
Dagosto, M. 1988. Implications of postcranial evidence for the origin of euprimates. *J. Hum. Evol.* **17:**35–56.
Dagosto, M. 1993. Postcranial anatomy and locomotor behavior in Eocene primates. In: D. L. Gebo (ed.), *Postcranial Adaptation in Nonhuman Primates,* pp. 199–219. Northern Illinois University Press, DeKalb.
Dutrillaux, B., and Rumpler, Y. 1988. Abence of chromosomal similarities between tarsiers (*Tarsius syrichta*) and other primates. *Fol. Primatol.* **50:**130–133.
Fleagle, J. G. 1986. The fossil record of early catarrhine evolution. In: B. Wood, L. Martin, and P. Andrews (eds.), *Major Topics in Primate and Human Evolution,* pp. 130–149. Cambridge University Press, Cambridge.
Fleagle, J. G. 1988. *Primate Adaptation and Evolution.* Academic Press, San Diego.
Fleagle, J. G., and Kay, R. F. 1983. New interpretations of the phyletic position of Oligocene hominoids. In: R. L. Ciochon and R. S. Corruccini (eds.), *New Interpretations of Ape and Human Ancestry,* pp. 181–210. Plenum Press, New York.
Fleagle, J. G., and Kay, R. F. 1985. The paleobiology of Catarrhines. In: E. Delson (ed.), *Ancestors: The Hard Evidence,* pp. 23–36. Alan R. Liss, New York.
Fleagle, J. G., and Kay, R. F. 1987. The phyletic position of the Parapithecidae. *J. Hum. Evol.* **16:**483–531.
Franzen, J. L. 1987. Ein neuer Primate aus dem Mitteleozän der Grube Messel (Deutschland, S-Hessen). *Cour. Forsch. Inst. Senckenberg* **91:**151–187.
Gheerbrant, E., and Thomas, H. 1992. The two first possible new adapids from the Arabian Peninsula (Taqah, early Oligocene of Sultanate of Oman). In: *Abstracts of the 14th Congress of the International Primatological Society,* pp. 349–350. Sicop Press, Strasbourg.
Gheerbrant, E. Thomas, H., Roger, J., Sen, S., and Al-Sulaimani, Z. 1994. Deux nouveaux primates dans l'Oligocene inferieur de Taqah (Sultanat d'Oman): premiers Adapiformes (?Anchomomyini) de la peninsula Arabique? *Palaeovertebrata* **22:**141–196.
Gingerich, P. D. 1977. Radiation of Eocene Adapidae in Europe. *Geobios Mem. Spéc.* **1:**165–182.
Gingerich, P. D. 1980. Eocene Adapidae: Paleobiogeography and the origin of South American Platyrrhini. In: R. L. Ciochon and A. B. Chiarelli (eds.), *Evolutionary Biology of the New World Monkeys and Continental Drift.* pp. 123–138. Plenum Press, New York.
Gingerich, P. D. 1981. Early Cenozoic Omomyidae and the evolutionary history of tarsiiform primates. *J. Hum. Evol.* **10:**345–374.
Gingerich, P. D. 1990. African dawn for primates. *Nature* **346:**411.
Gingerich, P. D. 1993.Oligocene age of the Gebel Qatrani Formation, Fayum, Egypt. *J. Hum. Evol.* **24:**207–218.
Gingerich, P. D., and Martin, R. 1981. Cranial morphology and adaptations in Eocene Adapidae, II: The Cambridge skull of *Adapis parisiensis. Am. J. Phys. Anthropol.* **56:**235–257.
Gingerich, P. D., and Sahni, A. 1984. Dentition of *Sivaladapis nagrii* (Adapidae) from the late Miocene of India. *Int. J. Primatol.* **5:**63–69.

Gingerich, P. D., and Simons, E. L. 1977. Systematics, phylogeny, and evolution of early Eocene Adapidae (Mammalia, Primates) in North America. *Contrib. Mus. Paleontol. Univ. Michigan* **24:**245–279.

Gingerich, P. D., Dashzeveg, D., and Russell, D. E. 1991. Dentition and systematic relationships of *Altanius orlovi* (Mammalia, Primates) from the early Eocene of Mongolia. *Geobios* **1991**(24)**:**637–646.

Godinot, M. 1984. Un nouveau gnere temoignant de la diversite des Adapines (Primates, Adapidae) a l'Eocene terminal. *C. R. Adad. Sci. Paris [II]* **299:**1291–1296.

Godinot, M. 1992. Early euprimate hands in evolutionary perspective. *J Hum. Evol.* **22:**267–284.

Godinot, M., and Mahboubi, M. 1992. Earliest known simian primate found in Algeria. *Nature* **357:**324–326.

Gregory, W. K. 1920. On the structure and relation of *Notharctus,* an American Eocene primate. *Mem. Am. Mus. Nat. Hist. N.S.* **351:**243.

Hartenberger, J.-L., and Marandat, B. 1992. A new genus and species of an early Eocene primate from North Africa. *Hum. Evol.* **7:**9–16.

Kappelman, J. 1992. The age of the Fayum primates as determined by paleomagnetic reversal stratigraphy. *J. Hum. Evol.* **22:**495–503.

Kay, R. F., and Simons, E. L. 1983. Dental formulae and dental eruption patterns in Parapithecidae (Primates, Anthropoidea). *Am. J. Phys. Anthrop.* **62:**363–375.

Kay, R. F., Thewissen, J. G. M., and Yoder, A. D. 1992. Cranial anatomy of *Ignacius graybullianus* and the affinities of the Plesiadapiformes. *Am. J. Phys. Anthrop.* **89:**477–498.

Krause, D. W., and Maas, M. 1990. The biogeographic origins of the late Paleocene–early Eocene mammalian immigrants to the western interior of North America. In: T. M. Bown and K. D. Rose (eds.), *Dawn of the Age of Mammals in the Northern Part of the Rocky Mountain Interior, North America,* pp. 71–105. Geological Society of America, Boulder.

Luckett, W. 1975. Ontogeny of the fetal membranes and placenta: Their bearing on primate phylogeny. In: W. Luckett and F. S. Szalay (eds.), *Phylogeny of the Primates: A Multidisciplinary Approach,* pp. 157–182. Plenum Press, New York.

Maas, M., Krause, D. W., and Strait, S. G. 1988. The decline and extinction of Plesiadapiformes (Mammalia: ?Primates) in North America: Displacement or replacement? *Paleobiology* **14:**410–431.

MacPhee, R. D. E. (ed.) 1993. *Primates and Their Relatives in Phylogenetic Perspective.* Plenum Press, New York.

MacPhee, R. D. E., and Cartmill, M. 1986. Basicranial structures and primate systematics. In: D. R. Swindler and J. Erwin (eds.), *Comparative Primate Biology, Vol. 1: Systematics, Evolution and Anatomy,* pp. 219–275. Alan R. Liss, New York.

Martin, R. D. 1990. *Primate Origins and Evolution: A Phylogenetic Reconstruction.* Princeton University Press, Princeton.

Miyamoto, M., and Goodman, M. 1990. DNA systematics and evolution of primates. *Ann. Rev. Ecol. Syst.* **21:**197–220.

Pocock, R. I. 1918. On the external characters of the lemurs and of *Tarsius. Proc. Zool. Soc. Lond.* **1918:**19–53.

Rasmussen, D. T. 1986. Anthropoid origins: A possible solution to the Adapidae–Omomyidae paradox. *J. Hum. Evol.* **15:**1–12.

Rasmussen, D T. 1990. The phylogenetic position of *Mahgarita stevensi:* protoanthropoid or lemuroid? *Int. J. Primatol.* **11:**437–467.

Rasmussen, D. T., and Simons, E. L. 1988. New specimens of *Oligopithecus savagei,* early Oligocene primate from the Fayum, Egypt. *Fol. Primatol.* **51:**182–208.

Rasmussen, D. T., and Simons, E. L. 1992. Paleobiology of the oligopithecines, the earliest known anthropoid primates. *Int. J. Primatol.* **13:**477–508.

Rasmussen, D. T., Bown, T. M., and Simons, E. L. 1992. The Eocene–Oligocene transition in continental Africa. In: D. Prothero and W. Berggren (eds.), *Eocene–Oligocene Climatic and Biotic Evolution,* pp. 548–566. Princeton University Press, Princeton.

Rose, K. D., and Bown, T. M. 1991. Additional fossil evidence on the differentiation of the earliest euprimates. *Proc. Natl. Acad. Sci. USA* **88:**98–101.
Rose, K. D., and Walker, A. 1985. The skeleton of early Eocene *Cantius,* oldest lemuriform primate. *Am. J. Phys. Anthrop.* **66:**73–89.
Rosenberger, A. L. 1985. In favor of the *Necrolemur*–tarsier hypothesis. *Fol. Primatol.* **45:**179–194.
Rosenberger, A. L. 1986. Platyrrhines, catarrhines, and the anthropoid transition. In: B. Wood, L. Martin, and P. Andrews (eds.), *Major Topics in Primate and Human Evolution,* pp. 66–88. Cambridge University Press, Cambridge.
Rosenberger, A. L., and Dagosto, M. 1992. New craniodental and postcranial evidence of fossil tarsiiforms. In: S. Matano, R. Tuttle, H. Ishida, and M. Goodman (eds.), *Topics in Primatology, Vol. 3: Evolutonary Biology, Reproductive Endocrinology, and Virology,* pp. 37–51. University of Tokyo Press, Tokyo.
Rosenberger, A. L., and Szalay, F. S. 1980. On the tarsiiform origins of Anthropoidea. In: R. L. Ciochon and A. B. Chiarelli (eds.), *Evolutionary Biology of the New World Monkeys and Continental Drift.* pp. 139–157. Plenum Press, New York.
Russell, D., and Gingerich, P. D. 1987. Nouveaux primates de l'Eocene du Pakistan. *C.R. Acad. Sci. Paris, [2]* **304:**209–214.
Sigé, B., Jaeger, J.-J., Sudre, J., and Vianey-Liaud, M. 1990. *Altiatlasius koulchii* n. gen. et sp., primate omomyide du Paleocene superieur de Maroc, et les origines des euprimates. *Paleontographica [A]* **214:**31–56.
Simmons, N. B., Novacek, M. J., and Baker, R. J. 1991. Approahes, methods, and the future of the Chiropteran monophyly controversy: A reply to JD Pettigrew. *Syst. Zool.* **40:**239–243.
Simons, E. L. 1961. The dentition of *Ourayia:* Its bearing on relationships of omomyid prosimians. *Postilla* **54:**1–20.
Simons, E. L. 1962. A new Eocene primate genus *Cantius,* and a revision of some allied European lemuroids. *Bull. Br. Mus. (Nat. Hist.) Geol.* **7:**1–30.
Simons, E. L. 1972. *Primate Evolution.* New York: MacMillan.
Simons, E. L. 1986. *Parapithecus grangeri* of the African Oligocene: An archaic catarrhine without lower incisors. *J. Hum. Evol.* 205–213.
Simons, E. L. 1989. Description of two genera and species of late Eocene arthropoidea from Egypt. *Proc. Natl. Acad. Sci. U.S.A.* **86:**9956–9960.
Simons, E. L. 1990. Discovery of the oldest known anthropoidean skull from the Paleogene of Egypt. *Science* **247:**1567–1569.
Simons, E. L. 1992. Diversity in the early Tertiary anthropoidean radiation in Africa. *Proc. Natl. Acad. Sci. USA* **89:**10743–10747.
Simons, E. L., and Rasmussen, D. T. 1989. Cranial morphology of *Aegyptopithecus* and *Tarsius* and the question of the tarsier–anthropoid clade. *Am. J. Phys. Anthrop.* **79:**1–23
Simons, E. L., and Russell, D. 1960. Notes on the cranial anatomy of *Necrolemur. Breviora* **127:**1–14.
Simons, E. L., Bown, T. M., and Rasmussen, D. T. 1986. Discovery of two additional prosimian primate families (Omomyidae, Lorisidae) in the African Oligocene. *J. Hum. Evol.* **15:**431–438.
Strait, S. G. 1991. *Dietary Reconstruction in Small-Bodied Fossil Primates.* Ph.D. Thesis, SUNY at Stony Brook.
Sudre, J. 1975. Un prosimien du Paleogene ancien du Sahara Nord-Occidental: *Azibius trerki* n.g.n.sp. *C.R. Acad. Sci. Paris [D]* **280:**1539–1542.
Szalay, F. S., and Dagosto, M. 1980. Locomotor adaptations as reflected in the humerus of Paleogene primates. *Fol. Primatol.* **34:**1–45.
Szalay, F. S., and Li, C.-K. 1986. Middle Paleocene euprimate from southern China and the distribution of primates in the Paleogene. *J. Hum. Evol.* **15:**387–397.
Tattersall, I., and Schwartz, J. 1985. Evolutionary relationships of living lemurs and lorises (Mammalia, Primates) and their potential affinities with European Eocene Adapidae. *Anthropol. Papers Am. Mus. Nat. Hist.* **60:**1–110.

Thomas, H., Roger, J., Sen, S., Bourdillon de Grissac, C., and Al-Sulaimani, Z. 1989. Découverte de vertébrés fossiles dans L'Oligocène inférieur du Khofar (Sultanat d'Oman). *Geobios* **22:**101–120.

Wible, J., and Covert, H. 1987. Primates: Cladistic diagnosis and relationships. *J. Hum. Evol.* **16:**1–22.

Williams, B. A. 1994. *Phylogeny of the Omomyidae and Implications for Anthropoid Origins.* PhD Dissertation, University of Colorado.

Williams, B. A., and Covert, H. H. 1994. An early Eocene anaptomorphine primate (Omomyidae) from the Washakie Basin, Wyoming, with comments on the phylogeny and paleobiology of anaptomorphines. *Am. J. Phys. Anthrop.* **93:**323–340.

Wilson, J. A., and Szalay, F. S. 1976. New adapid primate of European affinities from Texas. *Fol. Primatol.* **25:**294–312.

Systematic Index

Geological/Geographic Index

GPSR Compliance
The European Union's (EU) General Product Safety Regulation (GPSR) is a set of rules that requires consumer products to be safe and our obligations to ensure this.

If you have any concerns about our products, you can contact us on

ProductSafety@springernature.com

In case Publisher is established outside the EU, the EU authorized representative is:

Springer Nature Customer Service Center GmbH
Europaplatz 3
69115 Heidelberg, Germany

www.ingramcontent.com/pod-product-compliance
Ingram Content Group UK Ltd.
Pitfield, Milton Keynes, MK11 3LW, UK
UKHW051132260726
13967UKWH00010B/2994
* 9 7 8 1 4 7 5 7 9 1 9 8 3 *